Flora Australiensis

A Description of the Plants of the
Australian Territory

VOLUME 4: STYLIDIEAE TO PEDALINEAE

GEORGE BENTHAM
FERDINAND VON MUELLER

CAMBRIDGE
UNIVERSITY PRESS

CAMBRIDGE UNIVERSITY PRESS

Cambridge, New York, Melbourne, Madrid, Cape Town,
Singapore, São Paolo, Delhi, Tokyo, Mexico City

Published in the United States of America by Cambridge University Press, New York

www.cambridge.org
Information on this title: www.cambridge.org/9781108037419

© in this compilation Cambridge University Press 2011

This edition first published 1869
This digitally printed version 2011

ISBN 978-1-108-03741-9 Paperback

CAMBRIDGE LIBRARY COLLECTION

Books of enduring scholarly value

Life Sciences

Until the nineteenth century, the various subjects now known as the life sciences were regarded either as arcane studies which had little impact on ordinary daily life, or as a genteel hobby for the leisured classes. The increasing academic rigour and systematisation brought to the study of botany, zoology and other disciplines, and their adoption in university curricula, are reflected in the books reissued in this series.

Flora Australiensis

George Bentham (1800–84) was one of Britain's most influential botanists, whose own collection of plant specimens numbered more than 100,000. Although he donated his herbarium to the Royal Botanic Gardens, Kew in 1854, he continued to make significant contributions to the field, including this exhaustive, seven-volume work detailing the plant life of Australia, which was published from 1863 to 1878. It was part of a series of works commissioned by the British government to document the flora in its colonies. Using the extensive numbers of specimens at Kew – and with the help of Ferdinand Mueller (1825–96), a German botanist in Australia – Bentham was able to compile descriptions of more than 8,000 species of Australian plants, making these volumes the first completed compendium of the flora of any large continental area. Volume 4, published in 1869, describes 26 orders of dicotyledon flora in the subclass monopetalae.

Cambridge University Press has long been a pioneer in the reissuing of out-of-print titles from its own backlist, producing digital reprints of books that are still sought after by scholars and students but could not be reprinted economically using traditional technology. The Cambridge Library Collection extends this activity to a wider range of books which are still of importance to researchers and professionals, either for the source material they contain, or as landmarks in the history of their academic discipline.

Drawing from the world-renowned collections in the Cambridge University Library, and guided by the advice of experts in each subject area, Cambridge University Press is using state-of-the-art scanning machines in its own Printing House to capture the content of each book selected for inclusion. The files are processed to give a consistently clear, crisp image, and the books finished to the high quality standard for which the Press is recognised around the world. The latest print-on-demand technology ensures that the books will remain available indefinitely, and that orders for single or multiple copies can quickly be supplied.

The Cambridge Library Collection will bring back to life books of enduring scholarly value (including out-of-copyright works originally issued by other publishers) across a wide range of disciplines in the humanities and social sciences and in science and technology.

FLORA AUSTRALIENSIS.

FLORA AUSTRALIENSIS:

A DESCRIPTION

OF THE

PLANTS OF THE AUSTRALIAN TERRITORY.

BY

GEORGE BENTHAM, F.R.S., P.L.S.,

ASSISTED BY

FERDINAND MUELLER, M.D., F.R.S. & L S.,

GOVERNMENT BOTANIST, MELBOURNE, VICTORIA.

VOL. IV.

STYLIDIEÆ TO PEDALINEÆ.

PUBLISHED UNDER THE AUTHORITY OF THE SEVERAL GOVERNMENTS
OF THE AUSTRALIAN COLONIES.

LONDON:
L. REEVE & CO., 5, HENRIETTA STREET, COVENT GARDEN.
1869.

PRINTED BY TAYLOR AND CO.,
LITTLE QUEEN STREET, LINCOLN'S INN FIELDS.

CONTENTS.

CONSPECTUS OF THE ORDERS CONTAINED IN THE FOURTH VOLUME.

Class I. **DICOTYLEDONS.**

Subclass II. MONOPETALÆ.

(Continued from Vol. III.)

(*Ovary inferior in the first three Orders, in some* Ericaceæ, Myrsineæ, Styracaceæ, *and* Gesneriaceæ, *superior in the others.*)

LXIII. Stylidieæ. Herbs or rarely undershrubs. Leaves radical or scattered, or in whorl-like clusters. Flowers irregular or rarely regular, the fifth lobe of the corolla usually smaller or different from the other four. *Stamens 2, the filaments connate with the style in a column free from the corolla* (anthers sessile on the top of the style). Ovary inferior, wholly or partially 2-celled. Seeds albuminous.

LXIV. Goodenovieæ. Herbs or shrubs. Leaves alternate or radical. Flowers more or less irregular. Stamens 5. Ovary inferior, at least as to the corolla, 2-celled or rarely 1-celled. *Style with a cup-shaped or peltate indusium under the stigma.* Seeds albuminous.

LXV. Campanulaceæ. Herbs, usually with a milky juice. Leaves alternate. Flowers irregular or regular. Corolla-lobes valvate. Stamens usually free from the corolla, as many as its lobes and alternating with them. Ovary more or less inferior, with 2 to 5 many-ovulate cells. Seeds albuminous.

LXVI. Ericaceæ. Shrubs. Leaves alternate. Flowers regular. Stamens usually free from the corolla, twice as many as its lobes. *Anthers 2-celled, opening in terminal pores* (except in *Wittsteinia*). Ovary inferior or superior, with as many cells as corolla-lobes (fewer in *Wittsteinia*). Seeds albuminous.

LXVII. Epacrideæ. Shrubs. Leaves usually alternate, rigid and strict. Flowers regular. Stamens as many as corolla-lobes and alternate with them, or rarely fewer. *Anthers 1-celled.* Ovary superior, with 5 or fewer cells. Seeds albuminous.

LXVIII. Plumbagineæ. Herbs or rarely shrubs. Leaves radical or alternate. Flowers regular. Calyx tubular. Stamens 5, opposite to the corolla-lobes or petals. Ovary 1-celled, *with 1 ovule suspended from a free filiform placenta; styles or style-branches* 5. Seeds rarely albuminous.

LXIX. Primulaceæ. Herbs. Leaves radical or alternate, rarely whorled. Flowers regular. Stamens as many as corolla-lobes and opposite to them. Ovary 1-celled, with peltate ovules attached to a free central placenta. Fruit usually dehiscent. Seeds albuminous.

LXX. Myrsineæ. Trees or shrubs. Leaves alternate, usually dotted. Flowers regular. Stamens as many as corolla-lobes and opposite to them. Ovary 1-celled, with peltate ovules attached to a free central placenta. Fruit succulent or hard, usually indehiscent. Seeds rarely without albumen.

LXXI. Sapotaceæ. Trees or shrubs, the juice often milky. Leaves alternate. Flowers regular. Corolla-lobes as many or twice as many as calyx-segments. Stamens as many as corolla-lobes and opposite to them, or twice as many. Ovary 2- or more celled, with 1 ovule in each cell. Fruit succulent or hard, usually indehiscent. Seeds with or without albumen.

LXXII. EBENACEÆ. Trees or shrubs, not milky. Leaves alternate. Flowers regular, usually diœcious. Corolla-lobes 3 to 5. Stamens indefinite (few or many). Ovary 3- or more-celled, with 1 or 2 ovules in each cell. Fruit succulent, usually indehiscent. Seeds albuminous.

LXXIII. STYRACACEÆ. Trees or shrubs. Leaves alternate. Flowers regular, hermaphrodite. Corolla-lobes as many or twice as many as calyx-lobes. Stamens usually more than twice as many, rarely twice as many as corolla-lobes or fewer. Ovary, or at least the fruit, more or less inferior, 2- to 5-celled, with 2 or more ovules in each cell. Fruit usually succulent and indehiscent. Seeds albuminous.

LXXIV. JASMINEÆ. Trees shrubs or climbers. Leaves opposite or very rarely alternate. Flowers regular. Corolla with 4, 5 or more lobes rarely 2-petaled or none. *Stamens 2, alternating with the carpels.* Ovary 2-celled, with one or two ovules in each cell. Fruit succulent or capsular. Seeds with or without albumen.

LXXV. APOCYNEÆ. Trees shrubs or twiners, rarely perennial herbs. Leaves opposite or rarely scattered. Flowers regular. Stamens 5, alternate with the corolla-lobes; anthers connivent round the stigma. Ovary of two distinct carpels, the styles connected upwards, or rarely the carpels united from the base. Fruit of 1 or 2 follicles drupes or berries. Seeds usually albuminous.

LXXVI. ASCLEPIADEÆ. Twiners or rarely herbaceous perennials or shrubs. Leaves opposite. Flowers regular. Stamens 5, alternate with the corolla-lobes; anthers connate round the stigma, 2- or 4-celled; *pollen consolidated in 1 or 2 masses in each cell.* Ovary of 2 distinct carpels; the styles united upwards. Fruit follicular. Seeds with little albumen.

LXXVII. LOGANIACEÆ. Trees shrubs twiners or herbs. Leaves opposite, often connected by stipules or raised lines. Flowers regular. Stamens as many as corolla-lobes and alternate with them. Anthers free. Ovary usually 2-celled. Style single. Fruit a capsule or berry. Seeds albuminous.

LXXVIII. GENTIANEÆ. Herbs with a bitter taste. Leaves opposite or in the *Menyantheæ* alternate. Flowers regular. Stamens as many as corolla-lobes and alternate with them. Anthers free. Ovary 1-celled with 2 or rarely more parietal placentas rarely completely dividing it into 2 cells; ovules numerous and minute. Style single. Fruit a capsule, rarely indehiscent. Seeds albuminous.

LXXIX. HYDROPHYLLACEÆ. Herbs or rarely undershrubs. Leaves alternate or the lower ones opposite. Flowers regular in unilateral racemes or cymes. Stamens as many as corolla-lobes and alternate with them. Anthers free. Ovary 1-celled with 2 parietal placentas or rarely 2-celled. *Styles or style-branches* 2. Fruit a capsule. Seeds albuminous.

LXXX. BORAGINEÆ. Herbs usually coarsely hirsute or in drupaceous genera trees or shrubs. Leaves usually alternate. Flowers regular in cymes or unilateral racemes. Stamens as many as corolla-lobes and alternate with them or very rarely fewer. Ovary 2- or 4-celled with 1 ovule in each cell or 2-celled with 2 parallel ovules in each cell. Style single, entire or rarely forked. Fruit a drupe, or dry and separating into 2 or 4 nuts. Seeds with little or no albumen.

LXXXI. CONVOLVULACEÆ. Twiners or rarely erect herbs shrubs or trees. Leaves alternate. Flowers regular, usually axillary. Corolla-limb folded in the bud. Stamens 5, alternate with the corolla-lobes or angles. Ovary of 2 to 4 cells or carpels with 1 or 2 erect ovules in each. Style single and entire or 2-branched or 2 distinct styles. Fruit capsular or succulent and indehiscent. Seeds with little or no albumen; *cotyledons very much folded* (or inconspicuous in *Cuscuta*).

LXXXII. SOLANEÆ. Herbs shrubs or soft-wooded trees. Leaves alternate. Flowers regular or nearly so. Corolla-lobes folded or rarely imbricate in the bud. Stamens as many as corolla-lobes and alternate with them. Ovary 2-celled or spuriously 4-celled (rarely 3- or 4-celled), with several ovules in each cell. Style single. Fruit a berry or a capsule. Seeds albuminous, the embryo usually curved or annular.

LXXXIII. SCROPHULARINEÆ. Herbs or rarely shrubs or small trees. Leaves alternate or opposite. Flowers irregular with the corolla-lobes bilabiate or imbricate in the bud or rarely nearly regular with the corolla-lobes folded. Perfect stamens 4 in pairs or 2, the fifth rudimentary or wanting, or very rarely perfect. Ovary 2-celled with several ovules in each cell. Fruit a capsule or very rarely a berry. Seeds albuminous; embryo usually straight.

LXXXIV. LENTIBULARIEÆ. Herbs either aquatic with floating capillary-divided leaves or terrestrial with radical or without any leaves. Flowers irregular, the corolla 2-lipped. Stamens 2; anthers 1-celled. *Ovary 1-celled, with peltate ovules inserted on a free central placenta.* Fruit a capsule. Seeds small, without albumen.

LXXXV. OROBANCHACEÆ. Leafless herbs, not green, parasites on roots. Flowers irregular. Stamens 4, in pairs; anthers 2-celled. Ovary 1-celled, with 2 or 4 parietal placentas and very numerous ovules. Fruit capsular. Seeds albuminous.

LXXXVI. GESNERIACEÆ. Herbs or when shrubby often epiphytical or climbing, rarely erect shrubs. Leaves opposite. Flowers usually irregular. Perfect stamens 4 in pairs or rarely 2 only. Ovary 1-celled, with 2 parietal placentas and numerous ovules. Fruit a berry or capsule. Seeds with or without albumen.

LXXXVII. BIGNONIACEÆ. Woody climbers trees or shrubs. Leaves opposite, often compound. Flowers irregular. Perfect stamens 4 in pairs or 2 only. Ovary 2-celled with 2 distinct and sometimes distant placentæ on the dissepiment in each cell; ovules usually numerous. Fruit a capsule usually long and narrow. Seeds winged, without albumen.

LXXXVIII. ACANTHACEÆ. Herbs or shrubs rarely twiners. Leaves opposite. Flowers more or less irregular. Perfect stamens 4, in pairs or 2 only. Ovary 2-celled with 2 or more superposed ovules in each cell. Fruit a capsule opening elastically in 2 valves. Seeds without albumen, usually subtended by hooked or rarely cup-shaped or minute retinacula.

LXXXIX. PEDALINEÆ. Herbs. Leaves opposite. Flowers irregular. Perfect stamens 4 in pairs or rarely 2 only. Ovary composed of 2, rarely 3 or 4 carpels, but divided (at least after flowering) into twice as many cells by spurious dissepiments. Ovules 2 or more or rarely 1 only in each spurious cell (half-carpel). Fruit hard and indehiscent or capsular. Seeds without albumen.

(*Verbenaceæ*, *Labiatæ*, and *Plantaginæ*, completing the *Monopetalæ*, will be given in the fifth volume.)

FLORA AUSTRALIENSIS.

Order LXIII. STYLIDIEÆ.

Calyx-tube adnate to the ovary, the limb of 5 divisions, all free or more or less united in 2 lips, the upper one consisting of 3, the lower of 2. Corolla usually irregular, deeply divided into 5 lobes, of which one (the lowest), called the *labellum*, much smaller or very different from the others, or rarely the corolla as well as the calyx regularly 5- or 6-lobed. Stamens 2, the filaments connate with the style in a column free from the corolla; anthers sessile at the top of the column, 2-celled, the cells at length divaricate. Style or stigma terminal, entire or 2-lobed, concealed between the anthers or protruding from them. Ovary 2-celled or 1-celled except quite at the base, with many ovules attached to the centre of the dissepiment, surmounted frequently by 1 or 2 glands at the base of the style. Capsule opening from the top downwards in 2 valves parallel to the dissepiment. Seeds numerous or rarely solitary by abortion, very small, with a minute embryo in a fleshy albumen.—Herbs or rarely undershrubs. Leaves radical or scattered or collected in whorl-like tufts. Flowers hermaphrodite or very rarely unisexual, in terminal racemes or thyrsoid or corymbose panicles, rarely reduced to spikes or to single flowers, the primary inflorescence usually centripetal, the secondary often or sometimes the whole inflorescence centrifugal.

A small Order, chiefly Australian, a very few species being found in tropical Asia, or in New Zealand and Antarctic America, and these all belong to Australian genera except the two species of *Helophyllum*, Hook. f. The Order is very nearly allied to *Campanulaceæ*, and some species have quite the habit of some *Lobelias*, but they constantly differ in the close union of the filaments with the style.

Corolla irregular, the 5th lobe or labellum very different from the others.
  Column elongated and folded (usually elastic). Labellum small or narrow . . . . 1. STYLIDIUM.
  Column erect, usually short, not elastic. Labellum with a hood-shaped lamina covering the anthers or elastically reflexed . . . 2. LEVENHOOKIA.
Corolla regular or nearly so, the lobes all similar . . . . 3. FORSTERA.

1. STYLIDIUM, Swartz.

(Forsteropsis, *Sond.*)

Calyx-lobes 5, often more or less united in 2 lips. Corolla irregular, 1 of the lobes or labellum much smaller and turned down or rarely nearly as long and curved upwards, the other 4 ascending in pairs. Column elongated and bent down or folded, elastic in most of the species if not in all. Stigma undivided. Ovary 2-celled.—Habit and foliage those of the Order. Flowers in racemes, panicles or corymbose cymes on terminal peduncles or radical scapes.

A genus comprising nearly the whole Order, and entirely Australian, with the exception of one species extending into tropical Asia, and another East Indian species not yet identified with certainty with any Australian one. The majority of the species form a rosette or spreading tuft of radical leaves, from the midst of which springs the scape. Sometimes the following year the new leaves and scape are close upon the old ones, forming a dense, tufted stock, the bases of the leaves sometimes assuming a bulbous appearance; in others, one or two short stems are formed above the old tuft, each crowned by a new rosette and scape, and sometimes several successive tufts of leaves, separated by short stems or branches, may be observed; these are termed *proliferous* stems or branches, and occasionally emit adventitious roots from several of the lower tufts. In a few species the leaves are all, or only the small upper ones, in almost regular whorls; and in a few others they are alternate or scattered without forming tufts. The inflorescence in different species shows every gradation, from the simple raceme or raceme-like panicle to the corymb or to the dichotomous cyme with sessile or pedicellate flowers in the forks. The precise form of the corolla, the direction of its lobes in the expanded flower, and the small scales or glandular appendages in the throat or at the base of the labellum may be constant in many cases, and might serve for good specific characters; but these parts are so delicate that there is great uncertainty in describing them from dried specimens. Different botanists have described them differently in the same species, and I have myself found considerable discrepancies in this respect in different flowers even of the same specimens; the characters founded on them must therefore not be absolutely relied on. The colour of the flower is also said to be constant in some species, and has been made use of as a specific character since the time of Brown; but it appears to be variable in other species, and in most cases it is either unknown, or only given in vague and often contradictory notes of collectors. It is only a botanist resident on the spot that can complete the specific characters in the above respects.

SECT. I. **Tolypangium.**—*Capsule globular-ovoid, obovoid or oblong.*

SERIES I. **Squamosæ.**—*Stock tufted, rarely proliferous. Leaves radical, intermixed with lanceolate, scarious scales, which are wanting in all the following sections. Scapes leafless, except a few scattered bracts.—Western species.*

Inflorescence a long raceme, simple or slightly branched at the base.
  Leaves obovate or orbicular, not above 1 in. long. Scape
    glabrous . . . . 1. *S. carnosum.*
  Leaves linear or lanceolate, 4 in. to 1 ft. long. Scape hairy . 2. *S. pilosum.*
Inflorescence a loose thyrsoid panicle. Leaves linear. Scape hairy.
  Leaves usually glabrous. Scapes (with the inflorescence) much
    longer than the leaves . . . . 3. *S. reduplicatum.*
  Leaves pubescent. Scapes but slightly exceeding the leaves . . 4. *S. scabridum.*
Inflorescence short, compact and spike-like. Leaves linear.
  Spike oblong, hirsute as well as the scape. Bracts small . . . 5. *S. hirsutum.*
  Spike contracted into a depressed head. Bracts ½ in. long with
    scarious margins . . . . 6. *S. crossocephalum.*
  (In a few of the *Lineares* and other series the bases of the old leaves persist on the stock in the form of scales, always much shorter and more rigid than the true scales of the *Squamosæ.*)

SERIES II. **Peltigeræ.**—*Stock tufted or proliferous-branched. Leaves radical. Flowers in a dense spike or cluster with numerous bracts produced below their insertion, the scapes otherwise leafless.—Western species.*

Scapes erect and rush-like or long and twining. Spike oblong.
  Flowers almost sessile . . . . 7. *S. junceum.*

Scapes 1 to 2 in.. Flowers sessile in a terminal globular cluster of
  small, leaf-like bracts . . . . 8. *S. guttatum.*
Stock or stem proliferous-branched and rooting at the tufts.
  Flowers pedicellate in terminal, globular clusters of small, leaf-
  like bracts . . . . 9. *S. repens.*

SERIES III. **Lineares.**—*Perennials with a tufted or shortly proliferous stock or
rarely annuals. Leaves radical, linear or rarely linear-lanceolate. Scapes leafless except
a few scattered bracts. Inflorescence racemose, paniculate or corymbose.*

Calyx-lobes united in 2 lips. Inflorescence nearly simple, long and
  narrow.—Eastern species.
  Leaves linear or linear-lanceolate, 2 to 9 in. long. Fruit ovoid-
    oblong . . . . 10. *S. graminifolium.*
  Leaves linear-subulate, under 2 in. long. Fruit narrow-oblong.
    Flowers small . . . . 11. *S. lineare.*
Calyx-lobes all free. Inflorescence thyrsoid or racemose. Capsule
  ovoid or oblong.—Western species.
  Leaves 2 to 6 in. long. Scape hirsute. Inflorescence thyrsoid . 12. *S. elongatum.*
  Leaves under 2 in. long.
    Leaves very narrow linear, mucronate. Raceme simple, loose,
      glandular . . . . 13. *S. spinulosum.*
    Leaves linear but flat, obtuse or mucronate-acute, densely
      tufted. Raceme loose, usually simple . . . . 14. *S. cæspitosum.*
    Leaves linear or slightly cuneate, usually erect, obtuse or
      rather acute. Raceme simple or nearly so. Stock often
      slightly elongated.
      Raceme nearly glabrous. Flowers purple or rarely yellow . 15. *S. violaceum.*
      Raceme glandular-pubescent. Flowers usually yellow . . 16. *S. luteum.*
    Leaves densely tufted, linear or slightly cuneate, with a fine,
      usually hair-like point.
      Racemes mostly simple, slightly glandular-pubescent . . 17. *S. piliferum.*
      Panicle narrow thyrsoid or rarely almost a simple raceme,
        clothed with yellowish, glandular hairs . . . . . . 18. *S. ciliatum.*
Calyx-lobes free. Inflorescence corymbose-paniculate. Capsule
  ovoid or oblong.—Eastern species.
  Perennial. Leaves in a dense globular tuft, with fine, hair-like
    points. Capsule 2 to 2½ lines long . . . . 19. *S. sobuliferum.*
  Annual. Leaves linear-filiform. Capsule 1 to 1½ lines long . 20. *S. Floodii.*
Calyx-lobes free. Inflorescence corymbose-paniculate. Capsule
  globular. Annual. Western species. . . . . 21. *S. dispermum.*

SERIES IV. **Androsaceæ.**—*Small annuals, with radical rosulate leaves and few-
flowered, leafless scapes. Labellum long and ascending. Stigma stipitate between the
anthers (sessile in all other series). Capsule globular.—Southern species.*

Corolla-tube spurred . . . . 22. *S. calcaratum.*
Corolla without any spur . . . . 23. *S. perpusillum.*

SERIES V. **Spathulatæ.**—*Perennials with a tufted or proliferous stock or rarely
annuals. Leaves radical or terminal, from linear-spathulate to obovate. Scapes leafless
except a few scattered bracts.*

Tropical species. Leaves usually thin.
  Stock thick and woolly. Leaves pubescent, with a hair-like
    point. Panicle narrow-thyrsoid . . . . 24. *S. eriorhizum.*
  Annuals or with a slender stem below the terminal tuft of
    leaves.
    Raceme long and simple . . . . 25. *S. debile.*
    Panicle loose.
      Scape with long, spreading, glandless hairs . . . . . 26. *S. floribundum.*

Scape glabrous or with short, glandular hairs . . . . 27. *S. leptorhizum.*
Western species.   Perennials with firm leaves.
Leaves linear-cuneate, thick and very obtuse.   Panicle loosely
   thyrsoid, glandular-pubescent . . . . 28. *S. assimile.*
Leaves oblanceolate to obovate, spathulate, flat.   Raceme simple
   or nearly so.
Leaves pubescent.
Stock proliferous-branched.   Leaves under ½ in. long.
   Scapes short, few-flowered . . . . 29. *S. rupestre.*
Leaves radical, ½ to 1½. in. long.   Raceme long.
   Leaves entire . . . . 30. *S. spathulatum.*
   Leaves toothed . . . . 31. *S. Barleei.*
Leaves hirsute with long hairs.   Raceme long . . . . . 32. *S. lineatum.*
Leaves glabrous . . . . 33. *S. glaucum.*

SERIES VI. **Diversifoliæ.**—*Perennials with tufted or rosulate radical leaves.   Scapes with* 1, 2, 3, *or* 4 *whorls of linear leaves, much smaller than the radical ones.   Western species.*

Radical leaves spathulate, under 2 in.   Whorl-leaves very small
   and few.   Raceme simple or nearly so.
   Leaves obscurely striate . . . . 34. *S. amœnum.*
   Leaves with marked striæ . . . . 35. *S. striatum.*
Radical leaves obovate-orbicular, thick.   Whorl-leaves subulate . 36. *S. diversifolium.*
Radical leaves narrow-spathulate, 2 to 4 in.   Whorl-leaves narrow.
   Panicle compact, thyrsoid . . . . 37. *S. articulatum.*
Radical leaves linear or oblanceolate.
   Whorls several . . . . 38. *S. Brunonianum.*
   Whorls 1 or 2, very minute . . . . 16. *S. luteum.*
Radical leaves linear subulate.   Whorl-leaves similar but smaller . 39. *S. diuroides.*

SERIES VII. **Verticillatæ.**—*Stems elongated, simple or branched.   Leaves all linear in distant whorls, without larger radical ones.   Western species.*

Inflorescence glabrous or nearly so . . . . 40. *S. scandens.*
Inflorescence hirsute, with spreading hairs . . . . 41. *S. verticillatum.*

SERIES VIII. **Sparsifoliæ.**—*Undershrubs or shrubs, with linear spreading leaves scattered along the branches and not collected in radical or terminal tufts.*

Western species.   Branches covered with the adnate cartilaginous
   bases of the petioles . . . . 42. *S. glandulosum.*
Eastern species.   Petioles without adnate bases . . . . . 43. *S. laricifolium.*

SERIES IX. **Imbricatæ.**—*Stems branching, slender but hard, covered with small imbricate, almost scale-like leaves, not collected in radical or terminal tufts.   Western species.*

Leaves with scarious margins.   Flowers few, sessile within the last
   leaves.   Bracts transparent.   Corolla-tube short . . . . . 44. *S. Preissii.*
Leaves with scarious margins.   Flowers in a short terminal sessile
   spike or raceme.   Bracts leaf-like.   Corolla-tube exserted . . 45. *S. imbricatum.*
Leaves without scarious margins.   Flowers few, in a pedunculate
   head or short raceme.   Bracts leaf-like.   Corolla-tube short . 46. *S. adpressum.*

SECT. II. **Nitrangium.**—*Capsule linear or very narrow-oblong.*

SERIES X. **Tenellæ.**—*Slender annuals.   Leaves small or thin, alternate or scattered, the lower ones sometimes more crowded but not distinctly tufted or rosulate.*

Leaves linear, chiefly in the lower part of the stem.   Bracts very
   small.
Capsule 2 to 4 lines long.   Western species.
   Flowers pedicellate.

Corolla scarcely half as long again as the calyx-lobes. La-
bellum inappendiculate . . . . 47. *S. despectum.*
Corolla twice as long as the calyx-lobes. Labellum appen-
diculate . . . . 48. *S. utricularioides.*
Flowers sessile. Corolla small . . . . 49. *S. pygmæum.*
Capsule ½ to 1 in. long.
Flowers pedicellate. Stems erect. Western species . . . 50. *S. longitubum.*
Flowers sessile. Stems very slender or diffuse. Tropical
species.
Capsule ½ in. long, not beaked . . . . 51. *S. diffusum.*
Capsule ¾ to 1 in. long, more or less distinctly beaked . . 52. *•S. fissilobum.*
Leaves scattered along the stem, contracted at the base, the
lower ones not larger, the upper ones passing into the bracts.
Tropical species.
Floral leaves or bracts opposite. Capsule sessile . . . . . 53. *S. alsinoides.*
Floral leaves or bracts alternate. Capsule pedicellate . . . 54. *S. tenerrimum.*

SERIES XI. **Corymbulosæ.**—*Slender annuals or small perennials. Leaves radical,
rosulate or tufted. Scapes rarely exceeding 6 in. Flowers corymbose or sessile in the
forks or along the scape or its branches, or solitary.*

Leaves radical, very small, oblong-linear or spathulate, not form-
ing a bulb.
Flowers corymbose. Capsule oblong-linear. Western species . 55. *S. brachyphyllum.*
Flowers 1 or 2. Capsule very narrow-linear. Tropical species 56. *S. capillare.*
Leaves radical, not forming a bulb, the lamina thin, obovate-orbi-
cular. Tropical species.
Flowers in a loose, irregularly-corymbose panicle. Capsule
narrow-linear.
Calyx-lips undivided, short and broad. Corolla-lobes nearly
equal . . . . 57. *S. rotundifolium.*
Calyx lower lip 3-partite. Corolla with 2 lobes much longer
and bifid.
Very slender, slightly glandular. Flowers white . . . 58. *S. schizanthum.*
Rather more rigid and glandular. Flowers pink . . . 59. *S. lobuliflorum.*
Flowers sessile along the scape or its branches. Capsule shortly
linear . . . . 60. *S. uliginosum.*
Leaves radical, small, linear or spathulate, the dilated base of the
petioles forming a little brown bulb. Western species.
Flowers in a small regular corymb. Capsule linear . . . . 61. *S. pulchellum.*
Flowers in a loose, irregular corymb. Capsule oblong.
Scape without any or very few scattered bracts below the in-
florescence . . . . 62. *S. petiolare.*
Scape with a whorl of minute bracts below the inflorescence . 63. *S. emarginatum.*
Leaves radical, linear, densely tufted, not bulbous, ½ to 2 in. long.
Western species.
Flowers in a compact corymb, with the central one sessile.
Leaves not very narrow. Bracts oblong or ovate. Calyx-
lobes broad, very obtuse . . . . 64. *S. corymbosum.*
Leaves very narrow. Bracts and calyx-lobes linear . . 65. *S. lepidum.*
Flowers in a loose, divaricate, irregularly corymbose panicle . 66. *S. streptocarpum.*
Flowers solitary on each scape . . . . 67. *S. uniflorum.*
Stem shortly developed below the terminal tuft of leaves and pe-
duncles. Tropical species (except *S. brachyphyllum*).
Leaves linear-subulate. Peduncles 1-flowered . . . . . 68. *S. pedunculatum.*
Leaves oblanceolate or spathulate. Stem thick and hard.
Flowers corymbose . . . . 69. *S. pachyrhizum.*
Leaves petiolate, orbicular, membranous (½ to 1 in.).
Flowers sessile along the branches of the peduncles . . . 70. *S. muscicola.*

Leaves very small, narrow.   Flowers pedicellate . . . . 55. *S. brachyphyllum.*

(See also 77. *S. bulbiferum* and 78. *S. breviscapum,* which have the inflorescence sometimes almost corymbose, but a proliferous-branched stock.)

SERIES XII. **Thyrsiformes.**—*Perennials with a tufted or proliferous-branched stock
or stem, with radical or terminal tufts of leaves. Flowers in an oblong or elongated
thyrsoid panicle or raceme.*

Stock hard, at length horizontal.   Leaves radical, elongated, thick,
   oblanceolate or spathulate.
  Scape 1 to 2 ft. high, glabrous at the base.   Inflorescence long
    and narrow.   Capsule 5 to 8 lines long . . . . 71. *S. crassifolium.*
  Scape ½ to 1 ft., glandular-pubescent or villous.   Inflorescence
    thyrsoid.   Capsule 9 to 10 lines long . . . . 72. *S. pycnostachyum.*
Stock tufted.   Leaves linear.
  Leaves rather broad, with a very fine joint.   Inflorescence glan-
    dular-pubescent.   Capsule 4 to 6 lines long . . . . . . 73. *S. pubigerum.*
  Leaves narrow-linear, obtuse or shortly pointed.
    Inflorescence nearly glabrous.   Capsule 3 lines long . . . 74. *S. canaliculatum.*
    Inflorescence glandular-pubescent or villous.   Capsule 4 to 6
     lines . . . . 75. *S. leptophyllum.*
Stock or stem proliferous-branched.   Leaves narrow-linear.
  Inflorescence thyrsoid, many-flowered.   Upper leaves 1 in. long
    or more . . . . 76. *S. dichotomum.*
  Inflorescence short, few-flowered or compact.   Leaves under
    ½ in. long.
    Flowers few, loosely racemose.   Capsule long, linear . . . 77. *S. bulbiferum.*
    Flowers in a compact cluster.   Capsule lanceolate-linear . . 78. *S. breviscapum.*
  Inflorescence narrow-racemose.   Branches of the stem slender.
    Leaf-tufts woolly at the base . . . . 79. *S. eglandulosum.*

SECT. III. **Rhynchangium.**—*Capsule lanceolate or linear, contracted into a slender
beak.   Perennials.   Leaves linear, scattered along the stem, the upper ones usually
forming a terminal tuft.*

Both cells of the capsule equal and fertile . . . . 80. *S. fasciculatum.*
Upper cell of the capsule scarcely half as broad as the more per-
  fect one.   Capsule much falcate.   Raceme usually simple.
  Raceme several inches long.   Beak of the capsule short . . . 81. *S. falcatum.*
  Raceme sessile, scarcely exceeding the leaves.   Beak of the
    capsule long . . . . 82. *S. rhynchocarpum.*
Upper cell of the capsule reduced to a filiform rib . . . . . 83. *S. adnatum.*

(52. *S. fissilobum,* a slender annual, with very small leaves, has also the linear capsule
more or less distinctly beaked.)

SECT. I. TOLYPANGIUM, *Endl.*—Capsule globular, ovoid, obovoid or
oblong.

SERIES 1. SQUAMOSÆ.—Stock tufted, very rarely proliferous, the radical
leaves intermixed with lanceolate, scarious scales, often enclosing the young
shoots, and which appear to be abortive petioles.   Scapes leafless or with
a few small, scattered bracts.

1. **S. carnosum,** *Benth. in Hueg. Enum. 71.*   Stock tufted, at length
thick, but not proliferous, with a few linear-lanceolate, acuminate, scarious
scales both outside and inside the leaves.   Leaves all radical, from obovate
to narrow-lanceolate, obtuse, rather thick and glabrous, from ½ in. to above
1 in. long, besides the long petiole, which is more or less dilated and

scarious at the base.   Scape 1 to 2 ft. long, glabrous below the inflorescence, with a very few small scale-like leaves or bracts, the upper half occupied by the slender, glandular-pubescent raceme.   Flowers small, on short pedicels or the lower ones 2 or 3 together on a short peduncle.   Calyx about 2 lines long, the lobes free, lanceolate or linear, rather shorter than the tube. Corolla " whitish," the upper lobes nearly 3 lines long, the throat with small glandular appendages, the labellum small, ending in a fine point, without appendages at the base.   Capsule nearly globular, 1½ lines diameter.—DC. Prod. vii. 332 ; Sond. in Pl. Preiss. i. 370 ; *S. leptostachyum,* Lindl. Swan Riv. App. 28 ; Sond. l. c.

W. Australia. King George's Sound to Swan River, *Huegel, Drummond, 1st Coll. n.* 530 ; *Preiss, n.* 2233, 2234, *Oldfield.* The species has much of the aspect of *S. diversifolium,* from which it differs in the presence of the scarious scales on the stock, in the less bulbous appearance of the base of the leaves, and in the small leaves on the scape very few and not collected in whorls.

2. **S. pilosum,** *Labill. Pl. Nov. Holl.* ii. 63. *t.* 213.   Stock tufted, at length thick but not proliferous, with lanceolate, scarious scales, 1 to 2 in. long, intermixed with the leaves.   Leaves radical, broadly linear or lanceolate, acute, nearly flat, of a firm consistence and glabrous, from 4 or 5 in. to 1 ft. long, including the long petiole, which is not dilated at the base.   Scapes leafless, except the bracts, 1 to 2 ft. long, more or less clothed with long spreading hairs, intermixed in the inflorescence with short glandular ones, the upper moiety forming a long raceme, simple or shortly branched at the base.   Flowers rather large, all pedicellate.   Calyx about 3 lines long ; lobes shorter than the tube, free.   Corolla-lobes 3 to 4 lines long, the throat with clavate appendages ; labellum small, obtuse, with basal appendages. Capsule ovoid, glandular-villous, 4 to 6 lines long.—R. Br. Prod. 567 ; DC. Prod. vii. 332 ; *Candollea pilosa,* Labill. in Ann. Mus. Par. vi 453. t. 63 ; *S. longifolium,* Rich. in Pers. Syn. ii. 210 ; *S. plantagineum,* Sond. in Pl. Preiss. i. 371.

W. Australia. King George's Sound, *Labillardière, R. Brown,* and others ; *Preiss, n.* 2298 ; *Drummond, n.* 79, *2nd Coll. n.* 263, *5th Coll. n.* 350, also, perhaps, *n.* 351, in which the raceme is shorter, but it seems to belong to this rather than to *S. reduplicatum.*

3. **S. reduplicatum,** *R. Br. Prod.* 568. Stock thick, rarely proliferous, with broadly lanceolate scarious scales round the leaves.   Leaves all radical, linear, acutely acuminate, narrowed into a long petiole not dilated at the base, from rather broad and nearly flat to very narrow with revolute margins, glabrous or minutely glandular-pubescent, 3 or 4 in. to nearly 1 ft. long. Scapes leafless, ½ to 1 ft. or rarely 1½ ft. long, with more or less of spreading hairs intermixed on the inflorescence with glandular pubescence. Flowers of *S. pilosum* but usually larger, of a yellowish-white or pale pink, in a short loose raceme, sometimes reduced to 3 or 4 flowers, the lower ones on long pedicels, sometimes numerous in a shortly pyramidal or almost corymbose panicle.   Calyx of *S. pilosum.*   Corolla-lobes more unequal, the 2 larger ones from 6 to 9 lines long, connate to the middle.   Capsule ovoid, usually much smaller than in *S. pilosum* (2 to 3 lines long).—DC. Prod. vii. 332 ; *S. schœnoides,* DC. Prod. vii. 782 ; Sond. in Pl. Preiss. i. 372 ; *S. hebegynum,* DC. Prod. vii. 782 ; *S. caricifolium,* Lindl. Swan Riv. App. 28 ; Sond.

l. c. 372 : *S. pilosum*, Sond. l. c. 371; Lindl. Bot. Reg. 1842, t. 41, not of Labill.; *S. affine*, Sond. l. c. 371 ; *S. Drummondii*, Grah. in Edinb. New Phil. Journ. xxx. 208, and in Maund. Botanist, v. t. 213.

W. Australia. From King George's Sound to Swan River, *Baxter, Collie, Oldfield, Preiss, n.* 2291, 2292, *Drummond, n.* 7, 525, 526, *2nd Coll. n.* 276; and eastward to Lucky Bay, *R. Brown;* and Cape Le Grand and Cape Arid, *Maxwell.*—The Swan River specimens are usually larger, and have often broader and flatter leaves; those from the eastern drier parts of the area are usually smaller, with narrower, more revolute, almost terete leaves, and fewer larger flowers, but exceptions are found to both. In some eastern specimens there is a tendency to prolification of the stock. Some of the smaller specimens, when reduced to very few flowers, have almost the aspect of *S. hirsutum*, but with much larger flowers.

4. **S. scabridum,** *Lindl. Swan Riv. App.* 28. Very closely allied to the pubescent forms of *S. reduplicatum*, and perhaps a variety of that species. It is a smaller plant, the leaves rarely above 3 in. long, with shorter points and always pubescent; the scape sometimes not exceeding the leaves, rarely twice as long, with a loosely corymbose panicle, the pedicels and calyxes often very hirsute, the bracts and calyx-lobes narrow-linear or subulate ; the flowers smaller than in *S. reduplicatum*, and the capsule shorter.—Sond. in Pl. Preiss. i. 372 ; *S. laxiflorum*, DC. Prod. vii. 782 ; *S. leptocalyx*, Sond. in Pl. Preiss. i. 373.

W. Australia. Swan River, *Drummond, 1st Coll. n.* 532, 533; Princess Royal Harbour, *Preiss, n.* 2289 *(Sonder)* ; also *Drummond, n.* 402.

5. **S. hirsutum,** *R. Br. Prod.* 568. Stock tufted, not so thick as in *S. reduplicatum*, with narrow-lanceolate, scarious scales amongst the leaves. Leaves narrow-linear, acutely acuminate, sometimes all under 2 in., sometimes 6 to 8 in. long, glabrous or glandular-pubescent. Scapes 6 in. to above 1 ft. high, leafless, with spreading hairs often intermixed on the inflorescence with glandular pubescence. Flowers nearly sessile, forming a dense, oblong, spike-like raceme, rarely above 1 in. long and very hairy. Calyx about $3\frac{1}{2}$ lines long, the lobes free or the 2 lower broader ones united at the base. Corolla pink or red, the larger lobes nearly equal, the throat appendages conspicuous ; labellum lanceolate, with crisped margins and short appendages or sometimes none. Capsule ovoid, from 3 to 4 lines long.— DC. Prod. vii. 332 ; Sond. in Pl. Preiss. i. 372 ; Bot. Mag. t. 3194.

W. Australia. King George's Sound and adjoining districts, *R. Brown* and others, *Drummond, n.* 113, *3rd Coll. n.* 166, *5th Coll.* 353 (or 359 ?), *Preiss, n.* 2294; also Swan River, *Drummond, 1st Coll.*

6. **S. crossocephalum,** *F. Muell. Fragm.* vi. 5. Stock tufted, with a few long lanceolate scales intermixed with the leaves, either entirely scarious or with a leaf-like centre. Leaves radical, narrow-linear, acute or almost obtuse, scabrous-pubescent like those of *S. scabridum*, 2 to 3 or rarely 4 in. long. Scapes simple and leafless, quite glabrous, longer than the leaves. Flowers sessile, in a short, dense, depressed head, surrounded by lanceolate-acuminate bracts, about $\frac{1}{2}$ in. long, very fine-pointed, with rather broad, scarious, slightly-ciliate margins. Flowers only seen in bud, and then not so long as the bracts. Calyx-lobes linear, acuminate, with scarious margins, the 2 lower ones united to the middle, the others free. Corolla with broad

appendages to the throat opposite the larger lobes, and very small ones opposite to the smaller lobes ; labellum linear-lanceolate, with a few glandular hairs. Ovary 1-celled, with several ovules on a short, basal placenta.

W. Australia, *Drummond.*

SERIES 2. PELTIGERÆ.—Stock tufted or proliferous-branched. Leaves radical, linear. Flowers in dense spikes or clusters, the bracts numerous, and more or less produced below their insertion into a short spur or appendage.

7. **S. junceum,** *R. Br. Prod.* 569. Quite glabrous or with a very few glandular hairs on the ovary. Stock at length thick and almost woody. Leaves radical, linear, acute, 1 to 1¼ in. long, sometimes all withered away at the time of flowering. Stem or scape erect and rush-like, and ¼ to ½ ft. high, or more or less flexuose or twining, and then lengthening to 2 or 3 ft. or more, leafless except the bracts. Flowers small, nearly sessile in a short, dense, spike-like raceme. Bracts lanceolate, acute, with scarious margins, produced at the base below their insertion, imbricate before flowering. Calyx-lobes narrow and very acute, all free. Corolla white, pink or pale yellow, with small, glandular appendages to the throat, the labellum lanceolate, without appendages. Capsule ovoid, the calyx-lobes more deciduous than in most species, and with a prominent rib on each side at the suture of the valves.—DC. Prod. vii. 334 ; Sond. in Pl. Preiss. i. 377 ; *S. scariosum,* DC. Prod. vii. 783.

W. Australia. In very wet places common from King George's Sound to Swan River, *R. Brown, Huegel,* and others, *Drummond, 1st Coll. n.* 539, *3rd Col[?]. n.* 179, *also* 26, 41, 132, *and in some sets* 171, *Preiss, n.* 2264. To the eastward, near Lake Leven, *Maxwell,* these specimens more elongated and twining, with rather larger flowers than any others, although some of Drummond's are nearly as much so.

8. **S. guttatum,** *R. Br. Prod.* 571. Quite glabrous, except a few glandular hairs on the inflorescence. Stock tufted or very shortly proliferous. Leaves all radical, narrow-linear, acute or obtuse, ½ to 1 in. long. Scape 1½ to 2 in. long, leafless below the inflorescence, bearing a cluster or head of sessile flowers surrounded by and intermixed with numerous oblong-linear, leaf-like bracts, which are shorter than the calyx-tube, and shortly produced at the base below their insertion. Calyx about 2½ lines long, the lobes free, rather acute, as long as the tube. Corolla with appendages to the throat and to the very narrow, acute labellum. Capsule narrow-oblong, contracted at the top, about 2 lines long, but not seen quite ripe, and then perhaps broader.—DC. Prod. vii. 336 ; Sond. in Pl. Preiss. i. 387 ; *S. androsaceum,* DC. Prod. 783.

W. Australia. From King George's Sound, *R. Brown* and others, to Vasse river, *Preiss, n.* 2243, and Swan River, *Drummond, 1st Coll. n.* 540, *also n.* 36 *and* 40, and eastward to Cape Arid, *Maxwell.*

9. **S. repens,** *R. Br. Prod.* 571. Quite glabrous. Stock forming very slender proliferous branches, and often rooting at the tufts. Leaves at the base and ends of the branches densely tufted, linear, acute, under ¼ in. long, with a few also scattered along the branches, the floral ones or nearly all

shortly produced at the base below their insertion, and often with scarious margins. Flowers small, intermixed with the leaves or bracts of the terminal tufts, on slender pedicels of 2 to 4 lines. Calyx-lobes free, very acute. Corolla red or white, with appendages to the labellum and not to the throat in the specimens examined, but just the contrary as observed by Brown. Capsule narrow ovate, 1 line long.—DC. Prod. vii. 336 ; *S. radicans*, Sond. in Pl. Preiss. i. 381.

W. Australia. King George's Sound, *R. Brown* and others, to Swan River, *Preiss, n.* 2299, 2300, also *Drummond, 2nd Coll. n.* 272, 273, *3rd Coll. n.* 171.

SERIES 3. LINEARES.—Leaves linear or rarely linear-lanceolate, all radical or in a tuft at the end of a short, proliferous stem, without intermixture of scarious scales. Scape leafless, except the small bracts of the inflorescence, and sometimes a very few, small, bract-like leaves scattered below the inflorescence.

10. **S. graminifolium,** *Swartz ; DC. Prod.* vii. 333. Glabrous or nearly so, except the glandular-pubescent inflorescence. Stock tufted or shortly proliferous, rarely lengthening to 4 or 5 in. Leaves linear, rather rigid, acute or obtuse, nearly flat, varying much in breadth, the margins entire or with minute, cartilaginous denticulations, dilated and more or less scarious at the base, sometimes not exceeding 2 in., in luxuriant specimens 6 to 9 in. long. Scapes from ½ to 1½ ft., the upper ¼ to ½ occupied by a narrow, simple raceme or interrupted spike. Bracts small. Flowers nearly sessile or shortly pedicellate. Calyx 3 or 4 lines long, the lobes broad and obtuse, united nearly to the top into two lips. Corolla-lobes nearly equal, the labellum rather long, obtuse, the appendages variable. Capsule ovoid-oblong, ¼ to nearly ½ in. long.—R. Br. Prod. 568 ; Labill. Pl. Nov. Holl. ii. 65. t. 215 ; Hook. f. Fl. Tasm. i. 235 ; Bot. Reg. t. 90 ; Bot. Mag. t. 1918 ; *Candollea serrulata*, Labill. in Ann. Mus. Par. vi. 454. t. 64 ; *Stylidium serrulatum*, Rich. in Pers. Syn. ii. 210 ; *Ventenatia major*, Sm. Exot. Bot. ii. 13. t. 66 ; *S. canaliculatum*, Poir. Dict. Suppl. v. 412.

Queensland. Moreton Bay, *A. Cunningham.*
N. S. Wales. Common in grassy lands Port Jackson to the Blue Mountains, *R. Brown, Sieber, n.* 253, and others ; northward to Hastings and Clarence rivers, *Beckler ;* New England, *C. Stuart ;* and southward to Illawarra and Twofold Bay, *A. Cunningham.*
Victoria. From Port Phillip to the Australian Alps, *F. Mueller* and others ; Portland, *Allitt ;* Glenelg river, *Robertson ;* Wimmera, *Dallachy.*
Tasmania. Very abundant throughout the colony, from the level of the sea to an elevation of 4000 ft., *J. D. Hooker.*
S. Australia. Port Adelaide, *Blandowsky ;* foot of Mount Barker, *Whittaker ;* Lofty Range, *F. Mueller.*
S. armeria, Labill. Pl. Nov. Holl. ii. 66. t. 216 ; DC. Prod. vii. 333 ; Lem. Jard. Fleur. iii. t. 286 ; *Candollea armeria*, Labill. in Ann. Mus. Par. vi. 455, is the same plant with rather broader leaves. *S. melastachys*, R. Br. Prod. 568 ; DC. l. c. 333, separated at first on account of the leaves being almost entirely without the minute denticulations, is not distinguishable even as a variety, the leaves varying in this respect on the same specimen.
S. umbellatum, Labill. Pl. Nov. Holl. ii. 66. t. 217 ; DC. Prod. vii. 332 ; *Candollea umbellata*, Labill. in Ann. Mus. Par. vi. 456 ; *S. polystachyum*, Rich. in Pers. Syn. ii. 210, is probably, as suggested in Hook. f. Fl. Tasm. i. 235, an accidental monstrosity of *S. graminifolium.*

11. **S. lineare,** *Swartz ; DC. Prod.* vii. 333.　Glabrous, except the inflorescence, like *S. graminifolium,* and resembles that species in its linear leaves, dilated towards the base, racemose inflorescence and undivided calyx-lips, but it is a smaller species, the leaves very narrow, acute, usually incurved, not exceeding 2 in. and often much shorter.　Scapes slender, the raceme 2 to 4 in. long, and much looser than in *S. graminifolium.*　Flowers smaller, on pedicels of 1 to 2 lines, the calyx-tube linear, and the ripe capsule much narrower than in that species.—R. Br. Prod. 568 ; *Ventenatia minor,* Sm. Exot. Bot. ii. 15. t. 67 ; *S. planifolium,* Poir. Dict. Suppl. v. 412.

N. S. Wales. Marshes about Port Jackson, *R. Brown* and others ; Blue Mountains, *A. Cunningham.*　F. Mueller proposes to unite this with *S. graminifolium.*

If the calyx-lips of *S. setaceum,* Labill. Pl. Nov. Holl. ii. 65 ; DC. l. c. 333 ; *Candollea setacea,* Labill. in Ann. Mus. Par. vi. 455, from the Terre Van Leeuwin, are really undivided as described, they would indicate a species allied to *S. lineare ;* in other respects his description answers to *S. spinulosum.*

12. **S. elongatum,** *Benth.*　Stock tufted, the broad bases of the old leaves giving it at length a bulbous aspect.　Leaves all radical, linear, acute or nearly so, flaccid, 4 to 8 in. long, usually glabrous.　Scape rarely twice as long as the leaves, hirsute, with spreading hairs intermixed on the inflorescence with glandular pubescence.　Panicle long and narrow, but many-flowered, almost all the peduncles 3- to 7-flowered.　Calyx-lobes free.　Corolla without any appendages to the throat, but long ones to the labellum.　Capsule ovate.

W. Australia, *Drummond ;* Champion Bay, *Oldfield.　Drummond's 4th Coll. n.* 170, appears to be the same species, but with a shorter, looser panicle, almost like that of *S. assimile.*　In the ripe capsule there are usually only 1 or 2 perfect seeds in each cell.

13. **S. spinulosum,** *R. Br. Prod.* 569.　Stock tufted or shortly proliferous, more slender than in *S. cæspitosum.*　Leaves very narrow-linear, mucronate, bordered by minute, cartilaginous serratures or short hairs, mostly under 1 in. long.　Scapes about 4 to 8 in. high, sprinkled with a few glandular hairs from the base.　Racemes usually simple, loose, glandular-pubescent or sometimes almost villous, resembling those of *S. cæspitosum.*　Flowers white, with red streaks outside like those of *S. cæspitosum,* but the appendages of the throat of the corolla prominent, and none on the labellum according to Brown ; these characters, however, may not be constant.　Capsule rather narrower than in *S. cæspitosum.*—DC. Prod. vii. 333 ; Sond. in Pl. Preiss. i. 373.

W. Australia. King George's Sound and adjoining districts, *R. Brown, Baxter, Preiss, n.* 2279, *Drummond, 3rd Coll. n.* 176, *F. Mueller.*

S. aciculare, Sond. in Pl. Preiss. i. 373, from D'Urville's collection, which I have not seen, appears from the character given to be the same species.　The character given of *S. setaceum,* Labill., above quoted, appears also to answer to that of *S. spinulosum,* except that the calyx-lips are said to be undivided as in *S. graminifolium.*

14. **S. cæspitosum,** *R. Br. Prod.* 569.　Glabrous, except a few glandular hairs on the calyx and sometimes on the pedicels, more rarely on the foliage.　Stock tufted, scarcely proliferous.　Leaves narrow-linear, obtuse, acute or with a short point, which is fine but not hair-like as in *S. piliferum,* mostly

1 to 1½ in. long, more crowded and spreading than in *S. violaceum*, broader
and flatter than in *S. spinulosum*.   Scapes slender, 6 in. to nearly 1 ft. high,
the raceme loose, mostly simple, but the pedicels bracteate at or above the
middle, and sometimes 2- or 3-flowered.   Flowers small.   Calyx-lobes free,
obtuse.   Corolla "yellow or nearly white," the labellum with appendages,
the throat usually without.   Capsule ovoid, in the ordinary form 2 lines long.
—DC. Prod. vii. 333 ; Sond. in Pl. Preiss. i. 373.

W. Australia.   King George's Sound and adjoining districts, *R. Brown. A. Cunningham,* and others, *Preiss, n.* 2275 ; *Drummond, n.* 46, 49, 126, 132, *3rd Coll. n.* 168.

Var. ? *subbulbosum.*   Stock thicker, almost bulbous from the persistent bases of old
leaves.   Leaves usually ciliate, with a few long glandular hairs.   Raceme more stout.
Capsule ovoid-oblong, 3 lines long.   Swan River, *Drummond, 1st Coll.* and *3rd Coll. n.*
172.   *S. squamellosum,* DC. Prod. vii. 782, Sond. in Pl. Preiss. i. 377, is probably this
variety.

This and the two following species, which may be really very distinct and readily recognized in the fresh state by the colour and form of the corolla, are very difficult to characterize from dried specimens.

15. **S. violaceum,** *R. Br. Prod.* 569.   Quite, glabrous or with a few
glandular hairs on the calyx and pedicels.   Stock simple or tufted, but more
slender than in *S. cæspitosum,* and sometimes proliferous or slightly elongated
below the terminal tuft.   Leaves narrow-linear, obtuse or rather acute, but
without fine points, erect or slightly spreading, 1 to 2 in. long.   Scape often
above 1 ft. long, with or rarely without small, scattered, bract-like leaves.
Raceme loose, 2 to 6 in. long, simple or compound, the peduncles bearing
above the middle a small bract, and sometimes 2 or 3 pedicellate flowers.
Flowers small, violet-purple or rarely "variegated or yellow" (*Preiss*).
Calyx-lobes free, as long as the tube.   Corolla with small appendages to the
throat, but none to the labellum.   Capsule small, ovoid-globular.—DC.
Prod. vii. 334 ; Sond. in Pl. Preiss. i. 377 ; Bauer, Illustr. t. 5.

W. Australia.   King George's Sound and adjoining districts, *R. Brown* and many
others, *Drummond, 3rd Coll. n.* 173, 174, 175, and *Suppl. n.* 79, 81, 82.   It is possible
that the specimens said to have variegated flowers may be hybrids.

16. **S. luteum,** *R. Br. Prod.* 570.   Very near *S. violaceum,* with nearly
the same habit and foliage, but rather more slender, and the inflorescence
and sometimes also the base of the scapes and foliage more or less glandular-
pubescent ; the minute leaves on the scape sometimes but not always collected into 1 or 2 minute whorls, thus connecting this series with the next,
the flowers yellow and sometimes rather larger than in *S. violaceum,* the capsule also larger and glandular-villous.—DC. Prod. vii. 334.

W. Australia.   King George's Sound, *R. Brown, M'Lean, Drummond, n.* 112, *F.
Mueller.*   When the minute whorled leaves on the scape are present, this species may be
distinguished from *S. amœnum* by the narrow leaves, from *S. Brunonianum* by the minuteness of the whorl.   It will, however, probably prove to be not specifically distinct from *S.
violaceum,* and possibly the larger-fruited specimens which I have here included, may be
rather referable to *S. cæspitosum,* but the limits of this species can only be determined by
the examination of fresh flowers.

17. **S. piliferum,** *R. Br. Prod.* 569.   Glabrous or with short glandular hairs on the inflorescence, and sometimes on the margins of the leaves.
Stock tufted, not proliferous.   Leaves all radical in a dense tuft, which is

almost globular when they are small, linear or narrowly linear-spathulate,
ending in a fine often long and hair-like point, about ½ in. long in the smaller
specimens, 1½ in the luxuriant forms.   Scape from 3 or 4 in. to above 1 ft.
high, the upper portion occupied by a loose, simple or slight y-branched
raceme, the peduncles always bearing 1 or 2 bracts.   Calyx-lobes free.
Corolla variable in size, yellowish or according to others white tinged with
purple or pink, with appendages to the labellum but none to the throat.
Capsule obovoid, about 2 lines long.—DC. Prod. vii. 333 ; *S. bicolor,* Lindl.
Swan Riv. App. 28, Sond. in Pl. Preiss. i. 374.

W. Australia.   King George's Sound, *R. Brown, Baxter, A. Cunningham ;* Stirling
range and Upper Kalgan, *F. Mueller ;* Swan River, *Drummond, 1st Coll. n.* 545 in some
collections, 546 in others, *Preiss, n.* 2276.   *Drummond's 5th Coll. Suppl. n.* 97 is a rather
broader-leaved form, which we have also from Swan River, *Collie,* and *Drummond's 2nd Coll.
n.* 277, has a more branching raceme, like that of *S. ciliatum,* but nearly glabrous.

18. **S. ciliatum,** *Lindl. Swan Riv. App.* 28.   Resembles *S. piliferum,*
and perhaps a variety distinguished by the longer and more copious yellowish
glandular hairs which clothe the inflorescence and sometimes also the base of
the scape and foliage.   Stock tufted.   Leaves linear, 1 to 1½ in. long, terminating in a hair-like point.   Scape 6 in. to 1 ft. high.   Panicle or raceme
from shortly pyramidal to narrow and 3 or 4 in. long, the peduncles mostly
branched, but not really corymbose as in *S. soboliferum.*   Calyx-lobes free.
Corolla more or less yellow (or sometimes white or pink ?), variable in size,
the appendages of the throat and labellum small or wanting.   Capsule obovoid, turbinate, 2 to 2½ lines long.—Bot. Mag. t. 3883, Sond. in Pl. Preiss.
i. 374 ; *S. saxifragoides,* Lindl. Swan Riv. App. 28, Bot. Mag. t. 4529,
(copied in Lem. Jard. Fleur. t. 34), Sond. l.c. 374 (with rather broader leaves) ;
S. hispidum, Lindl. l. c. 29, Sond. l. c. 375 ; *S. setigerum,* DC. Prod. vii. 782.

W. Australia.   Swan River, *Drummond, 1st Coll. n.* 544 in some collections, 545 in
others; *Preiss, n.* 2269, 2277.   Tone and Murchison rivers, *Oldfield,* Mongerup and
eastward to Cape Le Grand, *Maxwell.*   There seems at first sight considerable difference
between the two forms figured in the 'Botanical Magazine,' but there are too many intermediates to admit them as distinct varieties, and, probably, the whole must be reduced to
S. piliferum.   The *S. pubigerum* much resembles this species, but may be readily known
by the linear ovary and capsule.

19. **S. soboliferum,** *F. Muell. in Hook. Kew Journ.* viii. 162, *and in
Trans. Vict. Inst.* 131.   Stock small, densely tufted.   Leaves all radical,
forming an almost globular tuft, linear or narrow linear-cuneate, rarely ½ in.
long, ending in a long, hair-like point, glabrous.   Scapes slender, leafless, 3
to 6 in. long, glabrous at the base or with a few glandular hairs.   Flowers
all pedicellate, in a loose, almost dichotomous, glandular-pubescent corymb,
reduced sometimes to 3 or 4 flowers.   Calyx 1½ lines long ; the lobes free,
obtuse, as long as the tube.   Corolla " pale pink or white," the throat naked
or with very small stipitate glands ; labellum obtuse, with small appendages.
Capsule ovoid, 2 to 2½ lines long.

Victoria.   Sandy, stony declivities of the Grampians, Serra, and Victoria ranges, *F.
Mueller* and others.   The foliage is that of the small specimens of *S. piliferum,* but the
inflorescence different.

20. **S. Floodii,** *F. Muell. Fragm.* i. 149.   A slender annual, some-

times shortly proliferous at the base. Leaves all radical or in a second tuft at the end of the short stem, very narrow-linear, almost filiform, obtuse or acute, $\frac{1}{2}$ to 1 in. long, glabrous or sprinkled with a few hairs. Scapes or peduncles usually several, leafless, slender, 3 to 6 in. high, usually sprinkled with a few hairs at the base, the inflorescence glandular-pubescent. Flowers very small, in loose, more or less corymbose panicles, all pedicellate. Calyx a little more than 1 line long, the lobes short and obtuse, free or the 2 lower ones united. Corolla "pink," the tube longer than the calyx-lobes, the lobes unequal, the throat with or without small appendages, the labellum without any. Capsule ovoid, 1 to $1\frac{1}{2}$ lines long.

N. Australia. Gravelly banks of the Upper Victoria and Roper rivers, *F. Mueller.*
Queensland. Sources of Cape river, *Bowman.*

21. **S. dispermum,** *F. Muell. Fragm.* iv. 93. Stock thick, crowned by a sort of bulb formed by the bases of old leaves. Leaves all radical, linear, mucronate, rather rigid, 1 to 2 in. long, glabrous and smooth or with scabrous margins. Scapes glandular-pubescent, 4 to 6 in. high, leafless except the bracts, branching from near the base into a broad, almost corymbose, glandular-pubescent or villous panicle. Flowers shortly pedicellate, very small and numerous. Calyx-lobes free, as long as the tube. Corolla with 2 prominent, fringed, glandular appendages, which appear to be on each side of the labellum (in the throat, *F. Mueller*). Capsule globular, about 1 line diameter. Ovules 6 to 8 in each cell of the ovary, crowded near the top of the dissepiment, but only 1 or 2 in each cell come to maturity.

W. Australia. Moist, gravelly places, Murchison river, *Oldfield.*

SERIES 4. ANDROSACEÆ. Slender annuals, with small, rosulate, radical leaves, and few-flowered, loosely-corymbose panicles. Labellum long, ascending. Stigma stipitate between the anthers. Capsule globular.

22. **S. calcaratum,** *R. Br. Prod.* 570. A slender annual, sprinkled with glandular hairs, especially on the inflorescence. Leaves all radical, rosulate, ovate or orbicular, 1 to 3 lines long, on short, slender petioles. Scape usually 3 to 4 in. high, but sometimes not 1 in., simple and 1-flowered or more frequently branched at the top into a very loose, irregular corymb of 5 to 9 pink flowers, with small bracts under and sometimes on the branches or long pedicels. Calyx-tube almost globular, about 1 line long, the lobes as long, free, oblong, obtuse. Corolla-tube very short, produced on the side opposite the labellum into a slightly incurved spur, usually nearly as long as the calyx-lobes, sometimes longer than them or very short and reduced to a small protuberance. Corolla-lobes nearly equal or 2 rather shorter; the labellum scarcely shorter, narrow, concave, ascending, more or less denticulate at the end or rarely quite entire, and no appendages to the throat. Stigma bearded and stipitate between the anthers. Capsule nearly globular.—DC. Prod. vii. 335 ; Bauer, Illustr. t. 15 ; *S. androsaceum*, Lindl. Swan Riv. App. 29 ; *S. Lindleyanum*, Sond. in Pl. Preiss. i. 390.

Victoria. Grampians, *F. Mueller.*
S. Australia. Near Echunga, *F. Mueller.*
W. Australia. King George's Sound and adjoining districts, *R. Brown, Oldfield, F. Mueller*, and others, to Swan River ; *Drummond, 1st Coll. n.* 511, 512, 513, 571, *2nd Coll. n.* 279, 280 ; *Preiss, n.* 2245.

23. **S. perpusillum,** *Hook. f. in Hook. Lond. Journ.* vi. 266, *and Fl. Tasm.* i. 235. A slender annual of 1 to 2 in., sprinkled with glandular hairs, closely resembling the smaller forms of *S. calcaratum,* but without any spur or protuberance to the tube of the corolla. Leaves rosulate, linear-spathulate or obovate, rarely above 1 line long. Flowers small, few, in a loose corymb or solitary. Calyx and capsule of *S. calcaratum* and corolla also the same, except the absence of the spur. Stigma bearded and stipitate.—*S. perminutum,* F. Muell. Fragm. vi. 78.

Victoria. Serra Ranges and Mount M'Ivor, *F. Mueller;* near Portland, *Allitt.*
Tasmania. Wet, sandy soil, Georgetown, *Gunn.*
W. Australia. Salt lake, Middle Island, *Maxwell;* sandy swamps and wet rocks between King George's Sound and Mount Perongerup, *F. Mueller.*

SERIES 5. SPATHULATÆ. Tufted or proliferous perennials or annuals. Leaves all radical or in terminal tufts, from linear-spathulate to obovate. Stigma concealed between the anthers. Capsule ovoid or oblong.

24. **S. eriorhizum,** *R. Br. Prod.* 569. Stock tufted, the base of the leaves when old enveloped in a short, dense wool. Leaves all radical, from obovate to narrow-spathulate, from 1 to 2 or rarely near 3 in. long, rounded at the end, but with a fine, sometimes hair-like point, edged with a short, woolly pubescence, otherwise glabrous or sprinkled with a few short glandular hairs. Scapes glandular-pubescent, 4 to 8 in. high, the greater part occupied by a loose but narrow panicle. Calyx-lobes cohering at the base into 2 lips. Corolla small, " pink or nearly white ;" the labellum convex with very short appendages, the throat without any. Capsule narrow-ovoid. —DC. Prod. vii. 334 ; F. Muell. Fragm. i. 147.

Queensland. Shoalwater Bay, *R. Brown ;* Suttor river, *F. Mueller;* Rockingham Bay, *Dallachy ;* Broad Sound, Connor and Isaacs rivers, *Bowman ;* Dogwood Creek, *Leichhardt.*

25. **S. debile,** *F. Muell. Fragm.* i. 149. Glabrous or with a very slight glandular pubescence on the inflorescence, and apparently annual or perhaps with a very slender rootstock. Stems slender, often 4 or 5 in. long, bearing a few leaves below the terminal tuft or rosette. Leaves obovate oblanceolate or spathulate, mucronate-acute, $\frac{1}{2}$ to 1 in. long, including the petiole, usually thin. Scape filiform, from a few in. to $1\frac{1}{2}$ ft. long, the greater part occupied by a slender raceme, quite simple, or the lower peduncles rarely 2-flowered, the pedicels all short, with 1 or 2 bracts. Calyx-lobes short and very obtuse, the 2 lower ones often united. Corolla-lobes unequal, undivided, without appendages to the throat but with small ones to the labellum. Capsule narrow-oblong, about 4 lines long.

Queensland. Brisbane river, Moreton Bay, *F. Mueller ;* Port Curtis, *M'Gillivray.*
N. S. Wales. Hastings river, *Beckler ;* Richmond river, *Fawcett ;* New England, *C. Stuart.*
The species has very much the aspect of a *Lobelia.*

26. **S. floribundum,** *R. Br. Prod.* 569. Apparently annual. Leaves radical, rosulate, petiolate, oblong oblanceolate or spathulate, acute or obtuse, $\frac{3}{4}$ to $1\frac{1}{2}$ in. long, thin and glabrous. Scape filiform, $\frac{1}{2}$ to 1 ft. long, clothed with long spreading glandless hairs, nearly simple or paniculately branched ;

leafless except the minute bracts. Flowers small, on slender pedicels. Calyx scarcely 2 lines long, the lobes free, narrow and very small. Corolla scarcely twice as long as the calyx-lobes; the labellum with 2 small appendages, but none to the throat. Capsule oblong-clavate, 2 lines long.—DC. Prod. vii. 334; F. Muell. Fragm. i. 148.

N. Australia. Islands of the Gulf of Carpentaria, *R. Brown;* moist, shady places at the sources of Wentworth river, and near M'Adam range, *F. Mueller.*

27. **S. leptorhizum,** *F. Muell. Fragm.* i. 148. Apparently annual, but the slender stem sometimes 1 to 3 in. long, bearing a few leaves below the terminal tuft or rosette, quite glabrous as well as the leaves, or sprinkled with a few short glandular hairs. Leaves rosulate, from oblanceolate to obovate or spathulate, obtuse or mucronate-acute, thin and membranous, often almost glaucous, $\frac{1}{2}$ to 1 or rarely 2 in. long, including the petiole. Scape slender, leafless, except the minute bracts, from 2 or 3 in. to $1\frac{1}{2}$ ft. long, glabrous or glandular hairy. Flowers small, in a slender, very loosely-branched panicle or raceme along its branches, with or without one in the fork, all on rather long slender pedicels without bracteoles. Calyx-lobes free, narrow and small. Corolla lobes unequal (the 2 uppermost more united?), the labellum with 2 appendages, but none to the throat, or, according to F. Mueller, the appendages to the throat and not to the labellum. Capsule oblong-clavate, 2 lines long.—*S. semipartitum,* F. Muell. Fragm. i. 147.

N. Australia. Greville's Island, Regent river, N. Coast, *A. Cunningham* (in Herb. R. Br.); dry pastures on the Victoria river and between M'Adam Range and Providence Hill, *F. Mueller;* Port Essington, *Armstrong.* The glandular hairs on the staminal column mentioned by F. Mueller appear to be very inconstant, sometimes rather copious, sometimes very few or none.

Var. *pilosum.* Leaves, scapes and inflorescence glandular-hairy.—Van Diemen's Gulf, N.W. Coast, *A. Cunningham.*

28. **S. assimile,** *R. Br. Prod.* 569. Glabrous, except the glandular pubescent inflorescence. Stock tufted, rather thick. Leaves all radical, oblong-spathulate to almost linear, very obtuse, thick and turning black when dry, often glaucous underneath. Scapes solitary or several together, from 3 or 4 in. to nearly 1 ft. high, divided from below the middle into a loose, pyramidal panicle. Flowers small. Calyx-lobes free. Corolla " with appendages to the labellum but none to the throat." Capsule ovoid, rather narrow.—DC. Prod. vii. 333; Sond. in Pl. Preiss. i. 377.

W. Australia. King George's Sound and adjoining districts, *R. Brown* and others, *Drummond, 5th Coll. n. 346 and Suppl. n.* 80, *Preiss, n.* 2272; Fitzgerald Ranges, Esperance Bay, and Cape Paisley, *Maxwell.*

29. **S. rupestre,** *Sond. in Pl. Preiss.* i. 375. Very closely allied to *S. spathulatum,* with the same pubescent, spathulate leaves, simple inflorescence and flowers, but the stock is always proliferous-branched, the leaves much smaller, not exceeding $\frac{1}{2}$ in. in length, in dense, rosulate, terminal tufts, and the racemes very loose, usually reduced to 2 or 3 flowers on long pedicels.—*S. glaucum,* R. Br. Prod. 569, not of Labill.; *S. glaucum, β?* *Brownei,* DC. Prod. vii. 334; *S. Lehmannianum,* Sond. in Pl. Preiss. i. 375.

W. Australia. King George's Sound, *Baxter, Preiss, n.* 2261, and eastward to Cape Riche, *Preiss, n.* 2262, *Drummond, 5th Coll. n.* 352; Lucky Bay, *R. Brown;* Cape Arid and Cape Paisley, *Maxwell.* This will probably prove to be a variety only of *S. spathulatum.*

30. **S. spathulatum,** *R. Br. Prod.* 569. Stock tufted or shortly proliferous. Leaves radical, rosulate, from obovate to oblong-spathulate, obtuse or acute, more or less pubescent or sprinkled with glandular hairs on both sides, $\frac{1}{2}$ to $1\frac{1}{2}$ in. long including the petiole. Scapes glabrous or pubescent, from a few inches to $1\frac{1}{2}$ ft. high, including the loose, elongated, simple raceme. Flowers small, pedicellate. Calyx-lobes free. Corolla pale yellow, with appendages usually both to the throat and the labellum. Capsule narrowly obovoid-oblong.—DC. Prod. vii. 333 ; *S. bellidifolium,* Sond. in Pl. Preiss. i. 376.

W. Australia. King George's Sound and adjoining districts, *R. Brown* and others, *Drummond, n.* 130, *2nd Coll. n.* 273, *3rd Coll. n.* 167, 177, *5th Coll. n.* 352, *Preiss, n.* 2259.

31. **S. Barleei,** *F. Muell. Fragm.* vi. 5. *t.* 69. Stock tufted or shortly proliferous. Leaves rosulate, ovate or spathulate, striate and prominently but irregularly toothed, glandular-pubescent on both sides, 1 to $1\frac{1}{2}$ in. long including the long petiole. Scape slender, often above 1 ft. long, glandular-pubescent, bearing occasionally 2 or 3 small, scale-like leaves below the inflorescence. Raceme long, loose, simple, with small, linear bracts, the pedicels longer than the calyx. Calyx not 3 lines long, the lobes free, as long as the tube. Corolla small, with glandular appendages to the throat ; labellum ending in a fine point, without appendages at the base. Capsule ovoid, often oblique, scarcely $1\frac{1}{2}$ lines long.

W. Australia, *Drummond, n.* 129, *2nd Coll. n.* 274. The toothed leaves, exceptional in the genus, if constant, readily distinguish this species, which is otherwise very near *S. spathulatum.*

32. **S. lineatum,** *Sond. in Pl. Preiss.* i. 376. Stock hard, tufted. Leaves rosulate, obovate or spathulate, obtuse, striate, sprinkled upon the upper or both surfaces with long, glandular hairs, rarely exceeding 1 in. including the petiole. Scape 1 to $1\frac{1}{2}$ ft. high, with a few, scattered, scale-like leaves below the inflorescence, glabrous or glandular-hairy. Racemes rather long and loose, simple or the lower peduncles with 2 or 3 flowers, all the pedicels longer than the calyx. Calyx-lobes free, longer than the tube. Corolla with very conspicuous appendages to the throat, but none to the small, convex labellum. Capsule ovoid, but not seen ripe.

W. Australia. Swan River, *Drummond, 1st Coll.*

33. **S. glaucum,** *Labill. Pl. Nov. Holl.* ii. 64. *t.* 214. Quite glabrous or very slightly glandular-pubescent, especially the inflorescence. Stock tufted or rarely lengthening out into a leafy stem of 2 or 3 in. Leaves radical and rosulate or tufted at the end of the stock, from obovate to oblong-spathulate, sometimes rather narrow, obtuse or mucronate-acute, often glaucous underneath, mostly 1 to $1\frac{1}{2}$ in. long including the petiole. Scape $\frac{1}{2}$ to about 1 ft. high, often with a few, very small, scale-like leaves below the inflorescence, but scattered and not in whorls. Flowers rather small, in a long, simple, loose raceme. Calyx-lobes free. Corolla with appendages

both to the throat and labellum. Capsule shortly ovoid.—DC. Prod. vii. 334 (excl. var. *β.*) ; *Candollea glauca*, Labill. in Ann. Mus. Par. vi. 454. t. 64 ; *S. nudum*, Lindl. Swan Riv. App. 29.

W. Australia. King George's Sound and adjoining districts, *A. Cunningham. Maclean, Harvey, Drummond*, 4*th Coll. n.* 174, 3*rd Coll. n.* 178. Our specimen of Preiss's n. 2237, referred by Sonder, Pl. Preiss. i. 378, to *S. amœnum*, has not the verticillate leaves of that species, and seems to belong to *S. glaucum*. This, the chief character which separates the two species, may not, however, be constant.

SERIES 6. DIVERSIFOLIÆ. Leaves rosulate or tufted on the stock or shortly proliferous stem, without intermixture of scarious scales, but with the addition of 1, 2 or more whorls of smaller narrow leaves on the scape below the inflorescence.

34. **S. amœnum,** *R. Br. Prod.* 570. Glabrous, except the glandular-pubescent inflorescence. Stock tufted or shortly proliferous. Radical leaves oblanceolate spathulate or almost obovate, mucronate-acuminate, ¾ to 1½ in. long including the petiole, rather thick, scarcely striate, often glaucous underneath, quite entire or slightly denticulate at the end. Scape ½ to 1½ or even 2 ft. high, with a single whorl of small, narrow, acute leaves above the middle. Raceme rather loose, 2 to 4 in. long, simple or very rarely the lower peduncles 2-flowered. Pedicels longer than the calyx. Calyx-lobes free. Corolla "bluish" or "pink," usually violet when dry, the throat with very few appendages, the labellum without any, but ending in a fine point. Capsule ovoid-globular, about 1½ lines long.—DC. Prod. vii. 334.

W. Australia. King George's Sound, *R. Brown, Baxter ;* Cape Naturaliste and Blackwood river, *Oldfield ;* Stirling Range and Toolbrunup Lake, *Maxwell. Drummond's* 3*rd Coll. n.* 178, seems also to be the same species, which scarcely differs from *S. glaucum*, except in the whorl of leaves on the scape, and even this is sometimes very minute so as to be easily overlooked.

Var. *caulescens.* Stock shortly proliferous or lengthening out below the terminal tuft into a leafy stem of 1 to 2 in.—*S. caulescens*, Lindl. Swan Riv. App. 29 ; DC. Prod. vii. 782 ; Sond. in Pl. Preiss. i. 378.—Swan River, *Drummond*, 1*st Coll. n.* 528, also *n.* 27.

35. **S. striatum,** *Lindl. Swan Riv. App.* 28. Glabrous or the inflorescence slightly glandular-pubescent. Stock densely tufted or slightly proliferous. Leaves oblanceolate or oblong-spathulate, acute, rather rigid, prominently striate and slightly glaucous in the typical form. Scapes 1 to 2 ft. high, usually with a single whorl of very small, oblong-linear leaves above the middle, which, however, is wanting in a few of the smaller specimens. Racemes 3 to 6 in. long, rather loose, simple or the lower peduncles slightly branched, the pedicels longer than the ovary. Calyx-lobes free, longer than the ovary. Corolla with appendages to the throat but none to the labellum. Capsule shortly ovate, but not seen ripe.

W. Australia. Swan River, *Drummond*, 1*st Coll. and n.* 125 ; *n.* 25 *and* 2*nd Coll. n.* 278 are larger specimens, often with 2 whorls on the scape.

Var. *glaucum.* Leaves less rigid, more obtuse, often longer, glaucous underneath, those on the scape broader than in the typical form, the raceme looser, often branched.—*S. striatum*, Sond. in Pl. Preiss. i. 379.—Swan River, *Preiss, n.* 2238 ; Swan and Vasse rivers, *Oldfield*, also *Drummond*, 5*th Coll. n.* 348, 349.

36. **S. diversifolium,** *R. Br. Prod.* 570. Glabrous, except the more

or less glandular-pubescent inflorescence. Stock tufted, usually crowned by the bulb-like, persistent bases of the leaves. Leaves radical, from broadly obovate or orbicular to ovate or spathulate, 1 to 2 in. long including the rather long petiole, obtuse, thick, often glaucous underneath. Scapes 1 to 2 ft. long, with 1, 2, 3 or rarely 4 whorls of small, narrow, acute leaves. Raceme slender, simple, 2 to 4 in. long, the pedicels all short and 1-flowered or rarely the lower ones 2-flowered. Calyx-lobes free, longer than the ovary. Corolla almost without appendage, the labellum ending in a fine point. Capsule nearly globular, not above 1½ lines long.—DC. Prod. vii. 334; *S. marginatum* and *S. pruinosum*, Sond. in Pl. Preiss. i. 379.

W. Australia. King George's Sound, *R. Brown, Baxter, Preiss, n.* 2236, to Swan River, *Drummond,* 1*st Coll. n.* 529, *Preiss, n.* 2232, and Vasse river, *Oldfield.* The species may be confounded at first sight with *S. carnosum,* but is readily distinguished by the absence of the scarious scales on the stock, and the presence of verticillate leaves on the scape.

37. **S. articulatum,** *R. Br. Prod.* 570. Glabrous, except the glandular-pubescent inflorescence. Stock tufted, often very thick or more or less proliferous. Radical leaves oblong-spathulate, obtuse or mucronate-acute, narrowed into a long petiole, attaining with the petiole 4 in. in the larger specimens, 2 in. in the smaller ones, rather thick, glaucous underneath. Scape varying from ½ to 1½ ft. high, but usually rather short, stout, bearing 1 or 2 whorls of small, linear or lanceolate leaves. Raceme or panicle dense, 2 to 4 in. long, the peduncles short, but the lower ones often with 2 or 3 flowers. Calyx-lobes free. Corolla rather large, the appendages of the throat very prominent, the labellum apparently without any. Capsule ovoid-oblong, fully 3 lines long.—DC. Prod. vii. 334.

W. Australia. King George's Sound, *R. Brown, Baxter, Drummond,* 4*th Coll. n.* 173. Readily known by its stout habit, short inflorescence, and large capsule. *S. robustum,* Sond. in Pl. Preiss. i. 378, from Sussex district, Preiss. n. 2235, which I have not seen, is probably, as he suggests, a very tall specimen of this species ; he describes the capsule, however, as only 2 lines long.

38. **S. Brunonianum,** *Benth. in Hueg. Enum.* 72. Glabrous and often glaucous, even the inflorescence scarcely glandular. Stock tufted or rarely shortly proliferous. Radical leaves from linear to oblanceolate, acute or rarely almost obtuse, 1 to 2 in. long or even longer, rather flaccid and scarcely striate. Scapes 1 to 1½ ft. long or rarely under 1 ft, with 2 to 5, usually 3 or 4, whorls of narrow, acute leaves. Raceme loose, 2 to 4 in. long, with numerous small flowers, the pedicels usually bracteate but rarely 2-flowered, the lowest often collected in whorls. Calyx-lobes free. Corolla with appendages to the throat, the labellum without any or with very small ones. Capsule small, globular.—DC. Prod. vii. 334 ; Sond. in Pl. Preiss. i. 380 ; Bot. Reg. 1842, t. 15 ; *S. compressum,* Lindl. Swan Riv. App. 29.

W. Australia. Swan River, *Drummond,* 1*st Coll. n.* 521 *or* 527, *also n.* 13 *and* 3*rd Coll. n.* 176 ; Murchison river, *Oldfield.*

Var. *minor.* Smaller, with small, very acute leaves, ½ to ¾ in. long.—*S. tenue,* Sond. in Pl. Preiss. i. 380.—Kalgan and Harvey rivers, *Oldfield ;* foot of Stirling Ranges, *F. Mueller,* also *Drummond,* (4*th Coll. ?*) *n.* 170. This variety almost connects the species with *S. diuroides.* The flowers are pink (*F. Mueller*).

39. **S. diuroides,** *Lindl. Swan Riv. App.* 29. Glabrous, except the slightly glandular inflorescence. Stock tufted or rarely shortly proliferous, densely covered with the persistent bases of old leaves. Radical leaves densely tufted, linear-subulate, acute or almost obtuse, the base scarcely dilated, usually about 1 in. long. Scapes 4 to 8 in. high, with a single whorl of setaceous leaves. Raceme loose, 1 to 1½ in. long, the lower pedicels rather long and bracteate but rarely 2-flowered. Flowers small. Calyx-lobes free, narrow. Corolla with short, broad appendages to the throat, but none to the acuminate labellum.—Sond. in Pl. Preiss. i. 380.

W. Australia. Swan River, *Drummond, 1st Coll. n.* 536, 537, *3rd Coll. n.* 170.

SERIES 7. VERTICILLATÆ. Stems elongated, simple or branched, the leaves all in distinct, whorl-like tufts, the lower ones not larger.

40. **S. scandens,** *R. Br. Prod.* 570. Quite glabrous, the stems in small specimens slender but nearly erect and simple, when luxuriant very flexuose or climbing to the height of 2 or 3 ft. and branching at some of the nodes. Leaves all collected in dense, whorl-like, distant tufts, without scattered ones between the tufts, linear, 1 to 2 in. or even longer when luxuriant, with a hooked or involute point. Racemes terminal, solitary or 2 or 3 together, shortly pedunculate, loose but few-flowered, and almost always simple. Pedicels bracteolate, longer than the calyx. Calyx-lobes free, as long as the ovary. Corolla pink, with more or less prominent appendages to the throat and labellum. Capsule broadly ovoid or globular.—DC. Prod. vii. 334 ; Sond. in Pl. Preiss. i. 381 ; Bot. Mag. t. 3136 ; Paxt. Mag. xv. 149, with a figure.

W. Australia. King George's Sound and adjoining districts, *R. Brown* and others, and thence to Vasse river, *Oldfield,* and eastward to Mount Bland, *Maxwell,* also *Drummond, n.* 5, 111, 123, *2nd Coll. n.* 275, *4th Coll. n.* 49, *Preiss, n.* 2295, 2296.

41. **S. verticillatum,** *F. Muell. Fragm.* iv. 94. Glabrous, except the inflorescence. Stems slender, elongated, flexuose or climbing and branching at the nodes as in *S. scandens.* Leaves all collected in dense, whorl-like, distant tufts without scattered ones between the tufts, narrow-linear, acute, not much exceeding ½ in. Peduncles hirsute with long, spreading, glandular hairs, bearing a loose cyme or short raceme of few flowers. Calyx-lobes free, as long as the tube, very narrow. Capsule ovoid, hispid, about 2 lines long.

W. Australia, *Drummond, n.* 93 ; Stirling Range, *Maxwell.* Scarcely differs from the smaller forms of *S. scandens,* except in the hirsute inflorescence.

SERIES 8. SPARSIFOLIÆ. Undershrubs or shrubs, with linear, spreading leaves scattered along the branches, and not collected in radical or terminal tufts.

42. **S. glandulosum,** *Salisb. Parad. Lond. t.* 77. An erect or spreading undershrub or shrub, ½ to 1½ ft. high, the branches covered with the cartilaginous, adnate bases of the petioles. Leaves scattered, but rather crowded along the branches, not collected in terminal tufts, linear, mucronate, under 1 in. long, glabrous or, especially the upper ones, slightly glan-

dular-pubescent. Panicles terminal, loosely branched, copiously glandular-pubescent, pedunculate or the lower branches proceeding from the base. Pedicels short. Calyx-lobes shorter than the ovary, free or shortly united in 2 lips. Corolla without appendages to the throat, but with linear ones to the labellum. Epigynous glands more prominent than in most species. Capsule ovoid, rather narrow, nearly 3 lines long.—*S. fruticosum*, R. Br. Prod. 570; DC. Prod. vii. 335.

W. Australia. Lucky Bay, *Brown, Baxter.*

43. **S. laricifolium,** *Rich. in Pers. Syn.* ii. 210. An undershrub with elongated leafy branches, rarely exceeding 1 ft. without the inflorescence, quite glabrous or the inflorescence sparingly glandular-pubescent, and sometimes a few hairs sprinkled on the foliage. Leaves scattered but rather crowded along the branches, not collected in terminal tufts, narrow linear, mucronate, usually ½ to 1 in. long, not leaving the adnate bases of those of *S. glandulosum.* Panicle or raceme loose, pedunculate, often above 6 in. long. Pedicels longer and the ovary more attenuate at both ends than in *S. glandulosum.* Calyx-lobes free, shorter than the ovary. Corolla with appendages to the labellum, but usually without any to the throat. Capsule oblong-turbinate, 4 to 6 lines long.—DC. Prod. vii. 335; Juss. in Ann. Mus. Par. xviii. 19. t. 3; Hook. Exot. Fl. t. 32; Bot. Reg. t. 550; *S. tenuifolium*, R. Br. Prod. 570; Link et Otto, Pl. Sel. t. 26; Bot. Mag. t. 2249.

N. S. Wales. Port Jackson to the Blue Mountains, *R. Brown, Sieber, n.* 172, and others; northward to New England, *C. Stuart;* southward to Illawarra, *Shepherd;* the latter luxuriant specimens with leaves 1 to 2 in. long.

SERIES 9. IMBRICATÆ. Stems slender, branching but hard, covered with small imbricate, almost scale-like leaves, not collected in terminal or radical tufts.

44. **S. Preissii,** *F. Muell. Fragm.* iii. 122. Stems simple or dichotomously branched, often flexuose, rarely above 3 in. high, completely covered with the closely-imbricated, scale-like leaves. These are ovate-lanceolate, acute, ½ to ¾ line long, with broad scarious, more or less ciliate margins and tips. Flowers 2 to 4, sessile within the last leaves, surrounded by scarious transparent bracts. Calyx 1½ lines long, the lobes broad and very obtuse, much longer than the tube, free or nearly so, but very much imbricate, the ends scarious and streaked with red. Corolla-tube shorter than the calyx-lobes, the throat without appendages, but the labellum fringed with long hairs. Capsule not seen.—*Forsteropsis Preissii*, Sond. in Pl. Preiss. i. 393.

W. Australia. *Drummond, 5th Coll. n.* 347; Cape Riche, *Preiss, n.* 438; from the Fitzgerald Range to Cape le Grand, *Maxwell.* This and the two .following species have a very peculiar habit, and Sonder proposed the present one as a distinct genus, characterizing it by the ovary and capsule 1-celled, with a central placentæ. The ovary appears to me, however, to be in fact 2-celled, but the very thin dissepiment splits very readily on each side of the placenta, which is thus left free as the capsule dries, but this is the case with several other *Stylidia;* and the structure of the flowers, as well as of the fruit of *Forsteropsis*, appear to me to be entirely those of *Stylidium.*

45. **S. imbricatum,** *Benth. in Hueg. Enum.* 73. Stems hard but

slender, simple or branched, erect or flexuose, ½ to 1½ ft. high, entirely covered with closely-imbricated, scale-like leaves. These are ovate, mucronate or aristate, with more or less scarious margins, ½ to 1 line long, the upper ones narrower and more acuminate. Raceme terminal, sessile, simple or scarcely compound, ½ to 1 in. long ; the almost sessile flowers intermixed with scale-like bracts similar to the leaves, but more acuminate, with usually a few woolly hairs on the rhachis and a few glandular ones on the calyx. Calyx 1½ lines long ; the lobes obtuse, free or nearly so, rather longer than the tube. Corolla-tube longer than the calyx-lobes, but not always twice as long ; the throat with small appendages ; the labellum ciliate. Capsule apparently ovoid, but not seen ripe.—DC. Prod. vii. 337.

W. Australia. King George's Sound, *Baxter, Huegel,* towards Cape Riche, *Harvey, Maxwell.*

46. **S. adpressum,** *Benth.* Stems simple or slightly branched, in our specimens 2 to 3 in. high, covered with the small imbricate leaves. These are lanceolate, concave or triangular, rigid, acute or mucronate, about 1 line long, glabrous or scarcely ciliate, less closely appressed than in the preceding two species. Peduncles terminal, glandular-hairy, 2 to 3 lines long, bearing a short compact raceme or head of 3 to 6 small flowers, surrounded by a few small rigid bracts. Calyx glandular-hispid, the lobes acute, free or nearly so, rather longer than the tube. Corolla-tube not exceeding the calyx-lobes, the appendages apparently as in *S. imbricatum,* but not seen very distinctly. Capsule ovoid, small, but not seen ripe.

W. Australia, *Drummond, 2nd Coll. n.* 38, *3rd Coll. n.* 182 (in some sets).

Sect. II. Nitrangium, *Endl.* Capsule linear or very narrow-oblong.

Series 10. Tenellæ. Slender annuals. Leaves small or thin, alternate or scattered, the lower ones sometimes more crowded, but not distinctly tufted or rosulate.

47. **S. despectum,** *R. Br. Prod.* 571. A little erect glabrous annual of 1 to 3 cr 4 in., sometimes very slender, sometimes rather firmer and branching in the upper part. Leaves small and scale-like, few and scattered, not rosulate or rarely a few longer narrow ones gathered together in a little loose tuft near the base, the others varying from ovate and under 1 line to lanceolate or linear and 2 or 3 lines long. Flowers pink, few, shortly pedicellate, forming an irregular corymb. Calyx about 2 lines long, the lobes very short, the 2 lower ones more or less united. Corolla scarcely exceeding the calyx-lobes, without appendages to the throat. Capsule linear, 3 lines long.—DC. Prod. vii. 336 ; Hook. f. Fl. Tasm. i. 235 ; *S. inundatum,* R. Br. Prod. 571 ; DC. Prod. vii. 336 ; Sond. in Pl. Preiss. i. 385.

Victoria. On the Yarra and Mount M'Ivor, *F. Mueller ;* Mount Emu, *Whan.*
Tasmania. Port Dalrymple, *R. Brown ;* Western Plains, *Backhouse ;* George Town, *Gunn ;* King's Island, *M'Gowan.*
S. Australia. Mount Muirhead and St. Vincent's Gulf, *F. Mueller.*
W. Australia. King George's Sound, *R. Brown, F. Mueller ;* Gordon, Murray, and Tweed rivers, *Oldfield.* The latter specimens belong to a long, slender variety. Brown's and Mueller's King George's Sound specimens have the same habit as the Port Dalrymple ones.

48. **S. utricularioides,** *Benth. in Hueg. Enum.* 73. A slender annual, glabrous or with a slight glandular pubescence on the inflorescence, 1 to 3 in. high, simple or scarcely branched at the top. Leaves scattered, linear or lanceolate, 1 to 2 lines long, the lower ones sometimes approximate but not rosulate. Flowers very few or solitary, pedicellate. Bracts small, setaceous. Calyx about 3½ lines long, the lobes narrow, much shorter than the tube, free or more frequently partially united in 2 lips. Corolla-lobes fully twice as long as the calyx-limb, without appendages to the throat or to the labellum. Capsule linear, 3 to 4 lines long.—DC. Prod. vii. 336 ; Sond. in Pl. Preiss. i. 386.

W. Australia. Swan River, *Huegel, Preiss, n.* 2246. Very near the slender forms of *S. despectum,* but the flowers are larger and the capsule longer.

49. **S. pygmæum,** *R. Br. Prod.* 571. A slender annual, glabrous or with a slight glandular pubescence on the inflorescence, 1 to 3 in. high, simple or scarcely branched at the top. Leaves scattered, linear or lanceolate, 1 to 2 lines long, the lower ones sometimes approximate but not rosulate. Flowers 1 to 3, sessile or very nearly so. Calyx about 2 lines long, the lobes free, narrow, about half as long as the tube. Corolla more than twice as long as the calyx-lobes, without appendages to the throat or labellum. Capsule oblong-linear, scarcely 2 lines long.—DC. Prod. vii. 336 ; Bauer, Illustr. t. 15 ; Sond. in Pl. Preiss. i. 387 ?

W. Australia. King George's Sound, *R. Brown, Harvey.* Very near the slenderest forms of *S. despectum,* but readily known by the sessile flowers and shorter broader capsule.

50. **S. longitubum,** *Benth. in Hueg. Enum.* 73. An erect glabrous annual, resembling the larger forms of *S. despectum,* but at once known by the larger flowers and longer capsules. Stems 3 to 5 in. high, paniculately branched above the middle. Leaves scattered, mostly near the base of the stem, but not rosulate, the lowest sometimes very short and ovate, the others oblong or linear, 1 to 3 lines long. Flowers all pedicellate, in a somewhat corymbose, loose panicle, the bracts very narrow and acute. Calyx-tube, at the time of flowering, 3 to 4 lines long, the lobes narrow, more or less united in 2 lips. Corolla at least twice as long as the calyx-lobes, with appendages to the throat. Capsule linear, 7 to 9 lines long.—DC. Prod. vii. 336 ; Sond. in Pl. Preiss. i. 386.

W. Australia. Swan River, *Drummond, 1st Coll. n.* 519 , Sussex district, *Preiss, n.* 2251; Serpentine river, *Oldfield.* This species closely resembles the E. Indian *S. tenellum,* Swartz, but the inflorescence is more corymbose, the flowers all pedicellate, the bracts broader, and the corolla larger.

51. **S. diffusum,** *R. Br. Prod.* 571. A very slender, usually branched annual, glabrous or the inflorescence slightly glandular, erect and only 1 to 2 in. high or much longer and diffuse. Leaves very small, the lower ones approximate but not rosulate, linear or oblong, 1 to 2 lines long, the upper ones still smaller and narrow. Flowers lateral and sessile or terminal, the small bracts not opposite. Calyx-tube 4 to 5 lines long, the lobes free, narrow, not 1 line long. Corolla scarcely exceeding the calyx-lobes, the 2 larger lobes bifid. Capsule linear, often incurved, ½ in. long or rather more.—DC. Prod. vii. 336.

Queensland. Shoalwater Bay, *R. Brown.* This may prove to be a very slender, usually diffuse variety of *S. tenellum,* Swartz.

52. **S. fissilobum,** *F. Muell. Fragm.* i. 154. A weak, filiform annual of ½ to 1½ ft., paniculately branched at the top, glabrous or with a few small glandular hairs on the inflorescence. Leaves very small and scattered, the lower ones more approximate but not rosulate, 1 to 1½ lines long, all linear-subulate. Panicle loose and somewhat corymbose. Flowers sessile. Calyx-tube long and filiform, the lobes linear-lanceolate, often partially united in 2 lips. Corolla twice as long as the calyx-lobes, the larger lobes bifid, with appendages both to the throat and labellum (*F. Mueller*). Capsule narrow-linear, ¾ to 1 in. long, very slender and beak-like at the end.

N. Australia. Grassy, inundated places on the Victoria river, between Main Camp and Steep Head, *F. Mueller.*

53. **S. alsinoides,** *R. Br. Prod.* 572. A glabrous and weak but usually erect and branching annual of ⅓ to nearly 1 ft. Leaves alternate or scattered below the inflorescence, shortly petiolate, broadly ovate, 2 to 4 lines long, obtuse and thin, the upper ones passing gradually into the narrow, acute, sessile floral leaves or bracts, which are almost always opposite. Flowers sessile in one axil of each pair of bracts. Calyx-tube linear, nearly ½ in. long, the lobes narrow, not 1 line long, the 2 lower ones more or less united. Corolla very small, the lobes united at the base in pairs (without appendages?). Capsule linear, 8 to 9 lines long.—DC. Prod. vii. 337; F. Muell. Fragm. i. 151.

N. Australia. Victoria river towards Stokes Range (starved, with narrow leaves) and Wickham river, Gulf of Carpentaria, *F. Mueller.*
Queensland. Endeavour river, *Banks and Solander, A. Cunningham;* Lizard Island, *M'Gillivray;* Rockingham Bay, *Dallachy.*

S. mitrasacmoides, F. Muell. Fragm. i. 150, from Palm Island, Victoria river, *Flood,* seems to be a small, starved specimen of *S. alsinoides;* the bracts, as in the typical form, are mostly opposite.

54. **S. tenerrimum,** *F. Muell. Fragm.* i. 150. A little glabrous annual, with weak, filiform, simple or slightly-branched stems of 1 to 3 in. Leaves scattered, lanceolate, acute, narrowed at the base, 1 to 1½ lines long, the lower ones not more approximate and usually smaller. Flowers small, on rather long axillary and terminal pedicels. Calyx about 2 lines long, the lobes free, shorter than the tube. Corolla small, white, the lobes entire, with a dark spot at the base of each (*F. Mueller*). Capsule narrow-oblong or linear, contracted at the base, 2 to 3 lines long.

N. Australia. Margins of swamps between M'Adam Range and Providence Hill, *F. Mueller.*

SERIES 11. CORYMBULOSÆ. Slender annuals or small perennials. Leaves radical or at the ends of the short stems, rosulate or tufted. Scapes leafless except the bracts, rarely exceeding 6 in. Flowers corymbose or sessile in the forks or along the scape or its branches, or solitary.

55. **S. brachyphyllum,** *Sond. in Pl. Preiss.* i. 386. A slender (annual?), glabrous except the slightly glandular inflorescence. Leaves rosulate or

forming a little tuft, but not bulbous, radical or at the top of a loosely-sheathed stem of 3 or 4 lines, linear, flaccid, 2 to 4 lines long. Scapes 2 to 4 in. high, branching upwards into a loose, irregular, almost corymbose panicle. Flowers all pedicellate. Calyx about 3 lines long, the lobes narrow, much shorter than the tube, the 2 upper ones sometimes united. Corolla twice as long as the calyx-lobes (without appendages ?), the tube very short. Capsule narrow, about 3 lines long.

W. Australia. Swan River, *Drummond, 1st Coll. n.* 518, 519, 523 (often mixed with *S. petiolare*), *Preiss, n.* 2248 (the specimen seen in Herb. F. Mueller, a very reduced slender form), Tone river, *Oldfield.*

56. **S. capillare,** *R. Br. Prod.* 570. A little slender, glabrous annual. Leaves radical, rosulate, like those of *S. brachyphyllum* or more spathulate, 2 to 3 lines long. Scapes capillary, 1 to 2 in. high, bearing 1 or 2 pedicellate flowers. Calyx-lobes free, very narrow. Corolla small, the 2 larger lobes more united, with appendages to the throat but none to the labellum (*R. Brown*). Capsule long and linear.—DC. Prod. vii. 335.

Queensland. Endeavour river, *Banks and Solander.*

57. **S. rotundifolium,** *R. Br. Prod.* 571. Glabrous, except a few glandular hairs on the inflorescence. Leaves radical, rosulate, obovate or orbicular, shortly petiolate, 2 to 3 lines long. Scapes filiform, 6 to 8 in. high, with a few, minute, scattered, narrow leaves, loosely corymbose at the top. Flowers small, shortly pedicellate or nearly sessile. Calyx-lobes united in 2 short, broad, entire lips. Corolla-lobes nearly equal, entire or emarginate, without appendages to the throat. Capsule linear, 5 to 8 lines long.—DC. Prod. vii. 335 ; F. Muell. Fragm. i. 151.

N. Australia. Plains at the foot of M'Adam Range, *F. Mueller* ; Hunter's River, York Sound, *A. Cunningham.*

Queensland. Endeavour river, *Banks and Solander* ; Shoalwater Bay, *R. Brown.*

The species is very near *S. uliginosum,* but the flowers are not so closely sessile, and the capsule much longer.

58. **S. schizanthum,** *F. Muell. Fragm.* i. 152. A slender annual, with rosulate, radical leaves, closely resembling *S. rotundifolium* in foliage, inflorescence, and in the slender, linear capsules, but the calyx-lobes are linear and free, at least the 3 upper ones, and the corolla-lobes are larger and more deeply bifid, the 2 larger ones more divided, and united at the base into a kind of lower lip. Capsule above ½ in. long.

N. Australia. Moist pastures on the Victoria river, *F. Mueller.* This species has the calyx and corolla nearly of *S. uliginosum* with the capsule of *S. rotundifolium.*

59. **S. lobuliflorum,** *F. Muell. Fragm.* i. 153. A slender, erect annual, with the habit of *S. schizanthum,* but rather more rigid and minutely glandular-pubescent from the base of the scape. Leaves petiolate, rosulate, broadly obovate or orbicular, ¼ to ½ in. long, glabrous, rather firmer than in *S. schizanthum.* Scapes about 6 in. high, branching in the upper part into a loose, almost corymbose panicle. Flowers nearly sessile in the forks or terminating the branches. Calyx-lobes linear, much shorter than the slender tube, the 2 upper ones more or less united, the others free. Corolla with the 2 larger lobes bifid. Capsule linear, ½ in. long or rather more.

N. Australia. Moist sandy pastures on the Victoria river, near Steep Head, *F. Mueller.* This seems to me to be a more glandular variety of *S. schizanthum.* I do not quite understand the differences described in the form of the corolla. As far as I can ascertain from the dried specimens, it seems to me to be the same in both.

60. **S. uliginosum,** *Swartz; DC. Prod.* vii. 336. A small, slender annual, glabrous or with a very few small, glandular hairs on the inflorescence. Leaves radical, ovate or orbicular, 2 to 5 lines long. Scape erect, 3 to 5 in. high, simple or slightly branched. Flowers sessile along the scape or its branches, each in the axil of a small bract. Calyx about 3 lines long, the lobes linear, free or shortly united in 2 lips, half as long as the tube. Corolla scarcely longer than the calyx-lobes, the upper lobes rather longer and bifid. Capsule linear, about 3 lines long.—Benth. Fl. Hongk. 195; *S. tenellum,* R. Br. Prod. 571, not of Swartz; *S. tenerum,* Spreng. Syst. iii. 749; DC. Prod. vii. 335.

Queensland. Endeavour river, *Banks and Solander;* Shoalwater Bay, *R. Brown.* This species is also in Ceylon and South China, and a rather more robust variety in Silhet, Chittagong, and the Malayan peninsula.

61. **S. pulchellum,** *Sond. in Pl. Preiss.* i. 381. Glabrous, except a minute and rare glandular pubescence on the inflorescence. Stock very small and slender, but slightly bulbous with the dilated, scarious, lanceolate bases of the petioles. Leaves few, radical, with an obovate, oval or oblong obtuse lamina of 2 or rarely 3 lines. Scape filiform, 2 to 4 in. high, leafless, bearing a loose, corymbose panicle of $\frac{1}{2}$ to 1 in., with small, narrow, obtuse bracts. Flowers pedicellate. Calyx 2 to $2\frac{1}{2}$ lines long, the lobes free, narrow, shorter than the ovary. Corolla-tube nearly as long as the calyx-lobes, without appendages to the throat or to the small,·obtuse labellum. Capsule linear, about 2 lines long, but not seen quite ripe.—*S. corymbosum,* Benth. in Hueg. Enum. 72, not of R. Br.

W. Australia. King George's Sound, *Baxter;* Swan River, *Huegel, Drummond,* 1*st Coll. n.* 520, 3*rd Coll. n.* 180; Bull's Creek, *Preiss, n.* 2242; Tone and Murray rivers, *Oldfield.* Very closely allied to *S. petiolare,* and perhaps a variety, but the flowers appear to be smaller and more regularly corymbose, and the capsule narrower.

62. **S. petiolare,** *Sond. in Pl. Preiss.* i. 382. Glabrous, except a very few small, glandular hairs on the inflorescence. Stock small and slender, but the dilated bases of the petioles form a little bulb, at first ovoid and pale brown, in the older plants much darker, and attaining 3 lines diameter. Leaves radical, the lamina from oblong-linear to ovate, obtuse, usually 2 to 3 lines long. Scape slender, 2 to 4 or rarely 5 in. high, leafless except the small bracts, branching in the upper part into an irregular corymb or rarely reduced to a single flower, the pedicels long and slender. Calyx $2\frac{1}{2}$ to nearly 3 lines long, the lobes much shorter than the tube, obtuse or acute. Corolla pink, larger than in the allied species, the tube as long as the calyx-lobes, the lobes entire or slightly emarginate, the throat with small appendages, the labellum without any. Capsule narrow-oblong, about 2 lines long.—*S. obtusatum,* Sond. l. c.

W. Australia. Swan River, *Drummond,* 1*st Coll. n.* 522, 523, 524 (sometimes mixed with *S. emarginatum*), *Preiss, n.* 2244; Dandenong, *Oldfield;* Stirling Range, *F. Mueller.* Drummond's 2nd Coll. n. 281 is perhaps the same, but the specimens are imperfect. Sonder distinguishes his two species chiefly by the form of the calyx-lobes, which are certainly in some specimens broad and very obtuse, in others narrow and almost acute, but there are many intermediates. In both, the labellum is said in the diagnosis to be inappendiculate, in the description to be appendiculate. I have not found any appendage to the labellum, but small, glandular ones to the throat.

63. **S. emarginatum,** *Sond. in Pl. Preiss.* i. 383. A small, slender plant, with the little, brown, bulbous stock and small, petiolate, radical leaves of *S. petiolare,* but readily known by a minute whorl of 3 or 4 leaves below the inflorescence. The scape, 2 to 3 in. high, bears usually only 1 or 2 flowers, and is more glandular than in *S. petiolare,* the calyx and corolla very nearly as in that species, of which this one may prove to be a variety only ; the lobes of the corolla are, however, said to be more deeply emarginate.

W. Australia. Swan River, *Drummond, n.* 521, 524 (partly) ; Victoria Plains, *Oldfield;* towards the Great Bight, *Maxwell* (with almost orbicular leaves).

64. **S. corymbosum,** *R. Br. Prod.* 571. Glabrous, except a few glandular hairs on the inflorescence. Stock densely tufted. Leaves radical, linear, sometimes rather broad, rigid and incurved as in *S. piliferum,* rarely above ½ in. long, terminating in a fine, almost hair-like point. Scape leafless, 3 to 6 in. high, bearing a dense, more or less compound corymb. Bracts small, rather thick, ovate or oblong as in *S. guttatum,* but not produced below their insertion, and sometimes narrow. Flowers sessile or nearly so. Calyx about 3 lines long, the lobes free, rather broad, very obtuse, scarcely half as long as the tube. Corolla-tube short, the throat without appendages, the labellum narrow, acuminate, ciliate and appendiculate. Capsule rather broadly linear, contracted at the top but not beaked, 4 to 5 lines long.– DC. Prod. vii. 335.

W. Australia. Lucky Bay, *R. Brown;* King George's Sound or to the eastward, *Baxter;* margins of swamps inland from Cape Le Grand, *Maxwell;* also *Drummond, 3rd Coll. n.* 165 (or 195 ?), 172.

Var. ? *proliferum.* Stock more or less proliferous.—Rocks of Mount Perongerup, *F. Mueller.* The specimens are not quite in flower, but probably belong to this species.

65. **S. lepidum,** *F. Muell. Herb.* Glabrous, except the slightly glandular-pubescent or hirsute inflorescence. Stock tufted. Leaves radical, linear, narrow but flat, with a short, fine point, ¼ to ½ in. long. Scape leafless, slender, almost filiform, 2 to 3 in. long, bearing a small cyme of 3 to 5 flowers, rarely reduced to a single flower. Bracts small, linear, mucronulate. Flowers sessile or nearly so, like those of *S. corymbosum,* but the calyx-lobes narrower. Capsule narrow-oblong, 3 or rarely 4 lines long.

W. Australia, *Drummond, n.* 114, *3rd Coll. n.* 181 ; Tone and Blackwood rivers, *Oldfield.* This differs from *S. corymbosum* chiefly in the small narrow leaves, the narrow acute bracts, and the narrow calyx-lobes.

66. **S. streptocarpum,** *Sond. in Pl. Preiss.* i. 385. Quite glabrous. Stock densely tufted or rarely shortly proliferous, or at length ¼ to 1 in. long below the tuft of leaves. Leaves radical, narrow-linear, 1 to 2 in. long, usually mucronate. Scapes several, divaricately branched from below the middle, forming a broad, more or less corymbose panicle, 6 to 8 in. high, in the smaller specimens reduced to a small, few-flowered cyme. Bracts small,

oblong, obtuse. Flowers nearly sessile. Calyx nearly ½ in. long, the lobes very short, broad and obtuse, cohering more or less in two lips. Corolla without appendages to the throat, the labellum very short, without any or with 2 short appendages. Capsule linear or linear-lanceolate, about ½ in. long or more.

W. Australia. Swan River, *Preiss, n.* 2273 (in fruit), *Drummond, n.* 29 (a single specimen in flower). Sonder describes the calyx-lobes as acute, but probably by mistake, as in both the above specimens they are remarkably obtuse.

Var. *tenellum.* Smaller and more slender, but equally divaricate and dichotomous.— *S. divaricatum,* Sond. in Pl. Preiss. i. 385.—Vasse river, *Preiss, n.* 2274.

Var.? *macrocarpum.* Panicle less divaricate and not so corymbose, the lower branches being sometimes shorter. Capsule sometimes ¾ in. long.—Swan River, *Collie ;* Murchison river, *Oldfield,* also *Drummond, n.* 131, *2nd Coll. n.* 271.

S. rigidulum, Sond. in Pl. Preiss. i. 389, which I have not seen, and of which the precise locality is not given, may be a small form of this species with the stem or stock 1 in. long below the leaves, and the flowers reduced to very few.

67. **S. uniflorum,** *Sond. in Pl. Preiss.* i. 381. Stock densely tufted or shortly proliferous. Leaves radical, very narrow-linear, 1 to 2 in. long, glabrous and smooth or with cartilaginous-serrulate margins. Scapes slender, 1-flowered, scarcely exceeding the leaves or rarely 3 or 4 in. high, ciliate-hirsute, leafless except a small bract under the flower. Calyx hairy, but scarcely glandular, about 4 lines long, the lobes free, short, and narrow. Corolla pale flesh-coloured, the tube very short; lobes unequal, with appendages to the labellum but scarcely any to the throat. Capsule linear-falcate, contracted towards the end, 5 to 6 lines long.

W. Australia. King George's Sound, *Drummond ;* Hay district, *Preiss, n.* 2253 ; Tone and Blackwood rivers, *Oldfield ;* Upper Kalgan river, *F. Mueller.*

68. **S. pedunculatum,** *R. Br. Prod.* 571. A perennial, sprinkled with a few hairs, forming a short, leafy stem of about ½ in. below the terminal tuft. Lower leaves rosulate at the base of the stem but often soon disappearing, oblong-lanceolate or almost ovate, 1 to 3 lines long, narrowed into a short petiole, those of the stem few, scattered and linear, those of the terminal tuft numerous, narrow-linear, terminating in long, hair-like points. Scapes or peduncles amongst the terminal leaves numerous, filiform, 1-flowered, 1 to 2 in. long. Calyx-lobes very small, the 2 lower ones united at first, all at length free. Corolla very small, the lobes unequal, without appendages to the throat or labellum (*R. Brown*). Capsule glabrous, linear, 3 to 4 lines long.—DC. Prod. vii. 337 ; *S. bryoides,* F. Muell. Fragm. vi. 91.

N. Australia. Port Essington, *Armstrong.*
Queensland. Endeavour river, *Banks and Solander, A. Cunningham ;* Rockingham Bay, *Dallachy.*

69. **S. pachyrhizum,** *F. Muell. Fragm.* i. 152. Glabrous. Stem or stock thick, erect, 1 to 2 in. high, branching at the top. Leaves few, scattered, more crowded under the scapes, petiolate, oblanceolate or spathulate, rather thick. Peduncles or scapes several, 3 to 6 in. long, branching from near the base into a loose, spreading, corymbose panicle. Flowers sessile in the forks or terminating the branches. Calyx-lobes linear, free or the lower ones more or less united, much shorter than the slender tube. Corolla

small, the lobes nearly equal, the throat with small appendages, but none to the oblong labellum.   Capsule linear, 4 to 6 lines long.

N. Australia. Pastures between Providence Hill and M'Adam Range, *F. Mueller.* The specimens not in good foliage.

70. **S. muscicola,** *F. Muell. Fragm.* i. 153.   Stems simple, erect, herbaceous but rather rigid, 1 to 4 in. high, with a few scattered leaves, and crowned by a spreading cluster of orbicular, membranous leaves on rather long petioles, the lamina $\frac{1}{2}$ to above 1 in. diameter, usually glabrous.   Peduncles several from amongst the terminal leaves and exceeding them, but flowering from near the base.   Flowers sessile, in the axils of minute bracts, forming interrupted, more or less glandular spikes.   Calyx-tube very slender, lobes small, narrow, the 2 lower ones more or less united.   Corolla very small, the 2 larger lobes emarginate, the throat without appendages.   Capsule very narrow-linear, 8 to 10 lines long.

N. Australia. In tufts of moss near springs and cataracts on the Upper Victoria river, *F. Mueller*, also from King's voyage in herb. R. Brown.   This species has nearly the inflorescence of *S. uliginosum*, with a very peculiar habit and foliage.

Series 12. THYRSIFORMES.   Perennials, with a tufted or proliferous-branched stock or stem, with radical or terminal tufts of leaves.   Flowers in an oblong or elongated thyrsoid panicle or raceme.

71. **S. crassifolium,** *R. Br. Prod.* 571.   Usually glabrous, except the glandular-pubescent inflorescence.   Stock short, thick and hard.   Leaves radical, dilated at the base, but not forming persistent bulbs, lanceolate or almost linear, 4 to 6 in. long, including the long petiole, rather thick and turning black in drying as in *S. pycnostachyum.* Scape 1 to 2 ft. long or even more, including the long raceme-like panicle.   Bracts small, lanceolate.   Peduncles very short, mostly 2- or 3-flowered.   Calyx-lobes short, acute, free or shortly united in 2 lips.   Corolla pink, without appendages or with very small ones to the labellum.   Capsule linear or oblong-linear, 5 to 8 lines long.—DC. Prod. vii. 335 ; Sond. in Pl. Preiss. i. 384 ; *S. leptobotrys,* DC. Prod. vii. 783 ; Sond. l. c. 384 ? ; *Dampiera ? inundata,* De Vr. in Pl. Preiss. i. 404.

W. Australia. King George's Sound, *R. Brown, Baxter* and others, to Swan and Vasse rivers, *Collie, Drummond, Oldfield, Preiss, n.* 1523, and others, eastward to Phillips Flats, *Maxwell.*

72. **S. pycnostachyum,** *Lindl. Swan Riv. App.* 29.   Stock tufted, the persistent bases of the petioles forming a bulb.   Leaves radical, from obovate-spathulate to oblanceolate, mostly 3 or 4 in. long, including the long petiole, more or less sprinkled with glandular hairs.   Scape ascending, glandular-pubescent or villous, 6 in. to nearly 1 ft. high, the thyrsoid panicle occupying about one-half.   Bracts lanceolate.   Pedicels very short.   Calyx-lobes linear, free.   Corolla without any appendages or only very small ones to the labellum.   Capsule linear, 9 to 10 lines long, slightly contracted at the top, but not seen quite ripe.—*S. thyrsiforme,* DC. Prod. vii. 733 ; Sond. in Pl. Preiss. i. 384.

W. Australia. Swan River, *Drummond, 1st Coll. n.* 531.

73. **S. pubigerum,** *Sond. in Pl. Preiss.* i. 383. Glabrous or nearly so, except the yellowish glandular-pubescent or villous inflorescence. Stock densely tufted or very shortly branched. Leaves radical, linear, flat, and often dilated upwards, rather rigid and curved inwards, ½ to 1 in. long, ending in a long hair-like point. Scapes 3 to 6 in. high, including the oblong thyrsoid panicle of 1 to 3 in. Bracts, small, narrow, and acute. Flowers shortly pedicellate. Calyx about 4 lines long, the lobes obtuse, not half so long as the tube, more or less united in 2 lips. Corolla-lobes nearly equal, without any appendages to the throat or labellum. Capsule linear, 5 or 6 lines long, but not seen quite ripe.

W. Australia. Swan River, *Drummond, 1st Coll. n.* 543, 546, also *n.* 25 ; *Preiss, n.* 2278. With the exception of the linear ovary and capsule, the species much resembles *S. ciliatum.*

74. **S. canaliculatum,** *Lindl. Swan Riv. App.* 29. Glabrous, except a very few small glandular hairs sprinkled on the inflorescence. Stock densely tufted, not proliferous, the base of the petioles somewhat bulbous. Leaves narrow-linear, scarcely pointed, rather flaccid, 1 to 2 or rarely 3 in. long. Scape ½ to nearly 1 ft. long, the panicle 1 to 3 in. long, loosely thyrsoid, most of the peduncles or branches 3-flowered. Bracts small, linear. Flowers all pedicellate. Calyx 2½ lines long, the narrow acute lobes as long as the tube and free. Corolla with very small appendages to the throat, the labellum narrow, acuminate, without appendages. Capsule oblong-linear, about 3 lines long.—*S. thesioides,* DC. Prod. vii. 783 ; Sond. in Pl. Preiss. i. 383.

W. Australia. Swan River, *Drummond, 1st Coll. n.* 538, also *n.* 28; *Preiss, n.* 2257, 2258.

75. **S. leptophyllum,** *DC. Prod.* vii. 783. Stock densely tufted. Leaves narrow-linear, flaccid, 1½ to 3 or even 4 in. long, shortly mucronate, glabrous or nearly so. Scapes 6 to 8 in. high, usually hirsute with spreading hairs from the base. Panicle thyrsoid or narrow and raceme-like, glandular-pubescent, the hairs often yellowish. Bracts narrow, small. Flowers very shortly pedicellate. Calyx about 3 lines long, the lobes free, narrow, obtuse, rather shorter than the tube. Corolla without appendages to the throat. Capsule linear, about 4 lines long.—Sond. in Pl. Preiss. i. 384.

W. Australia. Swan River, *Drummond, 1st Coll., Preiss, n.* 2254. Scarcely differs from *S. dichotomum,* except in the short not proliferous stock.

Var. *mucronifolium.* Leaves rather shorter, with a fine point.—Blackwood river, *Old-field;* Sterling and Plantagenet ranges and Phillips river, *Maxwell. S. mucronifolium,* Sond. in Pl. Preiss. i. 383, from Plantagenet, *Preiss, n.* 2256, is, from the description, evidently this variety, but I have not seen the specimen.

76. **S. dichotomum,** *DC. Prod.* vii. 783. Very closely allied on the one hand to *S. leptophyllum,* on the other to *S. bulbiferum,* differing from the former chiefly in the proliferous-branched stock, often 2 to 3 in. long, from *S. bulbiferum* in the longer leaves and more numerous flowers. Leaves crowded at the base and ends of the branches and scattered between the tufts, attaining sometimes 1 in. or more, narrow-linear and acute. Scapes 2 to 4 in. high, glandular-pubescent. Panicle or compound raceme more or

less thyrsoid and glandular-hairy like that of *S. leptophyllum*, the flowers rather numerous. Capsule linear, 3 to 4 lines long.—Sond. in Pl. Preiss. i. 387 ; *S. mucronifolium*, Hook. Bot. Mag. t. 4538 (copied in Lemaire, Jard. Fleur. t. 59, and in Fl. des Serres, vi. t. 606) ; *S. Hookeri*, Planch. in Fl. des Serres, vi. 229.

W. Australia. Swan River and adjoining districts, *Drummond*, 1*st Coll. n.* 534 (referred by Sonder to *S. leptophyllum*), 535, also *n.* 18, 19, *3rd Coll. n.* 169 ; *Preiss, n.* 2284, 2285 ; Phillips Range, *Maxwell*.

77. **S. bulbiferum,** *Benth. in Hueg. Enum.* 73. A small much-branched proliferous species, erect or rooting at the lower tufts, glabrous except the glandular-pubescent scapes and inflorescence. Leaves very narrow-linear, scarcely acute, from about ¼ to above ½ in. long, densely tufted at the ends and bases of the branches, with a few smaller intermediate ones, the bases of the old tufts often bulb-like. Scapes or peduncles ½ to about 2 in. long, with a loose almost corymbose raceme of 3 to 7 flowers, rarely reduced to a single one. Bracts short and very narrow. Calyx-lobes free, very obtuse, not ⅓ the length of the tube. Corolla without appendages to the throat. Capsule narrow, slightly contracted upwards, but not beaked, under ½ in. long in the normal form.—DC. Prod. vii. 336 ; Sond. in Pl. Preiss. i. 388 ; *S. proliferum*, DC. Prod. vii. 783.

W. Australia. Swan River and adjoining districts, *Drummond*, 1*st Coll.*, also *n.* 171 ; *Preiss, n.* 2281, 2283, *Oldfield*.

Var. *macrocarpum.* Capsule almost sessile, 8 to 9 lines long.—*S. recurvum*, Grah. in Bot. Mag. t. 3913.—Harvey river, *Oldfield*.

Var. *ciliatum*, Sond. Leaves ciliate-hirsute. Peduncles all or nearly all 1-flowered.— Swan River, *Drummond*, 1*st Coll. n.* 541.

78. **S. breviscapum,** *Br. Prod.* 572. A small, much-branched, proliferous species, the stock or stem and foliage glabrous. Leaves very narrow-linear, scarcely acute, about ½ in. long, densely tufted at the base and ends of the branches, with a few scattered ones between the tufts. Panicles ovoid and very compact, on very short scapes or peduncles clothed with spreading white hairs. Flowers in nearly sessile clusters, intermixed with small leaf-like bracts. Calyx-lobes very short, obtuse, with scarious margins. Corolla-lobes narrow, the throat without appendages, the labellum very small. Capsule linear-lanceolate, glandular-pubescent, 3 to 4 lines long, slightly contracted at the top, but not beaked.—DC. Prod. vii. 337 ; *S. eriopodum*, DC. Prod. vii. 784 ; Sond. in Pl. Preiss. i. 388.

W. Australia. Swan River, *Drummond*, 1*st Coll. n.* 547, also *n.* 21 ; Darling Range, *Collie* ; Gordon river, *Oldfield* ; near Maddington, *Preiss, n.* 2282 ; King George's Sound, *A. Cunningham* ; Lucky Bay, *R. Brown* (specimens in fruit and very near the following variety) ; base of Stirling Range, *F. Mueller*.

Var. *erythrocalyx.* Inflorescence not so dense. Peduncles or scapes less hairy. Calyx very red.—*S. involucratum*, F. Muell. Fragm. i. 154.—Fitzgerald Ranges and Cape Arid, *Maxwell*.

79. **S. eglandulosum,** *F. Muell. Fragm.* i. 150. Stems rather slender, more or less proliferous-branched, from a few inches to a foot long or more, glabrous, except a dense cottony wool about the old leaf-tufts, scarcely per-

ceptible on the young ones. Leaves narrow-linear, acute or mucronate, usually glabrous, crowded in dense tufts at the base and ends of the branches, with scattered intermediate ones. Racemes from 2 to 4 in. or rarely 5 or 6 in. long, including the very short peduncle, pubescent, but scarcely or not at all glandular, simple or nearly so, the pedicels all short, the lower ones rarely 2- or 3-flowered. Calyx-lobes free, narrow, not half so long as the tube. Corolla small, white with purple spots, the throat without appendages. Capsule oblong-linear, about 3 lines long.

Queensland. Arid hills between the Suttor, Belyando, Mackenzie, and Burnett rivers, *F. Mueller;* Alice river, *Mitchell;* Dogwood Creek, *Leichhardt.*

N. S. Wales. Darling Downs, *Woolls;* N.W. interior, probably on the Bogan, *Mitchell,* 1*st Expedition* (1831). These specimens were determined as a new species by Lindley and distributed as *S. laricifolium,* Lindl. (not of Rich.), but never described (the *S. laricifolium* attributed to Lindley, in Walp. Rep. ii. 704, being a misprint for *S. caricifolium*) ; without the basal woolly tufts the specimens are not unlike those of *S. laricifolium,* Rich., but readily distinguished by the short pedicels, narrow capsules, and almost total want of glandular hairs.

SECT. III. RHYNCHANGIUM.　Capsule lanceolate or linear, contracted into a slender beak. Perennials. Leaves linear, scattered along the proliferous-branched stock or stem, the upper ones usually crowded into terminal tufts.

80. **S. fasciculatum,** *R. Br. Prod.* 572. Glabrous or the inflorescence glandular-pubescent. Leafy stems usually elongated, sometimes attaining 1 ft., simple or slightly proliferous-branched. Leaves scattered along the stems, the upper ones collected in a terminal whorl-like tuft, linear, acute or almost obtuse, narrowed below the middle, the longer ones from 1 to 2 in. Spike-like panicles or compound racemes varying from almost sessile and 2 or 3 in. long to pedunculate and 10 in. long, the flowers more or less clustered along the rhachis on very short peduncles or almost sessile. Bracts small, lanceolate or linear. Calyx-tube long and linear, the lobes small and free. Capsule membranous, lanceolate or linear-lanceolate, straight or slightly falcate, but both valves perfect and nearly equal, ending in a slender beak, $\frac{1}{2}$ to nearly $\frac{3}{4}$ in. long, including the beak.—DC. Prod. vii. 337.

W. Australia. King George's Sound, *R. Brown,* also *Drummond, n.* 128, and 2*nd Coll. n.* 270, the typical form with short proliferous stems and short inflorescence.—*S. cicatricosum,* Sond. in Pl. Preiss. i. 390, is probably the typical *S. fasciculatum.* It cannot be *S. adnatum,* var. *propinquum,* for the capsule is described as "subæquivalvis."

Var. *elongatum.* Stems elongated ; racemes 6 to 10 in. long.—Flinders Bay, *Collie,* also *Drummond, n.* 127 and 2*nd Coll. n.* 269 (in some sets).

The species differs from *S. adnatum* only in the capsule.

81. **S. falcatum,** *R. Br. Prod.* 572. Very near *S. fasciculatum,* with the same linear leaves and elongated or proliferous stem, leafy below the terminal tuft ; but the stem as well as the rhachis of the inflorescence is pubescent. Raceme 4 to 6 in. long, shortly pedunculate or nearly sessile, simple or nearly so, the pedicels very short, and all 1-flowered or very rarely 2-flowered. Capsule lanceolate, falcate, curving downwards, the upper cell and valve much narrower than the lower one and semi-abortive, the beak short but slender.—DC. Prod. vii. 337.

W. Australia. King George's Sound, *R. Brown, Oldfield.*

S. Lessoni, DC. Prod. vii. 337, from the incomplete diagnoses given, and Sond. in Pl. Preiss. i. 388, from the more detailed character, seems to be this species, as suggested by Sonder.

82. **S. rhynchocarpum,** *Sond. in Pl. Preiss.* i. 389. This may be a variety of *S. falcatum* corresponding to the abbreviated form of *S. fascicula-tum* and *S. adnatum,* but I have seen no intermediates. Leafy stems rarely above 6 in. long, proliferous-branched, with the linear leaves of the allied species. Spike compound but reduced to a sessile cluster, concealed in the terminal tuft of leaves from which the summits of the flowers alone protrude. Capsule falcate as in *S. falcatum,* with the upper cell narrow and the lower one broad, but the slender beak is much longer than in that species.

W. Australia. *Drummond,* 1st *Coll. n.* 542, also *n.* 128.

83. **S. adnatum,** *R. Br. Prod.* 572. Glabrous, except the more or less glandular inflorescence. Leafy stems simple or proliferous-branched as in *S. fasciculatum,* varying from 2 or 3 in. to nearly 1 ft. long. Leaves scattered along the stem, the upper ones crowded in a terminal whorl-like tuft, linear but sometimes all very narrow, in other specimens all rather broad, and from $\frac{1}{2}$ to $1\frac{1}{2}$ in. long. Spike-like panicles or compound racemes dense, usually rather short and nearly sessile, but sometimes as long as in the long variety of *S. fasciculatum.* Flowers, as in that species, in nearly sessile clusters along the rhachis. The sole difference is in the capsule, which is lanceolate or linear, and beaked as in that species, but usually narrower and more falcate (curving upwards), and the upper cell is entirely abortive, the valve being reduced to a filiform rib on the upper side of the lower perfect cell. The length varies from about $\frac{1}{2}$ in., including the rather long beak, to nearly 1 in., including the short beak.—DC. Prod. vii. 337 ; *S. fascicula-tum,* Lindl. in Bot. Reg. t. 1459 ; Arn. in Bot. Mag. t. 3816, not of R. Br.

W. Australia. Goose Island Bay, *R. Brown ;* King George's Sound and adjoining districts, *Oldfield* and others, *Drummond, n.* 139 ; Salt Lake, Middle Island, *Maxwell.* These all belong to the commoner form with narrow leaves, the inflorescence from 3 to near 10 in. long.

Var. *abbreviatum.* Leaves narrow or broad. Inflorescence rarely above 2 in. long and very dense.—*S. propinquum,* R. Br. Prod. 572; DC. Prod. vii. 337 (with narrow leaves) ; *S. adnatum,* var. Br. in Bot. Mag. t. 2598, Bot. Reg. t. 914 (with broader leaves).—King George's Sound, *R. Brown* and others.

2. LEVENHOOKIA, R. Br.

(Coleostylis, *Sond.*)

Calyx-lobes 5. Corolla irregular, 4 lobes nearly equal, more or less con-tracted at the base into claws, the 5th or labellum usually shorter, with a very concave or hood-shaped lamina, enveloping the top of the column or elastically thrown back. Column short and erect, adnate at the base to one side of the calyx-tube. Stigma 2-lobed or undivided. Ovary 1-celled, with a basal placenta more or less connected with the sides by a short, incomplete dissepiment. Capsule globular, crowned by the calyx-lobes. Seeds few.— Small, erect, simple or corymbosely-branched annuals. Leaves small, alter-nate, not rosulate. Flowers crowded in short racemes at the ends of the stem or branches, each one pedicellate in the axil of a small, leaf-like bract. Capsule glandular-hispid in all the species.

The genus is limited to the southern colonies and chiefly West Australia. Although very near *Stylidium*, with the foliage and habit of some of the series *Tenellæ*, it is curiously distinguished by the irritability residing in the hood-shaped labellum, whilst the column is erect and immovable, whereas in *Stylidium* the irritability is in the long, folded column. De Candolle altered the name into *Leeuwenhoekia*, in conformity with the Dutch orthography of the name of the natural philosopher Leeuwenhoek, to whom it is dedicated ; but, in transferring that name into Latin, or rather in making a Latin word out of it, the modified spelling used by Brown for euphony's sake is in strict conformity with the practice of Linnæus and his immediate followers in similar cases, and gives a pronunciation much easier, and, at the same time, nearer to the true one, than if the original Dutch letters were preserved in languages where they have not the same value.

Labellum short, nearly sessile at the throat of the corolla or on a short, broad claw. Style without any (or rarely with a small) appendage under the anthers.
Corolla-tube much shorter than the calyx-lobes, and the lobes not as long again.
Glabrous except the calyx-tube. Flowers in a dense, leafy corymb.
Style-lobes 2, linear . . . . 1. *L. pusilla.*
Glandular-pubescent. Inflorescence simple or nearly so.
Labellum scarcely coloured. Style entire . . . . 2. *L. dubia.*
Labellum dark purple. Style 2-lobed . . . . 3. *L. Sonderi.*
Corolla-tube as long as or longer than the calyx-lobes.
Corolla-tube not much exceeding the calyx-lobes; lobes 3 lines long. Style 2-lobed . . . . 4. *L. pauciflora.*
Corolla-tube slender, twice or thrice as long as the calyx-lobes. Style entire . . . . 5. *L. leptantha.*
Labellum on a long claw and nearly as long as the other petals. Column with a linear appendage under the anthers.— COLEOSTYLES, *Sond.*
Corolla-tube very short . . . . 6. *L. stipitata.*
Corolla-tube from a little shorter to rather longer than the calyx-lobes . . . . 7. *L. Preissii.*

1. **L. pusilla,** *R. Br. Prod.* 573. Glabrous except the calyx-tube and capsule, 1 to 2 in. or rarely 3 in. high, densely corymbose in the upper part. Leaves obovate or oblong-spathulate, obtuse, rather thick, rarely above 3 lines long including the petiole, the leafy bracts similar or narrower, exceeding the flowers in some specimens, rather shorter in others. Pedicels short. Calyx rarely 1 line long, the lobes rather unequal, but variable in breadth and proportion. Corolla-tube very short, 4 lobes having each a short, linear claw and obovate white lamina scarcely above ¼ line diameter, the 5th or labellum purple, nearly as long as the other lobes, but the claw broad and concave, shorter than the hood-shaped lamina. Style with 2 linear lobes protruding from the anthers.—Bauer, Illustr. t. 15 ; DC. Prod. vii. 338 ; Sond. in Pl. Preiss. i. 392.

W. Australia. King George's Sound and adjoining districts, *R. Brown, Baxter,* and many others, and thence to Swan River, *Drummond, 1st Coll., Preiss, n.* 2241, 2247; eastward to Cape Arid, *Maxwell.*

2. **L. dubia,** *Sond. in Pl. Preiss.* i. 392. Glandular-pubescent or hirsute, 1 to 2 or rarely 3 in. high, simple or branched but not so corymbose as *L. pusilla.* Leaves obovate-oblong or ovate, 2 to 3 lines long including the

petiole, the leafy bracts similar or rather narrower. Pedicels short or scarcely exceeding the leaves. Calyx a little more than 1 line long. Corolla-tube shorter than the calyx-lobes (longer, *Sonder*); lobes obovate, contracted at the base but scarcely clawed, about 1 line long, equal (or 2 shorter, *Sonder*), apparently white, with a yellow, somewhat glandular throat; labellum on a short claw, the small, hood-shaped, pale or slightly-coloured lamina much shorter than the lobes. Column slender. Stigma obovate, entire (2-lobed, *F. Mueller*).—*L. creberrima*, F. Muell. Fragm. iii. 121.

Victoria. Very common in pastures, both in the plains and in the mountains, from the western limits of Gipps' Land, *F. Mueller.*

S. Australia. Very common round St. Vincent's Gulf, *F. Mueller.*

W. Australia. Swan River, *Drummond, 1st Coll. n.* 516, *Preiss, n.* 2252; Stirling Range, *F. Mueller.* I can discover no difference whatever between the W. Australian and the Victorian specimens, nor have I been able to trace the 2 lobes of the style described by F. Mueller in the latter. There may be, however, in this respect some slight sexual difference.

3. **L. Sonderi,** *F. Muell. Fragm.* i. 18. Glandular-pubescent, 1 to 2 in. high, simple in all the specimens seen. Leaves petiolate, orbicular, 1 to 1½ lines diameter, the leafy bracts rather smaller and narrower. Calyx about 1 line long. Corolla-tube shorter than the calyx-lobes, and the lobes but little exceeding them, obovate and nearly equal. Labellum with a very short claw, the hood-shaped lamina of a deep purple and nearly as long as the lobes. Style distinctly 2-lobed.—*Coleostyles Sonderi*, F. Muell. in Trans. Phil. Soc. Vict. i. 46.

Victoria. Violet Creek, *Wilhelmi.* Scarcely differs from *L. dubia* in the larger deep-coloured labellum.

4. **L. pauciflora,** *Benth. in Hueg. Enum.* 74. Glandular-pubescent, especially in the upper part, 2 to 3 in. high, simple in all the specimens seen. Leaves very few and small, petiolate, ovate or orbicular, the leafy bracts oblong-spathulate and still smaller. Flowers few, on rather long pedicels, in a very short, terminal raceme. Calyx nearly 2 lines long. Corolla-tube exceeding the calyx-lobes; lobes white, streaked with red, obovate-oblong or spathulate, 2 of them 3 lines long, the 2 others rather shorter; labellum on a short claw, the lamina much shorter than the other lobes, broadly hood-shaped, of a deep purple, with long, entire, uncoloured appendages at the base and a fringed appendage in the terminal notch. Style shortly 2-lobed or quite entire, the column usually with a membranous dilatation immediately under the anthers.— DC. Prod. vii. 338 ; *L. stylidioides*, F. Muell. Fragm. vi. 77.

W. Australia. King George's Sound to Stirling Range, *Huegel, Collie, Oldfield, F. Mueller.*

5. **L. leptantha,** *Benth.* Very slender, 1 to 3 in. high, usually simple and hirsute with a few small scattered hairs. Lower leaves obovate, upper ones lanceolate, 1 to 2 lines long, narrowed into a short petiole. Flowers very few, in a short, simple raceme or cluster. Calyx scarcely above ½ line long. Corolla-tube slender, nearly 2 lines long, the lobes obovate, about 1½ lines long, narrowed into a short claw; labellum very short, the purple hood-shaped lamina deeply notched. Column very short. Stigma entire, not exceeding the anthers.

W. Australia, *Drummond, n.* 128, 175, 282; Champion Bay and Murchison river, *Oldfield ;* also a few specĭmens mixed in *Preiss's n.* 2249 from Sussex district.

6. **L. stipitata,** *F. Muell. Fragm.* iv. 94. Glandular-pubescent or hirsute, 2 to 4 in. high, usually branched. Leaves linear or the lower ones obovate, the upper floral ones or bracts very narrow. Flowers small, numerous, on long and slender pedicels, the uppermost almost umbellate. Calyx scarcely 1 line long, the narrow lobes longer than the tube. Corolla-tube scarcely any ; lobes obovate, nearly 2 lines long, narrowed into claws ; labellum nearly as long, on a slender claw, the lamina broadly hood-shaped, with broad auricles or appendages at the base of the limb, and an appendage in the deep, terminal notch, sometimes short and purple like the limb, sometimes long and narrow. Column elongated but straight, with a reflexed, linear-lanceolate appendage immediately under the anthers. Stigma entire, very short, and rounded.—*Stylidium stipitatum*, Benth. in Hueg. Enum. 72 ; DC. Prod. vii. 335 ; *Coleostylis umbellulata*, Sond. in Pl. Preiss. i. 391.

W. Australia. Swan River, *Drummond*, 1st *Coll. n.* 514, *Preiss, n.* 2240 (partly) ; Mount Barker, *Oldfield ;* Champion Bay, *Walcott.* I cannot find, either in this or in *S. Preissii*, the adnate sheath at the base of the column described by Sonder, and from which he took the proposed generic name *Coleostylis.* In Huegel's 'Enumeratio' I had myself misdescribed the petals ; there are 4, not 2, with obovate lamina, and only 1, the labellum, with a slender claw and cordate (concave) lamina.

7. **L. Preissii,** *F. Muell. Fragm.* iv. 94. Pubescent or hirsute with short, glandular hairs, sometimes simple and 2 to 3 in. high, more frequently corymbosely-branched and 3 to 5 or even 6 in. high. Leaves from oblong-spathulate to linear, rarely exceeding $\frac{1}{2}$ in., and usually much shorter. Flowers pedicellate, in short, terminal racemes. Calyx a little more than 1 line long, with narrow lobes. Corolla-tube from nearly as long as to rather longer than the calyx-tube ; lobes obovate, about 2 lines long including the claw ; labellum as long as the lobes, with a similar claw, but the lamina broadly hood-shaped, with an oblong-spathulate appendage in the deep, terminal notch, and a yellow appendage on each side at the base of the lamina. Column slender, rather longer than the claws of the petals, but not folded, with a linear appendage immediately under the anthers. Stigma entire, very short, and rounded.—*Coleostylis Preissii*, Sond. in Pl. Preiss. i. 391.

W. Australia. Swan River and adjoining districts, *Drummond*, 1st *Coll. n.* 515, *Preiss, n.* 2249, 2250 (a few specimens of *L. leptantha* mixed under n. 2249).

3. **FORSTERA,** Linn. f.

Calyx-lobes 5, nearly equal, 2 sometimes more united. Corolla nearly regular, funnel-shaped or almost campanulate, the 5 lobes nearly equal or 2 rather larger. Column free from the calyx-tube. Ovary imperfectly 2-celled at the base. Stigma 2-lobed. Capsule 1-celled.—Small perennials. Leaves entire, crowded on the tufted stock or short stems, or imbricated along the densely tufted branches. Scapes or peduncles terminal, bearing 1 or rarely 2 or 3 flowers, which are often more or less unisexual.

The genus contains, besides the Australian species which is endemic, a few others from New Zealand and Antarctic America. With the peculiar genital column of *Stylidieæ*, it has the flowers otherwise nearer those of *Campanulaceæ*.

1. **F. bellidifolia,** *Hook. f. Ic. Pl. t.* 851 *and Fl. Tasm.* i. 236. Quite glabrous. Stock densely tufted. Leaves all radical¡ rosulate-obovate or oblong-spathulate, very obtuse, rather thick, rarely above ½ in. long. Scapes slender, 3 to 5 in. long, usually with 2 or 3 small, linear bracts under the flower. Calyx about 2 lines long, the linear-oblong lobes about as long as the globular tube. Corolla-tube about 1 line long, the lobes rather longer, obovate, 2 rather larger than the others, the throat often with small, glandular appendages. Stigma-lobes broad, shortly exceeding the anthers, which were not quite perfect in the flower examined, probably a partially female one. Capsule broadly ovoid, about 3 lines long.

Tasmania. Mount Sorell and summit of the ranges above Birch's Inlet, Macquarrie Harbour, *Milligan, Gunn ;* Mount Lapeyrouse, *C. Stuart.* The species comes nearest to the New Zealand *F. tenella,* Hook. f.

ORDER LXIV. **GOODENOVIEÆ.**

Calyx-tube adnate to the ovary or rarely free, the limb of 5 persistent lobes, sometimes very small, or united in a ring, or quite obsolete. Corolla irregular or rarely regular, with 5 lobes, valvate in the bud, their margins usually induplicate and expanding into glabrous *wings* as the flower opens. Stamens 5, alternate with the lobes of the corolla and inserted at the junction of the corolla-tube with the ovary or very rarely shortly adnate to the corolla-tube ; anthers 2-celled, the cells parallel, opening longitudinally, free or united in a ring round the style. Ovary wholly or partially inferior, at least as to the corolla, or rarely free within the corolla-tube, 1- or 2-celled ; ovules 1, 2 or more in each cell, erect or ascending (except in *Cctosperma*). Style simple and undivided (except in *Calogyne*), with a cup-shaped or 2-lipped dilatation, called *indusium*, at the top, enclosing the stigma. Fruit an indehiscent nut or drupe, or a capsule opening in 2 or 4 valves or rarely bursting irregularly or almost indehiscent. Seeds with a thin or crustaceous or thick and hard testa ; embryo straight in the centre of a somewhat fleshy albumen, which is rarely deficient ; radicle next the hilum.—Herbs undershrubs or rarely shrubs, the juice not milky. Leaves alternate or radical, very rarely irregularly opposite, entire, toothed or rarely pinnatifid Flowers hermaphrodite, axillary or in terminal spikes, racemes or panicles, the primary inflorescence centripetal, the secondary usually cymose and dichotomous. Bracteoles on the 1-flowered peduncles (when present) and bracts at the forks of the dichotomous cymes, opposite. Corolla yellow blue or white, rarely red or purple.

The Order is almost exclusively Australian, a very few species only of one genus (*Scævola*) being known from New Zealand, the Pacific islands, and the coasts of tropical and subtropical Africa, Asia, and America, and one of another genus (*Calogyne*), perhaps not different from an Australian one, extending to the coast of China. It is, no doubt, allied to *Campanulaceæ,* but, besides the differences in the ovules, fruit, and seeds, and the want of the milky juice of that Order, *Goodenovieæ* are readily known by the remarkable indusium, which evidently, together with the peculiar surrounding hairs of the style or of the corolla, acts a considerable part in assisting the proper impregnation of the stigma. The contrivances by which this impregnation is impeded, retarded, or promoted, appear to be very different in different genera, as, for instance, in *Goodenia, Leschenaultia,* and *Dampiera,* and are well worthy of careful observation on the part of botanists resident in Australia, where the flowers can be observed subject to the natural operation of insects, as well as of climatological and other external influeuces.

In the systematic arrangement and description of the genera and species, it is with regret that I have been unable to avail myself more largely of the elaborate monograph of De Vriese (' Goodenovieæ,' 4to, Haarlem, 1854), but the extraordinary confusion in both genera and species, as well of characters as of synonyms and identifications displayed in that work, and in the Hookerian and other herbaria that were placed at the author's disposal and which he has named, show that he could scarcely have formed any definite idea of his own genera or species, and that in working he must have generally contented himself with outer aspect, very rarely resorting to analysis. I feel therefore compelled to reject as doubtful those few of his species of which I have seen no tolerably authentic specimen.

Ovules 2 or more in each cell of the ovary or on each side of the imperfect or rudimentary dissepiment.
   Anthers connate round the style. Capsule linear, 4-valved at least at the base. Leaves narrow-linear or heath-like or reduced to scales.
      Indusium 2-lipped. Seeds hard, in 2 or 4 rows. Flowers solitary or in leafy corymbs . . . . 1. LESCHENAULTIA.
      Indusium cup-shaped. Seeds small and numerous. Flowers small in terminal clusters . . . . 2. ANTHOTIUM.
   Anthers free (when the flower is expanded). Capsule globular ovoid or oblong, opening from the top in 2 or 4 valves. Indusium cup-shaped.
      Calyx entirely free . . . . 3. VELLEIA.
      Calyx-tube adnate (sometimes exceedingly short), lobes free or adnate at the base.
         Style undivided . . . . 4. GOODENIA.
         Style 2- or 3-cleft . . . . 5. CALOGYNE.
   Anthers free. Fruit more or less succulent and indehiscent.
      Ovules several, erect or ascending . . . . 6. SELLIERA.
      Ovules 2 in each cell, *pendulous* . . . . 7. CATOSPERMA.
Ovules 1 or 2 in the whole ovary.
   Capsule 2-valved . . . . *Goodenia barbata.*
   Drupe or nut indehiscent.
      Calyx-tube adnate.
         Corolla-tube slit, lobes spreading, not auriculate. Ovules 2 (except in *S. fasciculata*). Anthers free . . . . . . 8. SCÆVOLA.
         Corolla-tube entire, lobes horizontally spreading. Ovules 2. Anthers free . . . . 9. DIASPASIS.
         Corolla-tube slit, upper lobes connivent, not auriculate. Ovule 1. Anthers free . . . . 10. VERREAUXIA.
         Corolla-tube slit, upper lobes auriculate. Ovule 1 (except the sect. *Dicœlia*). Anthers connate . . . . 11. DAMPIERA.
      Calyx and corolla-tubes almost closed over the ovary but free.
         Corolla-tube entire, lobes horizontally spreading. Ovule 1 . 12. BRUNONIA.

1. LESCHENAULTIA, R. Br.

(Latouria, De Vr.)

Calyx-tube linear, wholly adnate, lobes linear or lanceolate. Corolla oblique, the tube slit open to the base or rarely closed, the lobes all or partially erect and connivent or spreading. Anthers cohering round the style or rarely free. Ovary inferior, completely 2-celled, with several, sometimes numerous ovules ascending in 2 rows in each cell. Indusium broadly 2-lipped, the upper lip shorter, glandular inside and with a half-ring of short hairs on the outside at the base, the lower lip smooth or hairy inside ; stigma obsolete (or adnate to the upper lip?). Capsule linear, either entirely 4-valved, or seedless contracted and entire at the top and sometimes between the seeds. Seeds usually truncate, and more or less angular testa thick and hard, sometimes almost bony; embryo from half as long to nearly as long as the albumen.—Herbs undershrubs or shrubs. Leaves narrow-linear, entire, scattered or crowded and heath-like. Flowers either solitary and terminal or leaf-opposed or several in compact, leafy terminal corymbs, blue white yellow red or greenish.

The genus is limited to Australia, and is readily known as well by the habit as by the indusium, ovary, and fruit.

SECT. I. **Euleschenaultia.**—*Capsule seed-bearing and 4-valved to the top.*— *Flowers solitary, on short leafy branchlets or in compact leafy corymbs.*

Corolla (red yellow or greenish?), the 2 upper lobes connivent, the 3
  lower broad and very spreading.
    Corolla-lobes not longer than the cylindrical tube. Branchlets
    straight. Leaves rather crowded, heath-like.
      Upper corolla-lobes rather broad and erect . . . . 1. *L. formosa.*
      Upper corolla-lobes acuminate and recurved . . . . . . 2. *L. chlorantha.*
    Corolla-lobes very broad, longer than the short broad tube,
    branchlets usually curved or twisted. Leaves often ½ in. long . 3. *L. linarioides.*
Corolla (red yellow or greenish?), the tube long and cylindrical, the
  lobes short, connivent or slightly spreading.
  Leaves crowded or imbricate, almost acerose, 2 to 4 lines long.
    Corolla 6 to 8 lines long, the wings of the lobes united with
    the terminal mucro on the back . . . . 4. *L. tubiflora.*
  Leaves crowded, ½ to 1 in. long. Corolla nearly 1 in., the wings
    of the lobes spreading the terminal mucro between them . . . 5. *L. superba.*
  Leaves crowded, 2 to 4 lines long. Corolla nearly ¾ in., the lobes
    acuminate, recurved, not at all or scarcely winged . . . . . 6. *L. acutiloba.*
Corolla (red yellow or greenish?), the tube long and cylindrical, the
  lobes all very spreading and winged.
  Heath-like shrub. Leaves fine, under ½ in. long. Flowers nu-
    merous but scarcely corymbose . . . . 7. *L. laricina.*
  Undershrubs (or shrubs?), the stems (or branches?) corymbose
    at the top. Leaves scattered, often above 1 in. long.
    Very hispid. Flowers glandular-pubescent . . . . 8. *L. hirsuta.*
    Quite glabrous . . . . 9. *L. longiloba.*
Corolla (blue), the lobes all spreading, with broad, transversely-veined
  wings . . . . 10. *L. biloba.*
Corolla (blue lilac pale or white), lobes spreading, almost unilabiate,
  the wings of the upper or of all the lobes, narrow and veinless
  or nearly so.
  Corolla-lobes all with narrow wings. Flowers in densely leafy
    corymbs.
    Calyx-tube shorter than the floral leaves. Corolla 4 to 5 lines
    long . . . . 11. *L. expansa.*
    Calyx-tube longer than the floral leaves. Corolla 6 to 8 lines
    long . . . . 12. *L. floribunda.*
  Upper corolla-lobes lanceolate, arcuate, acute, not winged, lower
    ones with transversely veined wings . . . . 13. *L. heteromera.*

SECT. II. **Latouria.**—*Capsule ending in a slender seedless indehiscent teak. Leaves or scales distant; flowers solitary and terminal or leaf opposed.*

Leaves all reduced to small scales. Branches rigid, intricate, divaricate . . . . 14. *L. divaricata.*

Leaves and stems filiform. Capsule pedicellate . . . . 15. *L. filiformis.*
Leaves linear, acute, slightly flattened. Stems filiform. Capsule
   sessile . . . . 16. *L. agrostophylla.*

SECT. I. EULESCHENAULTIA. Capsule seed-bearing and 4-valved to the
top. Flowers solitary on short leafy branchlets or in compact leafy corymbs.

1. **L. formosa,** *R. Br. Prod.* 581. A weak, diffuse or spreading and
much-branched shrub, attaining sometimes 2 ft. Leaves rather loosely
scattered, obtuse or scarcely acute, 2 to 4 lines long. Flowers red, solitary,
terminating short leafy branchlets or becoming lateral by the growing out of
the upper axillary bud. Calyx-tube longer than the surrounding leaves;
lobes linear or linear-lanceolate. Corolla-tube 3 to 5 lines long, split to the
base; lower lobes large and spreading, but not longer than the tube; upper
ones broad rounded, erect and connivent, shorter than the lower ones. Capsule ½ to 1. in. long.—DC. Prod. vii. 519; De Vr. in Pl. Preiss. i. 414;
Sweet, Fl. Austral. t. 26; Bot. Reg. t. 916; Bot. Mag. t. 2600; *L. oblata,*
Sweet, Fl. Austral. t. 46; *L. multiflora,* Lodd. Bot. Cab. t. 1579; DC.
Prod. vii. 519; *L. Baxteri,* G. Don in Loud. Hort. Brit. 79.

W. Australia. King George's Sound and adjoining districts, *R. Brown, Drummond,*
n. 178, *Preiss, n.* 1457, 1458, *F. Mueller* and others; eastward to Stokes Inlet and Cape
Arid, *Maxwell.*

Sweet distinguishes *L. oblata* by the larger wings of the corolla-lobes and by the
pubescent filaments and corolla, but I find the wings sometimes quite as broad with perfectly
glabrous filaments, and the hairs, when present on the filaments and back of the corolla, are
but very few and small. F. Mueller sends some specimens which he considers as distinct
from *L. formosa* in having an erect stem; I can find no other differences.

2. **L. chlorantha,** *F. Muell. Fragm.* ii. 20. A low, diffuse, much-
branched shrub, with the habit of *L. formosa,* but the leaves are finer, ¼ to ½
in. long. Inflorescence and flowers the same, except that the two upper
connivent lobes of the corolla are acuminate and more or less recurved, and
the colour, according to Oldfield's label, is pale green, which does not show
on the specimen.

W. Australia. Murchison river, *Oldfield.*

Some specimens from the Kalgan river, *Oldfield, in Herb. Hook.,* have the fine leaves
and acuminate upper corolla-lobes of *L. chlorantha,* but the flowers are marked on Oldfield's
label as turquoise blue, the corresponding specimens in Herb. F. Mueller are marked as
having the flowers red. Perhaps the whole are a variety only of *L. formosa.* In the dried
state the flowers look red in all.

3. **L. linarioides,** *DC. Prod.* vii. 519. A prostrate or divaricately
branched shrub, attaining sometimes several feet, the branches often much
incurved at the end. Leaves not crowded, slender, acute, sometimes exceeding ½ in. Flowers rather large, terminating short branchlets and sessile
within the upper leaves, often several together in a terminal leafy corymb, the
flowering branchlets or even the calyx-tube itself often much incurved or
twisted. Calyx-lobes lanceolate. Corolla-tube short, broad, and gibbous,
he lower lobes often ¾ in. long, with very broad wings of a greenish-yellow,
the upper lobes reddish, oval-oblong, erect and connivent, much shorter than
the lower ones. Capsule curved, 1 to 1½ in. long.—*L. arcuata,* De Vr. in
Pl. Preiss. i. 416, Gooden. 186; Bot. Mag. t. 4265, copied into Fl. des

Serres, iii. t. 219.; Paxt. Mag. xiv. 245, with a fig. ; *Scævola grandiflora,*
Benth. in Hueg. Enum. 70 ; DC. Prod. vii. 512.

W. Australia. Swan River, *Fraser, Drummond, 1st Coll. n.* 186, also *n.* 147 ? ;
Preiss, n. 1465 ; Murchison river, *Oldfield;* Champion Bay, *Walcott;* Dirk Hartog's
Island, *Milne.*

4. **L. tubiflora,** *R. Br. Prod.* 581. A bushy shrub, sometimes low,
diffuse, and densely branched, sometimes more erect, 3 to 4 ft. high, with
divaricate or shortly virgate branches. Leaves crowded or densely imbricate,
2 to 4 lines long, rigid and almost acerose, with pellucid points. Flowers
from a greenish- to an orange-yellow or perhaps red, solitary on the short
branchlets, sessile within the terminal leaves, which are longer than the
ovary. Calyx-lobes like the leaves. Corolla 6 to 8 lines long, the tube cylindrical, at first closed, but splitting more or less on the upper side, the
lobes all shorter than the tube, connivent or scarcely spreading, the wings
short and united at the end, with the terminal mucro at the back. Capsule
narrow as in the other species, but not longer than the surrounding leaves.—
DC. Prod. vii. 519 ; De Vr. Gooden. 183, but only as to Brown's plant; *L.
pinastroides,* Lehm. Pl. Preiss. ii. 244 ; De Vr. Gooden. 133. t. 36.

W. Australia. King George's Sound and adjoining districts, *R. Brown, Drummond,
n.* 163, 404, *Preiss, n.* 430, and others ; eastward to Stokes Inlet and Fitzgerald Ranges,
Maxwell.

Preiss's specimens, n. 1458, referred here by De Vriese, belong to *L. formosa;* Drummond's n. 65, also quoted by him under *L. tubiflora,* is a species of *Eremophila* or some
allied genus.

5. **L. superba,** *F. Muell. Fragm.* vi. 10. A bushy shrub of 2 ft. with
virgate branches. Leaves crowded, $\frac{1}{2}$ to 1 in. long. Flowers large, yellow
(*Maxw.*), sessile in a cluster of small leaves, and often 2 or 3 together at the
ends of the branchlets. Calyx-lobes linear-subulate. Corolla nearly 1 in.
long, the tube cylindrical, slit on the upper side, the lobes short, nearly
equal, scarcely spreading, with rather broad, distinct wings, mucronate-
acuminate between them. Capsule $\frac{3}{4}$ to 1 in. long.

W. Australia, *Drummond, n.* 354 ; Phillips river and E. Mount Barren, *Maxwell.*

6. **L. acutiloba,** *Benth.* An erect or diffuse shrub. Leaves rather
crowded, and sometimes imbricate, obtuse or with short points, 2 to 3 lines
long. Flowers (red? or yellow?) solitary and terminal, sessile within the
last leaves, but the ovary usually exceeding them. Calyx-lobes acute, sometimes slightly lanceolate. Corolla-tube about $\frac{1}{2}$ in. long, cylindrical, slit to
the base; lobes short, erect or slightly spreading, acutely acuminate, not
winged.

W. Australia. Young river, *Maxwell.*

7. **L. laricina,** *Lindl. Swan Riv. App.* 27. A much branched, erect,
bushy shrub. Leaves rather crowded, usually fine, $\frac{1}{4}$ to $\frac{1}{2}$ in. long. Flowers
varying from white or lilac to the richest red (*Drummond*), sessile in the
upper axils, often numerous, the calyx-tube not exceeding the floral leaves.
Calyx-lobes like the leaves, but shorter than the corolla-tube. Corolla-tube
fully $\frac{1}{2}$ in. long, slit to the base, the lobes all similar, sometimes nearly as

long as the tube, but usually shorter, the wings rather broad, with a small
point between them.—De Vr. Gooden. 185.; *L. splendens*, Hook. Bot. Mag.
t. 4256, copied into Fl. des Serres, ii. t. 176; Paxt. Mag. xiv. 245, with
a figure.

W. Australia. Swan River, *Drummond, 1st Coll. ;* Darling Range, *Collie.*

L. parviflora, De Vr. in Pl. Preiss. i. 416, and *L. tenuifolia*, De Vr. l. c. 415, are both
referred by the author (Gooden. 185) to *L. laricina*, but the characters he gives are those
of *L. expansa.* I have not seen the specimens.

8. **L. hirsuta,** *F. Muell. Fragm.* vi. 9.   Apparently an undershrub,
hispid with rigid hairs.   Stems decumbent or erect, sometimes exceeding
1 ft., slightly branched towards the top.   Leaves linear, acute, often above
1 in. long.   Flowers large (red?), sessile in the upper axils.   Calyx-lobes
linear-subulate, acute.   Corolla about 1¼ in. long, glandular-pubescent outside ; tube narrow ; lobes rather shorter than the tube, all winged, and probably spreading.   Style long.   Capsule very long, but not seen ripe.

W. Australia. Between Moore and Murchison rivers, *Drummond, 6th Coll. n.* 145.
The habit is nearly that of *L. longiloba*, but the hairs are quite exceptional in the whole
genus.

9. **L. longiloba,** *F. Muell. Fragm.* vi. 10.   Apparently an undershrub,
with several decumbent or ascending stems under 1 ft. high, corymbosely
branched towards the end only.   Leaves not crowded, linear, mucronate-
acute, mostly above ½ in. long, sometimes rather broad and almost lanceolate,
the upper, especially the floral ones, irregularly opposite.   Flowers (red ?)
sessile in the upper axils.   Calyx-lobes usually long and acute.   Corolla
nearly 1 in. long, the tube narrow, less woolly inside than in most species,
the lobes all spreading and winged, mucronate-acuminate, as long as the
tube.   Capsule long, but not seen ripe.

W. Australia, *Drummond*, the typical specimens without n., but n. 179 of the 4th
Coll. appears to be the same.

10. **L. biloba,** *Lindl. Swan Riv. App.* 27 ; *Bot. Reg.* 1841, *t.* 2.   A
weak shrub of 2 or 3 ft.   Leaves rather slender, not very crowded, ¼ to
nearly ½ in. long.   Flowers blue, sessile in the upper axils, few together in
each branchlet, but sometimes forming a broad, leafy corymb.   Calyx-lobes
like the leaves.   Corolla varying from 8 lines to nearly 1 in. long, the lobes
longer than the tube, all spreading, more or less mucronate, with broad,
spreading, dark blue wings, marked with parallel, transverse veins like those
of *Scævola striata.*   Capsule ¾ to 1½ in. long.—Paxt. Mag. viii. 151 ; De
Vr. Gooden. 182. t. 35 ; *L. grandiflora*, DC. Prod. vii. 519 ; De Vr. Gooden.
181 ; *L. Drummondi*, De Vr. l. c. 182.

W. Australia. King George's Sound to Swan River, *Drummond, 1st Coll. n.* 3, *2nd
Coll. n.* 415, *Preiss, n.* 1463, 1466, *Harvey, Oldfield*, and others.

L. grandiflora, Lindl. Swan Riv. App. 26, is a large-flowered form of the same species.

11. **L. expansa,** *R. Br. Prod.* 581.   A weak branching or diffuse
shrub of 1 to 2 ft.   Leaves mostly obtuse, under ½ in. long.   Flowers small,
pale blue, yellowish or white, sessile in small, compact, leafy corymbs at the
ends of the branches.   Calyx-tube shorter than the surrounding leaves, the

lobes usually as long as or longer than the corolla-tube. Corolla 4 to 5 lines long, the lobes longer than the tube, all digitately spreading and winged, the wings undulate, not broad, irregularly and very sparingly veined. Capsule about ½ in. long.—DC. Prod. vii. 519.

W. Australia. King George's Sound, *R. Brown, Baxter,* and others ; Swan River, *Drummond, 1st Coll.* ; Flinders Bay, *Collie ;* Mount Barker, Tone river, Wilson's Inlet, *Oldfield.* The flowers are usually described as 2-bracteolate, but they are so in appearance only, the flower being between 2 nearly opposite leaves similar to the subtending one, but in the axil of one of them is another flower-bud, which, as it is developed, causes the branchlet to grow out, and the leaves become alternate.

12. **L. floribunda,** *Benth. in Hueg. Enum.* 70. A shrub of 2 or 3 ft., with spreading or rarely virgate branches. Leaves rather crowded but not imbricate, finely pointed or almost obtuse, 2 to 4 lines long. Flowers white, pale blue, lilac or yellowish, in terminal, leafy corymbs, but not so dense as in *L. expansa,* the flowers themselves larger, and the calyx-tube at the time of flowering always longer than the surrounding leaves. Corolla 6 to 8 lines long, the lobes nearly equal, all digitately spreading and longer than the tube, the wings not at all or very obscurely veined, the point of the lobe in the notch between them.—DC. Prod. vii. 519 ; De Vr. Gooden. 186 ; *L. glauca,* Lindl. Swan Riv. App. 27 ; De Vr. Gooden. 184 ; *L. pallescens,* De Vr. in Pl. Preiss. i. 415 ; *L. expansa,* De Vr. Gooden. 184, partly, not of R. Br.

W. Australia. Swan River, *Huegel, Drummond, 1st Coll., Preiss, n.* 1460, 1464, 1468 ; Murchison river, *Oldfield.*

13. **L. heteromera,** *Benth.* A shrub, with rather rigid branches. Leaves mucronate-acute, thicker and more rigid than in most species, usually under ½ in. long, not imbricate. Flowers apparently blue, few in the upper axils, forming a short, loose, leafy spike, the calyx-tube longer than the surrounding leaves. Calyx-lobes very acute. Corolla (pale blue?) 6 to 8 lines long, the 3 lower lobes with broad, blue, transversely-veined wings as in *L. biloba,* but the 2 upper ones shorter, linear-lanceolate, acute, with narrow, involute margins but not winged. Capsule about ¾ in. long.

W. Australia, *Drummond, n.* 142, *Mylne ;* E. Mount Barren, Moir's Inlet, Oldfield and Phillips rivers, *Maxwell.*

SECT. II. LATOURIA, *Endl.*—Capsule ending in a slender, seedless, indehiscent beak. Leaves or scales of the stem distant. Flowers solitary, terminal or leaf-opposed.

14. **L. divaricata,** *F. Muell. Fragm.* iii. 33, 167. Summits of the plant leafless, rigid, flexuose, intricately branched, the leaves replaced by small, oblong-linear, distant scales. Flowers sessile, terminal or opposite the scales. Calyx-tube 2 to 3 lines long, the lobes narrow, about half as long. Corolla yellow (*F. Muell.*), 6 to 8 lines long, the lobes about as long as the tube, all narrow and expanded, the 3 lower ones winged, the 2 upper ones lanceolate-falcate, not winged. Fruit often above 1 in. long, ripening but very few large, thick seeds, and contracted into a neck at the top, and also between the seeds.

S. Australia. Cooper's Creek, *Wheeler.*

15. **L. filiformis,** *R. Br. Prod.* 581. An annual or perennial, with filiform, slightly branched stems of ½ to 1 ft. Leaves distant, filiform, ¼ to 1 in. long. Flowers blue (*R. Br.*), terminal or leaf-opposed, very shortly pedicellate. Calyx-tube long and slender, the lobes short and subulate. Corolla ½ in. long or rather more, the 3 lower lobes as long as the tube, with broadly oblong, erect, almost parallel wings, leaving a deep sinus between them, the margins of the lobes very undulate below the wings, the 2 upper lobes separated much lower down, and winged on one side only. Capsule slender, about 1 in. long when perfect, the upper half consisting of a slender, filiform beak. Seeds cylindrical or angular, truncate, the testa not quite so hard as in other species.—DC. Prod. vii. 519; F. Muell. Fragm. vi. 9; *Latouria filiformis,* De Vr. Gooden. 187.

N. Australia. Islands of the Gulf of Carpentaria, *R. Brown.*
Queensland. Shoalwater Bay passage, *R. Brown;* Rockingham Bay, *Dallachy.*

16. **L. agrostophylla,** *F. Muell. Fragm.* vi. 8. An erect annual, very closely allied to *L. filiformis,* and probably only a variety, the leaves are rather broader and flatter, the capsule not so slender, and usually sessile.

N. Australia. Sandstone tableland of the Upper Victoria river and Macarthur river, Gulf of Carpentaria, *F. Mueller.* Without fuller sets of specimens, I have been unwilling to unite this with *L. filiformis;* but Brown's Carpentaria island specimens appear to me to be intermediate between F. Mueller's and those from Queensland.

2. ANTHOTIUM, R. Br.

Calyx-tube adnate; lobes 5, free. Corolla with the 2 upper lobes free to the base, erect, connivent, each of them winged on one side, with a broad, concave, inflexed auricle covering the indusium, the 3 lower lobes united to the middle. Anthers cohering round the style. Indusium cup-shaped, not ciliate, enclosing the stigma. Ovary entirely inferior, 2-celled; ovules numerous, in 2 rows in each cell, mostly ascending. Capsule opening laterally, in 4 valves (cohering at the top?). Seeds small, slightly compressed; testa crustaceous.—Glabrous perennials, with a tufted stock. Leaves radical, entire. Flowering stems leafless or nearly so, simple or branched. Flowers (probably varying from a yellow to a purplish-red) in terminal clusters or dense corymbs, rarely almost solitary.

The genus is limited to Australia. It approaches *Dampiera* in the flower, with the ovary of *Leschenaultia,* and a peculiar habit.

Radical leaves narrow-linear. Calyx-tube contracted to the base.
    Upper lip of the corolla exceeding the calyx-lobes . . . . . . 1. *A. humile.*
Radical leaves linear-cuneate, oblong or lanceolate. Calyx-tube rounded
    at the base. Upper lip of the corolla shorter than the calyx-lobes . 2. *A. rubriflorum.*

1. **A. humile,** *R. Br. Prod.* 582. Radical leaves linear-terete or very narrow and thick but flat, 1½ to 4 in. long, the petioles shortly dilated at the base. Stems sometimes scarcely exceeding the leaves, erect and simple, sometimes ½ to 1 ft. long, ascending, erect (or procumbent?), and more or less branched, leafless except short, linear bracts under the branches. Flowers small, in a terminal, compact corymb or head when the stem is simple, or in smaller clusters at the ends of the branches, sessile within very

short bracts. Calyx-tube linear-turbinate, 2 to 3 lines long; lobes lanceolate or linear, usually shorter than the tube. Upper lobes of the corolla exceeding the calyx-lobes, broad and concave at the base, slightly contracted under the auricle, lower lip about 3 lines long, the lobes oblong, with narrow wings, and somewhat concave at the end. Capsule 3 to 4 lines long, but not seen quite ripe. Seeds very small, the testa very minutely glandular-tuberculate or nearly smooth.—DC. Prod. vii. 520; De Vr. Gooden. 188. t. 37; *Leschenaultia humilis*, Spreng. Syst. i. 720; *Goodenia junciformis*, De Vr. in Pl. Preiss. i. 413; *G. geniculata*, De Vr. l. c. (altered afterwards l. c. ii. 244 to *G. genuflexa*); and *G. pygmæa*, De Vr. in Pl. Preiss..i. 413.

W. **Australia,** *Drummond, n.* 416, 181, 183; King George's Sound and Lucky Bay, *R. Brown, Baxter;* Plantagenet and Stirling Ranges, *Maxwell;* sandy plains inundated in winter, Swan River, *Preiss, n.* 1456, 1522.

De Vriese refers here also (Gooden. 188 and in Herb. Hook.) Drummond's n. 191, which is *Scævola tortuosa*, and n. 140, which I have not seen.

2. **A. rubriflorum,** *F. Muell. Herb.* Radical leaves oblong-spathulate, lanceolate or linear-lanceolate, $1\frac{1}{2}$ to 2 in. long, the petiole dilated and membranous at the base. Stems erect and simple or nearly so in the specimens seen, 3 to 6 in. high, leafless or with 1 or 2 erect, short, linear leaves. Flowers red (*F. Muell.*), in a compact, terminal corymb, sessile and surrounded by bracts as in *A. humile*, but the calyx-tube is rather shorter and broader, rounded at the base and prominently ribbed, the upper connivent lobes of the corolla are not contracted under the auricle, but broader and rounded at the top, not exceeding the calyx-lobes, and the lower lip is nearly twice as long. Ovules of *A. humile* or even more numerous. Fruit not seen.

W. **Australia,** *Drummond, n.* 180, *Maxwell.*

3. VELLEIA, Sm.

(Euthales, *R. Br.*)

Calyx free from the ovary, consisting of 3 or 5 sepals, either distinct or connate in a campanulate tube at the base. Corolla oblique, the tube adnate to the ovary at the base, with a hollow protuberance sometimes produced into a spur, the upper part split on the upper side nearly or quite to the ovary, the 2 upper lobes separate lower down, unequally winged and auriculate, or rarely all equal and equally winged. Stamens free. Ovary adnate to the corolla at the base, the summit free within the corolla-tube, nearly 1-celled, divided only at the very base or rarely to the middle into 2 imperfect cells. Style simple; indusium usually large, closed and almost folded when the flower expands. Capsule quite free from the calyx-lobes, equally 4-valved, or rarely 2-valved with entire or 2-cleft valves. Seeds of *Goodenia*, flat, with a callous or winged margin.—Herbs, with a short, thick stock and radical leaves, or in one species forming a thick, branching, leafy stem. Scapes (or peduncles in *V. macrophylla*) erect or ascending, dichotomously- or trichotomously-branched, many-flowered. Bracts opposite, free or connate. Flowers yellow, like those of *Goodenia*.

The species are all Australian. The genus is very nearly allied to *Goodenia*, but differs constantly in the free calyx and peculiar dichotomous inflorescence.

Bracts large, broad, connate.
  Sepals 5, lanceolate, nearly distinct . . . . 1. *V. panduriformis.*
  Sepals 5, ovate, connate at the base . . . . 2. *V. connata.*
  Sepals 3, orbicular-cordate . . . . 3. *V. perfoliata.*
Bracts distinct or very shortly united at the base.
  Sepals 5, united in a campanulate, 5-lobed cup.
    Stock tufted. Leaves all radical . . . . 4. *V. trinervis.*
    Stem erect, leafy . . . . 5. *V. macrophylla.*
  Sepals 5, distinct or nearly so.
    Flowers rather large. Sepals obtuse, 4 to 6 lines long. Co-
      rolla frequently spurred. Capsule not exceeding the calyx . 6. *V. paradoxa.*
    Flowers small. Sepals acute, about 2 lines long. Corolla not
      spurred. Capsule exceeding the calyx . . . . 7. *V. cycnopotamica.*
  Sepals 3, distinct or nearly so.
    Upper lobes of the corolla deeply separate. Scapes longer than
      the leaves.
      Glabrous. Sepals broadly cordate, almost orbicular . . . 8. *V. lyrata.*
      Glabrous. Sepals ovate-lanceolate, obliquely cordate and
        shortly decurrent . . . . 9. *V. macrocalyx.*
      Villous. Sepals broadly ovate, not cordate . . . . . . 10. *V. pubescens.*
      Glabrous. Sepals oblong-lanceolate . . . . 11. *V. spathulata.*
    Corolla-lobes all nearly equal. Scapes shorter than the leaves 12. *V. montana.*

1. **V. panduriformis,** *A. Cunn. Herb.* Glabrous and glaucous.
Radical leaves petiolate, obovate, toothed. Scapes or stems very tall, less
forked than in other species, with several pairs of large, broad, connate
bracts, entire or scarcely toothed, each one 1 to 2 in. diameter. Flowers in
dense cymes in one axil of each of the upper pairs of bracts (only one fork
of the primary inflorescence elongated) or the terminal cyme regularly dichotomous. Sepals 5, lanceolate, nearly equal, about ½ in. long, free or slightly
connate at the base. Corolla not seen. Capsule about as long as the calyx.
Seeds about 4 lines diameter, including the broad transparent wing.

N. Australia. Goodenough Bay and Point Cunningham, N.W. coast, *A. Cunningham.* The specimens seen of this and the two following species are very incomplete; the
radical leaves of *V. panduriformis* are described from the drawing of a plant formerly raised
in Kew Gardens from Cunningham's seeds.

2. **V. connata,** *F. Muell. in Hook. Kew Journ.* viii. 162, *and in Trans.
Phil. Soc. Vict.* i. 18. Glabrous and glaucous. Radical leaves petiolate,
obovate-oblong and toothed (*Herb. Hook.*) òr oblong-spathulate and entire
(*Herb. F. Muell.*), 2 to 3 in. long. Stem tall, dichotomous, with large broad
connate entire or toothed bracts at the forks. Sepals 5, ovate or ovate-
lanceolate, acuminate, the outer one fully 4 lines long, the others smaller,
usually connate at the base. Corolla 7 to 8 lines long, the lower lobes not
winged (or the wings destroyed in the specimens seen?), the upper ones
winged on one side and slightly ciliate. Capsule about 3 lines diameter.
Seeds about 1 line diameter, including the wing.

N. S. Wales. Scrubby sandhills towards the junction of the Murray and Murrumbidgee, *F. Mueller.*
S. Australia. Between the Bonney river and Mount Morphett, *M'Douall Stuart,* a
single specimen in Herb. F. Muell., with the calyx-lobes much more connate than in the
typical specimens.

3. **V. perfoliata,** *R. Br. Prod.* 581. Glabrous, except a little short

wool in the axils of the leaves and bracts. Radical leaves petiolate, obovate or
oblong, toothed or entire. Scapes or stems tall, glaucous, dichotomous, with
large broad connate entire or toothed bracts at the forks. Flowers shortly
pedicellate in the forks, the upper ones in a compact corymb. Sepals 3, orbicular-cordate, about 3 lines long, the 2 inner ones scarcely smaller than the
outer one. Corolla yellow, about ½ in. long, glabrous outside, the lower
lobes winged, the 2 upper ones winged on the outer side only, with a concave densely hairy auricle below the wing. Style densely hairy. Capsule
shorter than the calyx. Seeds very flat and broad, with a thickish margin
not winged.—DC. Prod. vii. 518.

N. S. Wales. Blue Mountains, *Gordon (Herb. R. Br.), Miss Atkinson.* In the
dried specimens the wings of the corolla-lobes sometimes disappear.

4. **V. trinervis,** *Labill. Pl. Nov. Holl.* i. 54. *t.* 77. Glabrous or
rarely sprinkled with appressed hairs or in one variety villous. Radical
leaves on long petioles, broadly or narrow oblong, entire or remotely toothed,
sometimes distinctly 3-nerved, but the veins usually very obscure. Scapes
dichotomous, sometimes as in *V. paradoxa,* above 1 ft. high, with erect
branches, sometimes low and ascending as in *V. pubescens,* the bracts under
the forks lanceolate or linear, free or slightly connate at the base. Flowers
small. Sepals united in a campanulate calyx, about 2½ to 3 lines long, with
a turbinate tube and 5 unequal lobes, the larger outer one usually longer
than the tube. Corolla 5 to 6 lines long, slightly pubescent outside, the
lobes all broadly winged, the 2 upper ones rather unequally so and separated
nearly to the base. Dissepiment of the capsule more prominent than in most
species, attaining nearly to the middle. Seeds flat, not winged.—*Goodenia
tenella,* Andr. Bot. Rep. t. 466; Bot. Mag. t. 1137, not of R. Br.; *Euthales trinervis,* R. Br. Prod. 580; DC. Prod. vii. 517; De Vr. Gooden.
169.

W. Australia. King George's Sound, *R. Brown, Preiss,* n. 1515, and many others,
and thence to Swan River, *Drummond, 1st Coll., 2nd Coll. n.* 400, 401; Blackwood and
Tweed rivers and Port Gregory, *Oldfield;* eastward to Cape Arid and Cape le Grand, *Maxwell.*

The genus *Euthales* was separated by Brown from *Velleia* solely on account of the
gamosepalous calyx, but that occurs also, though in a less degree, in *V. connata;* F. Mueller
proposes to join it rather with *Goodenia;* but besides the inflorescence and capsule, which
are entirely those of *Velleia,* the calyx is always free, whilst in *Goodenia* the calyx-tube is
entirely adnate, and where that is very short the lobes are also adnate at the base, and persist on the capsule.

Var. *villosa,* more or less villous, the leaves often very densely so on the under side, but
sometimes sprinkled only with a few hairs. Scapes numerous, ascending, rarely above 6 in.
high.—*Drummond, 4th Coll. n.* 188; King George's Sound, *Collie;* Vasse river, *Oldfield;*
Stirling Ranges, Don river, and Cape Arid, *Maxwell.*

Euthales pilosella, De Vr. in Pl. Preiss. i. 414, from King George's Sound, *Preiss, n.*
1438, which I have not seen, would appear, from the character given, to be this variety, although the author (Gooden. 174) refers it to *V. spathulata,* an Eastern species to which
the description does not at all apply.

5. **V. macrophylla,** *Benth.* Closely allied to the larger forms of *V.
trinervis,* but the stock grows out into an erect leafy branching stem, and,
including the large loose dichotomous panicles, attains 3 or 4 ft. The whole
plant glabrous. Stem-leaves in the ordinary form 2 to 6 in. long, toothed

and narrowed into a rather long petiole. Peduncles axillary, paniculate, with opposite bracts at the forks, precisely like the paniculate scapes of *V. trinervis*, but the flowers and capsules usually larger.—*Euthales macrophylla*, Lindl. Bot. Reg. 1840, Misc. 54, and 1841, t. 3 ; De Vr. Gooden. 170. t. 33 ; Maund. Botanist, t. 209 ; *Goodenia macrophylla*, F. Muell. Fragm. vi. 11.

W. Australia. *Drummond, n.* 141, *3rd Coll. n.* 189 ; Kalgan river, *Oldfield.* The station, " Sydney, *Clowes,*" given by De Vriese, is owing to a mistake in Herb. Hook.

Var. *foliosa.* Stems thick. Leaves crowded, obovate or broadly cuneate, coarsely toothed, scarcely petiolate, but contracted into a short, broad, stem-clasping base, the axils woolly.—*Drummond, n.* 182 ; summit of Stirling Range, *F. Mueller.*

Euthales filiformis, De Vr. in Pl. Preiss. i. 414 and Gooden. 171, from W. Australia, *Preiss, n.* 1889, described from imperfect specimens without flower, and which I have not seen, is not recognizable, and may belong to some very different genus.

6. V. paradoxa, *R. Br. Prod.* 580. Glabrous, pubescent or hirsute. Leaves radical, petiolate, from broadly obovate and under 2 in. to narrow-oblong and above 4 in. long, coarsely toothed or almost entire, sometimes almost lyrate. Scapes ascending or erect, ½ to 1½ ft. high, dichotomously or trichotomously branched. Bracts opposite at the forks, free, the lower ones sometimes ovate and deeply lobed at the base, the upper ones small, linear, and entire. Sepals 5, free, oblong-linear or lanceolate or the outer larger one ovate-lanceolate and sometimes 6 lines long, the others smaller. Corolla yellow, pubescent outside, the hollow protuberance of the tube usually produced into a spur, sometimes 4 lines long, sometimes very small or obsolete ; lobes all winged, the 2 upper ones separate much lower than the others. Indusium very large. Capsule shorter than the outer sepals. Seeds broadly winged.—DC. Prod. vii. 518 ; De Vr. Gooden. 172 ; Hook. f. Fl. Tasm. i. 233 ; Bot. Reg. t. 971.

Queensland. Dawson river, *F. Mueller ;* Port Curtis, *M'Gillivray ;* Plains of the Condamine, *Leichhardt ;* Ipswich, *Nernst ;* Warwick, *Beckler.*

N. S. Wales. Cow-pasture plains, *R. Brown ;* Mudgee, *Woolls ;* Nangas, *M'Arthur ;* from the Murray and Darling to the Barrier Range, *Victoria and other Expeditions,* and northward to Richmond river, *Fawcett,* New England, *C. Stuart.*

Victoria. Port Phillip, *R. Brown ;* on the Yarra, *F. Mueller ;* Glenelg river, *Robertson ;* Creswick, *Whan ;* Wimmera, *Dallachy.*

Tasmania. Port Dalrymple, *R. Brown ;* not uncommon in many parts of the island, *J. D. Hooker.*

S. Australia. Gulf of St. Vincent's, Holdfast Bay, Lofty Ranges, etc., *F. Mueller* and others.

V. arguta, R. Br. Prod. 580, DC. Prod. vii. 518 ; De Vr. Gooden. 173, from Spencer's Gulf, *R. Brown,* is a small form, with the leaves rather more sharply toothed than usual, but not otherwise different. The western stations given for the species by De Vriese are founded upon a mixture of specimens of *V. trinervis* and of *Goodenia filiformis,* which had been glued down upon the same sheet in the Hookerian Herbarium.

Var. *stenoptera,* F. Muell. More divaricate; flowers rather smaller ; spur very small or none ; seeds with a very narrow or scarcely any wing. To this belong the Queensland and the New England and Richmond river specimens. The spur of the corolla disappears also sometimes on the common Southern form, and is always variable in size.

7. V. cycnopotamica, *F. Muell. Fragm.* vi. 7. Sprinkled with a few rigid hairs or nearly glabrous. Radical leaves petiolate, oblanceolate, acutely

toothed, 1 to 2 in. long. Scapes slender, ascending, 6 to 8 in. high, dichotomous, with small opposite lanceolate free bracts at the forks. Flowers
small, yellow. Sepals 5, lanceolate, acute, about 2 lines long and nearly
equal or the outer one broader, usually hispid. Corolla not seen perfect but
apparently like those of other not spurred species. Capsule exceeding the
calyx, somewhat compressed, with the 2 valves not splitting so readily into 4
as in other species. Seeds orbicular, winged.

W. Australia. *Drummond* (*2nd Coll.?*), *n.* 402, 410. The Hookerian and other
specimens seen by De Vriese were referred by him to *V. lanceolata*, Lindl. (*Goodenia filiformis*), or to *V. lyrata*, which is entirely eastern.

8. **V. lyrata,** *R. Br. Prod.* 580. Glabrous except a few hairs
on the stock. Leaves radical, oblong-spathulate, deeply toothed below the
middle or lyrate-pinnatifid, often several in. long. Scapes erect or ascending,
6 in. to nearly 1 ft. high, dichotomous, with spreading branches, and small
opposite ovate-lanceolate or linear free bracts at the forks. Sepals 3, broadly
cordate, the outer ones 3 to 4 lines long and broad, the 2 others rather
smaller. Corolla yellow, about ½ in. long, glabrous outside, the lobes broadly
winged, the 2 upper ones separated low down, unequally winged, with a
concave inflexed auricle near the base. Capsule rather shorter than the calyx.
Seeds orbicular, flat, with a rather thick margin, not winged.—DC. Prod. vii.
518; De Vr. Gooden. 173; Bot. Reg. t. 551; Hook. Exot. Fl. t. 24;
Guillem. Ic. Pl. Austral. t. 4; *V. spathulata*, Juss. in Ann. Mus. Par. xviii.
17. t. 1, not of R. Br.

N. S. Wales. Marshy places about Port Jackson, *R. Brown, Sieber n.* 223, and
many others. The western specimens referred by De Vriese to *V. lyrata*, belong to *V.
cycnopotamica* and *V. trinervis*.

9. **V. macrocalyx,** *De Vr. Gooden.* 176. *t.* 34. Glabrous except a
little wool at the base of the leaves and the inflorescence sometimes slightly
pubescent. Rootstock often thick and elongated. Leaves radical, petiolate,
obovate-oblong or spathulate, often several in. long, obtuse, entire sinuate-
toothed or rarely almost lyrate. Scapes sometimes short and few-flowered,
but when full grown ½ to 1 ft. high, dichotomous, with spreading branches
and small narrow opposite bracts at the forks as in *V. lyrata*. Sepals 3,
ovate-lanceolate or almost ovate, mucronate-acute, obliquely cordate at the
base, with the margins slightly decurrent, the outer one often 5 lines long,
the others smaller. Corolla about 6 lines long, the 2 upper lobes separated
lower down than the others and winged on one side only. Capsule of *V.
lyrata*. Seeds not seen.

Queensland. Burdekin river, *F. Mueller.*
N. S. Wales. Upper Clarence river, *Herb. F. Mueller.*

10. **V. pubescens,** *R. Br. Prod.* 581. Softly pubescent or villous.
Leaves radical, petiolate, oblong-spathulate and 2 to 4 in. long or shorter and
almost obovate, irregularly toothed or lobed towards the base. Scapes
ascending, shortly exceeding the leaves, dichotomous, with small opposite
lanceolate bracts at the forks. Sepals 3, broadly ovate or the inner ones
ovate-lanceolate, sometimes shortly united at the base, not decurrent. Corolla
pubescent outside, 6 or 7 lines long, the lobes all broadly winged, but the

wings in some specimens soon disappearing, the 2 upper lobes separated almost to the base and auriculate on the outer side below the wing.—DC. Prod. vii. 518.

Queensland. Shoalwater Bay, *R. Brown ;* Burdekin river, *Dallachy* (a dwarf stunted state).

11. **V. spathulata,** *R Br. Prod.* 580. Glabrous except the woolly axils. Leaves radical, mostly oblong-spathulate and 2 to 3 in. long, but sometimes shorter and obovate or longer and narrow, entire or with minute distant teeth. Scapes slender, ascending or spreading, rarely above 6 in. long, sometimes very much branched and almost filiform, the bracts at the forks small, oblong or linear. Flowers small, yellow. Sepals 3, scarcely 3 lines long, oblong-lanceolate or the outer one almost ovate and slightly cordate. Corolla 4 to 5 lines long, the 2 upper lobes separated low down and winged on one side only. Seeds flat, with a thickish border, not winged or rarely with a very narrow wing.—DC. Prod. vii. 518 ; De Vr. Gooden. 174.

Queensland. Shoalwater and Keppel Bays, *R. Brown ;* Port Arthur, *M'Gillivray ;* Wide Bay, *Bidwill ;* Rockingham Bay, *Dallachy ;* Rockhampton, *Thozet ;* Brisbane river, Moreton Bay, *Leichhardt, F. Mueller.*

N. S. Wales. Newcastle, *R. Brown.*

The Western *Euthales pilosella,* De Vr. in Pl. Preiss. i. 414, referred here by De Vr. Gooden. 174, must surely be *V. trinervis,* var. *villosa.*

12. **V. montana,** *Hook. f. in Hook. Lond. Journ.* vi. 265 ; *Fl. Tasm.* 234. *t.* 68, B. More or less densely hirsute or nearly glabrous. Leaves radical, petiolate, mostly oblong-spathulate but varying from obovate to oblanceolate, obtuse, entire, rather thick, 1 to 2 in. long. Scapes very short, dichotomous, the whole inflorescence not exceeding the leaves ; bracts opposite at the forks, small and narrow. Sepals 3, lanceolate or oblong, about 3 lines long, but unequal. Corolla of a dingy reddish-yellow (*Gunn*), 4 to 5 lines long, pubescent outside, the lobes all short, nearly equal, and equally winged. Capsule small, nearly globular. Seeds flat, with a slightly thickened border, not winged.—De Vr. Gooden. 176.

Victoria. Snowy plains on Snowy River at an elevation of 4000 to 5000 feet, Mount Wellington (Gipps' Land) and Haidinger Range, *F. Mueller.*

Tasmania. Mountain districts, forming large patches, on Hampshire Hills, Western Mountains, etc., *J. D. Hooker.*

With the habit, inflorescence, calyx, ovary and fruit of *Velleia,* this differs from all the other species in the corolla, which is nearer to that of *Scævola* or of the section *Monochila* of *Goodenia.*

4. GOODENIA, Sm.

(Picrophyta, *F. Muell. ;* Stekhovia, Tetraphylax *and* Aillya, *De Vr.*)

Calyx-tube adnate, sometimes exceedingly short, and usually shorter than the ovary ; lobes 5, sometimes adnate at the base. Corolla-tube with the adnate part blended with the calyx-tube or produced above it or with a hollow protuberance on the under side between the calyx-lobes, sometimes forming a spur at the base of the calyx, slit on the upper side down to the adnate part, the limb oblique, the 2 upper lobes separated lower down, and often arching over the genitalia or rarely all 5 lobes nearly equal and digitately spreading, all equally winged or the upper lobes with the wing more decurrent on the outer side, and sometimes dilated into a concave, inflexed auricle, which is usually thinner and less coloured than in *Dampiera.* Stamens free. Ovary inferior (at least as to the corolla), except the convex summit more or less prominent within the corolla, more or less 2-celled, the dissepiment from almost rudimentary to reaching nearly to the top of the cavity, always with a curved notch at the top, leaving the summit of the ovary 1-celled. Ovules ascending, rarely solitary in each cell, usually several in 2 rows, sometimes blended into 1 or crowded in more than 2 rows. Style undivided. Indusium cup-shaped, enclosing the truncate or divaricately 2-lobed stigma. Capsule opening in 2 valves parallel to the dissepiment or rarely splitting into 4 valves. Seeds usually flat, with a callous or winged margin, rarely thicker, and not bordered. Embryo usually about half as long as the albumen or shorter.—Herbs undershrubs or rarely shrubs. Leaves alternate or radical. Peduncles either axillary or in terminal racemes or panicles, 1-flowered, with or without bracteoles or bearing a cyme or cluster of few flowers, which are usually pedicellate above the bracts or rarely sessile between them.

The species are all limited to Australia ; a few have the corolla, and some others the habit of *Scævola,* but the majority are different from that genus in both respects, and all are readily known by their dehiscent capsule as well as by the ovules ; in all, except *G. barbata,* more than one in each cell or on each side of the dissepiment.

Sect. I. **Monochila.**—*Corolla with the lobes all nearly equal and digitately spreading. Ovules either imbricate in two rows or few and erect from the base of the ovary.*

Flowers all axillary and nearly sessile. Branches all leafy.
Leaves entire, with revolute margins, white underneath. Flowers
in short, leafy heads or spikes. Dissepiment very short.
Ovules few . . . . 1. *G. phyllicoides.*
Leaves toothed, viscid-pubescent. Dissepiment attaining to or
exceeding the middle of the ovary.
Branches long, virgate. Leaves not ½ in. long, the floral ones
all similar . . . . 2. *G. viscida.*
Branches short. Leaves about 1 in. long, stem-clasping, the
floral ones small. Spikes terminal, leafy. Capsule cylindrical . . . . 3. *G. xanthotricha.*
Flowers in a long, leafless, clustered and interrupted spike or narrow panicle. Leaves crowded at the base of the stem or up to
the inflorescence . . . . 4. *G. scapigera.*

Sect. II. **Eugoodenia.**—*Corolla with the two upper lobes separated much lower than the others, and arching over the genitalia or rarely spreading. Ovules either imbricate in 1 or 2 rows in each cell of the ovary or few and erect from the base of the cavity.*

(Corolla-lobes less unequal in *G. Ramelii.*)

Series I. **Racemosæ.**—*Flowers (yellow, white or blue) in a long, terminal, leafless, interrupted spike raceme or panicle.*

Capsule oblong-cylindrical, about ½ in. long, at length 4-valved.
Raceme usually simple . . . . 5. *G. quadrilocularis.*
Capsule ovoid or ovoid-oblong, the 2 valves entire.
Peduncles all nearly equally cymose along the common rhachis.
Flowers blue . . . . 6. *G. Ramelii.*

E 2

Peduncles long, filiform, the upper ones 1-flowered, the lower
　　ones irregularly branched.　Flowers yellow.　Leaves linear-
　　subulate . . . .　7. *G. pinifolia.*
Peduncles short, the lower ones often several-flowered, the upper
　　1-flowered.　Flowers yellow.
　　Leaves oblong, toothed, decurrent . . . .　8. *G. decurrens.*
　　Leaves linear, entire, very narrowly decurrent . . . . .　9. *G. racemosa.*
Flowers (yellow) in sessile clusters along the common rhachis.
　　Leaves all or chiefly radical.
　　Leaves from obovate to narrow oblong-spathulate, mostly
　　　toothed . . . .　10. *G. bellidifolia.*
　　Leaves linear or slightly linear-cuneate, entire . . . . .　11. *G. stelligera.*

SERIES II. **Bracteolatæ.**—*Peduncles axillary or radical or the upper ones very
shortly racemose, bracteolate, 1-flowered, or when several-flowered the bracteoles at the base
of the pedicels.　Flowers yellow or white.*

Glabrous, viscid or rarely viscid-villous undershrubs or shrubs.
　　Leaves ovate, obovate, elliptical or lanceolate.　Common
　　peduncles very short, at least the lower ones, bearing several
　　flowers.
　　Leaves petiolate.　Capsule cylindrical, the dissepiment reaching
　　　far above the middle . . . .　12. *G. ovata.*
　　Leaves sessile, stem-clasping.　Capsule ovoid-oblong, the dissepi-
　　　ment reaching halfway . 　. . . . .　13. *G. amplexans.*
　　Leaves narrowed at the base.　Capsule ovoid or oblong.
　　　Dissepiment scarcely any.　Seeds 2 or 3, strophiolate, thick,
　　　　not bordered . . . .　14. *G. strophiolata.*
　　　Dissepiment half as long as the ovary.　Seeds several, flat,
　　　　with a distinct border.　Stem usually decumbent . . .　15. *G. varia.*
Glabrous or viscid-villous herbs.　Leaves linear.　Peduncles 1-
　　flowered.
　　Glabrous, decumbent.　Bracteoles very small.　Dissepiment half
　　　as long as the ovary.　Seeds several . . . .　16. *G. lævis.*
　　Glandular-pubescent or glabrous.　Bracteoles linear.　Dissepi-
　　　ment scarcely any.　Ovules 2.　Seeds scarcely flattened.
　　　Branching shrubs.　Leaves obovate to lanceolate, mostly
　　　　toothed . . . .　14. *G. strophiolata.*
　　　Undershrub.　Leaves oblong-linear, mostly entire, with revo-
　　　　lute margins . . . .　17. *G. barbata.*
　　Viscid-villous.　Bracteoles linear.　Dissepiment short.　Ovules
　　　few.　Seeds broad . . . .　18. *G. arthrotricha.*
　　Glabrous or hoary pubescent.　Bracteoles very small.　Dissepi-
　　　ment short.　Ovules few.　Seeds flat, oblong . . . . .　19. *G. disperma.*
Villous or rarely glabrous herbs, with tufted or rosulate, radical
　　leaves, and decumbent, prostrate or creeping stems (very rarely
　　nearly erect).　Peduncles 1-flowered, radical or axillary, often
　　long.
　　Calyx-lobes linear, obtuse.
　　Leaves linear-spathulate or narrow-oblong, entire or sinuate,
　　　all radical or stems shortly ascending . . . .　20. *G. geniculata.*
　　Leaves obovate ovate or lyrate.　Leafy stems procumbent .　21. *G. lanata.*
　　Calyx-lobes subulate or acuminate.
　　Bracteoles at a distance from the flowers.
　　　Pedicels filiform.
　　　Leaves all petiolate, orbicular obovate or cuneate.
　　　　Plant glabrous mealy or shortly pubescent . . . .　22. *G. hederacea.*
　　　　Plant hispid-villous . . . .　23. *G. hirsuta.*
　　　Leaves nearly sessile, ovate or lanceolate, deeply toothed .　24. *G. heterophylla.*

Pedicels mostly very short and rather thick. Stems long
  and flagellate. . . . .
  Plant glabrous . . . . 25. *G. glabra.*
  Plant very villous . . . . 26. *G. strongylophylla.*
  Bracteoles close to the calyx . . . . 27. *G. rotundifolia.*
(50. *G. corynocarpa*, with a cylindrical capsule, has sometimes small bracteoles close to
the calyx.)

SERIES III. **Cæruleæ.**—*Peduncles axillary, bracteolate (except in G.* Vilmoriniæ)
1-*flowered or the lower ones loosely cymose. Flowers blue.*

Bracteoles large, leafy, ovate or oblong. Plant glabrous . . . 28. *G. azurea.*
Bracteoles linear or small.
  Plant very viscous-villous. Cymes leafy . . . . 29. *G. scævolina.*
  Plant hoary or white, with woolly hairs . . . . 30. *G. incana.*
  Plant glabrous or nearly so or the inflorescence glandular.
    Stems procumbent, with small, ovate, stem-clasping leaves . 31. *G. leptoclada.*
    Stems erect, with linear or lanceolate leaves.
      Calyx-lobes acute. Lower corolla-lobes about as long as
        the entire part.
        Radical and lower leaves the longest. Calyx-lobes very
          narrow.
          Flowers about ¾ in. long. Stems rigid . . . . . 32. *G. cærulea.*
          Flowers 4 to 5 lines long. Stems slender. Peduncles
            filiform . . . . 33. *G. trichophylla.*
        Lowest leaves shorter and more distant than the succeed-
          ing ones. Calyx-lobes lanceolate . . . . 34. *G. Hassallii.*
      Calyx-lobes obtuse. Lower corolla-lobes much longer than
        the very short entire part . . . . 35. *G. pterigosperma.*
Bracteoles none . . . . 36. *G. Vilmoriniæ.*
(6. *G. Ramelii*, with a long, leafless panicle, has also the flowers blue.)

SERIES IV. **Foliosæ.**—*Erect or rarely decumbent herbs or undershrubs, usually glandular-pubescent or hirsute, with leafy stems. Peduncles axillary, 1-flowered, articulate
under the flower, without bracteoles (or very rarely a few peduncles have 1 or 2 minute
ones). Flowers yellow white or purple.*

Stem-leaves all broad, abruptly petiolate or pinnate. Plant glandular-pubescent or rarely glabrous, not villous.
  Leaves more or less pinnate.
    Protuberance of the corolla-tube produced into a spur . . . 37. *G. calcarata.*
    Protuberance of the corolla-tube decurrent but not spurred.
      Terminal lobe of the leaves usually cuneate at the base,
        lateral ones several.
        Flowers yellow . . . . 38. *G. Nicholsoni.*
        Flowers purple . . . . 39. *G. Macmillani.*
      Terminal lobe of the leaves truncate or cordate at the base,
        lateral ones few or none . . . . 40. *G. grandiflora.*
  Leaves without lateral segments on the petioles.
    Leaves broadly ovate or cordate, acute or acuminate. Flowers
      yellow . . . . 40. *G. grandiflora.*
    Leaves orbicular, petioles very short . . . . 41. *G. Chambersii.*
    Leaves ovate-lanceolate or lanceolate, narrowed at the base.
      Flowers white . . . . 42. *G. albiflora.*
Stem-leaves contracted at the base into a short petiole or almost
  sessile. Plant villous or silky-hairy.
  Very villous. Corolla-tube minutely spurred at the base of the
    calyx . . . . 44. *G. Mitchellii.*
  Villous, often viscid. Leaves toothed. No spur to the corolla.
    Calyx-lobes leafy. Corolla-lobes all winged . . . . . 46. *G. sepalosa.*

Calyx-lobes narrow. Upper lobes of the corolla much narrower
than the others, and scarcely winged . . . . 45. *G. heterochila.*
(See also 51. *G. mollissima.*)
Hairs appressed, almost silky. Leaves entire, long-lanceolate,
crowded . . . . 43. *G. Strangfordii.*
Stem-leaves sessile or stem-clasping. Annuals, with spreading
hairs.
Floral leaves sessile, narrow. Dissepiment of the capsule very
short.
Calyx-lobes lanceolate or linear, leafy. Leaves mostly toothed 46. *G. sepalosa.*
Calyx-lobes linear-subulate. Stem-leaves mostly entire, linear
or linear-lanceolate . . . . 47. *G. hispida.*
Floral leaves cordate, auriculate, not decurrent. Dissepiment of
the capsule exceedingly short . . . . 48. *G. auriculata.*
Floral leaves more or less decurrent. Dissepiment reaching to
the middle of the capsule . . . . 49. *G. Armstrongiana.*
Stem-leaves contracted into a petiole. Capsule linear, ½ to 1 in. long 50. *G. corynocarpa.*

(See also 54. *G. elongata*, 55. *G. pinnatifida*, 59. *G. glauca*, and 60. *G. filiformis*, with
the stems sometimes leafy with linear leaves.)

SERIES V. **Pedicellosæ.**—*Herbs with the leaves chiefly radical or tufted, the stem
leaves either few and distant or collected in terminal tufts, rarely scattered and linear or
reduced to bracts. Peduncles long, radical or in terminal tufts, or axillary, 1-flowered,
without bracteoles.*

Softly tomentose villous or hirsute. Leaves orbicular obovate or
broadly oblong, coarsely toothed.
Leaves hirsute. Dissepiment reaching nearly to the top of the
ovary . . . . 51. *G. mollissima.*
Leaves softly villous or tomentose. Dissepiment almost rudi-
mentary. Seeds winged . . . . 52. *G. cycloptera.*
Hispid with spreading hairs. Leaves narrow.
Stem-leaves linear or linear-lanceolate. Tropical species . . . 47. *G. hispida.*
Stem-leaves few and small. Western species . . . . . 60. *G. filiformis*, var.
Glabrous or with scattered or appressed or silky hairs.
Radical leaves obovate ovate or oblong, entire or sinuate.
Flowers small. Leaves all rosulate or in terminal tufts. Dis-
sepiment very short. Seeds broadly bordered . . . . 53. *G. tenella.*
Flowers large. Stem-leaves few but scattered. Dissepiment
reaching the middle of the ovary. Seeds scarcely bordered 54. *G. elongata.*
Radical leaves pinnatifid.
Flowers small. Leaves and pedicels all radical or in terminal
tufts. Dissepiment reaching above the middle of the ovary.
Seeds winged . . . . 57. *G. heteromera.*
Flowers rather large. Leaves sometimes scattered. Dissepi-
ment short. Seeds winged . . . . 55. *G. pinnatifida.*
Flowers small. Leaves sometimes scattered. Seeds broadly
winged . . . . 56. *G. coronopifolia.*
Radical leaves entire, linear or lanceolate.
Dissepiment reaching far above the middle of the ovary.
Leaves and peduncles all radical or in terminal tufts . . 57. *G. heteromera.*
Stems ascending, with scattered peduncles and floral leaves
or bracts . . . . 58. *G. concinna.*
Dissepiment short. Leaves and peduncles scattered. Seeds
winged.
Flowers rather arge, yellow. Indusium glabrous . . . 59. *G. glauca.*
Flowers small, yellow. Indusium pubescent . . . . . 60. *G. filiformis.*
Flowers small (purple?). Radical leaves often pinnatifid . 56. *G. coronopifolia.*
Doubtful, perhaps allied to *G. glauca*, but foliage unknown . . 61. *G. microptera.*

SECT. III. **Amphichila.**—*Corolla with the 2 upper lobes separated much lower down than the others. Ovules very numerous, closely packed in more than 2 rows in each cell of the ovary. Flowers small, in leafless panicles or on radical scapes.*

Flowering stems ½ to 2 ft. high, several times longer than the radical leaves.

Leaves in the lower part of the stem from obovate to lanceolate, mostly toothed . . . . 62. *G. paniculata.*

Leaves chiefly or entirely radical, linear or linear-lanceolate, entire.

Flowers purple . . . . 63. *G. purpurascens.*

Flowers yellow.

Panicle loose and moderately branched. Leaves not conspicuously veined . . . . 64. *G. gracilis.*

Panicle very much branched. Leaves rigid, conspicuously veined . . . . 65. *G. lamprosperma.*

Flowering stems shorter than or scarcely exceeding the linear or linear-lanceolate radical leaves . . . . 66. *G. humilis.*

Flowering stems 2 to 4 in. high, much exceeding the tufted or rosulate membranous radical leaves.

Leaves linear. Stock almost bulbous . . . . 67. *G. Leytoniana.*

Leaves ovate, rosulate . . . . 68. *G. bicolor.*

Dwarf creeping or stoloniferous plant. Leaves ovate, in rosulate tufts. Scapes or peduncles filiform, mostly 1-flowered . . . 69. *G. pumilio.*

G. stolonifera, De Vr. Gooden. 135, from Verreaux's collection, which I have not seen, may possibly from his description be the same as *G. tenella.*

G. lanceolata and *G. latifolia* are garden names taken up by Steudel, but which have never to my knowledge been described.

SECT. I. MONOCHILA, *G. Don.* Corolla (usually yellow or white) with the lobes all nearly equal and digitately spreading as in *Scævola.* Ovules either imbricate in two rows in each cell or few and erect from the base of the ovary.

1. **G. phylicoides,** *F. Muell. Fragm.* i. 206. An erect branching undershrub or shrub of 1 to 2 ft., clothed with a short white woolly tomentum. Leaves rather crowded, oblong-lanceolate or almost linear, obtuse, entire, with revolute margins, narrowed at the base, ½ to 1 in. or rarely 1½ in. long, coriaceous and becoming glabrous and shining above, white underneath. Flowers white, sessile in the upper axils, forming a short compact terminal leafy spike or head. Calyx-tube exceedingly short; lobes linear, softly ciliate, about 3 lines long. Corolla tomentose-pubescent outside, 6 to 7 lines long, the lobes all nearly equal and spreading. Ovary almost 1-celled, the dissepiment almost rudimentary in the bottom of the cavity; ovules about 5 or 6, erect. Indusium very shortly ciliate. Fruit not seen.—*Dampiera ? polygalacea,* De Vr. Gooden. 115.

W. Australia. *Drummond, n.* 856 ; Kalgán river, *Oldfield ;* Plantagenet and Stirling Ranges and scrubby plains towards West Mount Barren, *Maxwell.*

2. **G. viscida,** *R. Br. Prod.* 578. A glabrous usually viscid perennial, with a thick almost woody stock and erect virgate simple or slightly-branched leafy stems of ½ to 1½ ft. Leaves sessile, from broadly lanceolate to almost linear, rarely exceeding ½ in. and often much smaller, entire or slightly toothed. Flowers small, yellow, axillary, sessile or nearly so, and scarcely

exceeding the leaves. Bracteoles linear. Calyx-lobes linear, much longer than the tube. Corolla glabrous outside, 3 to 4 lines long, the lobes nearly equal, slightly and irregularly winged. Dissepiment of the ovary reaching to above the middle; ovules 5 or 6 in each cell. Indusium shortly 2-lobed, not at all or very minutely ciliate; stigma divaricately 2-lobed, but scarcely more so than in many other species. Fruit not seen.—DC. Prod. vii. 516; *Stekhovia viscida*, De Vr. Gooden. 168; *Goodenia spicata*, F. Muell. Fragm. iii. 35.

W. Australia. Lucky Bay, *R. Brown;* Oldfield river and moist flats west of Moir's Iulet, *Maxwell;* also *Drummond*, 3rd *Coll. n.* 164.

3. **G. xanthotricha,** *De Vr. Gooden.* 155. An erect glandular-pubescent and apparently viscid undershrub or shrub with leafy branches. Leaves sessile and stem-clasping, lanceolate or linear, with revolute margins, toothed or rarely entire, ¾ to 1½ in. long, the floral ones gradually smaller, linear and entire. Flowers sessile in dense terminal leafy spikes, at first very short, at length 3 or 4 in. long. Bracteoles linear, obtuse. Calyx-tube short; lobes linear, obtuse, 3 lines long or rather more. Corolla glandular-pubescent outside, 6 to 7 lines long, the lobes nearly equal and equally winged. Ovary adnate to the corolla-tube above the calyx-tube, the dissepiment reaching nearly to the top; ovules rather numerous, in 2 rows in each cell. Indusium very shortly ciliate. Capsule oblong-linear, 3 or 4 lines long, opening at length nearly to the base in 4 valves.—*G. leptotheca*, F. Muell. Fragm. vi. 13.

W. Australia. *Drummond*, 4th *Coll. n.* 195. The specimens are identified by De Vries in Herb. Hook. as his *G. xanthotricha,* which he at first intended to refer to *Dampiera,* and they agree with his description, excepting as to the seeds, said to be only 2 in each cell, which he must have taken from some other plant.

4. **G. scapigera,** *R. Br. Prod.* 578. A stout erect perennial or under-shrub of 1 to 2 ft., quite glabrous or with a very little wool in the lower axils. Leaves in the typical form crowded in the lower part of the stem, linear or lanceolate, thick, entire or rarely with a few minute remote teeth, 1½ to 3 in. long, narrowed into a petiole, usually dilated at the base. Flowers white, in a leafless narrow panicle, sometimes contracted into a short dense spike at the end of a long peduncle, sometimes 6 to 9 in. long, interrupted or with long branches at the base, the primary inflorescence racemose and the upper peduncles 1-flowered, the lower peduncles or branches irregularly several-flowered. Bracts and bracteoles narrow, subulate-acuminate. Calyx-lobes subulate. Corolla glabrous outside, 6 to 8 lines long, the adnate part of the tube with a saccate protuberance decurrent between the calyx-lobes, the lobes nearly equal and all winged, the throat with purple spots or streaks under each lobe. Ovary with a dissepiment reaching above the middle and rather numerous ovules in 2 rows in each cell. Indusium not at all or scarcely ciliate. Seeds broad, with a thickened and sometimes winged margin.—DC. Prod. vii. 516; F. Muell. Fragm. i. 114; *Scævola stricta*, De Vr. in Pl. Preiss. i. 408; *Stekhovia scapigera*, De Vr. Gooden. 167. t. 32.

W. Australia. Lucky Bay, *R. Brown;* Point Henry, *Oldfield;* Konkoberup hills, *Preiss, n.* 1511; cold summits of Stirling Range, *F. Mueller;* sandhills and rocks, E. Mount Barren to Cape le Grand, *Maxwell,* also *Drummond, n.* 403.

Var. (B. R. Br.) *foliosa*, F. Muell. Leaves sessile, from ovate to lanceolate, toothed, crowded on the stem up to the inflorescence.—With the typical form, *R. Brown, Maxwell.*

Var. *parviflora.* Leaves narrow, all from the stock. Spike at the end of a long leafless scape or peduncle, short and very compact. Corolla 4 to 5 lines long.—Plantagenet, Stirling, and Phillips Ranges, *Maxwell.*

Sect. II. EUGOODENIA. Corolla with the 2 upper lobes separated much lower than the others and arching over the genitalia or rarely spreading. Ovules either imbricate in 1 or 2 rows in each cell of the ovary or few and erect from the base of the cavity.

Series 1. RACEMOSÆ. Flowers (yellow white or blue) in a long terminal leafless interrupted spike raceme or panicle.

5. **G. quadrilocularis,** *R. Br. Prod.* 578. Glabrous, erect, with hard simple or slightly-branched stems, 1 to 1½ ft. high, leafy only in the lower half. Leaves petiolate, from obovate or oblong-spathulate to lanceolate, toothed, the larger ones often 2 to 3 in. long, the upper ones smaller and more sessile, the floral ones reduced to small linear bracts or the lowest rarely larger and more leafy. Flowers solitary under each bract in a terminal virgate interrupted raceme, the upper pedicels very short, with the linear bracteoles close under the flower, the lower ones sometimes very long, with the bracteoles distant. Calyx-tube nearly 3 lines long, the lobes shorter, lanceolate. Corolla slightly pubescent outside, ¾ in. long, the upper lobes separated low down, with a broad wing on the outside, forming a slightly concave auricle. Ovary 2-celled nearly to the summit, with numerous ovules in 2 rows in each cell. Capsule about ½ in. long, splitting at length to the base into 4 valves marked on the inside with the impressions of the ends of the seeds, which are orbicular, flat, with a thickened margin.—DC. Prod. vii. 515 ; F. Muell. Fragm. vi. 13 ; *G. Taylori,* F. Muell. Fragm. iii. 141 ; *Tetraphylax quadrilocularis,* De Vr. Gooden. 165 (but not *Dampiera Lindleyi* nor Preiss's n. 1474, quoted by De Vriese).

W. Australia. Lucky Bay, *R. Brown ;* King George's Sound or to the eastward, *Baxter ;* sand hills, Orleans Bay, *Maxwell.* The capsule is described as 4-celled, I have only found a slight protrusion of the placenta along the centre of the dissepiment, but nothing like a complete cross dissepiment, real or spurious.

6. **G. Ramelii,** *F. Muell. Fragm.* iii. 20. *t.* 17. Probably a tall plant, with the habit of *G. decurrens.* Stem-leaves lanceolate, entire, more or less decurrent. Inflorescence a foot long, the floral leaves reduced to very small bracts, the rhachis rigid and glabrous. Peduncles numerous, more equal in length than in most species, each one bearing a glandular-pubescent divaricate cyme of 3 to 7 blue flowers. Bracteoles minute. Calyx densely glandular-pubescent, the tube about 2 lines long, the lobes shorter, lanceolate. Corolla glandular-pubescent outside, fully ½ in. long, the upper lobes free almost to the base and unequally winged, but not so much so as in some species. Ovary divided up to about ¾, the ovules in 2 rows. Indusium ciliate. Capsule oblong, but not seen ripe.

N. Australia. Attack Creek, *M'Douall Stuart,* a single imperfect specimen (*Herb. F. Muell.*).

7. **G. pinifolia,** *De Vr. Gooden.* 157. *t.* 30.   An erect shrub, the short leafy branches clothed with a white tomentum, otherwise glabrous.   Leaves crowded, linear-terete, entire, ½ to 1 in. long.   Panicles terminal, loose and slender, about 6 to 9 in. long, the primary branches or peduncles filiform, the upper ones 1-flowered, the lower irregularly several-flowered.   Bracts and bracteoles minute.   Calyx-lobes linear, acute.   Corolla yellow, glabrous outside, nearly ½ in. long, the tube sometimes adnate higher up than the calyx, the upper lobes separated lower than the lower ones and rather unequally winged.   Dissepiment of the ovary exceedingly short; ovules 3 or 4.   Indusium not ciliate.   Capsule ovoid-globular, scarcely 2 lines long.   Seeds rather large broad and flat, with a thickish border.

W. Australia.   Swan River, *Drummond, 1st Coll.*

8. **G. decurrens,** *R. Br. Prod.* 575.   A rather rigid perennial, with ascending or erect stems of 1 to 2 ft., leafy only in the lower half, quite glabrous or the inflorescence slightly pubescent.   Leaves oblong or the lower ones rarely ovate, toothed or rarely entire, 1½ to 3 or even 4 in. long, the upper ones all more or less decurrent on the stem, the floral ones all reduced to small linear bracts.   Flowers rather large, yellow, in a terminal leafless irregular raceme or narrow panicle, the upper peduncles 1-flowered, the lower ones often bearing a loose cyme of 3 to 5 flowers.   Bracteoles very small.   Calyx-lobes linear or lanceolate.   Corolla about ¾ in. long, the upper lobes separate lower down than the others and unequally winged.   Dissepiment of the ovary reaching high up; ovules several in 2 rows in each cell.   Indusium ciliate.   Capsule 3 to 4 lines long.   Seeds flat with a thickened margin.—DC. Prod. vii. 513 ; De Vr. Gooden. 138.

N. S. Wales.   Common in the Blue Mountains, *R. Brown, Sieber, n.* 231, *Fraser,* and others.

9. **G. racemosa,** *F. Muell. Fragm.* i. 114.   A glabrous shrub or undershrub of 1½ to 3 ft., the branches angular, with raised lines decurrent from the leaves.   Leaves rather crowded, narrow-linear, entire, mostly 1 to 1½ in. long, the floral ones reduced to small linear bracts.   Flowers rather small, in a dense terminal raceme, the peduncles short, the lower ones often with a cluster of 2 or 3 flowers, the upper ones 1-flowered.   Calyx-lobes subulate.   Corolla slightly glandular-pubescent, about ½ in. long, the upper lobes separate lower down than the others.   Ovary very prominent within the corolla-tube, the dissepiment short; ovules not numerous, in 2 rows in each cell.   Capsule ovoid, about 3 lines long.   Seeds rather large, flat.

Queensland.   Arid ranges on the Burnett river, *F. Mueller.*

10. **G. bellidifolia,** *Sm. in Trans. Linn. Soc.* ii. 349.   A perennial with a tufted stock, glabrous except the inflorescence, or the base of the leaves also pubescent, the flowering stems leafless or nearly so, 1 to 1½ ft. high.   Leaves radical, petiolate, from broadly obovate and 1 to 1¾ in. long to narrow oblong-spathulate and 3 or 4 in. long, always broader than in *G. stelligera,* entire or irregularly toothed, rather thick; the stem ones few and very small or none besides the small floral bracts.   Flowers rather small, sessile or nearly so in little clusters along the rhachis of a long interrupted

spike, the upper ones usually solitary. Bracteoles small. Calyx-lobes linear or subulate. Corolla less villous outside than in *G. stelligera*, rarely ½ in. long, the upper lobes separate low down and unequally winged. Dissepiment of the ovary reaching high up ; ovules in 2 rows in each cell. Indusium ciliate. Capsule ovoid, about 2 to 2½ lines long. Seeds flat, with a thickish border.—R. Br. Prod. 575 ; DC. Prod. vii. 513 ; De Vr. Gooden. 122.

Queensland. Dawson river, *F. Mueller ;* near Brisbane, *Leichhardt.*
N. S. Wales. Port Jackson to the Blue Mountains, *R. Brown, Sieber n.* 230, *Fl. Mixt. n.* 619, and others ; Hastings and Clarence rivers and Mount Mitchell, *Beckler ;* New England, *C. Stuart.*

G. spathulata, De Vr. Gooden. 123, from Botany Bay, *Huegel*, must, from the character given, be the same species.

11. **G. stelligera,** *R. Br. Prod.* 575. A perennial, with a tufted stock and erect, almost leafless stems of 1 to 1½ ft., glabrous or the inflorescence pubescent. Radical leaves linear or slightly linear-cuneate, obtuse, rather thick, entire, sometimes 5 or 6 in. long but usually half that size ; stem-leaves very few and much shorter, floral ones reduced to linear bracts. Flowers yellow, sessile or nearly so, in distant clusters of 2 or 3, the upper ones solitary, in a long, interrupted spike. Calyx-lobes linear or linear-lanceolate. Corolla densely villous outside, with a glandular pubescence concealed under the longer hairs, 7 to 8 lines long, the upper lobes separated low down, and unequally winged. Dissepiment of the ovary reaching high up ; ovules in 2 rows in each cell. Indusium ciliate. Capsule ovoid-oblong, 3 to 4 lines long. Seeds flat, with a thickish border.—DC. Prod. vii. 513 ; De Vr. Gooden. 124 ; *S. armeriæfolia*, Sieb. ; DC. Prod. vii. 513 ; De Vr. Gooden. 129, as to the eastern plant ; *G. longifolia*, De Vr. Gooden. 127.

Queensland. Moreton Island, *M'Gillivray, F. Mueller.* Some of the more luxuriant specimens proliferous-branched, with tufts of leaves and a secondary flowering stem.
N. S. Wales. Port Jackson to the Blue Mountains, *R. Brown, Sieber n.* 229, and others ; northward to Hastings river, *Beckler ;* near Durval, *Leichhardt ;* and southward to Mount Imlay, *F. Mueller.*

De Vriese quotes Sieber's specimens both for *G. stelligera*, Goodenov. p. 124, and for *G. armeriæfolia*, p. 129, and mixes up with the latter Preiss's n. 2032, which is *Diaspasis fili-folia.*

SERIES 2. BRACTEOLATÆ. Peduncles axillary or radical or the upper ones very shortly racemose, bracteolate, 1-flowered or when several-flowered the bracteoles at the base of the pedicels. Flowers yellow or white.

12. **G. ovata,** *Sm. in Trans. Linn. Soc.* ii. 347. An erect, glabrous, often somewhat viscid shrub or undershrub of 2 to 4 ft. Leaves petiolate, from ovate to broadly lanceolate or the lower ones sometimes almost orbicular-cordate, denticulate, 1 to 2 in. long. Peduncles axillary, often 2 together or forked near the base, slender and often several-flowered, but rarely exceeding the leaves. Bracteoles very small, at a distance from the flower. Calyx-tube linear, lobes subulate. Corolla yellow, glabrous outside, about ½ in. long, the upper lobes deeply separate. Dissepiment reaching high up in the ovary. Indusium strongly ciliate. Capsule narrow, 4 to 6 lines long, slightly tapering at the base. Seeds flat, almost in a single row in each cell. —Cav. Ic. vi. t. 506 ; Vent. Jard. Cels. t. 3 ; Andr. Bot. Rep. t. 58 ; R. Br.

Prod. 576; DC. Prod. vii. 514; De Vr. Gooden. 141; Hook. f. Fl. Tasm. i. 232.

Queensland. Moreton Bay, *Fitzalan.*
N. S. Wales. Port Jackson to the Blue Mountains, *R. Brown, Sieber, n.* 232 *and Fl. Mixt. n.* 538, and others; Hastings river, *Beckler.*
Victoria. Port Phillip, *R. Brown;* near Melbourne, *Adamson, Robertson;* dry rocks near Morra-Morra and Mount Disappointment, *F. Mueller;* Creswick, *Whan.*
Tasmania. Kent's Group and Port Dalrymple, *R. Brown;* abundant in copse woods, etc., *J. D. Hooker.*
S. Australia. Onkaparinga, Torrens river, Lofty Range, etc., *F. Mueller* and others; Three-Well river, *Waterhouse.*

G. acuminata, R. Br. Prod. 575; DC. Prod. 513; De Vr. Gooden. 140, is a broadly lanceolate-leaved form, common in the Blue Mountains, passing gradually into the common broad-leaved form, and sometimes broad and narrow leaves may be seen on the same specimen.

13. **G. amplexans,** *F. Muell. in Trans. Phil. Inst. Vict.* ii. 70. An erect, glandular-pubescent or villous perennial or undershrub. Leaves sessile and stem-clasping, with broad auricles, ovate-lanceolate or oblong, denticulate, mostly 2 to 3 in. long, the upper ones gradually smaller, and in some small states all ovate and under 1 in. long. Flowers solitary or 2 or 3 together in the axils, the common peduncle exceedingly short or none, and the pedicels shorter than the calyx-tube, with very small bracteoles at their base. Calyx-lobes rather short. Corolla yellow, glandular outside, 6 to 8 lines long, the upper lobes separated much lower than the others. Capsule ovoid-oblong, the dissepiment reaching about halfway up. Seeds oval-oblong, flat, with a rather thick margin.

Victoria? Nile rivulet, *Herb. F. Mueller.*
S. Australia. Ridges and gullies near Adelaide, Holdfast Bay, Lofty Ranges, *F. Mueller.*

Var. *parvifolia.* Small, with leaves scarcely above ½ in. long.—Mount Arapiles, *Dallachy.*

14. **G. strophiolata,** *F. Muell. Fragm.* i. 119. A glabrous, viscid, branching shrub of 1 to 2 ft. or sometimes twice that height, erect or with straggling, divaricate branches. Lower leaves obovate, oblong-spathulate or elliptical, the upper ones or sometimes nearly all oblanceolate, acutely toothed or entire, mostly narrowed into a short petiole, in some specimens 1 to 1½ in. long, in others all under 1 in., the upper floral ones smaller. Peduncles in the upper axils nearly as long as or longer than the leaves, 1-flowered with linear bracteoles at a distance from the flower, or short and 2- or 3-flowered, the pedicels when the flower is solitary reflexed from the bracteoles after flowering as in *G. geniculata* and *G. heterophylla.* Calyx-lobes linear, acute, often free lower down than the corolla-tube. Corolla glabrous or minutely tomentose outside, ½ in. long or rather more, the upper lobes separated much lower down than the others, but nearly equally winged. Ovary with a short dissepiment, and 3 or 4 erect ovules on each side. Fruit ovoid-oblong, 2 to 3 lines long. Seeds usually only 2 perfect, oblong, thick and smooth, without any prominent margin, like those of *G. barbata,* but the funicle is expanded into a conspicuous, oblique strophiola.

W. Australia, *Drummond, n.* 196, 355; rocks and clay flats E. Mount Barren to

Phillips and Fitzgerald Ranges, *Maxwell.* The strophiola of the seed, already perceptible on the ovule, is, as far as hitherto observed, exceptional in the genus.

15. **G. varia,** *R. Br. Prod.* 576. A glabrous perennial or undershrub, the stems either long and prostrate or decumbent at the base, and ascending to the height of ½ to 1 ft. Lower leaves usually broad, obovate or orbicular, narrowed into a petiole, the upper ones ovate-oblong or sometimes in erect branches lanceolate or linear, all thick, coriaceous, toothed or rarely entire. Peduncles axillary, 1- to 3-flowered, shorter than the leaves. Bracteoles small, narrow. Calyx-tube short, lobes narrow. Corolla glabrous outside, 6 to 7 lines long, the tube adnate higher up than the calyx, but the summit of the ovary free, the upper lobes separated much lower than the others. Dissepiment reaching to about the middle of the ovary; ovules in 2 rows. Capsules 3 to 4 lines long. Seeds flat, oblong, not winged.—DC. Prod. vii. 514; De Vr. Gooden. 142; F. Muell. Fragm. i. 205.

S. Australia. Petrel Bay, Flinders Island, Memory Cove, *R. Brown;* rocky places and salt lagoons on the Murray and Flinders Range, St. Vincent's and Spencer's Gulfs, *F. Mueller* and others; Venus and Streaky Bays, *Warburton.*

G. marginata, De Vr. Gooden. 143, said by him to be near *G. varia,* may not be a *Goodenia* at all, as neither the flower nor the fruit has been seen.

16. **G. lævis,** *Benth.* Quite glabrous, procumbent or ascending, and branched. Lower leaves oblong-cuneate, obtuse, with 2 or 3 coarse teeth or lobes, narrowed into a short petiole, 1 to 1½ in. long, the upper ones narrow-linear, entire, all rather thick and smooth. Peduncles axillary, 1-flowered, rarely exceeding the leaves. Bracteoles linear, sometimes very small, close under the flower. Calyx-lobes linear. Corolla yellow, glabrous outside, about ½ in. long or rather more, the lobes all winged, the upper ones separated low down, and the wings unequal. Dissepiment reaching high up in the ovary; ovules numerous, in 2 rows. Style slightly pubescent. Indusium ciliate. Capsule ovoid-oblong, about 3 lines long. Seeds flat, not winged.

W. Australia. Phillips Ranges, *Maxwell.*

17. **G. barbata,** *R. Br. Prod.* 576. A perennial or sometimes undershrub, with erect branches, often long and virgate, more or less glandular-pubescent. Leaves oblong-linear or lanceolate, obtuse, with revolute margins, entire or slightly toothed, rarely above 1 in. long and usually shorter. Peduncles 1-flowered in the upper axils, rarely exceeding the leaves. Bracteoles linear, at a distance from the flowers, and the pedicels reflexed after flowering. Adnate calyx-tube very short; lobes linear or linear-lanceolate, 2 to 3 lines long, adnate at the base. Corolla about ¾ in. long, pubescent outside, with stellate mixed with glandular hairs, the tube with long, reflexed fringes inside descending from the margins of the lobes, the 2 upper lobes separate almost to the adnate part of the tube. Dissepiment of the ovary very short, with 1 erect ovule on each side. Capsule ovate. Seeds 2, oblong, less compressed than in most species, smooth and almost shining, without any prominent border.—DC. Prod. vii. 514; De Vr. Gooden. 145; *G. cistifolia,* A. Cunn.; DC. Prod. vii. 516; De Vr. Gooden. 150; F. Muell. Fragm. ii. 110, 176.

N. S. Wales. Blue Mountains, *Caley in Herb. Br.*, *A. Cunningham*, and others; Camden, *Leichhardt*; heathy ground and dry, stony ridges, Twofold Bay, *F. Mueller.* **Tasmania.** Port Dalrymple, *R. Brown.* A single, imperfect specimen.

18. **G. arthrotricha,** *F. Muell. Herb.* Herbaceous, apparently perennial, scabrous-pubescent or hirsute, the stems erect and branching, 1 to 2 ft. high. Leaves all linear and entire, the lower ones often 2 or 3 in. long. Peduncles axillary, 1-flowered, longer than the leaves, with 2 long, linear bracteoles or the lower ones branching out into a loose, dichotomous cyme. Calyx-lobes linear or linear-lanceolate, $\frac{1}{4}$ to $\frac{1}{2}$ in. long. Corolla $\frac{3}{4}$ to 1 in. long, the lobes broadly winged, the upper ones separated low down. Indusium ciliate. Dissepiment not reaching to the middle of the ovary; ovules few, in 2 rows on each side. Capsule ovoid, 3 to 4 lines long. Seeds broad and flat, not winged.

W. Australia, *Drummond, 4th Coll. n.* 190 *and* 197.

19. **G. disperma,** *F. Muell. Fragm.* i. 113. Herbaceous, erect, branching, rather slender, hoary-pubescent or nearly glabrous, the specimens seen all under 1 ft. high. Leaves linear, entire, 1 to 3 in. long, without larger radical ones. Peduncles short, axillary, 1-flowered or the flowers almost sessile. Bracts very small, setaceous, at a distance from the flower when pedunculate. Calyx-lobes almost setaceous. Corolla under $\frac{1}{2}$ in. long, pubescent outside, the upper lobes separated almost to the base and shorter than the others, broadly winged on the outer edge, but scarcely auriculate. Dissepiment of the ovary exceedingly short, with about 6 erect ovules. Capsule ovoid-oblong, about 3 lines long. Seeds oblong, flat, about 2 lines long, minutely granular, with scarcely any border.—*G. sessiliflora,* F. Muell. Fragm. iv. 145.

Queensland. Sandy plains between the Mackenzie, Dawson, and Burnett rivers, *F. Mueller;* Cape river, *Bowman.* Although only from 2 to 4 of the seeds usually come to maturity, there appear to be always at least 6 ovules.

20. **G. geniculata,** *R. Br. Prod.* 577. A perennial, with a tufted and often creeping rootstock, emitting occasionally short, decumbent or ascending leafy stems, rarely more vigorous and 6 to 9 in. high. Indumentum in some specimens consisting entirely of simple hairs, in others woolly at the base of the leaves or the whole plant cottony or rarely entirely clothed with a long, dense wool. Leaves chiefly radical, petiolate, from linear to obovate-oblong, obtuse, entire or slightly sinuate-toothed, varying from about 1 to above 3 in. long, those on the short stems more sessile. Scapes or peduncles 1-flowered, usually longer than the leaves. Bracteoles linear, at a distance from the flower, the pedicels bent back from the bracteoles after flowering. Calyx-lobes oblong or linear, rather obtuse. Corolla yellow, pubescent outside, $\frac{1}{2}$ to $\frac{3}{4}$ in. or rarely 1 in. long, the 2 upper lobes separated low down and unequally winged. Dissepiment of the ovary reaching to above the middle, the conical summit of the ovary free. Ovules varying from 7 or 8 to twice that number in each cell, in 2 rows. Capsule ovoid, 3 to 4 lines long. Seeds broad, flat, with a thick margin.—DC. Prod. vii. 514; De Vr. Gooden. 136; Hook. f. Fl. Tasm. i. 232.

N. S. Wales. Between the Murray and Darling rivers, *Victoria and other Expeditions.*

Victoria. Port Phillip, *R. Brown* and others ; near Melbourne, *F. Mueller, Adamson ;* Portland, *Allitt ;* Grampians, *Wilhelmi ;* Wimmera, *Dallachy.*

Tasmania. Rocky Cape, *Gunn.*

S. Australia. Port Lincoln, *R. Brown, Wilhelmi ;* from the Murray to St. Vincent's and Spencer's Gulf, *F. Mueller* and others.

W. Australia, *Drummond, n.* 405 ; Mount Barker, Bremer Bay to Phillips Flats, *Maxwell.*

In the Tasmanian specimens, and most of those from the neighbourhood of Melbourne, the indumentum consists entirely of simple hairs, or with a little wool at the base of the leaves. In the var. *primulacea* (*G. primulacea,* Schlecht. Linnæa, xx. 601 ; De Vr. Gooden. 158), there are a few, or scarcely any simple hairs, and the foliage is more or less clothed with a short and close or longer and looser intricate cottony wool. This is the commonest form in South Australia, N.W. Victoria, and on the Murray. In a third var., *eriophylla* (*Scævola geniculata,* De Vr. in Pl. Preiss. i. 404 ; *Goodenia affinis,* De Vr. Gooden. 137), the leaves are large, thick, and soft, and the whole plant is covered with a long, shaggy wool. To this belong the W. Australian and some of the S. Australian specimens. Some of Brown's from Port Lincoln pass into a fourth variety, *robusta,* with the wool of *eriophylla,* but with stout, almost erect stems ½ to nearly 1 ft. high, the scapes or peduncles both radical and axillary as in the other varieties. This, which, at first sight, would be taken for a very distinct species, we have from Wimmera, *Dallachy ;* Lake Koorong, *Herb. F. Mueller ;* and Marble Range, *Wilhelmi.*

21. **G. lanata,** *R. Br. Prod.* 577. Very near the var. *primulacea* of *G. geniculata,* with the same indumentum, peduncles, and flowers, but the rootstock appears to be less creeping, the stems usually elongated, prostrate, and often rooting at the nodes, and the leaves are much broader, mostly obovate, and more deeply toothed.—DC. Prod. vii. 514 ; De Vr. Gooden. 146. t. 26.

N. S. Wales. New England, *C. Stuart.*

Victoria. Glenelg river, Ballarat, Eureka, Bunip Creek, *F. Mueller ;* Creswick Diggings, *Whan.*

Tasmania. Port Dalrymple, *R. Brown ;* common in dry pastures, *J. D Hooker.*

J. D. Hooker (Fl. Tasm. i. 232) unites this with the northern *G. hederacea,* from which it appears to me to differ in the calyx and bracteoles as well as in indumentum. F. Mueller unites it with *G. geniculata,* to which it is certainly closely allied, but I have never seen any forms of that species with prostrate stems and broad leaves approaching those of *G. lanata.*

22. **G. hederacea,** *Sm. in Trans. Linn. Soc.* ii. 349. A perennial, with a thick hard often woody stock and long trailing rather slender stems, often rooting at the nodes, and sometimes ascending or nearly erect to the height of 6 in. or more, the whole plant sometimes clothed with a very close white tomentum, which is rarely wanting on the under side of the leaves, the upper side soon becoming glabrous. Leaves petiolate, obovate or orbicular, usually irregularly toothed, cuneate or cordate at the base, the upper floral ones sometimes narrow-ovate or spathulate and entire. Peduncles slender, axillary, usually exceeding the leaves, 1- to 3-flowered. Bracteoles small and narrow, at a distance from the flower. Calyx-tube very narrow-turbinate, prominently 5-ribbed, the lobes narrow and acute. Corolla slightly tomentose outside, ½ to ¾ in. long, the upper lobes separated low down. Dissepiment of the ovary reaching far above the middle. Capsule ovoid.

Seeds flat, not winged.—R. Br. Prod. 577 ; DC. Prod. vii. 514 ; De Vr.
Gooden. 147. t. 27.

Queensland. Burnett river, *F. Mueller ;* Moreton Bay, *C. Stuart.*
N. S. Wales. Port Jackson to the Blue Mountains, *R. Brown* and others; northward to Clarence river and Mount Mitchell, *Beckler ;* New England, *C. Stuart ;* Head of
the Gwydir river, etc., *Leichhardt,* and in the western interior between the upper Bogan
and Lachlan rivers, *L. Morton.*
Victoria. Upper Snowy River, Haidinger Range, Mount Buller, at an elevation of
5000 ft., *F. Mueller.*

The Queensland specimens have generally more erect and shorter stems, with the leaves
not so broad, but appear to belong to the same species, differing from *G. geniculata* in habit,
in the smaller, differently-shaped calyces with acute lobes, and in their slender peduncles
frequently 3-flowered, from *G. rotundifolia* easily distinguished by the bracteoles at a distance from the flower.

23. **G. hirsuta,** *F. Muell. Fragm.* iii. 35. The single fragment upon
which this species is founded shows the procumbent habit of *G. hederacea,*
but is hispid all over with rigid hairs. Radical leaves petiolate, obovate or
ovate, coarsely toothed, and sometimes with 1 or 2 additional lobes on the
petiole ; leaves on the procumbent stems shortly petiolate, broadly ovate or
orbicular, ½ to 1 in. long. Peduncles axillary, longer than the leaves, with
linear bracteoles about the middle. Flowers unknown. Capsule 3 to 4 lines
long, the dissepiment reaching at least to the middle, with several large flat
winged seeds.

Central Australia, *M'Douall Stuart (Herb. F. Muell.).*

24. **G. heterophylla,** *Sm. in Trans. Linn. Soc.* ii. 349. A pubescent
or hirsute perennial or perhaps sometimes annual, with decumbent ascending
or rarely erect rather slender leafy stems of ½ to 1 ft. or sometimes much
longer and procumbent or flexuose. Leaves ovate-lanceolate or almost linear,
truncate or contracted at the base, and often shortly petiolate, coarsely and
irregularly toothed or lobed or rarely entire, ½ to 1 in. long or rarely more.
Peduncles axillary, filiform, bearing 1 or rarely 2 or 3 rather small flowers ;
bracteoles setaceous, at a distance from the flower. Calyx-tube turbinate,
prominently 5-ribbed, the lobes setaceous. Corolla yellow, scarcely ¼ in.
long, glabrous or slightly pubescent outside ; the upper lobes separated low
down, unequally winged, with slightly concave auricles. Dissepiment of
the ovary reaching to about the middle. Ovules variable in number, in 2
rows. Capsule ovoid, about 2 lines long.—Cav. Ic. t. 508 ; R. Br. Prod.
576 ; DC. Prod. vii. 514 ; De Vr. Gooden. 144 ; *G. teucriifolia,* F. Muell.
in Trans. Vict. Inst. ii. 70.

Queensland. Glasshouse Mountains and Burnett river, *F. Mueller.*
N. S. Wales. Port Jackson to the Blue Mountains, *R. Brown, Sieber, n.* 228, and
others ; Newcastle, *Leichhardt.*

De Vriese refers to *G. heterophylla,* the *G. pubescens,* Sieb. Pl. Exs. n. 178, DC. Prod.
vii. 514. I have not seen the plant, but DC. says, " bracteis flori proximis," which would
rather refer it to *G. rotundifolia,* if it be a *Goodenia* at all, and not *Scævola microcarpa,*
as suggested by F. Mueller.

25. **G. glabra,** *R. Br. Prod.* 577. A glabrous perennial, with a tufted
stock, emitting several long simple prostrate stems, like stolons, but not

usually rooting. Radical leaves petiolate, from obovate to oblong-spathulate, rather thick, entire or minutely toothed, 1 to 3 in. long, those of the prostrate stems distant, sessile or nearly so, obovate or cuneate, rarely above ½ in. long. Flowers solitary, either radical on long peduncles or axillary on shorter, sometimes very short ones. Bracteoles very small, at a distance from the flower when the peduncle is elongated. Calyx-lobes subulate-acuminate. Corolla yellow, the upper lobes separated low down and unequally winged, with a large inflexed auricle on the outer side. Dissepiment reaching to about the middle of the ovary. Capsule ovoid or oblong. Seeds rather large, flat, granular-tuberculate, with a rather thick smooth border.—DC. Prod. vii. 515 ; *G. flagellifera,* De Vr. in Mitch. Trop. Austr. 378, and Gooden. 146. t. 25.

Queensland. Shoalwater Bay, *R. Brown ;* Port Curtis, *M'Gillivray ;* Rockhampton, *Thozet, O'Shanesy ;* Peak Downs, *F. Mueller ;* Bokhara Creek, Ottley's Station, etc., *Leichhardt ;* Armadillo, *W. Barton.*

26. **G. strongylophylla,** *F. Muell. Fragm.* vi. 12. The fragments upon which this species is founded appear to be procumbent or prostrate flagellate branches, like those of *G. glabra,* but densely hirsute with rather soft hairs. Leaves nearly sessile, very broadly ovate or orbicular, toothed, not exceeding ½ in. diameter. Flowers axillary. Bracts subulate at the base of the very short pedicels. Calyx very villous, with narrow lobes. Corolla hairy outside, 5 or 6 lines long, the upper lobes separate low down and auriculate. Dissepiment reaching to about the middle of the ovary, with several ovules in 2 rows in each cell. Capsule ovoid, about 3 lines long. Seeds few, large, flat, with a thickish margin, rarely bordered by a narrow wing.

Queensland. Princhester Creek, *Bowman.*

27. **G. rotundifolia,** *R. Br. Prod.* 576. A slightly pubescent or viscid perennial, the stems short and erect or long and trailing. Leaves petiolate, broadly ovate or orbicular, coarsely toothed, rounded or cordate at the base, ½ to 1 in. diameter or sometimes more, those of the smaller branches sometimes obovate and narrowed at the base. Peduncles slender, shorter than or slightly exceeding the leaves, bearing 1 or more frequently a cyme of 3 to 5 flowers. Bracteoles very narrow, close under the flower when solitary. Calyx-tube hispid ; lobes subulate, ciliate. Corolla pubescent outside, about ½ in. long, the upper lobes separate low down and unequally winged. Dissepiment of the ovary very short or rarely reaching to the middle. Ovules several in 2 rows in each cell. Capsule broad, about 2 lines long. Seeds several, flat, with a smooth margin, not winged.—DC. Prod. vii. 514 ; De Vr. Gooden. 145.

Queensland. Shoalwater Bay, *R. Brown ;* Brisbane river, Moreton Bay *F. Mueller, Leichhardt ;* Rockhampton, *Dallachy* and others.

N. S. Wales. Near Newcastle, *R. Brown ;* Hunter's River, *Oldfield ;* Clarence river, *Beckler ;* New England, *C. Stuart* (the latter with larger flowers and capsules).

The new species in Banks's collection mentioned by Gærtn. fil. Fruct. iii. 165, under the name of *G. intermedia,* is in all probability *G. rotundifolia.*

Series 3. Cæruleæ. Peduncles axillary, bracteolate (except in *G.*

Vilmoriniæ), 1-flowered or the lower ones loosely cymose. Flowers blue.

28. **G. azurea,** *F. Muell. Fragm.* i. 117. A glabrous glaucous undershrub of 1 to 2 ft., with rigid spreading branches. Lower leaves petiolate, obovate, sparingly toothed, 2 to 3 in. long; upper ones sessile, obovate ovate or oblong, very obtuse and entire, the uppermost under $\frac{1}{2}$ in. long. Peduncles axillary, rigid, longer than the leaves, 1-flowered or the lower ones bearing a cyme of 3 to 7 flowers. Bracteoles large, leafy, ovate or obovate. Calyx-tube glandular, lobes lanceolate. Corolla blue, 7 to 8 lines long, sparingly glandular-pubescent outside, with a saccate protuberance extending to the base of the calyx-tube, the upper lobes separate almost to the base, broadly winged on the outer side. Dissepiment of the ovary reaching far above the middle; ovules rather numerous, in 2 rows in each cell. Capsule 4 to 5 lines long. Seeds small, orbicular, flat, with a somewhat thickened margin.

N. Australia. Upper Victoria river and table land at the sources of Sturt's Creek, *F. Mueller.*

29. **G. scævolina,** *F. Muell. Fragm.* i. 118. An undershrub of 2 to 4 ft., viscid-pubescent or hirsute. Leaves obovate oblong or oblanceolate, acutely and coarsely toothed, narrowed at the base and often shortly decurrent, 2 to 3 in. long, the floral ones much smaller and less toothed or entire, the uppermost reduced to small bracts. Peduncles in the upper axils, bearing each a cyme of 3 to 7 flowers with leafy bracts at the forks, or the uppermost short and 1-flowered, forming a broadly pyramidal leafy viscidvillous panicle. Calyx-lobes linear-lanceolate. Corolla blue, $\frac{3}{4}$ to nearly 1 in. long, pubescent or almost hirsute outside, unilabiate (*F. Muell.*), but the upper (lateral) lobes separate lower down and unequally winged. Dissepiment reaching nearly to the top of the ovary; ovules rather numerous, in 2 rows. Capsule oblong, 4 to 6 lines long. Seeds flat, minutely granular, with a narrow smooth border.

N. Australia. Sandstone hills, upper Victoria river, *F. Mueller.*

30. **G. incana,** *R. Br. Prod.* 578. A tufted perennial, clothed all over with hoary or white woollen hairs, sometimes long and loose, sometimes short and close. Stems usually several, simple or nearly so, erect, 6 to 8 in. or rarely above 1 ft. high. Radical leaves petiolate, oblong-spathulate or oblanceolate, 1 to 2 in. high, stem-leaves smaller and linear or rarely lanceolate, all entire thick and soft. Peduncles axillary, 1-flowered, often all turned to one side, shorter or rarely longer than the leaves. Bracteoles linear, more or less distant from the flower. Calyx-lobes linear. Corolla blue, about $\frac{3}{4}$ in. long, pubescent or hirsute outside, the lobes all broadly and nearly equally winged, but the upper ones separated nearly to the base, and the outer wings fringed near the base. Anthers minutely pointed. Dissepiment of the ovary reaching far above the middle; ovules rather numerous, in 2 rows in each cell. Capsule ovoid or almost globular, about 3 lines long. Seeds flat, orbicular, bordered by a thin and irregular but sometimes broad wing.—DC. Prod. vii. 516; De Vr. Gooden. 152. t. 28; F. Muell. Fragm. iii. 141; *Scævola pterosperma,* De Vr. in Pl. Preiss. i. 408.

W. Australia. Lucky Bay, *R. Brown ;* King George's Sound to Cape Riche, *Baxter*, *Preiss, n.* 1499, 1501, also *Drummond, 1st Coll.* (*2nd Coll. ?*) *n.* 20, *3rd Coll. n.* 155 ; Plantagenet and Stirling Ranges to Stokes and Muir's Inlets and Fitzgerald Ranges, *Maxwell.*

31. **G. leptoclada,** *Benth.* A glabrous or minutely pubescent perennial, with slender but rigid procumbent or ascending scarcely branched stems, not exceeding 6 in. in our specimens. Radical leaves petiolate, oblong-cuneate or oblanceolate, entire or with a few prominent teeth, often above 1 in. long ; stem-leaves sessile, stem-clasping, ovate or oblong, under ½ in. long or the lower ones rarely longer, all thick and coriaceous. Peduncles 1-flowered in the upper axils, much longer than the leaves, with a pair of linear bracteoles at some distance from the flower. Calyx-lobes linear, rather long. Corolla apparently blue, pubescent outside, about ½ in. long, the 2 upper lobes deeply separated and rather unequally winged, but not auriculate. Dissepiment of the ovary very short ; ovules few. Capsule small, ovoid or globular. Seeds very few, orbicular, rather large, not winged.

W. Australia, *Drummond, n.* 188. The habit approaches that of *G. glabra* on a small scale, but the flowers appear to be blue, and the small stem-clasping leaves are very characteristic.

32. **G. cærulea,** *R. Br. Prod.* 578. A perennial, usually tufted, glabrous glandular-pubescent or rarely sprinkled with soft spreading hairs, the inflorescence or at least the calyx always glandular-pubescent. Stems usually several, angular, simple flexuose or branched, erect or ascending, often rigid, mostly ½ to 1 ft. high. Leaves linear, rigid, the radical ones sometimes slightly dilated above the middle and rarely with 1 or 2 teeth, all the others quite entire, the upper ones small and almost terete. Peduncles axillary, often longer than the leaves, spreading and turned to one side. Bracteoles very small, usually about the middle. Flowers blue, rather large, like those of *G. incana*, but glabrous or slightly glandular-pubescent outside. Capsule of *G. incana*, the seeds (in the very few specimens in which they have been seen ripe) broadly but irregularly winged or sometimes almost without wings.—DC. Prod. vii. 515 ; *G. rigida*, Benth. in Hueg. Enum. 71 ; DC. Prod. vii. 516 ; De Vr. Gooden. 156 ; *Scævola tenera*, De Vr. in Pl. Preiss. i. 409, partly ; *Goodenia pterygosperma*, De Vr. Gooden. 153, as to the King George's Sound and Swan River specimens, not of R. Br. ; *G. teretifolia*, De Vr. Gooden. 130 ; *G. Barilletii*, F. Muell. Fragm. iii. 140.

W. Australia. King George's Sound and adjoining districts, *R. Brown, Baxter,* and others ; thence to Swan River, *Huegel, Drummond, n.* 395 *and 2nd Coll. n.* 297, *Preiss, n.* 1482, and Murchison river, *Oldfield* ; eastward to Stokes Inlet, *Maxwell.*

The northern specimens are taller and more slender than the southern ones, but do not otherwise appear different.

33. **G. trichophylla,** *De Vr. in Herb. Hook.* A perennial, glabrous except the calyx, with a tufted stock and erect simple or branched stems of ½ to 1 ft., like *G. cærulea*, but very much more slender. Leaves very narrow-linear, the radical ones 2 to 4 in. long, dilated at the base, the stem ones almost filiform. Flowers small, on filiform pedicels longer than the leaves. Bracteoles minute, at a distance from the flowers or obsolete. Calyx slightly glandular-pubescent, 1½ lines long, the lobes linear, acute. Corolla (of a

purplish-blue ?) 4 to 5 lines long, glabrous or slightly glandular-pubescent outside, the lobes all nearly equally winged, but the upper ones separated much lower down. Dissepiment of the ovary reaching to above the middle; ovules several, in 2 rows in each cell. Capsule ovoid-oblong, small, but not seen ripe. Seeds orbicular, flat.

W. Australia, *Drummond, 2nd Coll. n.* 407, *3rd Coll. n.* 158.

34. **G. Hassallii,** *F. Muell. Fragm.* vi. 10. *t.* 51. A perennial, quite glabrous, except sometimes a little wool in the axils ; the stems erect, branching, leafy, about 1 ft. high, often several together, but not forming a tufted stock like *G. cærulea,* and the radical leaves not leaving the broad persistent bases of that species and its allies. Leaves linear or lanceolate, entire or with a few distant teeth or lobes, often 2 to 3 in. long, the upper floral ones much smaller and narrow-linear. Peduncles rarely exceeding the floral leaves or bracts, 1-flowered or rarely the lower ones 2- or 3-flowered. Bracteoles linear-terete, at a distance from the flower. Calyx-lobes lanceolate, acute. Corolla blue, 6 to 7 lines long, glabrous or sprinkled with a few hairs outside, the lobes all broadly and nearly equally winged, but the upper ones separated nearly to the base. Dissepiment of the ovary reaching to above the middle ; ovules several, in 2 rows in each cell. Capsule ovoid-oblong, about 3 lines long. Seeds flat, bordered by a smooth margin, but not at all or very narrowly winged.

W. Australia, *Drummond.*

35. **G. pterigosperma,** *R. Br. Prod.* 578. A glabrous perennial, with a tufted stock, and usually numerous, erect, scarcely branched stems of 1 ft. or more. Radical leaves and a few at the base of the stems petiolate, linear or linear-lanceolate, often 2 in. long or more, thick, bordered by a few small, distant teeth or entire, the other leaves few, small, and distant. Flowers of a deep blue, terminal or on long peduncles in the upper axils. Bracteoles very small and obtuse, at a distance from the flower. Calyx quite glabrous, the lobes linear and obtuse, shorter or scarcely longer than the tube. Corolla about ½ in. long, quite glabrous outside or rarely sprinkled with a few hairs, the lobes all broadly winged, and separated very low down, the 2 upper ones almost to the base, with the outer wings descending quite to the base. Dissepiment of the ovary reaching to above the middle ; ovules several, in 2 rows in each cell. Capsule ovoid or globular, 2 to 3 lines long. Seeds broadly winged.—DC. Prod. vii. 515 ; De Vr. Gooden. 153, partially ; *G. cyanea,* F. Muell. Fragm. i. 155.

W. Australia. Lucky Bay, *R. Brown,* and thence eastward to Stokes Inlet, Cape le Grand, Phillips Range, etc., *Maxwell.*

36. **G. Vilmoriniæ,** *F. Muell. Fragm.* iii. 19. *t.* 16. An erect herb, with a few woolly hairs in the lower part, especially about the axils, the inflorescence glandular-pubescent. Leaves linear, entire, 2 to 4 in. long, the floral ones much shorter. Peduncles axillary, longer than the leaves, 1-flowered, without bracteoles. Calyx-lobes rather short. Corolla blue (*F. Muell.*), 7 to 8 lines long, glandular-pubescent outside ; lobes broadly winged, the 2 upper ones separated lower down, and with a concave auricle

at the base of the wing on the outer side. Dissepiment of the ovary reaching
to above the middle. Ovules rather numerous, in 2 rows in each cell. Ripe
capsule not seen ; young seeds with rather broad wings.

N. Australia. Between Bonney river and Mount Morphett, *M'Douall Stuart* (*Herb.
F. Mueller*).

G. Bonneyana, F. Muell. Fragm. t. 53, is, as far as I am aware of, as yet undescribed, and
I find no specimen in F. Mueller's collection. From the plate it appears only to differ from
G. Vilmoriniæ in being more hairy, and in having long, linear bracteoles.

SERIES IV. FOLIOSÆ.—Erect or rarely decumbent herbs or undershrubs,
usually glandular-pubescent or hirsute, with leafy stems. Peduncles axillary,
1-flowered, articulate under the flower, without bracteoles or very rarely a
few peduncles have 1 or 2 minute ones. Flowers yellow white or purple.

The hollow protuberance at the base of the corolla-tube is more conspicuous in several of
the species of this series than in the others, and sometimes forms a more or less prominent
spur as in *Velleia*.

37. **G. calcarata,** *F. Muell. Fragm.* vi. 14. An erect, rather stout,
glabrous, and often glaucous annual of $\frac{1}{2}$ to 1 ft., the stem very angular.
Leaves deeply pinnatifid, with ovate-oblong or lanceolate, deeply-toothed segments, the terminal one sometimes large and the others small or all more
nearly of a size, the upper floral leaves reduced to small bracts. Peduncles
solitary in the axils, forming a terminal raceme. Bracteoles none or very
minute under the flowers. Calyx and corolla nearly those of *G. grandiflora*,
but the hollow protuberance of the adnate part produced into a small spur.
Capsule ovate, 4 to 5 lines long. Seeds of *G. grandiflora.*—*Picrophyta calcarata*, F. Muell. in Linnæa, xxv. 422.

N. S. Wales. Between the Darling river and the Barrier Range, *Panton.*
S. Australia. Stony declivities near Cudnaka, *F. Mueller ;* Lake Gillies, *Burkitt.*

38. **G. Nicholsoni,** *F. Muell. Fragm.* i. 203. *t.* 4. A shrub or
undershrub, softly but minutely glandular-pubescent or tomentose, very
closely allied to the pinnate-leaved state of *G. grandiflora*, differing only in
the terminal lobe of the leaves being more cuneate and lobed at the base, not
cordate, and the bracteoles more frequently present, either at the base of the
peduncles or higher up, although many peduncles are entirely without.
Flowers of *G. grandiflora* or the wings of the corolla-lobes rather narrower,
the hollow protuberance is prominent in some flowers, inconspicuous in
others.

S. Australia. N.W. interior, *M'Douall Stuart's Expedition.* Probably a variety
only of *G. grandiflora.*

39. **G. Macmillani,** *F. Muell. Fragm.* i. 119. *t.* 5. Very closely allied
to *G. grandiflora*, and perhaps a variety, with the same stature, inflorescence,
and size and structure of the flowers and fruit, but the leaves are more decidedly lyrate, with several segments, the terminal one larger, but not so
much so as in *G. grandiflora*, and the flowers are purple.

Victoria. Schistose valleys of the Macalister river, *F. Mueller.*

40. **G. grandiflora,** *Sims; Bot. Mag. t.* 890. Herbaceous, erect,

rather stout, more or less glandular-pubescent, attaining sometimes 3 or 4 ft. Leaves petiolate, from broadly ovate to ovate-lanceolate, truncate or cordate at the base, toothed, often above 2 in. long, with or without the addition of a few small segments along the petiole. Peduncles axillary, shorter than the leaves, 1-flowered, without any or with very minute bracteoles at the base, solitary or the lower ones sometimes 2 or 3 together on a very short, common peduncle. Flowers large, yellow, more or less streaked with purple. Calyx-lobes linear-lanceolate. Corolla glabrous or slightly pubescent outside, usually about 1 in. long but sometimes larger, the 2 upper lobes more deeply separated, the adnate part of the tube with a hollow protuberance, sometimes scarcely conspicuous, sometimes very prominent and reaching to the base of the calyx. Dissepiment of the ovary reaching far above the middle; ovules in 2 rows in each cell. Capsule ovoid-oblong, sometimes ½ in. long, but often smaller. Seeds broad, flat, with a thickish margin, not winged.—R. Br. Prod. 576; DC. Prod. vii. 514; Bonpl. Jard. Malm. t. 6; Bot. Reg. 1845. t. 29; De Vr. Gooden. 143; F. Muell. Fragm. i. 204; *G. appendiculata*, Jacq. Fragm. t. 92.

Queensland. Broad Sound, *R. Brown;* Wide Bay, *Bidwill;* Burnett and Burdekin rivers, *F. Mueller;* Rockhampton, *Dallachy* and others.

N. S. Wales. Port Jackson to the Blue Mountains, *R. Brown* and others; Richmond river, *Fawcett;* Macleay and Clarence rivers, *Beckler.*

S. (or N. ?) Australia. Mount Freeling, Central Australia, *M'Douall Stuart's Expedition.*

G. mollis, R. Br. Prod. 577; DC. Prod. vii. 515; De Vr. Gooden. 151, is a variety, or, perhaps, a state of the plant, with rather shorter capsules, and no small leaf-segments along the petiole. To this form belong the majority, but not all, of the Queensland specimens, and a few of the N. S. Wales ones.

41. **G. Chambersii,** *F. Muell. Fragm.* i. 204. A shrub or under-shrub, softly but minutely glandular-pubescent, very closely allied to *G. grandiflora*, and, judging from the fragmentary specimens preserved, perhaps a variety. Leaves smaller on shorter petioles, orbicular, coarsely toothed, ½ to ¾ in. diameter in our specimens, and without accessory segments. Flowers rather smaller than in *G. grandiflora*, the lobes rather narrower, and the upper ones less deeply separated than in that species, but the structure otherwise the same, and the saccate protuberance prominent.

S. (or N. ?) Australia. Mount Freeling, *M'Douall Stuart's Expedition.*

42. **G. albiflora,** *Schlecht. Linnæa*, xx. 599. An erect, glabrous perennial, with a thick, almost woody rootstock; stems angular, 1 ft. high or more. Leaves ovate-lanceolate or lanceolate, acutely toothed, 1½ to 3 in. long, narrowed into a petiole without accessory lobes, the upper floral ones very much smaller, soon passing into lanceolate, entire bracts. Flowers entirely of *G. grandiflora*, but white, and forming more of a terminal, leafy raceme, the saccate protuberance of the corolla-tube at least as prominent as in *G. grandiflora*.—*Picrophyta albiflora*, F. Muell. in Linnæa, xxv. 421.

S. Australia. Gawler river, *Behr;* common in dry, gravelly beds of streams, St. Vincent's Gulf, Encounter and Holdfast Bays, etc., more rare at the base of Lofty Range, *F. Mueller.*

43. **G. Strangfordii,** *F. Muell. Fragm.* vi. 11. *t.* 52. A perennial,

simple or branching, and almost woody at the base, more or less pubescent or almost silky, with a little wool in the axils of the leaves, the stems erect, leafy, under 1 ft. high. Leaves rather crowded, lanceolate, entire, 3 to 4 in. long, narrowed into a petiole. Peduncles slender, shorter than the leaves, 1-flowered, without bracteoles or rarely 2 linear bracteoles on the lower ones as represented in the plate, the flower readily disarticulating from the peduncle as in most of the ebracteolate species. Calyx-lobes linear. Corolla silky, pubescent outside, 7 to 8 lines long in most specimens, the upper lobes deeply separated, rather short, broadly auriculate. Dissepiment of the ovary reaching above the middle; ovules rather numerous. Seeds orbicular, flat, not winged.

N. Australia. Upper Victoria river, *F. Mueller.;* Elsey's Creek, *Herb. F. Mueller.*

Var. *grandiflora.* Flowers much larger, the lower peduncles long, with 2, long, linear bracteoles.—Flinders river, *Kennedy* (*Herb. F. Muell.*). This is the form figured.

44. **G. Mitchellii,** *Benth.* A densely villous-tomentose, rather coarse perennial, with decumbent or ascending stems, under 1 ft. in our specimens. Radical and lower leaves on long petioles, obovate-oblong, coarsely toothed or almost pinnatifid, thick and soft, 2 to 3 in. long, the upper ones small, the uppermost floral ones scarcely ½ in. long, but all more or less petiolate. Peduncles 1-flowered, without bracteoles. Calyx-lobes narrow-linear. Corolla about ¾ in. long, apparently yellow, hirsute outside, the upper lobes rather shorter, separated almost to the base, with concave auricles, the tube with a small, concave protuberance, forming a minute spur at the base of the calyx. Dissepiment of the ovary exceedingly short; ovules few. Capsule globular, about 3 lines diameter. Seeds usually 1 or 2 only perfect, large, flat, with a broad, thick margin, not winged.

Queensland. In the interior, *Mitchell.* The precise station not given.

45. **G. heterochila,** *F. Muell. Fragm.* iii. 142. Herbaceous, softly pubescent or villous. Leaves oval-oblong or lanceolate, entire or toothed, 1 to 2 in. long, contracted at the base, the lower ones not seen. Peduncles 1-flowered, axillary, slender, shorter than the leaves, articulate under the flower. Bracteoles none or very minute. Calyx-lobes linear or setaceous. Corolla under ½ in. long, the 3 lower lobes winged and truncate, the 2 upper ones separate lower down and shorter, narrow-lanceolate, acute, scarcely winged, but with an auricle on the outer side below the middle. Dissepiment of the ovary very short; ovules 4 or 5, large and flat.

N. Australia. Newcastle water and Burke river, *M'Douall Stuart's Expedition.*

Var. *? foliosa.* Stouter and more hirsute. Stems decumbent or erect, under 1 ft. high. Leaves rather larger than the other form, and the lobes of the corolla less dissimilar.—Victoria river, *F. Mueller.*

Var. *? racemosa.* Apparently annual and very hirsute, the upper peduncles longer than the very small floral leaves, the lobes of the corolla less dissimilar than in the typical specimens.—Camden Harbour, N.W. Australia, *Herb. F. Mueller.*

Var. *? runcinata.* Leaves deeply toothed or pinnatifid, otherwise like the typical form.—Arnhem's Land, *F. Mueller.*

The specimens of all the above forms, as well as of those included under the 2 following species (*G. sepalosa* and *G. hispida*) are too few and too imperfect to judge of their value as species or varieties. Amongst them is probably included *G. melanoptera*, F. Muell. Fragm.

i. 115, from Victoria river, which I have been unable to identify satisfactorily, and I find no specimens so named in his collections.

46. **G. sepalosa,** *F. Muell. Herb.* A low, branching, hispid or viscid-villous herb, apparently perennial, but flowering the first year, the stems decumbent or erect, not attaining 1 ft. in any specimens seen. Leaves oblong lanceolate or oblong-linear, coarsely and irregularly toothed or rarely entire, shortly petiolate or the upper ones only contracted at the base, the larger ones 2 to 3 in. long. Peduncles 1-flowered, axillary, without bracteoles, in some specimens very short, in others at least as long as the leaves. Calyx-lobes more or less lanceolate and leafy, acuminate. Corolla yellow, hairy outside, ¾ to 1 in. long, the lobes broadly winged, the 2 upper ones separate almost to the base, the wings unequal, the outer one almost auriculate. Dissepiment of the ovary exceedingly short, almost rudimentary. Capsule globular. Seeds few, flat, granular-tuberculate, with a narrow, smooth margin.

N. Australia. N.W. coast, *Bynoe;* Camden Harbour, *Martin;* Port Essington, *Armstrong.* F. Mueller, Fragm. vi. 12, refers this species to *G. hispida;* but the general aspect, as well as the more toothed, broader, usually more petiolate leaves and broader calyx-lobes appear to me too different to warrant the union without better specimens, most of those preserved being little more than fragmentary. The fragment alluded to by F. Mueller, as gathered by Kennedy on the Flinders river, has much more of the foliage, but not the indumentum of *G. hispida.*

Var. *brachypoda,* F. Muell. Herb. Much-branched and very leafy, the upper leaves small. Flowers small (the corolla about ½ in. long), on very short peduncles or almost sessile.—Victoria river, *F. Mueller.*

47. **G. hispida,** *R. Br. Prod.* 577. Apparently annual, hispid with rigid spreading hairs, which almost disappear from the old plants. Stems slender, erect, branching, ½ to 1 ft. long or rarely more. Radical leaves sometimes oblong and narrowed into a short petiole, all the others sessile, linear or narrow-lanceolate, 1 to 2 in. long, entire or with a very few acute prominent teeth, especially near the base. Peduncles in the upper axils often longer than the leaves, without bracteoles. Calyx-lobes subulate. Corolla hairy outside, about ½ in. long, the 2 upper lobes separated low down. Capsule globular or ovoid. Dissepiment exceedingly short. Seeds flat, not winged.—DC. Prod. vii. 515 ; De Vr. Gooden. 133.

N. Australia. Islands of the Gulf of Carpentaria, *R. Brown;* Copeland Island, *A. Cunningham;* Bowen's Straits, Port Essington, *Armstrong.* These specimens are much less hispid than Brown's, but much older and more elongated and probably belong to the same species.

48. **G. auriculata,** *Benth.* Apparently annual, sprinkled with spreading hairs or nearly glabrous. Stems weak, 1 ft. long or more. Lower leaves unknown ; floral leaves ovate-lanceolate, entire, cordate, sessile and stem-clasping with broad auricles. Peduncles 1-flowered, much longer than the leaves, without bracteoles. Flowers rather large, yellow. Calyx hispid, with linear acuminate lobes. Corolla ¾ in. long, pubescent outside. Dissepiment of the ovary exceedingly short. Capsule nearly globular. Seeds few, oval-oblong, flat, granular-tuberculate, with a somewhat thickened smooth border, not winged.

N. Australia. Depot Creek, Upper Victoria river, *F. Mueller.* As far as shown by the specimens, this species appears to be more nearly allied to *G. Armstrongiana* than to *G. hispida*, but the broad auriculate floral leaves are different from those of either.

49. **G. Armstrongiana,** *De. Vr. Gooden.* 138. *t.* 24. Apparently annual, glabrous or sprinkled with soft hairs. Stems slender, leafy, erect or flexuose, 1 to 2 ft. long. Leaves sessile and stem-clasping, the lower ones ovate and sometimes bordered by a few small teeth, the upper ones quite entire, lanceolate, more or less decurrent along the stem. Peduncles filiform, axillary, longer than the leaves, without bracteoles. Flowers small, yellow. Calyx-lobes setaceous. Corolla not $\frac{1}{2}$ in. long, glabrous or with a few rigid hairs outside, the lobes broadly winged. Capsule ovoid, about 3 lines long, the dissepiment reaching above the middle. Seeds flat, granular-rugose, with a rather thick smooth margin.

N. Australia, *R. Brown Herb.* (an imperfect specimen, without the precise station), Port Essington, *Armstrong;* Victoria river and grassy flats between Providence Hill and Macadam Range, *F. Mueller.*

50. **G. corynocarpa,** *F. Muell. Fragm.* ii. 16. Herbaceous and glabrous or sprinkled with appressed silky hairs. Stems ascending, sometimes above 2 ft. high. Leaves petiolate, ovate-lanceolate oblong or lanceolate, the radical ones sometimes many inches long and coarsely toothed, the stem ones smaller and distant, the upper floral ones reduced to small sessile bracts. Peduncles axillary, 1-flowered, forming a long loose leafy raceme. Bracteoles none or minute and close under the flower. Calyx-tube cylindrical, longer than the setaceous lobes. Corolla (yellow?) 7 to 8 lines long, very silky-hairy outside, the upper lobes separated lower down, with a broad concave auricle. Style usually very short. Dissepiment of the ovary reaching far above the middle. Capsule linear, about $\frac{1}{2}$ in. long in the common form. Seeds almost in a single row in each cell, flat, black, minutely granulate, with a paler coloured smooth border.

W. Australia. Murchison river, *Oldfield.* The inflorescence and the linear capsule give this species a very distinct aspect.

Var. *macrocarpa.* Bracteoles conspicuous, although very small. Capsule about 1 in. long.—Between Moore and Murchison rivers, *Drummond, 6th Coll. n.* 146.

SERIES 5. PEDICELLOSÆ. Herbs with leaves chiefly radical or tufted, the stem-leaves either few and distant or collected in terminal tufts, rarely scattered and linear or reduced to small bracts. Peduncles long, radical or in terminal tufts or axillary, 1-flowered, usually articulate under the flower, without bracteoles or rarely here and there with very minute ones.

51. ? **G. mollissima,** *F. Muell. Herb.* A small plant, very villous all over with soft hairs. Radical leaves petiolate, obovate, deeply crenate, 1 to $1\frac{1}{2}$ in. long. Stems short, decumbent, with a few small almost sessile orbicular or obovate leaves. Peduncles axillary, rather longer than the leaves, 1-flowered, without bracteoles. Flowers rather large, yellow. Calyx hispid, the lobes narrow and acute. Corolla pubescent outside, 7 to 8 lines long. Dissepiment of the ovary reaching far above the middle. Ovules numerous. Fruit not seen.

Queensland. Near Cooper's Creek, *Bowman.* Although the specimens are not developed enough to fix the affinities of this plant, it appears to be a very distinct species, differing from *G. lanata* in the absence of bracteoles, from *G. cycloptera* in the ovary.

52. **G.cycloptera,** *R. Br. in App. Sturt, Exped.* 20. Annual or with a perennial tufted or creeping rootstock, tomentose-pubescent or softly villous and often glandular. Radical leaves on long petioles, obovate or oblong, coarsely toothed, 1 to 2 or even 3 in. long. Stems either very short and erect or decumbent, or ascending and from 6 in. to 1 ft. long, with a few leaves usually crowded towards the end or scattered when the stems are much lengthened, all smaller, narrower, and more entire than the radical ones and nearly sessile. Peduncles in the upper axils or terminal tufts, 1-flowered, without bracteoles or rarely with very minute ones. Calyx hirsute, the lobes narrow. Corolla yellow, pubescent outside, 6 or 7 lines long, the tube sometimes forming a minute spur at the base of the calyx, the 2 upper lobes deeply separated, unequally winged and auriculate. Dissepiment of the ovary very short or almost rudimentary, with few ovules. Capsule orbicular or shortly ovate. Seeds large, orbicular, flat, usually black with a paler coloured rather thick marginal wing.

N. S. Wales. Darling river, *Dallachy, Mrs. Ford;* thence to the Lachlan, *Burkitt, Victorian Expedition.*

S. Australia. Scrub to the N.E. of Lake Gairdner, *Babbage;* Spencer's Gulf, *Warburton.* I have not seen Sturt's specimens described by Brown, but his character leaves no doubt as to the identity.

53. **G. tenella,** *R. Br. Prod.* 577. A small slender plant, appearing sometimes annual but evidently stoloniferous, glabrous or sprinkled with a few hairs and with a little wool in the axils of the leaves. Leaves radical or rosulate, ovate oblong or oblanceolate, entire or obscurely crenate, $\frac{1}{4}$ to $\frac{1}{2}$ in. long, besides the petiole. Peduncles filiform without bracteoles, 1-flowered, and longer than the leaves or sometimes bearing a tuft or umbel of pedicels, with a few leaves narrower than the radical ones. Flowers very small, yellow. Calyx-lobes linear. Corolla hairy outside, about 3 lines long, the upper lobes deeply separated and all broadly winged. Dissepiment of the ovary exceedingly short; ovules few. Capsule ovoid, about 2 lines long. Seeds few, large, flat, granular-tuberculate, with a rather broad smooth border, but not winged.—DC. Prod. vii. 515 ; De Vr. Gooden. 150.

W. Australia. King George's Sound, *Bauer, Drummond, n.* 184.

Var. *major.* Larger in every part, the leaves $\frac{1}{2}$ to 1 in. long, and the flowers larger.—Muddy bed of Don river, *Maxwell ;* Karri Dale, *Herb. F. Mueller.*

Scævola pusilla, De Vr. in Pl. Pr. i. 412 or *Goodenia pusilla,* De Vr. Gooden. 131, from Plantagenet district, *Preiss, n.* 1470, which I have not seen, appears from the description to be the same as *G. tenella.*

54. **G. elongata,** *Labill. Pl. Nov. Holl.* i. 52. *t.* 75. A perennial, sprinkled with simple appressed or spreading hairs or rarely glabrous, the rootstock tufted or slender and creeping. Radical leaves on long petioles, obovate or oblong-spathulate, entire or obscurely sinuate-toothed. Stems sometimes very short, but usually ascending or erect and weak, $\frac{1}{2}$ to above 1 ft. long, with a few distant oblong or lanceolate nearly sessile leaves. Peduncles 1-flowered, without bracteoles, long and slender, sometimes scapelike amongst the radical leaves, but mostly in the axils of the stem-leaves.

Flowers rather large, yellow. Calyx-lobes linear or linear-lanceolate. Corolla 7 to 9 lines long, the 2 upper lobes deeply separated, the wings unequal, with a concave auricle on the outer side. Dissepiment usually about half as long as the ovary but variable, with 4 to 6 ovules in each cell. Indusium usually hairy. Capsule ovoid, about 3 lines long. Seeds not so flat as in most species, the border scarcely prominent and not winged.—R. Br. Prod. 577; DC. Prod. vii. 515; De Vr. Gooden. 148 (as to Brown's plant, but not the western one associated with it, which is *G. filiformis*); Hook. f. Fl. Tasm. i. 232.

Victoria.   Ovens river, Buffalo Range, Dandenong mountains, *F. Mueller.*

Tasmania.   Port Dalrymple, *R. Brown*; not uncommon in marshy soil in various parts of the colony, *J. D. Hooker.*

55. **G. pinnatifida,** *Schlecht. Linnæa,* xxi. 440. A perennial, sprinkled with appressed hairs or nearly glabrous, the rootstock tufted or creeping. Radical leaves petiolate, oblong-spathulate or narrow, deeply toothed or pinnatifid or a few of the outermost obovate and crenate, often 2 to 3 in. long. Stems ascending or erect, $\frac{1}{2}$ to 1 ft. long, with narrow entire or pinnatifid leaves, few and distant or sometimes clustered at the ends of the stems. Peduncles very long, axillary, 1-flowered, without bracteoles or rarely with very minute ones at the base. Flowers rather large, yellow. Calyx-lobes linear or linear-lanceolate. Corolla sprinkled with rigid appressed hairs, $\frac{1}{2}$ to $\frac{3}{4}$ in. long in the smaller forms, $\frac{3}{4}$ to 1 in. in the larger ones, the lower lobes often ciliate below the wings, the upper ones deeply separated, with a broad concave auricle on the outer side. Dissepiment of the ovary short, but the ovules rather numerous. Capsule nearly orbicular, 3 to 4 lines long. Seeds few, large, very flat, black with a lighter-coloured border.

N. S. Wales.   Darling and Lachlan rivers, *Burkitt*, and thence to the Barrier Range, *Victorian and other Expeditions;* Nangas, *Macarthur;* New England, *C. Stuart;* Manilla river, *Leichhardt;* Darling Downs, *Law.*

Victoria.   Common on the Yarra, Forest Creek, towards Bacchus Marsh, *F. Mueller* and others; Wimmera, *Dallachy.*

S. Australia.   Holdfast Bay and Flinders Range, *F. Mueller;* Port Lincoln, *Wilhelmi.*

56. **G. coronopifolia,** *R. Br. Prod.* 577. Apparently annual although tufted, glabrous except a few long hairs towards the base of the leaves. Radical leaves linear, pinnatifid with linear lobes or rarely nearly entire, $1\frac{1}{2}$ to 3 in. long. Stems slender, erect or ascending, $\frac{1}{2}$ to nearly 1 ft. high, sometimes almost filiform, with a very few distant linear leaves. Peduncles 1-flowered, long and slender, without bracteoles. Flowers small, yellow (*R. Brown*), but assuming a purplish hue when dry. Calyx-lobes linear, short. Corolla glabrous outside, 3 to 4 lines long, the 2 upper lobes deeply separated and with broad auricles on the outer side. Dissepiment of the ovary short. Capsule orbicular or ovoid. Seeds few, large, flat, broadly winged.— DC. Prod. vii. 515; De Vr. Gooden. 149 (as to Brown's plant, but not Mitchell's nor the New Zealand one associated with it).

N. Australia.   Cambridge Gulf, N.W. coast, *A. Cunningham;* islands of the Gulf of Carpentaria, *R. Brown.*

57. G. heteromera, *F. Muell. Fragm.* i. 115. A glabrous or sparingly pubescent perennial, with a tufted stock emitting stems or stolons at first sometimes erect but usually prostrate, and all leafless except a terminal tuft, which often roots and becomes a new plant. Leaves radical or in the terminal tufts, petiolate, linear-spathulate or narrow-oblong, entire or rarely pinnatifid, often several inches long. Peduncles or scapes 1-flowered, slender, rarely as long as the leaves, without bracteoles or rarely with very minute ones at the base. Calyx-lobes variable, usually unequal. Corolla yellow, glabrous or hairy outside, under ½ in. long. Dissepiment of the ovary reaching above the middle. Capsule obovoid, about 3 lines long. Seeds several, flat, winged.

Queensland? In the interior, *Mitchell.*
N. S. Wales. Gwydir river, *Leichhardt;* Darling river, *Victorian and other Expeditions.*
Victoria. Murray and Avoca rivers, *F. Mueller.*

58. **G. concinna,** *Benth.* A perennial, either glabrous or sprinkled with appressed hairs or the inflorescence sometimes glandular-pubescent. Leaves radical or crowded on the short simple stock, petiolate, linear or linear-lanceolate, entire, 2 to 3 in. long. Scapes or flowering branches erect, ½ to nearly 1 ft. high, leafless or with 1 or 2 filiform leaves, besides small bracts under the peduncles. Flowers yellow, rather larger than in *G. filiformis,* on slender peduncles without bracteoles, the upper ones sometimes almost umbellate. Calyx-tube scarcely any, the lobes linear, adnate at the base. Corolla glabrous outside or slightly glandular, fully ½ in. long, the adnate tube with a slight saccate protuberance, the lobes broadly winged, the 2 upper ones unequally so and separated low down. Dissepiment of the ovary reaching far above the middle ; ovules numerous in 2 rows. Capsule ovoid-oblong, 2 to 3 lines long. Seeds flat, granular-tuberculate, with a rather broad smooth margin, but not winged.

W. Australia, *Drummond, n.* 360 ; Point Henry, *Oldfield ;* E. Mount Barren, Eyre's Relief, Phillips and Fitzgerald Ranges, *Maxwell.*

59. **G. glauca,** *F. Muell. in Trans. Vict. Inst.* 1855, 40. A perennial, glabrous and glaucous in the typical form, silky-hairy in a common variety, with a slender creeping rootstock and ascending or erect stems, scarcely exceeding 6 in. in most specimens, sometimes densely tufted. Radical leaves oblong-lanceolate or linear-lanceolate, entire or rarely with a few prominent teeth, 1 to 3 in. long; stem-leaves linear or linear-lanceolate, entire. Peduncles 1-flowered, longer than the leaves, without bracteoles. Flowers rather large, yellow. Calyx-lobes narrow, acute. Corolla nearly glabrous outside, nearly ¾ in. long, the 2 upper lobes separate low down, unequally winged and auriculate. Dissepiment of the ovary short. Indusium glabrous. Capsule globular, about 3 lines long. Seeds few, large, flat, with a broad not very thin wing.

N. Australia. Sturt's Creek, *F. Mueller.*
Queensland. In the interior, *Mitchell ;* plains of the Condamine, *Leichhardt ;* Dawson river, *F. Mueller ;* Armadillo, *W. Barton.*
N. S. Wales. Liverpool plains, *Leichhardt ;* Darling river to the Barrier Range, *Victorian and other Expeditions.*

Victoria. Murray and Avoca rivers, *F. Mueller.*
S. Australia. Murray river, *F. Mueller ;* Lake Torrens to Wills' ard Cooper's Creeks, *Howitt's Expedition.*
W. Australia ? Stokes Inlet, *Maxwell.* A single small specimen (in Hert. F. Muell.) with large flowers appears to be this species, but insufficient for positive identification.

This species has usually the large flowers of *G. elongata* and *G. pinnatifida,* from which it may be readily distinguished by the glabrous indusium, as pointed out by F. Mueller, as well as by the narrow leaves. It is very nearly allied to *G. coronopifolia,* differing chiefly in the large flowers. Some forms of the western *G. filiformis* come also very near to it, but besides the small flowers the seeds in that species are rarely winged and then irregularly so.

Var. *sericea.* Clothed all over or sprinkled with silky hairs. To this belong the northern and most of the N. S. Wales specimens.

Var. *glandulosa.* More or less glandular-pubescent. Darling river, *Dallachy ;* Goyinga mountains, *Victorian Expedition.*

60. **G. filiformis,** *R. Br. Prod.* 578. A slender annual or tufted perennial, glabrous or sprinkled with a few hairs in the typical form, the inflorescence sometimes glandular, and occasionally a little wool in the axils. Leaves in the typical form linear or filiform, but in some varieties more or less lanceolate, especially the radical ones, which are often 2 to 3 in. long, the stem ones few and smaller, all entire. Flowering stems ascending or erect, ½ to 1 ft. long or more, sometimes ending in an umbel-like tuft of leaves and peduncles. Flowers yellow, usually small. Peduncles 1-flowered, filiform, without bracteoles. Calyx-lobes linear. Corolla glabrous outside, rarely above ½ in. long, the adnate part of the tube with a hollow protuberance sometimes forming a short spur at the base of the calyx, the lobes all broadly winged, the 2 upper ones separated low down. Dissepiment short ; ovules not numerous. Capsule small. Seeds few, orbicular, with broad margins, sometimes irregularly winged.—DC. Prod. vii. 515 ; De Vi. Gooden. 134, but with a wrong station ; *Scævola umbellata,* De Vr. in Pl. Preiss. i. 411 ; *Aillya umbellata,* De Vr. Gooden. 76. t. 13 ; *G. elongata,* De Vr. in Pl. Preiss. i. 412, not of Labill.

W. Australia. King George's Sound, *R. Brown, Baxter,* and thence to Swan River, *Drummond, n.* 401, *Preiss, n.* 1428, 1429, 1430, 1435, 1450 ; Tone, Blackwood, and Murchison rivers, *Oldfield ;* Warrenup, *Maxwell.*

Var. *pulchella.* Glabrous, softly hairy or hispid. Leaves linear-spathulate or lanceolate. *G. pulchella,* Benth. in Hueg. Enum. 71 ; DC. Prod. vii. 515 ; *Velleia lanceolata,* Lindl. Swan Riv. App. 26 ; De Vr. Gooden. 175 (partly, for the specimens referred to in that work, and named *V. lanceolata* by De Vriese in the herbaria quoted, belong to eight different species, viz., *Goodenia cycloptera, G. glauca, G. hispida, G. filiformis, Calogyne Berardiana, Velleia cycnopotamica, Stylidium leptorhizum,* and *Vandellia pubescens*). The var. *pulchella* appears to be fully as common as the filiform one, and to pass gradually into it. Drummond's specimens, n. 185, are remarkably hirsute ; his n. 408 has very small flowers, but the specimens are imperfect.

61? **G. microptera,** *F. Muell. Fragm.* iii. 34. The specimens of this species consist of 2 small fragments, apparently very near to *G. glauca.* The leaves are wanting. The flowers are of the size of those of *G. glauca,* but the calyx-teeth are shorter, and the wings of the corolla-lobes are narrower, except the auricle of the upper lobes, which is broad. The dissepiment of the ovary reaches above the middle, and the capsule is rather longer than in

G. glauca, but the true value and affinities of the species cannot be correctly judged of without better specimens.

N. Australia. Nichol Bay, N.W. coast, *Walcott.*

Sect. III. Amphichila, *DC.* Corolla with the 2 upper lobes separated much lower down than the others. Ovules very numerous, closely packed in more than 2 rows in each cell of the ovary, which is 2-celled, except the very short, free summit. Capsule less coriaceous than in the other sections. Seeds numerous and small.—Flowers small, in leafless panicles or on radical scapes.

62. **G. paniculata,** *Sm. in Trans. Linn. Soc.* ii. 348. A perennial, more or less hirsute or rarely glabrous, with a tufted stock and erect stems of 1 to 2 ft. Radical leaves petiolate, from obovate to narrow-lanceolate, irregularly toothed or rarely entire, 1 to 2 in. long when broad, often twice that length when narrow; stem-leaves few and much smaller, the floral ones reduced to linear bracts. Flowers yellow, in a loose, terminal panicle, the primary branches racemose, the secondary cymose, the flowers all pedicellate within the small, linear bracts. Calyx-lobes subulate, short. Corolla about ½ in. long, glandular-pubescent outside, with longer simple hairs intermixed. Capsule about 2 lines long, the dissepiment reaching nearly to the top. Seeds numerous and very small, but orbicular and flattened.—R. Br. Prod. 575; Cav. Ic. t. 507; DC. Prod. vii. 513; De Vr. Gooden. 125; *G. flexuosa,* De Vr. Gooden. 126 (from the character given).

Queensland. Dawson, Mackenzie, and Burnett rivers, *F. Mueller;* Rockingham Bay, *Dallachy;* Rockhampton, *O'Shanesy.*

N. S. Wales. Port Jackson to the Blue Mountains, *R. Brown* and others; Hastings river, *Beckler;* Richmond river, *Fawcett.*

Victoria. Lake Wellington and seacoast near Snowy River, *F. Mueller.*

F. Mueller (Fragm. i. 116) refers to this species *G. gracilis* and *G. humilis,* which are certainly closely allied; but, if they are united, *G. lamprosperma* should also be included.

63. **G. purpurascens,** *R. Br. Prod.* 578. Glabrous, pubescent or the foliage almost woolly. Stock tufted; stems erect, 1 to 2 ft. high. Radical leaves linear or lanceolate, attaining sometimes 6 in. or more, entire or with a few remote teeth, the broader ones contracted into a long petiole; stem-leaves few and linear or none besides the small bracts. Flowers small, " of a purplish blue," usually numerous, in a loose panicle, the primary racemose branches bearing loose, irregular, dichotomous cymes. Calyx-tube very short; lobes linear. Corolla-tube adnate much higher up than the calyx; lobes broadly winged, the 2 upper ones separated low down, and the wings very unequal. Dissepiment of the ovary reaching to the summit or nearly so. Capsule rarely 2 lines long. Seeds exceedingly numerous and small, flat, with a wing-like margin of very variable breadth.—DC. Prod. vii. 515; De Vr. Gooden. 153; F. Muell. Fragm. i. 117.

N. Australia. Victoria river and Sturt's Creek, *F. Mueller;* islands of the Gulf of Carpentaria, *R. Brown;* Fitzmaurice river and near Providence Hill, *F. Mueller.*

Queensland. Rockingham Bay, *Dallachy;* Cleveland Bay, *Bowman.*

Var. ? *minima,* F. Muell. Very slender and small. Stems filiform, 3 to 4 in. high.

Panicle little branched, with few very small flowers.—Upper Victoria river, *F. Mueller.*
Perhaps a distinct species, but the specimens not perfect.

64. **G. gracilis,** *R. Br. Prod.* 575. A perennial, glabrous or with
more or less of wool about the stock. Radical leaves petiolate, linear or
linear-lanceolate, varying from 1 to 5 or 6 in. long, usually entire. Flowering stems always much longer than the radical leaves, and sometimes above
1 ft., with very few, linear leaves. Panicle loose as in *G. paniculata* and
G. purpurascens. Flowers yellow, and fruit entirely of *G. paniculata.* Seeds
very numerous, small, flat, smooth, and shining.—DC. Prod. vii. 513.

Queensland. Broad Sound, *R. Brown ;* Rockhampton and Keppel Bay, *Thozet.*
N. S. Wales. Aiton Plains, S. of Liverpool Plains, *A. Cunningham ;* Murray and
Darling rivers, *Dallachy.*
Victoria. Murray, Ovens, Broken, and King rivers, *F. Mueller.*

65. **G. lamprosperma,** *F. Muell. Fragm.* i. 116. Very closely allied
to *G. gracilis,* and perhaps a variety, more rigid and very much branched, 1
to 1½ ft. high, with very few stem-leaves. Radical leaves linear or lanceolate, but thicker and more rigid than in *G. gracilis,* prominently veined, often
rather broader and occasionally toothed. Flowers smaller than in some specimens of *G. gracilis,* but quite like those of other specimens, and in other
respects quite the same, as well as the capsule and seeds.

N. Australia. Victoria river, Sturt's Creek, and Macadam Range, *F. Mueller.*

66. **G. humilis,** *R. Br. Prod.* 575. A small, tufted perennial, glabrous,
except the inflorescence, which is pubescent or rarely the leaves and base of
the stem are slightly hirsute. Leaves radical, petiolate, linear-lanceolate or
oblong, sometimes all under 1 in., sometimes several inches long, quite entire
in all the specimens seen. Flower-stems or scapes almost leafless, paniculate, shorter than or scarcely exceeding the leaves, the primary branches
racemose, the secondary cymose, the flowers pedicellate above the small,
narrow bracts. Flowers yellow, and fruits of *G. paniculata* or the flowers
rather smaller. Ovules quite as numerous as in that species, but the seeds
sometimes fewer and rather larger.—DC. Prod. vii. 513 ; Hook. f. Fl. Tasm.
i. 233. t. 68 ; De Vr. Gooden. 131. t. 23 ; *G. graminifolia,* Hook. f. in Hook.
Lond. Journ. vi. 265 ; De Vr. Gooden. 133 ; *G. nana,* De Vr. Gooden.
132.

Victoria. Port Phillip, *R. Brown ;* Melbourne, *Adamson ;* Glenelg river, *Robertson ;*
Portland, *Allitt ;* Wannon, *Wilhemi.*
Tasmania. Northern parts of the island in marshy soil, *J. D. Hooker.*

67. **G. Laytoniana,** *F. Muell. Herb.* A small, glabrous perennial,
with a densely tufted stock. Radical leaves numerous, linear, obtuse, entire,
thin, from under 1 in. to above 2 in. long. Stems slender, erect, 2 to 6 in.
high, leafless except the bracts, divided into a loose 2- or 3-chotomous or
almost umbellate panicle of few flowers, with leaf-like linear bracts at the
forks. Pedicels filiform, without bracteoles. Flowers small, yellow. Calyx
about 2 lines long, the lobes as long as the tube. Corolla 3 to 4 lines long,
slightly hairy outside, the 2 upper lobes separated low down and unequally
winged. Indusium glabrous, ciliate. Dissepiment of the ovary reaching to

the top of the adnate part; ovules very numerous. Capsule narrow, about
2 lines long, opening in 4 valves. Seeds very small, flat, not winged.—
G. tenella, F. Muell. Fragm. ii. 111, not of R. Br.

W. Australia, *Drummond, 1st Coll. n. 42, also n.* 159, 406; swampy flats, Don
river, *Maxwell.*

68. **G. bicolor,** *F. Muell. Herb.* A small, tufted, pubescent plant,
apparently perennial. Leaves rosulate, petiolate, obovate or ovate, thin,
entire or minutely and remotely toothed, the larger ones above 1 in.
long. Scapes slender, erect, leafless or with very small, linear, bract-like
leaves, 2 to 4 in. high, bearing a loose raceme of very few small flowers.
Bracts small and narrow. Pedicel ¼ in. long, without bracteoles. Calyx
nearly 2 lines long, the lobes as long as the tube. Corolla 3 to 4 lines long,
pubescent outside, the upper lobes purple, the lower yellow (*F. Muell.*), the
upper ones separate to the base. Dissepiment reaching nearly to the top of
the ovary, with numerous ovules in several rows. Capsule oblong, about
2 lines long. Seeds very numerous and small, flat, orbicular, with a thickened margin.

N. Australia. Pastures between Macadam Range and Providence Hill, *F. Mueller.*
This appears to be the plant alluded to as near *G. paniculata* by F. Mueller, Fragm. i. 117.

69. **G. pumilio,** *R. Br. Prod.* 579. A small, shortly creeping or stoloniferous perennial, hoary with an irregularly stellate pubescence. Leaves
in rosulate tufts, petiolate, ovate or orbicular, obtuse, entire, 1 to 3 lines long.
Peduncles numerous in the tufts, slender, ½ to ¾ in. long, each with a single,
small flower, without bracteoles. Calyx not above 1 line long at the time of
flowering, the lobes short. Corolla only seen in bud, and then the 2 upper
lobes rather larger than the others. Anthers ovate. Ovary almost completely 2-celled; ovules in about 4 rows in each cell. Capsule oblong or
ovoid-oblong, 1½ to nearly 2 lines long. Seeds small, orbicular, flat, not
winged.—DC. Prod. vii. 516.

N. Australia. South Goulburn island, *A. Cunningham.*
Queensland. Endeavour river, *Banks and Solander.*

5. CALOGYNE, R. Br.

(Distylis, *Gaudich.*)

Calyx, corolla, stamens, ovary, capsule, and seeds of *Goodenia*. Style
deeply divided into 2 incurved and connivent branches, each with a dimidiate
indusium open on the inner edge, and enclosing a half-stigmate, with sometimes a third intermediate shorter branch, with a straight indusium open on
both edges, and enclosing the central portion of the stigmate.—Erect or diffuse annuals. Leaves alternate. Peduncles axillary, without bracteoles.
Flowers yellow.

Besides the two Australian species, a third has been published from the coasts of China.
It is, however, so near to the *C. pilosa* that it may, perhaps, prove to be a variety only.
The genus is very closely allied to *Goodenia*, with a precisely similar habit, differing in the
single character of the divided style.

Style with 2 branches. Dissepiment reaching at least the middle of
the ovary . . . . 1. *C. Berardiana.*

Style with 3 branches. Dissepiment exceedingly short, almost rudimentary . . . . 2. *C. pilosa.*

1. **C. Berardiana,** *F. Muell. Fragm.* vi. 7. An erect, glandular-pubescent or hirsute annual of ½ to 1 ft. Leaves lanceolate or almost linear, bordered by a few teeth or almost entire, the radical ones petiolate, in some specimens almost lobed and 2 to 3 in. long, in others few, small, and entire; the stem-leaves few and distant, gradually passing into narrow-linear bracts. Calyx-lobes linear. Corolla hairy outside, usually about ½ in., but in some specimens 8 to 9 lines long, the tube very shortly closed above the ovary, the upper lobes separated very low down and unequally winged, with an inflexed, concave auricle, the lower lobes equally winged. Anthers minutely mucronate. Dissepiment of the ovary reaching to about the middle, with 6 to 8 ovules in each cell. Style divided from halfway down to near the base into 2 branches, without the third intervening one of *C. pilosa.* Capsule ovoid, 3 to 4 lines long. Seeds flat, orbicular, winged.—*Distylis Berardiana,* Gaud. in Freyc. Voy. Bot. 460. t. 80 ; DC. Prod. vii. 517 ; *Calogyne distylis,* F. Muell. Fragm. vi. 6.

N. Australia. Dampier Archipelago, N.W. coast, *A. Cunningham* (in fruit only).
W. Australia. Swan River, *Drummond,* 1st *Coll. n.* 409, *and Suppl. n.* 27 ; Old-field river, *Maxwell ;* Basaltic Ranges, N. of Stirling Range, *F. Mueller :* Murchison river ? *Oldfield, Burges* (these specimens with larger flowers). The aspect of the plant is that of some specimens of *Goodenia glauca,* but with a different indumentum. De Vries did not recognize it, but named the specimens he had to examine from the Hookerian Herbarium either *Velleia lanceolata* or *Euthales trinervis.*

2. **C. pilosa,** *R. Br. Prod.* 579. An erect or branching and diffuse, more or less hispid annual of ½ to 1 ft. Leaves sessile or the lower ones petiolate, the upper ones often stem-clasping, lanceolate or almost linear, 1 to 2 in. long, marked with a few distant teeth, and the floral ones usually with 1 or 2 lobes on each side at the base. Calyx-lobes lanceolate, leafy, hirsute. Corolla slightly hispid outside, scarcely ½ in. long, the tube shortly closed above the ovary, the upper lobes separated very low, unequally winged, with an inflexed, concave auricle, lower lobes equally winged. Anthers mucronate-acuminate. Style with the 2 branches of *C. Berardiana,* and a third shorter intervening one. Ovary almost completely 1-celled, the very short dissepiment dividing the cavity into two, as in *Goodenia,* but sometimes almost rudimentary. Ovules about 6. Capsule nearly globular, 2 to 3 lines diameter. Seeds oval, flat, with a rather broad, thick border.—DC. Prod. vii. 517 ; De Vr. Gooden. 180, partly ; *Goodenia dubia,* Spreng. Syst. i. 721.

N. Australia. Arnhem Bays and islands of the Gulf of Carpentaria, *R. Brown.*
Queensland. Rockingham Bay, *Dallachy. C. chinensis,* Hance, from the coasts of China, is closely allied to this species, and may prove to be a variety only. The specimens from the N.W. coast and from Port Essington referred here by De Vriese belong to *Goodenia sepalosa* and *G. hispida.* In outward aspect *Calogyne pilosa* much resembles some specimens of *Goodenia hispida.*

6. SELLIERA, Cav.

Calyx-tube adnate; limb free, of 5 lobes or segments. Corolla-tube slit on the upper side to the ovary, the limb of 5 nearly equal lobes, at length

digitately spreading, the margins slightly inflexed or winged. Stamens free.
Ovary inferior, more or less 2-celled, with several erect ovules. Style undivided. Indusium cup-shaped, enclosing the short truncate stigma, the margin not at all or very minutely ciliate. Fruit more or less succulent, indehiscent. Seeds compressed or irregularly shaped. Embryo about half as
long as the albumen.—Small or creeping perennials. Leaves entire. Flowers
axillary, sessile or pedunculate.

Of the two species, both Australian, one is endemic, the other extends to New Zealand
and extratropical South America. The genus has been reduced by Brown and De Candolle
to *Goodenia*, the flowers being nearly those of the section *Monochila*, but the indehiscent
fruit accompanied by a different habit has induced me to follow J. D. Hooker in restoring it
as originally established by Cavanilles.

Stems creeping. Flowers pedicellate. Seeds rather numerous, flat . . 1. *S. radicans.*
Stems densely branched in tufts of 1 to 1½ in. Flowers sessile. Seeds
few, irregularly shaped . . . . 2. *S. exigua.*

1. **S. radicans,** *Cav. Ic.* v. 49. *t.* 474. A glabrous prostrate or creeping perennial, extending sometimes to several feet. Leaves clustered at the
nodes or sometimes solitary, linear oblong spathulate or rarely ovate, obtuse,
entire, thick, narrowed into a long petiole, from 1 to 2 in. long or when very
luxuriant twice that size. Peduncles 1-flowered, axillary, shorter than the
leaves. Bracteoles small, at a distance from the flower unless the pedicel is
very short. Calyx-lobes divided to the ovary, lanceolate or linear. Corolla
3 to 4 lines long, glabrous outside, the lobes not winged. Ovary 2-celled
nearly to the top, with numerous ovules in 2 rows in each cell. Fruit ovoid
or oblong, about 2 or rarely 3 lines long. Seeds compressed, with a distinct
margin, and sometimes irregularly winged.—Hook. f. Fl. Tasm. i. 231 ; De
Vr. Gooden. 163 ; *Goodenia radicans*, Pers. Syn. i. 195 ; DC. Prod. vii. 516;
G. littoralis, R. Br. in Flind. Voy. ii. 561; *G. repens*, Labill. Pl. Nov. Holl.
i. 53. t. 76 ; R. Br. Prod. 579 ; DC. Prod. vii. 516; *Selliera repens*, De Vr.
Gooden. 162 ; *S. herpystica*, Schlecht. Linnæa, xx. 598 ; De Vr. Gooden.
164.

Victoria. Common on the Yarra river, *F. Mueller* and others ; Darebin Creek, *F.
Mueller ;* Glenelg river, *Robertson ;* Little River, *Fullagar ;* Wimmera, *Dallachy* (in leaf
only).

Tasmania. King's Island and Port Dalrymple, *R. Brown.* Common in marshy
places near the sea, *J. D. Hooker.*

S. Australia. Port Lincoln, *R. Brown ;* Gawler river, Holdfast Bay, *F. Mueller.*

The species is also in New Zealand and in extratropical South America.

2. **S. exigua,** *F. Muell. Fragm.* iii. 142. A dwarf, procumbent, very
much-branched glabrous perennial, forming dense tufts of 1 to 1½ in. diameter, or the stock proliferous-branched, each branch terminating in a similar
tuft. Leaves oblong-lanceolate, spathulate or almost linear, thick, entire, 1
to 1½ lines long, on petioles of 2 to 4 lines. Flowers axillary, sessile.
Bracteoles small, linear. Calyx-tube turbinate, limb campanulate, deeply 5-
lobed, rather longer than the tube. Corolla glabrous outside, 3 lines long or
rather more, the lobes nearly equal and digitately spreading, all winged.
Ovary with 3 or 4 erect ovules. Indusium not ciliate. Fruit nearly globular, 1 to 1½ lines diameter, indehiscent or irregularly bursting. Seeds 3 in

the fruit examined, smooth and shining, irregularly shaped.—*Goodenia exigua*, F. Muell. l. c.

W. Australia. Margin of Moir's Inlet, *Maxwell.*

7. CATOSPERMA, Benth.

Calyx-tube adnate ; lobes 5, free. Corolla-tube slit on the upper side to the ovary, the limb of 5 nearly equal lobes, at length digitately spreading, the margins shortly winged. Stamens free. Ovary wholly inferior, 2-celled, with 2 ovules in each cell suspended from the top. Indusium cup-shaped, enclosing the short stigma, minutely ciliate. " Drupe 10-ribbed, 4-seeded, the cells imperfectly 2-locellate " (*F. Muell.*).—Glabrous herb. Leaves toothed. Flowers yellow, in axillary pedunculate cymes.

The genus is limited to a single species, endemic in Australia, with the flowers of some species of *Goodenia* or of *Scævola*, but differing remarkably from the whole Order in the insertion of the ovules.

1. **C. Muelleri,** *Benth. in Hook. Ic. Pl. t.* 1028. A glabrous perennial, the stems procumbent or ascending, 1 to 1½ ft. long. Leaves petiolate, ovate or obovate, irregularly toothed, the larger ones above 1 in. long without the petiole. Peduncles usually bearing 3 flowers on rather long pedicels, the central one without bracteoles, the lateral ones with two minute bracteoles below the middle. Calyx-tube about 1½ lines long ; lobes shorter, linear-lanceolate. Corolla about ½ in. long, glabrous outside, slightly pubescent inside. Style glabrous or nearly so. Fruit, according to F. Mueller, 3 to 4 lines long.—*Scævola goodeniacea*, F. Muell. Fragm. i. 121.

N. Australia. Gravelly banks of Victoria river, Hooker's and Sturt's Creeks, *F. Mueller.* The specimens I have seen have the fruit as yet very young, and the ripe seeds are unknown.

8. SCÆVOLA, Linn.

(Pogonetes, *Lindl.*, Temminckia, Kamphusia, Crossotoma, Molkenboeria, *and* Merkusia, *De Vr.*)

Calyx-tube adnate, limb usually very short, annular cup-shaped or of 5 distinct segments, sometimes obsolete. Corolla oblique, the tube slit open to the base on the upper side, the lobes nearly equal or the upper ones shorter, usually at length digitately expanding. Stamens free. Ovary wholly inferior or rarely the summit free, 2-celled with 1 erect ovule in each cell or 1-celled with 1 or 2 erect ovules. Style undivided ; indusium cup-shaped, enclosing the truncate or divaricately 2-lobed stigma. Fruit inde-hiscent, with a more or less succulent or thin and membranous exocarp, and a hard woody or bony rarely thin and crustaceous endocarp. Seeds 1 in each cell, erect, albuminous, embryo usually terete and as long as the albumen, or rarely the cotyledons broader than the radicle.—Herbs undershrubs or shrubs. Leaves alternate. Flowers solitary between 2 bracteoles, sessile or peduncu-late in the axils of the leaves or subtending bracts or the peduncles dichoto-mously branched, with a flower in each fork. Indumentum simple or stellate, the hairs outside the corolla usually reversed, the tube of the corolla always more or less villous inside, the lobes at their base sometimes fringed with long teeth or bristles often tipped with a minute white tuft, and continued more or less into the tube down the lines of junction of the petals.

Out of Australia there are several species in the islands of the Pacific, and two maritime ones (one of them Australian) are widely diffused over the warmer seacoasts of the new as well as the Old World; the remaining Australian species are all endemic.

SECT. I. **Sarcocarpæa.**—*Shrubs. Leaves usually large with woolly axils. Flowers in lateral or axillary dichotomous cymes. Exocarp usually very succulent.*

Leaves obovate.   Calyx-lobes oblong or linear   . . . . . 1. *S. Kœnigii.*

SECT. II. **Crossotoma.**—*Shrubs. Flowers solitary on short slender axillary peduncles.*

Bracteoles very small, free or none.   Leaves entire.
  Leaves linear-cuneate obovate or ovate, crowded or clustered on
    short branchlets or nodes. . . . .   2. *S. spinescens.*
  Leaves elliptical or ovate, scattered along the branches   . . .   3. *S. Grœneri.*
Bracteoles ovate or oblong, connected halfway up on the upper
    side.   Leaves sinuate or coarsely toothed.
  Plant densely stellate-tomentose . . . .   4. *S. tomentosa.*
  Plant minutely stellate-pubescent . . . .   5. *S. atriplicina.*

SECT. III. **Pogonanthera.**—*Herbs or undershrubs. Peduncles or pedicels axillary, 1 flowered or the lower ones bearing a dichotomous cyme of 3 or more flowers or rarely flowers sessile on leafless nodes. Anthers in the first 5 species tipped with a minute tuft of hairs.*

Decumbent diffuse or prostrate leafy herbs.   Leaves, at least the
    lower ones, toothed.
  Wings of the corolla-lobes marked with numerous transverse
      veins.   Anthers penicillate.
    Bracteoles ovate or oblong, leafy . . . .   6. *S. striata.*
    Bracteoles linear, usually very small . . . .   7. *S. phlebopetala.*
  Wings of the corolla-lobes veinless.
    Peduncles mostly as long as or longer than the leaves.   An-
        thers penicillate.
      Bracteoles ovate.   Western species . . . .   8. *S. pilosa.*
      Bracteoles linear.   Eastern species.
        Corolla-wings broad   . . . .   9. *S. hispida.*
        Corolla-wings very narrow or scarcely any . . . . . 10. *S. apterantha.*
    Peduncles shorter than the leaves.   Plant prostrate or creep-
        ing.   Anthers not penicillate   . . . . 11. *S. Hookeri.*

(The flowers are shortly pedicellate with large bracteoles in 35. *S. platyphylla,* but form a terminal leafy spike as in other *Xerocarpæas.*)

Erect or ascending undershrubs or rigid herbs.   Leaves narrow,
    entire or reduced to scales.   Anthers not penicillate.
  Stem-leaves very small and distant or none.
    Plant hirsute erect and much-branched . . . . 12. *S. parvifolia.*
    Branches glabrous elongate, rush-like.   Flowers sessile . . 13. *S. restiacea.*
    Branches glabrous elongate divaricate.   Flowers pedicellate . 14. *S. depauperata.*
    Stems dwarf erect tortuose . . . . 15. *S. tortuosa.*
  Leaves linear, mostly 1 in. long or more.   Plant glabrous.
    Peduncles elongated, mostly 3- or 5-flowered   . . . . . 16. *S. Cunninghamii.*
    Peduncles none.   Pedicels elongated, solitary or clustered with
        minute bracteoles at their base.   Fruit beaked . . . . 17. *S. collaris.*

SECT. IV. **Xerocarpæa.**—*Herbs undershrubs or shrubs. Flowers sessile or nearly so in the axils of floral leaves or bracts, all or the upper ones forming a terminal dense or interrupted and leafy spike.*

SERIES I. **Globuliferæ.**—*Ovary 2-celled. Hairs on the back of the indusium short or few or scattered. Flowers rather large (usually blue), with rows of soft bristles or subulate teeth tipped with minute white tufts descending in the throat from the margins of the lobes.*

Glabrous or rarely with long spreading hairs at the base of the
  stem.
　Leaves mostly sessile, ovate oblong or broadly lanceolate, entire 18. *S. angulata.*
　Leaves mostly petiolate, ovate or elliptical, toothed . . . . 19. *S. nitida.*
　Leaves all linear or lanceolate, entire or very rarely toothed.
　　Fruit ovoid, rugose . . . . 20. *S. globulifera.*
　　Fruit large, globular, smooth, striate . . . . 21. *S. porocarya.*
Whole plant or at least the inflorescence hairy.
　Lower leaves petiolate, broadly lanceolate, toothed. . . . . 22. *S. attenuata.*
　Rigidly hirsute. Leaves linear or rarely lanceolate, entire.
　　Corolla and inflorescence glandular pubescent. Fruit smooth 23. *S. glandulifera.*
　　No glandular hairs. Fruit very rugose . . . . . . 24. *S. anchusæfolia.*
　Densely tomentose-villous. Leaves lanceolate or linear, soft,
　　thick, entire . . . . 25. *S. holosericea.*
　Villous with appressed hairs or almost glabrous. Stems decum-
　　bent. Leaves thick, entire, the lower ones obovate . . . 26. *S. suaveolens.*

(A few of the penicillate cilia in the throat of the corolla are observable in some other series, especially in 42. *S. microcarpa* and others of the *Monospermæ*.)

SERIES II. **Macrostachyæ.**—*Ovary 2-celled. Hairs on the back of the indusium short or few or scattered. Flowers usually smaller than in the* Globuliferæ *(white ?), with very few or no penicillate bristles within the throat. Fruiting spikes usually continuous with small bracts.*

Branching tomentose-villous shrub. Leaves small, entire, mostly
　sessile . . . . 27. *S. revoluta.*
Glabrous or viscid shrubs. Leaves petiolate, broad, toothed.
　Leaves ovate or obovate. Bracts ovate or elliptical. Fruit
　　ovoid. Tropical species . . . . 28. *S. ovalifolia.*
　Leaves obovate or orbicular. Bracts linear or lanceolate. Fruit
　　globular or depressed . . . . 29. *S. crassifolia.*
Undershrubs or herbs. Leaves linear or lanceolate.
　Inflorescence or the whole plant hirsute. Fruit ovoid-oblong.
　　Upper leaves and bracts rigidly ciliate. Tropical species . . 30. *S. macrostachya.*
　　Bracts very hairy all over. Western species.
　　　Bracts lanceolate. Indusium ciliate . . . . . . 31. *S. longifolia.*
　　　Bracts ovate, acuminate. Indusium scarcely ciliate . . 32. *S. lanceolata.*
　Plant glabrous. Fruit very small, broader than long. Indusium
　　not ciliate . . . . 33. *S. thesioides.*

SERIES III. **Pogogyneæ.**—*Ovary 2-celled. Indusium with a dense tuft of hairs at the base on the back as long as the indusium itself.*

Leaves sessile, broadly stem-clasping.
　Leaves ovate-lanceolate, toothed. Bracts lanceolate. Spike
　　dense. Bracteoles linear-subulate . . . . 34. *S. macrophylla.*
　Leaves ovate or oblong, mostly entire, the floral ones large.
　　Bracteoles broad . . . . 35. *S. platyphylla.*
　Leaves membranous, broad, coarsely toothed, with toothed auri-
　　cles. Spike interrupted. Bracteoles oblong or lanceolate . 36. *S. auriculata.*
Leaves mostly petiolate or narrowed at the base.
　Lower leaves obovate, toothed. Stem usually decumbent . . 37. *S. æmula.*
　Lower leaves obovate, very acutely toothed . . . . . . 38. *S. humilis.*
　Leaves small, entire, thick. Small erect undershrub . . . . 39. *S. amblyanthera.*

(The dense tuft of hairs at the back of the indusium occurs also in some species of the section *Pogonanthera*, and in the first two species of the series *Monospermæ*.)

SERIES IV. **Monospermæ.**—*Ovary 1-celled, 2-ovulate. Fruit usually 1-seeded.*

Flowers in terminal interrupted leafy spikes.
  Indusium with a dense tuft of long hairs at the base on the back
    as in the *Pogogyneæ*.
    Leaves stem-clasping and auriculate . . . . 40. *S. microphylla.*
    Leaves contracted at the base or petiolate . . . . . 41. *S. cuneiformis.*
  Indusium glabrous or shortly hairy on the back.
    Leaves obovate or oblong, toothed . . . . 42. *S. microcarpa.*
    Leaves linear, revolute, entire . . . . 43. *S. linearis.*
Flowers in short dense leafy spikes or clusters, lateral or on short
  axillary branchlets. Leaves entire.
    Glabrous. Leaves linear or lanceolate . . . . . . 44. *S. Oldfieldii.*
    More or less hairy. Leaves linear or lanceolate . . . . 45. *S. paludosa.*
    Closely silky-tomentose or villous. Leaves elliptical, acute . 46. *S. sericophylla.*
    Densely tomentose or villous. Leaves oblong or linear, ob-
      tuse, thick and soft . . . . 47. *S. canescens.*
    Leaves narrow-linear, nearly glabrous.
      Axils very woolly-white . . . . 48. *S. humifusa.*
      Axils not woolly. . . . . 43. *S. linearis.*

SERIES V. **Parvifloræ.**—*Ovary 1-celled, 1- or 2-ovulate, the upper portion free. Style covered with short purple glandular hairs. Leaves very narrow. Flowers small and numerous in terminal spikes.*

Leaves short, fine, clustered and heath-like. Bracts nearly as long
  as the flowers. Corolla-tube nearly as long as the lobes. Ovules
  solitary . . . . 49. *S. fasciculata.*
Leaves long, crowded. Bracts very short. Corolla-tube much
  shorter than the lobes. Ovules 2 . . . . 50. *S. stenophylla.*

S. lyratifolia, De Vr. in Pl. Preiss. i. 405 ; *Merkusia lyratifolia,* De Vr. Gooden. 60, from the cataracts of Swan River, Preiss, n. 1485, which I have not seen, is probably, from the description given, not a *Scævola.*

SECT. I. SARCOCARPÆA, *G. Don.*—Shrubs. Leaves usually large, with woolly axils. Flowers in lateral or axillary, dichotomous cymes. Exocarp usually very succulent. Species mostly maritime or insular.

1. **S. Kœnigii,** *Vahl, Symb.* iii. 36. An erect shrub, with a thick, almost succulent stem, the branches, leaves, and inflorescence either silky-pubescent or nearly glabrous, but always with a tuft of long, silky or woolly hairs in the axils. Leaves obovate-oblong, 3 to 5 in. long, rounded and very obtuse at the top, entire or rarely broadly crenate, narrowed at the base into a very short petiole. Cymes axillary, very much shorter than the leaves. Bracts small. Calyx-lobes oblong-linear, very variable in length, but usually as long as or longer than the tube. Corolla about ¾ in. long, more or less pubescent outside, the wings of the lobes narrow. Ovary 2-celled. Drupe ovoid or nearly globular, the endocarp bony, near 4 lines long, the exocarp succulent.—R. Br. Prod. 583 ; DC. Prod. vii. 505 ; Hook. and Thoms. in Journ. Linn. Soc. ii. 8 ; Bot. Mag. t. 2732 ; *S. Taccada,* Roxb. ; Wight, Illustr. t. 137 ; *S. sericea,* Forst. ; Br. Prod. 583 ; DC. Prod. vii. 506 ; *S. Lobelia,* De Vr. in Kruidk. Arch. ii. 20 and Gooden. 20 ; Benth. Fl. Hongk. 198 ; *S. macrocalyx,* De Vr. Gooden. 26. t. 3 ; *S. chlorantha,* De Vr.

l. c. 27 ; *S. Lambertiana*, De Vr. l. c. 28 ; *S. montana*, Labill. Sert. Austr. Caled. 41. t. 42.

N. Australia. N. coast, *A. Cunningham ;* mouth of Victoria river, *F. Mueller.*
Queensland. Along the tropical seacoast ; Palm Island, *Banks and Solander ;* Endeavour Strait, Low Island, and Northumberland Islands (the latter with smaller leaves crenate at the end), *R. Brown ;* Escape Cliffs, *Hulls ;* Port Denison, *Fitzalan ;* Edgecombe Bay, *Dallachy.*

A common seacoast plant in the warmer parts of the Old World, and chiefly within the tropics, occurring occasionally also in the West Indies ; but there the most common species is *S. Plumieri,* distinguished by the truncate, annular, calyx-limb without the lobes of *S. Kœnigii,* and usually by thicker leaves. This *S. Plumieri* occurs also on the coasts of Africa, Ceylon, and other parts of Asia, but has not yet been found in Australia. De Vriese reunites it with *S. Kœnigii,* but that is owing to his having mistaken Vahl's plant. The distinction between the two species was well pointed out by Gærtner and by Vahl, and has been maintained by all subsequent botanists, however much they have multiplied species. Nevertheless, the specimens determined by De Vriese as *S. Plumieri* belong to *S. Kœnigii,* and he has distributed the true *S. Plumieri* (which he could not have mistaken if he had looked at Plumier's figure) in his *S. Macræi, S. senegalensis, S. Sieberi, S. Thunbergii,* and *S. uvifera.*

In the ' Flora Honkongensis ' I had followed De Vriese in calling the species *S. Lobelia,* Linn., but I cannot find that Linnæus ever gave it such a name in any published work. He published the genus in his ' Mantissa,' p. 145, without any specific name, having previously designated Plumier's and Jacquin's plant as *Lobelia Plumierii. Scævola Lobelia* first occurs in Gmel. Syst. ·Nat. Veg. 361, but that refers especially to *S. Plumieri,* as evidenced by the authority of Plumier's and Jacquin's figures quoted. He might, indeed, if he had known of it, have included also the *S. Kœnigii,* for he misquotes Gærtner's figure under a wrong name, but he never gave the name of *S. Lobelia* exclusively or specially to the latter species.

SECT. II. CROSSOTOMA, *G. Don.*—Shrubs. Flowers solitary, on short, slender, axillary peduncles. Fruit sometimes with as succulent an exocarp as in *Sarcocarpæa,* in other species nearly dry.

2. **S. spinescens,** *R. Br. Prod.* 586. A rigid, scrubby shrub of several feet, glabrous or hoary-tomentose, the short branchlets often, but not always converted into short, simple or branched spines. Leaves often clustered on short branchlets or nodes, obovate spathulate oblong or linear, obtuse, entire, thick, under ½ in. when broad, sometimes near 1 in. long when narrow. Flowers few or solitary in the clusters of leaves, the peduncles slender, but rarely as long as the leaves. Bracteoles small, narrow-linear. Calyx-limb exceedingly short, annular, truncate. Corolla white, 6 to 8 lines long, tomentose or glabrous outside. Indusium ciliate. Ovary 2-celled. Drupe ovoid, rather large, with a bony endocarp, and thick, succulent mesocarp. Seeds with a large embryo, and very little albumen.—DC. Prod. vii. 512 ; *S. oleoides* and *S. lycioides,* DC. l. c. ; *Pogonetes,* Lindl. Introd. Nat. Syst. ed. 2. 443 ; *Crossotoma spinescens, C. oleoides,* and *C. lycioides,* De Vr. Gooden. 36 to 38.

N. Australia. Dampier's Archipelago, *A. Cunningham.*
Queensland, *Bowman ;* Bokhara Creek, *Leichhardt ;* between Thomson and Flinders rivers, *Sutherland ;* on the Maranoa, *Mitchell ;* Armadillo, *W. Barton.*
N. S. Wales. N. extremity of Peel's Range, *A. Cunningham ;* between the Murray and Darling rivers, *Victorian and other Expeditions.*
S. Australia. Islands of Nuyt's Archipelago and Petrel Bay (the very spinescent

form, with small, broad leaves), *R. Brown;* Murray Scrub, Flinders Range, and Spencer's Gulf, *F. Mueller;* Lake Gillies, *Burkitt.*

W. Australia, *Drummond, n.* 413, *and Suppl. n.* 51 ; Murchison river, *Oldfield.*

Generally speaking the leaves are small, broad, and whitish tomentose on the most scrubby spinescent specimens, narrow, longer, and green on the luxuriant ones, but sometimes the two occur on different branches of the same specimen. The fruit described was only seen on one of Drummond's specimens, n. 51, without flowers. On other specimens it appears to be ovoid, rather small, dry, and rugose, but evidently not arrived at perfection, and without any seed.

3. **S. Grœneri,** *F. Muell. Fragm.* vi. 15.   An erect, glabrous, branching shrub.   Leaves numerous but not clustered, petiolate, ovate-elliptical oblong or lanceolate, coriaceous, acute or obtuse, entire, $\frac{1}{2}$ to 1 in. long. Pedicels axillary, 1-flowered, short but slender, the flowers slightly exceeding the leaves.   Bracteoles linear.   Calyx-lobes exceedingly small, ovate. Corolla slender, 7 to 9 lines long, glabrous outside ; lobes with narrow wings, the middle ones sometimes fringed at the base with a few long teeth, decurrent in the throat.   Ovary 2-celled.   Indusium glabrous, ciliate.   Style lobes rather long, and divaricate.   Fruit not seen.—*Merkusia myrtifolia,* De Vr. Gooden. 72.

W. Australia, *Drummond, n.* 363.

4. **S. tomentosa,** *Gaudich. in Freyc. Voy. Bot.* 460. *t.* 81.   An erect shrub, clothed all over with a close, stellate tomentum.   Leaves ovate, obovate or oblong, narrowed into a short petiole, coarsely sinuate-toothed, above 1 in. long.   Pedicels axillary, 1-flowered, much shorter than the leaves.   Bracteoles leafy, entire, ovate and obtusely acuminate or oblong, connate on the upper side to about the middle.   Calyx-limb a very short, ciliate ring.   Corolla 10 to 11 lines long, tomentose outside, the tube gibbous at the base on the lower side, the upper lobes fringed at the base. Ovary 2-celled.   Style slightly hairy.   Indusium glabrous, ciliate.   Fruit not seen.—DC. Prod. vii. 506 ; *Temminckia tomentosa,* De Vr. Gooden. 13.

W. Australia.   Sharks' Bay (*Gaudichaud*), *Maitland Brown* (a single specimen in Herb. F. Muell.).

5. **S. atriplicina,** *F. Muell. Fragm.* ii. 18.   A branching shrub, sprinkled with a minute, stellate pubescence.   Leaves often crowded on the short branchlets, ovate obovate or oblong, obtuse, entire or coarsely sinuate-toothed, 1 to 2 in. long.   Pedicels axillary, 1-flowered, slender, shorter than the leaves.   Bracteoles leafy, entire, ovate and obtusely acuminate or oblong, connate on the upper side to about the middle.   Calyx-limb a very short ring or quite obsolete.   Corolla about $\frac{3}{4}$ in. long, pubescent outside, the tube gibbous at the base on the lower side, the upper lobes often fringed at the base.   Filaments long.   Ovary 2-celled.   Style slightly hairy.   Indusium ciliate. Drupe small, ovoid, the endocarp bony and rugose, the exocarp more or less succulent.

W. Australia.   Port Gregory, *Oldfield;* also in *Drummond's* collection.   The specimens are very few, and not perfect.   It may prove to be a variety of *S. tomentosa.*

SECT. III. POGONANTHERA, *G. Don.*—Herbs or undershrubs. Peduncles or pedicels axillary, 1-flowered or the lower ones bearing a dichotomous cyme of 3 or more flowers, and not forming a terminal spike.   Anthers in the first 5 species tipped with a minute tuft of hair, but not in the others. Fruit usually hard, the exocarp scarcely succulent.

6. **S. striata,** *R. Br. Prod.* 586.   A scabrous-pubescent or hispid perennial, with diffuse, decumbent or ascending slightly-branched stems of ½ to 1½ ft., flowering sometimes the first year, and then more simple and erect. Leaves obovate or oblong-spathulate, coarsely toothed, the lower ones petiolate, the upper ones sessile but narrowed at the base, and all slightly stem-clasping, in elongated specimens the upper leaves small and distant.   Peduncles axillary, longer than the leaves, 1-flowered or the lower ones rarely branching out into a several-flowered cyme.   Bracteoles large, foliaceous, ovate-oblong or lanceolate.   Calyx-lobes linear, from scarcely longer than the tube to 3 or 4 times that length.   Corolla often above 1 in. long, the wings of the lobes broad and elegantly marked with transverse veins.   Anthers tipped with a tuft of short bristles.   Ovary 2-celled.   Indusium with a dense tuft of hairs on the back, the margin minutely ciliate.   Fruit ovoid or oblong, 3 to 3½ lines long, the drupe verrucose within the calyx-tube. Seeds oblong, the embryo terete, nearly as long as the albumen.—DC. Prod. vii. 511; *S. calliptera,* Benth. in Hueg. Enum. 70; DC. l. c. 511; *S. macropoda,* DC. l. c. 509; *S. Benthamia,* De Vr. in Pl. Preiss. i. 411; *Molkenboeria striata,* De Vr. Gooden. 42.

W. Australia. King George's Sound and neighbouring districts, *R. Brown* and others, and thence to Vasse and Swan Rivers, *Drummond, 1st Coll., 2nd Coll. n.* 18, *3rd Coll. n.* 392, *Preiss, n.* 1508, 1520, *Oldfield,* and others; Champion Bay, *Oldfield;* eastward to Lake Leven and Salt River, *Maxwell.*

S. macrodonta, DC. Prod. vii. 511; De Vr. in Pl. Preiss. i. 411, appears to me to be the same plant.   The wings of the corolla-lobes are said to be without veins, but that is not the case either in A. Cunningham's specimens from King George's Sound or in Preiss's specimens described by De Vriese.

7. **S. phlebopetala,** *F. Muell. Fragm.* ii. 18.   A scabrous-pubescent or hispid herb, with diffuse, procumbent or elongated and flexuose stems, closely allied to *S. striata,* with the same inflorescence, and transverse veins to the wings of the corolla-lobes, but the leaves are narrower, the lower ones sometimes obovate, the upper ones all narrow, oblong-lanceolate or linear, and the bracteoles are always linear, sometimes very small, sometimes as long as the calyx or longer.   Flowers rather smaller than in *S. striata,* and apparently of a deeper blue, the wings also deeply coloured.

W. Australia, *Drummond, n.* 189, 393, *or* 399; Murchison river, *Oldfield and Walcott.*   Probably a variety of *S. striata.*

8. **S. pilosa,** *Benth. in Hueg. Enum.* 69.   Herbaceous, but with a hard, almost woody rootstock or sometimes suffruticose, erect, 1 to 3 ft high, hispid with spreading hairs.   Lower leaves petiolate, obovate or oblong, coarsely toothed, 2 or 3 in. long, upper ones much smaller, sessile, and stem-clasping, from broadly oblong-cuneate to lanceolate.   Peduncles axillary, longer than the leaves, 1-flowered.   Bracteoles large, ovate or broadly lanceolate.   Calyx-tube oblong or obovoid, pubescent, the lobes either about half as long as the breadth of the tube or quite obsolete.   Corolla like that of

S. hispida, blue, the wings of the lobes without transverse veins, and the lobes more or less fringed at their junction. Anthers tipped with a minute tuft of hairs. Ovary 2-celled. Fruit not seen.—DC. Prod. vii. 511; *Molkenboeria pilosa*, De Vr. Gooden. 39. t. 7; *M. semiamplexicaulis*, De Vr. Gooden. t. 5, as to the plate, but not the description, p. 41.

W. Australia. Swan River, *Huegel, Drummond, 1st Coll. and n.* 393; Harvey river, *Oldfield.*

S. membranacea, Benth. in Hueg. Enum. 69; DC. Prod. vii. 511; *Molkenboeria membranacea*, De Vr. Gooden. 40, appears to be only a slight variety of *C. pilosa*, with broader and thinner leaves.

9. **S. hispida,** *Cav. Ic.* vi. 7. *t.* 510.   Herbaceous, apparently annual, scabrous-pubescent or hirsute with spreading hairs, the stems erect or branching at the base, 1 to 2 ft. high.   Leaves linear or lanceolate and sessile or the lower ones narrow-oblong, and contracted into a petiole, entire or the broader ones remotely toothed, the larger ones 2 or 3 in. long, the upper ones smaller.   Peduncles axillary, as long as or longer than the leaves, 1-flowered or the lower ones bearing 3 or sometimes a cyme of several flowers.   Bracteoles leafy, linear.   Calyx-lobes linear, much longer than the tube.   Corolla ¾ to 1 in. long, the wings of the lobes broad, but without transverse veins.   Anthers tipped with a tuft of minute bristles.   Indusium densely hairy on the back at the base, the margin ciliate.   Ovary 2-celled.   Drupe oblong, about 3 lines long.   Seeds oblong, the embryo nearly terete. —R. Br. Prod. 586; DC. Prod. vii. 511; *Goodenia ramosissima*, Sm. in Tr. Linn. Soc. ii. 349, and Specim. Bot. Nov. Holl. t. 5; *Merkusia hispida*, De Vr. Gooden. 62.

Queensland. Near Brisbane, *W. Hill.*
N. S. Wales. Port Jackson to the Blue Mountains, very common, *R. Brown, Sieber, n.* 225, and many others.
Victoria. Merriman's Creek, Gipps' Land, rare, *F. Mueller.*

10. **S. apterantha,** *F. Muell. Fragm.* i. 121.   An erect, hispid, branching herb, apparently annual, about 1 ft. high, closely resembling *S. hispida*.   Leaves linear or lanceolate, entire or sparingly toothed, the larger ones 2 or 3 in. long.   Inflorescence, bracteoles, and flowers of *S. hispida*, except that the corolla is rather smaller, and the wings of the lobes are exceedingly narrow or scarcely developed.   Anthers tipped with minute bristles.   Indusium nearly glabrous, the margin ciliate.   Ovary 2-celled.

Victoria. Ranges beyond the Snowy River, *F. Mueller.*   This may prove to be a variety of *S. hispida.*

11. **S. Hookeri,** *F. Muell.; Hook. f. Fl. Tasm.* i. 231. *t.* 67.   Prostrate or creeping, much branched, rooting at the nodes, more or less hirsute or rarely almost glabrous.   Leaves ovate obovate spathulate or oblong, irregularly toothed or rarely entire, narrowed at the base or shortly petiolate, in many specimens scarcely ½ in. long, in luxuriant ones 1 in. or more. Pedicels 1-flowered, axillary, rarely as long as the leaves.   Bracteoles leafy, ovate or oblong.   Calyx-lobes obsolete or very rarely one of them elongated. Corolla of a dirty white, 3 or 4 lines long, pubescent outside.   Anthers without terminal tufts.   Ovary 2-celled.   Style hairy.   Indusium not at all

or very minutely ciliate. Fruit ovoid, slightly rugose, about 1½ lines long.
—*Merkusia Hookeri,* De Vr. Gooden. 56. t. 12.

N. S. Wales. Bogs, Blue Mountains, *A. and R. Cunningham.*

Victoria. Marshy places, mouth of Albert river (very luxuriant), Australian Alps at an elevation of 4000 to 6000 ft., snowy plains at the sources of the Yarra, summit of Mount Useful, Munyong Mountains, etc., *F. Mueller.*

Tasmania. Marshy places, Rocky Cape, Mount Wellington, Hampshire Hills, etc., *J. D. Hooker.*

12. **S. parvifolia,** *F. Muell. Herb.* An erect, much branched, villous herb or undershrub, the specimens all under 1 ft. high. Leaves lanceolate or oblong-linear, obtuse, the longest under ½ in., and mostly reduced to small bracts. Flowers solitary on rigid, axillary peduncles of ½ in. or more. Bracteoles very small. Calyx-lobes lanceolate, shorter than the tube. Corolla about ¼ in. long, hairy outside; wings of the lobes narrow. Anthers with a small, glabrous point. Ovary 2-celled. Style hairy; indusium nearly glabrous, ciliate.

N. Australia. Hooker's Creek, *F. Mueller.*

13. **S. restiacea,** *Benth.* Glabrous or nearly so. Stems elongated, rigid, terete, striate, rush-like, but branching. Leaves all reduced to minute, distant, acute scales. Flowers solitary and sessile at the nodes, with minute, scale-like bracts at their base. Calyx-lobes linear, rigid, as long as the tube. Corolla 6 to 7 lines long, bearing rigid, appressed hairs outside, the lobes all equally winged and apparently nearly equal. Ovary 2-celled. Indusium not seen. Fruit oblong, about 2 lines long, rugose and hispid with short, incurved hairs.

W. Australia, *Drummond, 5th Coll. n.* 169. I have seen several specimens, but with only very few scattered fruits, and on one in the Hookerian herbarium are two much-injured flowers.

14. **S. depauperata,** *R. Br. App. Sturt's Exped.* 20. A glabrous herb or undershrub, probably tall, with long, spreading, rigid branches. Lower leaves unknown, upper ones reduced to small, distant, linear, recurved, rigid bracts. Flowers solitary or few, in irregular cymes, on long, rigid, axillary, divaricate peduncles (ultimate branches of R. Br.), and pedicellate above the small bracteoles. Calyx-limb campanulate, 5-lobed, at least as long as the adnate tube. Corolla about ¾ in. long, pubescent outside, the tube very villous inside, the wings of the lobes ciliate. Ovary 2-celled. Style nearly glabrous, except a dense tuft of hairs on the back of the indusium at its base, the margin of the indusium ciliate. Fruit unknown.—*Merkusia ? depauperata,* De Vr. Gooden. 74 ; *Scævola patens,* F. Muell. Fragm. iii. 33.

S. Australia. Cooper's Creek, *Wheeler* (a single specimen in Herb. Mueller) ; salt ground in lat. 26°, *D. Sturt.* I have not seen this specimen, but Brown's description leaves no doubt as to its identity.

15. **S. tortuosa,** *Benth.* A small perennial, with rigid, tortuous, erect stems of 3 or 4 in. Radical leaves petiolate, oblong or linear, entire or rarely minutely toothed, pubescent, about 1 in. long, those of the stem all reduced to minute scales or one or two at the base more developed. Peduncles

1-flowered, radical or in the lower axils, rigid, 1 to 2 in. long, with a pair of small, linear bracteoles close under the flower. Calyx-lobes lanceolate, nearly as long as the tube. Corolla hairy outside, nearly ¾ in. long. Ovary 2-celled. Indusium with a tuft of rather long hairs on the back at the base. Fruit unknown.

W. Australia, *Drummond, 4th Coll. n.* 191. These specimens were referred by De Vriese in Herb. Hook. and Gooden. 189 to *Anthotium humile,* to which they bear no resemblance whatever.

16. **S. Cunninghamii,** *DC. Prod.* vii. 508. Apparently an undershrub or shrub, the typical form glabrous or nearly so. Leaves linear or linear-lanceolate, entire, the lower ones narrowed into a petiole and 1½ to 2 in. long, the upper ones passing gradually into small bracts. Peduncles axillary, mostly nearly as long as the leaves, and bearing a cyme of 3 flowers, the central one sessile, a few of the upper ones short and 1-flowered. Calyx-lobes broadly ovate, very obtuse, shorter than the tube. Corolla 6 to 7 lines long, nearly glabrous outside, the tube hairy inside; lobes very narrow. Ovary 2-celled. Style slightly hairy. Indusium ciliate. Unripe fruit nearly globular.—*S. Maitlandi,* F. Muell. in Trans. Edinb. Bot. Soc. vii. 497.

N. Australia. Dampier's Archipelago, *A. Cunningham ;* Nichol Bay, *Gregory's Expedition.*

Var. *hispida.* Villous all over with rigid hairs, the other characters the same.—Nichol Bay, *Gregory's Expedition;* Depuech Island, *Bynoe.*

17. **S. collaris,** *F. Muell. Rep. Babb. Exped.* 15, Quite glabrous and smooth, apparently a shrub or undershrub. Leaves linear or linear-lanceolate, entire, thick, 1 to 2 in. long. Flowers axillary, solitary or clustered in the lower axils, the common peduncle not developed, but the pedicels elongated, with minute bracteoles at their base. Calyx-tube elongated, contracted at the top, the lobes very small, acute. Corolla (yellowish, *F. Muell.*) about 5 lines long, glabrous outside, the tube slightly pubescent inside. Style hairy. Drupe ovoid, 2-celled, about ½ in. long, including the long neck into which it is contracted, endocarp hard, prominently ribbed, exocarp succulent, not thick. Seeds oblong, the embryo nearly terete.

S. Australia. Sand ridges near Wonnomulla in the N. interior, *Babbage's Expedition,* also from *M'Douall Stuart's Expedition.* The specimens are in fruit, and I have only seen the fragments of the flower described by *F. Mueller,* who found the indusium without cilia.

SECT. IV. XEROCARPÆA, *Don.*—Herbs undershrubs or shrubs. Flowers sessile or nearly so in the axils of floral leaves or bracts, solitary or the lower ones very rarely in clusters of 2 or 3, the upper ones forming a terminal spike, either dense with short bracts or interrupted and leafy.

SERIES 1. GLOBULIFERÆ.—Ovary 2-celled. Hairs on the back of the indusium short or few and scattered. Flowers rather large (usually blue), with rows of soft bristles or subulate teeth, tipped with minute, white tufts descending in the throat of the corolla from the margins of the lobes.

18. **S. angulata,** *R. Br. Prod.* 586. Tall, erect and branching, gla-

brous or hirsute with long, spreading hairs. Lower leaves ovate-lanceolate, acute, entire or with a few coarse teeth, usually contracted at the base, but almost sessile, 1 to 2 in. long, upper ones sessile, from ovate to lanceolate, entire, rounded or cordate at the base, under 1 in. long, the floral ones or bracts almost as long as the flowers. Flowers almost sessile, distant, forming an interrupted, leafy spike, the lower ones or nearly all distant. Bracteoles lanceolate, often as long as the bracts. Calyx-limb campanulate, lobes lanceolate. Corolla glabrous outside, about 7 lines long or rather more, with long, subulate teeth or soft bristles descending in the throat from the margins of the lobes. Style glabrous or hairy; indusium ciliate. Ovary 2-celled.—DC. Prod. vii. 511; *Merkusia angulata*, De Vr. Gooden. 69.

N. Australia. South Goulburn Island, *A. Cunningham;* islands of the Gulf of Carpentaria, *R. Brown.*

19. **S. nitida,** *R. Br. Prod.* 584. An erect undershrub or shrub, attaining 3 or 4 ft., quite glabrous, and sometimes viscid. Leaves mostly petiolate, oval-elliptical or lanceolate, 2 to 3 in. long, more or less toothed, the upper ones smaller and narrower, the floral ones all but the lowest reduced to lanceolate or linear, entire bracts, rarely exceeding 6 lines. Flowers sessile, in rather dense spikes of 2 to 6 in. Bracteoles small, narrow-linear. Calyx-limb a very short, truncate ring. Corolla (white? or pale blue?) 6 to 8 lines long, glabrous outside, with rather numerous, long teeth or soft bristles, tipped with minute tufts, descending in the throat from the margins of the lobes. Ovary 2-celled. Style slightly hairy. Indusium ciliate. Seeds ovoid-oblong, about 2 lines long.—DC. Prod. vii. 509; *Merkusia? nitida*, De Vr. Gooden. 73; *Scævola multiflora*, Lindl. Swan Riv. App. 26; *Merkusia multiflora*, De Vr. Gooden. 48; *Scævola Drummondii*, DC. Prod. vii. 508.

W. Australia. King George's Sound, *R. Brown, A. Cunningham,* and others, to Swan River, *Fraser, Drummond, 1st Coll.,* and others; Géographe Bay, Gordon, Chapman, and Blackwood rivers, *Oldfield.*

S. fastigiata, De Vr. in Pl. Preiss. i. 406, or *Merkusia fastigiata,* De Vr. Gooden. 48, from Bald Head, *Preiss, n.* 1491, which I have not seen, is, from the short characters given, most likely to be *S. nitida.*

20. **S. globulifera,** *Labill. Pl. Nov. Holl.* i. 55. *t.* 78. A glabrous undershrub of 1 to 2 ft., growing sometimes into a shrub of twice that height. Leaves linear or lanceolate, rather thick, entire or when broad slightly toothed, narrowed below the middle but shortly dilated and stem-clasping at the base, the longer ones 2 to 3 in. long, the floral ones or bracts linear-lanceolate, the upper ones shorter than the flowers, glabrous or ciliate with a few hairs. Flowers sessile in a terminal spike, at length long and interrupted with the lower ones distant. Calyx-limb broadly cup-shaped, sinuate-toothed, nearly half as long as the tube. Corolla (blue?) 8 to 9 lines long, glabrous outside, with long subulate teeth or bristles tipped with a minute white tuft descending in the throat from the margins of the lobes. Ovary 2-celled. Style slightly hairy, especially at the top. Indusium glabrous, ciliate. Fruit ovoid, rugose, the mesocarp having (at least in the dried state) 2 spurious empty cells often as large as the real ones and alternating with them, but sometimes irregular.—R. Br. Prod. 584; DC.

Prod. vii. 508; *Merkusia globulifera*, De Vr. Gooden. 52; *Scævola cæspitosa*, R. Br. Prod. 585; DC. Prod. vii. 510; *Merkusia cæspitosa*, De Vr. Gooden. 63 (partly, for *Goodenia barbata* and several other plants are so named by him in Herb. Hook.).

W. Australia. King George's Sound and adjoining districts, *R. Brown, Preiss, n.* 509; eastward to Cape Knobb and Cape le Grand, *Maxwell.*

Var.? *humilis.* More branching and slightly hairy, the specimens 6 to 8 in. high. Leaves smaller and occasionally toothed, but otherwise apparently the same.—W. Australia, *Drummond.*

21. **S. porocarya,** *F. Muell. Fragm.* ii. 19. A single specimen in fruit only, closely resembling some specimens of *S. globulifera*, but the fruit is much larger (about 4 lines diameter), globular, very smooth but marked with longitudinal striæ, the endocarp thick and hard, with 2 or more irregular empty cavities besides the 2 perfect cells.

W. Australia. Murchison river, *Oldfield* (*Herb. F. Mueller*).

22. **S. attenuata,** *R. Br. Prod.* 583. An erect shrub or undershrub of $1\frac{1}{2}$ to 2 ft., hirsute with rigid scattered hairs. Leaves petiolate, the larger ones broadly lanceolate, bordered by a few acute teeth, strongly veined, 2 to 3 in. long, the upper ones linear or linear-lanceolate, mostly entire. Flowers sessile in terminal leafy spikes, at length long and interrupted, with the lower ones distant. Bracts linear, the lower ones exceeding the flowers, the upper ones small; bracteoles linear, small. Calyx-limb prominent, annular, sinuate. Corolla blue, about $\frac{3}{4}$ in. long, hairy inside, with rather numerous subulate teeth or soft bristles tipped with minute white tufts descending in the throat from the margins of the lobes. Ovary 2-celled. Style very hairy. Indusium ciliate.—DC. Prod. vii. 508; Bot. Mag. t. 4196; *Merkusia attenuata*, De Vr. Gooden. 61.

W. Australia. King George's Sound, *R. Brown, Baxter.* A specimen from Cape Naturaliste, *Oldfield* (*Herb. F. Muell.*), appears to be the same.

23. **S. glandulifera,** *DC. Prod.* vii. 510. An erect rigid herb or undershrub, with the scabrous rigid hairs, foliage, and inflorescence of *S. anchusæfolia*, but the flowers are larger (the corolla $\frac{3}{4}$ to 1 in. long) and the whole inflorescence as well as the outside of the corolla are pubescent with glandular hairs intermixed with the rigid ones, and the drupe is prominently ribbed, but otherwise smooth or nearly so.—*S. rufa*, De Vr. in Pl. Preiss. i. 405; *Merkusia glandulifera*, De Vr. Gooden. 67.

W. Australia. Swan River, *Drummond, 1st Coll.* often mixed with *S. anchusæfolia* and *S. longifolia*, also n. 39; Kalgan, Tone, and Gordon rivers, *Oldfield*; Mount Barker, *F. Mueller*; Mount Barren flats and Salt River, *Maxwell.* The character derived from the fruit has only been observed in very few specimens, the glandular pubescence appears to be constant.

24. **S. anchusæfolia,** *Benth. in Hueg. Enum.* 68. An erect or more rarely procumbent or prostrate herb or undershrub, scabrous and hirsute with spreading hairs. Leaves linear or oblanceolate, entire or coarsely toothed when broad, 1 to 2 in. long or sometimes more, mostly narrowed into a petiole and partly dilated at the base, the floral ones or bracts linear,

under ¼ in. long. Flowers blue, sessile or nearly so in a terminal leafy spike. Bracteoles linear, much smaller than the bracts. Calyx-limb obsolete. Corolla ½ in. long or rather longer, more or less hairy outside but not glandular, with subulate teeth or soft bristles tipped with minute tufts descending in the throat from the margins of the lobes. Ovary 2-celled. Style hairy. Indusium pubescent, ciliate. Fruit ovoid, very rugose, 2 to 3 lines long, the thick hard pericarp showing 2 or more empty cavities nearly as large as the true cells.—DC. Prod. vii. 510 ; *Merkusia anchusæfolia,* De Vr. Gooden. 65 (partly, several other plants being so named by him in Herb. Hook.).

W. Australia. Swan River, *Huegel, Drummond, 1st Coll.*

25. **S. holosericea,** *De Vr. in Pl. Preiss.* i. 408. An erect undershrub of 1 to 2 ft., densely villous with rather soft or more rigid hairs. Leaves petiolate, lanceolate, entire or remotely toothed, rather thick, the larger ones 2 in. long or more, passing into the linear or linear-lanceolate floral leaves or bracts of which the uppermost are shorter than the flowers. Flowers blue, sessile in a terminal leafy spike, at length much elongated. Bracteoles linear. Calyx-limb broadly cup-shaped, truncate or sinuate-toothed. Corolla nearly ¾ in. long, hairy outside, with numerous subulate teeth or soft bristles tipped with minute tufts descending in the throat from the margins of the lobes. Ovary 2-celled. Style very villous. Indusium ciliate. Fruit not seen.

W. Australia. Swan River, *Preiss, n.* 1478, 1512 ; Coogee, *Oldfield.* This species is closely allied on the one hand to *S. anchusæfolia,* on the other to *S. suaveolens,* but can scarcely be united with either. *S. sphærocarpa,* De Vr. in Pl. Preiss. i. 409, from Swan River, *Preiss, n.* 1512, of which I have only seen an imperfect specimen, appears to be the same.

26. **S. suaveolens,** *R. Br. Prod.* 585. A prostrate or decumbent hard perennial or undershrub, extending sometimes in dense masses for several feet, more or less clothed with appressed silky hairs or rarely glabrous. Leaves petiolate, from obovate to oblong-spathulate, quite entire, thick, the larger ones 2 to 3 in. long, the upper ones smaller or linear when on elongated branches. Flowers blue, sessile in interrupted terminal hirsute spikes, the bracts oblong-linear, shorter than the flowers, the bracteoles still smaller and linear. Calyx-limb broadly cup-shaped, ciliate, otherwise entire or unequally 5-lobed. Corolla 7 to 8 lines long, villous or rarely nearly glabrous outside, with numerous subulate teeth or soft bristles tipped with minute tufts descending in the throat from the margins of the lobes. Ovary 2-celled. Style slightly hairy. Indusium ciliate, but often on one side only. Drupe sometimes ovoid, rugose, 3 lines long and nearly dry but then evidently unripe ; when fully formed, it appears to be much larger with a very succulent exocarp as in *S. Kœnigii.*—DC. Prod. vii. 510 ; *Goodenia calendulacea,* Andr. Bot. Rep. t. 22 ; *Merkusia suaveolens,* De Vr. Gooden. 62.

Queensland. Near the sea, Sandy Cape, Keppel Bay, Broad Sound, etc., *R. Brown ;* Curtis' Island, *Henne, Thozet ;* Burdekin Expedition, *Fitzalan ;* Moreton Island, *F. Mueller, M'Gillivray.*

N. S. Wales. Seacoast, Botany Bay, *R. Brown ;* Manly Beach, *Woolls ;* northward to Clarence river, *Henderson ;* Richmond river, *Wilcox ;* southward to Kiama, *Harvey ;* Gabo island, *Mapleton.*

Victoria. Port Phillip, *R. Brown;* from the mouth of the Glenelg, *Allitt,* to Wilson's Promontory, *F. Mueller.*

S. Australia. Rivoli Bay and Lake Alexandrina, *F. Mueller.*

SERIES 2. MACROSTACHYÆ. Ovary 2-celled. Hairs on the back of the indusium short or few and scattered. Flowers usually smaller than in the *Globuliferæ* (white ?), with very few or no penicillate bristles in the throat of the corolla. Fruiting spikes usually continuous or only the lower flowers distant, with small bracts.

27. **S. revoluta,** *R. Br. Prod.* 586. An undershrub or spreading shrub of 2 to 3 ft., softly villous all over and sometimes silky. Leaves from obovate-oblong to cuneate or lanceolate, mostly obtuse, flat or with revolute margins, entire, thick and soft, rarely exceeding ½ in. and mostly shorter, the upper ones passing gradually into oblong-linear bracts, either as long as or shorter than the flowers. Flowers sessile in a terminal leafy spike, sometimes short and interrupted at the base, sometimes very much elongated. Calyx-limb obsolete. Corolla nearly ½ in. long, hairy outside. Ovary 2-celled. Style slightly hairy. Indusium ciliate. Fruit small, oblong, rugose, but not seen ripe.—DC. Prod. vii. 510; *Merkusia revoluta,* De Vr. Gooden. 64, partly.

N. Australia. Islands of the Gulf of Carpentaria, *R. Brown;* Sea Range and Upper Victoria river, *F. Mueller.*

28. **S. ovalifolia,** *R. Br. Prod.* 584. An erect branching perennial or undershrub, about 1 ft. high, usually pubescent, sometimes viscid, rarely almost glabrous, of a pale green. Lower leaves petiolate, ovate or obovate, coarsely toothed or sometimes entire, 1 to 2 in. long besides the petiole, upper ones smaller and more sessile; floral leaves or bracts ovate-lanceolate or elliptical, leafy but mostly under ½ in. long, entire or the lower ones toothed. Flowers sessile, in an interrupted leafy spike. Bracteoles linear or linear-lanceolate. Calyx-lobes exceedingly short, rounded. Corolla about ¾ in. long, glabrous or pubescent outside. Ovary 2-celled, but one ovule sometimes abortive. Style slightly hairy. Indusium ciliate. Fruit small, ovoid-oblong.—DC. Prod. vii. 509; *Merkusia ovalifolia,* De Vr. Gooden. 50.

N. Australia. Sandstone table-land, Upper Victoria river, *F. Mueller;* islands of the Gulf of Carpentaria, *R. Brown;* Sweers Island, *Henne.*

Queensland. Flinders river, *Bowman.*

N. S. Wales. Darling river, *Nielsen;* Mount Goningbery, *Victorian Expedition.*

S. Australia. Cooper's Creek, *Howitt's Expedition.*

29. **S. crassifolia,** *Labill. Pl. Nov. Holl.* i. 56. *t.* 79. A glabrous shrub, either low and decumbent or very divaricate and attaining 2 or 3 ft. Leaves on rather long petioles, obovate orbicular or spathulate, 1 to 2 in. long, thick and rigid, bordered by small teeth. Flowers sessile or nearly so, in rather dense spikes of 1 to 3 in., of which several together form sometimes a terminal panicle. Bracts linear-lanceolate, entire, rarely above ¼ in. long, the bracteoles much smaller. Calyx-limb short, broad, and truncate. Corolla about ½ in. long, glabrous outside. Ovary 2-celled. Style nearly glabrous. Indusium sparingly ciliate. Fruit small, globular or depressed,

and usually slightly compressed, hard, and almost woody.—R. Br. Prod. 584;
DC. Prod. vii. 508; *Merkusia crassifolia*, De Vr. Gooden. 46.

S. Australia. S. coast, *R. Brown;* Holdfast Bay, *F. Mueller*; near Adelaide, *Blandowsky;* Spencer's Gulf, *F. Mueller;* Port Lincoln, *Wilhelmi;* Streaky Bay, *Warburton.*
W. Australia. King George's Sound to Cape Riche, *Baxter, Harvey, Preiss, n.* 1486,
and others; Moir's Inlet, *Maxwell;* Swan River, *Fraser;* estuary of Murchison river,
Oldfield.

30. **S. macrostachya,** *Benth.* A shrub or undershrub, glabrous or
sprinkled with rigid hairs, the branches virgate. Leaves linear or lanceolate,
rarely above 1 in. long, ciliate with rigid hairs, passing into the floral leaves
or bracts, which are mostly nearly ½ in. long, and all rigidly ciliate. Flowers
sessile, in long, rather dense, leafy spikes; bracteoles linear, smaller than
the bracts. Calyx-limb obsolete. Corolla nearly ½ in. long, hairy outside.
Ovary 2-celled. Indusium scarcely ciliate. Fruit oblong, rugose, 1½ lines
long and dry, but containing perfect seeds.—*Merkusia macrostachya*, De Vr.
Gooden. 51.

N. Australia. Lacrosse Island, Cambridge Gulf, and Regent's River, *A. Cunningham;*
Usborne's Harbour, *Bynoe.*

31. **S. longifolia,** *De Vr. in Pl. Preiss.* i. 410. A perennial, with a
tufted stock or a very short, woody stem, slightly hairy or glabrous, except
the inflorescence, which is always very hirsute. Flowering stems more or
less leafy, erect or ascending, attaining usually about 1 ft. Leaves linear or
lanceolate, thick, with revolute margins, the lower ones often 4 to 6 in. long,
narrowed below the middle, and shortly dilated and sheathing at the base,
the upper ones linear-lanceolate, passing into the small, lanceolate bracts.
Flowers "dirty white," rather small, sessile in a terminal spike, at first
dense, afterwards long and interrupted. Bracteoles linear or linear-lanceolate. Calyx-lobes very short and broad, almost united in a ciliate ring.
Corolla about ½ in. long, hairy outside. Ovary 2-celled. Style slightly
hirsute upwards; indusium glabrous, ciliate. Fruit not seen.—*Merkusia
longifolia*, De Vr. Gooden. 69.

W. Australia. Swan River, *Drummond, 1st Coll. n.* 411; Vasse river, *Preiss, n.*
1472, 1483; Harvey river, *Oldfield.*
S. depressa, De Vr. in Pl. Preiss. i. 410; *Merkusia depressa*, De Vr. Gooden. 70, from
W. Australia, Preiss, n. 1502, which I have not seen, may be the same as *β. longifolia.*
The description given answers except as to the calyx-lobes, said to be linear, which is not
the case with any species of the group; but, perhaps without examination, De Vriese may
have confounded the bracts with the calyx-lobes.

32. **S. lanceolata,** *Benth. in Hueg. Enum.* 69. A perennial, sometimes low and tufted, sometimes erect, branched, and 1 ft. or more high,
more or less hirsute, especially the inflorescence. Leaves linear or linear-lanceolate, entire or very rarely with a few remote teeth, the radical ones
sometimes above 2 in. long, the stem ones much shorter, dilated and stem-clasping at the base, most of them nearly glabrous and rather thick, the
lowest floral ones linear, with rather broad, lanceolate bases, but mostly reduced to ovate or broadly lanceolate, acuminate bracts, nearly as long as the
flowers. Flowers nearly sessile, forming a terminal spike, either dense or at

length long and interrupted. Bracteoles linear. Calyx-limb annular or very shortly cup-shaped, sinuate. Corolla about 4 lines long, more or less hairy outside, sometimes with a few penicillate soft bristles at the base of the lobes. Ovary 2-celled. Style shortly hirsute. Indusium shortly ciliate or quite without cilia, at least on one side. Fruit ovoid, ribbed and furrowed, about 2 lines long, but not seen ripe.—DC. Prod. vii. 510 ; *S. lasiantha,* F. Muell. Fragm. i. 207.

W. Australia. Swan River, *Huegel;* wet places, Kalgan, Gordon, and Vasse rivers and Champion Bay, *Oldfield;* Phillips river, *Maxwell.* Very near *S. longifolia,* differing chiefly in the bracts, and in the shorter cilia of the indusium. From *S. thesioides* it differs in the hairiness, in the bracts, and in the shape of the fruit.

33. **S. thesioides,** *Benth. in Hueg. Enum.* 68. An erect, glabrous undershrub, attaining 2 or 3 ft. Leaves linear, not crowded, the upper ones very narrow, the lower ones sometimes lanceolate, all obtuse and entire or rarely with 1 or 2 minute teeth, rather thick, 1 to 2 in. long, the floral ones reduced to linear bracts, dilated at the base, 3 or 4 lines long. Flowers, small, sessile, in terminal, leafy spikes. Bracteoles linear, dilated at the base, nearly as long as the subtending bract. Calyx-tube shortly pubescent, the limb a scarcely prominent ring, sometimes obscurely lobed. Corolla about 4 lines long, glabrous outside. Ovary 2-celled. Style covered with short hairs. Indusium glabrous, not ciliate. Fruit small, compressed, broader than longer, with a thick, somewhat corky pericarp.—DC. Prod. vii. 508 ; *S. squarrosa,* Lindl. Swan Riv. App. 26 ; *S. polystachya,* DC. Prod. vii. 508 ; *S. paniculata* and *S. flaccida,* De Vr. in Pl. Preiss. i. 407 ; *Merkusia thesioides,* De Vr. Gooden. 53. t. 11.

W. Australia. Swan River, *Huegel, Drummond, Preiss, n.* 1516, 1521 ; Port Gregory and South Hutt river, *Oldfield ;* Oldfield and Phillips rivers, *Maxwell.*

Series 3. POGOGYNEÆ.—Ovary 2-celled. Indusium with a dense tuft of hairs at the base on the back, as long as the indusium itself, and often purple coloured.

34. **S. macrophylla,** *Benth.* Erect (from a woody stock ?), very hispid, 1 to 2 ft. high. Leaves ovate or ovate-lanceolate, acute, mostly toothed, stem-clasping at the base, 1 to 1½ in. long in some specimens, much smaller in others, the upper ones passing gradually into lanceolate bracts. Flowers almost sessile, in a terminal, leafy spike, at first very dense, but afterwards lengthening. Bracteoles linear-subulate. Calyx-lobes exceedingly small, ovate, ciliate. Corolla ¾ to 1 in. long, silky hairy or hispid outside. Style flattened. Indusium almost surrounded at the base by a dense tuft of purplish hairs as long as itself, the margin ciliate with white hairs. Ovary 2-celled. Drupe small, oblong, rugose.—*Molkenboeria macrophylla,* De Vr. Gooden. 44. t. 8.

W. Australia, *Drummond, 5th Coll. n.* 362 ; Cape Riche, *Maxwell.*

35. **S. platyphylla,** *Lindl. Swan Riv. App.* 26. Erect, woody at the base, with rigid, herbaceous branches, hispid with spreading hairs or almost glabrous. Leaves sessile and stem-clasping, ovate obovate or oblong, entire or with a few coarse teeth, 1 to 1½ or rarely 2 in. long, the upper floral

ones gradually smaller. Flowers large, sessile or on a very short pedicel, in a terminal, leafy spike. Bracteoles large and leafy. Calyx-lobes very small, ovate, obtuse. Corolla above 1 in. long, silky hairy, the lobes acuminate, winged. Indusium with a very dense tuft of long hairs on the back. Ovary 2-celled. Fruit not seen.—*S. semiamplexicaulis*, DC. Prod. vii. 509 ; *S. Candollei*, De Vr. in Pl. Preiss. i. 405 ; *Molkenboeria platyphylla*, De Vr. Gooden. 43. t. 6.

W. Australia. Swan River, *Drummond, 1st Coll. also n.* 15, 391, *Preiss, n.* 1497. Although the flowers are very shortly pedunculate, yet the general inflorescence is much more that of *Xerocarpæa* than of *Pogonanthera.*

36. **S. auriculata,** *Benth.* Herbaceous (or suffruticose?), pubescent or hirsute, the stems procumbent or ascending, and often very long. Leaves ovate or obovate, irregularly and coarsely toothed or rarely entire, the lower ones petiolate, the succeeding ones contracted below the middle, but all clasping the stem with broad toothed auricles, the lower floral ones broadly cordate-ovate, the upper ones much smaller, gradually reduced to small bracts. Flowers sessile or the lower ones very shortly pedicellate, the upper ones forming a long, interrupted, leafy spike. Bracteoles leafy. Calyx-limb a minute ring or quite obsolete. Corolla slightly pubescent outside, about ¾ in. long or sometimes smaller. Ovary 2-celled. Indusium with a very dense tuft of long hairs on the back at the base. Fruit ovoid-oblong, about 2 lines long.—*M. semiamplexicaulis*, De Vr. Gooden. 41, as to the character and reference to Drummond, but neither the plate (t. 5) nor the synonym of DC. nor Preiss's plant.

W. Australia, *Drummond, 3rd Coll. n.* 153. Perongerup and Plantagenet and Stirling Ranges, *Maxwell.* It was by mistake that in Hueg. Enum. I placed this in *Pogonanthera* : the anthers are not tipped with hairs, and the inflorescence is that of *Xerocarpæa.*

37. **S. æmula,** *R. Br. Prod.* 584. Herbaceous, diffuse, ascending or rarely erect, rather coarse, clothed with rigid, mostly appressed hairs or nearly glabrous. Leaves petiolate, obovate or cuneate, coarsely toothed, the lower ones sometimes 2 or 3 in. long, the upper ones smaller, the floral ones sessile, ovate-oblong or almost lanceolate, entire or with a few teeth. Flowers sessile, in a spike often 8 to 10 in. long, and much interrupted. Bracteoles linear or linear-lanceolate. Calyx-lobes exceedingly short. Corolla hairy outside, 8 to 10 lines or sometimes nearly 1 in. long. Ovary 2-celled. Style glabrous or hairy, but always with a dense tuft of rigid, often purple hairs at the top as long as the indusium. Fruit small, ovoid, rugose.— DC. Prod. vii. 509 ; *S. sinuata*, R. Br. Prod. 584 ; DC. Prod. vii. 509 ; *Merkusia sinuata*, De Vr. Gooden. 58 ; and *M. ? æmula*, De Vr. l. c. 74.

N. S. Wales. Bittangabee Flats, *Mossman.*
Victoria. Yowaka river and Bunyip Creek, *F. Mueller ;* mouth of the Glenelg river, *Allitt ;* Wimmera, *Dallachy ;* Grampians, *Wilhelmi.*
Tasmania, *from Herb. Lindley.*
S. Australia. Port Lincoln, *R. Brown ;* Rivoli and Guichen Bay, Flinders and Elder's Ranges and towards Mount Gambier, *F. Mueller ;* Mount Seal, *Warburton ;* Lake Gillies, *Burkitt.*
W. Australia. Goose Island Bay, *R. Brown ;* towards Cape Riche, *Drummond, 5th Coll. n.* 361 ; Point Henry, *Oldfield ;* Cape Arid, *Maxwell.*

The smaller specimens may sometimes be mistaken for *S. microcarpa*, but may be readily known by the tuft of hairs at the back of the indusium, and by the 2-celled ovary.

38. **S. humilis,** *R. Br. Prod.* 585. A low, branching, diffuse perennial, more or less pubescent. Leaves obovate or cuneate, acutely and prominently toothed, the lower ones 1 to 1½ in. long including the petiole, the upper ones passing into the oblong or lanceolate, acute, entire or toothed, sessile bracts. Flowers sessile in a leafy spike, short but interrupted. Bracteoles linear or lanceolate, ciliate. Calyx-lobes small, ovate. Corolla pubescent outside, about ½ in. long. Ovary 2-celled. Style more or less hairy. Indusium with a dense tuft of hairs on the back at the base, as long as the indusium itself, the margin densely ciliate. Fruit ovoid-oblong, about 2 lines long.—DC. Prod. vii. 509 ; *Merkusia humilis*, De Vr. Gooden. 59.

S. Australia. Spencer's Gulf, *R. Brown ;* Wonnomulla, *Babbage's Expedition.* The latter specimen agrees in every respect with Brown's, except that the tuft of hairs at the back of the indusium is not quite so long. F. Mueller, in Rep. Babb. Exped. 15, refers it to *S. microcarpa ;* but, besides the difference in foliage, I have always found the ovary of the latter species, as described by Brown, 1-celled, without any trace of dissepiment between the two ovules. *S. humilis* is much more nearly allied to *S. æmula*, and perhaps a variety only.

39. **S. amblyanthera,** *F. Muell. Fragm.* i. 121. A small, erect, branching perennial, shortly villous all over like *S. revoluta.* Leaves obovate-cuneate, entire, rather thick and soft, all under ½ in. long in the single specimen seen, the floral ones similar, but smaller. Flowers sessile. Bracteoles very small. Calyx-limb obsolete. Corolla pubescent outside, nearly ½ in. long. Ovary 2-celled. Style with a dense tuft of long, purplish hairs behind the indusium, which is ciliate with long hairs. Fruit small, ovoid-oblong, slightly tuberculate.

N. Australia. Granite valleys of the Upper Nicholson river, Gulf of Carpentaria, *F. Mueller.* The fragmentary specimens preserved have much the aspect of *S. revoluta*, but with smaller leaves, and readily distinguished by the tuft of hairs at the back of the indusium.

Series 4. Monospermæ.—Ovary 1-celled, with 2 ovules often so closely appressed as to appear at first like a single one. Fruit usually small and one-seeded, or rarely both seeds ripen.

40. **S. microphylla,** *Benth.* A diffuse, prostrate or ascending, pubescent or hirsute perennial, but apparently flowering also the first year, the stems and branches slender, but often above 1 ft. long. Lower leaves petiolate, obovate or oblong, 1 to 1½ in. long, the others smaller, sessile, and clasping the stem with broad auricles, all coarsely toothed, the floral ones cordate-ovate or ovate-lanceolate, entire or nearly so, mostly about ¼ in. long. Flowers almost sessile, in a long, interrupted, leafy spike. Bracteoles leafy, ovate-lanceolate. Calyx-lobes small, ovate. Corolla sparingly pubescent outside, under ½ in. long. Ovary 1-celled, with 2 ovules. Style with a dense tuft of hairs at the back of the indusium, as long as the indusium itself, which is ciliate. Fruit ovoid-oblong, about 1 line long, smooth, but not seen quite ripe.—*Molkenboeria microphylla*, De Vr. Gooden. 44. t. 9.

W. Australia. Swan River, *Fraser, Drummond, n.* 187, 190, Karri Dale, *Walcott.*
In many respects near *S. auriculata,* but the leaves and flowers smaller, and the ovary, in
numerous specimens examined, never showed any trace of dissepiment between the ovules.
The "semen longitudine sectum," f. 10 of De Vriese's plate is evidently one of the unripe
seeds separated from the other, and showing at the base the minute placenta, which the
author appears to have mistaken for the embryo.

41. **S. cuneiformis,** *Labill. Pl. Nov. Holl.* i. 56. *t.* 80.　Herbaceous,
apparently annual, clothed with short, appressed hairs or nearly glabrous.
Stems erect or ascending, ½ to 1 ft. high.　Leaves petiolate, the lower ones
obovate, attaining often 2 in., the upper ones oblong-cuneate, passing into
the sessile, ovate-oblong or broadly lanceolate bracts, which are mostly under
½ in. long and quite entire.　Flowers sessile, in a long, interrupted spike.
Bracteoles broadly lanceolate.　Calyx-lobes very small, ovate.　Corolla
slightly pubescent outside, 7 to 8 lines long.　Ovary 1-celled, with 2 ovules.
Indusium with a dense tuft of usually purplish hairs at the base on the back,
as long as the indusium itself.　Fruit small.—R. Br. Prod. 584; DC. Prod.
vii. 509; *Merkusia cuneiformis,* De Vr. Gooden. 54.

W. Australia, *Labillardière.*　Cape Arid and Cape le Grand, *Maxwell.*　Although
I have not seen Labillardière's specimens, his figure and description leave no doubt as to the
identity of the species.　The leaves are nearly those of *S. æmula,* but more glabrous.　The
ovary is always 1-celled, as in *S. microphylla,* but the foliage is very different.

42. **S. microcarpa,** *Cav. Ic.* vi. 6. *t.* 509.　A procumbent, diffuse or
ascending, very rarely almost erect perennial, more or less pubescent.　Leaves
petiolate, obovate ovate or cuneate, coarsely toothed, the lower ones often 1
to 1½ in. long, the upper ones smaller, passing into the sessile, ovate-oblong
or lanceolate, entire or toothed floral leaves or bracts, which are mostly shorter
than the flowers.　Spike usually long and interrupted.　Bracteoles linear.
Calyx-lobes small, ovate.　Corolla hairy outside, 7 to 9 lines long in the
typical form, the lobes fringed at the base with a few long cilia, sometimes penicillate, and descending into the throat as in the *Globuliferæ.*　Ovary 1-celled,
with 2 ovules.　Style more or less hairy.　Indusium glabrous, ciliate.　Fruit
small, usually 1-seeded.—R. Br. Prod. 585; DC. Prod. vii. 509; *Goodenia
albida,* Sm. in Trans. Linn. Soc. ii. 348; *G. lævigata,* Curt. Bot. Mag.
t. 287; *Merkusia microcarpa,* De Vr. Gooden. 55.

N. S. Wales.　Port Jackson to the Blue Mountains, *R. Brown* and others; Glendon
and Newcastle, *Leichhardt,* Macleay and Clarence rivers, *Beckler;* New England, *C.
Stuart.*
Victoria.　Portland forest, *Herb. F. Mueller.*
S. Australia.　Burra-Burra, *Hinteracker;* Port Adelaide, *Blandowsky.*

Var. *pallida.*　More diffuse and smaller, the flowers smaller, the corolla under ½ in. long,
and nearly glabrous outside; style glabrous or nearly so.—*S. pallida,* Br. Prod. 585; DC.
Prod. vii. 510; *Merkusia pallida,* De Vr. Gooden. 56 —Port Phillip, *R. Brown;* seacoast
from Wilson's Promontory to the Glenelg, *F. Mueller* and others; Rivoli Bay and Onkaparinga, *F. Mueller.*　Compared with the common large-flowered N. S. Wales form, this
plant looks very different, but the flowers vary much in size, and the character indicated of
the glabrous styles is not constant, for some of the S. Australian small-flowered specimens
(*e. g.* from Holdfast Bay, *Whittaker*) have the flowers of *S. pallida,* with the very hairy
style of *S. microcarpa.*
Goodenia pubescens, Sieb. Fl. Mixt. n. 608, is referred by F. Mueller, and probably correctly, to *S. microcarpa,* but the specimens I have seen have no flowers.

43. **S. linearis,** *R. Br. Prod.* 586. A diffuse undershrub or spreading shrub, more or less villous. Leaves sessile, linear or linear-lanceolate, obtuse, with revolute margins, the larger ones 1 in. long, the axils not woolly, the floral ones similar or more lanceolate and smaller. Flowers sessile in the upper axils, forming an interrupted, leafy spike. Bracteoles linear. Calyx-limb very short, annular, sinuate. Corolla pubescent outside, 6 to 7 lines long, with the penicillate cilia in the throat of *S. microcarpa.* Ovary 1-celled, with 2 ovules. Style slightly hairy. Indusium shortly ciliate. Fruit oblong, about 1½ lines long.—DC. Prod. vii. 510; *Merkusia linearis,* De Vr. Gooden. 67.

S. Australia. Port Lincoln, *R. Brown, Wilhelmi;* Encounter Bay, *F. Mueller;* Kangaroo Island, *Waterhouse.* In this species and in *S. paludosa* Brown describes the ovary as monospermous. I have always found in the ovary 2 ovules closely appressed to each other without any trace of dissepiment, but only one of them appears to enlarge, so as to leave the fruit monospermous.

44. **S. Oldfieldii,** *F. Muell. Fragm.* ii. 19. An erect or divaricate shrub of several ft·, closely allied to *S. paludosa,* but quite glabrous except the inflorescence. Leaves from oblong-lanceolate to linear-lanceolate, acute, narrowed into a long petiole, the larger ones 2 to 3 in. long, a few of the lower ones sometimes smaller and obovate, the floral ones small, narrow, with broad, sheathing bases, all entire or very rarely when broad with a few teeth. Spikes dense, axillary, and leafy, always shorter than the subtending leaves, the flowers sessile. Bracteoles linear, dilated and sheathing at the base. Calyx-limb exceedingly short and truncate. Corolla 6 to 8 lines long, silky-pubescent outside. Ovary 1-celled, with 2 ovules. Indusium glabrous, shortly ciliate. Fruit small, oblong.

W. Australia. Murchison river, *Oldfield.* Very closely allied to *S. paludosa,* and perhaps only a large-flowered, glabrous variety.

45. **S. paludosa,** *R. Br. Prod.* 586. A spreading decumbent or prostrate hard perennial or undershrub, more or less hirsute with appressed hairs or rarely nearly glabrous. Leaves from linear-lanceolate to oblong-lanceolate, acute, narrowed into a long petiole, the larger ones 2 to 3 in. long, the floral ones much smaller and dilated at the base, all entire. Spikes dense, axillary, leafy, always shorter than the subtending leaves. Flowers sessile. Bracteoles linear with dilated sheathing bases. Calyx-limb exceedingly short and truncate. Corolla under ½ in. long, pubescent outside. Ovary 1-celled, with 2 ovules. Indusium glabrous, shortly ciliate. Fruit small.—DC. Prod. vii. 511; *Merkusia paludosa,* De Vr. Gooden. 68.

W. Australia King George's Sound, *R. Brown.*

Var. *prostrata.* Stems long and prostrate.—*S. repens,* De Vr. in Pl. Preiss. i. 406; *Dampiera repens,* De Vr. Gooden. 114.—Swan River, *Preiss, n.* 1519; *Drummond,* 1st Coll.

46. **S. sericophylla,** *F. Muell. Herb.* An erect shrub of 2 to 3 ft., the branches and foliage hoary or silvery with a very close silky tomentum. Leaves crowded on the short branchlets, obovate-oblong or oblanceolate, obtuse or softly mucronate, narrowed below the middle, rather thick, ¾ to 1 in. long. Flowers sessile in the axils, scarcely so long as the

leaves. Bracteoles linear, sheathing and dilated at the base. Calyx-limb almost obsolete. Corolla 6 to 8 lines long, densely tomentose outside, the lobes narrow, winged only in the upper half. Indusium with a tuft of long hairs outside, the margin ciliate. Ovary 1-celled, with 2 ovules.

W. Australia. Murchison river, *Oldfield.* F. Mueller (Fragm. ii. 19) considers this as a variety of *S. Oldfieldii,* but it appears to me to be different in foliage and indumentum, as well as in the tuft of hairs on the back of the indusium.

47. **S. canescens,** *Benth. in Hueg. Enum.* 69. A much-branched shrub or undershrub, densely clothed with a soft stellate tomentum often intermixed with long silky or rarely spreading hairs. Leaves linear-oblong or broadly lanceolate, obtuse, entire, soft and thick, the larger ones 2 or 3 in. long and narrowed into a long petiole dilated and stem-clasping at the base, but mostly shortly narrowed at the base. Flowers in short dense spikes or clusters, sessile in the axils or terminating short axillary branches, the floral leaves or bracts linear, soft, often as long as the flowers. Bracteoles similar to the bracts but smaller. Calyx-limb broadly annular or almost obsolete. Corolla ½ in. long, hairy outside. Ovary 1-celled, with 2 ovules. Style nearly glabrous or with short hairs at the back of the indusium. Fruit small.—DC. Prod. vii. 510 ; *S. trinervis,* De Vr. in Pl. Preiss. i. 407 ; *S. glaucescens,* De Vr. l. c. 410 ; *Dampiera canescens,* De Vr. Gooden. 114. t. 19.

W. Australia. Swan River, *Huegel, Drummond, 1st Coll. n.* 412, also *n.* 24 *and* 25, *Fraser, Preiss, n.* 1477, 1479.

One specimen of Preiss's *S. trinervis* is less tomentose, with broader thinner leaves and more developed flowering branchlets, perhaps grown in a less exposed situation ; another is like the common form. Some imperfect specimens from Murchison river, *Oldfield,* appear to be a variety with stout, very villous-tomentose branches, the axils very woolly, and with smaller flowers.

48. **S. humifusa,** *De Vr. in Pl. Preiss.* i. 410. A diffuse or prostrate much-branched undershrub or small shrub, slightly pubescent besides the white-woolly axils. Leaves linear with revolute margins, the longer ones above 1 in. long, narrowed below the middle and dilated at the base, but most of them much shorter and densely tufted in the axils of the older ones, sometimes not 3 lines long and buried in the wool, all quite entire and rather thick. Flowers sessile in short dense leafy spikes. Bracteoles linear-lanceolate, with woolly axils. Calyx-limb reduced to a minute ciliate ring or quite obsolete. Corolla 4 to 5 lines long, slightly hairy outside. Ovary 1-celled, with 2 erect ovules. Style slightly hairy. Indusium scarcely ciliate. Fruit small.—*Merkusia humifusa,* De Vr. Gooden. 70.

W. Australia, *Drummond*; plains of the Avon, *Preiss, n.* 1480 ; Port Gregory and Murchison river, *Oldfield.* I presume this to be the plant intended by De Vriese, from the specimen with Preiss's above-quoted number in F. Mueller's Herbarium, and De Vriese's description agrees well in many points, but in others it is totally at variance, taken perhaps from some different species confounded with it. *Merkusia molluginea,* De Vr. Gooden. 71, from Swan River, *Drummond,* is a bad specimen apparently of *S. humifusa.*

SERIES 5. PARVIFLORÆ. Ovary 1-celled, 1- or 2-ovulate, the convex summit free. Style covered with short purple glandular hairs. Leaves very narrow. Flowers small and numerous in terminal spikes.

The two following species have both been published as *Goodenias*, to which the only approach seems to me to be the free summit of the ovary. The ovules limited to 2 or 1, and the fruit, a hard indehiscent nut, are entirely those of *Scævola.*

49. **S. fasciculata,** *Benth. in Hueg. Enum.* 68. An erect shrub of 2 or 3 ft., nearly glabrous except some long woolly hairs at the base of the leaves. Branches virgate. Leaves linear almost setaceous, clustered at the nodes, from 2 to 4 lines or rarely ½ in. long, the upper and floral ones similar but less clustered. Flowers small, numerous, sessile in the upper axils, forming a terminal leafy spike. Bracteoles linear, usually unequal. Calyx-lobes linear, as long as the 5-ribbed tube. Corolla about 4 lines long, white, often with dark spots in the throat, glabrous outside, the lobes about as long as the tube, their margins induplicate but not so distinctly winged as in most species. Ovary 1-celled, with a single ovule, the convex summit shortly free within the corolla. Style covered with short dark glandular hairs. Fruit ovoid, 1 to 1½ lines long, rather more than half inferior.—DC. Prod. vii. 508 ; *Goodenia squarrosa,* De Vr. in Pl. Preiss. i. 413, Gooden. 154. t. 29.

W. Australia. Swan River, *Huegel, Drummond, 1st Coll. and n.* 33, *Preiss, n.* 1467 ; Harvey river, *Oldfield.*

50. **S. stenophylla,** *Benth.* An erect glabrous shrub with numerous virgate branches. Leaves rather crowded but not clustered, very narrow linear, almost terete, ½ to 1 in. long or sometimes more, quite entire, the floral ones reduced to short bracts, all but the lower ones scarcely exceeding the calyx. Flowers numerous, sessile or nearly so in terminal spikes. Bracteoles very small. Calyx-tube scarcely ½ line long, the lobes oblong, erect. Corolla glabrous outside, about 5 lines long, the entire part much shorter than the lobes, the adnate part of the tube with a slight concave protuberance on the lower side. Ovary 1-celled, with 2 ovules, the very convex summit free within the corolla. Style hispid with short purple glandular hairs. Indusium very deep, not ciliate. Fruit ovoid, 1 to 1½ lines long, rather more than half inferior. Seeds 1 or 2, oval-oblong, more or less flattened, especially when both ripen, not bordered.—*Goodenia stenophylla,* F. Muell. Fragm. i. 113.

W. Australia, *Drummond, n.* 365 *and 3rd Coll. Suppl. n.* 54 ; Young river, Middle Mount Barren, Phillips Range, etc., *Maxwell.*

9. DIASPASIS, R. Br.

Calyx-tube adnate, limb of 5 distinct segments. Corolla oblique, the tube entire, the lobes nearly equal, spreading. Stamens free, entirely included as well as the style in the tube of the corolla. Indusium cup-shaped, entire. Ovary wholly inferior, 1-celled, with 2 erect ovules. Fruit small, ovoid, dry or nearly so, but indehiscent. Seeds of *Scævola.*—Herb with linear leaves. Flowers axillary, pedunculate, solitary between 2 bracteoles.

The genus is limited to a single species, endemic in Australia, closely allied to *Scævola,* differing in the nearly regular corolla enclosing the genitalia in its tube.

1. **D. filifolia,** *R. Br. Prod.* 587. A perennial, glabrous or sprinkled

with short hairs, with a hard stock and erect or ascending wiry stems slightly branched, sometimes all under 1 ft. high, sometimes weaker and 2 ft. long. Leaves linear, almost terete, ½ to 1½ in. long, entire, or the lower ones rarely dilated and bordered by a few teeth. Peduncles rather shorter or longer than the leaves, the linear-terete bracteoles close under the flower. Calyx-lobes ovate, very much shorter than the tube. Corolla ½ to ¾ in. long, the lobes longer than the tube, their wings broad and veinless. Indusium more or less hairy, but the margin not ciliate.—DC. Prod. vii. 505 ; De Vr. Gooden. 178 ; *Goodenia armeriæfolia*, De Vr. in Pl. Preiss. i. 412, not of DC., *G. glandulifera*, De Vr. Gooden. 129 ; *Scævola clandestina*, F. Muell. Fragm. i. 206.

W. Australia. King George's Sound and adjoining districts, *R. Brown, Drummond, n.* 148, 187, *Preiss, n.* 2032, and many others.

10. VERREAUXIA, Benth.

Calyx-tube adnate ; lobes 5, free. Corolla-tube slit on the upper side to the ovary, the lobes nearly equal, the 2 upper ones separated rather lower down (arching over the style ?), not auriculate, the 3 lower ones spreading, all equally winged. Stamens free. Ovary wholly inferior, 1-celled, with 1 ovule erect from the base. Indusium cup-shaped, enclosing the stigma, the margin ciliate. Fruit small, more or less flattened, indehiscent, usually crowned by the annular persistent base of the corolla. Seed flat, with a thin crustaceous testa. Embryo terete in the centre of the albumen.—Herbaceous or suffruticose plants, stellate-tomentose or woolly. Leaves entire. Flowers in terminal leafless panicles or spikes.

The genus is limited to Australia. With the flowers and flat seeds of *Goodenia*, it has the ovary and fruit of *Dampiera*.

Leaves all radical. Panicle long loose and slender, with filiform cymose branches . . . . 1. *V. paniculata.*
Stems leafy. Flowers clustered along a long simple rhachis . . . 2. *V. Reinwardtii.*

1. **V. paniculata,** *Benth.* A perennial, with a thick tufted and densely woolly stock. Leaves all radical, obovate or oval-oblong, obtuse, entire, narrowed into a petiole, thick and softly tomentose or woolly on both sides, 1½ to 3 in. long. Scapes erect, slender, 1 to 1½ ft. high, becoming quite glabrous, leafless or with 1 or 2 very small linear cottony leaves, the branches or peduncles filiform, bearing each a loose cyme of 3 to 7 small (yellow?) flowers, or the uppermost 1-flowered. Bracts minute. Calyx-tube hispid with long purplish hairs intermixed with a close cottony wool; lobes linear-lanceolate nearly as long as the tube. Corolla 3 to 4 lines long, hispid outside with short simple hairs intermixed with a glandular pubescence. Fruit only seen young, but then appears flattened with 1 broad erect flat seed.—*Dampiera Verreauxii*, De Vr. Gooden. 118. t. 20.

W. Australia, *Drummond, 4th Coll. n.* 186 ; sandy plain of Quangen, *Preiss, n.* 1454. In Drummond's specimens, the only ones I could analyse, the flowers are much injured, and I do not feel certain whether the form and position of the upper lobes of the corolla are the same as in *V. Reinwardtii* or not.

2. **V. Reinwardtii,** *Benth.* An undershrub or shrub of 2 to 3 ft.,

clothed with a close white cottony wool.   Leaves obovate to oblong, obtuse, entire, narrowed into a petiole, thick and soft, 1 to 2 in. long or the lower ones larger on long petioles.   Flowers yellow, clustered along the rhachis of long interrupted leafless and cottony terminal spikes or the uppermost solitary, all sessile or nearly so.   Bracts very small, obtuse.   Calyx in some specimens densely cottony, in others hirsute with purplish hairs ; lobes linear, shorter than the tube.   Corolla 4 to 5 lines long, cottony or hirsute outside. Fruit ovate, flattened, about 2½ lines long, softly tomentose, crowned by the calyx-lobes, with a hard endocarp (the pericarp separating from the calyx-tube ?).   Seed flat, filling the cavity of the fruit.—*Scævola Reinwardtii*, De Vr. in Pl. Preiss. i. 409 ; *Dampiera Reinwardtii*, De Vr. Gooden. 97. t. 15.

W. Australia.   *Drummond, 2nd Coll. n.* 404.   Champion Bay and Murchison river, *Oldfield and Walcott.*   The specimens with the two kinds of indumentum on the flowers are mixed in Oldfield's collections, and I can find no other differences between them.

11. DAMPIERA, R. Br.

(Linschotenia, *De Vr.*)

Calyx-tube adnate to the ovary ; lobes 5, very small, often concealed under the indumentum or quite obsolete.   Corolla-tube deeply slit on the upper side, but usually entire and persistent at the base, the remainder circumsciss and deciduous, 2 upper lobes deeply separated, unequally winged, erect and connivent, enclosing the summit of the style in two thick concave auricles, one on the outer side of each lobe below the wing, the 3 lower lobes broadly winged and spreading.   Anthers cohering in a tube round the style.   Ovary 1-celled, with 1 erect or ascending straight or recurved ovule, rarely 2-celled with 1 erect ovule in each cell ; indusium somewhat 2-lipped, not ciliate. Fruit small and indehiscent.   Seed variously shaped ; testa rather thin ; embryo in the centre of the albumen.—Herbs undershrubs or shrubs, the indumentum usually stellate or branched, cottony or woolly.   Leaves entire or obtusely toothed or sinuate.   Flowers purple blue or white, rarely yellow, the margins of the corolla-lobes undulate below the wings and forming prominent lines decurrent inside the tube.   Peduncles simple or irregularly (mostly cymosely) branched, solitary or clustered in the upper axils or the nearly sessile flowers forming terminal spikes.

The genus is limited to Australia.   It is a natural one, readily known by the peculiar auricles of the upper corolla-lobes and coherent anthers combined with the solitary ovules. Several *Goodenias* have indeed auriculate upper corolla-lobes, but the auricles are never so conspicuously concave and thick as in *Dampiera*, and they have always free anthers and a capsular fruit.   One species of *Scævola* has an uniovulate ovary, and in two *Dampieræ* it is 2-celled and 2-ovulate, but the two genera are very distinct in their corolla and anthers.

The indumentum in *Dampiera* is almost always more or less stellate, and in many species normally so.   Where the hairs appear long and simple they are usually stellate or with short crowded branches at the base, with one branch long and simple; where they are long and plumose, the branches are scattered along the main ones, but yet often more crowded at the base ; where the hairs appear strigose and appressed, the branches are few, stellate in principle, but parallel and divaricate in opposite directions or reduced to a single centrally attached (2-branched) hair.

SECT. I. **Linschotenia.**—*Flowers sessile or shortly pedicellate, in terminal, leafless, simple or branched spikes or racemes.   Ovary 1-celled, with 1 oblong ovule, laterally attached above the base.—Plants tomentose or woolly.*

Spikes branched. Leaves oblong or lanceolate. Eastern species . 1. *D. Linschotenii.*
Spikes or racemes long and simple. Western species.
    Leaves obovate or cuneate . . . . 2. *D. spicigera.*
    Leaves linear-terete. Flowers small . . . . 3. *D. teres.*

SECT. II. **Dicœlia.**—*Peduncles axillary, loosely and irregularly cymose. Ovary 2-celled, with 1 erect, linear ovule in each cell. Plants glabrous or nearly so, with angular or winged stems.*

Leaves lanceolate or linear . . . . 4. *D. trigona.*
Leaves ovate . . . . 5. *D. prostrata.*

SECT. III. **Camptospora.**—*Peduncles axillary or terminal, 1-flowered or irregularly cymose. Ovary (often oblique or gibbous) 1-celled, with 1 vertical, horseshoe-shaped ovule. (Seeds usually recurved over a spurious, partial dissepiment.)*

Glabrous or nearly so. Branches 2- or 3-winged. Upper leaves
    few and small.
    Leaves oblong or lanceolate. Calyx-tube globular ; lobes con-
      cealed in the indumentum . . . . 6. *D. alata.*
    Leaves oblong-cuneate. Calyx-tube gibbous ; lobes prominent . 7. *D. corcnata.*
Glabrous or nearly so. Branches terete or sulcate. Leaves few,
    linear-terete. Calyx-tube very gibbous.
    Flowers terminal, solitary . . . . 8. *D. carinata.*
    Peduncles axillary, 2- or 3-flowered . . . . 9. *D. sacculata.*
Hoary-tomentose or woolly. Branches terete. Leaves cuneate.
    Calyx-tube rather oblique . . . . 10. *D. inccna.*

SECT. IV. **Eudampiera.**—*Peduncles axillary or terminal, solitary or clustered, 1-flowered or irregularly cymose. Ovary 1-celled, with 1 linear or oblong ovule, erect from the base.*

Leaves flat or with recurved margins, stellate-tomentose under-
    neath or rarely glabrous. Hairs of the flowers plumose.
    Weak, trailing herb. Western species . . . . 11. *D. hederacea.*
    Erect shrubs or undershrubs. Eastern species.
      Leaves orbicular, ovate or ovate-lanceolate.
        Bracts ovate, leafy . . . . 12. *D. ferruginea.*
        Bracts minute . . . . 13. *D. Brownii.*
      Leaves oblong-lanceolate or almost linear . . . . 14. *D. lanceolata.*
Leaves flat or with revolute margins, stellate-tomentose under-
    neath or on both sides. Flowers stellate-tomentose.
    Leaves nearly flat. Peduncles cymose, with leafy bracts.
      Leaves cuneate or narrowed into a petiole. Cymes loose . . 15. *D. altissima.*
      Leaves ovate to oblong-elliptical, sessile. Cymes very short . 16. *D. marifolia.*
    Leaves mostly narrow, and much revolute.
      Peduncles mostly very short. Bracteoles small. Eastern
        species . . . . 17. *D. rosmarinifolia.*
      Peduncles mostly longer than the leaves. Bracteoles linear.
        Western species . . . . 18. *D. lavandulacea.*
Leaves rigid, not revolute, glabrous when full grown or none.
    Hairs of the flowers plumose.
    Stems terete.
      Leaves none or small and linear-terete . . . . 19. *D. juncea.*
      Leaves few, small, and flat . . . . 20. *D. oligophylla.*
    Stems very angular. Leaves oblong or cuneate, flat . . . . 21. *D. loranthifolia.*
Leaves rigid, flat, glabrous when full grown (except *D. triloba*).
    Hairs of the flowers rigidly appressed, with parallel branches.
    Stems angular. Leaves sessile.
      Calyx-teeth prominent. Leaves oblong or linear.

SECT. V. **Cephalantha.**—*Leaves radical. Scapes simple or branched. Flowers
in dense, terminal heads. Ovary 1-celled, with 1 linear-oblong ovule, erect from the
base* . . . . 34. *D. eriocephala.*

SECT. I. LINSCHOTENIA.—Flowers sessile or shortly pedicellate, in terminal, leafless, simple or branched spikes or racemes. Ovary 1-celled, with
1 oblong ovule, laterally attached above the base. Plants tomentose or
woolly.

1. **D. Linschotenii,** *F. Muell. Fragm.* vi. 28. Probably a tall undershrub, erect, clothed with a soft, white cotton, very dense on the stems and
under side of the leaves, disappearing on their upper surface, mixed on the
inflorescence with rather longer hairs. Branches terete. Leaves shortly
petiolate, oblong or lanceolate, thick, entire, flat or concave, the lower ones
above 2 in. long, the upper ones smaller. Flowers nearly sessile, solitary
within each bract, forming long spikes, branching at the base into a pyramidal, leafless panicle. Bracts small, linear or lanceolate; bracteoles very
small. Calyx-tube densely covered with long hairs, the lobes obsolete.
Corolla 4 to 5 lines long, covered outside with a stellate tomentum, mixed
with longer hairs. Upper lobes of the corolla shorter than the others.
Ovary 1-celled, with 1 oblong ovule, erect, but attached laterally a little
above the base.—*Linschotenia discolor,* De Vr. in Mitch. Trop. Austr. 346 ;
Gooden. 120. t. 22.

Queensland. Near Mount Pluto and Mount Faraday, *Mitchell.* The " paracorollæ
cuculliformes," by which De Vriese proposed to distinguish this plant generically from

Dampiera, are nothing but the auricles of the upper corolla-lobes, one of the principal characters of the whole genus *Dampiera*.

2. **D. spicigera,** *Benth.* An undershrub, with erect, simple or branched stems of $\frac{1}{2}$ to 1 ft., hoary as well as the foliage with a close, stellate or intricate tomentum, disappearing with age from the upper side of the leaves. Leaves obovate-oblong or cuneate, very obtuse, entire or obtusely toothed at the end, coriaceous, flat or concave, under $\frac{1}{2}$ in. long when broad, nearly 1 in. when narrow, the floral ones all reduced to very small bracts. Flowers solitary within each bract, and sessile or nearly so, forming long, terminal, simple spikes. Bracteoles minute or none. Calyx-lobes very small, ovate, almost concealed in the tomentum. Corolla about $\frac{1}{2}$ in. long, clothed with a stellate tomentum, mixed with a few longer, simple hairs. Ovary 1-celled; ovule erect, straight or slightly incurved at the end, laterally attached a little above the base.

W. Australia, *Drummond*, 3rd *Coll. n.* 154. Included by De Vries in *D. lavandulacea*, by F. Mueller in *D. incana*.

Var. *lanata.* Stouter, more shrubby, attaining 1 to 2 ft., and very woolly, especially the branches. Flowers rather larger.—Murchison river, *Oldfield.*

3. **D. teres,** *Lindl. Swan Riv. App.* 27. An erect, branching, rather slender undershrub, our specimens not exceeding 1 ft., hoary all over, with a close, minute, stellate tomentum. Leaves linear-terete, obtuse, entire, $\frac{1}{4}$ to $\frac{1}{2}$ in. long. Flowers small, blue, in loose, terminal, slender spikes or racemes of 2 to 4 in., each one on a short pedicel in the axil of a small, linear bract, with 2 smaller bracteoles close under the flower. Calyx-lobes erect, small but conspicuous. Corolla 5 to 6 lines long, with the same close tomentum as the rest of the plant. Ovary 1-celled; ovule erect, straight, laterally attached a little above the base.—De Vr. Gooden. 96.

W. Australia. Swan River, *Drummond*, 1st *Coll. n.* 12.

SECT. II. DICŒLIA.—Peduncles in the upper axils loosely and irregularly cymose. Ovary 2-celled, with 1 erect, linear ovule in each cell. Plants glabrous or nearly so, with angular or winged stems.

4. **D. trigona,** *De Vr. in Pl. Preiss.* i. 401. Herbaceous and glabrous. Stems diffuse ascending or erect, very angular, slender and weak or rarely rigid, and almost winged. Leaves sessile or petiolate, lanceolate or almost linear, entire or rarely toothed, 1 to 2 in. long, the lower ones sometimes shorter and almost ovate, the upper floral ones small and very narrow. Flowers blue, rather large, on slender, flexuose, branching peduncles in the upper axils, forming a very loose, terminal, panicle. Calyx-tube contracted at the top, and apparently continuous with the persistent base of the corolla, without any perceptible lobes. Corolla 7 to 8 lines long, glabrous or sprinkled with appressed hairs. Ovary 2-celled, with 1 linear ovule erect from the base in each cell. Fruit oblong, about 2 lines long, crowned by the persistent base of the corolla. Seeds nearly terete.—Hook. Ic. Pl. t. 1026; *D. biloculata*, F. Muell. Fragm. ii. 17.

W. Australia. Swan River, *Drummond*, 1st *Coll.*; near Maddington, *Preiss, n.*

1471 ; Vasse and Blackburn rivers and Cape Leschenault, *Oldfield;* Phillips Flats, *Maxwell;* King George's Sound, *F. Mueller.*

Var. *tenuis.* Very slender, with smaller, slender flowers.—*Drummond, 4th Coll. n.* 192. De Vriese, Gooden. p. 113, reduces this species to *D. coronata,* Lindl., which is, however, widely distinct in habit and structure.

5. **D. prostrata,** *De Vr. Gooden.* 83 (*not of Pl. Preiss.*). Glabrous or nearly so except the inflorescence. Stems broadly and acutely 3-angled, almost winged. Leaves sessile, with a broad base, ovate, acute, acutely toothed or almost lobed, coriaceous and rigid, 1 to 1½ in. long, the lower ones unknown. Peduncles in the upper axils branching and several-flowered, with small, linear bracts. Flowers (blue ?) sprinkled with appressed, simple hairs. Calyx-teeth quite obsolete. Corolla fully ½ in. long. Ovary 2-celled, with 1 linear ovule erect from the base in each cell. Fruit not seen ripe, but apparently the same as in *D. trigona.*

W. Australia, *Drummond, n.* 364; Cheynes Beach, *Maxwell.*

SECT. III. CAMPTOSPORA.—Peduncles axillary or terminal, 1-flowered or irregularly cymose. Ovary (often oblique or gibbous) 1-celled, with 1 vertical ovule erect from the base, but recurved into a horseshoe-shape or almost annular. Seed where known recurved over a spurious semidissepiment.

6. **D. alata,** *Lindl. Swan Riv. App.* 27. Glabrous except the inflorescence or slightly silky-pubescent. Stems erect or ascending, not much branched, 1 to 2 ft. high, with 2 or 3 very much raised angles or wings decurrent from the leaves, sometimes 2 or 3 lines broad. Leaves coriaceous, sometimes oblong or broadly lanceolate, entire or toothed, and above 1 in. long, sometimes very small or linear or reduced to minute scales. Peduncles in the upper axils solitary or 2 together, 1-flowered or loosely 2- or 3-flowered, with minute bracteoles close under the flower. Calyx-tube nearly globular, oblique, the lobes very small, almost concealed in the indumentum. Corolla 6 to 9 lines long, clothed with appressed or rather loose hairs. Ovary 1-celled; ovule vertical, horseshoe-shaped or almost annular. Fruit about 2 lines diameter, crustaceous, separable from the herbaceous calyx-tube. Seed horseshoe-shaped, curved over the spurious semidissepiment.—De Vr. Gooden. 112; Hook. Ic. Pl. t. 1027; *D. cauloptera,* DC. Prod. vii. 504; De Vr. in Pl. Preiss. i. 402, and Gooden. 111. t. 18 ; *D. trialata, D. epiphylloidea,* and *D. Lindleyi,* De Vr. in Pl. Preiss. i. 401, 402.

W. Australia. King George's Sound to Swan River, *Drummond, 1st Coll., 2nd Coll. n.* 426, *Suppl. n.* 10; *Preiss, n.* 1444, 1476, 1494, 1514, *Oldfield,* and others; Stirling Range and Salt River, *Maxwell;* Murchison river, *Oldfield.*

7. **D. coronata,** *Lindl. Swan Riv. App.* 27. A perennial, glabrous except the flowers, with erect, simple or branched stems of 1 to 1½ ft., with 2 or 3 raised angles or wings decurrent from the leaves. Lower leaves petiolate, oblong-cuneate or almost obovate, coarsely toothed, 1 to 2 in. long, the upper ones small, narrow, and sessile. Peduncles in the upper axils bearing usually few, shortly-pedicellate flowers, forming altogether a loose, terminal, leafy panicle. Bracts minute. Flowers blue, covered with appressed, dark-coloured hairs. Calyx-tube oblique and very gibbous, the lobes broad and

obtuse, very small, but prominent. Corolla about $\frac{1}{2}$ in. long. Ovary 1-celled, with 1 erect, horseshoe-shaped ovule. Fruit not seen.

W. Australia. Swan River, *Collie, Drummond,* 1*st Coll.,* 2*nd Coll. n.* 396.

8. **D. carinata,** *Benth.* A small perennial, the stock tufted and woolly, the rest of the plant glabrous except the flowers. Stems erect, branching, terete, rigid, almost leafless, the specimens seen not exceeding 6 in. Leaves few, small, linear-terete or reduced to minute scales. Flowers solitary and terminal, about 4 lines long, silky-white with appressed and stellate hairs. Calyx-tube short, with a broad, semicircular, keel-like appendage or gibbosity on one side, the lobes ovate-lanceolate as long as the tube, Ovary 1-celled, with 1 erect, horseshoe-shaped ovule.

W. Australia, *Drummond,* 2*nd Coll. n.* 397 (*Herb. F. Muell.*).

9. **D. sacculata,** *F. Muell. Herb.* Herbaceous, glabrous except the flowers, the stems rush-like, terete or scarcely angular, simple or branched, sometimes erect and 1 to 1$\frac{1}{2}$ ft. high, sometimes longer and flexuose (or almost scandent?). Leaves linear, very narrow and thick, obtuse, entire, the lower ones sometimes 1 to 2 in. long, and occasionally rather broad and flat, the upper ones distant, very small, and almost terete. Peduncles in the upper axils mostly 2 or 3 together, 1-flowered or rarely 2-flowered. Calyx-tube very short, oblique and gibbous, the lobes minute, concealed under the indumentum. Corolla about $\frac{1}{2}$ in. long, of a deep purple-blue clothed as well as the calyx with black, appressed hairs. Ovary 1-celled, with 1 erect, horseshoe-shaped or almost annular ovule. Fruit not seen.

W. Australia. Blackwood and Upper Kalgan rivers, *Oldfield;* Stirling Range, *Maxwell, F. Mueller.*

10. **D. incana,** *R. Br. Prod.* 588. A diffuse or divaricately-branched shrub, hoary in every part with a close intricate or stellate tomentum. Leaves obovate or oblong-cuneate, the lower ones shortly petiolate and above 1 in. long, the upper ones smaller and sessile but narrowed at the base, all quite entire or very rarely slightly angular. Flowers solitary or 2 together terminating short leafy branches and forming an irregularly corymbose leafy panicle. Bracts and bracteoles small, linear, very obtuse. Calyx-tube very short, the minute lobes concealed under the indumentum. Corolla about 4 lines long, densely tomentose outside. Ovary 1-celled, with 1 erect horseshoe-shaped ovule. Fruit not seen.—DC. Prod. vii. 503; De Vr. Gooden. 95.

W. Australia. Sharks' Bay, *Gaudichaud, A. Cunningham;* Dirk Hartog's Island, *A. Cunningham;* Murchison river, *Oldfield.*

Var. *fuscescens.* Tomentum of the flower looser, of a more leaden colour.—Murchison river, *Oldfield.*

F. Mueller, Fragm. ii. 17, says that *D. incana* extends over the desert land of the interior to S. Australia and Victoria, varying much in inflorescence, etc., but he includes in the species *D. spicigera* and *D. marifolia,* both of which appear to me to differ too much in the ovule as well as in the inflorescence to be placed even in the same section as *D. incana.*

Sect. IV. Eudampiera. Peduncles axillary or terminal, solitary or clustered, 1-flowered or irregularly cymose. Ovary 1-celled with 1 linear or oblong ovule erect from the base.

11. **D. hederacea,** *R. Br. Prod.* 588. Herbaceous, with weak diffuse or trailing stems, the whole plant clothed with stellate or plumose rather loose and sometimes rigid hairs. Leaves petiolate, broadly cordate ovate or ovate-lanceolate, the lower ones 1 to 1½ in. long, angular or almost lobed, the upper ones small and entire, all rather thin, loosely woolly underneath, becoming nearly glabrous above when old. Peduncles in the upper axils rather slender, bearing each an irregular cyme of blue flowers. Bracts small, linear. Calyx-lobes lanceolate, nearly as long as the tube. Corolla 5 to 6 lines long, clothed with dense dark plumose hairs. Ovary 1-celled, with 1 oval-oblong flattened ovule. Fruit not seen.—DC. Prod. vii. 503; De Vr. Gooden. 79.

W. Australia. King George's Sound and adjoining districts, *R. Brown, A. Cunningham, Drummond, n.* 143 *and 3rd Coll. n.* 188; Franklin river, *Maxwell;* Upper Kalgan river, *F. Mueller.*

12. **D. ferruginea,** *R. Br. Prod.* 588. An erect undershrub of 1 to 2 ft., clothed with a loose stellate tomentum intermixed with long hairs shortly plumose at the base ; branches terete, sulcate. Leaves very shortly petiolate, ovate orbicular or rhomboidal, rather rigid, often 3-nerved, entire or coarsely toothed, ¾ to 1½ in. long, becoming glabrous when old, the upper ones sessile and smaller. Peduncles in the upper axils several-flowered, with ovate leaf-like bracts or the peduncles growing out into leafy branches. Flowers blue, almost sessile in the axils of the leafy bracts, densely clothed with loosely stellate hairs and long ones plumose at the base. Calyx-lobes obsolete. Corolla about ½ in. long. Ovary 1-celled, with 1 erect narrow ovule.—DC. Prod. vii. 503 ; De Vr. Gooden. 93. t. 14.

Queensland. Shoalwater Bay, *R. Brown ;* dry ridges, Burnett river, *F. Mueller ;* Magnetic Island, *Burdekin Expedition,* and Port Denison, *Fitzalan ;* Rockingham Bay, *Dallachy.*

13. **D. Brownii,** *F. Muell. Fragm.* vi. 29. A tall shrub, more or less scabrous-pubescent hirsute or almost woolly with stellate hairs. Leaves petiolate, orbicular oval or rarely ovate-lanceolate, thick and often undulate or coarsely sinuate-toothed in the broad-leaved forms, usually quite entire in the oval-leaved varieties, usually from ½ to 1 in. long, but twice as much when very luxuriant. Peduncles solitary or clustered in the upper axils, sometimes much shorter than the leaves, sometimes especially in the oval-leaved varieties much elongated, and usually bearing an irregular cyme of 3 or more sessile flowers. Bracts minute. Flowers purple or blue, densely clothed with dark-coloured plumose hairs, sometimes very long and spreading, especially in the broad-leaved forms. Calyx-lobes minute and concealed under the indumentum or quite obsolete. Corolla usually about ½ in. long. Ovary 1-celled, with 1 erect narrow ovule. Fruit oblong, nearly 2 lines long, transversely rugose. Seed nearly terete.—*D. undulata, D. rotundifolia, D. ovalifolia,* and *D. purpurea,* R. Br. Prod. 587, 588 ; DC. Prod. vii. 503 ; De Vr. Gooden. 84 to 86, 93 ; *D. omissa,* De Vr. in Ned. Kruidk. Arch. ii. 10 ; *D. melanopogon,* De Vr. Gooden. 87 ; *D. nervosa,* De Vr. in Ned. Kruidk.

Arch. ii. 12, Gooden. 92 ; *D. bicolor*, De Vr. Gooden. 89, from the characters in Ned. Kruidk. Arch. ii. 11.

N. S. Wales. Port Jackson and the Blue Mountains, very abundant, *R. Brown* and many others; to the N.W. of Bathurst, *Fraser, A. Cunningham ;* New England, *C. Stuart.*

Victoria. Rocky mountains on the Macalister river at an elevation of 2000 to 3000 ft., *F. Mueller.*

R. Brown had already indicated that the four species he proposed were very closely connected with each other, and F. Mueller appears to have been quite right in uniting them. The broad-leaved forms, *D. undulata* and *D. rotundifolia*, run so closely into each other, in the numerous specimens before me, as to be inseparable even as varieties. *D. ovalifolia* and *D. purpurea*, also inseparable from each other, have smaller, more entire, more oval leaves, more narrowed at the base and sometimes almost obovate-cuneate, and generally a looser inflorescence and shorter indumentum on the flowers, especially in the southern specimens, but many specimens pass into the broad-leaved form. Both appear to be very abundant in the Blue Mountains. Sieber's n. 227 belongs to the oval-leaved variety. Some of the northern specimens have the leaves almost ovate-lanceolate.

14. **D. lanceolata,** *A. Cunn. ; DC. Prod.* vii. 503. An erect or diffuse undershrub or shrub nearly allied to *D. rosmarinifolia*, but approaching in some respects to *D. Brownii.* Branches usually sulcate, scabrous or stellate-hairy. Leaves elliptical oblong-linear or lanceolate, entire or with a few teeth, mostly larger and less revolute than in *D. rosmarinifolia*, much narrower than in *D. Brownii*, glabrous but scabrous above when full grown, slightly hoary-tomentose underneath. Peduncles in the upper axils solitary or clustered, mostly longer than the floral leaves, 1- or few-flowered. Bracts very small. Flowers clothed with a loose stellate tomentum intermixed with longer plumose hairs. Calyx-teeth very small, almost concealed in the indumentum. Corolla about ½ in. long. Ovary 1-celled, with 1 erect narrow ovule.—*D. Cunninghamii*, De Vr. Gooden. 91 ; *D. adpressa*, De Vr. Gooden. 100. t. 16. f. 1, not of A. Cunn.

N. S. Wales. Peel's Range and Wellington Valley near Croker's Range, *A. Cunningham ;* between the Upper Bogan and Lachlan rivers, *L. Morton.*

Victoria. Murray river, *Dallachy ;* near Lake Koorong, *Herb. F. Mueller.*

In the above-quoted plate, De Vr. Gooden. t. 16. f. 2, appears to represent a very different plant ; perhaps *D. glabrescens.*

15. **D. altissima,** *F. Muell. Herb.* A branching shrub of 4 to 6 ft. (*Oldfield*), more or less hoary or white with a loose stellate tomentum which disappears from the older leaves. Leaves mostly oblong-cuneate, ½ to 1 in. long, entire or slightly toothed, flat and rather soft, the lower ones somewhat larger obovate and petiolate, the upper ones small narrow and entire. Peduncles in the upper axils branching out into loose dichotomous cymes with opposite leafy bracts at the forks or degenerating into more normal leafy branches. Bracteoles linear. Flowers blue, rather large, covered with a densely intricate white stellate tomentum. Calyx-lobes small, almost concealed in the tomentum. Corolla 7 to 8 lines long. Ovary 1-celled, with 1 narrow ovule erect from the base.

W. Australia. White Peak, Murchison river, *Oldfield.* Like *D. marifolia*, this has at first sight some resemblance to *D. incana*, but with a very different ovary.

Var. ? *dura.* Stature lower. Leaves more rigid, mostly cuneate and often almost lobed at the end, the indumentum looser, mixed on the flowers with a few longer hairs plumose at the base only.—*Drummond, 5th Coll. n.* 71.

16. **D. marifolia,** *Benth.* A diffuse or erect much-branched shrub or undershrub, under 1 ft. high in all the specimens seen, hoary all over with a close stellate tomentum mixed with longer rather rigid more or less plumose hairs. Leaves sessile, ovate to oblong-elliptical, obtuse, entire, with slightly revolute margins, usually about ½ in. long and hoary on both sides, sometimes on luxuriant shoots nearly 1 in. long and glabrous above. Peduncles in the upper axils nearly as long as or longer than the leaves, 1- to 3-flowered, with linear leafy bracts or bracteoles. Flowers blue, stellate-tomentose outside. Calyx-lobes ovate or ovate-lanceolate, almost concealed in the indumentum, although nearly as long as the tube. Corolla 4 to 5 lines long. Ovary 1-celled, with 1 ovoid-oblong straight ovule erect from the base.

N. S. Wales. Confluence of the Murray and Murrumbidgee, *F. Mueller.*
Victoria. Wimmera, *Dallachy.*

These specimens are referred by F. Mueller, Fragm. ii. 17, to *D. incana,* of which they have the inflorescence and nearly the indumentum, but neither the foliage nor the ovule.

17. **D. rosmarinifolia,** *Schlecht. Linnæa,* xx. 603. An erect or rarely diffuse or prostrate undershrub or much-branched shrub of ½ to 1½ ft., the branches, underside of the leaves, and young shoots more or less hoary or white with a close stellate or intricate tomentum. Leaves sessile or nearly so, oblong or almost linear, very obtuse, the margins much revolute, thick, smooth and shining on the upper surface, mostly about ½ in. long and quite entire; the lower ones sometimes 1 in. long and occasionally broader, slightly angular and narrowed into a short petiole. Peduncles in the upper axils solitary or clustered, exceedingly short or rarely as long as the leaves. Bracts small, linear. Flowers blue (or white or red according to Behr), densely stellate-tomentose, with a few longer hairs. Calyx-lobes very small, almost concealed in the indumentum. Corolla about ½ in. long. Ovary 1-celled, with 1 narrow erect ovule.

Victoria. Murray desert and Wimmera, *Dallachy;* near Lake Koorong, *Herb. F. Mueller.*
S. Australia. From the Murray to St. Vincent's Gulf, *F. Mueller* and others; Lake Gillies, *Burkitt;* near Spencer's Gulf, *F. Mueller.*

Var. *dysantha.* Indumentum of the flowers much longer and looser, as in *D. lanceolata,* but the upper leaves much revolute and white underneath, as in *D. rosmarinifolia.*—Grampians, *Wilhelmi;* St. Vincent's Gulf, *F. Mueller.*

18. **D. lavandulacea,** *Lindl. Swan Riv. App.* 27. A much-branched erect or diffuse rigid perennial, not exceeding 1 ft. in any specimens seen, the young parts clothed with a white cotton which disappears at least from the upper side of the leaves. Leaves oblong lanceolate or almost linear, the larger ones ¾ in., but mostly ½ in. long or less, sessile or the lower ones shortly petiolate, coriaceous, with recurved or revolute margins, entire or obscurely toothed. Peduncles in the upper axils usually clustered, as long as or longer than the leaves, mostly 1-flowered. Bracteoles linear. Flowers blue, covered with short stellate or branched cottony hairs. Calyx-lobes short, almost concealed in the indumentum. Corolla under ½ in. long.

Ovary 1-celled, with 1 short oblong erect ovule.—*D. repens*, DC. Prod. vii. 503.

W. Australia. Swan River, *Drummond, 1st Coll. n.* 13, *also n.* 398.   Murchison river, sandy plains Kalgan river, Stirling Range, *Oldfield ;* Oldfield river, *Maxwell* (with shorter peduncles).   This species is sometimes very near *D. rosmarinifolia,* but the pedun-cles are generally longer, the bracteoles much more developed, the ovule shorter, etc.

D. Preissii, De Vr. in Pl. Preiss. i. 403, from York district, *Preiss, n.* 1431, which I have not seen, is probably a more tomentose-woolly variety of the same species.

19. **D. juncea,** *Benth.*   Herbaceous, glabrous except the flowers or the young shoots white-tomentose.   Stems terete or slightly angular, rush-like, erect and about 1 ft. high or longer and flexuose (or scandent ?).   Leaves none or very small and linear-terete or reduced to minute scales.   Flowers rather large, blue, solitary or clustered at the upper nodes, clothed with a short dense intricate plumose tomentum of a leaden colour, the short thick pedicels continuous with the calyx and almost as thick, with a minute scale-like bract about halfway up.   Calyx-lobes very short and broad, scarcely distinguishable in the dense indumentum.   Corolla nearly ¾ in. long.   Ovary 1-celled, with 1 linear ovule erect from the base.

W. Australia. Swan River, *Drummond, 1st Coll.,* also *5th Coll. n.* 168.   Plantage-net and Stirling Ranges, Dillon Bay, Cape Knobb, *Maxwell.*   Drummond's specimens were included by De Vriese in *D. parvifolia.*

20. **D. oligophylla,** *Benth.*   A perennial or undershrub, hoary when young with a close stellate tomentum, becoming glabrous with age.   Stems numerous, erect, slender but rigid, simple or branched, terete and sulcate, under 1 ft. high.   Leaves few and distant, oval or oblong, flat, coriaceous, entire, under ½ in. long and sometimes none of them half that size.   Pedun-cles mostly 1-flowered, solitary or 2 together in the upper axils, scarcely ex-ceeding the leaves.   Indumentum of the flowers stellate, lead-coloured, often mixed with longer slightly plumose hairs.   Calyx-lobes obsolete.   Corolla about ½ in. long.   Ovary 1-celled, with 1 oblong erect ovule.

W. Australia. *Drummond, 4th Coll. n.* 193 ; Gordon and Kalgan rivers, *Oldfield.*

21. **D. loranthifolia,** *F. Muell. Herb.*   Very closely allied to the narrow-leaved forms of *D. fasciculata,* with the same rigid angular stems, sessile rigid oblanceolate flat leaves and clustered peduncles, and similar flowers ex-cept that their indumentum consists of spreading plumose hairs, snow-white or somewhat brown, and that the calyx-teeth are quite obsolete.

W. Australia. Among rocks, Phillips river (with white indumentum on the flowers), and above Middle Mount Barren (with a brownish indumentum), *Maxwell.*

22. **D. stricta,** *R. Br. Prod.* 589.   A rigid perennial, nearly glabrous except the inflorescence, the young shoots rarely stellate-pubescent.   Stems broadly angular or compressed, erect or rarely decumbent.   Leaves sessile, the lower ones sometimes broadly obovate or cuneate, but mostly ob-long or linear, from ½ to 1½ in. long, flat, rigid, entire or coarsely angular-toothed.   Flowers blue, solitary or irregularly clustered in the upper axils, densely covered with appressed usually rust-coloured hairs apparently simple but really branched or stellate at the base, with divaricate and parallel

branches. Pedicels usually very short, the bracts and bracteoles rigid, linear or lanceolate. Calyx-lobes ovate or ovate-lanceolate, generally prominent, although covered with the hairs of the tube. Corolla rarely under ½ in. long and sometimes 7 or 8 lines, the persistent base very short. Ovary 1-celled, with 1 straight ovule erect from the base.—DC. Prod. vii. 504; Hook. f. Fl. Tasm. i. 230; De Vr. Gooden. 109 (but in all these only the Eastern plant); *D. fasciculata*, DC. Prod. vii. 504, De Vr. Gooden. 105 (as to the Port Jackson plant); *Goodenia stricta*, Sm. in Trans. Linn. Soc. ii. 349.

Queensland. Glasshouse Mountains, *F. Mueller*.

N. S. Wales. Port Jackson to the Blue Mountains, *R. Brown, Sieber, n.* 224, 226, and others; Hastings river, *Beckler*; Illawarra, *A. Cunningham*; Berrima, *M'Arthur*.

Victoria. Mount Macedon, Bunip Creek, Plenty Range, *F. Mueller*.

Tasmania. Flinders Island and Cape Barren Island, *Gunn*; South Esk river, *Strelecky*; towards George Bay, *Bissill*.

Var. *laxa*. Decumbent? Leaves broadly cuneate, coarsely toothed. Peduncles elongated.—To this belong the Victoria specimens.

Var.? *oblongata*. Leaves usually oblong, entire or .nearly so, 1 to 2 in. long. Indumentum of the flowers looser and darker.—*D. oblongata*, R. Br. Prod. 588; DC. Prod. vii. 504; De Vr. Gooden. 106, partly. To this form belong the Queensland specimens and those from Hastings river. I follow J. D. Hooker and F. Mueller in uniting it with *D. stricta*, but it has a somewhat different aspect. Possibly, however, the western *D. leptoclada*, *D. fasciculata*, and even *D. loranthifolia*, may prove to be varieties only of *D. stricta*, which it will be then very difficult to define.

23. **D. leptoclada,** *Benth.* A perennial, glabrous except the flowers, with decumbent, ascending or erect stems of 1 to 1½ ft., often broadly triangular in the lower part, the branches slender, elongated, less prominently angled. Leaves sessile, lanceolate or oblong, linear, obtuse, entire or with 1 or 2 prominent teeth on each side, thick and flat, the larger ones 1 to 2 in. long but mostly smaller, the floral ones narrow. Peduncles solitary or 2 or 3 together in the upper axils, rarely exceeding the leaves, rather slender, 1- to 3-flowered. Bracts minute. Flowers blue, clothed with lead-coloured or blackish, appressed, parallel-branched hairs. Calyx-lobes ¼ to ½ line long, but usually concealed under the indumentum. Corolla 5 to 7 lines long, the wings of the lobes broad. Ovary 1-celled, with 1 oblong-linear ovule, erect from the base.

W. Australia. King George's Sound and adjoining districts, *R. Brown, Oldfield, Maxwell, F. Mueller*; perhaps also Champion Bay, *Oldfield*. R. Brown referred his specimens doubtfully to *D. stricta*, of which this species may possibly be a Western form, but the habit is much weaker, the deep blue wings of the corolla-lobes broader, the indumentum of the flowers of a different colour, the bracts very minute, etc. Drummond's specimens, n. 161, may be the same plant in a more advanced state, with the peduncles longer than the leaves, but they are very imperfect.

Var. *parviflora*. Still more slender, with smaller flowers.—Cape Arid, *Maxwell (Herb. F. Muell.)*.

24. **D. fasciculata,** *R. Br. Prod.* 588. A rigid perennial or undershrub, nearly glabrous except the flower, or with more or less of stellate tomentum or clustered hairs, especially on the young shoots and about the inflorescence. Stems angular or compressed, erect or decumbent, not much branched, 2 to 3 ft. high (*Oldfield*). Leaves sessile, obovate, cuneate-oblong or rarely oblanceolate, entire or coarsely angular-toothed, thick, and coria-

ceous, mostly 1 to 2 in. long, the upper ones often irregularly opposite or
whorled. .Peduncles usually clustered in the upper axils, 1- or rarely 2-
flowered, shorter or scarcely longer than the leaves, mostly stellate-tomentose.
Flowers blue, clothed with appressed, parallel-branched hairs. Calyx-teeth
much smaller than in *D. stricta*, sometimes scarcely conspicuous, and usually
concealed under the indumentum. Corolla ½ in. long or rather more. Ovary
1-celled, with 1 straight ovule erect from the base.—DC. Prod. vii. 504 ;
De Vr. Gooden. 105 (both as to the Western plant only) ; *D. subverticillata*,
De Vr. in Pl. Preiss. i. 403 ; Gooden. 108.

W. Australia. King George's Sound and adjoining districts, *R. Brown, Baxter, Oldfield, Maxwell, Drummond, n.* 359, *Preiss, n.* 1510, and thence eastward to Cape Arid,
Maxwell.

Var. *angustifolia.* Leaves all lanceolate or the larger ones oblong-cuneate.—Cape Arid,
Maxwell.

25. **D. subspicata,** *Benth.* Glabrous except the flowers. Branches
erect, rigid, 3-angled. Leaves sessile, oblong-lanceolate, obtuse, entire,
thick, flat, coriaceous, 1 to 2 in. long, the floral ones much smaller. Flowers
blue, nearly sessile, solitary or 2 or 3 together, in the axils of the floral
leaves or leafy bracts, on pedicels shorter than the leaves, forming long, virgate, flowering branches or leafy, interrupted spikes. Bracts or bracteoles
minute. Calyx-tube narrow, 5-ribbed, glabrous, the lobes minute. Corolla about ½ in. long, clothed with appressed, parallel-branched hairs.
Ovary 1-celled, with 1 erect, linear ovule.

W. Australia. Near the base of Mount Bland, *Maxwell.* There are but very few
specimens, remarkable for their peculiar inflorescence and glabrous calyx, but possibly
further materials may show their connection with some one of the foregoing species.

26. **D. triloba,** *Lindl. in Swan Riv. App.* 27. A perennial or undershrub, with a thick, woody stock, and numerous ascending or erect, slightly
angular stems, not exceeding 1 ft. in any of our specimens, the whole plant
clothed with a stellate tomentum, either loose and floccose or dense and ferruginous, often disappearing from the old leaves and branches. Leaves
from broadly obovate to oblong-cuneate, angular-toothed or entire, coriaceous, the lower ones petiolate and often 2 in. long, the upper ones smaller,
contracted at the base, but sessile or nearly so, the floral ones sometimes
irregularly verticillate. Peduncles clustered in the upper axils, and usually
longer than the leaves, few-flowered. Bracts very small and narrow.
Flowers rather small. Calyx densely stellate-tomentose, the lobes obsolete.
Corolla about 4 lines long, clothed with appressed, parallel-branched hairs.
Ovary 1-celled, with 1 erect, linear ovule.—De Vr. Gooden. 80 ; *D. repanda*,
De Vr. in Pl. Preiss. i. 400 ; Gooden. 80 ; *D. Drummondi* and *D. hæmato-
tricha*, De Vr. Gooden. 82, 94.

W. Australia, *Drummond, 2nd Coll. n.* 105, *3rd Coll. Suppl. n.* 57 ; Swan River,
Preiss, n. 1518.

27. **D. linearis,** *R. Br. Prod.* 588. A rigid herb or undershrub, glabrous except the inflorescence or sprinkled with irregularly stellate hairs, the
stems erect or diffuse, scarcely angular, sometimes all under 1 ft., sometimes
exceeding 1½ ft. Leaves sessile or the lower ones contracted into a short

petiole, linear-oblong or spathulate, rarely obovate-oblong, entire or with a few teeth, coriaceous and rigid, $\frac{3}{4}$ to $1\frac{1}{2}$ in. long. Peduncles in the upper axils usually longer than the floral leaves and several-flowered, but sometimes short. Bracteoles linear. Flowers blue, covered with rather long hairs, usually loosely spreading but sometimes almost appressed and irregularly branched or stellate at the base. Calyx-lobes quite obsolete. Corolla about $\frac{1}{2}$ in. long. Ovary 1-celled, with 1 ovule erect from the base.—DC. Prod. vii. 504; De Vr. Gooden. 104; *D. azurea*, De Vr. in Pl. Preiss. i. 400; Gooden. 103; *D. eriophora*, De Vr. in Pl. Preiss. i. 400; *D. erecta*, De Vr. in Pl. Preiss. i. 401, according to De Vr. Gooden. 104.

W. Australia. King George's Sound, *R. Brown*, and thence to Swan and Vasse Rivers, *Drummond, 1st Coll. n. 17, Preiss, n.* 1475, 1500, *Huegel, Oldfield*, and others; Mount Manypeak, *Maxwell*.

The broad-leaved forms, to which this specific name is scarcely applicable, may still be distinguished from *D. cuneata* by their shape as well as by the larger, more loosely villous flowers, less rigid bracts, etc.

28. **D. cuneata,** *R. Br. Prod.* 588. A perennial, with erect and virgate or diffuse and more branching stems, sprinkled with short, stellate hairs, and a few longer ones soft and simple or shortly plumose at the base, the inflorescence more covered with long, soft hairs; branches angular. Leaves sessile, obovate or oblong-cuneate, entire or angular-toothed, coriaceous, nearly glabrous, mostly about $\frac{1}{2}$ in. long, but on luxuriant shoots twice that size. Peduncles in the upper axils longer than the leaves, bearing each an irregular cyme of 3 or more flowers, with narrow, rigid, leafy bracts of 3 to 4 lines, the lower peduncles sometimes replaced by slender, leafy branches, with flowers solitary in the axils. Calyx-lobes quite obsolete. Corolla 4 to 5 lines long, covered with long hairs, intermixed with smaller, stellate ones. Ovary 1-celled, with 1 erect, oblong ovule. Fruit oblong, 1 to $1\frac{1}{2}$ lines long. —DC. Prod. vii. 504; De Vr. Gooden. 102; *D. lanuginosa*, De Vr. Gooden. 81.

W. Australia. King George's Sound, *R. Brown* and many others, *Drummond, n.* 140 (*or* 40?) *and* 157 (*or* 127?). There are also many specimens without flowers of Drummond's n. 27, which may be this species.

29. **D. sericantha,** *F. Muell. Herb.* A small perennial, glabrous except the flowers. Stems slender but rigid, erect or ascending, branching, under 1 ft. high. Leaves sessile, but narrowed at the base, oblong-cuneate, entire or angular-toothed, mostly small but occasionally nearly 1 in. long, obtuse, thick, and flat. Flowers small, deep blue, 2 or 3 together on rather long peduncles in the upper axils. Bracts linear, rigid. Calyx-lobes obsolete. Corolla about 4 lines long, silky-white outside, with long, appressed hairs. Ovary 1-celled, with 1 erect, linear ovule.

W. Australia. Lucky Bay, *Maxwell.* This may possibly prove to be an anomalous form of *D. parvifolia*, but it is much less rigid, and the inflorescence is different.

30. **D. parvifolia,** *R. Br. Prod.* 589. Herbaceous, rigid, and glabrous when full grown except the flower, the stock and axils sometimes woolly, and the young shoots hoary-tomentose. Leaves thick, a few of the lower ones sometimes obovate or cuneate, slightly toothed at the end, and 1

or even 2 in. long, the others quite entire, small and narrow or a few of the larger ones 1 in. long. Flowers sessile, solitary or few together in clusters towards the ends of the branches, forming a leafy panicle. Bracts dry, rigid, acute, more or less imbricate, usually exceeding the calyx. Calyx-lobes obsolete. Corolla under ½ in. long, covered with simple or scarcely plumose, silky hairs. Ovary 1-celled, with 1 erect, narrow-oblong ovule.— DC. Prod. vii. 504; De Vr. Gooden. 110 (only as to Brown's plant).

W. Australia. Lucky Bay, *R. Brown,* and thence to Orleans Bay and Cape Paisley, *Maxwell.*

31. **D. glabrescens,** *Benth.* Probably an undershrub, closely resembling the Eastern *D. adpressa* in aspect and indumentum, but apparently more branched. Leaves sessile, the larger ones lanceolate, entire or minutely and remotely toothed, sometimes 3-nerved, 1½ to 2 in. long, those of the branchlets oblong, entire, about ½ in. long, all obtuse, flat, and quite smooth, loosely and almost silky tomentose when young, glabrous when full grown. Peduncles in the upper axils shorter than the leaves, 1- to 3-flowered. Bracts small, linear. Flowers covered with a white, close, stellate tomentum, mixed with a few, long, soft, almost simple hairs. Calyx-lobes quite obsolete. Ovary 1-celled, with 1 oblong, erect ovule.

W. Australia, *Drummond, 4th Coll. n.* 194; South Hutt and Murchison rivers, *Oldfield.* The plant figured by De Vriese, Gooden. t. 16. f. 2, as a second specimen may, perhaps, belong to this species, of some forms of which it is a fair representation.

32. **D. adpressa,** *A. Cunn.; DC. Prod.* vii. 503. An undershrub, with a thick stock, and several erect, simple or slightly-branched stems of 1 to 2 ft., clothed when young as well as the leaves with a close, white, almost floccose tomentum, which disappears from the adult leaves. Leaves very shortly petiolate, ovate or lanceolate, acute or rarely obtuse, entire or obscurely toothed, coriaceous, rather thick, smooth, and quite flat, ½ to 1 in. long. Peduncles axillary, rarely exceeding the leaves, 2- or 3-flowered. Bracts small, linear. Flowers rather small, clothed with a dense, stellate or woolly tomentum, mixed with long, almost simple hairs. Calyx-lobes oblong-linear, sometimes rather long, but apparently deciduous, and much concealed by the indumentum. Ovary 1-celled, with 1 straight ovule erect from the base.—*D. lanceolata,* De Vr. Gooden. 101. t. 17.

Queensland, *Mitchell;* Cape river, *Bowman;* Thermometer Creek, *Leichhardt.*
N. S. Wales. Croker's Range, *A. Cunningham.*

By some mistake, De Vriese has interchanged the names of *D. lanceolata* and *D. adpressa,* although he had Cunningham's named specimens before him.

33. **D. diversifolia,** *De Vr. in Pl. Preiss.* i. 403; *Gooden.* 117. A prostrate perennial or undershrub, extending sometimes to a considerable breadth, with numerous short, dense, leafy branches, quite glabrous in every part. Radical leaves oblong-spathulate or oblanceolate, 1 to 1½ in. long, all the others lanceolate, oblanceolate or linear, rarely above ½ in. long, coriaceous, acute, entire or minutely toothed, often concave like those of *Epacrideæ.* Flowers blue, differing from all other species in being quite glabrous outside, on short, axillary peduncles or branchlets, with 1 or 2 leaves and a pair of bracteoles close under the flower. Calyx-lobes ovate-triangular, shorter than

the persistent base of the corolla. Corolla about ½ in. long. Ovary 1-celled, with 1 straight ovule erect from the base.—*Scævola prostrata,* De Vr. in Pl. Preiss. i. 406.

W. Australia, *Drummond, n.* 358, *3rd Coll. n.* 160 ; Mount Barker, Gordon and Kalgan rivers, Stirling Range, *Oldfield ;* Gardiner river, *Maxwell.*

Sect. 5. Cephalantha.—Leaves radical. Scapes simple or branched. Flowers in dense, terminal heads. Ovary 1-celled, with 1 linear-oblong, straight ovule erect from the base.

34. **D. eriocephala,** *De Vr. Gooden.* 118. *t.* 21. A perennial, with a thick, tufted stock. Leaves radical, petiolate, obovate oval or oblong, obtuse, entire or sinuate-toothed, rather thick, often several inches long, glabrous above, clothed underneath in the typical form with a close, white tomentum. Scapes erect, woolly-tomentose, 1 to 2 ft. high, leafless, and simple below the flower-head or more or less corymbosely branched, with sessile, oblong or linear leaves or bracts subtending the branches. Flowers blue, in dense, villous heads rarely growing out into short spikes, at the ends of the stem or branches. Bracts lanceolate, ciliate, otherwise glabrous or silky-villous. Calyx-lobes small, and almost concealed in the indumentum. Corolla 6 to 8 lines long, clothed with long, silky, appressed or spreading hairs. Ovary 1-celled, with 1 erect ovule.

W. Australia, *Drummond, 5th Coll. n.* 69, 70 ; rocks, Mongerup, Stirling Range, *Maxwell.* The simple-scaped specimens have very much the aspect of *Brunonia.*

Var. ? *concolor,* F. Muell. Leaves thick, glabrous on both sides. Flowers much smaller ; the corolla 4 to 5 lines long, with narrow lobes.—*Drummond, 4th Coll. n.* 162.

12. BRUNONIA, Sm.

Calyx-tube free but contracted over the ovary ; lobes 5. Corolla nearly regular, inserted at the base of the calyx-tube, the tube cylindrical ; lobes 5, valvate, spreading, the 2 upper ones separated rather lower down. Stamens 5, inserted at the base of the corolla-tube, the filaments cohering upwards, the anthers cohering in a ring round the style. Ovary free, but enclosed in the calyx-tube, 1-celled, with a single erect anatropous ovule. Style simple ; stigma shortly 2-lobed, enclosed in a cup-shaped indusium. Fruit a small nut enclosed in the hardened calyx-tube. Seed erect, without albumen. Embryo straight, cotyledons ovate, radicle short, inferior.—Silky-hairy perennial. Leaves radical. Flowers in a dense head, intermixed with bracts, on a leafless scape.

The genus is limited to a single species endemic in Australia. R. Brown appears to me to have been quite right in including it in *Goodenovieæ,* of which it has the remarkable indusium. It has since been raised into an independent Order, on account of the free ovary, regular flowers, and exalbuminous seed, and has even been removed far away from *Goodenovieæ* to the neighbourhood of *Plumbagineæ.* The ovary and fruit are, however, so completely enclosed in the constricted calyx-tube as to be really less free than in *Lobelia xalapensis,* the exceptionally regular flowers are but little more so than in *Diaspasis,* and in some species of *Scævola,* especially *S. spinescens,* I have found the albumen much reduced. The habit of *Brunonia* is also so little different from that of *Dampiera eriocephala,* that I have seen the latter placed in covers of *Brunonia* as a new species.

1. **B. australis,** *Sm. in Trans. Linn. Soc.* x. 367. *t.* 28.　A tufted perennial, clothed in every part with long silky hairs closely appressed in some specimens, more frequently more or less spreading.　Leaves radical, from obovate to linear-cuneate, quite entire, softly mucronate, contracted into a petiole, mostly 2 to 4 in. long.　Scapes 6 in. to above 1 ft. high, bearing a dense globular or hemispherical flower-head of ½ to ¾ in. diameter.　Flowers numerous, sessile, intermixed with bracts of which a few of the outer ones are broad and leafy though not longer than the flowers, forming a kind of involucre, the inner ones small and narrow, ciliate with long hairs ; there are also close around each flower 3 or 4 concave truncate but jagged and ciliate bracts of which at least the 2 innermost are scarious and transparent. Calyx-tube very short, the lobes 1½ to 2 lines long, plumose-ciliate, almost always tipped with a glabrous pedicellate gland.　Corolla blue　the tube linear, hirsute, shorter than the calyx-lobes, the lobes oblong, glabrous, about as long as the tube.　Fruit small.—R. Br. Prod. 590 ; A. DC. Prod. xii. 616 ; Hook. f. Fl. Tasm. i. 229 ; Bot. Reg. t. 1833 ; Paxt. Mag. vii. 267, with a fig. ; *B. sericea,* Sm. in Trans. Linn. Soc. x. 367. *t.* 29 ; R. Br. Prod. 590 ; A. DC. Prod. xii. 616 ; *B. simplex,* Lindl. in Mitch. Trop. Austr. 82.

Queensland.　Shoalwater Bay, *R. Brown ;* Keppel Islands, *M'Gillivray ;* Rockhampton, *Thozet ;* Darling Downs, *H. Law ;* Mount Pluto, *Mitchell.*

N. S. Wales.　Peel's Range and Lachlan river, *A. Cunningham ;* between the Upper Bogan and Lachlan rivers, *L. Morton.*

Victoria.　Port Phillip, *R. Brown ;* dry pastures from Melbourne to the mouth of the Glenelg and towards Lake Hindmarsh, *F. Mueller, Adamson,* and others.

Tasmania.　Port Dalrymple, *R. Brown ;* common in dry pastures in several parts of the colony, but local, *J. D. Hooker.*

S. Australia.　Bugle Range, Holdfast Bay, Black Forest, *F. Mueller ;* St. Vincent's Gulf, *Blandowsky.*

W. Australia.　Swan River, *Drummond, 1st Coll. n.* 417 ; Murchison river and Champion Bay, *Oldfield.*

The two forms commonly distinguished as species pass into one another very gradually. Where the indumentum is more silky and shorter, the glabrous tips of the calyx-lobes are prominent, and these specimens have usually smaller flower-heads.　Where the hairs are longer the tips are concealed amongst them, and perhaps sometimes, but very rarely, disappear altogether ; the former state is most common in the northern districts, the latter in the southern ones, but intermediate ones are also very frequent.

ORDER LXV. **CAMPANULACEÆ.**

Calyx-tube adnate to the ovary, the limb of 3 to 10, usually 5, persistent lobes.　Corolla regular or irregular, with 3 to 10, usually 5, lobes, valvate in the bud, the margins often induplicate.　Stamens as many as the lobes of the corolla, alternate with them, inserted at the base of the corolla-tube, but free from or very rarely more or less adnate to it.　Anthers opening longitudinally, free or united in a ring round the style.　Ovary inferior or rarely semisuperior or free except the broad base, 2- or more-celled, with numerous ovules in each cell.　Style simple, entire or divided at the top into as many stigmatic lobes or branches as there are cells to the ovary.　Fruit usually a capsule, opening either in short valves at the top or in lateral pores or slits, rarely an indehiscent berry.　Seeds numerous, small.　Embryo straight, often very small, in a fleshy albumen.—Herbs or very rarely shrubs, with a

juice usually milky.   Leaves alternate or very rarely opposite, usually undivided and toothed, rarely deeply pinnatifid, without stipules.   Flowers hermaphrodite or very rarely unisexual, either axillary solitary or clustered or in
terminal spikes racemes or leafy panicles.   Corolla frequently blue or white,
more rarely purple or red, very rarely yellow.

A considerable Order, most abundant in the temperate regions of the northern hemisphere and in S. Africa, but extending also over the tropics both in the New and the Old
World.   The limits of the genera are as yet very unsatisfactorily determined.   Of the four
Australian ones as at present constituted, two extend over nearly the whole range of the
Order, a third, *Pratia*, is limited to the extratropical regions of the southern hemisphere,
unless it be deemed to include the tropical Asiatic *Piddingtonia*, the fourth, *Isotoma*, is
either strictly Australian or may be extended to include a very few South African as well as
European species.

Corolla usually irregular.   Anthers united round the style.   Ovary
    2-celled.   (**Lobelieæ.**)
    Corolla-tube slit open to the base, the limb very irregular, 2-
      lipped.
      Capsule opening between the calyx-lobes in 2 loculicidal valves .   1. Lobelia.
      Fruit, often succulent, either indehiscent or bursting irregularly
        below the calyx-lobes . . . .   2. Pratia.
    Corolla-tube entire or very shortly slit, the limb spreading, nearly
      regular or oblique . . . .   3. Isotoma.
Corolla regular, campanulate.   Anthers free.   Ovary 3- to 5-celled .   4. Wahlenbergia.

1. LOBELIA, Linn.

(Rapuntium *and* Grammatotheca, *Presl ;* Holostigma, *Don.*)

Calyx-tube hemispherical, turbinate ovoid oblong or rarely linear, limb of
5 lobes, open or reduplicate-valvate in the bud.   Corolla slit open on the
upper side to the base, 5-lobed, the 2 upper lobes usually shorter, more deeply
separated and erect or curved upwards, forming a more or less distinct upper
lip, the 3 lower spreading in a lower 3-lobed lip.   Stamens inserted at the
base of the corolla, sometimes very shortly adnate to it, the filaments often
united above the middle; anthers united in an oblique or slightly incurved
ring round the style.   Ovary 2-celled.   Stigma broadly 2-lobed and often
surrounded by a ring of retractile hairs.   Capsule opening loculicidally
within the calyx-lobes in 2 valves, rarely splitting also longitudinally below
the calyx-lobes when old.—Herbs, often acrid with a milky juice, the Australian ones either annual or creeping and rooting at the base.   Pedicels 1-
flowered, either axillary or terminal or in terminal racemes, sometimes bearing
2 small bracteoles, which are never constant in the same species.   Flowers
in a few species diœcious by the abortion or sterility of the anthers in the
females, and the imperfection of the undivided stigma and abortion of the
ovules in the males.

The genus is numerous in species and widely spread over the greater part of the globe,
but chiefly abundant in North America, South Africa, and Australia.   There are also several
species within the tropics both in the Old and in the New World, but none in Northern Asia ;
and in Europe the few species known are strictly western or Mediterranean.   Of the 18
Australian species two are also in S. Africa, one of them, a maritime one, extending to New
Zealand and extratropical South America, the remaining 16 are all endemic.

Sect. I. **Holopogon.**—*All the anthers bearded (tipped with a tuft of short rigid* *hairs or bristles).*   *Flowers terminal or in terminal racemes (in* L. Bergiana *the lower ones axillary).*

Annuals either erect or the lateral stems shortly decumbent at the base.
  Flowers in a one-sided terminal raceme.   Capsule gibbous.
    Lower leaves mostly pinnatifid.   Middle lower lobe of the corolla broad.   Seeds 3-winged . . . . 1. *L. heterophylla.*
    Leaves linear, entire or rarely toothed.   Middle lower lobe of the corolla narrow.   Seeds not winged . . . . . . 2. *L. gibbosa.*
    Lower leaves ovate, cut.   Racemes very loose.   Seeds small, not winged.
      Middle lower lobe of the corolla narrow . . . . . . 3. *L. dentata.*
      Middle lower lobe of the corolla broad . . . . . . 4. *L. gracilis.*
  Flowers terminating the branches or long branch-like peduncles.
    Capsule twice as long as broad, rather oblique.   Leaves narrow, the lower ones usually pinnatifid.
      Seeds tubercular-rugose on the back, smooth but not shining in front with a prominent rib . . . . 5. *L. rhytidosperma.*
      Seeds compressed, very smooth and shining . . . . . 6. *L. tenuior.*
    Capsule scarcely longer than broad, very gibbous.   Seeds very small.
      Lower leaves ovate or obovate, cut . . . . 7. *L. rhombifolia.*
      Leaves all small linear and entire or nearly so . . . . 8. *L. parvifolia.*
Stem creeping or procumbent at the base; branches ascending.
  Leaves all petiolate, cordate.   Flowers in a loose terminal leafless raceme.   Capsule globular . . . . 9. *L. trigonocaulis.*
  Leaves scarcely petiolate, obovate lanceolate or linear.   Flowers axillary or in a terminal raceme leafy at the base.   Capsule linear . . . . 10. *L. Bergiana.*

(In 11. *L. anceps* and a few others of the following section there are short rigid hairs at the back of the upper anthers near the top, but not forming a terminal tuft.)

SECT. II. **Hemipogon.**—*Two lower anthers tipped with tufts of short bristles or with single bristles or points, three upper ones without any.   Flowers solitary on axillary pedicels, the uppermost rarely forming a leafy raceme.*

Flowers hermaphrodite (anthers, stigma, and ovules, all perfect).
  Lower leaves cuneate or obovate, rather thick, entire or obscurely toothed.   Pedicels rarely much longer than the leaves.
    Stems ascending, upper leaves narrow.   Pedicels short.   Capsule oblong or linear . . . . 11. *L. anceps.*
    Stems short, creeping.   Pedicels longer than the obovoid capsule . . . . 12. *L. surrepens.*
  Leaves toothed.   Pedicels slender, much longer than the leaves.
    Leaves ovate or orbicular, mostly petiolate, thin . . . . 13. *L. membranacea.*
    Leaves sessile, linear or linear-lanceolate . . . . 14. *L. stenophylla.*
    Leaves sessile, ovate or orbicular, prominently toothed . . 15. *L. quadrangularis.*
Flowers more or less diœcious, the males with an entire stigma and very short ovary with abortive ovules ; females with rudimentary stamens or abortive anthers.
  Slender annual.   Stamens in the female rudimentary . . . 16. *L. dioica.*
  Perennials, decumbent or creeping at the base.   Anthers in the female without pollen or quite abortive.
    Glabrous.   Leaves oblong or lanceolate, toothed.   Pedicels long . . . . 17. *L. purpurascens.*
    Pubescent.   Leaves linear or oblong, entire or toothed.   Pedicels shorter or not much longer than the leaves . . . 18. *L. pratioides.*

L. erinus, Linn., A. DC. Prod. vii. 370, a decumbent plant with something of the habit of *L. anceps*, but the stems not winged, usually hirsute at the base, and much larger blue flowers on longer pedicels, a S. African species much cu tivated for ornament, has established itself in the vicinity of gardens about Melbourne (*F. Mueller*).

L. longiscapa, De Vr. in Pl. Preiss. i. 398, from the interior of S.W. Australia, Preiss, n. 1435, is unknown to me, but, from the character given, it seems very doubtful whether it belongs to the genus.

L. dubia, De Vr. in Pl. Preiss. ii. 242, from Swan River, *Preiss, n.* 1440, is still more doubtful and not described so as to be capable of identification.

SECT. I. HOLOPOGON.—All the anthers bearded, that is, tipped with a tuft of short, rigid hairs or bristles. Annuals (except the last two species), either erect or the lateral branches shortly decumbent. Flowers either terminal and solitary or in terminal racemes, or in *L. Bergiana* the lower ones axillary.

1. **L. heterophylla**, *Labill. Pl. Nov. Holl.* i. 52. *t.* 74. An erect annual, simple or slightly branched, more or less pubescent or hirsute with short hairs, rarely almost glabrous, ⅓ to 1½ ft. high. Lower leaves usually pinnatifid, with few, narrow-linear lobes or the radical ones sometimes obovate and deeply cut, the upper ones small, linear and entire. Flowers rather large, in a loose, one-sided raceme, the pedicels mostly longer than the calyx-tube, with 1 or 2 linear bracteoles at their base, which, however, are sometimes wanting. Calyx-lobes almost subulate, longer than the tube. Corolla ¾ to nearly 1 in. long, of a deep blue, the middle lower lobe very broadly obovate or obcordate, the lateral ones shorter and oblong or obovate, the upper ones still shorter and narrower. Anthers all tipped with a dense tuft of short bristles. Capsule broadly obovoid, usually about 3 lines long, oblique, but less so than in *L. gibbosa.* Seeds much longer than in *L. gibbosa*, the 3 angles bordered by scarious, transparent wings.—R. Br. Prod. 564; A. DC. Prod. vii. 359; Bot. Reg. t. 2014; Paxt. Mag. ix. 101, with a figure (a weak, elongated, almost twining form).

W. Australia. Swan River, *Huegel, Drummond*, 1*st Coll.* (*n.* 420?), *Oldfield;* Murchison river and Gordon river, *Oldfield;* eastward from King George's Sound, *R. Brown,* to Cape le Grand, *Maxwell.* The Tasmanian station given by A. De Candolle is taken from the 'Botanical Register,' where, through a nurseryman's mistake, the plant is said to have been raised from Tasmanian seeds. Drummond's very imperfect specimens, n. 184, appear to be an almost glabrous form, with oblong-linear, obtuse, almost entire leaves. In the species generally, the form and proportion of the lobes of the corolla and the hairs of the upper ones appear to be somewhat variable.

2. **L. gibbosa**, *Labill. Pl. Nov. Holl.* i. 50. *t.* 71. An erect, glabrous annual, simple or with a few, erect branches, ½ to 1½ ft. high. Leaves not numerous, linear, entire or with a few small teeth or rarely the lower ones broader and deeply toothed. Flowers of a deep blue, in a terminal, one-sided raceme, the pedicels usually short, between 2 short, linear bracteoles, the subtending bract often wanting, and sometimes the bracteoles also. Calyx-lobes very narrow, about as long as the short, broad tube. Corolla variable in size, often above ½ in. long, the 3 lower lobes oblong, obtuse or acute, the 2 upper ones shorter, incurved, acute, glabrous or with a few long hairs. Anthers all tipped with a tuft of short bristles. Capsule from

¼ to nearly ½ in. long, broadly and very obliquely obovate, gibbous on the upper side. Seeds very numerous, variable in size, always much smaller than in *L. heterophylla*, with 3 more or less prominent angles, but not winged.— R. Br. Prod. 564; A. DC. Prod. vii. 358; Hook. f. Fl. Tasm. i. 238; *L. simplicicaulis*, R. Br. l. c.; A. DC. l. c.; *L. stricta*, R. Br. l. c.; *L. Browniana*, Rœm. et Schult. Syst. v. 71; A. DC. Prod. vii. 359.

Queensland. Shoalwater Bay, *R. Brown ;* Burdekin Expedition, *Fitzalan ;* Port Denison, *Bowman ;* Moreton Island, *F. Mueller.*

N. S. Wales. Port Jackson to the Blue Mountains, *R. Brown, A. Cunningham, Woolls ;* New England, *C. Stuart ;* Hastings river, *Beckler.*

Victoria. Common in dry pastures from the western limits to Wilson's Promontory, Dandenong Ranges, etc., *F. Mueller* and others.

Tasmania. Mount Wellington and Port Dalrymple, *R. Brown ;* abundant in light, sandy soil, *J. D. Hooker.*

S. Australia. Memory Cove, *R. Brown* ; Lofty Ranges, *F. Mueller ;* Boston Point, Port Lincoln, etc., *Wilhelmi ;* Kangaroo Island, *Waterhouse.*

W. Australia, *Drummond, n.* 177 ; Murchison river, *Oldfield ;* Swan River, *Fraser* and others ; Salt River, *Maxwell.*

The differences in the hairiness of the upper lobes of the corolla, as already observed by J. D. Hooker, appear far too variable to serve for specific characters. The Northern specimens are generally more slender, with smaller flowers, than the Southern ones, but I can find no other difference.

3. **L. dentata,** *Cav. Ic.* vi. 14. *t.* 522. Very nearly allied to, and probably a variety of *L. gibbosa*, but weaker, and not so erect, sometimes very scrambling and almost twining, approaching in foliage and inflorescence the *L. gracilis*. Lower leaves small, ovate, deeply cut, the others few, narrow, pinnatifid or toothed. Raceme very loose, the flowers fewer, on longer pedicels than in *L. gibbosa*, but as large as in that species, the middle lower lobe oblong, not obovate as in *L. gracilis*. Capsule obovoid, broad, very oblique, and gibbous on the upper side. Seeds small, ovoid or 3-angled, smooth.—R. Br. Prod. 564; A. DC. Prod. vii. 364.

N. S. Wales. Port Jackson to the Blue Mountains, *R. Brown* and others ; Clarence river, *Beckler.*

4. **L. gracilis,** *Andr. Bot. Rep. t.* 340. A glabrous annual, erect, or branching and shortly decumbent at the base, more slender than *L. gibbosa*, and usually under 1 ft. high. Leaves small, the lowest ones ovate and deeply cut, the others lanceolate and pinnatifid, or linear and toothed or entire, all narrowed at the base. Flowers rather smaller than in *L. gibbosa,* on long pedicels, in a very loose, unilateral raceme, the lower subtending bracts often more or less leaf-like. Calyx of *L. gibbosa*. Corolla with the middle lower lobe obovate, the lateral ones oblong or obovate-oblong, the 2 upper ones incurved, hairy or glabrous. Capsule rarely 2 lines diameter, as broad as long, very oblique, gibbous on the upper side. Seeds very small.—R. Br. Prod. 563 ; A. DC. Prod. vii. 364 ; Bot. Mag. t. 741 ; *L. dentata*, Sieb. Pl. Exs. not of Cav.

N. S. Wales. Port Jackson to the Blue Mountains, *R. Brown, Sieber, n.* 179, and others.

Var. *major.* Larger and more luxuriant, with larger, deeply-toothed leaves, the flowers larger, but with the broad lower corolla-lobe of *L. gracilis.—L. trigonocaulis,* Hook. Bot. Mag. t. 5088, not of F. Muell.—Crevices of rocks, Mount Lindsay, *W. Hill.*

5. **L. ᵣhytidosperma,** *Benth.* An erect annual, simple or slightly branched, glabrous or very sparingly and minutely pubescent, ½ to 1 ft. high. Radical leaves obovate, toothed, lower stem ones pinnatifid, upper ones small, linear, and scarcely toothed. Flowers large, singly terminating the stems and long branch-like peduncles. Calyx-tube narrow, the lobes often 3 to 4 lines long, but variable. Corolla nearly of *L. heterophylla,* the middle lower lobe broadly obovate, the lateral ones smaller, the 2 upper smaller, incurved, and usually hairy. Capsule slightly oblique but not gibbous, when full grown 6 to 8 lines long and 2 to 2½ lines broad, tapering at the base. Seeds small, ovate, convex and prominently tubercular-rugose on the back, smooth but opaque and with a prominent rib on the inner face.—*L. simplicicaulis,* Benth. in Hueg. Enum. 74, not of R. Br.

W. Australia. Swan River, *Huegel, Drummond, 1st Coll. n.* 419, 421. Very near the more simple, erect forms of *L. tenuior,* but the flowers are larger, and the seeds very different.

6. **L. tenuior,** *R. Br. Prod.* 564. An annual, more or less pubescent or hirsute, with short hairs or rarely nearly glabrous, branching and slightly decumbent at the base, with erect or ascending stems, often 1 ft. high or more, each with a single terminal flower or branching into few, long, 1-flowered peduncles. Radical leaves usually small, obovate, and deeply toothed; stem leaves linear, the lower ones pinnatifid, the upper small, linear, and entire or toothed. Flowers large, like those of *L. heterophylla,* but the calyx-tube narrow. Middle lower lobe of the corolla broadly obovate, the lateral ones also obovate, the 2 upper much smaller, incurved. Capsule when full grown 6 to 8 lines long and 2 to 2½ lines broad, tapering at the base, scarcely oblique. Seed small, compressed, very smooth and shining, without prominent angles.—*L. ramosa,* Benth. in Maund. Botanist, ii. t. 93 ; A. DC. Prod. vii. 359 ; *L. longepedunculata,* De Vr. in Pl. Preiss. i. 394, erroneously referred to *L. simplicicaulis* in Pl. Preiss. ii. 242 ; *L. adscendens,* De Vr. in Pl. Preiss. i. 395 ; *L. heterophylla,* Hook. Bot. Mag. t. 3784, not of Labill.; Paxt. Mag. vi. 197, with a figure.

W. Australia. King George's Sound, *R. Brown, F. Mueller ;* Swan River, *Huegel, Drummond, 1st Coll.,* also *Suppl. n.* 4, *Preiss, n.* 1425, 1452. *L. ciliata,* De Vr. in Pl. Preiss. i. 397, from the same locality, appears to be the same species, but the seeds are not yet sufficiently formed to be certain. I find I had correctly identified this species in describing Huegel's plants at Vienna, but afterwards, owing to a misprint in Nees' edition of Brown's ' Prodromus,' where (T) is given for this plant instead of (M), I fancied I must have been wrong in identifying a south-western with a tropical species, and described it as new under the name of *L. ramosa.*

7. **L. rhombifolia,** *De Vr. in Pl. Preiss.* i. 397. A glabrous annual, branching and decumbent at the base, with usually numerous, erect or ascending branches, from 3 or 4 in. to nearly 1 ft. high. Lower leaves obovate or cuneate, and deeply cut, ½ to ¾ in. long, the others small, few, lanceolate or almost linear, mostly with a few deep teeth or lobes. Flowers resembling those of *L. gracilis,* but on long, slender peduncles, terminating the stems and branches. Calyx-lobes very narrow, longer than the tube. Corolla about ½ in. long or rather more, the middle lobe narrow-obovate. Anthers all tipped with tufts of short bristles. Capsule very obliquely

obconical, and gibbous on the upper side, about 3 lines long.   Seeds very small and numerous, smooth, but opaque.

S. Australia. Onkaparinga and Encounter Bay, *F. Mueller;* Mount Jagged and Mount Barker, *Whittaker.*

W. Australia. King George's Sound, *Wakefield;* Kalgan and Vasse rivers, *Oldfield;* Albany, Upper Kalgan, and Hay rivers, *F. Mueller;* Swan River, *Drummond, 1st Coll. n.* 422.

F. Mueller believes this to be a variety of *L. parvifolia,* of which it has many characters, but the foliage is more that of *L. gracilis.* I refer it to De Vriese's *L. rhombifolia* from the diagnosis given, for I have not seen Preiss's specimens, n. 1439, from Mount Elphinstone and Swan River.

8. **L. parvifolia,** *R. Br. Prod.* 564.   An erect, glabrous annual, ½ to 1 ft. high or rarely more, simple or slightly branched.   Leaves few and very small, lanceolate or linear, sessile or slightly decurrent, entire or scarcely toothed.   Flowers few, rather large, terminal or on long, branch-like peduncles, forming sometimes a loose panicle, but not racemose.   Calyx-lobes narrow, longer than the short tube.   Corolla 6 to 8 lines long, the lower middle lobe obovate, the lateral ones narrower, the upper ones smaller and incurved.   Anthers all tipped with tufts of short bristles.   Capsule about 3 lines long, very obliquely obovoid, and gibbous on the upper side.   Seeds very small, compressed, smooth, but opaque.—A. DC. Prod. vii. 359.

W. Australia. King George's Sound and adjoining districts, *R. Brown Drummond,* n. 58, *Oldfield;* eastward to Cape le Grand, *Maxwell.*

9. **L. trigonocaulis,** *F. Muell. Fragm.* i. 18.   Nearly glabrous or sprinkled with a few, small hairs.   Stems (from a perennial rootstock?) creeping at the base, weak, and ascending to 1 ft. or more, somewhat 3-angled.   Leaves all on rather long petioles, ovate-cordate or the lower ones almost orbicular, coarsely toothed or crenate, ½ to above 1 in. diameter.   Flowers few and distant, in a terminal raceme, on short pedicels, in the axils of very small, linear bracts.   Calyx-tube broadly turbinate, the lobes narrow and rather longer.   Corolla fully ½ in. long, the lower lobes oblong, rather acute, the 2 upper ones shorter and incurved.   Filaments hairy at the base ; anthers all tipped with tufts of short bristles.   Capsule almost globular or broader than long, scarcely oblique.   Seeds ovoid, rather large, minutely foveolate.

Queensland. Shady forests, Brisbane river, Moreton Bay, *F. Mueller.*
N. S. Wales. Hastings and Macleay rivers, *Beckler;* New England, *C. Stuart.*

10. **L. Bergiana,** *Cham. in Linnæa,* viii. 217.   A glabrous perennial, with a slender, creeping rootstock (or sometimes annual?), the stems procumbent ascending or erect, often 2 to 3 feet long, slightly 3-angular or compressed.   Lower leaves often petiolate, obovate or cuneate, the others sessile or nearly so, oblong lanceolate or linear, entire or slightly-toothed, passing gradually into the narrower floral leaves, the uppermost reduced to small bracts.   Flowers sessile in the axils of the floral leaves or bracts, between 2 linear bracteoles.   Calyx-tube linear, the lobes narrow, acute, serrulate or ciliate.   Corolla about 4 lines long, the 3 lower lobes united in a broad, 3-lobed lip, the 2 upper ones narrower, curved, and ascending.

Anthers all tipped with a tuft of short bristles, otherwise glabrous. Capsule linear, ¾ to 1 in. long, the conical summit opening loculicidally in 2 valves, as in other *Lobelias,* the adnate tube usually splitting when old into 2 valves parallel to the dissepiment, but remaining entire at the top and the base.— *Grammatotheca* (the whole genus), Presl, Prod. Mon. Lob. 43 ; A. DC. Prod. vii. 348 ; *G. erinoides,* Sond. Fl. Cap. iii. 532, with the synonyms adduced ; *G. Dregeana,* Presl, l. c. ; A. DC. Prod. vii. 348, 784 ; Deless. Ic. Sel. v. t. 6 ; F. Muell. Fragm. iv. 171 ; *Lobelia macrocarpa,* De Vr. in Pl. Preiss. i. 396 ; *L. amplexicaulis,* De Vr. l. c. 397 ; *L. stenotheca,* F. Muell. Fragm. ii. 20.

W. Australia, *Drummond, n.* 43, 156, 179; wet, swampy places or growing in water, *Gordon ;* Toue, Moore, and Murchison rivers, *Oldfield ;* borders of Lake Keiermula and Sussex district, *Preiss, n.* 1453, 1443. The species is also common in S. Africa.

The proposed genus *Grammatotheca* (of which Sonder is quite correct in reducing the supposed species to a single one) was founded on the capsule said to be triquetrous, 1-celled, with parietal placentas, and opening laterally in 3 valves, of which 2 bear the placeutas ; an extraordinary structure, copied from Presl by A. De Candolle and by Sonder without verification, and for which I cannot trace the slightest foundation. Probably in a very careless, superficial observation Presl mistook the dissepiment for a third valve, and imagined the rest from Chamisso's having associated the plant with *Clintonia* on account of the linear capsule. The true structure, already alluded to by R. Brown (Prod. p. 562) is very well represented in the above-quoted 'Icones' of Delessert. This splitting of the capsule can scarcely be called a dehiscence, which is on the summit as in other *Lobelias,* and the habit is too near that of *L. anceps,* in which the capsule is almost linear, to admit of its generic separation. I have not taken up the specific name *erinoides* as being that given by Thunberg, for, although this plant may be so named in Thunberg's herbarium, he evidently had chiefly in view Linnæus's *L. erinoides,* as he quotes the synonym from Willdenow, where the capsule is described as obovate.

SECT. II. HEMIPOGON.—Two lower anthers tipped with tufts of short bristles or with single bristles or points, sometimes very minute, 3 upper ones without terminal bristles, but sometimes with short hairs on the back. Stems usually prostrate or creeping, or the branches ascending or erect from a decumbent base. Flowers solitary, on axillary pedicels, the uppermost rarely forming a leafy raceme.

11. **L. anceps,** *Thunb. ; DC. Prod.* vii. 375. A glabrous perennial, the rootstock often shortly creeping ; stems decumbent, ascending or erect, from a few in. to 1½ ft. long, angular or more or less winged by the decurrent leaves. Lower leaves petiolate, obovate or cuneate, sometimes 2 in. long but usually smaller, the larger stem-leaves oblong-spathulate, lanceolate or almost linear, passing gradually into narrow, sessile floral leaves, the uppermost reduced to very small bracts, all entire or sparingly toothed. Flowers small, on very small, axillary pedicels, the upper ones forming a slender raceme. Calyx-lobes broadly lanceolate, acute, not half so long as the linear-cuneate tube. Corolla blue or almost white, the lower lobes forming a broadly obovate, 3-fid lip, the 2 upper ones small, acute, and falcate. Anthers with a few short hairs on the back, the 2 lower ones tipped with a tuft of short bristles. Capsule oblong-linear, 3 to 4 lines or in luxuriant specimens 5 lines long, contracted at the base.—Hook. f. Fl. Tasm. i. 237 ; *L. decumbens,* Sims, Bot. Mag. t. 2277 ; *L. rhizophyta,* Schult. ; Sims, Bot.

Mag. t. 2519; *L. alata*, Labill. Pl. Nov. Holl. i. 51. t. 72; R. Br. Prod.
562; De Vriese in Pl. Preiss. i. 395; *L. cuneiformis*, Labill. Pl. Nov. Holl.
i. 51. t. 73; *L. uncinata* and *L. stricta*, De Vr. in Pl. Preiss. i. 396.

Queensland. Port Curtis, *M'Gillivray.*

N. S. Wales. Port Jackson, *R. Brown, Woolls;* Hastings and Clarence rivers,
Beckler; Twofold Bay, *F. Mueller;* Lord Howe's Island, *M'Gillivray.*

Victoria. Port Phillip, *R. Brown;* chiefly near the sea, from Glenelg river to Wilson's
Promontory, *F. Mueller* and others.

Tasmania. Common in marshy places, especially near the sea, *J. D. Hooker.*

S. Australia. Memory Cove, *R. Brown;* around St. Vincent's Gulf, *F. Mueller;*
Kangaroo Island, *Waterhouse.*

W. Australia. From King George's Sound and adjoining districts, *R. Brown, A.
Cunningham, Preiss, n.* 1496, and others, to Swan River, *Fraser, Huegel, Preiss, n.* 1431,
1443, and Murchison river, *Oldfield* (with longer pedicels), and eastward to Middle Island,
Maxwell.

The species is also in New Zealand, South Africa, and extratropical South America.

On the immediate seacoast the leaves are often larger, firmer, and more obovate, but this
form, which is the *L. cuneiformis*, Labill., can scarcely be called a distinct variety, for many
specimens show this broad foliage at the base, whilst some branches grow out with the narrower leaves of the ordinary form. To this obovate form belongs also probably *L. saxicola*,
De Vr. in Pl. Preiss. i. 398, from the rocks on Mistaken Island, Preiss, n. 498, which,
however, I have not seen.

L. erecta, De Vr. l. c. 395, from Swan River, *Preiss, n.* 1447, which I have not seen,
must, from his description, be a luxuriant form of the same species, said to attain 6 to 8 ft.

L. angustifolia, Benth. in Hueg. Enum. 74; A. DC. Prod. vii. 358, was described from
imperfect specimens, which, on further examination, appear to be the summits of a very
luxuriant plant of *L. anceps*, with narrow-linear floral leaves 2 to 4 in. long.

12. **L. surrepens,** *Hook. f. Fl. Tasm.* i. 237. t. 69 A.   A small glabrous creeping or prostrate perennial, the branching stems from 1 or 2 in. to
near 6 in. long.   Leaves obovate or oblong-cuneate, obtuse, quite entire or
rarely obscurely toothed, $\frac{1}{4}$ to near 1 in. long.   Pedicels axillary, rarely as
long as the leaves.   Calyx-tube narrow-turbinate, the lobes short broad and
obtuse.   Corolla 3 to 4 lines long, the lobes oblong, nearly equal but oblique.
Anthers glabrous, the lower ones tipped each with 1 or 2 rigid flat bristles
or points.   Fruit not seen ripe, but apparently capsular, the summit being
conical as in *Lobelia.*

Tasmania. Marshy ground in alpine places at an elevation of 3000 to 4000 ft., *J. D.
Hooker.* This resembles in some respects *Pratia platycalyx*, but the leaves are broader,
the flowers twice as large, showing no signs of unisexuality, and the young fruit is that of
Lobelia. From *Isotoma fluviatilis* it differs in the larger entire leaves, and in the corolla
split to the base; from *L. anceps* in the shape of the capsule.

13. **L. membranacea,** *R. Br. Prod.* 563.   Glabrous, with long procumbent filiform stems, often rooting at the lower nodes.   Leaves petiolate,
broadly ovate-cordate or orbicular, thin and membranous, mostly sinuate-
toothed, rarely much exceeding $\frac{1}{2}$ in. diameter, a few of the uppermost more
sessile and ovate.   Flowers on long filiform axillary pedicels.   Calyx-tube
turbinate, the lobes small and narrow.   Corolla 4 to 5 lines long, the lower
lobes oblong, the upper ones narrower, more acute, and incurved.   Upper
anthers hirsute on the back without terminal tufts, lower ones with a single
small bristle or point on each.   Capsule obovate-turbinate, straight, about 2
lines long.   Seeds very small.—A. DC. Prod. vii. 365.

Queensland. Bustard Bay, *Banks and Solander* (with particularly large and thin leaves) ; Port Curtis, *M'Gillivray ;* Rockingham Bay, *Dallachy ;* Moreton Bay, *F. Mueller.*

Victoria. Some specimens from Fitzroy river, *Robertson,* appear to belong to this species, though with rather smaller leaves.

14. **L. stenophylla,** *Benth.* Glabrous. Stems (from a creeping base ?) very slender, ascending and often above 1 ft. long. Leaves sessile or nearly so, linear or linear-lanceolate, minutely and remotely toothed or entire, often above 1 in. long. Pedicels axillary, usually more than twice as long as the leaves. Calyx-lobes as long as or longer than the turbinate tube. Corolla about 4 lines long, the lobes nearly of equal length, the 3 lower ob-ovate-oblong, the 2 upper narrower, more acute, and curved upwards. Anthers glabrous, without terminal tufts, but the 2 lower tipped each with a small bristle. Capsule obovate, 2 to $2\frac{1}{2}$ lines long, scarcely oblique.

N. Australia. Port Essington, *Armstrong.*
Queensland. Burnett river, *F. Mueller ;* Rockhampton, *O'Shanesy ;* Broadsound, *Herb. F. Muell. ;* Brisbane river, Moreton Bay, *Backhouse, F. Mueller.*

15. **L. quadrangularis,** *R. Br. Prod.* 563 ?. A glabrous or rarely pubescent perennial, the stems much-branched, prostrate (or ascending?), angular. Leaves nearly sessile, ovate or orbicular, prominently toothed, rather thin, rarely exceeding $\frac{1}{2}$ in. Pedicels axillary, long and slender. Calyx-lobes lanceolate, about as long as the narrow turbinate tube. Corolla about 3 lines long, the lower lobes oblong, the 2 upper ones rather shorter, narrower and acute. Anthers glabrous or minutely pubescent on the back, without terminal tufts, but the lower ones tipped with 2 or 3 very small bristles. Capsule obovate-oblong, slightly oblique, about 3 lines long. Seeds very small and numerous.—A. DC. Prod. vii. 365 ; F. Muell. Fragm. iv. 182.

N. Australia. Banks of brooks and streams, Victoria and Fitzmaurice rivers, *F. Mueller.* Described from F. Mueller's specimens, at first distributed as *L. humistrata,* F. Muell., but afterwards referred by him, and probably correctly so, to *L. quadrangularis,* which was described by R. Brown from a specimen of Bauer's which I have not seen. It is not in Brown's herbarium.

16. **L. dioica,** *R. Br. Prod.* 565. A slender much-branched annual of 3 to 6 in., glabrous or sprinkled with a few short spreading hairs. Leaves sessile, oblong or lanceolate, bordered by a few small teeth or entire, 3 to 6 lines long, the floral ones gradually smaller. Flowers diœcious, on very slender pedicels in the axils of the upper leaves and much longer than them, forming a terminal leafy raceme, numerous and crowded at the ends of the branches in the males, fewer and more distant in the females. Calyx-lobes linear, about $\frac{1}{2}$ line long, the males without any tube, the females with an adnate tube of about $\frac{1}{2}$ line. Corolla about $1\frac{1}{2}$ lines long, the tube slit open on the upper side, the lobes obovate-oblong, nearly equal, but oblique. Anthers in the males slightly pubescent, the upper ones without terminal tufts, the 2 lower tipped by a very few small bristles ; the stamens rudimentary in the females. Stigma in the males small and entire, and the ovules abortive, but in the females the stigma is broadly 2-lobed as in other *Lobelias.* Capsule obovoid-oblong, about 2 lines long, scarcely oblique.—*Monopsis*

dioica, Presl, Prod. Mon. Lob. 11 ; *Holostigma dioicum*, G. Don, Gen. Syst. iii. 716, A. DC. Prod. vii. 352.

N. Australia. Gulf of Carpentaria, opposite Groote Island, *R. Brown ;* Roper river and near Macadam Range, *F. Mueller.*

17. **L. purpurascens,** *R. Br. Prod.* 563. Quite glabrous. Stems, from a perennial rootstock, branched, procumbent or ascending, angular, often above 1 ft. long, but sometimes very small and short. Leaves shortly petiolate or almost sessile, from ovate to oblong-lanceolate, toothed, rather firm, usually from ½ to 1 in. long. Pedicels axillary, much longer than the leaves, often reflexed after flowering. Flowers (in all the specimens seen) dioecious. Calyx-lobes narrow, acute, the tube very short or scarcely any in the males, obconical and rather narrow in the females. Corolla 4 to 5 lines long, the lower lobes oblong, obtuse, the 2 upper ones rather shorter narrower more acute and incurved. Anthers in the males glabrous, without terminal tufts, but the 2 lower tipped with 1 or 2 small bristles, smaller and without pollen or abortive in the females. Stigma small and undivided, and ovules abortive in the males, the stigma broadly 2-lobed in the females. Capsule narrow-ovoid, fully 3 lines long, the conical summit opening in 2 valves as in *Lobelia ;* the seeds rather large and often flattened as in *Pratia.*—A. DC. Prod. vii. 365.

Queensland. Southern tributaries of the Burnett river, *F. Mueller.*
N. S. Wales. Port Jackson to the Blue Mountains, *R. Brown* and others ; New England, *C. Stuart ;* Hastings and Macleay rivers, *Beckler.*
Victoria. Along brooks near the mouth of the Snowy River, *F. Mueller ;* Grampians, *Wilhelmi* (the latter a very small form).

The species has some resemblance with *Pratia Cunninghamii,* but is at once known by the long usually recurved pedicels.

18. **L. pratioides,** *Benth.* A slender creeping or prostrate much-branched perennial, more or less hoary with a minute pubescence. Leaves linear oblong or lanceolate, bordered by small distant teeth or rarely entire, mostly sessile but narrowed at the base, ¼ to ½ in. long, the lower ones rarely obovate. Flowers small, on axillary pedicels shorter or rather longer than the leaves, more or less dioecious. Calyx-tube scarcely any in the males, turbinate in the females. Corolla about 4 lines long, the lobes oblong, nearly equal but oblique. Anthers in the males glabrous, the upper ones without terminal tufts, the 2 lower tipped with a tuft of short bristles, glabrous and empty or abortive in the females. Capsule obliquely obovate, 2¼ lines long, the conical summit opening in 2 valves as in other *Lobelias.*

Victoria. Yarra river, Forest Creek, and Station Peak, *F. Mueller ;* Wendu Vale, *Robertson ;* Australian Pyrenees, *Wilhelmi.*
Tasmania. South Esk river near Perth, *C. Stuart.*

This species has some resemblance with *Pratia puberula,* but the leaves are much narrower and the fruit that of a *Lobelia.*

2. PRATIA, Gaudich.

Calyx of *Lobelia.* Corolla slit open on the upper side, 5-lobed, the lobes nearly equal but very oblique. Stamens of *Lobelia.* Ovary inferior, 2-celled,

the summit between the calyx-lobes nearly flat.   Style and stigma of
Lobelia. Fruit ovoid or globular, crowned by the calyx-lobes, indehiscent,
the pericarp usually succulent.—Herbs, with the habit of the section *Hemipogon* of *Lobelia,* usually creeping at the base, the branches sometimes
ascending. Flowers white (or blue ?) on axillary pedicels, and in most of the
Australian species more or less dioecious, by the abortion or sterility of the
anthers in the females, and of the ovary, ovules, and stigma in the males.

A small genus, containing, besides the Australian species which are all endemic, five or
six others dispersed over New Zealand, antarctic and extratropical South America. It
should probably include *Piddingtonia,* which differs in the greater irregularity of the corolla,
and both are only artificially distinguished from *Lobelia* by the indehiscent more or less
succulent fruit.

Glabrous. Leaves entire or obscurely toothed.
  Flowers and fruits almost sessile.
    Leaves ovate or orbicular, 1 to 2 lines long . . . . 1. *P. irrigua.*
    Leaves linear, 3 to 4 lines long. . . . . 2. *P. gelida.*
  Pedicels at least as long as the fruit.   Leaves linear-cuneate or ob-
      long-spathulate . . . . 3. *P. platycalyx.*
Glabrous.   Leaves ovate or oblong, toothed, ½ to 1 in. long.   Pedi-
  cels shorter or scarcely longer than the leaves . . . . . . 4. *P. erecta.*
Pubescent. Leaves ovate or orbicular.
  Pedicels shorter or scarcely longer than the leaves. Fruit 2 to 3
    lines diameter . . . . 5. *P. puberula.*
  Pedicels much longer than the very small leaves. Fruit 1 to 1½
    lines diameter . . . . 6. *P. pedunculata.*

1. **P. irrigua,** *Benth.* A little creeping glabrous perennial, scarcely
rising from the ground but forming intricate masses of several inches diameter.
Leaves ovate or orbicular, entire or scarcely toothed, 1 to 2 lines diameter.
Flowers very small, on exceedingly short axillary pedicels or almost sessile.
Calyx-lobes lanceolate.   Corolla scarcely more than 1 line long, the lobes
narrow, nearly equal, the 2 upper ones ascending and more acute.   Stamens
glabrous, with a minute bristle on each of the 2 lower anthers.   Fruit globular, indehiscent, about 1½ lines diameter, with a thin pericarp. Seeds
numerous, but rather large for the fruit, globular, smooth.—*Lobelia irrigua,*
R. Br. Prod. 563 ; A. DC. Prod. vii. 367.

Tasmania. Kent's group, Bass's Straits, *R. Brown (Herb. R. Br.).*

2. **P. gelida,** *Benth.* A little creeping glabrous perennial, like *P. irrigua,*
except that the leaves are linear, obtuse, entire and mostly 3 to 4 lines long.
Flowers, as in that species, very small and almost sessile. Stamens the
same with a minute bristle on each of the 2 lower anthers. Fruit globular,
indehiscent, about 1½ lines diameter, with a thin pericarp and rather large
globular seeds.—*Lobelia gelida,* F. Muell. Fragm. iv. 183.

Victoria. Summit of Haidinger Range, at an altitude of 5000 to 6000 ft., *F. Mueller.*
I do not feel certain that this and the preceding species are dioecious, as in the other Australian *Pratias,* but I think they are.   The fruit is certainly that of *Pratia,* globular, without the conical 2-valved apex of the *Lobelia* capsule.

3. **P. platycalyx,** *Benth.* A small glabrous creeping or prostrate perennial.   Leaves from obovate-cuneate and under ½ in. long to oblong or
linear and above 1 in. long, obtuse, quite entire, more or less narrowed into
a petiole, thick and smooth. Flowers small, on very short axillary pedicels.

Calyx-tube hemispherical in the males, ovoid in the females, the lobes short and broad, almost triangular. Corolla a little more than 1 line long, the lobes nearly equal, but oblique. Anthers in the males glabrous, the lower ones tipped with very minute points, and enclosing the small globular stigma; in the females the anthers are empty or abortive, with the rather large 2-lobed stigma protruding. Fruit ovoid-globular. Seeds rather large, ovate, compressed.—*Laurentia platycalyx*, F. Muell. in Trans. Vict. Inst. 1855, 39; *Lobelia platycalyx*, F. Muell. Fragm. iv. 183.

Victoria. In moist subsaline pastures from Port Phillip westward, *F. Mueller, Adamson;* Queenscliff and near Station Peak, *F. Mueller.*

4. **P. erecta,** *Gaudich. in Freyc. Voy. Bot.* 456. Glabrous, with a perennial rootstock and branching prostrate ascending or erect stems from a few in. to nearly 1 ft. long, but usually under 6 in. Leaves sessile or nearly so, ovate oblong or lanceolate, serrate, ½ to 1 in. long, rather firm. Flowers axillary, the pedicels either very short or rarely as long as the leaves. Calyx-tube in the males very shortly turbinate or scarcely any, ovoid in the females, the lobes lanceolate, ½ to ¾ line long. Corolla 2½ to 3 lines long, the lobes lanceolate, nearly equal, but the 2 upper more deeply separate. Anthers in the males glabrous, the 2 lower ones tipped with a tuft of minute bristles, all empty or abortive in the females. Fruit nearly globular, 3 to 4 lines diameter, slightly succulent.—*Lobelia concolor*, R. Br. Prod. 563; *Isolobus concolor* and *I. Cunninghamii*, A. DC. Prod. vii. 354; *Pratia Cunninghamii*, Hook. f. Fl. Antarct. i. 42.

Queensland. Suttor river, *F. Mueller;* Rockhampton and Bowen river, *Bowman;* in the interior, *Mitchell.*
N. S. Wales. Paterson's River, *R. Brown;* inundated banks of the Lachlan and Macquarrie rivers, *A. Cunningham, Fraser.*
Victoria. Borders of stagnant waters, Avoca and Bremer rivers, *F. Mueller.*

5. **P. puberula,** *Benth.* A small creeping or prostrate perennial, more or less pubescent. Leaves sessile, ovate or almost orbicular, toothed or almost entire, mostly 2 to 4 lines, rarely ½ in. long. Flowers axillary, the pedicels shorter or rarely longer than the leaves. Calyx-tube in the males very short or almost none, in the females ovoid, the lobes lanceolate, acute. Corolla about 3 lines long, the lobes nearly equal, but oblique and almost acute. Anthers in the males glabrous, the 2 lower ones each with a single minute bristle, empty or abortive in the females. Fruit globular, about 3 lines diameter, slightly succulent. Seeds rather large, ovoid or compressed.

N. S. Wales. Glendon, *Leichhardt.*
Victoria. Moist, grassy, and marshy places at Cobra and Mount Barkly, *F. Mueller.*
S. Australia. Cooper's Creek, *Bowman.*
This species bears some general resemblance to *Lobelia pratioides* and *Isotoma fluviatilis*, but is readily distinguished by the generic characters.

6. **P. pedunculata,** *Benth.* A very slender and slightly pubescent perennial, the intricate filiform creeping or prostrate stems extending to broad patches, with very shortly ascending flowering branches. Leaves almost sessile, ovate or orbicular, with few prominent teeth, 2 to 3 lines diameter. Flowers small, on slender axillary pedicels considerably longer than the

leaves. Calyx-tube very short in the males, obconical in the females, the lobes narrow-lanceolate, obtuse. Corolla 2 to 2½ lines long, the lobes nearly equal but very oblique. Fruit very small, pubescent, globular, not seen quite ripe, but evidently without the conical 2-valved summit of *Lobelia*.— *Lobelia pedunculata*, R. Br. Prod. 563 ; A. DC. Prod. vii. 367 ; Hook. f. Fl. Tasm. i. 237. t. 69 B.

N. S. Wales. Hunter's River, *R. Brown.*
Victoria. Ballan, Cape Otway, Apollo Bay, *F. Mueller;* Portland, *Allitt;* Emu Creek, *Whan.*
Tasmania. Not uncommon in good soil where damp, on the margins of rivers, and occasionally on mountains, *J. D. Hooker.* In the excellent plate in the 'Flora Tasmanica,' the whole specimen represented is a male ; fig. 3 is a female, but the artist has inserted the anthers from a male, thinking, no doubt, that those he found in the flower drawn were accidentally not normally imperfect.

3. ISOTOMA, Lindl.

(Lobelia, *sect.* Isotoma, *R. Br.* ; Enchysia (*partly*), *Presl;* Laurentia (*partly*), *A. DC.*)

Calyx of *Lobelia.* Corolla-tube cylindrical, entire or rarely very shortly slit on the upper side ; lobes 5, nearly equal, spreading, either quite horizontal or very shortly and obliquely campanulate at the base. Stamens inserted near the summit of the corolla-tube. Anthers of *Lobelia*, the upper ones without terminal tufts. Pistil, capsule, and seeds of *Lobelia.*—Herbs, with the habit of various species of *Lobelia.* Flowers axillary or in terminal racemes or solitary, on long scapes or peduncles, hermaphrodite in all the species known.

A small genus, perhaps too artificially distinguished from *Lobelia*, but the entire tube and epicorolline stamens are so exceptional in *Lobelieæ* that there may be a convenience in keeping it up. It is here characterized from the Australian species alone, which are all endemic. How far the extra-Australian species associated with it by Endlicher under the common name of *Laurentia*, and distributed into various genera by Presl and Alph. De Candolle, should or should not be considered as congeners, is a question requiring much further investigation. The precise form of the corolla, never perfectly regular, but more or less oblique, can scarcely be made use of for generic distinction, for, as far as can be judged of from dried specimens, it varies from species to species.

Erect and nearly simple. Flowers in a terminal, unilateral raceme . . 1. *I. Brownii.*
Erect and branching. Flowers on long, axillary pedicels.
  Leaves deeply toothed or cut. Flowers large.
    Leaves linear, pinnatifid . . . . 2. *I. axillaris.*
    Leaves ovate or lanceolate, with linear teeth or lobes . . . . . 3. *I. petræa.*
  Leaves oblong, small. Slender annual, with small flowers . . . . 4. *I. pusilla.*
Stems short or none. Leaves radical or nearly so. Scapes or erect
  pedicels long and slender . . . . 5. *I. scapigera.*
Creeping or prostrate perennial. Leaves ovate. Pedicels axillary . . 6. *I. fluviatilis.*

I. Baueri, Presl, Prod. Mon. Lob. 42 ; A. DC. Prod. vii. 412, described as having linear-lanceolate, serrate leaves, the upper ones ternately verticillate, and the corolla with ovate lobes twice as long as the entire tube, if an *Isotoma* at all, must be totally different from any species I have seen. No station but the general one of New Holland is given.

1. I. Brownii, *G. Don, Gen. Syst.* iii. 716. A glabrous, erect, simple or slightly branched annual, from 6 in. to 1½ ft. high. Leaves narrow-linear, entire, mostly from ½ to 1 in. long. Flowers often numerous, in a loose, unilateral, terminal raceme of 6 to 8 in., in the smaller specimens reduced to a very few flowers or to a single one. Calyx like that of *Lobelia heterophylla*, the subulate lobes at least as long as the very oblique tube. Corolla-tube cylindrical, 6 to 8 lines long; lobes broadly spathulate, slightly unequal, spreading horizontally to a diameter of about ¾ in. Anthers glabrous, the 2 lower ones each with a single flat bristle or point. Capsule in some specimens very obliquely obovoid, gibbous on the upper side, and about 4 lines long, in others less oblique, oblong, and 5 to 6 lines long. Seeds small, ovoid or 3-angled, very smooth and shining.—A. DC. Prod. vii. 412; *Lobelia hypocrateriformis*, R. Br. Prod. 565; Bot. Mag. t. 3075; *Isotoma brevifolia*, Presl, Prod. Mon. Lob. 43 (from the character given); A. DC. Prod. vii. 412; *Lobelia Lehmanni*, De Vr. in Pl. Preiss. i. 394.

W. Australia. King George's Sound and adjoining districts, *R. Brown* and others, and thence to Swan and Murchison rivers, *Oldfield, Drummond, n.* 423, *Preiss, n.* 1426.

2. **I. axillaris,** *Lindl. Bot. Reg. t.* 964. A glabrous perennial, flowering the first year so as to appear annual, but forming at length a hard rootstock, erect, with few, spreading branches, ½ to 1 ft. high. Leaves linear, irregularly pinnatifid, often 2 to 3 in. long, the lobes linear or linear-lanceolate. Pedicels axillary, 2 to 6 in. long. Flowers large, of a bluish-purple, very pale or with a yellowish-green tint outside. Calyx-lobes linear, rigid, as long as the oblong tube. Corolla-tube ¾ to 1 in. long, somewhat incurved and broader upwards; lobes spreading to a diameter of ¾ to 1 in., narrow, mucronate, slightly unequal, and not quite so flat as in *I. Brownii*. Anthers glabrous, the 2 lower ones with single, rigid bristles. Capsule cylindrical, tapering and slightly oblique at the base, 6 to 8 lines long, 2 to 3 lines broad. Seeds small, very minutely foveolate.—Gaudich. in Freyc. Voy. 455. t. 70; *Lobelia senecioides*, A. Cunn. in Bot. Mag. t. 2702; *Isotoma senecioides*, A. DC. Prod. vii. 412; Bot. Mag. t. 5073.

Queensland. Rocks in most exposed situations, Rockhampton, *Dallachy*.
N. S. Wales, *Caley* (*Herb. R. Br.*); Mudgee, *Woolls*; barren granite rocks near Bathurst, *Fraser, A. Cunningham*; between the Lachlan and Darling rivers, *Neilson*; New England, *C. Stuart*; Mount Mitchell, *Beckler*.
Victoria. Fissures of granite rocks, between Ten-mile Creek and Broken River, Mount Hope, Buffalo Ranges, *F. Mueller*.

3. **I. petræa,** *F. Muell. in Linnæa*, xxv. 420. Very closely allied to *I. axillaris*, and I should have proposed reducing it to a variety of that species but that, among numerous specimens from various localities, I have seen no intermediates. Habit, stature, inflorescence, flowers, and fruit the same, but the leaves are all ovate-oblong or elliptical, bordered by irregular linear or lanceolate teeth or lobes, but never longer than the breadth of the entire central part.

N. S. Wales. Goyinga Mountains, *Victorian Expedition*; between Stokes Range and Cooper's Creek, *Wheeler*.
S. Australia. Crystal Brook, *F. Mueller*; Purdie's Ponds, *Waterhouse*; Lake Gillies, *Burkitt*; Hugh's River, *M'Douall Stuart's Expedition*.
W. Australia. Between Moore and Murchison rivers, *Drummond, 6th Coll. n.* 47 or 67.

4. **I. pusilla,** *Benth. in Hueg. Enum.* 75. A glabrous annual of 2 to 4 in., usually erect and branching. Leaves oblong or lanceolate, mostly

sessile, obtuse, obscurely toothed, ¼ to ½ in. long. Peduncles in the upper
axils 1 to 2 in. long, slender, bearing each a small flower. Calyx-lobes
lanceolate, longer than the tube. Corolla-tube slender, 1½ lines long ; lobes
obovate, about as long as the tube, equally spreading, but the 2 upper ones
rather smaller. Anthers usually glabrous, the 2 lower ones tipped each with
a small bristle. Capsule narrow-turbinate, slightly oblique, about 2½ lines
long. Seeds small, ovate, smooth, and almost shining.—*Laurentia pusilla,*
A. DC. Prod. vii. 411; *Lobelia elegans,* De Vr. in Pl. Preiss. i. 396.

W. Australia. Swan River, *Huegel, Drummond, 1st Coll. n.* 424 *and n.* 81; *Preiss,*
n. 1434, 1436.

5. **I. scapigera,** *G. Don, Gen. Syst.* iii. 716. A slender, glabrous
annual. Leaves either radical and rosulate or on a stem of ½ to 1 in. long,
petiolate, obovate or oblong, slightly toothed, rarely exceeding ½ in., and
often much smaller. Scapes or pedicels erect, leafless, slender, 3 to 6 in.
long, 1-flowered. Calyx-lobes linear, narrow or broad, about as long as the
tube. Corolla 3 to 4 lines long, nearly regular, but somewhat oblique, the
tube either quite entire or at length very shortly slit open on the upper side,
the lobes short, obovate-oblong, and spreading, but the limb apparently
more campanulate than in the preceding species, and a white spot at the
base of the lower lobes, said to give it a bilabiate look (*Oldfield*). Anthers
glabrous, the 2 lower ones tipped each with a rigid bristle, surrounded by a
tuft of small ones. Capsule obliquely obovoid, often gibbous, sometimes
scarcely 3 lines long and almost as broad, sometimes 5 to 6 lines long and not
3 lines broad. Seeds very numerous and small, ovoid, smooth.—*Lobelia scapigera,* R. Br. Prod. 565 ; *Enchysia scapigera,* Presl, Prod. Mon. Lob. 41;
A. DC. Prod. vii. 409.

W. Australia. King George's Sound and adjoining districts, *R. Brown, F. Mueller,*
Drummond, 5th Coll. n. 68 ; Tone and Gordon rivers, *Oldfield ;* Swan River, *Drummond,*
1st Coll. n. 418 ; eastward to Esperance Bay and Middle Island, *Maxwell.*

Var. *pusilla,* R. Br. Very small, with single scapes of ½ to 1 in.—Goose Island Bay,
R. Brown ; east of King George's Sound, *Maxwell.*

Lobelia ophiocephala, De Vr. in Pl. Preiss. i. 397, from the borders of Lake Keiermula,
Preiss, n. 1446, and *Lobelia monanthus,* De Vr. l. c. 398, from Swan River, *Preiss, n.*
1432, which I have not seen, are probably this species.

6. **I. fluviatilis,** *F. Muell. Herb.* A small, prostrate or creeping perennial, usually pubescent, with the habit of some species of *Pratia.* Leaves
in the typical form oblong or almost linear, or the lower ones ovate or obovate, mostly 3 to 4 lines long, slightly toothed, shortly petiolate or the
upper ones sessile. Flowers on axillary pedicels, varying from the length of
the leaves to twice as long. Calyx-tube narrow, turbinate, pubescent; lobes
short, lanceolate. Corolla usually 5 to 6 lines long, but sometimes much
smaller, the entire tube longer than the calyx-lobes, the lobes oblong, almost
acute, nearly equal but oblique, and the 2 upper separated rather lower down.
Anthers glabrous, the 2 lower ones tipped each with one rigid broad bristle
and several smaller ones. Capsule about 2 lines long. Seeds ovoid, smooth.
—*Lobelia fluviatilis,* R. Br. Prod. 563 ; A. DC. Prod. vii. 366.

N. S. Wales. Banks of the Nepean, *R. Brown ;* near Bathurst, *Vicary ;* near Goulburn, *F. Mueller.*

Var. *inundata.* Leaves mostly ovate or orbicular.—*Lobelia inundata,* R. Br. Prod. 563 ; A. DC. Prod. vii. 367.

N. S. Wales. Port Jackson to the Blue Mountains, *R. Brown, A. Cunringham,* and others ; Paramatta, *Woolls* (with very small flowers).

Victoria. . Inundated places and borders of lagoons on the Yarra and Goulburn rivers, Ovens river, and to the western frontier, *F. Mueller* and others; Grampians, *Wilhelmi, F. Mueller.*

Enchysia Lessonii, Presl, Prod. Mon. Lob. 40 ; A. DC. Prod. vii. 409 ; *E. Baueri* and *E. Gaudichaudii,* Presl, l. c. 41 ; *Laurentia Baueri* and *L. Gaudichaudii,* A. DC. l. c. vii. 411, all known only from Presl's diagnoses, belong probably to this species, which connects the genus with true *Lobelias,* and has the habit of *L. pratioides* and *Pratia puberula,* but the insertion of the stamens near the orifice of the corolla-tube induces its reference to *Isotoma.*

4. **WAHLENBERGIA,** Schrad.

Calyx 5- or rarely 4-lobed or in abnormal flowers 6- or 7-lobed. Corolla regular, campanulate or more or less tubular at the base, with as many valvate lobes as calyx-lobes. Stamens free. Ovary 3- to 5-celled or rarely 2-celled. Style with as many stigmatic lobes as ovary-cells. Capsule opening at the top loculicidally within the calyx-teeth, in as many valves as cells.— Herbs. Leaves alternate or very rarely opposite or whorled. Peduncles terminal or in the upper axils, often forming loose, terminal, dichotomous, leafy panicles. Flowers usually blue.

A considerable genus, dispersed over various parts of the world, most abundant in Southern Africa. The two Australian species are both in New Zealand, and one appears to be the same as a common one in tropical Asia.

Stems leafy, simple or branched. Leaves sometimes crowded but not
  rosulate . . . . 1. *W. gracilis.*
Leaves all radical or crowded on very short, tufted stems. Scapes leafless 2. *W. saxicola.*

1. **W. gracilis,** *A. DC. Monogr. Camp.* 142 ; *Prod.* vii. 435. An exceedingly variable plant in stature, duration, and size of the flowers, glabrous or more or less clothed in the lower part with rigid hairs, sometimes a slender, simple or branched annual of 6 in. to 1½ ft., sometimes forming a perennial, almost woody rootstock, with numerous ascending or erect, simple or slightly branched stems, leafy chiefly in the lower part. Lower leaves from obovate, and under ½ in. long, to lanceolate or almost linear, and 1 in. long or even much more when very narrow, the upper ones fewer and narrower, and in slender varieties, nearly all linear-subulate or filiform. Flowers solitary, on long, terminal peduncles, without bracts, usually 5-merous, sometimes (said to be the early flowers) 4-merous, very rarely 6-merous or even 7-merous. Calyx-tube from ovoid to narrow-obconical, the lobes from broadly lanceolate and shorter than the tube to linear-subulate and twice as long. Corolla campanulate, more or less expanded, varying in size from ¼ in. to above 1 in. diameter. Filaments shortly dilated at the base. Ovary 3-celled or very rarely 2-celled.—Hook. f. Fl. Tasm. i. 239 ; *Campanula gracilis,* Forst. ; Br. Prod. 561 ; Sm. Exot. Bot. t. 45 ; Bot. Mag. t. 691 ; *C. vincæflora,* Vent. Jard. Malm. t. 12 ; *C. littoralis,* Labill. Pl. Nov. Holl. i. 49. t. 70 ; *C. capillaris,* Lodd. Bot. Cab. t. 1406 ; *C. quadrifida,* R. Br. Prod. 561 ; *Wahlenbergia quadrifida,* A. DC. Mon. Camp. 144 ; Prod. vii. 433 ; *W. Sieberi,*

A. DC. ll. cc.; *W. multicaulis*, Benth. in Hueg. Enum. 75; A. DC. Prod. vii. 433; *W. simplicicaulis*, De Vr. in Pl. Preiss. ii. 241.

N. Australia. Victoria river, *F. Mueller;* Port Essington, *Armstrong.*

Queensland, *R. Brown;* Albany Island, *F. Mueller;* Cape York, *E. Daemel;* Port Curtis, *M'Gillivray;* Rockingham Bay, *Dallachy;* Rockhampton, *Thozet;* Moreton Bay, *F. Mueller;* in the interior, *Mitchell;* plains of the Condamine, *Leichhardt.*

N. S. Wales. Port Jackson to the Blue Mountains, *R. Brown* and others; in the interior to the Lachlan and Darling and to the Barrier Range, *Victorian and other Expeditions;* New England, *C. Stuart ;* Hastings and Clarence rivers, *Beckler.*

Victoria. Common from the coast to the mountains, *F. Mueller* and others; in the Haidinger Range to an elevation of 5000 to 6000 ft., *F. Mueller;* Wimmera, *Dallachy.*

Tasmania, *R. Brown;* abundant in dry places throughout the island, *J. D. Hooker.*

S. Australia. Around St. Vincent's and Spencer's Gulf, *F. Mueller* and others; in the interior to Lake Gillies, *Burkitt;* Cooper's Creek, *Howitt's Expedition.*

W. Australia. King George's Sound, *R. Brown* and others, and thence to Swan River, *Drummond, n.* 151, 153, 164, 185, 421, 425, *Preiss, n.* 1883, 1884, 1886, 1887; Murchison river, *Oldfield;* eastward to Cape Arid, *Maxwell.*

The species is also in New Zealand, in the Eastern Archipelago, and extends over the East Indies if *W. agrestis,* A. DC.; Hook. and Thoms. in Journ. Linn. Soc. ii. 21, be, as is probable, the same species. Several distinct varieties have been enumerated by various authors, but they run so variously one into another that they would require to be differently defined for every separate collection of specimens.

Campanula Preissii, De Vr. in Pl. Preiss. ii. 241, from Preiss's specimens, n. 1892, and *Wahlenbergia Preissii,* De Vr. l. c., from the same collector's n. 1890 are unknown to me. If real *Wahlenbergias* they probably both belong to *W. gracilis,* which is the only species I have met with among the numerous W. Australian specimens I have seen.

2. **W. saxicola,** *A. DC. Monogr. Camp.* 144; *Prod.* vii. 433. A glabrous perennial, with a tufted or shortly creeping stock, rarely lengthening out into leafy branches of 1 in. or rather more. Leaves radical and rosulate or crowded on the short stems, petiolate, from obovate or spathulate to almost linear, entire or obscurely crenate, ½ to 1 in. long in the Tasmanian specimens, longer in some New Zealand ones. Scapes leafless, 1-flowered, 2 to 6 in. high. Flowers 5-merous, sometimes like those of *W. gracilis,* but usually more oblique and 1 or 2 of the anthers tipped with a small point. Ovary 2- or 3-celled.—Hook. f. Fl. Tasm. i. 239. t. 71; Handb. N. Zeal. Fl. 170; *Campanula saxicola,* R. Br. Prod. 561; *Wahlenbergia albomarginata,* Hook. Ic. Pl. t. 818; *Streleskia montana,* Hook. f. in Hook. Lond. Journ. vi. 267.

Tasmania. Summit of Mount Wellington (Table Mountain), *R. Brown, Gunn.* Also in New Zealand. Mr. Archer writes to me that the Table Mountain of R. Brown (frequently quoted in the former volumes of this Flora) is the one now known by the name of Mount Wellington. I learn from Dr. Hooker that it was still frequently called Table Mountain when he was in the island.

Order LXVI. ERICACEÆ.

Calyx more or less deeply divided into 4 or 5 teeth or lobes, the tube adnate to the ovary or quite free, sometimes exceedingly short. Corolla inferior or superior, the tube ovoid globular elongated or campanulate, the lobes spreading, valvate or imbricate in the bud, or (in a very few species not Australian) the petals distinct. Stamens twice as many or rarely of the same number as the corolla-lobes, inserted within the corolla but free from it. Anthers 2-celled, opening at the top by 2 separate pores or oblong slits or rarely by 2 slits extending their whole length. Hypogynous disk very small or none. Ovary having usually as many cells as lobes of the corolla, rarely (in genera not Australian) apparently twice as many or reduced to 3 or 2, with 1 or several ovules in each cell, the placentas attached to the axis. Fruit either capsular or succulent and indehiscent. Seeds very small, with a fleshy albumen, the embryo straight, often small.—Shrubs, sometimes very low creeping and almost herbaceous, more frequently erect and bushy or growing up into small trees, very rarely, in species not Australian, true herbs. Leaves entire or toothed, undivided, usually alternate, penninerved or 3-nerved. Flowers either axillary and solitary, or in short clusters or heads, or forming terminal racemes corymbs clusters or heads.

This large Order is widely spread over the whole world, especially in the temperate and colder regions, but not uncommon also in hilly districts within the tropics. In Australia, however, it is very much restricted both in area and in numbers, being only known in Tasmania and in the mountains of Victoria and New England. The three Australian genera belong to three of the great suborders (or, according to some, distinct Orders) into which *Ericaceæ* have been divided. One, *Wittsteinia*, is endemic; another, *Pernettya*, extends to New Zealand and South America; the third, *Gaultheria*, has a wide range over the hilly regions of Asia and America.

TRIBE I. **Vacciniese.**—*Ovary inferior. Fruit succulent, indehiscent.*
Anther-cells opening to the base in longitudinal slits . . . . . . 1. WITTSTEINIA.

TRIBE II. **Arbuteæ.**—*Ovary superior. Fruit succulent, indehiscent.*
Anthers with 2 awns to each cell, rarely none. Berry smooth . . . 2. PERNETTYA.

TRIBE III. **Andromedeæ.**—*Ovary superior. Fruit a capsule, opening loculicidally.*
Calyx more or less enlarged after flowering, often succulent and berry-
like enclosing the fruit . . . . 3. GAULTHERIA.

TRIBE 1. VACCINIEÆ. Ovary inferior. Fruit succulent, indehiscent.

1. WITTSTEINIA, F. Muell.

Calyx-tube adnate; lobes 5. Corolla campanulate; lobes 5, short, spreading, valvate or slightly induplicate in the bud. Stamens inserted at the base of the corolla; anthers versatile, the cells opening to the base in longitudinal slits. Ovary inferior, 2- or 3-celled, with several ovules in each cell. Fruit pulpy, crowned by the persistent calyx-lobes, the dissepiments obliterated. Seeds several, slightly flattened. Embryo minute, near the base of the albumen.—Prostrate or creeping shrub. Leaves coarsely toothed. Flowers axillary, solitary.

The genus consists of a single species, endemic in Australia and exceptional in the tribe and almost so in the whole Order in the dehiscence of the anthers, the embryo is also much smaller than in the majority of the genera.

1. **W. vacciniacea,** *F. Muell. Fragm.* ii. 136, iii. 166; *Pl. Vict.* ii. *t.* 51. Stems prostrate or creeping, with ascending branches of 6 in. to 1 ft., usually slightly pubescent. Leaves scattered or approximate in clusters of 2 or 3, obovate-oblong, obtuse, coarsely toothed, contracted into a short petiole, rather thick, penniveined, glabrous, pale or glaucous underneath, mostly

about 1 in. long or rather more. Flowers yellowish-green or reddish, pendulous, solitary in the axils on glabrous peduncles of 2 to 3 lines. Bracts 2 or 3, scattered, small and narrow. Calyx-lobes lanceolate, obtuse, nearly 1½ lines long, slightly imbricate in the bud. Corolla about 4 lines long. Stamens falling off with the corolla, but almost free from it; filaments as long as the corolla-tube; anthers ovate. Ovary 2-celled in the flowers I examined, 3-celled in those analysed by F. Mueller. Style rather thick, with a thick peltate stigma. Berry globular, greenish yellow or reddish. Seeds orbicular.

Victoria. Crevices of rocks and rocky summits of the Baw-Baw mountains, more rarely in the Albert Range and sources of the Yarra, at an elevation of 3500 to 5000 ft., *F. Mueller.*

TRIBE 2. ARBUTEÆ. Ovary superior. Fruit succulent, indehiscent.

2. PERNETTYA, Gaudich.

Calyx free, deeply divided into 5 segments. Corolla urceolate or campanulate; lobes 5, short, spreading, imbricate in the bud. Stamens 10, hypogynous, included in the corolla-tube; anther-cells opening in a large terminal or oblique foramen, each with 2 erect awns (4 to the anther) or rarely without any. Ovary 5-celled, with several ovules in each cell; style inserted in a central depression; stigma capitate or peltate. Fruit a globular indehiscent berry. Seeds small, embryo cylindrical in the centre of the albumen.—Low creeping or bushy shrubs. Leaves small, penniveined, entire or toothed. Flowers axillary, solitary.

The genus extends over the Andes of America from Mexico to Cape Horn, and thence to the Antarctic Islands and New Zealand, the only Australian species being endemic in Tasmania. As a genus, *Pernettya* differs slightly from the northern *Arbutus*, in the anthers with the awns when present 2 to each cell and erect as in *Gaultheria*, not solitary and reflexed as in *Arbutus*, and in the smooth not granular ovary and fruit.

1. **P. Tasmanica,** *Hook. f. in Hook. Lond. Journ.* vi. 268; *Fl. Tasm.* i. 242. *t.* 73 B. A small creeping shrub, the branches ascending to very few inches, usually glabrous. Leaves very shortly petiolate, oblong-elliptical or almost lanceolate, rather obtuse, coriaceous, entire or obscurely toothed, 2 to 3 lines long. Flowers solitary in the upper axils, on pedicels of 1 to 2 lines, with several bracts at the base or below the middle. Calyx-segments ovate, about 1 line long. Corolla urceolate-campanulate, nearly 2 lines long. Filaments dilated at the base; anther-cells without awns, the foramen extending nearly down to the base. Hypogynous disk short, undulate. Berry yellow red or cream-coloured, pulpy, 3 to 4 lines diameter.

Tasmania. On all the mountains, especially on a granite soil, forming large green cushions, *J. D. Hooker.* With much the aspect of *P. empetrifolia*, Gaudich., but with a more creeping habit, this is at once distinguished by the absence of any awns to the anthers.

TRIBE 3. ANDROMEDEÆ. Ovary superior. Fruit a capsule, opening loculicidally.

3. GAULTHERIA, Linn.

Calyx free, deeply divided into 5 segments, enlarged under or round the fruit and then often succulent.　Corolla urceolate ; lobes 5, short, spreading, imbricate in the bud.　Stamens 10, hypogynous, included in the corolla-tube ; anther-cells opening in a terminal or oblique foramen, each with 2 erect awns (4 to the anther).　Ovary 5-celled, with several ovules in each cell ; style inserted in a central depression ; stigma capitate or peltate. Fruit a globular capsule, opening loculicidally in 5 valves, more or less en-closed in the enlarged usually succulent and berry-like calyx.　Seeds small ; embryo cylindrical in the centre of the albumen.—Erect and bushy or low and creeping shrubs, often hispid with rigid hairs.　Leaves penniveined, en-tire or toothed.　Flowers in simple terminal or axillary racemes, each one pedicellate within a bract and 2 bracteoles, or solitary in the axils of the stem-leaves.

The genus is chiefly spread over the mountain regions of America from the Oregon to Cape Horn and eastward to Brazil, represented by a few species in Japan and the mountains of tropical Asia, and to the south extending through the Antarctic Islands and New Zea-land to eastern extratropical Australia.　Of the three Australian species, two are endemic, the third is a common New Zealand one.　The genus is chiefly distinguished by its berry-like calyx, the real fruit inside being capsular, but some southern species with the calyx occasionally scarcely enlarged and the capsule more or less succulent, closely connect the genus with *Pernettya*.

Erect shrub of 2 to 5 ft.　Leaves oblong or lanceolate.　Racemes ter-
　　minal or axillary with membranous bracts . . . .　1. *G. hispida.*
Diffuse or bushy shrub not above 1 ft.　Leaves oblong or lanceolate.
Flowers axillary, forming a terminal leafy raceme . . . . . .　2. *G. lanceolata.*
Small prostrate or depressed shrub.　Leaves orbicular or ovate.　Flowers
　　all axillary . . . .　3. *G. antipoda.*

1. **G. hispida,** *R. Br. Prod.* 559.　An erect spreading shrub, usually 2 to 3 ft. high but attaining 4 or 5 ft., the branches and often also the midrib of the leaves hispid with rigid spreading or appressed hairs.　Leaves shortly petiolate, lanceolate to elliptical-oblong, with obtuse or callous serratures, 1 to 2 in. long or rarely more.　Flowers in dense racemes, terminal or in the upper axils, shorter than the leaves.　Bracts membranous, broad, concave, 1 to 1½ lines diameter ; bracteoles smaller.　Calyx-segments at the time of flowering acute and not above 1 line long but soon enlarging.　Corolla broadly urceolate, about 2 lines long.　Fruiting-calyx depressed globular, succulent and berry-like, snow-white, surrounding or enclosing the fruit.— DC. Prod. vii. 594 ; Hook. f. Fl. Tasm. i. 241 ; A. Rich. Sert. Astrolab. t. 30.

N. S. Wales.　Summits of snowy mountains at the head of Bellenger river, at an elevation of 4000 ft., *C. Moore.*
Victoria.　Ranges of the Australian Alps at an elevation of 4000 to 6000 ft., rarely descending to 3000 ft., *F. Mueller.*
Tasmania.　Derwent river and Mount Wellington, *R. Brown ;* common or the moun-tains throughout the island at an elevation of 2000 to 4000 ft., *J. D. Hooker.*

2. **G. lanceolata,** *Hook. f. in Hook. Lond. Journ.* vi. 267 ; *Fl. Tasm.* i. 241. *t.* 72.　Stems, from a thick, woody base, diffuse or erect and bushy, not above 1 ft. high, more or less hispid with short stiff hairs or bristles. Leaves elliptical-oblong or lanceolate, rather acute, with callous serratures, ½ to 1 in. long.　Flowers on pedicels of 1 to 2 lines, solitary in the upper axils, but forming a short, terminal, leafy raceme, the floral leaves usually.

much smaller than the others. Bracts several, at the base of the pedicels, the uppermost about 1 line long. Calyx at the time of flowering fully half as long as the corolla. Corolla urceolate, about 2 lines long. Stigma peltate, 5-lobed. Fruiting-calyx berry-like, red, more or less enclosing the fruit.

Tasmania. Summits of Ben Lomond, the Western mountains, etc., *Gunn.* In the figure above quoted, by a mistake of the artist, the anthers are represented as having only 1 instead of 2 awns to each cell.

3. **G. antipoda,** *Forst.; DC. Prod.* vii. 594. The Tasmanian form of this very variable species is a small, depressed or prostrate shrub, with the branches, and sometimes also the margins of the leaves, hispid with stiff hairs or bristles. Leaves shortly petiolate, broadly ovate or orbicular, crenulate, 3 to 4 lines long. Flowers solitary, on very short, axillary peduncles, which, however, bear several small bracts. Fruit enclosed in a berry-like calyx, much larger than that of *G. hispida.*—Hook. f. Fl. Tasm. i. 241. t. 73A; *G. depressa,* Hook. f. in Hook. Lond. Journ. vi. 267.

Tasmania. Summits of Mount Olympus, Ben Lomond, Mount Lapeyrouse, etc., *Gunn* and others. The species is common in New Zealand, where it is usually a bushy shrub attaining 4 or 5 ft., as figured by A. Rich. Fl. N. Zel. t. 28; but on the higher mountains it is there also sometimes reduced to the small, prostrate state in which it is found in Tasmania.

ORDER LXVII. **EPACRIDEÆ.**

Calyx of 5 rarely 4 distinct sepals, much imbricate in the bud. Corolla regular, with a cylindrical, urceolate or campanulate tube, and 5 rarely 4 lobes, valvate or variously imbricate in the bud, more or less spreading or rarely cohering in a calyptra, or rarely the whole corolla separating into distinct petals. Stamens as many as corolla-lobes or rarely fewer, hypogynous, and free or more or less adnate to (inserted in) the corolla-tube; anthers versatile or rarely adnate, 1-celled (more or less perfectly 2-celled before opening), opening by a single longitudinal slit in 2 valves, leaving no longitudinal dissepiment or only a thin and slightly prominent one. Hypogynous disk annular or cup-shaped, entire, 5-lobed or consisting of 5 distinct scales, rarely deficient. Ovary superior, with 5 or sometimes fewer, rarely 6 to 10 cells; ovules solitary in each cell and pendulous, or several in each cell, the placenta proceeding from the axis immediately under the attachment of the style. Style simple and undivided, terminal in the uniovulate genera, inserted in a central, tubular depression of the ovary so as to be lateral (with reference to the carpels) or almost basal in the pluriovulate genera; stigma terminal, small, capitate or peltate, and sometimes slightly lobed. Fruit indehiscent, and more or less drupaceous in the uniovulate genera, capsular, and loculicidally dehiscent in the pluriovulate ones. Seeds with a thin, rarely almost crustaceous testa; embryo straight, much shorter than the albumen, terete or nearly so, the radicle next the hilum.—Shrubs or rarely trees. Leaves alternate or very rarely opposite, often crowded or imbricate, rigid, entire or scarcely denticulate, with several longitudinal, simple or forked nerves, sometimes prominent underneath, sometimes very fine and numerous or very obscure. Flowers axillary or terminal, either solitary and terminating peduncles more or less covered with imbricately scale-like or leaf-like bracts, or in spikes or racemes, each flower between 2 bracteoles in the axil of a subtending bract, the common peduncle usually ending in a small rudimentary flower with its subtending bract, the peduncles or spikes solitary or rarely several in a terminal panicle. Sepals usually finely marked with parallel or diverging veins. Corolla white or of various shades of red, rarely blue, green or yellowish.

The Order is almost confined to Australia, New Caledonia, New Zealand, and the Antarctic Islands, a few species spread over the islands of the Pacific and the Indian Archipelago, and a single one representing it in the mountains of extratropical South America; the extra Australian species belong to 4 out of the 24 Australian genera, except the South American species, and 1 or 2 from New Caledonia, which have been referred to genera not quite identical with Australian ones.

The division of the Order into two suborders, tribes or comprehensive genera, is remarkably clear and definite, and has been admitted by all; but the characters hitherto found available for their subdivision into lower groups have in some cases proved inconstant, and in others have been pronounced as of little value from à priori considerations; and many genera proposed by Brown, and generally adopted by all, have, nevertheless, been rejected, first by Poiret and Sprengel, and recently again by F. Mueller. The species, however, are so numerous that subdivision, whether into genera or into sections, is necessary, and the following appear to me to be the characters the most available for the purpose. The foliage, in many respects uniform and characteristic of the Order, divides nevertheless the tribe *Epacreæ* into three natural groups. The inflorescence and bracts, although, perhaps, less diversified in principle than was formerly supposed, offer still some modifications, which are constant in some genera, and very general in others. The calyx is remarkably uniform in the whole Order. The corolla, its shape, the æstivation of its lobes, and the arrangement of certain tufts of hairs it often bears (which probably take some part in aid of fertilization), has been made much use of by Brown and others for the distinction of genera, but is now almost entirely rejected by F. Mueller. It appears to me, however, to afford often most useful characters, although not always quite absolute. The filaments adnate to the corolla or free, flat or terete, the anthers connate or free, exserted or included, entire or 2-lobed, may be in some cases generic differences, but in others are specific only; the reduced number of stamens only serves to separate the monotypic genus *Oligarrhena*. The hypogynous disk, its presence or absence, and integrity or separation into scales, has completely broken down as a generic character, and may not always be constant in species. The ovary is nearly uniform in each of the great tribes, varying in Styphelieæ only in the number of cells, the differences being more frequently specific than generic, and in Epacreæ presenting only one marked modification of the placenta, which, however, neatly separates *Richea* and *Dracophyllum* from the rest of the tribe. The fruit is uniform in Epacreæ; in Styphelieæ the greater or lesser degree of succulence in the mesocarp, and of consolidation in the endocarp, is occasionally of considerable generic value, notwithstanding its vagueness. In the structure of the seeds I have not detected any differences of any importance. Their number and position either corresponds with that of the ovules; or, if altered in the course of growth, the consequent modifications do not appear to give any generic indications.

TRIBE I. **Styphelieæ.**—*Ovules solitary in each cell of the ovary, pendulous from the summit of the cavity. Style terminal. Fruit indehiscent, usually drupaceous.*

Anthers exserted. (Corolla-lobes revolute so as completely to expose
    the erect summits of the filaments and the anthers.)
Anthers free. Filaments glabrous.
    Corolla-tube long or slender. Fruit a 5-celled drupe . . . . 1. STYPHELIA.
    Corolla-tube very short. Fruit 5-pyrened and berry-like . . 6. PENTACHONDRA.
Anthers connate in a cone round the style. Filaments glabrous . 2. COLEANTHERA.
Anthers connivent or connate, enveloped with the filaments in a
    dense wool . . . . 3. ASTROLOMA.

Anthers wholly or partially enclosed in the corolla-tube or in the erect
base of the lobes, or rarely recurved with the lobes.

Corolla-tube (usually long) with 5 tufts of hairs or hairy scales or
a dense ring of hairs inside below the middle. Filaments usually
flat . . . . 3. ASTROLOMA.

Corolla-tube short, with 5 glandular scales inside below the middle 5. MELICHRUS.

Corolla-tube conical in the upper portion, with minute, erect
lobes. Anthers 2-lobed . . . . 4. CONOSTEPHIUM.

Corolla-tube without scales or hairy tufts below the middle, the
lobes spreading at least at the end. Filaments terete or
nearly so.

Corolla-lobes more or less imbricate in the bud, glabrous, the
throat closed with reflexed hairs or scales . . . . 9. BRACHYLOMA.

Corolla-lobes broadly induplicate in the bud, glabrous . . . . 10. NEEDHAMIA.

Corolla-lobes valvate in the bud, glabrous.

Corolla-tube cylindrical or urceolate.

Drupe with a several-celled, hard nucleus, the mesocarp
very pulpy. Flowers usually solitary, with imbricate
bracts . . . . 8. CYATHODES.

Drupe with a several-celled nucleus, the mesocarp moderately pulpy. Flowers in spikes or racemes, the bracts
and bracteoles distinct . . . . 11. LISSANTHE.

Drupe with 10 more or less separable pyrenes. Flowers
in spikes or clusters, the bracts and bracteoles distinct . 7. TROCHOCARPA.

Drupe berry-like, very pulpy, with 5 distinct pyrenes.
Flowers solitary or in spikes, the bracts and bracteoles
distinct . . . . 6. PENTACHONDRA.

Corolla-tube short and campanulate. Flowers very small.

Stamens and corolla-lobes 4 or 5. Ovary 1- or 2-celled . 14. MONOTOCA.

Stamens 2. Corolla-lobes 4. Ovary 2-celled . . . . 15. OLIGARRHENA.

Corolla-lobes valvate, with a reflexed beard at the tip, and reflexed hairs along the throat . . . . 13. ACROTRICHE.

Corolla-lobes valvate in the bud, bearded inside.

Drupe with several-celled, hard nucleus, the mesocarp very
pulpy. Flowers usually solitary, with imbricate bracts . 8. CYATHODES.

Drupe with a several-celled, rarely 1-celled, hard or thin
nucleus, the mesocarp moderately pulpy or dry. Flowers
in spikes or solitary, the bracts and bracteoles distinct . . 12. LEUCOPOGON.

Drupe with 10 (or fewer by abortion) separable pyrenes.
Flowers in spikes or clusters, the bracts and bracteoles
distinct . . . . 7. TROCHOCARPA.

TRIBE II. **Epacreæ.**—*Ovules several in each cell of the ovary. Style inserted in a
central tubular depression, so as to be lateral or basal. Capsule loculicidally dehiscent.*

Leaves petiolate, sessile or stem-clasping, not sheathing. Placentas
sessile or nearly so.

Bracts imbricate on the calyx, passing into the sepals.

Corolla-lobes quincuncially imbricate . . . . 16. EPACRIS.

Corolla-lobes contorted-imbricate . . . . 17. LYSINEMA.

Bracts or bracteoles at a distance from or scarcely reaching the
calyx, and very different from the sepals.

Stamens adnate to (inserted in) the corolla-tube . . . . . 18. ARCHERIA.

Stamens hypogynous, free . . . . 19. PRIONOTIS.

Leaves with an adnate, sheathing base, which falls off with the leaf,
leaving the denuded branches smooth and scarless. Placentas
sessile or nearly so.

Stamens adnate to (inserted in) the corolla-tube . . . . . 20. COSMELIA.

Stamens hypogynous, free. Corolla scarcely exceeding the calyx.

Corolla-tube very short; lobes very spreading, glabrous, more
 or less imbricate . . . . 21. SPRENGELIA.
Corolla-tube cylindrical; lobes erect, recurved or revolute,
 bearded inside, valvate in the bud . . . . 22. ANDERSONIA.
Leaves with an adnate, sheathing base, which falls off with the leaf,
 leaving annular scars on the denuded branches. Placentas de-
 pending from an ascending, recurved stipes.
Corolla circumsciss near the base, calyptriform, the lobes not
 opening . . . . 23. RICHEA.
Corolla not circumsciss, the lobes spreading . . . . 24. DRACOPHYLLUM.

TRIBE I. STYPHELIEÆ.—Ovules solitary in each cell of the ovary, and pendulous from the summit of the cavity. Style terminal. Fruit indehiscent, usually drupaceous.

1. **STYPHELIA**, Sm.

(Soleniscia, *DC.*)

Corolla-tube elongated, cylindrical or slightly ventricose, hairy inside at the throat, and with 5 tufts of hairs, sometimes confluent in a ring below the middle or rarely glabrous; lobes linear, bearded inside, much revolute, exposing the stamens. Filaments free from the throat, filiform, glabrous; anthers exserted, free, linear, 1-celled, attached about the middle. Hypogynous scales distinct or united in a cup-shaped disk. Ovary 5-celled, with 1 ovule in each cell. Style filiform, longer than the corolla-tube; stigma small. Fruit a drupe, with a dry or slightly pulpy mesocarp, and a hard, bony endocarp, with 5 cells and seeds or fewer by abortion.—Leaves sessile or scarcely petiolate. Flowers axillary, solitary, with the rudiment of a second or very rarely 2 or 3, on a very short peduncle. Bracts several, 1 or 2 of the uppermost more or less enlarged, and embracing the base of the calyx as well as the still larger bracteoles. Calyx usually coloured.

The genus is limited to Australia. The very much revolute corolla-lobes and exserted stamens distinguish it from all its allies except *Coleanthera*, which is readily known by the small flowers and connate anthers.

Styphelia was originally intended by Smith to comprehend the few drupaceous Epacrideæ then known. R. Brown, in adding a vast number of species, divided it into several groups which he proposed as distinct genera, considering the original genus as a tribe, but he stated at the same time that others might prefer to treat them as sections of one comprehensive genus. Sprengel attempted to consolidate some of these genera, but, by rearranging the species upon technical characters taken from books, without actual observation, he created nothing but confusion. De Candolle, Endlicher, Lindley, Sonder, and others, who have more or less studied the Order, have adopted the views of Brown, the excellence of whose groups has, with few minor exceptions, been fully confirmed by the additions since made. F. Mueller, after a careful study of a large number of species, which has supplied us with numerous valuable observations (Fragm. iv. and vi.), returns to the idea of one large genus divided into sections, justifying the change on the observation that some of the characters relied upon for the distinction of genera have failed, that there are frequently intermediate species connecting the several groups, and that *Styphelia*, if retained in its original comprehensive sense, is not, after all, so numerous in species as *Acacia* and others, which no one attempts to break up. Fully admitting the correctness of these statements, it may, however, be observed that, in the whole Vegetable Kingdom, there are few, if any, large genera which are not more or less connected with others by intermediate species; that in such genera as *Acacia* or *Eugenia*, for instance, throughout the whole of the 500 odd species known there is the greatest uniformity in the structure of the flowers; whilst in Stypheliæ there is considerable diversity, as well in the corolla as in the stamens and pistil; and that, if in the characters hitherto given some have failed, others have been brought forward in their support. F. Mueller, indeed, does not propose to remodel Brown's groups, but only to reduce their value in the systematic scale, of which the principal result is an altered nomenclature. It appears to me, therefore, to be a question more of convenience than of observation of fact, whether we should describe Stypheliæ as one tribe with several genera, or as one genus with several subgenera or sections. Had the genus remained hitherto undivided I might, perhaps, with F. Mueller, have adopted the latter view; but Brown's genera have now been so long and so generally recognized by botanists and horticulturists, that it appears to me that the proposed change, without really advancing the cause of science, would rather lead to practical confusion, as it has already added more than half a hundred superfluous names to the overloaded synonymy of the tribe.

SECT. I. **Eustyphelia.**—*Corolla-tube with* 5 *dense tufts of hairs below the middle, sometimes confluent in a ring.*

Diffuse or prostrate. Hypogynous scales quite distinct, lanceolate . 1. *S. adscendens.*
Erect shrubs. Hypogynous scales more or less cohering in a ring
   or cup, at least at the base.
   Branches pubescent-hirsute. Leaves concave.
     Leaves long-lanceolate, tapering into a fine point, 1 to 2 in.
      long, very concave. Corolla-tube about 1 in. . . . . . 2. *S. longifolia.*
     Leaves ovate to lanceolate, slightly concave, rarely exceeding 1
      in. Corolla-tube about ½ in. . . . . 3. *S. læta.*
   Branches glabrous or with a minute scarcely visible down.
     Leaves obovate-oblong to oblong-lanceolate, shortly tapering
      into a rigid point.
      Leaves concave or nearly flat. Sepals obtuse . . . . . 4. *S. triflora.*
      Leaves flat or slightly convex. Sepals acute. Corolla green 5. *S. viridis.*
     Leaves short, obovate or oblong, obtuse, flat or slightly convex.
      Western species . . . . 6. *S. Hainesii.*
     Leaves oblong-linear, abruptly mucronate, with recurved or re-
      volute margins . . . . 7. *S. tubiflora.*

SECT. II. **Soleniscia.**—*Corolla-tube slender, quite glabrous inside except a few hairs in the throat.*

Leaves ovate to lanceolate, flat or concave, about ½ in. long. Corolla-
   tube 1 in. long. Disk annular . . . . 8. *S. tenuiflora.*
Leaves ovate or broadly cordate, flat or concave, 3 to 5 lines long.
   Sepals acuminate. Corolla-tube 3 lines long. Disk annular . . 9. *S. melaleucoides.*
Leaves ovate, flat or concave, about 3 lines long. Sepals obtuse.
   Corolla-tube 1½ lines long. Disk annular . . . . 10. *S. pusilliflora.*
Leaves broad, concave, recurved at the end, not 2 lines long. Corolla-
   tube about 2 lines. Hypogynous scales free, lanceolate, acuminate 11. *S. leucopogon.*

SECT. I. EUSTYPHELIA. Corolla-tube with 5 dense tufts of hairs below the middle between the stamens, sometimes confluent in a ring.

1. **S. adscendens,** *R. Br. Prod.* 537. A much-branched diffuse and rigid shrub, forming broad matted patches, the branches prostrate or shortly ascending, often pubescent. Leaves rather crowded, lanceolate, 6 to 9 lines long, with a fine pungent point, the margins minutely scabrous-ciliate. Flowers solitary in the axils, almost sessile, surrounded by about 4 bracts gradually enlarged, embracing the calyx, and passing into the bracteoles which are about 2 lines long, obtuse. Sepals oblong, obtuse, about 4 lines long, slightly coloured at the tips. Corolla yellowish, more or less green at the tip, the tube about 6 lines long, with 5 dense tufts of hairs almost confluent inside below the middle and densely hairy from thence to the throat;

lobes linear-lanceolate, acute, densely bearded, scarcely shorter than the tube. Drupe ovoid, 5-ribbed, about as long as the calyx.—DC. Prod. vii. 735 ; Hook. f. Fl. Tasm. i. 243 ; F. Muell. Fragm. vi. 36.

Victoria. Stony places in the Grampians, *F. Mueller ;* Wimmera, *Dallachy.*

Tasmania. Port Dalrymple, *R. Brown ;* dry pastures, heaths, etc., near Hobarton, Circular Head, etc., common, *J. D. Hooker.*

2. **S. longifolia,** *R. Br. Prod.* 537. An erect shrub, with virgate softly pubescent branches. Leaves long-lanceolate, gradually tapering into a fine rigid point, concave, 1 to 2 in. long or the lower ones still longer. Flowers green, solitary in the axils, nearly sessile. Bracteoles about 3 lines long, rather obtuse. Sepals 7 to 9 lines long, acute. Corolla-tube nearly 1 in. long, with 5 dense confluent tufts of hairs above the base, less hairy in the upper part ; lobes linear, bearded inside. Hypogynous scales broad, more or less cohering at the base, free and spreading at the top. Drupe ovoid, 5-angled, 3 or 4 lines long.—DC. Prod. vii. 735 ; Bot. Reg. t. 24 ; Lodd. Bot. Cab. t. 1583.

N. S. Wales. Port Jackson, *R. Brown, Sieber, n.* 77, and others.

3. **S. læta,** *R. Br. Prod.* 537. An erect shrub, with spreading softly pubescent branches. Leaves in the typical form ovate ovate-lanceolate or broadly oblong, shortly tapering into a rigid point, under 1 in. long, flat or concave. Flowers pale red ?, very shortly pedicellate, solitary in the axils. Bracteoles obtuse, about 2 lines long. Sepals about 5 lines long, rather acute. Corolla-tube $\frac{1}{2}$ to $\frac{3}{4}$ in. long, with 5 dense tufts of hairs inside above the base, and hairy towards the throat ; lobes long, revolute, bearded inside. Hypogynous scales cohering in a scarcely lobed cup, but readily separating. —DC. Prod. vii. 735 ; *S. latifolia*, Sieb. Pl. Exs.

N. S. Wales. Port Jackson to the Blue Mountains, *R. Brown, Sieber, n.* 80, and others.

Var. *latifolia*. Leaves short, very broadly ovate and very concave.—*S. latifolia*, R. Br. Prod. 537 ; DC. Prod. vii. 735.—Hawkesbury river, *R. Brown.*

Var. *angustifolia*, Benth. in Hueg. Enum. 70. Leaves narrow-lanceolate, $\frac{1}{4}$ to nearly 1 in. long.—*S. angustifolia*, DC. Prod. vii. 735 ; *S. læta*, Reichenb. Icon. Exot. t 99. Port Jackson and Blue Mountains, *Sieber, n.* 79, and others.

4. **S. triflora,** *Andr. Bot. Rep. t.* 72. A tall shrub, quite glabrous or the branches very minutely pubescent. Leaves from obovate-oblong to oblong-lanceolate, very shortly tapering into a rigid point, flat or more or less concave, rarely exceeding 1 in., and the broad ones at the base of the shoots often very short. Flowers pale pink and yellow, very shortly-pedicellate, solitary or very rarely 2 (or 3 ?) together in the lower axils and often appearing clustered at the base of the shoot, especially when some of the floral leaves are small or abortive. Bracteoles obtuse, about 2 lines long. Sepals obtuse or almost acute, about 5 lines long. Corolla varying in size, but the tube usually about $\frac{3}{4}$ in., with 5 dense tufts of hairs inside above the base. Hypogynous scales more or less cohering in a truncate or shortly lobed cup. —R. Br. Prod. 537 ; DC. Prod. vii. 735 ; Bot. Mag. t. 1297 ; Lodd. Bot. Cab. t. 426 ; *S. glaucescens*, Sieb. Pl. Exs.

N. S. Wales. Port Jackson to the Blue Mountains, *R. Brown, Sieber, n.* 75, 86, and others. The name, as observed by Loddiges, is not well chosen, for the flowers are almost always solitary, and I have never seen them as figured by Andrews, probably from an over-luxuriant garden specimen. The species is very closely allied both to *S. læta* and *S. viridis*, differing from the former in the glabrous branches and clustered inflorescence, from the latter in the colour of the flowers as well as in the leaves rather concave than convex.

5. **S. viridis,** *Andr. Bot. Rep. t.* 312. An erect shrub, with spreading branches, quite glabrous or with a minute scarcely perceptible down. Leaves oblong lanceolate or obovate-oblong, abruptly narrowed into a short rigid point, flat or slightly convex, under 1 in. long. Flowers green, solitary in the axils, nearly sessile. Bracteoles broad and obtuse, under 2 lines long. Sepals almost acute, about $\frac{1}{2}$ in. Corolla-tube nearly $\frac{3}{4}$ in. long in the normal form, with 5 dense tufts of hairs inside above the base. Hypogynous scales broad, obtuse, free or slightly cohering.—DC. Prod. vii. 735 ; *S. viridiflora*, R. Br. Prod. 537 ; Sweet, Fl. Austral. t. 50.

Queensland. Moreton island, *F. Mueller.*
N. S. Wales. Port Jackson, *R. Brown, Sieber, n.* 78, and others ; Hunter's River, *A. Cunningham ;* Miall Brush, *Leichhardt ;* Darling Downs, *F. Law.*

Var. ? *breviflora.* Leaves narrower. Sepals more obtuse, about 4 lines long. Corolla-tube about $\frac{1}{2}$ in. long. To this belong the specimens from Queensland and from the northern parts of N. S. Wales.

6. **S. Hainesii,** *F. Muell. Fragm.* iv. 96. *t.* 28. A glabrous bushy shrub of about 2 ft. Leaves obovate-oblong, obtuse or rarely with a minute point, flat or slightly convex, 3 to 5 lines long. Flowers solitary in the axils, nearly sessile. Bracteoles very obtuse and broad, about $\frac{1}{2}$ line long. Sepals obtuse, about 2 lines. Corolla-tube about $\frac{3}{4}$ in. long with 5 dense tufts of hairs inside above the base. Hypogynous scales united in a short truncate cup.

W. Australia. Limestone cliffs and sand-drift hummocks, Eyre's Relief, and along the coast to Cape Paisley, *Maxwell.* Although the foliage and habit are different from those of the Eastern species, it quite agrees with them in the hairy tufts of the corolla-tube, as well as in the general shape and structure of the flowers.

7. **S. tubiflora,** *Sm. Bot. Nov. Holl.* 45. *t.* 14. An erect glabrous much-branched shrub. Leaves oblong-linear, sometimes slightly cuneate, abruptly mucronate, with revolute margins, mostly about $\frac{1}{4}$ in. long. Flowers red, solitary in the axils, nearly sessile or very shortly pedicellate. Bracteoles scarcely 1$\frac{1}{2}$ lines long, very broad, mucronate-acute. Sepals nearly 4 lines long, mucronate-acute. Corolla-tube nearly 1 in. long, the revolute lobes long and narrow. Hypogynous disk cup-shaped, truncate, scarcely lobed.—R. Br. Prod. 537 ; DC. Prod. vii. 735 ; Lodd. Bot. Cab. t. 1938 ; Paxt. Mag. xii. 29, with a fig. ; Maund. Botanist, t. 142.

N. S. Wales. Port Jackson, *R. Brown, Sieber, n.* 76, and many others.

Sect. II. Soleniscia.—Corolla-tube very slender, quite glabrous inside.

8. **S. tenuiflora** (*misprinted* tenuifolia), *Lindl. Swan Riv. App.* 25. An erect bushy rigid glabrous shrub of 2 to 3 ft. Leaves nearly sessile, from broadly ovate to lanceolate, mucronate-acuminate, flat or concave, mostly about $\frac{1}{2}$ in. long. Flowers solitary in the axils, sessile. Bracteoles scarcely 1 line long, very broad and obtuse, the outer bracts all minute or the upper-most scarcely enlarged. Sepals obtuse, smooth, about 2$\frac{1}{2}$ lines long. Co-

rolla-tube very slender, fully 1 in. long, without either tufts of hairs or scales inside ; the lobes short and thinly hairy inside, but linear and much revolute as in other *Styphelia.* Hypogynous disk short, 5-toothed. Drupe ovoid, nearly twice as long as the calyx.—*Soleniscia elegans*, DC. Prod. vii. 738 ; Deless. Ic. Sel. v. 21 ; *Styphelia elegans*, Sond. in Pl. Preiss. i. 29€.

W. Australia. From King George's Sound to Swan River, *Drummond, 1st Coll. also n.* 15, 481, *Preiss, n.* 468, 469, *Harvey, Maxwell.*

9. **S. melaleucoides,** *F. Muell. Fragm.* iv. 97. vi. 30. A shrub of 2 to 3 ft., glabrous or the branchlets minutely hoary-pubescent. Leaves in the typical form broadly ovate-cordate or almost orbicular, mucronate-acute, rigid, flat or concave, 3 to 5 lines long. Flowers solitary or 2 together on very short peduncles with a terminal rudiment, the bracts very small ; bracteoles scarcely ½ line long, very broad and obtuse. Sepals 1 to 1¼ line long, very obtuse. Corolla-tube slender, about 3 lines long, slightly hairy in the throat, otherwise glabrous inside and without scales ; lobes as long as the tube, closely revolute, bearded inside. Anthers linear. Hypogynous disk short, truncate or sinuate-toothed. Ovary 5-celled, tapering into the slender style.

W. Australia. In the interior from Eagle Hawk Camp, *Maxwell.*

Var. *ovata*, F. Muell. Leaves all ovate, not cordate, with a longer rigid point.—Israelite Bay, *Maxwell.*

10. **S. pusilliflora,** *F. Muell. Fragm.* iv. 105. Branchlets minutely pubescent. Leaves ovate, acute, with a fine rigid point, about 3 lines long, shining and rigid like those of *S. melaleucoides*, but rather convex than concave. Flowers solitary or 2 together on a very short peduncle with a terminal rudiment. Bracts very small ; bracteoles half as long as the calyx, and sepals 1½ lines long, all acutely acuminate. Corolla-tube scarcely exceeding the calyx, glabrous inside except a few hairs in the throat ; lobes revolute, as long as the tube, bearded inside. Anthers oblong-linear. Hypogynous disk truncate or sinuate-toothed. Ovary 5-celled, tapering into the slender style. —*Leucopogon exarrhenus*, F. Muell. Fragm. i. 178 ; *Styphelia exarrhena*, F. Muell. Fragm. vi. 31.

S. Australia. Near Penola in the Tattiara country, *Wood*, a single small specimen in Herb. F. Mueller, evidently allied to *S. melaleucoides*, but readily distinguished by the short corolla-tube and acutely acuminate bracts and sepals.

11. **S. leucopogon,** *F. Muell. Fragm.* iv. 97. vi. 31. An erect shrub of 1 to 2 ft., the branches rather slender, glabrous or minutely pubescent. Leaves broadly ovate, shortly acuminate, contracted into a short petiole, very concave and almost conduplicate but recurved at the end, not 2 lines long. Flowers shortly pedicellate, with very small bracts, the bracteoles not ½ line long. Sepals a little more than 1 line long, rather narrow but obtuse, smooth. Corolla-tube slender and cylindrical, but scarcely exceeding 2 lines, quite glabrous inside ; lobes nearly as long, bearded inside, much revolute. Hypogynous scales free, lanceolate, acuminate. Anthers rather small ; stamens otherwise, as well as the ovary and style, entirely those of the genus.— *Soleniscia pulchella*, Stschegl. in Bull. Mosc. 1859, i. 3 ; *Leucopogon exsertus*, F. Muell. Fragm. iii. 143.

W. Australia. *Drummond*, 5*th Coll. n.* 327; Phillips Range and Eyre's Relief, *Maxwell.*

2. COLEANTHERA, Stschegl.

(Michiea, *F. Muell.*)

Sepals small. Corolla-tube short, bearded inside at the throat, otherwise glabrous; lobes long, linear, valvate in the bud, more or less bearded inside, much revolute exposing the stamens. Filaments inserted in the throat; anthers exserted, linear, 1-celled, attached by the middle, cohering above the middle in a cone round the style. Hypogynous disk none or very obscure. Ovary 5-celled, with 1 ovule in each cell. Style filiform, longer than the corolla-tube; stigma small. Fruit a drupe with a dry or scarcely pulpy mesocarp and a hard endocarp, with 5 cells and seeds, or fewer by abortion.— Leaves flat or concave. Flowers small, 2 or 3 together on a short axillary peduncle, or solitary with a minute rudiment of another. Bracts few, small, the subtending one slightly enlarged and embracing the base of the calyx as well as the larger bracteoles.

The genus is limited to S.W. Australia. It is nearly allied to *Styphelia*, with which F. Mueller has recently united it (Fragm. vi. 80), but the small flowers give it the aspect of *Leucopogon*, and the exserted connate anthers readily distinguish it from both.

Leaves strongly veined underneath, ovate or lanceolate usually hairy .   1. *C. cœlophylla.*
Leaves smooth, the veins very fine and scarcely prominent, usually glabrous.
    Leaves ovate or orbicular. Corolla-lobes bearded to the end . . . 2. *C. myrtoides.*
    Leaves linear or linear-lanceolate. Corolla-lobes bearded at the base
    only . . . . 3. *C. virgata.*

1. C. cœlophylla, *Benth.* An erect bushy shrub of 1 to 2 ft., more or less hirsute with soft hairs on the branches and margins and bases of the leaves. Leaves ovate to lanceolate, obtuse or with a small callous point, concave, strongly striate underneath, $\frac{1}{4}$ to $\frac{1}{2}$ in. long. Flowers solitary or rarely 2 together on a very short peduncle, all axillary. Bracts very small; bracteoles broad, obtuse, not $\frac{1}{2}$ line long. Sepals about 1 line long, obtuse, ciliate. Corolla-tube shorter than the calyx, the lobes longer than the tube and bearded as in *C. myrtoides;* stamens and style the same as in that species.—*Leucopogon cœlophyllus*, A. Cunn.; DC. Prod. vii. 753.

W. Australia. Eastward of King George's Sound, *Baxter.* The strongly ribbed foliage gives this plant a very different aspect from that of *C. myrtoides*, independently of the hairs.

2. C. myrtoides, *Stschegl. in Bull. Mosc.* 1859, i. 4. An erect bushy shrub, attaining 4 to 5 ft., glabrous or the branches slightly pubescent. Leaves very shortly petiolate, often clustered towards the end of each annual shoot, ovate or almost orbicular, obtuse or with a short callous point, concave or nearly flat, the flabellate veins often forked but fine and scarcely conspicuous, 3 to 4 lines long. Peduncles very short, 1- to 3-flowered, those at the base of the shoots often without any subtending leaf. Bracts and bracteoles very small. Sepals ovate, obtuse, about 1 line long. Corolla-tube shorter than the calyx; lobes longer than the tube, bearded the whole length. Filaments filiform, glabrous, shorter than the lobes; anthers linear, 2 lines long. Stigma enclosed within the anther-cone or shortly protruding beyond it.—*Michiea symphyanthera,* F. Muell. Fragm. iv. 96. t. 27; *Styphelia Michiei,* F. Muell. Fragm. vi. 80.

W. Australia. *Drummond, 4th Coll. n.* 154, 5*th Coll. n.* 302; dry rocky ridges, Stirling Range, Salt River, Gardner Range, *Maxwell.*

3. **C. virgata,** *Stschegl. in Bull. Mosc.* 1859, i. 5. A shrub with elongated virgate or flexuose branches, usually loosely pubescent. Leaves shortly petiolate, linear or linear-lanceolate, obtuse, concave, glabrous or loosely pubescent, about ½ in. long on the main branches, shorter on the branchlets. Peduncles very short. Bracts and bracteoles of *C. myrtoides,* and flowers also the same, except that the corolla-lobes are bearded at the base only, and as well as the anthers are rather longer than in that species.

W. Australia, *Drummond, 5th Coll. n.* 303.

3. ASTROLOMA, R. Br.

(Ventenatia, *Cav.,* Stenanthera, *R. Br.,* Stomarrhena, *DC.,* Pentataphrus, *Schlecht.,* Mesotriche, *Stschegl.*)

Corolla-tube elongated, cylindrical or slightly ventricose, either with 5 tufts of hairs or densely hairy scales inside above the base or rarely without either; lobes linear or lanceolate, bearded inside or rarely glabrous, valvate, erect at the base round the anthers, spreading or recurved at the top only (more spreading in *A. stomarrhena*). Filaments short, often much flattened, inserted in the throat; anthers oblong or linear, 1-celled, attached above the middle or almost at the top. Hypogynous disk cup-shaped or annular, truncate or obscurely lobed. Ovary 5-celled, with 1 ovule in each cell; style filiform, as long as or longer than the corolla-tube; stigma capitate, often large and hairy, rarely 5-lobed. Fruit a drupe, with a dry or slightly pulpy mesocarp and a hard bony endocarp, with 5 cells and seeds, or fewer by abortion.—Leaves sessile or scarcely petiolate. Flowers solitary in the axils on very short pedicels or almost sessile, surrounded by several bracts of which from 2 to 4 of the innermost are gradually enlarged and as well as the still larger bracteoles embrace the base of the calyx.

The genus is limited to Australia. It is readily distinguished from *Styphelia* by the anthers more or less concealed within the corolla-tube or the erect base of its lobes, and from the larger flowered species of *Leucopogon* by the inflorescence, by the tufts of hairs in the tube, or by the dilated filaments, and in the great majority of species by all three characters.

Sect. I. **Stomarrhena.**—*Corolla-tubes with* 5 *tufts of hairs inside below the middle, alternating with the stamens and sometimes confluent in a ring.*

Filaments terete or scarcely flattened. Leaves convolute or concave, entire or minutely scabrous-denticulate.
 Plant hairy. Stamens protruding beyond the somewhat revolute lobes in a dense woolly mass . . . . 1. *A. stomarrhena.*
 Plant glabrous or the branches slightly pubescent. Stamens enclosed in the erect base of the lobes.
  Leaves narrow-lanceolate. Sepals acute, 7 to 8 lines long . 2. *A. macrocalyx.*
  Leaves narrow-lanceolate. Sepals almost obtuse, about 4 lines long . . . . 3. *A. xerophyllum.*

Filaments much flattened.
  Leaves concave, entire or scarcely scabrous-denticulate.
    Leaves spreading, ovate or lanceolate, 1 to 1½ lines long . .   4. *A. microphyllum.*
    Leaves erect, linear or linear-lanceolate, 4 to 6 lines long . 5. *A. prostratum.*
    Leaves erect, ovate or broadly lanceolate, above ½ in. long . 6. *A. tectum.*
  Leaves concave flat or slightly convex, denticulate-ciliate.
    Leaves ovate to broadly lanceolate, concave. Shrub erect . 7. *A. Candolleanum.*
    Leaves broadly lanceolate, flat. Shrub erect . . . . . 8. *A. microdonta.*
    Leaves lanceolate or oblanceolate, concave. Shrub low or
      diffuse . . . . 9. *A. pallidum.*
    Leaves cuneate obovate or oblanceolate, nearly flat. Shrub
      prostrate . . . . 10. *A. compactum.*
    Leaves lanceolate linear or almost subulate, tapering into a
      pungent point, the margins sometimes recurved . . . . 11. *A. humifusum.*
  Leaves narrow with recurved or revolute margins, entire or
    scarcely scabrous-denticulate.
    Leaves very spreading, lanceolate or linear-lanceolate, with
      rigid or pungent points . . . . 12. *A. divaricatum.*
    Leaves linear, with short points not pungent, usually hoary or
      glaucous underneath.
      Calyx fully 3 lines long. Plant usually softly pubescent . 13. *A. Drummondii.*
      Calyx 2 to 2½ lines long. Plant usually glabrous or mi-
      nutely pubescent . . . . 14. *A. microcalyx.*

SECT. II. **Pentataphrus.**—*Corolla-tube with 5 deflexed fringed scales inside below the middle alternating with the stamens. Filaments much flattened. Leaves linear with revolute margins. Bracts and sepals large.*

Calyx 3 to 4 lines long. Corolla-tube exserted. Anthers acuminate . . . . 15. *A. Baxteri.*
Calyx 6 to 8 lines long. Corolla-tube scarcely longer. Anthers
  very obtuse . . . . 16. *A. conostephioides.*

SECT. III. **Stenanthera.**—*Corolla-tube without tufts of hairs or scales inside below the middle. Filaments much flattened. Leaves linear, with revolute margins.*

Calyx 3 to 3½ lines long. Bracts small. Corolla-tube slender, ¾
  in., lobes bearded at the base . . . . 17. *A. longiflorum.*
Calyx 5 to 6 lines long. Bracts large. Corolla-tube not much
  exserted, lobes bearded at the end . . . . 18. *A. pinifolium.*

SECT. I. STOMARRHENA. Corolla-tube with 5 tufts of hairs inside below the middle, alternating with the stamens and sometimes confluent in a dense ring.

1. **A. stomarrhena,** *Sond. in Pl. Preiss.* i. 301. Stems usually several from a thick trunk, erect, simple or branched, ¼ to 1 ft. high, the whole plant more or less hirsute with long spreading hairs. Leaves sessile, erect, lanceolate, concave, tapering into a pungent point, the margins quite entire, ½ to 1 in. long, strongly striate. Flowers scarcely exceeding the leaves, nearly sessile. Upper bracts above 1 line; bracteoles about 2 lines long. Sepals 4 to 5 lines long, mucronate-acute, with the same spreading hairs as the rest of the plant. Corolla-tube about 6 lines long, with 5 tufts of hairs inside above the base and very hairy above them; lobes linear, acute, more revolute than in the other species but less so than in *Styphelia* and erect at the base, glabrous inside but hairy outside towards the top. Filaments not flattened, more than half as long as the lobes, very hairy, the anthers attached

near the top and connivent or slightly cohering round the style in a dense woolly mass exserted from the corolla-tube. Hypogynous disk truncate. Style filiform, longer than the corolla-tube; stigma distinctly 5-lobed.—*Styphelia lasionema* or *Astroloma lasionema*, F. Muell. Fragm. vi. 40.

W. Australia. Swan River, *Drummond, 1st Coll. n.* 475, *Preiss, n.* 410; Hamden, *Clarke.*

2. **A. macrocalyx,** *Sond. in Pl. Preiss.* i. 301. A glabrous erect bushy shrub of 1 to 3 ft. Leaves crowded, narrow-lanceolate, tapering into a pungent point, entire, concave or almost convolute, strongly striate, 1 to 1½ in. long. Flowers scarcely exceeding the leaves. Bracteoles shortly mucronate, 2 to 3 lines long. Sepals acute, 7 to 8 lines long. Corolla about as long as the calyx, the tube with 5 tufts of hairs inside above the base and hairy at the throat; lobes bearded, acuminate. Filaments short, scarcely flattened; anthers attached above the middle, 2-lobed at the top. Hypogynous disk short, truncate. Stigma capitate. Drupe ovoid, much shorter than the calyx.—*Styphelia macrocalyx*, F. Muell. Fragm. vi. 37.

W. Australia. Swan River, *Drummond, 1st Coll. n.* 477, *Oldfield;* near Pine Apple, *Preiss, n.* 413.

3. **A. xerophyllum,** *Sond. in Pl. Preiss.* i. 301. An erect shrub of 2 to 3 ft., the branches often elongated, minutely pubescent. Leaves narrow-lanceolate, tapering into a pungent point, not ciliate, very concave almost convolute, strongly striate, mostly ½ to ¾ in. long. Flowers white (*Oldfield*) scarcely exceeding the leaves or shorter. Bracteoles 1½ to 2 lines long. Sepals about 4 lines, obtuse or scarcely pointed. Corolla-tube broad, ventricose, scarcely or not at all exceeding the calyx, glabrous inside below the middle or with an obscure ring or tufts of few small hairs, very hairy towards the throat; lobes lanceolate, 2 to 2½ lines long, erect and bearded inside at the base, glabrous pointed and spreading at the tip. Filaments very short, not dilated, immersed in the wool of the lobes; anthers attached above the middle. Hypogynous disk truncate. Stigma globular and hairy.—*Stomarrhena xerophylla*, DC. Prod. vii. 738; *Styphelia xerophylla*, F. Muell. Fragm. vi. 38.

W. Australia. Swan River, *Drummond, 1st Coll. n.* 476, *Preiss, n.* 407; Murchison river, *Oldfield.* This species is anomalous in the want or minuteness of the tufts of hairs near the base of the corolla, but the inflorescence (without any rudimentary second flower), the habit, large flower, and other characters, clearly place it in *Astroloma* and not in *Leucopogon.*

4. **A. microphyllum,** *Stschegl. in Bull. Mosc.* 1859, i. 7. Branches apparently spreading or diffuse, with numerous erect or ascending slightly pubescent branchlets. Leaves rather crowded, spreading, ovate lanceolate or lanceolate-linear, minutely pointed, flat, rather striate, 1 to 1½ lines long, entire or minutely scabrous-denticulate. Flowers sessile in the lower axils or from leafless nodes at the base of the shoots. Bracts very small; bracteoles under 1 line long. Sepals smooth, coloured, about 2 lines long. Corolla-tube about 3 lines long, with 5 tufts of hairs forming a dense ring inside above the base, the throat hairy; lobes erect and bearded inside, with acutely acuminate spreading tips. Filaments short, flat; anthers long,

attached near the top. Hypogynous disk truncate. Style filiform, with a small capitate stigma.

W. Australia, *Drummond, 5th Coll. n.* 298. *A. juniperinum* or *Styphelia pentapogona,* F. Muell. Fragm. vi. 36, from gravelly places, Phillips and Fitzgerald Ranges, *Maxwell,* appears to me to be a form of the same species, with the leaves narrower than in Drummond's specimens.

5. **A. prostratum,** *R. Br. Prod.* 538. A low diffuse or prostrate shrub, with short ascending or erect branches, minutely pubescent. Leaves linear or linear-lanceolate, tapering into a short fine almost pungent point, flat or slightly concave, entire or minutely scabrous-denticulate, under $\frac{1}{4}$ in. long. Flowers red, nearly sessile. Bracteoles under 1 line long. Sepals 2 to nearly 3 lines, almost obtuse. Corolla-tube about 4 lines long, with 5 tufts of hairs inside near the base, the throat slightly hairy; lobes about 2 lines long, bearded inside, very acute. Filaments short and broad; anthers attached near the top, very obtuse or emarginate. Hypogynous disk truncate. Stigma capitate.—DC. Prod. vii. 738.

W. Australia. Lucky Bay, *R. Brown,* and probably the same locality, *Baxter;* South-west Bay, *Maxwell.*

6. **A. tectum,** *R. Br. Prod.* 538. The specimens usually show a thick, woody trunk or stock, with numerous thick, erect, simple or branched stems from 6 in. to nearly 1 ft. high. Leaves from almost ovate to broadly lanceolate, acute, with a short pungent point, minutely and obtusely denticulate, concave or nearly flat, usually erect and almost imbricate, $\frac{1}{2}$ to $\frac{3}{4}$ in. long. Flowers scarcely exceeding the leaves. Bracteoles nearly 2 lines. Sepals 4 to 5 lines long. Corolla-tube 6 to 7 lines long, with 5 dense tufts of hairs inside above the base, and villous above them; lobes bearded, acuminate. Filaments short, flattened; anthers attached above the middle. Hypogynous disk, short, truncate.—DC. Prod. vii. 739; *Styphelia tecta,* Spreng. Syst. i. 657; *Astroloma latifolium,* Sond. in Pl. Preiss. i. 302; *Styphelia platyphylla,* F. Muell. Fragm. vi. 37.

W. Australia, *Drummond, 4th Coll. n.* 149; Lucky Bay, *R. Brown;* and probably the same locality, *Baxter;* Mount Wuljenup, *Preiss, n.* 411; Stirling Range, *F. Mueller.*

7. **A. Candolleanum,** *Sond. in Pl. Preiss.* i. 302. An erect, bushy shrub of 1 to 2 ft., glabrous or the branches minutely pubescent. Leaves from broadly ovate-cordate to ovate or broadly lanceolate, tapering into a pungent point, more prominently denticulate-ciliate than in other species or rarely almost entire, concave or almost conduplicate, erect and imbricate or the end spreading or recurved, under $\frac{1}{2}$ in. long. Flowers red, solitary in the axils. Peduncles sometimes very short, sometimes 1 line long, with minute bracts, the 2 uppermost under the calyx $\frac{1}{2}$ line long; bracteoles 1 to $1\frac{1}{2}$ lines. Sepals about 3 lines long, coloured, obtuse, and minutely mucronate. Corolla-tube 4 to 5 lines long, the 5 tufts of hairs inside about or below the middle forming a dense ring; lobes 2 to $2\frac{1}{2}$ lines long, bearded, with very acute, glabrous tips. Filaments short, flat. Anthers attached near the top. Stigma large. Hypogynous disk truncate.—*Stomarrhena serratifolia,* DC. Prod. vii. 738; Deless. Ic. Sel. v. t. 22 (incorrect as to the corolla); *Styphelia Candolleana,* F. Muell. Fragm. vi. 38.

W. Australia, *Drummond, 1st Coll. n.* 471, *3rd Coll. n.* 192 *or* 194; Mount Bake-well, *Preiss, n.* 466; Middle Mount Barren, *Maxwell.*

8. **A. microdonta,** *F. Muell. Herb.* An erect shrub of 2 to 3 ft., with virgate branches, glabrous or minutely pubescent. Leaves lanceolate, taper-ing into a pungent point, denticulate-ciliate, flat, mostly about ½ in. long or rather more. Flowers red, nearly sessile. Bracts small; bracteoles scarcely above 1 line long. Sepals about 3 lines long, deeply coloured, obtuse, striate. Corolla-tube scarcely exceeding the calyx, with 5 dense tufts of hairs inside below the middle, the throat slightly hairy; lobes as long as the tube, densely bearded inside, less acuminate than in other species. Filaments short, flat; anthers attached above the middle. Hypogynous disk truncate. Stigma globular.

W. Australia. Murchison river, *Oldfield, Drummond, 6th Coll. n.* 121. Near *A. pallidum,* but more erect, the leaves larger and flatter, the corolla-tube shorter, and the lobes less mucronate.

9. **A. pallidum,** *R. Br. Prod.* 538. Diffuse or prostrate when grow-ing in sand, forming a tufted shrub of 1 ft. when in chinks of rocks, the numerous branchlets shortly pubescent. Leaves crowded, sessile or nearly so, lanceolate or oblanceolate, tapering into a pungent point, denticulate-ciliate, concave, rarely above ½ in. long. Flowers nearly sessile, usually white or flesh-coloured (sometimes dark red?). Bracteoles 1 to 1¼ lines long. Sepals 3 to 4 lines long, obtuse or with a minute point. Corolla-tube nearly twice as long as the calyx, with 5 dense tufts of hairs inside near the base, and the throat hairy; lobes lanceolate, about 3 lines long, bearded inside, with acute, glabrous points. Filaments short, flat; anthers attached near the top. Hypogynous disk truncate. Stigma large. Fruit ovoid or almost globular, about 3 lines diameter, often ripening only a single seed.—DC. Prod. vii. 739; Sond. in Pl. Preiss. i. 300; *Leucopogon blepharodes,* DC. Prod. vii 753; *Styphelia pallida,* Spreng. Syst. i. 658; F. Muell. Fragm. vi. 37.

W. Australia. King George's Sound, *R. Brown;* Swan and Canning rivers, *Preiss, n.* 424; Swan and Blackwood rivers, Cape Naturaliste, and King George's Sound, *Oldfield;* Lucky Bay, *Maxwell;* Stirling Range, *F. Mueller.*

10. **A. compactum,** *R. Br. Prod.* 538. Diffuse or prostrate, and much branched, glabrous or the branches slightly pubescent. Leaves oblan-ceolate cuneate or almost obovate, mucronate-acute, denticulate-ciliate, tapering into a more or less distinct petiole, flat or slightly concave or recurved and convex at the end, sometimes undulate, 3 to 5 lines long. Flowers red, axillary or from leafless nodes at the base of the shoots. Pedi-cels from very short to nearly 2 lines long. Bracteoles 1 line long or rather more. Sepals about 3 lines long, minutely mucronate. Corolla-tube longer than the calyx and sometimes nearly twice as long, with 5 tufts of hairs in-side, forming a ring about or a little below the middle; lobes densely bearded at the base, with glabrous, mucronate-acute tips. Filaments short, flat; anthers attached near the top. Hypogynous disk short, truncate. Stigma large.—DC. Prod. vii. 739; Sond. in Pl. Preiss. i. 300; *Styphelia compacta,* Spreng. Syst. i. 657; F. Muell. Fragm. vi. 38; *Astroloma cunei-*

folium, Sond. in Pl. Preiss. i. 300 ; *Styphelia cuneifolia,* F. Muell. Fragm. vi. 37.

W. Australia. Lucky Bay, *R. Brown ;* King George's Sound and adjoining districts, *Baxter, Oldfield,* and others, and thence to Swan River, *Drummond, 1st Coll. n.* 473, *5th Coll. n.* 300, *Preiss, n.* 421, 422, 423, and Murchison river, *Oldfield ;* eastward to Salt and Phillips rivers, *Maxwell.*

11. **A. humifusum,** *R. Br. Prod.* 538. Diffuse or prostrate and much branched, glabrous or the branches pubescent. Leaves narrow-lanceolate to linear or almost subulate, tapering into a pungent point, minutely denticulate-ciliate, flat or very slightly concave or convex, under $\frac{1}{2}$ in. long. Flowers red, axillary, on very short pedicels. Bracteoles mucronate, above 1 line long. Sepals mucronate, $2\frac{1}{2}$ to 3 lines long. Corolla-tube not twice as long as the calyx, with 5 tufts of hairs inside, forming a ring a little below the middle, and more or less hairy at the throat ; lobes about 3 lines long, bearded inside, with glabrous, mucronate-acute tips. Filaments short, flat ; anthers attached near the top. Hypogynous disk truncate. Stigma large, globular.—DC. Prod. vii. 738 ; Hook. f. Fl. Tasm. i. 244 ; Bot. Mag. t. 1439 ; Lodd. Bot. Cab. t. 1554 ; *Ventenatia humifusa,* Cav. Ic. iv. 28. t. 348 ; *Styphelia humifusa,* Pers. Syn. i. 174 ; F. Muell. Fragm. vi. 37 ; *Astroloma pallidum,* Sond. in Linnæa, xxvi. 246, not of R. Br.

N. S. Wales. Port Jackson, *R. Brown, Sieber, n.* 65, and others ; near Mount Aiton, *A. Cunningham ?* (not in flower).

Victoria. Near Melbourne, *Adamson ;* Wilson's Promontory, *F. Mueller ;* near Skipton, *Whan.*

Tasmania. Derwent river, *R. Brown ;* abundant on sandy and stony heaths, *J. D. Hooker.*

S. Australia. Rivoli Bay, Flinders and Mount Lofty Ranges, Torrens river, Kangaroo Island, *F. Mueller ;* Port Lincoln, *Wilhelmi.*

W. Australia. King George's Sound, *R. Brown,* or to the eastward, *Baxter.*

A. denticulatum, R. Br. Prod. 538 ; DC. Prod. vii. 739. *Styphelia denticulata,* Spreng. Syst. i. 658, from Memory Cove, *R. Brown,* Port Lincoln, *Wilhelmi,* etc., differs slightly from the common *A. humifusum* in its broad leaves, but many of the Tasmanian specimens have them almost, if not quite as broad.

12. **A. divaricatum,** *Sond. in Pl. Preiss.* i. 299. A straggling, divaricately-branched shrub of 1 to 2 ft., the branchlets minutely pubescent. Leaves lanceolate or linear-lanceolate, spreading or reflexed, mucronate-acute or tapering into a pungent point, very convex, and usually shining above, the margins revolute and entire, the larger ones above $\frac{1}{2}$ in. but mostly shorter. Flowers (red) solitary in the axils, besides a small rudiment in the upper bract. Bracteoles obtuse, about 1 line long. Sepals mucronate, 3 to 4 lines long. Corolla-tube not much longer than the calyx, with 5 tufts of hairs inside in a ring above the base, and the throat hairy ; lobes $2\frac{1}{4}$ lines long, bearded, with very acute, glabrous tips. Filaments short, flat ; anthers oblong, attached above the middle. Hypogynous disk truncate. Stigma globular.—*Cyathodes Baxteri,* DC. Prod. vii. 741 ; *Leucopogon epacridis,* DC. Prod. vii. 754 ; *Styphelia epacridis,* F. Muell. Fragm. vi. 38 ; *A. pungens,* Stschegl. in Bull. Mosc. 1859. i. 8 (with rather broad leaves) ; *A. splendens,* Planch. in Fl. des Serres. x. 129. t. 1018 (from the figure and description).

W. Australia. Swan River, *Drummond, 1st Coll. n.* 468, 469, *4th Coll. n.* 136, *5th Coll. n.* 296 ; gravelly places, York district, *Preiss, n.* 467 ; clay flats, Blackwood river, and sandstone rocks, Doubtful Island Bay, *Oldfield ;* Lucky Bay, *Baxter ;* near Cape Riche, *Harvey ;* Kalgan Ranges, Brewer Bay, and Cape Arid, *Maxwell ;* Stirling Range, *F. Mueller.*

A. marginatum, Sond. in˙ Pl. Preiss. i. 299, described from Preiss's specimens, n. 471, without flowers or fruit, if an *Astroloma* at all, may be this species. The leaves, however, are flatter.

13. **A. Drummondii,** *Sond. in Pl. Preiss.* i. 299. Stems from a thick, woody trunk, erect or ascending, simple or branched, often virgate, pubescent hirsute or rarely almost glabrous. Leaves linear, tapering into a short point, erect or rarely spreading, convex, with entire, recurved or revolute margins, often hoary underneath, and sometimes pubescent on both sides, rarely above ½ in. long, and mostly shorter. Flowers red, nearly sessile. Bracteoles above 1 line long. Sepals acute or almost obtuse, glabrous or pubescent, about 3 lines long. Corolla-tube 4 to 5 lines long, with 5 dense tufts of hairs inside above the base ; lobes about 2 lines long, bearded inside with acutely acuminate, glabrous tips. Filaments short, very flat ; anthers obtuse, inserted above the middle.—*Styphelia Drummondii,* F. Muell. Fragm. vi. 37 ; *Astroloma hirsutum,* Stschegl. in Bull. Mosc. 1859. i. 7.

W. Australia. Between King George's Sound and Swan River, *Harvey, Drummond, 4th Coll. n.* 135, 148 ; Vasse river, *Mrs. Molloy ;* Hay river, Esperance Bay, *Maxwell.*

Drummond's 3rd Coll. Suppl. n. 73, in bud only, appears to be this species ; n. 72 of the same Coll., in still younger bud, may be either this or *A. microcalyx.*

14. **A. microcalyx,** *Sond. in Pl. Preiss.* i. 298. A much branched, erect or diffuse shrub of 1 to 2 ft., the branchlets minutely pubescent. Leaves usually spreading, linear or narrow oblong, mucronate-acute or obtuse, minutely denticulate-ciliate or almost entire, convex with recurved margins, and often glaucous underneath, 3 to 4 lines or rarely ½ in. long. Flowers nearly sessile. Bracteoles under 1 line long. Sepals minutely mucronate, 2 to 2½ lines long, often pubescent. Corolla-tube 4 to 5 lines long, with 5 dense tufts of hairs inside above the base, slightly hairy at the throat ; lobes erect, with pointed tips, bearded inside towards the end. Filaments short, flat. Hypogynous disk very short, truncate. Fruit scarcely so long as the calyx.—*Styphelia microcalyx,* F. Muell. Fragm. vi. 37.

W. Australia. Swan River, *Drummond, 1st Coll. n.* 470, *Preiss, n.* 470 Possibly a small-flowered variety of *A. Drummondii.*

A. glaucescens, Sond. in Pl. Preiss. i. 298, from Swan River, Drummond, n. 475 in some collections, 478 in others, only known in fruit, appears to me to be the same as *A. microcalyx.* The leaves are obtuse or with a much smaller point than in the common form, but otherwise similar.

SECT. II. PENTATAPHRUS.—Corolla-tube with 5 deflexed, fringed scales inside below the middle, alternating with the stamens. Filaments much flattened.

15. **A. Baxteri,** *DC. Prod.* vii. 739. An erect or diffuse shrub, attaining 2 or 3 ft., the branches pubescent. Leaves linear, mucronate-acute, minutely serrulate-ciliate, convex, with recurved margins, rarely ½ in.

lo:ng. Flowers nearly sessile. Bracteoles nearly 2 lines long, and the larger bracts sometimes 1 line. Sepals coloured, 3 to 4 lines long, acute, scarcely striate. Corolla-tube ½ in. long, with 5 ciliate or fringed, reflexed scales inside below the middle alternating with the stamens, otherwise glabrous; lobes linear, erect, glabrous, 3 lines long. Filaments short, dilated upwards, and almost as broad as the anthers; anthers attached above the middle, acuminate. Hypogynous disk short, truncate. Fruit globular, shorter than the calyx.—*Stenanthera squamuligera*, F. Muell. Fragm. iv. 97 ; *Styphelia Baxteri*, F. Muell. Fragm. vi. 35.

W. Australia. King George's Sound and adjoining districts, *Fraser, Baxter, Drummond, Oldfield, F. Mueller;* along the coast to Cape le Grand and Cape Arid, *Maxwell;* known as "native Sarsaparilla," *Oldfield.*

16. **A. conostephioides,** *F. Muell. Herb.* An erect shrub, with spreading, pubescent branches. Leaves sessile, erect or spreading, linear or linear-lanceolate, rigid, tapering into a pungent point, with revolute margins, ½ to ¾ in. long or nearly 1 in. when very luxuriant. Flowers sessile. Bracts several, rather large, coloured like the calyx and passing into the bracteoles, which are not much shorter than the sepals, all acute and smooth. Sepals 6 to 8 lines long. Corolla-tube scarcely exceeding the calyx, with 5 ciliate or fringed scales inside above the base as in *A. Baxteri*, and not bearded at the throat. Filaments short, very flat ; anthers broad, very obtuse, attached near the top.—*Stenanthera conostephioides*, Sond. in Pl. Preiss. i. 296 ; *Pentataphrus Behrii*, Schlecht. Linnæa, xx. 618 ; *Styphelia Sonderi*, F. Muell. Fragm. vi. 36.

Victoria. Grampians, *F. Mueller;* Murray river, *Dallachy;* Skipton, *Whan;* Portland and Glenelg rivers, *Robertson, Allitt.*
S. Australia. Sandy and stony places near Adelaide, *Behr, Blandowsky;* Mount Lofty, *Whittaker;* Encounter Bay and Kangaroo Island, *F. Mueller.*

Sect. III. Stenanthera.—Corolla-tube without tufts of hairs or fringed scales inside. Filaments much flattened. Leaves linear, with revolute margins.

17. **A. longiflorum,** *Sond. in Pl. Preiss.* i. 297. Stems prostrate or diffuse, with numerous shortly ascending branches or rarely more erect and bushy, the branches usually pubescent. Leaves spreading, linear, tapering into a short point, serrulate-ciliate, convex, with recurved margins, sometimes much crowded very narrow and under 3 lines long, sometimes more distant ½ in. long and broader in proportion. Flowers almost sessile. Bracts very small, the upper ones passing into the bracteoles, which are unequal, the innermost about 1 line long, very broad and obtuse. Sepals about 3 lines long or rather more, obtuse, less rigid than in most species. Corolla-tube nearly or quite ¾ in. long, glabrous inside; lobes lanceolate, bearded at the base only. Filaments short, very flat ; anthers attached above the middle, very obtuse. Disk short. Fruit about as long as the calyx.—*Stenanthera ciliata*, Lindl. Swan Riv. App. 25 ; *Mesotriche longiflora*, Stschegl. in Bull. Mosc. 1859. i. 9 ; *Astroloma discolor*, Sond. in Pl. Preiss. i. 298 ; *Mesotriche discolor*, Stschegl. l. c.

W. Australia. Swan River, *Drummond,* 1*st Coll. n.* 472, 474, *Preiss, n.* 419, 420 ; Kalgan river, *Oldfield, F. Mueller.*

A. foliosum, Sond. in Pl. Preiss. i. 297, very remarkable for its crowded, short, and very narrow leaves, is, however, scarcely a variety, for the two states occur on different branches of the same specimen.

Var.? *dilatatum.* Leaves slightly dilated above the middle, and abruptly contracted into a pungent point. Calyx pubescent. Corolla, etc., of *A. longiflorum.—A. dilatatum,* Sond. in Pl. Preiss. i. 298.—Swan River, *Drummond.*

18. **A. pinifolium,** *Benth.* A rigid, much-branched shrub, sometimes small or diffuse, sometimes erect and 2 to 3 ft. high, the branchlets usually pubescent. Leaves crowded, very narrow linear, rigidly pointed, with revolute, scabrous margins, about ½ in. long. Flowers sessile and solitary in each axil, but often crowded at the base of the branchlets. Bracts several, the inner ones embracing the calyx, and passing into the bracteoles, which are 3 to 4 lines long, broad, and obtuse. Sepals 5 to 6 lines long, broad, obtuse, thin, scarcely striate. Corolla about ¾ in. long, reddish at the base, passing into yellow with green tips, the tube without any tufts of hairs inside near the base, but slightly hairy above the middle ; lobes lanceolate or almost linear, bearded inside towards the end. Filaments short and very flat ; anthers attached near the top, very obtuse. Hypogynous disk truncate or shortly lobed. Stigma small. Fruit globular, enclosed within the somewhat enlarged calyx.—*Stenanthera pinifolia,* R. Br. Prod. 538 ; DC. Prod. vii. 739 ; Hook. f. Fl. Tasm. i. 244 ; Bot. Reg. t. 218 ; *Styphelia pinifolia,* Spreng. Syst. i. 659 ; F. Muell. Fragm. vi. 36.

N. S. Wales. Port Jackson to the Blue Mountains, *R. Brown, Sieber, n.* 70, and others.

Victoria. Mount William, Grampians, up to 5000 ft. elevation, and Lake King in Gipps' Land, *F. Mueller.*

Tasmania. Circular Head, *Gunn ;* Launceston, *Laurence ;* St. Paul's River, *C. Stuart.*

4. CONOSTEPHIUM, Benth.

(Conostephiopsis, *Stschegl.*)

Corolla-tube enclosed in or scarcely protruding from the calyx, more or less conical in the upper part, without scales or tufts of hairs inside, but usually hairy towards the throat ; lobes very small, acute, valvate in the bud. Filaments very short, inserted at the base of or below the cone of the corolla ; anthers included in the cone, elongated, deeply divided into 2 lobes, joined by a short connectivum. Hypogynous disk of 5 distinct scales or none. Ovary 5-celled, with 1 ovule in each cell. Style usually slender, with a small stigma. Fruit a nearly dry drupe, enclosed in the calyx, the endocarp hard, with 5 cells and seeds or fewer by abortion.—Shrubs, with the habit and foliage of some species of *Astroloma.* Flowers solitary in the axils, usually pendulous. Pedicels with several bracts, of which 2 to 6 of the enlarged inner ones pass into the bracteoles, and with them embrace the base of the calyx.

The genus is limited to S.W. Australia. It is allied to *Astroloma,* but readily known by the anthers and by the corolla. The conical portion of the latter represents probably the erect portion of the lobes in *Astroloma ;* but whilst in that genus these lobes, though connivent at the base, are always distinct (at least, I have never seen them connate as figured,

probably by a mistake of the artist, in Delessert's plate of *Astroloma Candolleanum*), they are in *Conostephium* perfectly concrete, so as to form part of the tube.
Corolla-cone thicker at the base than the cylindrical lower part. Leaves convex or with recurved margins.
Leaves oblong-linear, ¾ to 1 in. long.   Flowers ½ in. long, on long
peduncles. Hypogynous scales 5.   Ovary glabrous   . . . .   1. *C. pendulum.*
Leaves narrow-linear, ½ to ¾ in. long. Flowers 4 lines, on short
peduncles. Disk none. Ovary pubescent at the top . . . .   2. *C. minus.*
Leaves ovate obovate or linear-cuneate. Flowers 4 to 4½ lines, on
short peduncles. Disk none. Corolla-tube shorter than the
cone. Ovary glabrous. . . . .   3. *C. Roei.*
Corolla cylindrical in the lower half, and gradually tapering at or towards the top.
Leaves obovate-oblong to narrow-oblong, mostly convex, obtuse or
mucronate   . . . .   4. *C. Preissii.*
Leaves ovate-lanceolate to linear-lanceolate, flat or concave, acute .   5. *C. planifolium.*

1. **C. pendulum,** *Benth. in Hueg. Enum.* 76.   An erect, branching, glabrous shrub of ½ to 1½ ft.   Leaves linear-oblong, acute or obtuse, with a short, pungent point, convex or with recurved margins, mostly ¾ to 1 in. long.   Peduncles recurved, 2 to 4 lines long.   Bracts numerous, 4 to 6 of the upper ones embracing the calyx besides the bracteoles, which are about 3 lines long.   Sepals about 4 lines long, broad, acute or almost obtuse.   Corolla about 6 lines long, narrow at the base, very much dilated above the middle, then conical, with minute lobes, hairy inside, especially in the cone and near the base.   Stamens inserted below the cone.   Hypogynous scales 5, narrow, acuminate, distant from each other.   Ovary and style glabrous. Fruit enclosed in the calyx.—DC. Prod. vii. 739 (from the synonym, but the plant figured in Deless. Ic. is *C. minus*) ; Sond. in Pl. Preiss. i. 303 ; *Styphelia conostephium,* F. Muell. Fragm. vi. 40.

W. Australia. King George's Sound to Swan River, *Huegel, Harvey, Drummond, n. 466, Preiss, n. 414, Oldfield.*

2. **C. minus,** *Lindl. Swan Riv. App.* 25.   An erect, branching shrub, usually under 1 ft. high.   Leaves linear, with a small, callous point, and closely revolute margins, ½ to ¾ in. long.   Flowers about 4 lines long, on peduncles of scarcely more than 1 line, at first erect, but at length usually recurved.   Bracts several, but not so numerous as in *C. pendulum*, the bracteoles nearly as long as the calyx.   Sepals scarcely 3 lines long, obtuse, broad, thin, shining.   Corolla glabrous outside or scarcely pubescent, the lower half narrow, thin, and glabrous, the cone much broader at the base, hairy inside.   Stamens inserted at the base of the cone.   No hypogynous disk.   Ovary obovoid, 5-furrowed, the upper half pubescent.   Style glabrous or slightly hairy.—Sond. in Pl. Preiss. i. 303 ; *Conostephium pendulum,* Deless. Ic. Sel. v. t. 23, not of Benth. ; *Conostephiopsis minor,* Stschegl. in Bull. Mosc. 1859, i. 6 ; *Styphelia Lindleyi,* F. Muell. Fragm. vi. 40.

W. Australia. Swan River, *Drummond, 1st Coll. n. 467, Preiss, n. 408.*

3. **C. Roei,** *Benth.*   An erect, branching shrub.   Leaves ovate, obovate, oblong or linear-cuneate, obtuse or with a short, callous point, convex, with recurved margins or the short ones flat, mostly 3 to 4 lines long.

Flowers 4 to 4½ lines long, almost sessile or at length pedicellate and reflexed. Bracteoles nearly as long as the calyx. Sepals scarcely 3 lines long, almost acute. Corolla with the thin, narrow part tapering to the base, and shorter than the broad, thick cone, which is hairy inside towards the top. Filaments inserted at the base of the cone and somewhat flattened; anther-lobes acuminate. No hypogynous disk. Ovary glabrous.

W. Australia. In the interior, *J. S. Roe.*

4. **C. Preissii,** *Sond. in Pl. Preiss.* i. 304. An erect, branching shrub, attaining sometimes 4 or 5 ft., but usually lower. Leaves from obovate-oblong and scarcely ½ in., to narrow-oblong and ¾ in. long, obtuse or with a small callous point, flat or the margins slightly recurved. Flowers about 4 lines long, on recurved peduncles of 1 to 3 lines. Bracts numerous; bracteoles nearly as long as the calyx. Sepals about 3 lines long, very broad and obtuse, shining. Corolla with the enclosed part pubescent outside, almost cylindrical, the short exserted portion alone more conical and glabrous, the real tube being very short, all the rest a long cone, of which more than half is cylindrical. Stamens inserted near the base of the corolla, the filaments slightly flattened; anthers long, less lobed than in the other species. Hypogynous disk very minute or none. Ovary oblong, glabrous; style shortly hairy.—*Conostephiopsis Preissii,* Stschegl. in Bull. Mosc. 1859. i. 6; *Styphelia Preissii,* F. Muell. Fragm. vi. 40.

W. Australia. Swan River, *Fraser, Preiss, n.* 416, *Drummond,* 3rd *Coll. n.* 183, *Harvey*; Murchison river, *Oldfield.*

5. **C. planifolium,** *F. Muell. Fragm.* vi. 30. An erect, branching shrub of 2 or 3 ft. Leaves ovate-lanceolate to lanceolate or oblong-linear, acute or almost pungent, flat or concave, rigid, prominently striate, 3 to 4 lines long. Flowers fully 4 lines long, on recurved peduncles of 1 line or more. Bracts numerous; bracteoles nearly as long as the calyx. Sepals scarcely 3 lines long, obtuse, coloured, pubescent. Corolla cylindrical and pubescent in the lower part, and not narrower than the glabrous cone. Filaments short and flat, inserted at the base of the cone; anther-lobes long, terminating in a hooked point. Hypogynous disk none. Ovary oblong, slightly hairy at the top; style hairy.—*Conostephiopsis Drummondii,* Stschegl. in Bull. Mosc. 1859. i. 6; *Styphelia conantha,* F. Muell. Fragm. vi. 30.

W. Australia, *Drummond,* 5th *Coll. n.* 299; sand hills between Cape Malcolm and Point Culver, *Maxwell.*

Stschegleev, whose determinations and descriptions of *Epacrideæ* are very accurate, appears to have relied too much for the establishment of *Conostephiopsis* and other new genera on modifications of the hypogynous disk, a character very rarely more than specific in *Epacrideæ.*

5. MELICHRUS, R. Br.

Corolla-tube short and broad, with 5 densely glandular scales inside alternating with the stamens; lobes longer than the tube, valvate in the bud, bearded or glabrous inside, spreading. Stamens inserted in the tube; filaments exceedingly short; anthers oblong, 1-celled. Hypogynous disk short, truncate. Ovary 5-celled, with 1 ovule in each cell; style very short or

reduced to a small cone ; stigma terminal, small. Fruit a drupe, with a dry
or slightly pulpy mesocarp and a hard, bony endocarp, with 5 cells and seeds
or fewer by abortion.—Leaves sessile, lanceolate. Flowers solitary in the
axils, sessile, surrounded by several bracts, of which the 2 innermost are enlarged, and, as well as the still larger bracteoles, embrace the base of the
calyx.

The genus is limited to Eastern Australia. It is allied to *Astroloma* and to *Leucopogon*,
distinguished from the former by the shape of the corolla, from the latter by its larger
flowers and more developed bracts, from both by the glandular scales inside the corolla-
tube.

Leaves ciliate, sprinkled with long hairs. Calyx (when open) broadly
  campanulate. Corolla shorter than the calyx, the tube very short . 1. *M. rotatus.*
Leaves without long hairs. Calyx ovoid. Corolla exceeding the calyx,
  the tube nearly as long as the lobes . . . . 2. *M. urceolatus.*

1. **M. rotatus,** *R. Br. Prod.* 539. A low, procumbent shrub, with
short, ascending branches. Leaves crowded, sessile, lanceolate, tapering into
a long and fine but not pungent point, flat, ciliate with long, soft hairs, and
hairy on both sides or at length glabrous and shining above, ½ to ¾ or sometimes almost 1 in. long. Bracteoles about 2 lines long. Calyx ovoid in
bud, but broadly campanulate when the flower is expanded, softly pubescent,
the sepals ovate, acute, ciliate, about 3 lines long. Corolla scarcely so long
as the calyx, the tube exceedingly short, with 5 large scales densely covered
with prominent glands, alternating with the stamens inside ; lobes broadly
lanceolate, glabrous except a few long hairs at the tip, expanded into a
rotate limb when the flower is open. Hypogynous disk very short, thick,
and fleshy.—DC. Prod. vii. 740 ; *Ventenatia procumbens,* Cav. Ic. iv. 28. t.
349. f. 1 (partly); *Styphelia procumbens,* Pers. Syn. i. 174 ; *Styphelia rotata,*
F. Muell. Fragm. vi. 38.

Queensland. Sandy Cape, *R. Brown.*
N. S. Wales. Port Jackson to the Blue Mountains, *R. Brown, Sieber, n.* 64, and
others ; Newcastle, *Leichhardt ;* New England, *C. Stuart.* Cavanilles' figure is a good
general representation of the plant, but the flower and the analysis must have been taken
from some *Astroloma.*

2. **M. urceolatus,** *R. Br. Prod.* 539. An erect shrub of 2 to 3 ft. or
rarely low and diffuse like *M. rotatus.* Leaves crowded or densely imbricate,
or especially the floral ones loosely spreading, lanceolate, rigid, tapering into
a fine pungent point, glabrous or rarely pubescent, from under ¼ in. to
nearly 1 in. long. Bracteoles about 1½ lines long. Calyx ovoid, glabrous
or pubescent, the sepals rigid, obtuse, nearly 3 lines long, often coloured.
Corolla-tube broad, shorter than the calyx, with the same glandular scales
inside as in *M. rotatus ;* lobes lanceolate, rather longer than the tube and
exceeding the calyx, recurved at the end but not rotate, bearded or nearly
glabrous inside. Hypogynous disk short, thin, truncate.—DC. Prod. vii.
740 ; *M. medius, M. erubescens,* and *M. adpressus,* A. Cunn. in DC. l. c.;
Styphelia urceolata, F. Muell. Fragm. vi. 38.

Queensland, *W. Hill ;* near-Warwick, *Herb. F. Mueller ;* ridges on the Burnett
river, *F. Mueller.*
N. S. Wales. Newcastle, *R. Brown ;* in the interior about Bathurst, Liverpool
Plains, etc., *A. Cunningham ;* head of the Gwydir, Boyd river, *Leichhardt ;* between the

Upper Bogan and Lachlan rivers, *L. Morton;* Mudgee, *Woolls;* New England, *C. Stuart;* Clarence river, *Beckler.*

Victoria. Forest Creek, Ovens river, Delatite river, *F. Mueller.*

Beckler's and Hill's (single) specimens (*Styphelia Cunninghamii,* F. Muell. Fragm. vi. 39) have large, very crowded, densely imbricated leaves but no flowers, another specimen has the floral leaves equally large but spreading, others are intermediate between them and the common form. Cunningham's four species, all from the neighbourhood of Bathurst, do not appear to me to be distinguishable even as varieties; the *M. adpressus* has not the leaves larger than the others. The colour of the flowers is stated by F. Mueller to be pale yellow; by A. Cunningham and Fraser it is given as white, pale pink or deep red, upon different specimens otherwise exactly alike.

6. PENTACHONDRA, R. Br.

Corolla-tube very short, or cylindrical and exceeding the calyx; lobes valvate in the bud, recurved or revolute, bearded inside. Filaments inserted at the top of the tube, rather long and erect with the anthers exserted, or short with the anthers more or less enclosed in the tube, or recurved with the lobes. Hypogynous disk consisting of scales either entirely distinct or more or less cohering. Ovary 5-celled; style long or short, with a small stigma. Fruit a baccate drupe, the mesocarp very pulpy, with 5 distinct pyrenes or fewer by abortion.—Diffuse or prostrate shrubs. Leaves usually crowded. Flowers (except in *P. verticillata*) solitary or 2 or 3 together at the ends of the branches, each one solitary in the axil of one of the last leaves on a short peduncle. Bracts several, small, the uppermost (above the one subtending the flower) with the rudiment of a second flower; bracteoles close under the calyx.

The genus is limited to the mountains of Tasmania, Victoria, and New Zealand, one Australian species being the same as a New Zealand one, the three others are endemic. It is united by F. Mueller with *Trochocarpa,* but the three genuine species have a different habit and inflorescence; the corolla-lobes are always bearded, and the fruit (which I have myself seen in one species only) is much more berry-like, with the pyrenes much more distinct, and five only in number, not ten. The fourth species, of which the fruit is unknown, is anomalous in inflorescence, and may possibly prove to be a *Cyathodes.*

Corolla nearly ½ in. long, the lobes much longer than the tube and re-
volute, exposing the erect stamens . . . . 1. *P. involucrata.*
Corolla not exceeding ¼ in., the lobes shorter than the tube, the anthers
wholly or partially included.
Flowers solitary, at the ends of the branches or in the last axils.
Leaves ovate or oblong, 1 to 2 lines long . . . . 2. *P. pumila.*
Leaves linear or lanceolate, 2 to 3 lines long . . . . . . 3. *P. ericæfolia.*
Flowers 2 or 3, in a short, terminal spike. Leaves linear, 2 to 3
lines long, crowded at the end of each year's shoot, with scarious
scales between each cluster or false-whorl . . . . 4. *P. verticillata.*

1. **P. involucrata,** *R. Br. Prod.* 549. A diffuse or prostrate shrub, with ascending or sometimes erect stems of 6 in. to 1 ft., the branches and sometimes the foliage pubescent or villous. Leaves nearly sessile, elliptical or lanceolate, acute or rather obtuse, flat or slightly concave, the margins softly ciliate, finely but prominently veined underneath or on both sides, ¼ to ½ in. or on barren branches ¾ in. long. Flowers 1, 2 or 3 together at the ends of the branchlets, each one solitary in one of the last leaves under the terminal bud. Bracts several, very small, the uppermost with a small rudi-

ment, the second subtending the flower. Bracteoles about half as long as the calyx. Sepals 1 line long, obtuse, ciliate. Corolla-tube scarcely exceeding the calyx; lobes about 4 lines long, very spreading or revolute, exposing the erect, glabrous, filiform filaments and wholly exserted anthers. Hypogynous scales distinct or slightly cohering. Ovary pubescent; style long.—DC. Prod. vii. 759; Hook. f. Fl. Tasm. i. 255; *Styphelia involucrata*, Spreng. Syst. i. 655; *Trochocarpa involucrata* or *Decaspora involucrata*, F. Muell. Fragm. vi. 57.

Tasmania. Summit of Mount Wellington, *R. Brown, J. D. Hooker.*

2. **P. pumila,** *R. Br. Prod.* 549. A small diffuse or prostrate shrub, the numerous branchlets ascending to a few inches, usually glabrous. Leaves crowded, ovate or oblong, obtuse or with a callous point, slightly concave, striate, 1 to 2 lines long. Flowers almost sessile, solitary at the ends of the short branchlets. Bracts several, very small, the terminal one with a small rudiment. Bracteoles fully half as long as the calyx. Sepals obtuse, ciliolate, about ¾ line long. Corolla-tube cylindrical, about 2 lines long, glabrous; lobes short, recurved, bearded inside. Anthers half included in the corolla-tube. Hypogynous scales distinct. Fruit very pulpy, 3 to 4 lines diameter, the pyrenes small and quite separate.—DC. Prod. vii. 759; Hook. f. Fl. Tasm. i. 255; *Epacris pumila*, Forst. Prod. 13; *Styphelia pumila*, Spreng. Syst. i. 656; *Leucopogon vaccinioides*, Sond. in Pl. Preiss. i. 325; *Pentachondra vaccinioides*, Sond. in Linnæa, xxvi. 252; *Trochocarpa pumila* or *Decaspora pumila*, F. Muell. Fragm. vi. 57.

Victoria. Mountains of Munyong, Baw-Baw, Mitta-Mitta, sources of the Yarra, at an elevation of 4500 to 6000 ft., *F. Mueller.*

Tasmania. Derwent river, *R. Brown;* summits of all the mountains above 3000 or 4000 ft., *J. D. Hooker.*

The species is also in New Zealand.

3. **P. ericæfolia,** *Hook. f. in Hook. Lond. Journ.* vi. 271; *Fl. Tasm.* i. 255. *t.* 77 A. A densely branched, diffuse or prostrate, heath-like shrub, extending to above a foot, with numerous, shortly ascending, glabrous or pubescent branchlets. Leaves erect and often closely appressed, linear or lanceolate, obtuse or with a callous point, concave, prominently 1- or 3-ribbed, 2 or rarely 3 lines long. Flowers 1, 2 or 3 together at the ends of the branches, each one solitary in the axil of one of the last leaves. Bracts several, very small, the two uppermost rather longer, one with a minute rudiment. Bracteoles about half as long as the calyx, ciliate. Sepals nearly 1 line long, obtuse, ciliolate. Corolla-tube about 2 lines long, pubescent or nearly glabrous outside, and slightly hairy inside; lobes shorter than the tube, bearded inside. Anthers half included in the corolla-tube. Hypogynous disk lobed, readily separating into distinct scales. Style long.

Tasmania. Abundant in the alpine districts between Marlborough and Lake St. Clair, *Gunn.*

4. **P. ? verticillata,** *Hook. f. Fl. Tasm.* i. 256. *t.* 77 B. A low, diffuse or prostrate shrub, extending to 1 ft. or more, the branches shortly ascending, covered with small, acuminate, almost scarious scales, which appear to be the persistent leaf-bud scales or abortive leaves at the base of each year's

shoot, which exist in most *Epacrideæ*, but are almost always very ceciduous. Perfect leaves clustered at the end of each year's shoot, as in *Cyathodes glauca* and *C. straminea*, petiolate, linear, shortly mucronate, with thick revolute margins, 2 to 3 lines long. Flowers 2 or 3 together besides the rudiment, in a short spike, at first terminal, but becoming lateral by the development of the new leaf-bud. Bracts and bracteoles acuminate, cil.ate, more than half as long, and sometimes nearly as long, as the calyx. Sepals oblong, $1\frac{1}{2}$ lines long, ciliolate, fringed. Corolla-tube scarcely so long as the calyx ; lobes nearly 1 line long, bearded inside. Anthers included in the tube. Hypogynous scales distinct. Ovary 5-celled ; style very short. Fruit unknown.

Tasmania. Mount Sorrell, Macquarrie Harbour, *Milligan.* Until the nature of the fruit is ascertained, it remains in some measure doubtful whether this should be referred to *Pentachondra*, to *Leucopogon*, or to *Cyathodes*.

7. TROCHOCARPA, R. Br.

(Decaspora, *R. Br.*)

Corolla-tube cylindrical or campanulate, glabrous or with reflexed hairs inside at the top ; lobes usually shorter than the tube, recurved, glabrous or bearded inside. Filaments inserted in the top of the tube, short, filiform ; anthers attached at or near the top, partially included in the tube or recurved with the lobes. Hypogynous disk truncate, lobed or separating into distinct scales. Ovary 10-celled, with 1 ovule in each cell ; style rather thick, usually short; stigma small. Fruit a globular or depressed drupe, the mesocarp pulpy, the endocarp separating or separable into 10 (or fewer by abortion) distinct pyrenes.—Shrubs. Leaves usually petiolate, flat or convex. Flowers several together in spikes, either terminal or in the axils of the previous year's leaves, or lateral on the old wood, each flower sessile within the small subtending bract and two bracteoles.

The genus is limited to Australia. It differs from *Leucopogon* in the separable pyrenes of the fruit ; the corolla-lobes are also in some species beardless, and the increased number of cells of the ovary is very rare in other *Stypheliæ*. I have followed F. Mueller in uniting the two genera, the close affinity of which Brown himself had pointed out, notwithstanding the different aspect which the looser inflorescence and large leaves give to the *T. laurina*. The three species since added form a third group, as different from the two others as these are from each other.

(The ovary is 6- to 10-celled also in *Cyathodes glauca*, in *Leucopogon pluriloculatus* and *L. pleiospermus*, and in *Acrotriche aggregata*.)

1. **T. laurina,** *R. Br. Prod.* 548.   A tree of 20 to 30 or even 40 ft., quite glabrous.   Leaves usually clustered at the ends of each year's shoots, so as to appear almost verticillate, petiolate, broadly oval or elliptical, acuminate, shining, 5- to 7-nerved on both sides, mostly $1\frac{1}{2}$ to 2 in. long. Flowers small, white, in terminal, solitary or clustered, interrupted spikes, $\frac{3}{4}$ to 1 in. long.   Bracts small; bracteoles obtuse, not half so long as the calyx.   Sepals $\frac{1}{2}$ line long or rather more, obtuse, striate.   Corolla-tube about 1 line long; lobes shorter than the tube, bearded to the middle as well as the upper part of the tube with reflexed hairs.   Hypogynous disk shortly lobed.   Ovary tapering into a short style.   Drupe depressed-globular, 3 to 4 lines diameter, the pyrenes less readily separable than in the other species.—DC. Prod. vii. 758; Bot. Mag. t. 3324; F. Muell. Fragm. vi. 57; *Cyathodes laurina,* Rudge in Trans. Linn. Soc. viii. 293. t. 9 (*Styphelia cornifolia* on the plate).

Queensland.   Brisbane river, Moreton Bay, *Fraser, F. Mueller.*
N. S. Wales.   Port Jackson, *R. Brown* and others; northward to Hastings and Clarence rivers, *Beckler, Wilcox;* New England, *C. Stuart;* Head of Bellinger river (a small and narrow-leaved, stunted variety), *C. Moore;* southward to Illawarra, *Shepherd.*

2. **T. disticha,** *Spreng. Syst.* i. 660.   A tall shrub, with slender branches, quite glabrous in the original form.   Leaves very shortly petiolate, spreading and somewhat distichous, from broadly ovate-lanceolate to narrow oblong-lanceolate, acute or almost obtuse, shining above, 3- or 5-nerved underneath, $\frac{1}{2}$ to 1 in. long.   Flowers red, in dense, terminal, recurved spikes of $\frac{1}{2}$ to $\frac{3}{4}$ in.   Bracts very small; bracteoles about half as long as the calyx. Sepals about 1 line long, broad, striate.   Corolla campanulate, the tube about 2 lines long, the lobes short, recurved, with a dense tuft of long hairs at the base reflexed into the tube, otherwise glabrous.   Hypogynous disk shortly lobed.   Fruit bluish-purple, succulent, with 10 distinct pyrenes.— *Cyathodes disticha,* Labill. Pl. Nov. Holl. i. 58. t. 82; *Decaspora disticha,* R. Br. Prod. 548; DC. Prod. vii. 758; Hook. f. Fl. Tasm. i. 254.

Tasmania.   Recherche Bay, *Labillardière* and others; South Port, *C. Stuart.*

Var. *Cunninghamii.*   Branches hirsute.   Leaves usually but not always smaller.—*Decaspora Cunninghamii,* DC. Prod. vii. 758; Deless. Ic. Sel. v. t. 25; Hook. f. Fl. Tasm. i. 254. —Macquarrie Harbour, *A. Cunningham;* Fagus forest, S.W. of Lake St. Clair, and Mount Olympus, *Gunn.*

3. **T. thymifolia,** *Spreng. Syst.* i. 660.   A low and diffuse or bushy shrub of 1 to 2 ft., the branches usually pubescent.   Leaves more petiolate than in other species, ovate to broadly oblong, obtuse or with a short callous point, convex, rather thick, obscurely veined or with the midrib prominent underneath, 1 to 2 lines long.   Flowers in dense, terminal, cylindrical, recurved spikes, of $\frac{1}{2}$ to $\frac{3}{4}$ in.   Bracts and bracteoles very short, broad, ciliolate. Sepals about 1 line long, broad, obtuse, coloured.   Corolla-tube campanulate, slightly exceeding the calyx; lobes shorter than the tube, bearded inside. Hypogynous scales distinct or slightly cohering.—*Decaspora thymifolia,* R. Br. Prod. 548; DC. Prod. vii. 758; Hook. f. Fl. Tasm. i. 254.

Tasmania.   Summit of Mount Wellington, *R. Brown, J. D. Hooker;* Western Mountains, *C. Stuart.*

4. **T. Clarkei,** *F. Muell. Fragm.* vi. 57. A small, diffuse shrub, glabrous or the branches slightly pubescent. Leaves elliptical-oblong to oblonglanceolate, obtuse or rarely almost acute, 3- or 5-nerved, 3 to 5 lines long. Flowers not very numerous, in dense globular heads, nearly sessile at the ends of the branches or in the axils of the leaves on the previous year's shoots, otherwise precisely as in *T. disticha.* Corolla campanulate, as in that species, and of the same size, with similar long tufts of hairs descending into the tube from the base of the lobes. Fruit much larger, fully 4 lines diameter, blue or of a bluish-purple, very pulpy.—*Decaspora Clarkei,* F. Muell. in Trans. Phil. Soc. Vict. i. 106, and in Hook. Kew Journ. viii. 163.

Victoria. Shady ravines of Mount Wellington, Gipps' Land, Baw-Baw mountains, Mount Barkly, and others of the Australian Alps, abundant at an elevation of 4000 to 5000 ft., *F. Mueller.*

5. **T. Gunnii,** *Benth.* A tall, densely-branched shrub, attaining sometimes 10 to 12 ft., glabrous or the branchlets pubescent. Leaves from ovalelliptical to oblong, obtuse or with a short callous point, strongly ribbed underneath, 3 to 4 lines long when broad, nearly $\frac{1}{2}$ in. when narrow. Flowers white, few together, in short, nearly globular spikes, terminating short, leafy branches, or sessile in the axils of the leaves of the previous year's shoots. Bracts and bracteoles about half as long as the calyx. Sepals broad, obtuse, about $\frac{3}{4}$ line long. Corolla-tube campanulate, shortly exceeding the calyx ; lobes short, the whole corolla glabrous inside and out. Hypogynous disk short, truncate. Fruit globular, succulent, 3 or 4 lines diameter, purple or violet (*Gunn*) or orange (*Oldfield*).—*Decaspora Gunnii,* Hook. f. in Hook. Lond. Journ. vi. 270, and Fl. Tasm. i. 254. t. 76.

Tasmania. Dense humid forests, S.W. of Lake St. Clair, and Hampshire Hills, *Gunn ;* Foot of Mount Lapeyrouse, *Oldfield.*

6. **T. parviflora,** *Benth.* Apparently a somewhat spreading, bushy shrub. Leaves oblong-elliptical or narrow-ovate, obtuse, thick, somewhat glaucous and finely veined underneath, 3 to 4 lines long. Flowers few together, in little clusters, almost sessile in the axils of the older leaves, apparently small, but only seen in young bud. Bracts and bracteoles very broad. Sepals quite those of the genus. Corolla as yet very small in the specimens, the lobes valvate and showing as yet no hairs, although those in the throat are already prominent. Fruits clustered, depressed-globular, about 2 lines diameter, with 10 distinct pyrenes, their outer edges prominent from the desiccation of the pulp.—*Decaspora parviflora,* Stschegl. in Bull. Mosc. 1859, i. 10.

W. Australia, *Drummond, 4th Coll. n.* 157. The aspect and inflorescence are nearly those of some forms of *Acrotriche ovalifolia,* but the fruit is totally different.

8. CYATHODES, Labill.

(Ardisia, *Gærtn.*)

Corolla-tube longer or rarely shorter than the calyx, cylindrical or contracted at the throat, glabrous or hairy inside above the middle, without tufts of hairs or scales below the middle ; lobes valvate in the bud, spreading or recurved towards the end, glabrous or bearded inside. Filaments inserted at the top of the tube, short, filiform or somewhat thickened; anthers wholly or partially enclosed in the tube or the erect base of the corolla-lobes. Hypogynous disk cup-shaped or annular, truncate, 5-lobed or consisting of 5 distinct scales. Ovary 5-celled or in one species 8- to 10-celled; style not exceeding the corolla-tube; stigma small. Fruit a baccate drupe, the mesocarp pulpy, the endocarp hard and bony, with 5 cells and seeds or fewer by abortion, or in one species 6 to 10.—Shrubs, usually much branched low and prostrate, but sometimes tall and almost arborescent. Leaves in most species white or hoary underneath. Flowers small, solitary in the axils, terminating short peduncles, with several imbricate bracts, the uppermost gradually enlarged and embracing the base of the calyx.

The genus extends over eastern Australia, New Zealand, the eastern Archipelago, and the Pacific Islands. Of the eight Australian species, one only, a maritime one, is also in New Zealand, the others are all endemic. The technical characters are very nearly those of *Leucopogon*, but the corolla-lobes are less bearded or glabrous, and the fruit much more pulpy; and the genus may, as far as I have observed, be easily determined by the inflorescence, except in one ambiguous species. The flower terminates the peduncle, the bracts being gradually enlarged to the 2 uppermost, which, although unequal and not different in insertion from the lower ones, must probably, nevertheless, be considered as bracteoles; and (except in *C. adscendens*) I have never found the rudimentary flower which in *Leucopogon* terminates the spike or raceme above the last flower and its subtending bract.

Leaves from oval-elliptical to oblong-linear, obtuse or minutely mucronate, white underneath.
  Tall shrub or tree. Leaves mostly above ½ in. long. Ovary 8- to 10-celled . . . . 1. *C. glauca.*
  Bushy or prostrate shrubs. Leaves mostly under ½ in. long. Ovary 5-celled.
    Leaves mostly 5- or 7-nerved underneath.
      Erect, bushy shrub. Corolla-lobes bearded along the centre. Disk truncate . . . . 2. *C. straminea.*
      Diffuse shrub. Corolla-lobes bearded above the middle. Disk of 5 distinct scales . . . . 3. *C. adscendens.*
    Leaves mostly 1- or 3-nerved underneath. Prostrate shrub. Disk 5-lobed . . . . 4. *C. dealbata.*
Leaves linear-lanceolate, green on both sides, very rigid, with a short, hard, not pungent point . . . . 5. *C. abietina.*
Leaves linear or lanceolate-subulate, tapering into a pungent point.
  Leaves mostly about ½ in. long. Peduncles very short. Corolla-lobes glabrous . . . . 6. *C. acerosa.*
  Leaves ¼ to ½ in. long. Peduncles 1 to 2 lines. Corolla-lobes hairy inside . . . . 7. *C. divaricata.*
  Leaves mostly ¼ in. long. Peduncles very short. Corolla-lobes glabrous . . . . 8. *C. parvifolia.*

1. **C. glauca,** *Labill. Pl. Nov. Holl.* i. 57. *t.* 81. Usually a weak straggling shrub or small tree, but attaining sometimes 30 to 40 ft. (*Labillardière, C. Stuart*), glabrous or the branchlets minutely pubescent. Leaves mostly clustered at the ends of the year's shoots so as to appear almost whorled, oblong-linear, minutely mucronate, entire, slightly convex, glaucous underneath, mostly ¾ to 1 in. long, but a few occasionally much shorter and broader. Flowers almost sessile, clustered with the floral leaves, and much shorter than them. Sepals nearly 2 lines long, obtuse, and minutely ciliate as well as the bracts. Corolla 3 to 3½ lines long, the tube very shortly ex-

ceeding the calyx; lobes shorter than the tube, bearded along the centre, the lower hairs longer and reflexed into the tube. Filaments rather thick; anthers linear, attached near the top. Disk truncate. Ovary 8- to 10-celled; style rigid, included in the corolla-tube. Fruit rather large, very pulpy.—R. Br. Prod. 539; DC. Prod. vii. 740; Hook. f. Fl. Tasm. i. 245; *Trochocarpa glauca*, Spreng. Syst. i. 660; *Styphelia Billardieri*, F. Muell. Fragm. vi. 43.

Tasmania. Derwent River, *R. Brown;* common in the mountainous parts of the island at an elevation of from 1000 to 3000 ft., *J. D. Hooker.*

2. **C. straminea,** *R. Br. Prod.* 539. A bushy shrub of 2 to 3 ft. Leaves usually crowded towards the ends of the year's shoots nearly as in *C. glauca,* but sometimes rather more scattered, narrow-oblong, obtuse or with a small callous point, slightly concave, glaucous underneath, rarely exceeding ½ in. or sometimes all under that length, and broadly oblong. Flowers nearly sessile. Sepals 2 to nearly 3 lines long, obtuse, and minutely ciliate as well as the bracts. Corolla from 4 to nearly 6 lines long, the tube exceeding the calyx, hairy inside towards the throat; lobes not half so long as the tube, slightly bearded inside along the centre. Filaments rather thick; anthers attached near the top. Hypogynous disk truncate. Ovary 5-celled; style shorter than the corolla-tube. Fruit very pulpy.—DC. Prod. vii. 741; Hook. f. Fl. Tasm. i. 245; *Styphelia straminea*, Spreng. Syst. i. 656; F. Muell. Fragm. vi. 43; *Cyathodes macrantha*, Hook. f. Fl. Tasm. i. 245.

Tasmania. Mount Wellington, at an elevation of 4000 ft., *R. Brown, J. D. Hooker;* sides of Mount Olympus, at 4000 to 5000 ft., *Gunn;* Meanaii Falls, *Archer;* western mountains, at 3000 to 4000 ft., *C. Stuart.*

Hooker's *C. macrantha* has narrower leaves and much larger flowers than the typical *C. straminea,* but C. Stuart's specimens are quite intermediate between the two.

3. **C. adscendens,** *Hook. f. in Hook. Lond. Journ.* vi. 263, *and Fl. Tasm.* i. 245. *t.* 74 A. A stout, diffuse shrub, with numerous ascending branches of 6 to 8 in. Leaves crowded along the branches, oblong or almost ovate, obtuse or with a short callous point, flat or slightly incurved, glaucous and striate underneath, ¼ in. long or rather more. Flowers sessile, 2 or 3 together in the same axil or sometimes solitary, but always with an additional rudiment. Bracteoles much shorter than the calyx. Sepals rather above 1 line long, obtuse, striate. Corolla-tube not 1½ lines long, nearly glabrous inside; lobes short, recurved, densely bearded above the middle. Filaments inserted below the throat; anthers attached near the top, almost entirely included in the tube. Hypogynous disk separating into distinct truncate scales. Ovary 5-celled; style short. Fruit small.—*Styphelia Hookeri*, F. Muell. Fragm. vi. 44; *Leucopogon petiolaris*, DC. Prod. vii. 753 (from the character given).

Tasmania. Summit of Mount Wellington, *J. D. Hooker* and others; western mountains, *C. Stuart.* This species has the inflorescence of *Leucopogon,* and ought, perhaps, to be transferred to that genus; but the fruit is much more pulpy than in any *Leucopogon,* and the aspect and foliage are quite those of *Cyathodes.*

4. **C. dealbata,** *R. Br. Prod.* 539. A small, diffuse or prostrate, much-branched shrub. Leaves crowded along the branches, oblong-linear, obtuse,

with a small, rigid, deciduous point, incurved, white underneath, with 1 or sometimes 3, very rarely 5, prominent ribs. Flowers very shortly pedicellate, solitary in the axils. Bracts several, gradually enlarged. Sepals $1\frac{1}{2}$ lines long, obtuse, scarcely ciliate. Corolla-tube $2\frac{1}{2}$ lines long, slightly hairy inside above the middle; lobes small, bearded inside. Filaments inserted below the top of the tube; anthers attached above the middle, almost entirely included in the tube. Hypogynous disk obtusely 5-lobed. Ovary 5-celled; style short. Fruit small, globular, with a pulpy mesocarp.—DC. Prod. vii. 741; Hook. f. Fl. Tasm. i. 245; *Styphelia dealbata*, Spreng. Syst. i. 659; F. Muell. Fragm. vi. 43.

Tasmania. Summit of Mount Wellington, *R. Brown, Gunn.*

5. **C. abietina,** *R. Br. Prod.* 540. A stout, rigid, erect, bushy shrub of 1 to 2 ft. Leaves crowded along the branches, erect or spreading, linear-lanceolate, very rigid, flat, with a short, hard, but not pungent point, $\frac{1}{2}$ to $\frac{3}{4}$ in. long. Flowers solitary in the axils, on very short pedicels. Bracts numerous. Sepals broad, obtuse, striate, rather unequal, the inner ones fully 1 line long. Corolla-tube rather broad, nearly $1\frac{1}{2}$ lines long, villous inside between the stamens to below the middle; lobes nearly 1 line long, spreading, bearded inside. Filaments filiform, half as long as the lobes; anthers attached near the top, half-exserted. Hypogynous scales irregularly lanceolate, free or united at the base. Ovary 5-celled. Drupe large, red, very pulpy.—DC. Prod. vii. 741; Hook. f. Fl. Tasm. i. 247; *Styphelia abietina*, Labill. Pl. Nov. Holl. i. 48. t. 68; F. Muell. Fragm. vi. 43.

Tasmania, *Labillardière;* D'Entrecasteaux Channel, *Gunn;* South Port Island, *C. Stuart.*

6. **C. acerosa,** *R. Br. Prod.* 539 *and* 540 *in the obs.* A shrub of several feet or a small tree, with spreading branches, or rarely low and diffuse. Leaves scattered, spreading or reflexed, linear or lanceolate-subulate, rigid, and tapering into a pungent point, with recurved margins, mostly about $\frac{1}{2}$ in. long. Flowers solitary in the upper axils, on short, recurved pedicels. Bracts several. Sepals very obtuse, scarcely above 1 line long. Corolla-tube rather broad, usually nearly 2 lines long, and glabrous inside, the lobes short, recurved, glabrous (or rarely sprinkled with a few long hairs?). Filaments very short, inserted just below the top of the tube; anthers half-exserted. Hypogynous scales broad, quite distinct or more or less united. Ovary 5-celled; style very short. Drupe rather large, pulpy. —DC. Prod. vii. 741; Hook. f. Handb. N. Zeal. Fl. 176; *Ardisia acerosa*, Gærtn. Fruct. ii. 78. t. 94; *Styphelia oxycedrus*, Labill. Pl. Nov. Holl. i. 49. t. 69; F. Muell. Fragm. vi. 43; *Cyathodes oxycedrus*, R. Br. Prod. 540; DC. Prod. vii. 741; Hook. f. Fl. Tasm. i. 246; *Lissanthe acerosa* and *L. oxycedrus*, Spreng. Syst. i. 660.

Victoria. Maritime rocks and sands, Sealer's Cove, Wilson's Promontory, Phillip's Island, *F. Mueller;* Rabbit Island, *J. Bosisto..*

Tasmania. Islands of Bass's Straits, *R. Brown;* Circular Head, Recherche Bay, and other parts of the island, *J. D. Hooker.*

This species is also in New Zealand.

7. **C. divaricata,** *Hook. f. Fl. Tasm.* i. 246. t. 74 B, *partly.* A rigid,

bushy, juniper-like shrub, closely resembling *C. acerosa*, but with smaller and finer leaves although equally rigid and pungent, and the flowers on recurved pedicels of 1 to 2 lines, with the corolla-lobes bearded inside with long, scattered hairs. Bracts very small, upper ones less enlarged than in the other species, and only 1 or 2 (bracteoles?) reach the calyx, which at first induced the placing the plant in *Lissanthe*, but the flower appears to be constantly terminal and solitary without the additional rudimentary one of *Lissanthe* and *Leucopogon.*—*Lissanthe divaricata*, Hook. f. in Hook. Lond. Journ. vi. 269.

Tasmania. Foot of Mount Wellington and other hills near Hobarton, *J. D. Hooker;* Swan Port, *Gunn;* elevated places among rocks, *Story.*

F. Mueller unites this species with *L. acerosa,* and, as it would appear, correctly so as to the seacoast and Victorian plant, from which the principal figure in Hooker's plate was taken, but the specimens from the above-quoted Tasmanian localities show differences, which can scarcely be pronounced insufficient for distinguishing a species without further investigation.

8. **C. parvifolia,** *R. Br. Prod.* 540. A rigid shrub of 2 to 4 ft. Leaves linear or lanceolate-subulate, tapering into a pungent point, rigid, with recurved margins, glaucous or white underneath, mostly about ¼ in. long. Flowers solitary in the upper axils, on exceedingly short, recurved peduncles. Bracts few. Sepals not 1 line long, ovate, obtuse. Corolla-tube about 1½ lines long, the lobes short, all quite glabrous. Filaments short, inserted at the throat. Hypogynous disk short, sinuate-toothed. Ovary 5-celled. Drupe pulpy.—DC. Prod. vii. 741; Hook. f. Fl. Tasm. i. 246; *Lissanthe parvifolia,* Spreng. Syst. i. 660.

Tasmania. Derwent River, *R. Brown;* abundant, especially in hilly parts of the island, ascending to 3000 ft., *J. D. Hooker.*

9. BRACHYLOMA, Sond.

(Lobopogon, *Schlecht.*)

Corolla-tube short, glabrous inside, but a ring of long hairs descending into it from tufts or fringed scales at the base of each lobe; lobes more or less imbricate in the bud (almost valvate in *B. daphnoides*), spreading, glabrous or slightly bearded. Filaments very short, inserted near the top of the tube; anthers 1-celled, attached above the middle, wholly or partially included in the tube. Hypogynous disk truncate or 5-lobed, readily separating into distinct scales. Ovary 5-celled, with 1 ovule in each cell. Style short; stigma small. Fruit a small depressed or globular drupe, the mesocarp somewhat pulpy.—Shrubs, with the foliage of some species of *Cyathodes* and *Leucopogon.* Flowers small, solitary in the axils. Pedicels short, with very small bracts or none besides the bracteoles.

The genus is limited to Australia. It is easily distinguished, as well by the imbricate æstivation of the corolla-lobes, which, although slight in one species, is very conspicuous in the others, as by the long hairs descending into the tube, which occur only in some species of *Leucopogon,* otherwise distinguishable by their inflorescence.

SECT. I. **Lobopogon.**—*Corolla-lobes obtuse. Bracts several.*

Leaves flat or slightly convex, pale underneath . . . . 1. *B. Preissii.*
Leaves slightly concave or almost flat, the two surfaces of the same colour.

Leaves oblong or lanceolate, with a rigid point. Corolla-lobes glabrous (besides the reflexed hairs of the base) . . . . 2. *B. concolor.*
Leaves linear or oblong-linear. Corolla-lobes bearded in the middle . 3. *B. ericoides.*

SECT. II. **Lissanthoides.**—*Corolla-lobes acutely acuminate.*   Bracts *none besides
the* 2 *bracteoles.*

Leaves lanceolate, tapering into a long, pungent point. Corolla-lobes
  much imbricate . . . . 4. *B.depressum.*
Leaves ovate-elliptical or oblong-lanceolate, with a short pungent point.
  Corolla-lobes moderately imbricate . . . . 5. *B. ciliatum.*
Leaves ovate elliptical or oblong-lanceolate, obtuse, with a minute callous
  point. Corolla-lobes slightly imbricate at the base . . . . . . 6. *B.daphnoides.*

SECT. I. LOBOPOGON.—Corolla-lobes obtuse.   Bracts several.

1. **B. Preissii,** *Sond. in Pl. Preiss.* i. 305.   An erect, bushy shrub,
the branches slightly hairy.   Leaves linear-oblong, obtuse but often mucronate, minutely denticulate-ciliate, flat or convex, pale or hoary underneath,
mostly about ½ in. long.   Peduncles axillary, 2 to 3 lines long, with several
bracts ; bracteoles at least half as long as the calyx.   Sepals 2 lines long or
rather more, almost scarious, shortly mucronate.   Corolla-tube shorter than
the calyx, rather broad, with reflexed scales in the throat copiously fringed
with long hairs ; lobes as long as the tube, obtuse, much imbricate in the
bud.   Filaments short, slightly dilated upwards ; anthers attached above the
middle.   Hypogynous disk truncate, minutely 5-toothed.   Ovary 5-celled.
Style short.   Fruit enclosed in the calyx, depressed-globular, remarkably
furrowed.—*Styphelia brachyloma,* F. Muell. Fragm. vi. 39.

W. Australia. Swan River, *Drummond, 1st Coll. n.* 480 ; Bull's Creek, *Preiss, n.*
426 ; Hamden, *Clarke ;* Harvey river, *Oldfield.* Drummond's specimens, n. 138 and 264,
appear to be the same species, but the remains of flowers are too‚imperfect to determine.

2. **B. concolor,** *F. Muell. Fragm.* vi. 39.   An erect, bushy shrub, glabrous or the branches slightly pubescent.   Leaves petiolate, oblong or oblanceolate, with a short fine point, slightly concave, thick and nearly as
smooth and shining underneath as above, mostly 3 to 4 lines long.   Peduncles under 1 line long, the bracts minute ; bracteoles very small.   Sepals ¾
line long.   Corolla nearly 2 lines long, quite glabrous, except the long hairs
reflexed into the tube from the base of the lobes ; lobes as long as the tube,
much imbricate.   Filaments exceedingly short, flat ; anthers attached near
the top, slightly cohering in a ring.   Hypogynous disk truncate, readily
separating into scales.   Ovary 5-celled ; style short.—*Styphelia geissoloma* or
Stenanthera brachyloma, F. Muell. Fragm. vi. 39.

W. Australia. To the eastward of King George's Sound, *Baxter, Maxwell.*

3. **B. ericoides,** *Sond. in Linnæa,* xxvi. 247.   A low, bushy or diffuse
shrub, the branches shortly pubescent.   Leaves erect or spreading, linear or
oblong-linear, with a fine rigid point, minutely denticulate-ciliate, flat or
slightly concave, mostly 3 to 4 lines long.   Flowers nearly sessile,   Bracts
few ; bracteoles half as long as the calyx.   Sepals about 2 lines long, very
obtuse and almost scarious.   Corolla about 3 lines long, the tube shorter
than the calyx, with reflexed scales in the throat fringed with long hairs
descending into the tube ; lobes rather shorter than the tube, ovate, obtuse,

imbricate in the bud, bearded in the centre.   Filaments very short; anthers
obtuse, attached near the top.   Hypogynous disk truncate.   Ovary 5-celled;
style longer than in the preceding species.   Fruit as long as the calyx, globular, with 5 raised lines.—Hook. Ic. Pl. t. 1038 ; *Lobopogon ericoides,*
Schlecht. Linnæa, xx. 620 ; *Stenanthera ericoides,* F. Muell. Fragm. iv. 98 ;
Styphelia lobopogona, F. Muell. Fragm. vi. 39.

Victoria. Grampians, *F. Mueller ;* Wimmera, *Dallachy.*
S. Australia. Murray desert, near Tornunda, Kangaroo Island, *F. Mueller ;* Encounter Bay, *Whittaker.*

SECT. II. LISSANTHOIDES.—Corolla-lobes acutely acuminate, more or
less dilated and overlapping at least at the base.   Bracts none besides the 2
bracteoles.

4. **B. depressum,** *Benth.* A very rigid, diffuse, much-branched shrub,
spreading sometimes to several ft.   Leaves sessile or nearly so, lanceolate or
linear-lanceolate, rigid, tapering into a long pungent point, the larger ones
½ in. long or rather more.   Flowers solitary at each node, but clustered at
the base of the short flowering branches, the subtending leaves mostly abortive.   Peduncles very short, without bracts ; bracteoles acuminate.   Sepals
lanceolate, acutely acuminate, about 1 line long.   Corolla-tube about 1 line
long, with a ring of long reflexed hairs in the throat ; lobes broad, acuminate,
much imbricate in the bud, quite glabrous.   Anthers attached near the top
and nearly sessile.   Hypogynous disk truncate, readily separating into distinct scales.   Ovary 5-celled ; style short.—*Lissanthe depressa,* F. Muell.
Fragm. i. 36 ; *Styphelia depressa,* F. Muell. Fragm. vi. 42 (not of l. c. 44).

Victoria. Mount Sturgeon, Grampians, *F. Mueller ;* Wimmera, *Dallachy.*
Tasmania. Granite rocks at Bicheno, *Story.*

5. **B. ciliatum,** *Benth.* A low, diffuse or prostrate shrub, with ascending
branches, glabrous or slightly pubescent.   Leaves shortly petiolate, ovate-
elliptical or oblong-lanceolate, with a short, pungent point, flat, often minutely
denticulate-ciliate, 3 to 4 lines long.   Peduncles very short, without any
bracts except the 2 bracteoles embracing the calyx.   Sepals about ¾ line long.
Corolla-tube longer than the calyx, rather broad, with a ring of long hairs
reflexed from the throat ; lobes acute, imbricate in the bud, but not so much so
as in *B. depressum,* usually shortly bearded inside.   Filaments short ; anthers
attached above the middle.   Hypogynous disk 5-lobed.   Ovary 5-celled ;
style short.—*Lissanthe ciliata,* R. Br. Prod. 541 ; DC. Prod. vii. 743 ;
Hook. f. Fl. Tasm. i. 248 ; *Styphelia ciliata,* F. Muell. Fragm. vi. 42.

Victoria. Arid plains near Mount Abrupt, and Victoria Range, *F. Mueller ;* Creswick
Creek, *Whan ;* heaths near Portland, *Robertson, Allitt.*
Tasmania. Port Dalrymple, *R. Brown ;* northern parts of the island, *J. D. Hooker.*

The species is very near *B. daphnoides,* differing chiefly in the pungent point to the
leaves, in the shorter, broader corolla-tube, and the depressed habit.

6. **B. daphnoides,** *Benth.* An erect, bushy shrub, glabrous or the
branchlets pubescent.   Leaves sessile or shortly petiolate, ovate-elliptical or
oblong-lanceolate, obtuse or with a short callous point, flat or slightly
concave, 3 to 4 lines long.   Flowers solitary in the axils or at the nodes,

but often several at the base of the shoots, without subtending leaves. Pedicels very short, without any bracts except the 2 unequal bracteoles embracing the calyx. Sepals ⅓ to ¾ line long, usually ciliate. Corolla-tube 1½
to nearly 2 lines long, with a ring of long hairs inside reflexed from the
throat; lobes narrow, much shorter than the tube, acute and nearly valvate,
but 2 of them at least slightly dilated and overlapping the intermediate one
at the base, glabrous or very shortly bearded inside. Filaments very short;
anthers attached above the middle. Hypogynous disk 5-lobed. Ovary 5-
celled; style short. Fruit small, globular.—*Styphelia daphnoides*, Sm. Bot.
Nov. Holl. 48; F. Muell. Fragm. vi. 42; *Lissanthe daphnoides*, R. Br. Prod.
541; DC. Prod. vii. 743; Lodd. Bot. Cab. t. 466; Hook. f. Fl. Tasm. i.
248; *L. Cunninghamii*, DC. Prod. vii. 743; *L. stellata*, Knowles and Westc.
Fl. Cab. iii. 79 (from the description given).

Queensland. Sandy Cape, *R. Brown;* Moreton island, *F. Mueller;* Mount Mitchell,
Beckler.

N. S. Wales. Port Jackson, *R. Brown, Sieber, n.* 100, and others; near Bathurst,
Woolls; St. George's Range, *A. Cunningham;* New England, *C. Stuart;* sandy ridges,
Cape Byron, *C. Moore;* and southward to Twofold Bay, *F. Mueller.*

Victoria. Grampians, Forest Creek, rocks on the Macalister river, and elsewhere in
the Australian Alps, ascending to 4000 ft., *F. Mueller;* Wendu vale, *Robertson.*

Tasmania? *R. Brown* (Prod.). I have seen no Tasmanian specimens; there are none
in Brown's Herbarium, and Scott's authority, quoted by J. D. Hooker, is founded on a specimen evidently incorrectly labelled in the Hookerian Herbarium.

S. Australia. Tattiara country, *Wood.*

10. NEEDHAMIA, R. Br.

Corolla-tube cylindrical; limb spreading, deeply divided into 5 lobes,
induplicate-valvate in the bud, with inflexed tips, not bearded. Stamens
inserted below the middle of the corolla-tube; filaments very short; anthers
entirely included. Hypogynous disk cup-shaped. Ovary 2-celled; style
short; stigma capitate. Fruit a small, dry drupe, 1- or 2-seeded.—A small
shrub. Leaves small, opposite or alternate. Flowers solitary in the upper
axils, without other subtending bracts than the floral leaf and 2 bracteoles.

The single species is limited to W. Australia. With the habit of some of the smaller
species of *Leucopogon*, it is distinguished by the æstivation of the corolla-lobes and their
want of beards.

1. **N. pumilio,** *R. Br. Prod.* 549. An erect or diffuse shrub, with
slender branches not exceeding 6 in. Leaves erect, often closely appressed,
linear or linear-lanceolate, acute or almost obtuse, concave or keeled, 1 to 2
lines long. Flowers at first forming a short terminal spike, the leaf-like
bracts either smaller and more ciliate than the upper stem-leaves, and some
of them coloured, others quite like the stem-leaves, and the axis ultimately
grows out into a prolongation of the branch, with the flowers axillary at its
base. Bracteoles linear, ciliate, as long as the calyx or even longer. Sepals
thin, ciliate, scarcely above ½ line long. Corolla-tube 1 to 1¼ lines long, the
lobes broad, shorter than the tube, the whole corolla glabrous, inside and
out. Hypogynous disk shortly lobed. Ovary pubescent; style exceedingly
short.—DC. Prod. vii. 759; Deless. Ic. Sel. v. t. 26; Sond. in Pl. Preiss.
i. 326; *Monotoca pumilio*, Spreng. Syst. i. 654.

W. Australia. King George's Sound, *R. Brown*, and neighbouring districts, *Huegel*, *Drummond*, 2nd *Coll. n.* 259, 5th *Coll. n.* 334; *Preiss, n.* 436, 437, and many others; Swan River, *Drummond*, 1st *Coll.* ; between Eyre's Range and Oldfield river, *Maxwell*. The station N. S. Wales, *Fraser*, given in the 'Plantæ Preissianæ' for a form of this species, is a mistake. Fraser's specimens were from King George's Sound.

11. LISSANTHE, R. Br.

Corolla-tube longer or shorter than the calyx, glabrous or hairy inside above the middle, without tufts of hairs or scales below the middle; lobes valvate in the bud, spreading or recurved upwards, glabrous inside. Filaments inserted at the top of the tube, short, filiform; anthers wholly or partially enclosed in the tube or erect base of the corolla-lobes Hypogynous disk cup-shaped, sinuate or 5-toothed. Ovary 5-celled; style not exceeding the corolla-tube; stigma small. Fruit a baccate drupe, the mesocarp pulpy, the endocarp hard and bony, with 5 cells and seeds or fewer by abortion.—Shrubs with the habit of *Leucopogon*. Flowers small, in small spikes or racemes, the terminal ones several-flowered, the axillary ones reduced to 2 or 3 or even a single one, but always ending with the rudiment of an additional flower, with 1 subtending bract and 2 bracteoles under each flower.

The genus is limited to Australia. It is here reduced to the first two sections of Brown, the third has the inflorescence and more or less imbricate corolla-lobes of *Brachyloma*, and is therefore referred to that genus. As thus defined, *Lissanthe* differs from *Cyathodes* in the inflorescence and in the less pulpy fruit, from *Leucopogon* solely in the want of the hairs or beards of the lobes of the corolla so universal in that genus.

Flowers racemose (pedicellate within the bracts and bracteoles).
  Leaves linear or linear-lanceolate with recurved margins and short rigid
    points ¾ to 1 in. long. Fruit mealy-pulpy . . . . 1. *L. sapida*.
  Leaves linear, tapering into a pungent point, under ½ in. long. Fruit
    small, nearly dry . . . . 2. *L. strigosa*.
Flowers spicate (sessile within the bracts). Leaves oblong-linear, obtuse . 3. *L. montana*.

1. **L. sapida,** *R. Br. Prod.* 540. An erect shrub of 2 to 3 ft, with spreading branches, glabrous or very minutely pubescent. Leaves linear-oblong or linear-lanceolate, with a very short rigid point, the margins recurved or revolute, white underneath, ¾ to 1 in. long. Flowers white, in loose racemes, either axillary and 2- or 3-flowered or terminal with more numerous flowers. Bracts and bracteoles small, at the base of the pedicels. Sepals very broad and obtuse, about 1 line long, the 2 outer ones usually thickened or produced at the base. Corolla-tube 2 to 2½ lines long, hairy inside above the middle; lobes about as long as the tube, spreading towards the end, quite glabrous. Anthers obtuse, attached rather above the middle. Hypogynous disk slightly toothed. Ovary 5-celled, pubescent on the top; style rather thick, as long as the corolla tube. Fruit red, with a mealy-pulpy mesocarp.—DC. Prod. vii. 742; Bot. Reg. t. 1275; Bot. Mag. t. 3147; *Styphelia sapida*, F. Muell. Fragm. vi. 42.

N. S. Wales. Grose river, *R. Brown*; various localities in the Blue Mountains, *Sieber, n.* 95, *Woolls*, and others; Bargo Brush, *Macarthur*, but apparently not generally common.

2. **L. strigosa,** *R. Br. Prod.* 540. A bushy shrub, sometimes low and

spreading, but usually erect and attaining about 2 ft., the branches glabrous or pubescent. Leaves linear, rigid, tapering into a pungent point, under ½ in. long. Flowers white or more or less pink, crowded in short racemes, 2 or 3 in the axillary ones, more numerous in the terminal one, each flower very shortly pedicellate within the bract and bracteoles. Sepals ovate, obtuse, about 1 line long. Corolla-tube about 1½ lines long, more or less hairy inside above the middle; lobes much shorter, glabrous. Anthers attached above the middle. Hypogynous disk shortly 5-lobed; style rather thick, pubescent at the base, shorter than the corolla-tube. Fruit small, globular.—DC. Prod. vii. 742; Hook. f. Fl. Tasm. i. 247; *Styphelia strigosa,* Sm. Bot. N. Holl. 49; F. Muell. Fragm. vi. 42; *Lissanthe subulata,* R. Br. Prod. 540; DC. Prod. vii. 742; *L. intermedia,* A. Cunn.; DC. l. c.; *L. rigida,* Benth. in Hueg. Enum. 76; DC. l. c.

N. S. Wales. Port Jackson and Blue Mountains, *R. Brown, Sieber, n.* 104, and others.
Victoria. Dry stony hills, Snowy River, Bendigo diggings, Nangatta river, Station Peak, etc., *F. Mueller;* Creswick, *Whan.*
Tasmania. Derwent river, *R. Brown;* abundant throughout the island in dry clayey or gravelly places, *J. D. Hooker.*
S. Australia. Onkaparinga river, *F. Mueller.*
In Brown's herbarium, small-leaved specimens from Port Jackson and from Tasmania represent *L. strigosa,* and his *L. subulata* from Grose river has much larger leaves, but in a large number of specimens from various localities the two can no longer be separated even as marked varieties. *L. propinqua,* A. Cunn. Herb., quoted by De Candolle is *Leucopogon juniperinus.*

3. **L. montana,** *R. Br. Prod.* 540. A small erect shrub of 6 in. to 1 ft. Leaves oblong-linear, obtuse or with an obscure callous point, flat, few-nerved, rarely exceeding ½ in. Flowers few together in short dense terminal spikes, each flower sessile within the subtending bract, with 2 broad bracteoles half as long as the calyx. Sepals very obtuse, about ½ line long. Corolla almost campanulate, about 1 line long, quite glabrous, the lobes rather longer than the tube. Hypogynous disk sinuate or shortly lobed. Ovary 5-celled; style short. Fruit small, white, pulpy.—DC. Prod. vii. 743; Hook. f. Fl. Tasm. i. 247.

Victoria. Munyong Mountains at an elevation of 5000 to 6000 ft., *F. Mueller.*
Tasmania. Mount Wellington, *R. Brown;* towards the summits of Mounts Wellington and others, *J. D. Hooker.*
F. Mueller, Fragm. vi. 45, under the name of *Styphelia montana,* unites this with *Leucopogon Hookeri,* and there is no doubt that the two plants closely resemble each other in foliage and general aspect, as already pointed out by Hooker, but independently of the want of the beard on the corolla-lobes, the difference in the size and shape of the flower appears to be constant, and, according to Gunn, the fruit has a clear translucent pulp in *L. montana,* whilst it is thick and opaque in *L. Hookeri.* The flowers appear to be partially diœcious in both species as in a few other small-flowered *Styphelieæ.*
Lissanthe mucronata, DC. Prod. vii. 743, from the east coast, Herb. Mus. Par., is distinguished by De Candolle from *L. montana* by the leaves strongly mucronate. I know of no such plant, possibly there may have been some mistake as to the genus.

12. LEUCOPOGON, R. Br.

(Perojoa, *Cav.,* Phanerandra, *Stschegl.*)

Corolla-tube longer or shorter than the calyx, glabrous or hairy inside above the middle, without tufts of hairs or scales below the middle ; lobes valvate in the bud, spreading or recurved in the upper portion, the whole inner surface, or rarely the lower portion only, densely bearded.　Filaments inserted at the top of the tube, short, filiform ; anthers wholly or partially enclosed in the tube or erect base of the corolla-lobes.　Hypogynous disk cup-shaped, truncate, 5-toothed, 5-lobed or formed of 5 distinct scales (none in *L. esquamatus*).　Ovary usually 2- 3- or 5-celled (in one species 1-celled and two others often 6- to 10-celled) ; style from very short to longer than the corolla-tube ; stigma small, rarely larger and peltate.　Fruit a drupe, the mesocarp sometimes pulpy but usually thin, the endocarp crustaceous or hard, with as many cells and seeds as in the ovary or fewer by abortion.—Shrubs of various habit, rarely rising into small trees.　Flowers small, rarely $\frac{1}{2}$ in. long, in small spikes (or very rarely racemes), terminal or axillary, sometimes many-flowered, sometimes reduced to few or a single one, but the rhachis always ending in the rudiment of an additional one ; each flower sessile or rarely pedicellate within a subtending bract (deficient in *L. flavescens*), with 2 bracteoles close under or rarely at a little distance from the calyx.

The genus, chiefly Australian, is represented also by a few species in New Zealand and some islands of the Malayan Archipelago and South Pacific.　Like *Lissanthe* it is only to be distinguished from *Cyathodes* either by the compound inflorescence or by the less pulpy fruit or by both characters, and would perhaps have been better united with that genus, were it not that the change would now produce so much inconvenience.　From *Lissanthe* it is still more artificially separated by the bearded corolla-lobes.

The sterile tips of the anthers, by which the first section of the genus is chiefly characterized, are sometimes exceedingly short but usually of a paler colour than the rest of the anther.　In a few species there is no discoloration, but they form a perfectly rounded extremity to the anther, not truncate or emarginate as is the open anther in the other sections. The characters derived from the hypogynous disk are not always very constant in species, especially where there is any approach to unisexuality as in *L. Hookeri* and a very few other species, and generally disks which are at first entire tend to separate into distinct scales as the flowering advances.　The number of cells of the ovary is more constant, but not strictly so ; the 2-celled species may in some flowers (though very rarely) add a third cell, and the 5-celled ones are occasionally reduced to 4, or the 3-celled ones may add a fourth.　In the fruit the number of cells is very frequently reduced by abortion.

The so-called beard which lines the corolla-lobes and suggested the generic name, is particularly conspicuous in nearly the whole of the first section.　It is then very white and dense, consisting on the upper portion of the lobes of long straight hairs inflexed in the bud, but erect when the flower expands, below this the hairs are woolly and intricate, and those at the base of the lobe are sometimes more or less reflexed into the tube as in *Brachyloma*.　In several species of the second and third sections, and a very few of the first, the beard is less dense, less woolly, and not so white ; in others of the second and third sections it is quite as in the first ; in *L. verticillatus* it occupies only the lower half of the lobes.

Cavanilles' generic name of *Perojoa*, contracted by Persoon to *Peroa*, is undoubtedly prior to *Leucopogon*, and it does not appear for what reason it was rejected by Brown, but the substitution of *Leucopogon* has now been so long and so universally adopted, that the resumption of *Perojoa*, except as sectional, would only create confusion without a single practical advantage.

Sect. I. **Perojoa.**—*Inflorescence chiefly or entirely terminal, rarely all axillary, and then the spikes long, slender and interrupted.　Anthers with sterile tips.　Style very short.*

Series I. **Psilostachyæ.**—*Spikes slender, interrupted, axillary or terminal, usually as long as the leaves.　Leaves flat or convex.*

Leaves (large) crowded at the ends of each year's shoots, so as to
   appear verticillate. Ovary 5- or 4-celled.
  Leaves usually 2 to 5 in. long. Spikes axillary or below the
   leaves. Corolla-tube longer than the calyx, lobes bearded in
   the lower half . . . . 1. *L. verticillatus.*
  Leaves about 1 in. long. Spikes terminal. Corolla-tube shorter
   than the calyx, lobes bearded to the end . . . . 2. *L. interruptus.*
Leaves scattered. Ovary 2-celled.
  Leaves broad, cordate, auriculate and stem-clasping.
   Leaves mostly ½ in. long or more. Eastern species . . . 3. *L. amplexicaulis.*
   Leaves under ¼ in. long. Stems filiform. Western species . 4. *L. alternifolius.*
  Leaves lanceolate. Fruit ovoid, red . . . . 5. *L. lanceolatus.*
Leaves scattered. Ovary 5-celled (see Ser. 2).
Leaves concave. Ovary 5-celled (see Ser. 7).

SERIES II. **Australes.**—*Spikes all terminal or also in the upper axils, short and
dense, or cylindrical and rather dense (interrupted in* L. distans). *Leaves flat or convex,
with recurved or revolute margins. Ovary 5-celled or rarely 4- or 3-celled.*

Spikes cylindrical, rather dense. Leaves oblong or lanceolate,
  finely veined. Ovary 5- or rarely 4-celled.
  Leaves mostly ½ to 1 in. long. Fruit ovoid-globose . . . . 6. *L. Richei.*
  Leaves mostly 1 to 2 in. long. Fruit depressed-globose · . . . 7. *L. australis.*
Spikes short and dense (except *L. distans*).
  Leaves oblong-linear or lanceolate (¼ to ½ in. long), erect or
   spreading. Ovary 5-celled.
   Leaves tapering into a rigid almost pungent point . . . . 8. *L. capitellatus.*
   Leaves obtuse or with a small callous point.
    Sepals obtuse, 1 to 1¼ line long. Anthers attached above
     the middle, with prominent sterile tips . . . . . . 9. *L. revolutus.*
    Sepals acute, 2 to 3 lines long. Anthers attached at or
     below the middle, tapering and recurved at the end, but
     2-valved to the tip . . . . 10. *L. atherolepis.*
  Leaves small (under ¼ in.), very convex, very spreading or re-
   flexed.
   Ovary 5-celled.
    Spikes short and dense. Flowers under 2 lines long.
     Leaves mostly ovate and 1 line to lanceolate and 2 lines
      long . . . . 11. *L. reflexus.*
     Leaves mostly narrow-linear, 2 to 3 lines long . . . 12. *L. corifolius.*
    Spikes interrupted or rarely dense. Flowers above 2 lines
     long. . . . . 13. *L. distans.*
   Ovary 3- or rarely 4-celled.
    Leaves orbicular, about 1 line diameter. Western species . 14. *L. gibbosus.*
    Leaves oblong or lanceolate, hirsute, 2 to 4 or 5 lines long.
     Eastern species . . . . 15. *L. thymifolius.*
  Leaves small (under ¼ in.), spreading nearly flat.
   Leaves scarcely petiolate, ovate-cordate lanceolate or linear.
    Ovary usually 5-celled . . . . 16. *L. cordatus.*
   Leaves on long petioles, orbicular. Ovary usually 4-celled . 17. *L. Bossiæa.*

SERIES III. **Collinæ.**—*Spikes all terminal or also in the uppermost axils, short and
dense, or cylindrical and rather dense. Leaves flat or convex with recurved or revolute
or thickened margins. Ovary 2-celled.*

Spikes short, dense or few-flowered. Bracts small.
  Leaves oblong or linear, obtuse or with a callous point (mostly
   3 to 6 lines).
   Trailing or procumbent and hirsute. Flowers very small . 18. *L. hirsutus.*
   Erect or diffuse and much branched.

Sepals obtuse . . . . 19. *L. collinus.*
Sepals acuminate . . . . 21. *L. compactus.*
Leaves (3 to 4 lines) linear-lanceolate, acuminate, twisted . . 20. *L. glacialis.*
Leaves mostly broad and scarcely exceeding 2 lines, occasionally
  linear and 3 lines long in the first 2 species.
  Leaves, at least the upper ones, mucronate-acute and squarrose.
    Sepals acute . . . . 22. *L. squarrosus.*
  Leaves all obtuse. Lower bracts leaf-like.
    Sepals narrow-acuminate. Corolla shortly exceeding the
      calyx. Eastern species . . . . 23. *L. microphyllus.*
    Sepals rather obtuse. Corolla twice as long as the calyx.
      Western species . . . . 24. *L. tetragonus.*
Spikes cylindrical. Bracts, at least the lower ones, leaf-like.
  Leaves very spreading or reflexed.
    Leaves orbicular-cordate, obtuse, 3 lines diameter . . . . 25. *L. phyllostachys.*
    Leaves cordate, acuminate, not 2 lines broad . . . . . . 26. *L. glabellus.*
  Leaves erect or spreading, ovate and 1 line to lanceolate and 3
    lines long, not cordate, nearly flat . . . . 27. *L. elatior.*

SERIES IV. **Striatæ.**—*Spikes all terminal or also in the upper axils, short and dense, or cylindrical and rather dense. Leaves nearly flat, the margins neither thickened nor recurved, usually small and obtuse, either strongly ribbed or the midrib prominently at the end. Ovary 2-celled.*

Leaves ovate, very flat, about 1 line long, scarcely ribbed . . . 28. *L. florulentus.*
Leaves ovate or ovate-lanceolate, 2 to 3 lines long, strongly ribbed.
  Spikes short . . . . 29. *L. striatus.*
Leaves ovate-lanceolate, 2 to 4 lines long, strongly ribbed. Spikes
  long. Plant softly villous . . . . 30. *L. lasiostachyus.*
Leaves oblong or oblong-lanceolate, 2 to 4 lines long, the midrib at
  least prominent. Spikes dense . . . . 31. *L. carinatus.*
Leaves narrow-lanceolate. Spikes cylindrical, many-flowered . . 37. *L. tenuis.*

SERIES V. **Oppositifoliæ.**—*Inflorescence and ovary of the* Striatæ. *Leaves small, erect, linear, obtuse, all opposite.*

Leaves with thickened margins, 2-furrowed underneath . . . . 32. *L. opponens.*
Leaves convex and smooth underneath . . . . 33. *L. oppositifolius.*

(27. *L. elatior*, 45. *L. lasiophyllus*, 57. *L. fimbriatus*, and perhaps a few other small-leaved species, have the leaves here and there opposite, especially on the sterile branches, but never all opposite as in the present series.)

SERIES VI. **Concurvæ.**—*Spikes all terminal or also in the upper axils, short and dense, or cylindrical and rather dense. Leaves more or less concave or keeled. Ovary 2-celled.*

Leaves ovate to broadly lanceolate, under ¼ in. long.
  Leaves sessile. Spikes cylindrical, many-flowered.
    Leaves acuminate and exceeding 1 line or obtuse and under 1
      line. Bracteoles small, acuminate . . . . 34. *L. tamariscinus.*
    Leaves obtuse, ciliate with fine hairs. Bracteoles acute, nearly
      as long as the calyx . . . . 35. *L. bracteolaris.*
  Leaves shortly petiolate. Spikes short and few-flowered. Brac-
    teoles small, obtuse . . . . 36. *L. elegans.*
Leaves lanceolate, mostly 4 to 6 lines long.
  Spikes all terminal, cylindrical, elongated. Leaves rather narrow 37. *L. tenuis.*
  Spikes all terminal, short and dense. Leaves rather broad, im-
    bricate, obtuse . . . . 38. *L. gnaphalioides.*
  Spikes terminal and in the upper axils, short and dense but
    many-flowered. Leaves distant.

N 2

Leaves broadly lanceolate. Eastern species . . . . . . 39. *L. concurvus.*
Leaves narrow lanceolate. Western species . . . . . . 40. *L. Gilbertii.*
Leaves linear-lanceolate or linear, mostly under 4 lines long.
  Leaves (usually glabrous) linear-lanceolate, 2 to 4 lines long,
    almost acute. Sepals acute . . . . 41. *L. gracilis.*
  Leaves linear-lanceolate, 2 to 4 lines long, tapering into a fine
    point. Sepals very acute.
      Corolla longer than the calyx. Plant hairy . . . . . . 42. *L. acicularis.*
      Corolla shorter than the calyx. Foliage glabrous . . . . 43. *L. cryptanthus.*
  Leaves (usually glabrous) linear, obtuse, keeled, 1 to 2 lines
    long. Sepals obtuse . . . . 44. *L. gracillimus.*
  Leaves (usually hairy) linear-lanceolate, 2 to 3, rarely 4 lines
    long, tapering as well as the bracteoles and sepals into obtuse,
    convolute points . . . . 45. *L. lasiophyllus.*
  Leaves (glabrous or hairy) about 2 lines long, very acute. Sepals
    very acute. Ovary surrounded by a ring of hairs . . . . 46. *L. cymbiformis.*

SERIES VII. **Virgatæ.**—*Leaves erect, concave. Ovary 5- or 3-celled (or exceptionally 4-celled). Western species except L. virgatus.*

Spikes rather loose, oblong or cylindrical. Leaves (mostly ½ in.)
    tapering to a point, finely veined.
  Leaves oblong-lanceolate or elliptical, with a rigid point. Ovary
    5-celled . . . . 47. *L. apiculatus.*
  Leaves linear-lanceolate, with a very short, callous point. Ovary
    usually 3-celled . . . . 48. *L. polystachyus.*
Spikes short and dense.
  Leaves (under ⅓ in.) tapering into a rigid point, finely veined.
    Eastern species . . . . 49. *L. virgatus.*
  Leaves (2 to 3 lines) linear, obtuse, keeled or obscurely veined . 50. *L. pulchellus.*
  Leaves ovate-lanceolate to linear-lanceolate or oblong-obtuse,
    strongly veined.
      Ovary glabrous, usually 5-celled.
        Leaves mostly narrow, scattered, ¼ in. long or under. . . 51. *L. polymorphus.*
        Leaves crowded, thick, rather broad, ¼ to ½ in. long . . . 52. *L. assimilis.*
      Ovary densely setose, 3-celled. Leaves 2 to 4 lines . . . 53. *L. Oldfieldii.*
  Leaves short, the larger ones very concave, and embracing the
    stem below the middle, spreading upwards. Ovary 5-celled.
      Leaves mostly 2 to 3 lines long . . . . 54. *L. cucullatus.*
      Leaves mostly 1 to 1½ lines long . . . . 55. *L. sprengelioides.*
  Leaves very small. Ovary 3-celled.
      Leaves (about 1 line) more or less spreading at least at the
        end . . . . 56. *L. obtusatus.*
      Leaves closely appressed to the end.
        Ovary glabrous. Leaves about 1 line long, slightly striate . 57. *L. fimbriatus.*
        Ovary densely hirsute. Leaves 1 to 2 lines long, strongly
          striate . . . . 58. *L. ozothamnoides.*

SECT. II. **Heteranthesis.**—*Spikes or clusters terminal, and sometimes also in the uppermost axils. Anthers obtuse or emarginate (in* L. atherolepis *attenuate towards the end) without sterile tips.*

Leaves linear, mucronate, with revolute margins. Flowers large.
    Anthers attached below the middle, tapering to the end . . . 10. *L. atherolepis.*
Leaves obtuse, flat or slightly convex. Anthers attached above the
    middle, and not tapering.
      Ovary 2-celled.
        Pubescent. Spikes terminal, capitate. Sepals with long,
          plumose points . . . . 59. *L. plumuliflorus.*

Glabrous. Spikes loose, with few pendulous flowers, terminal
  and in the upper axils. Sepals acute, glabrous . . . . 60. *L. unilateralis.*
Ovary 5-celled.
  Leaves shortly petiolate, rather thin, several-nerved.
    Leaves glaucous and striate underneath . . . . 61. *L. Hookeri.*
    Leaves green, and shining on both sides . . . . . . 62. *L. Macræi.*
    Leaves on long petioles, small, thick, with 2 furrows under-
     neath . . . . 63. *L. pleurandroides.*
  Ovary 4-celled. Leaves distinctly petiolate, small, orbicular . 17. *L. Bossiæa.*
Leaves linear, tapering into a pungent point.
  Leaves oblong-linear. Flowers sessile within the subtending
    bract. Ovary 5-celled . . . . 64. *L. melaleucoides.*
  Leaves narrow-linear. Flowers pedicellate within the bract and
    bracteoles.
    Leaves with revolute margins. Corolla-tube shortly exserted.
     Ovary 5- to 7-celled . . . . 65. *L. pluriloculatus.*
    Leaves concave. Corolla-tube twice as long as the calyx, the
     exserted part campanulate. Ovary 7- to 10-celled . . . 66. *L. pleiospermus.*
    Leaves with thickened margins and midrib. Corolla-tube
     cylindrical, 3 or 4 times as long as the calyx. Ovary 3-
     celled . . . . 67. *L. rubicundus.*
  Leaves lanceolate-linear. Flowers sessile. Corolla shorter than
    the calyx. Ovary 2-celled. Small, slender plant . . . . 43. *L. cryptanthus.*

SECT. III. **Pleuranthus.**—*Spikes all axillary, few-flowered or reduced to a single
flower besides the rudiment, the common peduncle very short or rarely as long as the leaves.
Anthers obtuse or emarginate, without sterile tips. Style usually slender and elongated,
rarely very short.*

SERIES I. **Confertæ.**—*Leaves small (1 to 2 lines), with recurved margins. Flowers
mostly solitary besides the rudiment, and nearly sessile, often forming dense leafy spikes.
Ovary 5-celled. Eastern species.*

Sepals obtuse . . . . 68. *L. attenuatus.*
Sepals acute . . . . 69. *L. confertus.*

SERIES II. **Ericoideæ.**—*Leaves narrow or rarely ovate, $\frac{1}{8}$ to 1 in. (except L. concinnus), with recurved or revolute margins. Flowers 2 or more together, in sessile or
shortly pedunculate clusters. Ovary 5- or 3-celled.*

Corolla-tube not exceeding the calyx. Flowers erect, at least at
  first.
  Leaves (under $\frac{1}{2}$ in.) obtuse, or with a small, callous point.
    Eastern species . . . . 70. *L. muticus.*
  Leaves (under $\frac{1}{2}$ in.) very shortly mucronate. Sepals under 1
    line. Eastern species . . . . 71. *L. ericoides.*
  Leaves (about $\frac{1}{2}$ in.) very shortly mucronate. Sepals about 2
    lines. Western species . . . . 72. *L. brevicuspis.*
  Leaves ($\frac{1}{4}$ to 1 in.) rigidly mucronate-acute. Sepals 1$\frac{1}{2}$ line,
    very acute. Western species . . . . 73. *L. propinquus.*
  Leaves (under $\frac{1}{2}$ in.) very spreading, rigidly mucronate. Sepals
    at least 1 line, obtuse. Western species . . . . . . 74. *L. insularis.*
Corolla-tube much longer than the small calyx. Flowers erect or
  pendulous. Western species.
  Leaves ovate-lanceolate . . . . 75. *L. Allittii.*
  Leaves oblong-linear . . . . 76. *L. racemulosus.*
Corolla-tube not exceeding the calyx. Flowers pendulous from the
  first. Fruit oblong.
  Leaves oblong-linear, erect or spreading (2 to 4 lines). Ovary
    usually 5-celled . . . . 77. *L. pendulus.*

Leaves ovate or oblong, very spreading or reflexed (1 to 2 lines).
Ovary usually 3-celled . . . . 78. *L. concinnus.*

SERIES III. **Micranthæ** —*Leaves oblong or lanceolate, nearly flat (or with recurved margins). Ovary 2-celled. Flowers small and nearly sessile.*

Leaves scattered, not mucronate, with recurved margins.
Leaves oblong-linear, ¼ to ½ in. long. Eastern species . . . 79. *L. margarodes.*
Leaves ovate or oblong, 1 to 2 lines long. Western species . 78. *L. concinnus.*
Leaves imbricate, flat or slightly concave.
Leaves obtuse (under ½ in.). Flowers solitary. No bracts besides the bracteoles . . . . 80. *L. flavescens.*
Leaves tapering into a short, fine point (3 to 4 lines). Hypogynous scales fringed with long hairs . . . . 81. *L. blepharolepis.*
Leaves tapering into a fine point (½ in. or more). No hypogynous disk or scales . . . . 82. *L. esquamatus.*

SERIES IV. **Planifoliæ.**—*Leaves flat or slightly convex or concave, rigid, usually shining above, the veins fine or inconspicuous. Ovary 5-celled.*

Corolla-tube shorter than the calyx.
Leaves obtuse or with a callous point. Flowers nearly sessile. Southern or western species.
Leaves sessile, orbicular-cordate, very spreading, often convex 83. *L. cordifolius.*
Leaves petiolate, obovate-cuneate or orbicular, often slightly concave . . . . 84. *L. rotundifolius.*
Leaves oblong or oblanceolate, nearly sessile, sometimes slightly convex . . . . 85. *L. planifolius.*
Leaves rigidly mucronate. Flowers nearly sessile or on a short peduncle, erect or at length spreading. North-eastern species.
Leaves obovate to oblong, often imbricate, with a very short point. Flowers about 2 lines long . . . . . . 86. *L. ruscifolius.*
Leaves obovate-oblong, imbricate, with a long, rigid point. Flowers about 3 lines long . . . . 87. *L. imbricatus.*
Leaves narrow-oblong, abruptly contracted into a long, rigid point. Flowers about 2 lines long . . . . . . 88. *L. cuspidatus.*
Leaves oblong-linear or lanceolate, tapering into a short, almost callous point. Flowers about 2 lines long . . . 89. *L. leptospermoides.*
Leaves linear-lanceolate, tapering into a fine point.
Leaves not twisted . . . . 90. *L. acuminatus.*
Leaves much twisted . . . . 91. *L. flexifolius.*
Leaves narrow, rigidly mucronate. Flowers pendulous. Eastern species.
Flowers nearly sessile. Sepals rather acute . . . . . 92. *L. biflorus.*
Flowers distinctly pedunculate. Sepals very acute.
Corolla glabrous outside . . . . 93. *L. setiger.*
Corolla pubescent outside . . . . 94. *L. exolasius.*
Corolla-tube shortly or scarcely exceeding the calyx. Leaves mucronate, the margins often recurved. Flowers erect.
Glabrous or minutely pubescent. Diffuse or prostrate shrub . 95. *L. Fraseri.*
Stem and foliage hirsute . . . . 96. *L. hirtellus.*
Corolla-tube considerably longer than the calyx.
Flowers pendulous in pairs. Leaves very shortly mucronate.
Leaves very spreading or reflexed, convex, 2 to 3 lines long. Flowers under 3 lines . . . . 97. *L. ovalifolius.*
Leaves erect or spreading, flat, 3 to 6 lines long. Flowers above 4 lines . . . . 98. *L. oxycedrus.*
Flowers erect.

Leaves obovate or cuneate, obtuse or with a minute, callous
point . . . . 99. *L. cuneifolius.*
Leaves oblong-lanceolate, with a rigid point.
 Sepals (about 1 line) less than half the corolla-tube.
  Western species . . . . 100. *L. strictus.*
 Sepals (about 2 lines) more than half the corolla-tube.
  Eastern species . . . . 101. *L. Mitchellii.*
Leaves linear, with a fine rigid point. Sepals very small.
 Corolla long and slender. Eastern species . . . . . 102. *L. juniperinus.*

SERIES V. **Concavæ.**—*Leaves concave or keeled.*

Leaves rigidly mucronate.
 Ovary 3- or 2-celled. Leaves lanceolate.
  Leaves tapering into a long, rigid point.
   Flowers erect, 3 to 4 lines long. Leaves broad or narrow.
    Eastern species . . . . 103. *L rufus.*
   Flowers pendulous, nearly 3 lines. Leaves rather narrow.
    Western species . . . . 104. *L. conostephioides.*
  Leaves narrow, very shortly pointed. Flowers erect, not 2
  lines long. Eastern species . . . . 105. *L. deformis.*
  (*Astroloma xerophyllum* is in technical characters very near *L. conostephioides*, but with
  much larger flowers, a 5-celled ovary, etc.)
 Ovary 5-celled.
  Leaves ovate or ovate-lanceolate, strongly striate, 2 to 3
  lines long. Corolla with the long dense white beards of
  Perojoa . . . . 106. *L. pogonocalyx.*
  Leaves finely veined or veinless. Corolla beards short, and
  not very white.
   Leaves narrow (2 to 4 lines), shortly mucronate.
    Flowers about 2 lines long ; leaves not keeled, finely
    veined underneath . . . . 107. *L. breviflorus.*
    Flowers about 1½ lines long. Leaves keeled but scarcely
    veined . . . . 108. *L. durus.*
   Leaves lanceolate or linear (about ½ in.), tapering into a
   pungent point . . . . 109. *L. multiflorus.*
   Leaves (2 to 4 lines) rather broad and very concave.
   Eastern species.
    Sepals 1 line long, obtuse. Corolla very little longer . 110. *L. appressus.*
    Sepals nearly 2 lines, very acute. Corolla 3 to 4 lines
    long . . . . 111. *L. neoanglicus.*
Leaves as broad as long, very obtuse or minutely mucronate, very
 concave, and embracing the stem.
 Leaves sessile, cordate, imbricate, 4 to 6 lines diameter, con-
 cealing the flowers . . . . 112. *L. obtectus.*
 Leaves sessile but narrowed at the base, cucullate, 1 to 1½ lines
 diameter. Flowers 3 lines long . . . . 113. *L crassiflorus.*
 Leaves petiolate, obovate-orbicular, 1 to 2 lines diameter.
 Flowers about 2 lines long . . . . 114. *L. strongylophyllus.*
Leaves oblong or almost ovate, obtuse or with a minute, callous
 point.
 Leaves 2 to 4 lines long. Corolla 1½ lines, the tube as long
 as the calyx. Fruit erect . . . . 115. *L. crassifolius.*
 Leaves 2 to 4 lines long. Corolla 2 lines, the tube very short,
 the lobes scarcely exceeding the calyx. Fruit erect . . . 116. *L. corynocarpus.*
 Leaves 1 to 2 lines long. Corolla 2 lines, the tube equal to the
 calyx, the lobes erect and connivent at the base. Fruit pendulous . . . . 117. *L. Woodsii.*
 Leaves 2 to 3 lines long. Corolla 3 lines, the slender tube at
 least twice as long as the calyx . . . . 118. *L. leptanthus.*

Sect. I. Perojoa.—Inflorescence chiefly or entirely terminal, rarely all axillary, and then the spikes long, slender, and interrupted. Anthers with sterile tips, which are either conspicuous and pale coloured or sometimes very small and recurved. Style very short.

Series 1. Psilostachyæ.—Spikes slender, interrupted, axillary or terminal, usually as long as the leaves. Leaves flat or convex.

1. **L. verticillatus,** *R. Br. Prod.* 541. A tall, erect shrub, quite glabrous. Leaves mostly crowded at the end of each year's shoot, so as to appear verticillate, broadly lanceolate, mostly 2 to 4 in. long, rarely only 1 in. and sometimes 5 or even 6 in., obtuse or with a callous or almost acute but not pungent point, flat or convex, with the margins recurved, the veins fine. Spikes slender, with small, distant, reddish flowers, axillary or below the leaves and usually as long as them. Bracts and bracteoles broad, not half so long as the calyx. Sepals obtuse, not quite a line long. Corolla-tube twice as long as the calyx; lobes half as long as the tube, bearded in the lower half only. Anthers linear, attached about the middle, the upper end forming a clavate, sterile tip, but of the same colour as the rest. Hypogynous disk truncate. Ovary 5- or rarely 4-celled. Style short. Fruit ovoid-oblong, 2 to 2½ lines long, with a hard endocarp.—DC. Prod. vii. 745; Sond. in Pl. Preiss. i. 307; F. Muell. Fragm. iv. 122; *Styphelia verticillata,* Spreng. Syst. i. 656; F. Muell. Fragm. vi. 43; *Lissanthe verticillata,* Lindl. Swan Riv. App. 25; *Leucopogon glaucescens,* DC. Prod. vii. 745.

W. Australia. King George's Sound and adjoining districts, *R. Brown, A. Cunningham, Preiss, n.* 431, *Maxwell, F. Mueller;* Gordon, Harvey, and Tone rivers, *Oldfield;* Swan River (?), *Drummond, 1st Coll., 2nd Coll. n.* 266.

2. **L. interruptus,** *R. Br. Prod.* 541. A glabrous shrub, with erect branches. Leaves mostly crowded at the end of each year's shoot, so as to appear verticillate, from almost oval to oblong-elliptical, obtuse or with a minute callous point, flat or nearly so, finely nerved, from ¾ to 1 in. long or rarely rather more. Spikes slender and interrupted, but not exceeding the leaves, solitary or 2 or 3 together at the ends of the branches. Flowers small, rather numerous. Bracts and bracteoles broad, not half so long as the calyx. Sepals obtuse, under 1 line long. Corolla-tube rather shorter than the calyx; lobes as long as the tube, bearded to the end. Anthers inserted below the linear, sterile tips. Hypogynous disk truncate. Ovary 5-celled; style short.—DC. Prod. vii. 745; *Styphelia interrupta,* Spreng. Syst. i. 656.

W. Australia. Goose Island Bay, *R. Brown;* King George's Sound, or to the eastward, *Baxter.*

3. **L. amplexicaulis,** *R. Br. Prod.* 543. An erect shrub, with long, straggling branches, hirsute with long soft hairs or nearly glabrous. Leaves sessile and clasping the stem with rounded auricles, spreading, cordate-ovate, acute, convex or with recurved margins, ciliate, striate on both sides, ½ to 1 in. long. Spikes slender, interrupted, terminal and in the upper axils, longer than the leaves. Bracts lanceolate, leaf-like and nearly as long as the flowers, or the upper ones or nearly all small and subulate; bracteoles ovate, acuminate, about half as long as the calyx. Sepals acutely acuminate, about 1 line

long. Corolla-tube much shorter than the calyx; lobes twice as long as the tube. Anthers attached below the rather long, sterile tips. Hypogynous disk very short, truncate. Ovary 2-celled, tapering into a short style. Fruit ovate, rather longer than the calyx.—DC. Prod. vii. 748 : *Styphelia amplexicaulis*, Rudge in Trans. Linn. Soc. viii. 292. t. 8 ; F. Muell. Fragm. vi. 44.

N. S. Wales. Port Jackson, *R. Brown, Sieber, n.* 92, and many others.

4. **L. alternifolius,** *R. Br. Prod.* 543. A little glabrous shrub, with very slender, often filiform, ascending or erect stems, of 6 in. to 1 ft. Leaves sessile and clasping the stem with rounded auricles, spreading, broadly cordate-ovate, obtuse or with a small callous point, flat or slightly convex, rarely above 2 lines and often not above 1 line long. Spikes in the upper axils slender, interrupted, rather longer than the leaves. Bracts scarcely above $\frac{1}{2}$ line long ; bracteoles still smaller, sepals about $\frac{3}{4}$ line, all rather acute. Corolla not seen. Fruit ovate, obtuse, 2-celled, rather longer than the calyx. —DC. Prod. vii. 748 ; *Styphelia alternifolia*, Spreng. Syst. i. 655.

W. Australia. Heaths about W. Cape Howe, *R. Brown.* This elegant little species, which I have not seen in any other collection, is like a miniature *L. amplexicaulis.*

5. **L. lanceolatus,** *R. Br. Prod.* 541. Usually a tall shrub or small tree, quite glabrous, but some varieties low and diffuse and others pubescent, the branchlets rather slender. Leaves erect or spreading, lanceolate, tapering at both ends, obtuse or with a callous point, flat, with fine nerves, in some specimens rarely exceeding 1 in., in others attaining 2 in. Spikes slender, interrupted, solitary in the upper axils or clustered at the ends of the branches, often exceeding the leaves. Bracts and bracteoles striate, fully half as long as the calyx. Sepals scarcely 1 line long, usually obtuse, but narrower than in *L. australis*, and sometimes almost acute. Corolla-tube not exceeding the calyx ; lobes as long as the tube. Anthers attached immediately under the prominent sterile tips. Hypogynous disk truncate, readily separating into distinct scales. Ovary 2-celled, tapering into the rather short style. Fruit ovate-globose, red, shortly exceeding the calyx.—DC. Prod. vii. 744 ; F. Muell. Fragm. iv. 124 ; Sweet, Fl. Austral. t. 47 ; Bot. Mag. t. 3162 ; *Styphelia lanceolata*, Sm. Bot. N. Holl. 49 (partly) ; F. Muell. Fragm. vi. 43 ; *L. australis*, Sieb. Pl. Exsicc. (not of R. Br.) ; *L. Cunninghamii*, DC. Prod. vii. 745 ; *L. affinis*, R. Br. Prod. 541 ; DC. Prod. vii. 745 ; *Styphelia affinis*, Spreng. Syst. i. 658.

Queensland. Stradbroke Island, *A. Cunningham ;* Port Macquarrie, *Backhouse ;* Moreton Bay, *Fitzalan.*

N. S. Wales. Port Jackson to the Blue Mountains, *R. Brown, Sieber, n.* 103, *Fl. Mixt. n.* 490, 496, and many others ; northward to Hastings and Macleay rivers, *Beckler ;* New England, *C. Stuart, Leichhardt ;* Mount Lindsay, *W. Hill ;* southward to Illawarra, *A. Cunningham ;* Twofold Bay, *F. Mueller ;* Gabo Island, *Maplestone.*

Victoria. Snowy River and Mitta-Mitta, *F. Mueller.*

Tasmania. Port Dalrymple, *R. Brown.* These specimens, upon which R. Brown founded his *L. affinis*, have certainly the flat leaves, the 2-locular, compressed fruit, and other characters of *L. lanceolatus.*

Var. *gracilis.* Branchlets very slender, more or less pubescent. Leaves and flowers small.—*L. pimeleoides*, A. Cunn. ; DC. Prod. vii. 744. To this variety belong all, or nearly all, the Queensland specimens.

Var. *gelidus*, F. Muell. Low and bushy. Leaves small and crowded. Spikes shorter, with the flowers less distant.—Barkly Range, Mount Baw-Baw, sources of the Yarra, Cobberas Mountains, etc., at an elevation of 3000 to 5000 ft., *F. Mueller.*

Var. ? *alpestris*, F. Muell. in Herb. Hook. Like the var. *gelidus*, but leaves more rigid and acute, with more prominent nerves. Ovary sometimes 3-celled, though usually 2-celled, as in the other forms.—*L. neurophyllus*, F. Muell. Fragm. i. 37, referred doubtfully to *L. australis* in F. Muell. Fragm. iv. 123. Summit of Mount William, in the Grampians, *F. Mueller.*

SERIES 2. AUSTRALES.—Spikes all terminal or also in the upper axils, short and dense or cylindrical and rather dense (interrupted in *L. distans*). Leaves nearly flat or convex, with recurved or revolute margins (not concave). Ovary 5-celled, rarely 4- or 3-celled.

6. **L. Richei,** *R. Br. Prod.* 541, *and in Bot. Mag. t.* 3251. A tall shrub or small tree, quite glabrous or the branches slightly pubescent. Leaves oblong lanceolate or oblanceolate, obtuse or with a small callous point, slightly convex or with recurved margins, finely veined, rarely exceeding 1 in. and often all under ¾ in. long. Spikes terminal and in the upper axils, cylindrical, often rather long but dense. Bracts and bracteoles striate, about half as long as the calyx. Sepals scarcely 1 line long, obtuse. Corolla-tube rather shorter than the calyx; lobes as long as the tube. Anthers attached above the middle, with more or less prominent sterile tips. Hypogynous disk 5-lobed. Ovary 5- or 4-celled. Style short. Fruit white, ovoid-globose.—DC. Prod. vii. 744 ; Sond. in Pl. Preiss. i. 305 ; Hook. f. Fl. Tasm. i. 249 ; F. Muell. Fragm. iv. 123 ; *Styphelia Richei*, Labill. Pl. N. Holl. i. 44. t. 60 ; F. Muell. Fragm. vi. 42 ; *Styphelia parviflora*, Andr. Bot. Rep. t. 287 ; *Leucopogon parviflorus*, Lindl. Bot. Reg. t. 1560 ; DC. Prod. vii. 745 ; Sond. in Pl. Preiss. i. 305 ; *Styphelia gnidium*, Vent. Jard. Malm. t. 23 ; *Leucopogon polystachyus*, Lodd. Bot. Cab. t. 1436, not of R. Br. ; *L. lanceolatus*, Sieb. Pl. Exs., not of R. Br.

Queensland. Moreton Island, *F. Mueller.*

N. S. Wales. Seashore, Port Jackson, *R. Brown, Sieber, n.* 102, and others ; Gabo Island, *Maplestone ;* Hastings river, *Beckler.*

Victoria. Port Phillip, *R. Brown ;* sand and rocky seacoasts, common, *F. Mueller* and others.

Tasmania. Islands of Bass's Straits and Storm Bay Passage, *R. Brown ;* abundant on sand-hills on all the coasts, *J. D. Hooker.*

S. Australia. Along the coast, St. Vincent's and Spencer's Gulfs and Kangaroo Island, *F. Mueller.*

W. Australia. King George's Sound, *R. Brown* and others, and thence to the eastward, *Maxwell,* and to Vasse and Swan rivers, *Drummond,* 1*st Coll. n.* 464, *Preiss, n.* 365, 370, 372, and others.

Var. ? *acutifolius.* Leaves lanceolate, very acute.—Stirling range, *F. Mueller.* The species is also on Chatham Island.

7. **L. australis,** *R. Br. Prod.* 541. A tall, bushy shrub, with erect, glabrous branches, very closely allied to *L. Richei*, differing chiefly in the longer leaves, rather longer spikes, and in the fruit. Leaves lanceolate, obtuse or with a callous point, mostly 1 to 2 in. long and the lower ones sometimes 3 in., convex or with recurved margins, finely veined. Spikes cylindrical, rather dense, but not so much so as in *L. Richei.* Bracteoles about half as

long as the calyx. Sepals about 1 line long, obtuse and rather broad.
Corolla-tube not exceeding the calyx ; lobes as long as the tube. Anthers
attached above the middle, with more or less prominent sterile tips. Hypo-
gynous disk lobed. Ovary short, 5-celled, abruptly contracted into a short
style. Drupe depressed-globular, yellow or white.—DC. Prod. vii. 744 ;
Hook. f. Fl. Tasm. i. 249 ; F. Muell. Fragm. iv. 123 ; *Styphelia australis*,
F. Muell. Fragm. vi. 43 ; *Leucopogon Drummondii*, DC. Prod. vii. 745 ;
Sond. in Pl. Preiss. i. 306 ; *L. paniculatus*, Sond. l. c.

Victoria. In moist heaths, and wet rocks, open woods, valleys, and shallow marshes,
always far from the sea, Bunip Creek, Corner Inlet, etc., *F. Mueller.*

Tasmania. Port Dalrymple, *R. Brown ;* in poor soil, generally near the sea, *J. D.
Hooker.*

S. Australia. Lofty Ranges, Rivoli Bay, *F. Mueller.*

W. Australia. King George's Sound, *R. Brown, Preiss, n.* 367 ; near the sea and
Tone river, *Oldfield ;* Swan River, *Drummond, 1st Coll., Preiss, n.* 368.

8. **L. capitellatus,** *DC. Prod.* vii. 747. An erect shrub, of 2 to 3 ft.,
glabrous or the branches slightly hirsute. Leaves linear or linear-lanceolate,
tapering into a short, fine, almost pungent point, rigid, convex or with
slightly recurved margins, about ½ in. long or rather more. Spikes short,
dense, terminal or in the uppermost axils, or on short, axillary, leafy
branchlets. Bracts short, obtuse, striate ; bracteoles very obtuse, not half so
long as the calyx. Sepals broad, obtuse, ciliolate, scarcely 1 line long.
Corolla nearly 2 lines, the lobes longer than the tube. Anthers attached
above the middle, with short, sterile tips. Disk truncate or sinuate-lobed.
Ovary short, broad, 5-celled ; style very short.—Sond. in Pl. Preiss. i. 311 ;
Styphelia capitellata, F. Muell. Fragm. vi. 31.

W. Australia. Swan River, *Drummond, 1st Coll. n.* 462, *Oldfield ;* Mahogany
Creek, *Preiss, n.* 371 ; Gordon and Salt rivers, *Maxwell.*

The var. *sparsiflorus*, Sond. l. c., from St. Ronan's Well, *Preiss, n.* 427, is an old fruiting
state, with a few of the remains of inflorescence, apparently (but not really) axillary from
the growth of the lateral shoot.

9. **L. revolutus,** *R. Br. Prod.* 542. An erect, bushy or rarely straggling shrub, attaining several ft., glabrous or more or less pubescent. Leaves
nearly sessile, linear or oblong, obtuse or with a small, callous point, convex,
with recurved or more frequently revolute margins, ¼ to ½ in. long or rarely
more. Spikes dense, terminal or in the uppermost axils, usually clustered.
Bracts small ; bracteoles about half as long as the calyx, with a prominent
midrib. Sepals obtuse, 1 to 1¼ lines long, minutely hoary-pubescent and
ciliolate as well as the bracteoles, or quite glabrous. Corolla 2 to 2½ lines
long, the lobes rather longer than the tube. Anthers attached above the
middle, with prominent sterile tips. Hypogynous disk truncate. Ovary
broad, 5-celled ; style very short. Fruit yellowish, nearly globular, scarcely
exceeding the calyx.—DC. Prod. vii. 746 ; Sond. in Pl. Preiss. i. 310 ; *Styphelia revoluta*, Spreng. Syst. i. 657 ; *Styphelia obovata*, F. Muell. Fragm.
vi. 31 ; *Leucopogon angustatus*, Benth. in Hueg. Enum. 77 ; Sond. in Pl.
Preiss. i. 311 ; DC. Prod. viii. 748.

W. Australia. King George's Sound and Goose Island Bay, *R. Brown ;* King
George's Sound to Cape Riche and neighbouring districts, *A. Cunningham, Drummond, 2nd*

Coll. n. 250, 5*th Coll. n.* 315, 317, *Preiss, n.* 393, 394, and many others; eastward to Eyre's Relief and Cape le Grand, *Maxwell.*

The species is closely allied to *L. collinus,* but has the ovary 5-celled. *L. rubricaulis,* R. Br. Prod. 542 ; DC. Prod. vii. 746 (*Styphelia rubricaulis,* Spreng. Syst. i. 656), is a glabrous form, with rather short and broad leaves and a spreading habit, but passes very gradually into the common one. Of *L. villosus,* R. Br. l. c.; DC. l. c. (*Styphelia villosa,* Spreng. Syst. i. 657), there is but a single specimen in Brown's herbarium, which only appears to differ from the common form in being rather more pubescent, as in Sonder's variety *hirsutus* of *L. angustatus.* The plants mistaken for *L. villosus* by Lindley and others are quite different. *Styphelia obovata,* Labill. Pl. N. Holl. i. 48. t. 67 (*Leucopogon obovatus,* R. Br. l. c.; DC. l. c.), is evidently figured from an abnormally broad-leaved specimen of this species, but even in that the leaves can scarcely be said to be obovate, and the name, as observed by F. Mueller, is so totally inappropriate that it is better to retain Brown's, which has, moreover, been generally adopted.

10. **L. atherolepis,** *Stschegl. in Bull. Mosc.* 1859, i. 13.   An erect shrub, with virgate branches, glabrous or pubescent. Leaves linear, erect or spreading, with a small callous point, the margins closely revolute, rarely exceeding ½ in. and mostly under that. Flowers large for the genus, in short spikes of 2 to 4, terminal or in the uppermost axils, forming usually a compact leafy head or oblong spike-like panicle. Bracts, bracteoles, and sepals rigid, acutely acuminate, minutely pubescent and striate, the sepals nearly 3 lines long and the bracteoles about 2 lines. Corolla-tube very short; lobes nearly 3 lines long. Filaments thick; anthers linear, attached about the middle, tapering to the recurved end, but obtuse and 2-valved to the end. Hypogynous disk of 5 short broad gland-like scales. Ovary 5-furrowed, 5-celled. Style short.—*L. grandiusculus* or *Styphelia grandiuscula,* F. Muell. Fragm. vi. 47.

W. Australia, *Drummond,* 5*th Coll. n.* 305 (in some sets, 306 in others). This species has the habit and very white corolla-beards of the *Perojoœ,* and in the anthers connects the sections *Perojoa* and *Heteranthesis.*

Var. *densiflorus.* Flowers rather smaller and more crowded, but much larger than in *L. revolutus,* with the acute sepals and the anthers of *L. atherolepis.*—Stirling Range, *F. Mueller.*

11. **L. reflexus,** *R. Br. Prod.* 544.   An erect shrub, with slender virgate branches, our specimens glabrous or the young shoots cottony and the branchlets slightly pubescent. Leaves closely reflexed or very spreading, from ovate and under 1 line long to lanceolate and nearly 2 lines, obtuse or with a minute callous point, very convex. Spikes short and dense, terminal or in the uppermost axils. Bracts small ; bracteoles acute, about half as long as the calyx. Sepals lanceolate, acute, 1½ lines long. Corolla-tube shorter than the calyx ; lobes about 1 line long. Anthers attached above the middle, with sterile tips. Hypogynous disk short, truncate. Ovary depressed, glabrous, 5-angled, 5-celled, style short.—DC. Prod. vii. 749 ; Sond. in Pl. Preiss. i. 314 ; *Styphelia reflexa,* Spreng. Syst. i. 655, not of Rudge ; *S. Brownii,* Spreng. Syst. (Index) v. 683 ; F. Muell. Fragm. vi. 32.

W. Australia. King George's Sound and adjoining districts, *R. Brown, A. Cunningham, Preiss, n.* 398, and others.

12. **L. corifolius,** *Endl. Nov. Stirp. Dec.* 15.   An erect shrub, more or less pubescent. Leaves linear or lanceolate, very spreading or reflexed,

rather acute or obtuse, with closely revolute margins, $1\frac{1}{2}$ to 3 lines long.
Spikes terminal, short dense and few-flowered.　Bracts obtuse, short ; bracteoles half as long as the calyx, acute, pubescent.　Sepals about 1 line long,
rather acute, pubescent, ciliate.　Corolla-tube very short ; lobes twice as
long as the tube.　Anthers attached below the prominent sterile tips.
Ovary, small, broad, 5-celled ; style short.—Sond. in Pl. Preiss. i. 317.

W. Australia.　Peaty soil amidst dense thickets near Albany, *Preiss, n.* 399 (Herb.
Sonder).　This species requires further confirmation from more perfect specimens.　It may
prove to be a variety of *L. reflexus.*

13. **L. distans,** *R. Br. Prod.* 544.　An erect shrub of 3 or 4 ft., with
slender elongated branches, glabrous or slightly hairy.　Leaves ovate-cordate
to ovate-lanceolate, obtuse or with a minute callous point, spreading or reflexed, very convex or with revolute margins, the nerves impressed on the
upper surface, 1 to 2 lines (mostly $1\frac{1}{2}$ lines) long.　Spikes terminal and in
the uppermost axils, 1 to $1\frac{1}{2}$ in. long in the normal form, with distant
flowers and a flexuose rhachis, forming a short terminal panicle, usually hoary-
pubescent.　Bracts small, striate ; bracteoles smooth, very broad and obtuse,
not half as long as the calyx.　Sepals about 2 lines long, hoary-pubescent
and ciliate, rather broad but usually acute.　Corolla-tube nearly as long as
the calyx ; lobes twice as long as the tube, the beard very long and dense.
Anthers linear, attached above the middle, with prominent sterile tips.　Hypogynous disk truncate, readily separating into distinct scales.　Ovary 5-
celled ; style short.　Fruit broad and flat, not exceeding the calyx.—DC.
Prod. vii. 748 ; Sond. in Pl. Preiss. i. 313 ; *Styphelia distans,* Spreng. Syst.
i. 655 ; F. Muell. Fragm. vi. 32.

W. Australia.　King George's Sound and adjoining districts, *Menzies, Baxter,
Preiss, n.* 390, *Drummond,* 3rd *Coll. n.* 190, 5th *Coll. n.* 316, and others.　In some of
Drummond's specimens the fruit is obovoid and much exceeding the calyx, but apparently
in a monstrous state without seed.

Var. *contractus.*　Spikes short, with the flowers near together.　Sepals rather more obtuse.　Anther-tips shorter.—*L. penicillatus,* Stschegl. in Bull. Mosc. 1859, i. 12.—*Drummond,* 5th *Coll. n.* 314.　Stirling Range, *F. Mueller.*—Intermediate, as it were, between
L. distans and *L. reflexus,* with the foliage and larger flowers of the former, but the inflorescence almost as much contracted as in the latter.

14. **L. gibbosus,** *Stschegl. in Bull. Mosc.* 1859, i. 12.　An erect bushy
shrub of 1 to 2 ft., the branchlets pubescent.　Leaves broadly orbicular,
obtuse or with a small reflexed point, very convex with recurved margins,
mostly reflexed and rarely above 1 line diameter.　Spikes short and dense,
terminal or in the upper axils.　Bracteoles thin, broad, ciliate, hirsute, more
than half as long as the calyx.　Sepals ciliate and hirsute, almost acute,
about 1 line long.　Corolla-tube very short ; lobes about 1 line long.　Anthers attached above the middle, with very short sterile tips.　Hypogynous
scales truncate, distinct or slightly cohering.　Ovary in all the flowers examined 3-celled ; style short.

W. Australia, *Drummond,* 3rd *Coll. n.* 74, 5th *Coll. n.* 310 ; limestone hills and
sands, Point Irwin, Doubtful Island Bay, *Oldfield* ; Cape Riche, *Harvey;* Kojonup, Middle
Mount Barren, Fitzgerald and Phillips rivers, *Maxwell ;* Stirling Range, *F. Mueller*

15. **L. thymifolius,** *Lindl. ms.*　An erect shrub, hirsute all over with

short spreading hairs. Leaves oblong or lanceolate, with a small soft recurved point and revolute margins, 2 to 4 lines long or rarely narrower and longer. Flowers in short spikes, terminal or in the upper axils. Bracts about half as long as the calyx. Sepals broadly lanceolate, acute, coloured, hirsute and ciliate, about 1 line long. Corolla-tube very short; lobes about 1 line long. Hypogynous disk truncate. Ovary broad, flat-topped, obtusely 3- or 4-angled and 3- or 4-celled; style short. Fruit obovoid, longer than the calyx, with 3 or rarely 4 broad prominent obtuse angles, and as many cells and seeds.

Victoria. Grampians, *Mitchell;* Victoria Range, *F. Mueller.* Determined at first by Lindley as a distinct species, but afterwards referred by him (Mitch. Three Exped.) to *L. villosus,* Br. (*L. revolutus,* var.), and united by F. Mueller with *L. collinus;* it is, however, intermediate between the two as to the number of carpels, and differs from both in being much more hirsute, and especially in the remarkably prominent broad angles of the ovary and fruit.

16. **L. cordatus,** *Sond. in Pl. Preiss.* i. 313. An erect bushy shrub of 3 to 4 ft., with the foliage and aspect of *L. squarrosus,* usually glabrous. Leaves ovate-lanceolate, scarcely acute, spreading or recurved, flat or convex, rigid, 1 to 2 or rarely 3 lines long. Spikes short, dense, often reduced to 1 or 2 flowers, terminal or in the uppermost axils, forming short dense leafy corymbs. Bracts leaf-like, obtuse or scarcely acute; bracteoles about half as long as the calyx. Sepals 1 line long or rather more, obtuse, not ciliate. Corolla-tube rather shorter than the calyx, marked inside about the middle with a prominent ring; lobes $1\frac{1}{2}$ lines long, the white beards remarkably long. Anthers attached above the middle, with recurved sterile tips. Hypogynous disk 5-lobed. Ovary depressed, pubescent, 5-celled. Fruit rather longer than the calyx, truncate and pubescent at the top.

W. Australia, *Preiss, n.* 388; sands, Géographe Bay, *Oldfield.*

17? **L. Bossiæa,** *F. Muell. Fragm.* vi. 47. An erect glabrous shrub attaining 3 or 4 ft. (*Maxwell*). Leaves distinctly petiolate, orbicular-cordate, obtuse, flat, rigid, prominently veined, 1 to 2 lines diameter. Flowers small, few together in short dense terminal spikes. Bracts very small; bracteoles not half so long as the calyx, obtuse, striate. Sepals $\frac{3}{4}$ line long, obtuse, slightly coloured, not ciliate. Corolla nearly $1\frac{1}{2}$ lines long, the lobes as long as the tube. Anthers attached below the very short recurved sterile tips, which are sometimes scarcely conspicuous. Hypogynous disk sinuate or obtusely lobed. Ovary short, usually 4-celled, but perhaps sometimes 5-celled; sty e short.—*Styphelia Bossiæa,* F. Muell. l. c.

W. Australia. Termination granite rocks towards the Great Bight, *Maxwell.* This species appears to have no immediate affinities; the comparatively long petioles are those of *L. pleurandroides,* the flowers and inflorescence are those of some of the small-leaved species with 2-celled ovaries, but in all the flowers examined I have found 4 cells.

SERIES 3. COLLINÆ. Spikes all terminal or also in the uppermost axils, short and dense or cylindrical and rather dense. Leaves flat or convex with recurved or revolute or thickened margins. Ovary 2-celled.

18. **L. hirsutus,** *Sond. in Pl. Preiss.* i. 310. A slender procumbent or trailing shrub, the branches and foliage hirsute. Leaves spreading, scattered,

oblong, obtuse, slightly convex or with recurved margins, 2 to 4 lines long.
Flowers minute, in short spikes, terminal or in the upper axils. Bracts and
bracteoles obtuse, about half the length of the calyx. Sepals scarcely above
½ line long, obtuse, ciliolate. Corolla ¾ line long, the lobes as long as the
tube. Anthers attached below the short hooked sterile tips. Hypogynous
disk sinuate-toothed. Ovary 2-celled, but 1 ovule often already abortive at
the time of flowering. Fruit compressed-globular, oblique, rugose when
dry, about 1 line diameter, with a single seed.—*Styphelia hirsuta*, F. Muell.
Fragm. vi. 31.

W. Australia, *Drummond, 1st Coll. n.* 465 ; south-west side of Mount Clarence,
Preiss, n. 464.

19. **L. collinus,** *R. Br. Prod.* 543. A shrub, sometimes erect and 3
to 4 ft. high with virgate branches, more rarely low and diffuse, nearly glabrous or softly pubescent. Leaves usually oblong or linear, obtuse or with
a short point, from very broad and scarcely 2 lines long to narrow and ½ in.
long, the margins recurved or revolute or probably nearly flat when fresh.
Spikes short and dense, terminal or in the uppermost axils or terminating
short leafy axillary branches. Bracts and bracteoles small. Sepals rather
broad, obtuse, about 1 line long. Corolla rarely 2 lines long, the lobes about
as long as the tube. Anthers attached under the more or less prominent
sterile tips. Hypogynous disk truncate. Ovary 2-celled, tapering into a
very short style. Fruit very small, 1-seeded by abortion or rarely both seeds
perfected.—DC. Prod. vii. 748 ; Hook. f. Fl. Tasm. i. 250 ; *Styphelia collina*, Labill. Pl. N. Holl. i. 47. t. 65 ; F. Muell. Fragm. vi. 45 ; *Leucopogon
ciliatus*, A. Cunn. ; DC. Prod. vii. 746 ; Hook. f. Fl. Tasm. i. 251. t. 75 A.

N. S. Wales (?). Mount Imlay, *L. Morton.* The specimens are very young and
somewhat doubtful. The ovary appears to be that of *L. collinus,* but the sepals are more
acute.
Victoria. Avon river; snowy plains, Mitta-Mitta, and Cabonga Mountains, *F.
Mueller.*
Tasmania, *Labillardière ;* Port Dalrymple and Derwent river, *R. Brown ;* abundant
throughout the island in dry gravelly places, etc., *J. D. Hooker.*
The commonest form in Tasmania is erect, shrubby, and slightly pubescent. From
mountain grassy situations the specimens show a small diffuse plant with slender branches
and small almost flat leaves. Cunningham's specimens of *L. ciliatus* have more the habit
and longer leaves of the common form, but they are nearly flat. They answer very well to
Labillardière's own specimens.
Styphelia reflexa, Rudge in Trans. Linn. Soc. x. 296. t. 19 ; DC. Prod. vii. 736, of
which I have been unable to find the original specimen, appears, from the plate and description, to be *Leucopogon collinus.*

20. **L. glacialis,** *Lindl. in Mitch. Three Exped.* ii. 127. A small erect
or diffuse shrub with pubescent branches. Leaves often crowded, erect or
spreading, linear or linear-lanceolate, acuminate, ciliate-denticulate, much
twisted in the dried state, 3 to 4 lines long. Spikes short, terminal and in
the uppermost axils. Bracts strongly striate, and bracteoles about half as
long as the calyx. Sepals about 1 line long, narrow but obtuse, ciliate.
Corolla-tube very short; lobes about 1 line long. Anthers attached below
the very short sterile tips. Hypogynous disk truncate or obtusely lobed.
Ovary 2-celled ; style very short.

Victoria. Summit of Mount William, Grampians, *Mitchell, Wilhelmi*; Serra Range, *F. Mueller*; Heaths near Portland, *Robertson, Allitt.* This is certainly near *L. collinus*, but the acuminate twisted leaves (as in *L. flexifolius*) give it a very different aspect, and the sepals are much narrower.

21. **L. compactus,** *Stschegl. in Bull. Mosc.* 1859, i. 13.   An erect shrub, hoary-villous all over with short hairs.   Leaves oblong-linear, obtuse or with a callous point, with revolute margins, mostly under $\frac{1}{2}$ in. long. Spikes short, densely clustered, terminal and in the upper axils.   Bracts and bracteoles acuminate, at least half as long as the calyx.   Sepals $1\frac{1}{4}$ lines long, acuminate, ciliate with long hairs as well as the bracts.   Corolla about 2 lines long, the lobes rather longer than the tube.   Anthers attached below the short sterile tips.   Hypogynous disk lobed, readily separating into distinct scales.   Ovary 2-celled ; style short.

W. Australia. *Drummond, 5th Coll. n.* 226.   Near *L. revolutus* and *L. collinus*, but much more villous than either, and differs from the former in the ovary, from both in the bracts and sepals.

22. **L. squarrosus,** *Benth. in Hueg. Enum.* 77.   An erect shrub, attaining 3 or 4 ft., much branched at the top, the branchlets usually pubescent. Lower leaves often linear-lanceolate erect and 3 lines long, but the greater number and always the upper ones ovate or ovate-lanceolate, scarcely exceeding 2 lines, with spreading or recurved tips or rarely spreading from the base, all rigid, acuminate or acute, flat or convex and mostly ciliate.   Spikes short dense and few-flowered, terminal or in the uppermost axils forming short leafy corymbs or clusters.   Bracts leaf-like, acuminate, longer than the bracteoles ; bracteoles about half as long as the calyx.   Sepals about $1\frac{1}{4}$ lines long, acuminate.   Corolla shortly exceeding the calyx, the tube very short.   Anthers attached below the small recurved sterile tips.   Hypogynous disk truncate.   Ovary 2-celled ; style short.—DC. Prod. vii. 750 ; Sond. in Pl. Preiss. i. 317 ; *Styphelia squarrosa*, F. Muell. Fragm. vi. 31.

W. Australia. Swan River and towards King George's Sound, *Huegel, Drummond, 1st Coll. n.* 463, *3rd Coll. n.* 186, *Preiss, n.* 403, and others.   Allied to *L. microphyllus*, but readily known by the acuminate squarrose leaves.   *L. cordatus* has a similar aspect, but a pubescent 5-celled ovary.

23. **L. microphyllus,** *R. Br. Prod.* 544.   An erect or straggling shrub with rather slender often twiggy branches, more or less pubescent.   Leaves ovate-oblong lanceolate or almost linear, obtuse, flat, erect or recurved, 1 to 2 lines long or sometimes all under 1 line, very rarely a few exceeding 2 lines.   Spikes terminal, very short dense and few-flowered, clustered so as to form little leafy heads at the ends of the branches.   Lower bracts leaf-like, acuminate ; bracteoles half as long as the calyx.   Sepals about 1 line long, narrow, acuminate and acute or rarely almost obtuse.   Corolla under $1\frac{1}{4}$ lines long, the lobes as long as the tube.   Anthers attached below the short sterile tips.   Hypogynous disk truncate or shortly lobed.   Ovary 2-celled, tapering into the style.   Fruit small, oblong, usually 1-seeded.—DC. Prod. vii. 749 ; *Perojoa microphylla*, Cav. Ic. iv. 29, t. 349 ; *Peroa microphylla*, Pers. Syn. i. 174 ; *Styphelia microphylla*, Spreng. Syst. i. 656 ; F. Muell. Fragm. vi. 45 ; *Leucopogon denudatus*, Sieb. ; DC. Prod. vii. 749 ; *Styphelia*

denudata, Spreng. Syst. Cur. Post. 67 ; *Leucopogon fraternus*, DC. Prod. vii. 749.

N. S. Wales. Port Jackson to the Blue Mountains, *R. Brown, Sieber, n.* 106, 109, *Fl. Mixt. n.* 498, and many others. The leaves are very variable in size, always larger on the main branches, and it is impossible to fix any limits so as to separate as varieties *L. denudatus* and *L. microphyllus.*

Var. *pilibundus.* More hairy. Leaves narrow, often almost acute. Inflorescence and flowers precisely the same.—*L. pilibundus*, A. Cunn. ; DC. Prod. vii. 746.—Near Bathurst, N. S. Wales, *A. Cunningham.* De Candolle places this amongst the *Axilliflora*, but the spikes only appear axillary from the shortness of the flowering branches, which, although axillary, are always leafy at the base.

Some specimens in the Hookerian as well as in Cunningham's Herbarium are marked " Point Possession, *Collie*," and " South Coast, *Baxter*," but I doubt much whether there is not here some error.

24. **L. tetragonus,** *Sond. in Pl. Preiss.* i. 317. An erect shrub, of 2 or 3 ft., the branches and often the leaves pubescent. Leaves crowded on the branchlets, often decussate and here and there irregularly opposite, oblong-lanceolate or almost ovate, obtuse, with recurved margins, usually ciliate with long hairs, mostly about 1 line long, or nearly 2 lines when narrow. Spikes short, dense, and few-flowered, terminal. Lower bracts leaf-like, concave, striate, longer than the bracteoles, upper ones smaller ; bracteoles broad, acuminate, about half as long as the calyx. Sepals about 1 line long, lanceolate, rather obtuse, coloured, ciliate. Corolla scarcely 2 lines long, the lobes longer than the tube. Anthers attached above the middle, with very short sterile tips. Hypogynous scales distinct, obovate. Ovary 2-celled ; style short.—*Styphelia tetragona*, F. Muell. Fragm. vi. 31.

W. Australia. Near Cape Riche, *Harvey, Drummond, 4th Coll. n.* 151 ; Konkoberup hills, *Preiss, n.* 387 ; Mount Bland, *Maxwell.*

25. **L. phyllostachys,** *Benth.* Erect and slightly branched, 1 to 1½ ft. high and quite glabrous. Leaves spreading or reflexed, a few of the lowest sessile, ovate or ovate-lanceolate, under 3 lines long, all the others broadly cordate-ovate or almost reniform, often above 3 lines diameter, obtuse or with a minute callous point, convex or nearly flat. Spikes cylindrical, many-flowered, terminal or in the uppermost axils, the upper broadly cordate leaves passing gradually into floral leaves or bracts similar but smaller, the uppermost scarcely above ½ line diameter and shortly acute. Bracteoles less than half as long as the calyx. Sepals about 1 line long, rather narrow, almost acute, smooth. Corolla scarcely 2 lines long, the lobes as long as the tube. Anthers attached near the top, with very small sterile tips. Hypogynous scales small, ovate, distinct or scarcely cohering. Ovary 2-celled ; style short. Drupe very small, oblong, 1- or 2-seeded.—*Styphelia glabella*, F. Muell. Fragm. vi. 32, not of Spreng., nor *Leucopogon glabellus*, Br.

W. Australia, *Drummond, 5th Coll. n.* 311.

26. **L. glabellus,** *R. Br. Prod.* 544. An erect shrub, of 2 to 3 ft., with rather slender branches, and the typical form quite glabrous. Leaves spreading or somewhat reflexed, varying from cordate-ovate to lanceolate, acute, flat or convex, 1 to 2 lines long or rarely 3 lines when narrow. Spikes

many-flowered, cylindrical, terminal or terminating very short axillary branch-lets, the upper leaves small and passing gradually into the lower bracts, which are like the leaves, but smaller. Bracteoles acute, not half so long as the calyx. Sepals not exceeding 1 line, rather narrow, acutely acuminate, often ciliate. Corolla about 1½ lines long, the lobes longer than the tube. Anthers attached near the top, with small sterile tips. Hypogynous scales ovate, distinct or slightly cohering. Ovary 2-celled; style short. Drupes small, oblong, 1- or 2-seeded.—DC. Prod. vii. 749; *Styphelia glabella*, Spreng. Syst. i. 655; *Leucopogon variifolius*, Sond. in Pl. Preiss. i. 314; *Styphelia variifolia*, F. Muell. Fragm. vi. 32, but not *L. elatior*, Sond.

W. Australia. King George's Sound and adjoining districts, *R. Brown, Drummond,* n. 29, *5th Coll. n.* 319, *Preiss, n.* 404, and others.

Var. *pubescens.* Branches, foliage, and inflorescence pubescent, in other respects pre-cisely as in the common form.—*L. lanigerus* (originally written *L. canigerus*), A. Cunn.; DC. Prod. vii. 749. King George's Sound or to the eastward, *Baxter.*

27. **L. elatior,** *Sond. in Pl. Preiss.* i. 314. An erect shrub, attaining 3 ft., with rather slender, glabrous branches. Leaves spreading, in some spe-cimens broadly ovate, obtuse, almost cordate at the base, rigid, flat or slightly convex, prominently ribbed, about 1 line long, occasionally but rarely oppo-site, in other specimens ovate and acute, and passing into ovate-lanceolate or even narrow-lanceolate and acute, and then 2 or even 3 lines long. Spikes terminal, cylindrical, rather dense, often many-flowered. Lower bracts like the leaves, but concave and gradually smaller; bracteoles about one-third as long as the calyx. Sepals scarcely above 1 line long, narrow, obtuse, slightly striate. Corolla under 2 lines long, the lobes rather longer than the tube. Anthers attached below the very short sterile tips. Hypogynous scales distinct. Ovary 2-celled; style short.—*L. decussatus*, Stschegl. in Bull. Mosc. 1859, i. 11; *L. semioppositus* or *Styphelia semiopposita*, F. Muell. Fragm. vi. 49.

W. Australia, *Drummond, 5th Coll. n.* 328; Vasse river, *Mrs. Molloy;* near Bus-selton, Sussex district, *Preiss, n.* 391.

SERIES 4. STRIATÆ.—Spikes all terminal or also in the uppermost axils, short and dense or cylindrical and rather dense. Leaves nearly flat, the margins neither thickened nor recurved, usually small and obtuse, either strongly ribbed or the midrib prominent at the end. Ovary 2-celled.

28. **L. florulentus,** *Benth.* An erect, branching shrub, our specimen quite glabrous. Leaves crowded, erect, ovate or obovate-oblong, very obtuse, thick, flat or slightly concave, sometimes slightly ciliate, with scarcely prominent veins, 1 to 1½ lines long. Spikes solitary, dense and many-flowered, termi-nating the numerous branchlets. Bracts ovate, obtuse, striate; bracteoles ovate-lanceolate, not half so long as the calyx. Sepals ¾ line long, rather obtuse, whitish when dry. Corolla 1 to 1½ lines long, the lobes about as long as the tube. Anthers attached below the very short sterile tips. Hy-pogynous scales distinct. Ovary 2-celled; style short.

W. Australia. Between King George's Sound and Swan River, *Harvey.* Although we have but a single specimen of this plant, it is a good one, covered with a profusion of small flowers, and which I am unable to refer to any other species, although allied on the one hand to *L. elatior* and on the other to *L. striatus.*

29. **L. striatus,** *R. Br. Prod.* 544. A spreading, much-branched shrub, usually low, but sometimes attaining 2 ft., glabrous or with pubescent branchlets. Leaves ovate or ovate-lanceolate, obtuse, thick, strongly ribbed underneath, and sometimes the midrib especially prominent above the middle as in *L. carinatus,* mostly 1½ to 3 lines long, flat and spreading or the upper ones erect and slightly concave, whilst the lowest are more spreading and slightly convex. Spikes dense, terminal or in the uppermost axils, ¼ to ½ in. long. Bracts and bracteoles about half as long as the calyx. Sepals about 1 line long, obtuse as well as the bracts, rather thick and rigid. Corolla under 2 lines long, the lobes about as long as the tube. Anthers attached near the top, with very small sterile tips. Hypogynous disk readily separating into distinct scales. Ovary 2-celled, with a broad, 5-furrowed summit; style very short.—DC. Prod. vii. 750; *Styphelia striata,* Spreng. Syst. i. 656; *L. nervosus,* R. Br. Prod. 544; DC. Prod. vii. 750; *L. rupestris,* Sond. in Pl. Preiss. i. 315.

W. Australia. Lucky Bay, *R. Brown, Baxter;* Konkoberup hills, Cape Riche, *Preiss, n.* 406 (partly); Mount Bland, *Maxwell.* R. Brown's specimens of his *L. striatus* and *L. nervosus* are, as observed in his 'Prodromus,' very nearly allied, and several of Baxter's connect them too closely to separate them as varieties.

30. **L. lasiostachyus,** *Stschegl. in Bull. Mosc.* 1859, i. 11. An erect shrub, with virgate branches, minutely but softly pubescent as well as the foliage. Leaves from ovate to lanceolate, obtuse or with a callous point, 2 to 4 or rarely 5 lines long when narrow, erect or spreading, flat or slightly concave, strongly ribbed underneath. Spikes cylindrical, dense and many-flowered, terminal and in the uppermost axils, ½ to 1 in. long, softly pubescent or villous. Bracts lanceolate or linear; bracteoles ovate-lanceolate, acuminate, more than half as long as the calyx. Sepals lanceolate, acuminate, 1½ lines long, softly villous as well as the bracts. Corolla-tube much shorter than the calyx; lobes twice as long as the tube. Anthers attached above the middle, with small, recurved, sterile tips. Hypogynous disk toothed, at length separable into distinct scales. Ovary 2-celled, but often broad and 5-angled at the top; style short.

W. Australia, *Drummond, 5th Coll. n.* 304; Stirling range, *F. Mueller.*

31. **L. carinatus,** *R. Br. Prod.* 545. An erect or spreading, much-branched shrub, of 1 to 2 ft., glabrous or with minutely pubescent branchlets. Leaves oval-oblong, lanceolate or almost linear, obtuse, thick and rigid, the midrib underneath prominent above the middle and sometimes almost as much ribbed as in *L. striatus,* mostly 2 to 3 lines long but in some specimens longer. Spikes short, dense, terminal or in the uppermost axils. Bracts and bracteoles scarcely half as long as the calyx, rather acute. Sepals 1 to 1¼ line long, rather rigid and acute. Corolla 1½ to 2 lines long, the lobes longer than the tube. Anthers attached above the middle, with short, sometimes very small, recurved, sterile tips. Hypogynous scales distinct. Ovary 2-celled, the summit usually broad.—*L. tectus,* Sond. in Pl. Preiss. i. 318; *Styphelia carinata,* Spreng. Syst. i. 658 (partly).

W. Australia. Lucky Bay, *R. Brown, Baxter;* Konkoberup hills, Cape Riche, *Preiss, n.* 406; Mount Melville, *F. Mueller;* towards Cape Riche and round Cape Arid to Cape Paisley, *Maxwell.*

Drummond's specimens, 5th Coll. n. 325, with narrow leaves, 3 to 5 lines long, seem almost to connect this with *L. tenuis;* on the other hand, those with short, broad leaves come near to *L. striatus. L. ovatus,* Sond. in Pl. Preiss. i. 319, from Preiss's collection, n. 375, seems to be a small-leaved state of the same species, but the specimen I have seen is a mere fragment.

SERIES 5. OPPOSITIFOLIÆ.—Spikes all terminal or also in the uppermost axils, short and dense. Leaves all opposite, small, erect, linear, obtuse. Ovary 2-celled.

32. **L. opponens,** *F. Muell. Fragm.* vi. 48. An erect shrub, of about 2 ft., with slender, virgate branches, glabrous or sprinkled as well as the foliage with short, spreading hairs. Leaves distinctly petiolate, all opposite, erect, obtuse, with thickened or recurved margins, so as to be 2-furrowed underneath, mostly 2 to 3 lines long. Spikes short, dense, terminal or in the uppermost axils. Bracts broad, concave, mostly opposite, shorter than the bracteoles; bracteoles about half as long as the calyx, almost acute, keeled. Sepals 1¼ to 1½ lines long, rather acute, scarcely coloured, minutely ciliolate. Corolla scarcely 2 lines long, the lobes rather longer than the tube. Anthers attached below the short sterile tips. Hypogynous disk obtusely lobed. Ovary 2-celled; style very short.—*Styphelia opponens,* F. Muell. l. c.

W. Australia. Sandy places, Phillips river, *Maxwell.*

33. **L. oppositifolius,** *Sond. in Pl. Preiss.* i. 316. An erect, heath-like shrub of 1 ft. or more, with slender branches, glabrous or pubescent as well as the foliage. Leaves all opposite, erect, narrow-linear or almost linear-lanceolate, obtuse, concave and keeled, 1 to 2 lines long. Flowers in very short terminal spikes. Bracts like the leaves, but smaller, and the lower ones opposite; bracteoles narrow, obtuse, about half as long as the calyx. Sepals 1 line long, lanceolate, rather obtuse, usually coloured. Corolla about 2 lines long, the lobes as long as the tube. Anthers attached under the prominent sterile tips. Hypogynous disk truncate or lobed, and readily separating into distinct scales. Ovary small, 2-celled; style very short.—*Styphelia oppositifolia,* F. Muell. Fragm. vi. 32.

W. Australia. King George's Sound, *Preiss, n.* 380, 400; Stirling range, *F. Mueller;* also in Maxwell's collection, without the precise station. The leaves in all the specimens seen are constantly opposite, but possibly *L. lasiophyllus* may be only a variety with larger leaves mostly alternate.

SERIES 6. CONCURVÆ.—Spikes all terminal or also in the upper axils, short and dense or cylindrical and rather dense. Leaves more or less concave or keeled. Ovary 2-celled.

34. **L. tamariscinus,** *R. Br. Prod.* 544. An erect shrub of 2 to 3 ft., with virgate branches and often numerous short bránchlets, glabrous or sprinkled with a few hairs. Leaves erect, ovate or lanceolate, acuminate, concave, often dilated near the base and almost embracing the stem, the larger ones on the main branches often 2 to 3 lines long, those on the smaller, slender

branchlets under 1 line, all usually turning black in drying, and the upper ones passing into the bracts. Spikes terminal, cylindrical and slender, many-flowered, ½ to 1 in. long. Bracts like the stem-leaves, but smaller ; bracteoles broad, shortly acuminate, not half so long as the calyx. Sepals under 1 line long, rather acute. Corolla 1¼ to nearly 1½ lines long, the lobes equal to or rather longer than the tube. Anthers attached below the minute sterile tips. Hypogynous disk obtusely lobed or the scales quite distinct. Ovary 2-celled ; style short.—DC. Prod. vii. 749 ; *Styphelia tamariscina,* Spreng. Syst. i. 656 ; *Leucopogon parvifolius,* DC. Prod. vii. 752 ; *L. vaginans,* Sond. in Pl. Preiss. i. 315.

W. Australia. King George's Sound, *R. Brown, A. Cunningham, Baxter, Drummond ;* Mount Melville, *F. Mueller ;* near Cape Riche, *Harvey ;* Phillips river, *Maxwell ;* also from Roe's collection, *Preiss, n.* 382.

35. **L. bracteolaris,** *Benth.* An erect shrub, with glabrous or pubescent branches. Leaves erect, ovate to broadly oblanceolate, obtuse or with an obtuse callous point, concave and often embracing the stem, ciliate with fine hairs, the veins fine or the midrib more prominent at the top, mostly about 2 lines long. Spikes terminal, cylindrical, rather dense, many-flowered. Bracts leaf-life, lanceolate, ciliate, concave, as long as the flowers ; bracteoles similar or more acute, almost or quite as long as the calyx. Sepals lanceolate, acute, rigid, ciliate, about 1 line long. Corolla but very little longer, the lobes rather longer than the tube. Anthers attached under the short sterile tips. Hypogynous scales small, distinct. Ovary 2-celled ; style very short.

W. Australia. King George's Sound or to the eastward, *M'Lean* (or rather *Baxter ?*).

36. **L. elegans,** *Sond. in Pl. Preiss.* i. 318. A slender shrub, of 3 to 4 ft., the branches more or less hirsute with long, soft, spreading hairs, or rarely glabrous. Leaves ovate to lanceolate, almost acute, fringed with long, soft hairs, concave, prominently striate, 2 to 3 lines long. Spikes terminal or very rarely in the uppermost axils, short and few-flowered. Lower or nearly all the bracts leaf-like, shortly petiolate, lanceolate, nearly as long as the flowers ; bracteoles broad, obtuse, about one-third as long as the calyx. Sepals obtuse or almost acute, smooth, ciliolate, about 1 line long. Corolla-tube about as long as the calyx ; lobes as long as the tube. Anthers attached below the rather long sterile tips. Hypogynous disk toothed. Ovary 2-celled ; style exceedingly short.—*Styphelia blepharophylla,* F. Muell. Fragm. vi. 34.

W. Australia, *Drummond, 4th Coll. n.* 150, *5th Coll. n.* 308 ; Konkoberup hills Cape Riche, *Preiss, n.* 378 ; Kalgan river, *Oldfield ;* Warricup hills, *Maxwell.*

37. **L. tenuis,** *DC. Prod.* vii. 744. A shrub, with erect, virgate, rather slender branches, usually glabrous. Leaves linear or narrow-lanceolate, obtuse or with a short, callous point, narrowed at the base but scarcely petiolate, erect, flat or slightly concave, the veins, especially the midrib towards the top, rather prominent, under ½ in. long except when very luxuriant, the upper ones smaller, and passing into the bracts. Spikes terminal, cylin-

drical, many-flowered. Lower or nearly all the bracts lanceolate and leaf-like, the upper ones short, broad, acuminate, and strongly ribbed; bracteoles not half so long as the calyx.   Sepals 1¼ lines long, narrow, acute, smooth like the bracteoles.   Corolla under 2 lines long, the lobes much longer than the tube.   Anthers attached below the short, recurved, sterile tips.   Hypogynous scales distinct, broadly obovate.   Ovary 2-celled, with a broad top; style short.   Young fruit oblong.

W. Australia.   Swan River, *Drummond, n.* 25, *5th Coll. n.* 325.

38. **L. gnaphalioides,** *Stschegl. in Bull. Mosc.* 1859. i. 14.   A stout, erect, bushy shrub, the branches and outside of the leaves softly pubescent or villous.   Leaves crowded, erect, ovate-lanceolate, obtuse, concave, rigid, prominently striate, mostly 3 to 4 lines long.   Spikes short and dense, terminal or in the uppermost axils, forming dense terminal heads.   Bracts and bracteoles about half as long as the calyx, obtuse, ciliate.   Sepals narrow, almost acute, 1½ to nearly 2 lines long.   Corolla 2½ lines long, the lobes as long as the tube.   Anthers attached a little below the rather long sterile tips.   Hypogynous disk toothed.   Ovary 2-celled, short, not dilated; style very short.

W. Australia, *Drummond, 4th Coll. n.* 152, *5th Coll. n.* 318.

39. **L. concurvus,** *F. Muell. Fragm.* iii. 144.   A small, decumbent shrub, with the elongated wiry branches, general habit, and inflorescence of *L. virgatus,* glabrous or hirsute with long, soft, spreading hairs.   Leaves ovate-lanceolate to lanceolate, acute or the lower ones obtuse, concave, the larger ones above ½ in. long, but mostly under that.   Flowers small, in short, dense spikes, terminal or in the upper axils, or the clusters appearing axillary from the shortness of the flowering branches.   Bracteoles nearly half as long as the calyx.   Sepals lanceolate, thin, almost acute, nearly 1¼ lines long.   Corolla-tube half as long as the calyx, the lobes longer than the tube.   Anthers inserted below the rather long sterile tips.   Hypogynous disk very short.   Ovary 2-celled; style very short.—*L. apiculatus,* Sond. in Linnæa, xxvi. 248, not of R. Br.; *Styphelia concurva,* F. Muell. Fragm. vi. 36.

S. Australia.   Encounter Bay and Onkaparinga river, *F. Mueller.*

40. **L. Gilbertii,** *Stschegl. in Bull. Mosc.* 1859. i. 15.   Stems slender, erect and virgate, 1 to 2 ft. high, quite glabrous.   Leaves erect, linear-lanceolate or lanceolate, tapering into a short, callous point, concave, very finely veined, the larger ones above ½ in. long, but mostly under that.   Flowers small, in short, dense spikes terminal or in the uppermost axils, or the clusters of spikes appearing axillary from the shortness of the flowering branchlets.   Bracts small; bracteoles about half as long as the calyx, all obtuse, striate.   Sepals ¾ line long, oblong, obtuse, coloured, ciliate.   Corolla 1¼ lines long, the lobes longer than the tube.   Anthers attached below the prominent sterile tips.   Hypogynous disk sinuate or lobed.   Ovary 2-celled, flat-topped; style very short.

W. Australia, *Drummond, 2nd Coll. n.* 263, *4th Coll. n.* 134, *Gilbert, n.* 41, *Oldfield;* Upper Hay river, *F. Mueller.*   The species is very near *L. tenuis,* but more slender, the flowers much smaller, in more compact spikes, and a different calyx.   It is also near

L. gracilis, and has precisely the inflorescence of *L. concurvus.* F. Mueller (Fragm. vi. 32) refers it to *L. multiflorus,* Br., which, however, belongs to the section *Pleurœnthus.*

41. **L. gracilis,** *R. Br. Prod.* 544. An erect or diffuse shrub of 1 to 1½ ft., with slender and wiry or rarely more rigid branches, usually glabrous. Leaves erect, linear-lanceolate, scarcely acute, concave, prominently 3- or 5-ribbed, 2 to 3 or rarely 4 lines long. Spikes short, dense, terminal, solitary or clustered. Bracts small, membranous, but not leafy; bracteoles lanceolate, obtuse, about half as long as the calyx. Sepals scarcely 1 line long, lanceolate, rather acute, minutely ciliolate. Corolla about 2 lines long, the lobes as long as the tube. Anthers attached below the prominent sterile tips. Hypogynous disk usually lobed. Ovary 2-celled; style very short.— DC. Prod. vii. 749; Sond. in Pl. Preiss. i. 316; *Styphelia gracilis,* Spreng. Syst. i. 658.

W. Australia. King George's Sound, *R. Brown, A. Cunningham, Preiss, n.* 381, 389.

42. **L. acicularis,** *Benth.* A small, heath-like shrub, with slender, erect branches, more or less hirsute with spreading hairs. Leaves erect, linear-lanceolate, tapering into a fine point, concave, prominently keeled or 3-nerved, 2 to 4 lines long. Flowers small, in short, dense spikes terminating the branches or the short axillary branchlets. Bracts small, green, the lower ones lanceolate, the upper ones ovate, acute. Bracteoles not half so long as the calyx, broad, ciliolate, rather acute. Sepals thin, narrow, acute, under 1 line long. Corolla about 1½ lines long, the lobes as long as the tube. Anthers attached below the very short sterile tips. Hypogynous scales distinct. Ovary 2-celled; style short.

W. Australia. Fitzgerald Range, *Maxwell.* Although I have seen but a single specimen it is a good one, and does not agree with any species known to me.

43 ? **L. cryptanthus,** *Benth.* A slender, much-branched. apparently diffuse shrub, not exceeding 6 in., the branches pubescent. Leaves erect, linear or linear-lanceolate, tapering into a pungent point, rigid, concave, prominently ribbed, 1 to 3 lines long. Flowers few, very small and inconspicuous, in short spikes, solitary or clustered at the ends of the branches, forming little leafy cymes. Bracts similar to the leaves, and mostly exceeding the flowers; bracteoles acutely acuminate, more than half as long as the calyx. Sepals acutely acuminate, under 1 line long. Corolla rather shorter than the calyx, the lobes as long as the tube. Anthers attached by the middle, oblong, obtuse, with very minute sterile tips or sometimes none. Hypogynous disk sinuate-lobed. Ovary 2-celled; style very short.

W. Australia. Swan River, *Drummond, 1st Coll. n.* 12. The specimens are numerous, and all small though complete with the root, the ends of the branches generally recurved. The species should, perhaps, technically belong to the section *Heteranthesis,* but the general affinities are with *Perojoa,* and the tips of the anthers have sometimes appeared to me to be closed and sterile.

44. **L. gracillimus,** *DC. Prod.* vii. 747. An erect, heath-like shrub of 1 to 3 ft., glabrous or the very slender branches minutely pubescent. Leaves erect, appressed or imbricate, narrow-linear, obtuse, concave, about 1 line or on the main branches 2 lines long. Flowers small, in short, terminal

spikes. Bracts, at least the lower ones, like the stem leaves, but smaller; bracteoles very obtuse, not half so long as the calyx. Sepals about $\frac{3}{4}$ line long, obtuse. Corolla $1\frac{1}{2}$ to nearly 2 lines long, the lobes longer than the tube. Anthers inserted below the prominent sterile tips. Hypogynous disk sinuate-toothed. Ovary 2-celled (or sometimes 3-celled?); style very short. —Sond. in Pl. Preiss. i. 312; *Styphelia gracillima*, F. Muell. Fragm. vi. 34.

W. Australia. Swan River, *Drummond, 1st Coll. n.* 23; Mount Bakewell, *Preiss, n* 395. Drummond's specimens, n. 75, are referred here by F. Mueller, but they have no flowers, and must therefore be doubtful.

45. **L. lasiophyllus,** *Stschegl. in Bull. Mosc.* 1859. i. 16. An erect shrub with virgate branches, our specimens under 1 ft. high, the branches and foliage pubescent, with short, rigid hairs. Leaves erect, sessile, linear or linear-lanceolate, rather obtuse, concave, prominently ribbed, 2 to 4 lines long, occasionally opposite. Flowers few, in short, dense spikes, terminal or in the uppermost axils. Bracts, at least the lower ones, like the leaves, but smaller; bracteoles acuminate, at least half as long as the calyx, and some-times almost passing into the sepals. Sepals $1\frac{1}{4}$ to $1\frac{1}{2}$ lines long, narrow, coloured, ciliate, and pubescent, with convolute, rather obtuse tips. Corolla about 2 lines long, the lobes as long as the tube. Anthers attached below the prominent sterile tips. Hypogynous disk obtusely lobed. Ovary gla-brous, 2-celled. Style very short.

W. Australia, *Drummond, 5th Coll. n.* 329; Stirling Range, *F. Mueller.* This plant is referred by F. Mueller, Fragm. vi. 32, to *L. oppositifolius*, from which it only ap-pears to differ in the rather longer, more pubescent leaves, mostly alternate.

46. **L. cymbiformis,** *A. Cunn.;. DC. Prod.* vii. 750. A bushy or wiry shrub of 1 to $1\frac{1}{2}$ ft., glabrous or the branches scarcely pubescent. Leaves erect, lanceolate or linear-lanceolate, tapering into a short, rigid point, concave, and usually keeled, 1 to 2 or rarely 3 lines long. Spikes very short, dense, and often only 2- or 3-flowered, terminal or in the upper-most axils, but sometimes becoming lateral by the elongation of a branch from one of the uppermost axils. Bracts lanceolate, acute, leaf-like; brac-teoles very acute, half as long as the sepals. Sepals $1\frac{1}{4}$ to $1\frac{1}{2}$ lines long, very acute, and sometimes greenish below the points. Corolla-tube shortly exceeding the calyx; lobes shorter than the tube. Anthers attached imme-diately under the exceedingly short, recurved, sterile tips. Hypogynous scales ovate, obtuse, usually distinct, with a dense ring of almost chaffy hairs or bristles within them round the ovary. Ovary elongated, tapering into the scarcely distinct style, angular, 2-celled.—Sond. in Pl. Preiss. i. 318; *Styphelia cymbiformis*, F. Muell. Fragm. vi. 34.

W. Australia, *Drummond, 3rd Coll. n.* 182, 188, *5th Coll. n.* 323; King George's Sound, *A. Cunningham;* Cape Riche, *Harvey;* Gordon river, *Preiss, n.* 385; Kalgan and Tone rivers, *Oldfield.* I have had the greatest difficulty in ascertaining the structure of the ovary. From more than a dozen different specimens I always found it diseased and black inside. In one specimen, however, it was distinctly 2-celled, but the ovules were still im-perfect; and were it not for the great uniformity in specimens gathered by different collectors in different localities, I should have suspected that the frequently abnormal inflorescences, the peculiar hairs round the ovary, and the unusual form of the latter organ, had been alike the effects of disease.

SERIES 7. VIRGATÆ.—Spikes all terminal or also in the upper axils, short and dense, or cylindrical and rather loose. Leaves erect, concave. Ovary 5- or 3-celled (or exceptionally 4-celled).

47. L. apiculatus, *R. Br. Prod.* 542. An erect shrub of 2 to 4 ft., glabrous or softly pubescent. Leaves erect or spreading, oblong-lanceolate or almost elliptical, acute, but with a callous point, concave or nearly flat, ½ to ¾ in. long. Spikes terminal and in the uppermost axils, loose and longer than the leaves. Bracts narrow-lanceolate, acute; bracteoles acuminate, fully half as long as the calyx. Sepals 1½ lines long, shortly acuminate, often coloured. Corolla-tube nearly as long as the calyx; lobes as long as the tube. Anthers linear, attached about the middle, with recurved sterile tips. Hypogynous disk sinuate-toothed, readily separable into distinct scales. Ovary depressed, 4- or 5-celled. Fruit much depressed, not exceeding the calyx.—DC. Prod. vii. 745; F. Muell. Fragm. iv. 105; *Styphelia apiculata*, Spreng. Syst. i. 656; F. Muell. Fragm. vi. 31; *L. Shuttleworthii*, Sond. in Pl. Preiss. i. 307.

W. Australia. Lucky Bay and Goose Island Bay, *R. Brown;* Lucky Bay ?, *Baxter;* Cape le Grand and along the coast to Cape Arid, *Maxwell.* The locality, N. S. Wales, given by Sonder was owing to a mistake in the label of the specimens he examined.

48. L. polystachyus, *R. Br. Prod.* 542. An erect shrub, attaining several feet, with slender, virgate branches usually glabrous. Leaves linear to lanceolate, tapering into a callous point, rigid, concave, erect or scarcely spreading, 4 to 8 lines long. Spikes rather short, dense, terminal or in the uppermost axils, usually crowded into an ovoid, terminal, leafy head. Bracts small, ovate, obtuse; bracteoles about half as long as the calyx, rather obtuse. Sepals 1½ lines long or rather more, thin, but rather rigid, and often pale pink. Corolla 2 lines long or rather more, the lobes longer than the tube. Anthers attached below the recurved sterile tips. Hypogynous disk sinuate or obtusely 5-lobed. Ovary (always?) 3-celled, although 5-angled at the base; style very short. Fruit as long as the calyx, truncate.—DC. Prod. vii. 746; Sond. in Pl. Preiss. i. 307; *Styphelia polystachya,* Spreng. Syst. i. 659; F. Muell. Fragm. vi. 31.

W. Australia. King George's Sound and neighbouring districts, *R. Brown, Drummond, 3rd Coll. n.* 189, *Preiss, n.* 363, 364, and many others.

49. L. virgatus, *R. Br. Prod.* 543. A low, decumbent or diffuse shrub, with ascending or erect wiry, glabrous branches of ½ to 1 ft. or rarely more rigid and bushy. Leaves lanceolate or linear-lanceolate, tapering into a rigid but not pungent point, concave, minutely ciliate, under ⅞ in. long. Spikes short, dense, terminal or in the uppermost axils, or appearing axillary from the shortness of the flowering branchlets. Bracts small; bracteoles half as long as the calyx. Sepals about 1 line long, obtuse, scarcely coloured. Corolla-tube rather shorter than the calyx; lobes longer than the tube. Anthers attached a little below the short sterile tips. Hypogynous disk truncate or obtusely lobed. Ovary broad, 5-celled; style short.—DC. Prod. vii. 748; Hook. f. Fl. Tasm. i. 249; *Styphelia virgata,* Labill. Pl. Nov. Holl. i. 46. t. 64; F. Muell. Fragm. vi. 42 (but not *L. glacialis,* Lindl.).

N. S. Wales. Port Jackson to the Blue Mountains, *R. Brown, Sieber, n.* 107, and others.

Victoria. Common in heathy ground and sterile regions, ascending into the mountains to an elevation of 4000 ft., *F. Mueller* and others.

Tasmania. Port Dalrymple and Derwent river, *R. Brown ;* abundant in dry, gravelly or sandy places throughout the island, *J. D. Hooker.*

S. Australia. Murray Desert, Onkaparinga and Torrens rivers, Lofty Range, *F. Mueller.*

Var. *brevifolius.* Leaves from ovate to lanceolate, mostly under ¼ in. long, passing sometimes into the common form on the longer branches.—Mount William in the Grampians, *F. Mueller ;* Wimmera, *Dallachy.*

50. **L. pulchellus,** *Sond. in Pl. Preiss.* i. 310.   An erect rather slender shrub, attaining 3 or 4 ft., the branches glabrous or pubescent.   Leaves erect, linear, obtuse, rather thick, and slightly concave, obscurely ribbed underneath, glabrous or slightly hairy, mostly 2 to 3 lines long.   Spikes short and dense, terminal or in the uppermost axils.   Bracts leaf-like but small ; bracteoles very obtuse, not half so long as the calyx.   Sepals about 1¼ lines long, obtuse, rather thin, ciliate and often pubescent.   Corolla 2 lines long or rather more, the tube very short, the lobes much longer.   Anthers attached below the prominent sterile tips.   Hypogynous disk short.   Ovary short and broad, 5-celled ; style very short.   Fruit small, truncate.—*L. triqueter,* Stschegl. in Bull. Mosc. 1859, i. 15 ; *Styphelia pulchella,* F. Muell. Fragm. vi. 34.

W. Australia. Swan River and adjoining districts, *Drummond,* 1*st Coll. n.* 29 and 460, *Preiss, n.* 396 and 401 ; Toodjay, *Gilbert, n.* 15 ; Tone river, *Oldfield ;* Dillon Bay, Phillips Ranges, *Maxwell;* also a glabrous form, with the sepals scarcely ciliate, Stirling Ranges, *Maxwell.*

51. **L. polymorphus,** *Sond. in Pl. Preiss.* i. 309.   An erect but often weak shrub of 1 to 3 ft., glabrous or more frequently the foliage as well as the branches pubescent or hairy.   Leaves erect, from ovate-lanceolate and scarcely 2 lines long to narrow-lanceolate or almost linear and 2 to 4 lines long, obtuse, rigid, concave, prominently ribbed underneath.   Spikes short, dense, terminal or in the uppermost axils.   Bracts and bracteoles obtusely acuminate and usually pubescent, not half so long as the calyx.   Sepals about 1¼ lines long, obtuse, ciliate, usually pubescent.   Corolla 2 to 2½ lines long, the lobes longer than the tube.   Anthers attached below the prominent sterile tips.   Hypogynous disk obtusely lobed.   Ovary 5-celled ; style very short.—*Styphelia polymorpha,* F. Muell. Fragm. vi. 31.

W. Australia. Swan River, *Preiss, n.* 392, 402 ; near Hampden, *Oldfield ;* Stirling Range, *F. Mueller.*   Sonder refers here Huegel's specimens from King George's Sound, which, however, on re-examination appear to me to be the true *L. assimilis,* and, on the other hand, Preiss's specimens, n. 383 and 384, from Swan River, referred by Sonder to *L. brachycephalus* (*L. cucullatus,* var.), appear to me to be a rather broader-leaved form of *L. polymorphus.*

52. **L. assimilis,** *R. Br. Prod.* 545.   An erect rigid shrub of 2 to 4 ft., glabrous or the branches minutely pubescent.   Leaves erect and often imbricate, in the typical form linear or lanceolate, obtuse, rigid, flat or concave, often prominently ribbed, ¼ to ½ in. long.   Spikes short and dense, terminal or in the uppermost axils.   Bracts and bracteoles obtuse, striate, about half

as long as the calyx.　Sepals 1 to 1¼ lines long, rather thin, obtuse, often coloured at the end.　Corolla 2 to 2½ lines long, the lobes rather longer than the tube.　Anthers attached a little below the prominent sterile tips.　Hypogynous disk truncate but readily separating into distinct scales.　Ovary depressed, 5-celled ; style short.　Fruit flat-topped, ribbed, about as long as the calyx or shortly exceeding it.—DC. Prod. vii. 750 ; *L. carinatus,* DC. l. c. not of R. Br. ; *L. vitellinus,* Sond. in Pl. Preiss. i. 309.

W. Australia. Lucky Bay, *R. Brown, Baxter ;* King George's Sound and neighbourhood, *A. Cunningham, Huegel, Oldfield, Preiss, n.* 363.

Var. *rudis.* Rather stouter.　Leaves broader, from oval-oblong to lanceolate, thick and strongly ribbed.　*L. rudis,* F. Muell. Fragm. iv. 106; *Styphelia rudis,* F. Muell. Fragm. vi. 32.

W. Australia. Fitzgerald Range, *Herb. Oldfield ;* Bald Island, *Oldfield in Herb. F. Mueller ;* Cape le Grand, *Maxwell.*

Both *L. assimilis* and *L. polymorphus* are allied in foliage to *L. striatus* and *L. carinatus,* but the latter two species have smaller and differently shaped flowers, and the ovary always 2-celled.

53. **L. Oldfieldii,** *Benth.*　An erect shrub of 2 to 3 ft., the branches and foliage softly pubescent.　Leaves erect, lanceolate, obtuse, rigid, concave, prominently ribbed, 2 to 4 lines long.　Spikes short, dense, terminal or in the uppermost axils.　Lower bracts almost leaf-like, longer than the bracteoles, upper ones smaller ; bracteoles acuminate, fully half as long as the calyx.　Sepals nearly 1½ lines long, thin, ciliate, rather broad but almost acute.　Corolla 2 lines long or rather more, the lobes 2 or 3 times as long as the very short tube.　Anthers attached below the prominent sterile tips.　Hypogynous disk truncate or sinuate.　Ovary densely hispid with white almost chaffy hairs, 3-celled ; style short.

W. Australia. Darling Range, *Oldfield.* The foliage is nearly that of the coarse varieties of *L. polymorphus.* but the ovary is very different.

54. **L. cucullatus,** *R. Br. Prod.* 545.　An erect rigid shrub, glabrous or with minutely pubescent branches.　Leaves crowded, broadly ovate or almost orbicular-cordate, obtuse or with a small callous point, rarely obtusely acuminate or those at the base of the branches passing into lanceolate, all very concave, almost conduplicate, smooth or finely striate, 2 to 3 lines long or nearly 4 when narrow.　Spikes short dense and few-flowered, terminal or in the uppermost axils.　Bracts like the leaves but very much smaller ; bracteoles strongly keeled but scarcely acute, about half as long as the calyx.　Sepals 1½ to nearly 2 lines long, rather obtuse, coloured at the end, minutely ciliate.　Corolla nearly 3 lines long, the lobes longer than the tube.　Anthers attached a little below the prominent recurved sterile tips.　Hypogynous disk short, truncate.　Ovary prominently ribbed, 5-celled or rarely 4-celled. —DC. Prod. vii. 750 ; Sond. in Pl. Preiss. i. 320 ; *Styphelia cucullata,* Spreng. Syst. i. 656 ; F. Muell. Fragm. vi. 32 ; *Leucopogon brachycephalus,* DC. Prod. vii. 746 ; Sond. in Pl. Preiss. i. 308 (partly ?).

W. Australia. King George's Sound, *R. Brown, Harvey, Preiss, n.* 377 ; Swan River, *F. Mueller, Drummond, 1st Coll. n.* 463.

Preiss's specimens, n. 377, have remarkably large, broad, concave leaves ; in Drummond's n. 461, they are remarkably small, but still very concave ; in Preiss's n. 383 and 384, they are small and much flatter but broader than in *E. polymorphus*, to which species they ought, perhaps, to be referred as a variety. Some of Drummond's larger Swan River specimens have large and small, more or less concave, leaves on different branches, and they are generally more acuminate than in those from King George's Sound. F. Mueller, Fragm. vi. 34, includes, under *Styphelia brachycephala*, specimens which I should refer to this and the three following species.

55. **L. sprengelioides,** *Sond. in Pl. Preiss.* i. 319. An erect shrub, with rather slender glabrous or scarcely pubescent branches. Leaves erect, sessile, ovate obovate or lanceolate, the larger ones very concave, embracing the stem to the middle, shortly spreading upwards and nearly 2 lines long, but the greater number crowded, very obtuse and scarcely 1 line long. Spikes short, dense, terminal or in the uppermost axils. Lower bracts leaf-like, the upper ones small; bracteoles very obtuse, about half as long as the calyx. Sepals 1 line long or rather more, obtuse, minutely ciliate and sometimes pubescent. Corolla about 2 lines long, the lobes longer than the tube. Anthers attached below the prominent sterile tips. Hypogynous disk truncate or sinuate-lobed. Ovary glabrous, 5-celled or rarely 4-celled; style very short.

W. Australia. Swan River, *Drummond, 1st Coll. n.* 27 *and* 461; York district, *Preiss, n.* 397. Very near *L. cucullatus,* differing chiefly in the much smaller leaves.

L. parvifolius, Sond. in Pl. Preiss. i. 319, but not of DC., appears to me to be the same as *L. sprengelioides,* differing from *L. parvifolius,* DC. (which is *L. tamariscinus*), in the leaves not acuminate, the short spikes, and the 3-celled not 2-celled ovary.

56. **L. obtusatus,** *Sond. in Pl. Preiss.* i. 313. Erect with numerous short erect pubescent branchlets. Leaves sessile, erect, mostly imbricate, ovate-oblong, very obtuse, thick, concave, the veins not prominent, about 1 line long. Flowers few in short dense terminal spikes or rarely also in the uppermost axils. Lower bracts like the leaves but smaller; bracteoles broad, obtuse, not half so long as the calyx. Sepals about 1 line long, obtuse, with scarious margins. Corolla about 2 lines long, the lobes longer than the tube. Anthers attached under the rather long sterile tips. Hypogynous disk shortly lobed. Ovary glabrous, 3-celled; style very short.—*L. brevifolius,* Stschegl. in Bull. Mosc. 1859, i. 17.

W. Australia, *Drummond, 5th Coll. n.* 322; Mount Bakewell, *Preiss, n.* 395 (partly).

Var. *elachophyllus,* F. Muell. Leaves narrow, more conspicuously ribbed underneath, the flowers rather smaller and the sepals and bracts usually ciliate.—Near Israelite Bay, *Maxwell,* also in Drummond's collection.

57. **L. fimbriatus,** *Stschegl. in Bull. Mosc.* 1859, i. 17. An erect shrub of 1 to 2 ft., the branches pubescent. Leaves erect and closely appressed, ovate or oblong, very obtuse, concave and almost embracing the stems, usually ciliate, striate but not so prominently as in *L. ozothamnoides,* rarely exceeding 1 line, occasionally opposite. Spikes terminal, short, dense and few-flowered. Bracts leaf-like but shorter and broader than the stem-leaves; bracteoles very broad with prominent keels, fully half as long as the calyx. Sepals about 1 line long, obtuse, with scarious ciliate margins. Corolla about 2 lines long, the lobes longer than the tube. Anthers attached below the short sterile tips. Hypogynous disk truncate. Ovary short, glabrous, 3-celled. Style very short.

W. Australia, *Drummond,* 3*rd Coll. n.* 187. This and the preceding species are referred by F. Mueller, Fragm. vi. 34, but with doubt, to *L. brachycephalus,* DC. (*L. cucullatus,* Br.).

58. **L. ozothamnoides,** *F. Muell. Herb.* Very near *L. fimbriatus,* and perhaps a variety. Leaves ovate, closely appressed and embracing the stem and occasionally opposite, as in that species, but larger, more prominently striate, 1 to 2 lines long. Inflorescence, flowers, anthers, and disk the same. Ovary similarly 3-celled, but densely hirsute with long hairs.

W. Australia. Dry sandy situations near Kinderup, *Oldfield.*

SECT. II. HETERANTHESIS. Spikes or clusters terminal and sometimes also in the uppermost axils. Anthers obtuse or emarginate, without sterile tips.

In this section the inflorescence is nearly that of *Perojoa,* whilst the anthers are those of *Pleuranthus.* The first rather anomalous species connects the section with *Perojoa,* several of the others have the aspect of *Lissanthe* or of *Cyathodes.*

59. **L. plumuliflorus,** *F. Muell. Fragm.* vi. 29. A weak shrub of 1 to 2 ft., the branches and foliage more or less pubescent and hirsute with rigid hairs. Leaves broadly ovate-cordate, very obtuse or with a minute callous recurved point, convex with recurved margins, under $\frac{1}{2}$ in. long. Spikes solitary, terminal, contracted into an almost globular plumose head. Bracts small, ovate, membranous; bracteoles ovate-lanceolate, acute, $\frac{3}{4}$ line long, ciliate-hirsute. Sepals $2\frac{1}{2}$ lines long, the lower part lanceolate, the rest narrow-linear and plumose-hirsute. Corolla nearly $1\frac{1}{2}$ lines long, the lobes shorter than the tube. Anthers oblong, attached near the top, emarginate, without sterile tips. Hypogynous disk lobed. Ovary 2-celled; style very short.—*Styphelia plumuliflora,* F. Muell. Fragm. vi. 29.

W. Australia. Between Moore and Murchison rivers, *Drummond,* 6*th Coll. n.* 122.

60. **L. unilateralis,** *Stschegl. in Bull. Mosc.* 1859, i. 19. A tall and erect or low and spreading but neat-looking shrub, glabrous or the branches and sometimes the foliage minutely pubescent. Leaves oblong-linear, obtuse or with a small callous point, flat or slightly convex, finely nerved underneath, $\frac{1}{4}$ to $\frac{1}{2}$ in. long. Spikes or racemes loose but few-flowered, terminal and in the uppermost axils, all turned to one side and more or less pendulous, the flowers very shortly pedicellate within the small broad bracts. Bracteoles not half so long as the calyx, obtuse or almost acute. Sepals 1 to $1\frac{1}{4}$ lines long, acute. Corolla $2\frac{1}{2}$ to nearly 3 lines long, the tube longer than the calyx, the lobes shorter. Anthers attached above the middle, obtuse without sterile tips. Hypogynous disk undulate or obtusely lobed. Ovary 2-celled, glabrous; style rather long.—*L. acutiflorus,* Stschegl. in Bull. Mosc. 1859, i. 18.

W. Australia, *Drummond,* 5*th Coll. n.* 305, 306 (these numbers applied to other plants in some sets); near the top of Mount Bland, *Maxwell;* Stirling Range, *F. Mueller.* The two forms distinguished by Stschegleev differ slightly in the size of the flower; the other characters given by him do not hold good in our specimen. F. Mueller refers the species as a variety to *L. pendulus.*

61. **L. Hookeri,** *Sond. in Linnæa,* xxvi. 248. A low diffuse or bushy

shrub. Leaves oblong-linear, obtuse or with an obscure callous point, shortly petiolate, flat or with recurved margins, few-nerved and glaucous underneath, rarely $\frac{1}{2}$ in. long and often not above $\frac{1}{4}$ in. Flowers few, in short terminal spikes. Bracts and bracteoles broad, about half as long as the calyx. Sepals very obtuse, $\frac{1}{2}$ to $\frac{3}{4}$ line long. Corolla about $1\frac{1}{2}$ lines long, the lobes about as long as the tube. Anthers attached above the middle, obtuse, without sterile tips. Hypogynous disk shortly lobed. Ovary 5-celled; style very short. Fruit small, nearly globular.—Hook. f. Fl. Tasm. i. 251. t. 75 B; *L. obtusatus*, Hook. f. in Hook. Lond. Journ. vi. 269, not of Sond.

N. S. Wales. Ben Lomond, New England, *Beckler ;* Upper Hastings river, *Moore and Carron.*

Victoria. Common in the Haidinger Range, Mount Barkley, Mount Buller, and others of the Australian Alps, *F. Mueller.*

Tasmania. Abundant in alpine situations throughout the island, *J. D. Hooker.*

The flowers are often partially diœcious, the males having longer anthers, a less perfect ovary, and a more developed disk than the females. In this respect, as well as in foliage and inflorescence, the species closely resembles *Lissanthe montana,* with which F. Mueller unites it under the name of *Styphelia montana,* Fragm. vi. 45.

62. **L. Macræi,** *F. Muell. in Trans. Phil. Soc. Vict.* i. 106, *and in Hook. Kew Journ.* viii. 163. An erect rigid shrub of 6 to 8 ft., with pubescent branches. Leaves rather crowded, spreading, petiolate, ovate to ovate-lanceolate, obtuse or with a small callous point, flat, green on both sides, shining above, mostly about $\frac{1}{4}$ in. long. Spikes few-flowered, but often exceeding the leaves, terminal or in the uppermost axils. Bracts and bracteoles very obtuse, often ciliolate, nearly half as long as the calyx. Sepals about $1\frac{1}{4}$ lines long, often but not always ciliolate. Corolla-tube rather broad, scarcely exceeding the calyx ; lobes rather shorter. Anthers attached near the top, obtuse, without sterile tips. Hypogynous disk broad, shortly lobed. Ovary 5-celled; style rather short, thickened towards the base. Fruit small, nearly globular.—*Styphelia Macræi,* F. Muell. Fragm. vi. 46.

Victoria. In valleys at the sources of the Mitta-Mitta, near Mount Hotham, Mount Latrobe, along torrents in the Cobberas mountains, at an elevation of 5000 to 6000 ft., in the Baw-Baw Range descending to 3500 ft., *F. Mueller.*

63. **L. pleurandroides,** *F. Muell. Fragm.* iii. 143. A stout, spreading, scrubby shrub of about 1 ft., glabrous or minutely pubescent. Leaves on slender petioles of from $\frac{1}{2}$ to 1 line, broadly oblong, very obtuse at both ends, thick, convex above, with 2 longitudinal furrows underneath, mostly about 2 lines long. Flowers few together, in dense, terminal spikes or clusters, scarcely exceeding the leaves, and the rhachis shorter than the petioles. Bracts very small ; bracteoles not half so long as the calyx. Sepals broad, obtuse, coloured at the end, the margins ciliate, about 1 line long. Corolla-tube about 2 lines long ; lobes not above 1 line. Anthers attached above the middle, oblong, obtuse, without sterile tips. Hypogynous disk short, sinuate. Ovary very hairy, 1-celled, with a single ovule suspended from a slender funicle ; style long and slender, sprinkled with a few long hairs.—*Styphelia pleurandroides,* F. Muell. Fragm. vi. 32.

W. Australia. Moir's Inlet, *Maxwell.* I have dissected several buds as well as open flowers without ever finding more than a single ovule, but the corolla and habit are entirely those of *Leucopogon*, not of *Monotoca*.

64. **L. melaleucoides,** *A. Cunn. ; DC. Prod.* vii. 750. An erect, robust shrub of several feet, the branches usually minutely pubescent. Leaves oblong-linear, tapering into a pungent point, flat or nearly so, smooth and shining, rarely exceeding ½ in. Spikes short, terminal, with occasionally a few single flowers in the uppermost axils. Bracts and bracteoles very small. Sepals obtuse or mucronate, rather above 1 line long. Corolla-tube shorter than the calyx ; lobes as long as the tube. Anthers attached at the top, very obtuse, without sterile tips. Hypogynous disk 5-lobed. Ovary 5-angled, 5-celled.—*L. linifolius*, A. Cunn. ; DC. Prod. vii. 747 ; *Styphelia linifolia*, F. Muell. Fragm. vi. 36.

Queensland· Barren heaths near Redcliffe point, Moreton Bay, *A. Cunningham ;* Mount Lindsay, *Fraser.*

N. S. Wales. Hunter's River, *A. Cunningham ;* Hastings and Macleay rivers, *Beckler ;* New England, *C. Stuart.*

This species has the foliage nearly of *L. leptospermoides*, but it is readily distinguished by the inflorescence.

65. **L. pluriloculatus,** *F. Muell. Fragm.* i. 37. A small but robust, erect, bushy shrub, the branchlets pubescent. Leaves crowded, linear, rigid, tapering into a pungent point, with revolute margins, mostly about ½ in. long. Flowers in short, terminal spikes or racemes, with a few occasionally in the uppermost axils, each flower shortly pedicellate within the small, subtending bract and bracteoles. Sepals obtuse, ¾ to nearly 1 line long. Corolla-tube shortly exceeding the calyx ; lobes nearly as long as the tube. Anthers attached near the top, oblong, very obtuse, without sterile tips. Hypogynous disk lobed. Ovary short, hairy, 5- to 7-celled. Fruit small, depressed-globular.—*Styphelia pluriloculata*, F. Muell. Fragm. vi. 32.

Queensland. Burnett river, *F. Mueller ;* in the interior, *Mitchell.*
N. S. Wales. Near Camden, *Leichhardt.*

66. **L. pleiospermus,** *F. Muell. Fragm.* vi. 41. An apparently erect, bushy shrub, with pubescent branches. Leaves oblong-linear, rather obtuse, with a short, rigid point, concave, finely veined underneath, mostly about ½ in. long. Racemes terminal or in the uppermost axils, rather loose, but rarely exceeding the leaves, the flowers small and pedicellate. Bracts and bracteoles small, broad, obtuse, striate, all at the base of the pedicels. Sepals about ½ line long, broad, obtuse, striate. Corolla 1½ lines long, the tube twice as long as the calyx, with the exserted part campanulate, the lobes short. Anthers attached near the top, obtuse, without sterile tips. Hypogynous disk truncate. Ovary short, broad, 7- to 10-celled ; style short. Fruit depressed-globular, smooth, about 2 lines diameter.—*Styphelia pleiosperma*, F. Muell. Fragm. vi. 41.

N. S. Wales, *Leichhardt ;* Darling Downs, *F. Law.*

67. **L. rubicundus,** *F. Muell. Herb.* (*not of Fragm.* iv. 99). A glabrous shrub of about 1 ft., with rather slender, but rigid branches. Leaves petiolate, linear or linear-lanceolate, tapering into a pungent point, rigid,

convex, with the margins and midrib thickened so as to be 2-furrowed underneath, mostly about ¼ in. long. Flowers apparently red, erect, in very short, terminal spikes or racemes, on pedicels of about ½ line, the bracts and bracteoles at the base of the pedicels not above half so long. Sepals broadly ovate, very obtuse, coloured, about ½ line long. Corolla-tube cylindrical, about 2 lines long, hairy inside about the middle, glabrous at the base and in the throat, the lobes bearded above the middle, more than 1 line long. Anthers attached above the middle, oblong, obtuse, without sterile tips. Hypogynous disk sinuate or obtusely lobed. Ovary slightly hairy, 3-celled (or 4-celled?); style slender, hairy.—*Cyathodes rubicunda,* F. Muell. Fragm. iv. 99; *Styphelia rubicunda,* F. Muell. Fragm. vi. 31.

W. Australia. Sandy plains between Point Malcolm and Point Culver, *Maxwell.* This and the preceding two species, with their pedicellate flowers, come very near to the first two species of *Lissanthe,* but with bearded corolla-lobes and drier fruits. In transferring this species from *Cyathodes* to *Leucopogon* I have been able to retain F. Mueller's specific name, as the other species of *Leucopogon* to which he had given it proves to be Sonder's *L. oxycedrus.*

SECT. III. PLEURANTHUS.—Spikes all axillary, few-flowered or reduced to a single flower besides the rudimentary one, the common peduncle very short or rarely as long as the leaves. Anthers obtuse or emarginate, without sterile tips. Style usually slender and elongated, rarely very short.

SERIES 1. CONFERTÆ.—Leaves small (1 to 2 lines long), with recurved margins. Flowers mostly solitary besides the rudiment, and nearly sessile, but often forming dense, leafy spikes along the branchlets. Ovary 5-celled.

68. **L. attenuatus,** *A. Cunn. in Field. N. S. Wales,* 341. A shrub, with long, spreading branches and numerous short branchlets, usually minutely pubescent as well as the foliage or sometimes hoary-villous. Leaves from ovate to oblong or lanceolate, mucronate, flat or with recurved margins, mostly about 1 line, or more rarely 2 lines long. Flowers solitary in each axil, with the rudiment of a second, or rarely 2, and often so crowded along the branchlets as to form leafy spikes. Peduncles exceedingly short. Bracts minute; bracteoles not half so long as the calyx. Sepals obtuse, 1 to 1¼ line long, obtuse. Corolla 1½ to 2 lines, the lobes as long as the tube. Hypogynous disk small, lobed. Anthers attached near the top, obtuse, without sterile tips. Ovary 5- or rarely 4-celled, striate; style short. Fruit striate, scarcely exceeding the calyx, often ripening only a single seed.— DC. Prod. vii. 752; *L. reclinatus,* A. Cunn.; DC. l. c.; *L. recurvatus,* A. Cunn.; DC. l. c. 754 and (the more villous specimens) *L. mucronatus,* DC. l. c. 751, and *L. ramulosus,* A. Cunn.; DC. l. c. 753.

N. S. Wales. Barren hills in the interior, near Cox's River, Daly's Plains, between the Lachlan and Macquarrie rivers, *A. Cunningham.* The dense white beards of the corolla and crowded flowers give this plant often the aspect of the section *Perojoa,* but the flowers are really all axillary, not in terminal spikes, and the anthers without sterile tips.

69. **L. confertus,** *Benth.* A shrub, with the habit, foliage, and inflorescence of *L. attenuatus,* and softly pubescent as in some varieties of that species. Leaves oblong or oblong-linear, with a very fine, rigid point, the

margins recurved, 1 to 2 lines long. Flowers rather larger than in *L. attenutus*, and as in that species solitary or 2 together in each axil, but crowded so as to form leafy spikes. Bracteoles mucronate-acute, more than half as long as the calyx. Sepals narrow, very acute, nearly 2 lines long. Corolla fully 2 lines long, the tube short, the lobes longer, with very white beards. Anthers attached near the top, obtuse, without sterile tips. Hypogynous disk very short. Ovary 5-celled, tapering into the style.

N. S. Wales. New England, *C. Stuart.*

SERIES 2. ERICOIDEÆ.—Leaves narrow ($\frac{1}{8}$ to 1 in.) except in *L. concinnus*, with recurved or revolute margins. Flowers 2 or more together, in sessile or shortly pedunculate axillary clusters. Ovary 5- or 3-celled.

70. **L. muticus,** *R. Br. Prod.* 543. An erect shrub, glabrous or the branches minutely pubescent. Leaves oblong-linear or oblanceolate, obtuse or with a minute, callous point, flat or the margins slightly recurved, mostly about $\frac{1}{2}$ in. long, the lower ones sometimes short and broad. Spikes axillary, few-flowered, shorter than the leaves. Bracts and bracteoles broad, obtuse, about half as long as the calyx. Sepals about 1 line long or scarcely more, rather narrow, obtuse. Corolla-tube about as long as the sepals; lobes as long as the tube. Anthers attached near the top, obtuse, without sterile tips. Hypogynous disk separable into truncate scales. Ovary oblong, 5-celled, tapering into the rather long style. Drupe oblong, glabrous or hairy, twice as long as the calyx, very obtuse, prominently 5-angled.—DC. Prod. vii. 747 ; *L. appressus,* Sieb. Pl. Exs., not of R. Br. ; *Styphelia mutica,* F. Muell. Fragm. vi. 45.

N. S. Wales. Port Jackson and Blue Mountains, *R. Brown, Sieber, n.* 101, *Fl. Mixt. n.* 495, and many others. There appears to be a slight degree of unisexuality in the flowers.

71. **L. ericoides,** *R. Br. Prod.* 543. A heath-like shrub, sometimes low and diffuse, sometimes erect and attaining several feet, glabrous or the branches and even the foliage more or less pubescent. Leaves mostly oblong-linear, obtuse, mucronate, $\frac{1}{4}$ to $\frac{1}{2}$ in. long, but sometimes narrow and acute, or, especially the lower ones, small broadly oblong or even ovate, always with recurved or revolute margins. Flowers few together, in close axillary clusters or spikes, rarely exceeding the leaves, but the spikes sometimes so numerous as to form long, dense, leafy racemes. Bracteoles about half as long as the calyx. Sepals scarcely 1 line long and sometimes shorter, narrow, mucronulate. Corolla variable in size, usually about 2 lines long, the lobes rather longer than the tube. Anthers linear, attached near the top, obtuse, without sterile tips. Hypogynous scales free or slightly cohering. Ovary 5-celled, pubescent or hairy, or rarely glabrous; style rather long. Fruit small, ovoid-oblong, often curved when partially abortive.—DC. Prod. vii. 747; Hook. f. Fl. Tasm. i. 250 ; *Styphelia ericoides,* Sm. Pl. N. Holl. 48 ; F. Muell. Fragm. vi. 45 ; *Epacris spuria,* Cav. Ic. iv. 27. t. 347 (not good) ; *Styphelia spuria,* Poir. Dict. vii. 485 ; *S. trichocarpa,* Labill. Pl. Nov. Holl. i. 47. t. 66 (not good) ; *Leucopogon trichocarpus,* R. Br. Prod. 543 ; DC. Prod. vii. 747.

Queensland. Moreton Island, *F. Mueller.*
N. S. Wales. Port Jackson to the Blue Mountains, *R. Brown, Sieber, n.* 105, *Fl. Mixt. n.* 499, and others ; brushy forest land in the interior, noith of Bathurst, *A. Cunningham ;* near Berrima, *Woolls.*
Victoria. Common on dry hills, from the Glenelg, *Robertson,* to Wilson's Promontory, *F. Mueller ;* in the interior on Victoria Range and the Grampians, and ascending Mount Cobberas to 6000 ft., *F. Mueller ;* Wimmera, *Dallachy.*
Tasmania. Port Dalrymple and Derwent river, *R. Brown ;* very abundant throughout the island in dry heaths, etc., *J. D. Hooker.*
S. Australia. Near Penola, *Woods ;* near Mount Gambier, *F. Mueller (Fragm. l. c.).*

72. **L. brevicuspis,** *Benth.* An erect, bushy shrub, glabrous or the branches and foliage more or less pubescent. Leaves broadly oblong or almost obovate-oblong, minutely but rigidly mucronate, convex, with recurved margins or nearly flat, mostly about $\frac{1}{2}$ in. long. Spikes axillary, very short, 2- or 3-flowered, erect or at length recurved. Bracts small, minutely mucronate. Bracteoles mucronate-acuminate, about half as long as the calyx. Sepals 2 lines long, softly pubescent, mucronate-acute. Corolla-tube as long as the calyx ; lobes rather shorter, erect at the base. Anthers obtuse, without sterile tips. Hypogynous disk large, with acuminate lobes. Ovary glabrous, 5-celled. Fruit ovoid or ovoid-oblong, about 3 lines long, the endocarp very hard.

W. Australia, *Drummond,* 2nd *Coll. n.* 249 ; Stirling Range, *F. Mueller.* Very near *L. propinquus,* but with the leaves almost of *L. Richei.*

73. **L. propinquus,** *R. Br. Prod.* 543. An erect, rigid shrub of 3 or 4 ft., glabrous or the branches minutely pubescent. Leaves linear, rigid, with a short, almost pungent point, convex, with recurved margins, $\frac{1}{2}$ to 1 in. long. Peduncles axillary, very short, erect or scarcely spreading, bearing usually 2 or 3 flowers, but sometimes 4 or 5, or only 1 besides the rudiment. Bracts small ; bracteoles not half so long as the calyx, all with a fine, rigid point. Sepals nearly $1\frac{1}{2}$ lines long, dry, mucronate. Corolla-tube shorter than the calyx ; lobes longer than the tube, much revolute. Anthers attached above the middle, obtuse or notched, without sterile tips. Hypogynous disk toothed, readily separating into distinct scales. Ovary 5-celled ; style rather long, with a broad stigma. Fruit from nearly globular to ovoid-oblong, 4 lines long, with a thick, hard endocarp.—DC. Prod. vii. 748 ; *Styphelia propinqua,* Spreng. Syst. i. 658 ; *L. pungens,* Sond. in Pl. Preiss. i. 324 ; *Styphelia pungens,* F. Muell. Fragm. vi. 34.

W. Australia. King George's Sound, *R. Brown, A. Cunningham,* and others ; Torbay and Cape Riche, *Oldfield ;* Gordon river, *Maxwell ;* Cape Naturaliste, *Collie ;* Swan River, *Fraser, Oldfield, Preiss, n.* 366, 373.

This has the foliage and nearly the inflorescence of *L. racemulosus,* but is readily known by the more compact spikes, more erect flowers, and short corolla-tube. The var. *abbreviata,* indicated by F. Mueller with short leaves and obtuse bracts, appears to me to be the *L. insularis.*

74. **L. insularis,** *A. Cunn. ; DC. Prod.* vii. 754. A rigid, scrubby, much-branched shrub of about 2 ft., glabrous or the branches pubescent. Leaves very spreading or reflexed, linear or rarely oblong, rigid, with a pungent point, the margins revolute, mostly 3 or 4 lines long. Peduncles axillary, very short, bearing 1 or 2 or rarely 3 to 5 flowers, erect or scarcely

spreading. Bracts very small ; bracteoles not half so long as the calyx, all obtuse. Sepals scarcely above 1 line long, dry, obtuse or rarely with a minute not pungent point. Corolla-tube about as long as the calyx ; lobes as long as or rather longer than the tube, sometimes cohering at the base so as to look like part of the tube till the flowering is advanced. Anthers attached above the middle, obtuse, without sterile tips. Hypogynous disk shortly lobed, readily separating into distinct scales. Ovary 5-celled ; style slender.—*L. subulatus*, F. Muell. Fragm. iv. 103 ; *Styphelia subulifolia*, F. Muell. Fragm. vi. 33 ; *L. oblongifolius*, Sond. in Pl. Preiss. i. 323 (with rather broader leaves).

W. Australia. Rottenest Island, *A. Cunningham ;* towards the Great Bight, *Maxwell ;* also in Drummond's and Preiss's collections. The flowers are generally solitary or 2 together in Cunningham's, Preiss's, and Maxwell's specimens, often 3 to 5 in Drummond's (which are in bud only). Maxwell's have the leaves more closely revolute than the others. All are very near *L. propinquus*, and perhaps a variety, with shorter, more sessile, more revolute, and more pungent leaves, the bracteoles and sepals, on the contrary, much more obtuse.

75. **L. Allittii,** *F. Muell. Fragm.* iv. 103. A rather stout, rigid, glabrous shrub, about 1 ft. high and not much branched (*Oldfield*). Leaves sessile, from ovate to ovate-lanceolate or lanceolate, with a short, almost pungent point, the margins much revolute, mostly about ½ in. long. Peduncles axillary, short, at length recurved, bearing 2 to 4 flowers, shortly pedicellate within the small bracts. Bracteoles broad, obtuse, not half so long as the calyx. Sepals scarcely above 1 line long, dry, obtuse. Corolla nearly 4 lines long, the lobes as long as the tube. Anthers obtuse, without sterile tips. Hypogynous disk obscurely lobed or crenulate, readily separating into distinct scales. Ovary 5-celled ; style elongated.—*Styphelia Allittii*, F. Muell. Fragm. vi. 34.

W. Australia. Murchison river, *Oldfield.* Differs from *L. racemulosus* chiefly in the breadth of the leaves.

76. **L. racemulosus,** *DC. Prod.* vii. 747. An erect, rigid shrub, sometimes low, with almost simple, erect stems, sometimes several feet high and branching, usually glabrous. Leaves linear or linear-lanceolate, rather rigid, with a short, usually pungent point, the margins revolute, ½ to 1 in. long. Peduncles axillary, very short, often at length recurved, bearing 2 to 5 flowers, shortly pedicellate within the small bracts. Bracteoles not one-third so long as the calyx. Sepals scarcely 1 line long, dry, rather narrow, but not acute. Corolla-tube fully 2 lines long, or, including the erect base of the lobes, which have the appearance of a continuation of the tube, 3 lines long, the lobes only shortly spreading above that. Anthers attached above the middle, obtuse, without sterile tips. Hypogynous scales ovate, acuminate. Ovary 5-celled ; style elongated. Fruit hard, globular, about 2 lines diameter.—Sond. in Pl. Preiss. i. 312 (excl. the var. *β*) ; *Styphelia racemulosa*, F. Muell. Fragm. vi. 33.

W. Australia. Swan River and thence to King George's Sound, *Fraser, Drummond, 1st Coll., 2nd Coll. n.* 267, *Preiss, n.* 369, and several others ; between Swan and Murchison rivers, *Oldfield.* The long falcate fruits described by F. Mueller from some of Oldfield's specimens (all probably from one bush) appear to me to be monstrous ; the hard portion contains the 5 consolidated abortive cells, and the lateral cavity on the convex side is outside of them. I find it always occupied by a long, black, loose body, apparently without any remaining organization, probably a decayed grub.

77. **L. pendulus,** *R. Br. Prod.* 545. An erect, heath-like, bushy shrub, attaining sometimes 3 or 4 ft., with numerous, slender, glabrous or minutely pubescent branches. Leaves erect or scarcely spreading, oblong-linear, obtuse or with a short, callous point, convex or with revolute margins, 2 to 4 lines or rarely nearly ½ in. long. Peduncles axillary, 1 to 2 lines long, recurved, 1- or 2-flowered. Bracts minute; bracteoles not half so long as the calyx, very obtuse. Sepals about 1 line long, dry, obtuse, not very broad. Corolla-tube about as long as the calyx; lobes twice as long, almost cohering at the base into a campanulate throat, with recurved ends. Anthers attached above the middle, linear, obtuse, without sterile tips, long in some specimens, short in others. Hypogynous disk toothed. Ovary oblong, 5-celled; style rather long. Fruit pendulous, ovoid-oblong, twice or three times as long as the calyx.—DC. Prod. vii. 751; *Styphelia pendula,* Spreng. Syst. i. 657; F. Muell. Fragm. vi. 33 (partly); *L. secundiflorus,* Sond. in Pl. Preiss. i. 320.

W. Australia. King George's Sound, *R. Brown,* and thence along the range to Cape Riche, *Drummond, 2nd Coll. n. 268, Preiss, Oldfield, Maxwell, F. Mueller.*

Var. *cuspidatus,* F. Muell. Leaves with a more prominent, rigid point. Flowers rather smaller.—*L. psilopus,* Stschegl. in Bull. Mosc. 1859, i. 19. *Drummond, 5th Coll. n.* 313, the flowers not yet expanded.

F. Mueller includes, as a pluriflorous variety, *L. unilateralis,* which appears to me to be well characterized by the inflorescence as well as by other peculiarities.

78. **L. concinnus,** *Benth.* A low, erect, very much-branched shrub, glabrous or the branches slightly pubescent. Leaves very spreading or reflexed, ovate or oblong, obtuse or with a small point, convex, shining, 1 to 2 lines long. Peduncles axillary, very short, recurved, 1- or 2-flowered. Bracts very small; bracteoles not half so long as the calyx. Sepals under 1 line long, obtuse or almost acute. Corolla-tube about 1 line long; lobes about as long as the tube. Anthers attached above the middle, linear, obtuse, without sterile tips. Hypogynous disk 5-lobed, separable into distinct scales. Ovary 3-celled or sometimes 2-celled; style slender.

W. Australia. Between King George's Sound and Swan River, *Harvey, Drummond;* near Albany, Kojonup, and table land in the interior from Eyre's Relief, *Maxwell.* Included by F. Mueller in *L. pendulus;* but, besides the difference in the foliage, I have never found more than 3 cells to the ovary, and in some specimens, especially of Harvey's and Drummond's, only 2 cells.

Series 3. Micranthæ.—Leaves oblong or lanceolate, nearly flat or with recurved margins. Ovary 2-celled. Flowers small and nearly sessile.

79. **L. margarodes,** *R. Br. Prod.* 542. A weak, straggling shrub, with pubescent branches. Leaves oblong-linear or oblanceolate, obtuse or with a minute callous point, the margins recurved, rarely exceeding ½ in. and mostly under that length. Peduncles axillary, very short, bearing 2 or 3 flowers or sometimes only 1 besides the rudiment. Bracts very small; bracteoles broad, obtuse, scarcely half as long as the calyx. Sepals about 1 line long, acute. Corolla-tube much shorter than the calyx; lobes narrow, twice as long as the tube. Anthers attached near the top, obtuse, without

sterile tips. Hypogynous disk rather long. Ovary oblong, compressed, 2-celled, tapering into a rather long style. Fruit oblong, obtuse, nearly 3 lines long, succulent angular and sterile at the base, the remaining seed-bearing portion striate and compressed.—DC. Prod. vii. 747 ; *Styphelia margarodes,* Spreng. Syst. i. 657 ; F. Muell. Fragm. vi. 36.

Queensland. Sandy Cape, Harvey Bay, *R. Brown ;* Stradbrooke Island, *Fraser ;* Moreton Island, *F. Mueller.*

N. S. Wales. Near Newcastle, *Leichhardt.*

80. **L. flavescens,** *Sond. in Pl. Preiss.* i. 322. A shrub of 2 to 4 ft., with erect branches, minutely pubescent, the foliage of a pale yellowish hue when dry. Leaves rather crowded, erect, oblong-linear, obtuse or with a minute callous point, flat or nearly so, contracted into a very short petiole, $\frac{1}{4}$ to $\frac{1}{2}$ in. long. Flowers axillary, solitary on an exceedingly short pedicel, the subtending bracts and rudimentary flower of the other species entirely wanting. Bracteoles broad, ciliate, about half as long as the calyx. Sepals under 1 line long, almost acute, with thin, ciliate margins. Anthers attached above the middle, obtuse, without sterile tips. Hypogynous scales shortly acuminate, free or slightly cohering. Ovary oblong, 2-celled ; style rather long. Fruit oblong, flat, nearly 3 lines long, with 3 to 5 raised ribs on each side, and contracted into a stipes at least as long as the calyx.—F. Muell. Fragm. iv. 100 ; *Styphelia flavescens,* F. Muell. Fragm. vi. 33.

W. Australia. King George's Sound or adjoining districts, *Baxter, Preiss, n.* 379, *Drummond, 5th Coll. n.* 312.

Var. *brevifolius.* Leaves 2 to 3 lines long, thicker and more striate, the floral ones not exceeding the flowers.—*Drummond, n.* 153 ; near Mount Bland, *Maxwell.*

The subtending bracts in this species appear to be entirely deficient, as in the section *Lissanthoides* of *Brachyloma,* but the corolla-lobes are strictly valvate and bearded. The foliage and general aspect of the plant is nearly that of *L. crassifolius,* but the want of bracts, the 2-celled ovary, and flat fruit, are very different.

81. **L. blepharolepis,** *F. Muell. Fragm.* vi. 48. An erect shrub, with virgate branches, glabrous or minutely pubescent. Leaves sessile, erect, oblong-lanceolate, shortly tapering into a fine point, flat or slightly concave, of a pale colour, the veins fine and not prominent, under $\frac{1}{2}$ in. long. Flowers very small, in axillary racemes of 2 to 5. Bracts very small ; bracteoles about half as long as the calyx. Sepals under 1 line long, thin, obtuse, ciliolate. Corolla almost urceolate, nearly 2 lines long, the lobes rather longer than the tube. Anthers attached near the top, oblong, obtuse, without sterile tips. Hypogynous scales distinct, ovate, fringed with a few long hairs. Ovary oblong, truncate, 2-celled ; style rather long.—*Styphelia blepharolepis,* F. Muell. Fragm. vi. 48.

W. Australia. Towards the Great Bight, *Maxwell.* The fringed or ciliate hypogynous scales are exceptional in the genus.

82. **L. esquamatus,** *R. Br. Prod.* 546. A bushy shrub, with erect branches, glabrous or minutely pubescent. Leaves erect, sometimes imbricate, lanceolate, rigid, tapering into a fine point, flat, mostly $\frac{1}{2}$ to $\frac{3}{4}$ in. long. Peduncles axillary, very short, bearing 1 or 2 flowers besides the rudiment. Bracts and bracteoles not half so long as the calyx. Sepals dry, obtuse,

above 1 line long.   Corolla-tube very short ; lobes exceeding the calyx, more
than twice as long as the tube.   Anthers attached a little above the middle,
oblong, obtuse.   Hypogynous disk entirely wanting.   Ovary oblong, 2-
celled ; style filiform.   Fruit oblong, compressed, nearly 2 lines long.—DC.
Prod. vii. 754 ; *Styphelia esquamata,* Spreng. Syst. i. 658 ; *L. fastigiatus,*
Sieb. ; G. Don, Gen. Syst. iii. 779 ; *Styphelia fastigiata,* Spreng. Syst. Cur.
Post. 67 ; *Leucopogon appressus,* DC. Prod. vii. 754, not of R. Br. ; *Phanerandra esquamata,* Stschegl. in Bull. Mosc. 1859, i. 20.

N. S. Wales.   Port Jackson, *R. Brown, Sieber, n.* 108, and several others.

SERIES 4. PLANIFOLIÆ.—Leaves flat or slightly concave or convex, rigid,
usually shining above, the veins fine or inconspicuous. Ovary 5-celled.

83. **L. cordifolius,** *Lindl. in Mitch. Three Exped.* ii. 122.   A tall,
bushy or spreading, much-branched shrub, glabrous or the branches pubescent.   Leaves nearly sessile, spreading or reflexed, broadly ovate or orbicular,
with a small rigid point or rarely quite obtuse, more or less cordate at the
base, thick and rigid, flat or somewhat convex, finely veined underneath,
mostly about ¼ in. diameter.   Peduncles axillary, very short, with few, frequently only 2, flowers or a single one besides the rudiment.   Bracts very
small ; bracteoles very obtuse, not half so long as the calyx.   Sepals 1½ to 1¾
lines long, dry, obtuse.   Corolla-tube as long as the calyx ; lobes about as long
as the tube.   Anthers attached above the middle, obtuse, without sterile tips,
at length often reversed so as to appear quite exserted.   Hypogynous disk
large, lobed.   Ovary 5-celled ; style rather long.—*L. rotundifolius,* Sond. in
Pl. Preiss. i. 323, not of R. Br. ; *Styphelia rotundifolia,* F. Muell. Fragm.
vi. 45 ; *Acrotriche ? latifolia,* A. Cunn. ; DC. Prod. vii. 757.

N. S. Wales.   Between Murray and Lachlan rivers, *Herb. F. Mueller.*
Victoria.   On the Murray, *F. Mueller ;* Wimmera, *Dallachy.*
S. Australia.   Friendly Bay, Boston Point, *Wilhelmi,* from the Murray to St. Vin-
cent's Gulf, *F. Mueller.*
W. Australia.   Murchison river, *Oldfield.*

84. **L. rotundifolius,** *R. Br. Prod.* 546.   A stout, erect, bushy shrub,
glabrous or the branches minutely pubescent.   Leaves erect or spreading,
obovate or almost orbicular, obtuse or with a small callous point, tapering
into a distinct petiole, flat or slightly concave, mostly 3 to 4 lines long.
Peduncles axillary, exceedingly short, bearing 2 or 3 flowers or a single one
besides the rudiment.   Bracts minute ; bracteoles very obtuse, not half so
long as the calyx.   Sepals 1 to 1¼ lines long, dry, obtuse.   Corolla nearly 3
lines long, the lobes longer than the tube, at length revolute down to the
calyx.   Anthers attached above the middle, obtuse, without sterile tips.
Hypogynous disk rather long, crenulate.   Ovary 5-celled, tapering into a rather long style.—DC. Prod. vii. 752 ; *Styphelia rotundifolia,* Spreng. Syst.
i. 655.

W. Australia.   Lucky Bay and Goose Island Bay, *R. Brown,* and (probably Lucky
Bay), *Baxter.*—*L. rotundifolius,* var. *oblongatus,* Sond. in Pl. Preiss. i. 324, may possibly
be the short-leaved variety of *L. oxycedrus ;* the buds in Preiss's specimens are too young
to ascertain the proportion of the parts.

85. **L. planifolius,** *Sond. in Pl. Preiss.* i. 322.   A bushy shrub, gla-

brous or the branches minutely pubescent. Leaves narrow-oblong or oblan-ceolate, with a short, callous point, contracted at the base or very shortly petiolate, flat or slightly convex, often glaucous underneath, in some speci-mens 3 to 4 lines, in others about ½ in. long. Peduncles axillary, very short, erect, bearing 2 or 3 flowers. Bracts very small; bracteoles obtuse, not half so long as the calyx. Sepals 1 to 1¼ lines long, dry, almost acute. Corolla-tube rather shorter than the calyx; lobes scarcely so long as the tube. Anthers attached above the middle, obtuse, without sterile tips Hypogy-nous scales narrow, acuminate, free (at least in old flowers). Ovary 5-celled; style rather long. Fruit obovoid or oblong, about 3 lines long.—*L. mega-carpus,* F. Muell. Fragm. iv. 102; *Styphelia megacarpa,* F. Muell. Fragm. vi. 32.

W. Australia. Swan River, *Drummond, n.* 30, *Preiss, n.* 415; Murchison river *Oldfield.*

86. **L. ruscifolius,** *R. Br. Prod.* 545. Erect and bushy, glabrous or nearly so. Leaves from broadly obovate to oblong-elliptical, obtuse or acute, but always with a short, rigid point, contracted at the base, slightly concave, smooth and shining, ¼ to ½ in. long. Peduncles axillary, very short, bearing 1 or 2 flowers besides the rudiment. Bracts very small; bracteoles broad, truncate with a minute point, not half so long as the calyx. Sepals nearly 1½ lines long, broad but almost acute, striate. Corolla about 2 lines long, the lobes very acute, longer than the tube. Anthers attached near the top, ob-tuse, without sterile tips. Hypogynous disk large, truncate. Ovary broad, flat-topped, 5-angled or almost 10-ribbed, 5-celled; style short. Fruit ovoid-oblong, twice as long as the calyx.—DC. Prod. vii. 752; *Styphelia ruscifolia,* Spreng. Syst. i. 656.

Queensland. Cape York and Lizard Island, *M'Gillivray;* Endeavour river, *Banks and Solander;* Port Bowen, Percy Island, *A. Cunningham.*

87. **L. imbricatus,** *R. Br. Prod.* 545. An erect shrub, of about 1½ ft., with divaricate branches, usually glabrous. Leaves crowded, erect and often imbricate, sessile but often contracted at the base, obovate-oblong, obtuse, but with a fine rigid point, slightly concave, under ½ in. long. Peduncles axillary, very short, bearing 1 or 2 flowers besides the rudiment. Bracts very small; bracteoles very broad, obtuse, not half so long as the calyx. Sepals 1½ lines long, dry, obtuse. Corolla-tube as long as the calyx; lobes as long as the tube. Anthers attached above the middle, obtuse, without sterile tips. Hypogynous disk crenate. Ovary 5- or sometimes 4-celled.—DC. Prod. vii. 752; *Styphelia imbricata,* Spreng. Syst. i. 656.

Queensland. Northumberland Island, *R. Brown (Herb. R. Brown).* This species has much larger flowers than *L. ruscifolius* and *L. cuspidatus,* which it otherwise resem-bles. *L. dasystylis,* Sond. in Pl. Preiss. i. 325, from the Paris Herbarium, appears to be the same as *L. imbricatus,* but the specimens I have seen are but fragments.

88. **L. cuspidatus,** *R. Br. Prod.* 545. An erect or spreading much-branched shrub, from under 1 to 3 or 4 ft. high, glabrous or the branches minutely pubescent. Leaves oblong or oblanceolate, contracted at the base and almost petiolate, shortly tapering into a fine rigid point, flat, shining, 3 to 4 lines long. Peduncles axillary, very short, bearing 1 or 2 small flowers

besides the rudiment. Bracts very small; bracteoles broad, obtuse, not half so long as the calyx. Sepals 1 to 1½ lines long, dry, acute or mucronate. Corolla nearly 3 lines long, the tube shorter than the calyx, the lobes longer than the tube. Anthers obtuse, without sterile tips. Ovary 5-celled; style rather long. Fruit ovoid, shortly exceeding the calyx.—DC. Prod. vii. 751; *Styphelia cuspidata,* Spreng. Syst. i. 657; *Acrotriche aristata,* Benth. in Hueg. Enum. 76; DC. Prod. vii. 757.

Queensland. Northumberland Islands, *R. Brown ;* Percy Island, *A. Cunningham ;* Rockhampton, *O'Shanesy, Dallachy* ; Warwick, *Beckler.*—Near *L. leptospermoides,* but readily known by the long fine point of the leaves and much longer flowers.

89. **L. leptospermoides,** *R. Br. Prod.* 546. An erect bushy shrub of 2 or 3 ft., the branches hoary-pubescent or rarely glabrous. Leaves oblong-linear to linear-lanceolate, tapering into a short rigid or callous point, flat or nearly so, shining, mostly about ½ in. long. Peduncles axillary, very short, bearing 1 or 2 flowers besides the rudiment. Bracts small; bracteoles about half as long as the calyx. Sepals about 1½ lines long, mucronate but scarcely acute. Corolla scarcely above 2 lines long, the lobes as long as the tube. Anthers attached near the top, obtuse, without sterile tips. Hypogynous disk truncate, readily separable into distinct scales. Ovary 5-angled, 5-celled; style short.—DC. Prod. vii. 751; *Styphelia leptospermoides,* Spreng. Syst. i. 659.

Queensland. Harvey Bay, Sandy Cape, *R. Brown ;* Moreton Island, *A. Cunningham, M'Gillivray, F. Mueller ;* towards Durval, *Leichhardt.* Also apparently a variety, with very obtuse sepals, but the specimens in bud only from Rockhampton, *Dallachy.* The species closely resembles *L. melaleucoides* in foliage, but the inflorescence is axillary, not terminal. *L. pauciflorus,* R. Br. Prod. 546; DC. Prod. vii. 752; *Styphelia pauciflora,* Spreng. Syst. i. 658, appears to be a depauperated state of *L. leptospermoides.*

90. **L. acuminatus,** *R. Br. Prod.* 545. A low spreading shrub, glabrous or the branches minutely pubescent. Leaves crowded, erect or spreading, linear-lanceolate, tapering into a fine pungent point and contracted into a short petiole, flat, under ⅓ in. long. Peduncles axillary, very short, bearing 1 or 2 small flowers besides the rudiment. Bracts very small; bracteoles broad, not half so long as the calyx. Sepals scarcely 1 line long, minutely mucronate, dry, finely striate as in *L. ruscifolius.* Corolla-tube shorter than the calyx, the lobes slightly exceeding it. Anthers obtuse, without sterile tips. Ovary 5-celled. Fruit obovoid, almost truncate, twice as long as the calyx.—DC. Prod. vii. 751; *Styphelia acuminata,* Spreng. Syst. i. 659.

N. Australia. North coast, *R. Brown,* the precise station not recorded (*Herb. R. Brown*). The specimens are not very good. It may prove to be a narrow-leaved variety of *L. ruscifolius ;* some specimens of a *Leucopogon* from the Moluccas appear to connect the two.

91. **L. flexifolius,** *R. Br. Prod.* 546. A rigid shrub of 1 to 2 ft., with numerous erect branches minutely pubescent. Leaves very crowded, linear or linear-lanceolate, tapering into a fine point, flat or concave, but much twisted when dry as in *L. glacialis,* 2 to 4 lines long. Peduncles axillary, exceedingly short, bearing 1 or 2 very small flowers besides the rudiment. Bracts very small; bracteoles about ⅓ as long as the calyx, broad,

obtuse, minutely ciliate. Sepals scarcely above ¾ line long, obtuse, minutely
ciliate. Corolla about 1 line long, the lobes as long as the tube. Anthers
attached at the top, obtuse, without sterile tips. Hypogynous disk sinuate.
Ovary 5-celled.—DC. Prod. vii. 754 ; *Styphelia flexifolia*, Spreng. Syst. i.
659.

Queensland. Shoalwater Bay, *R. Brown.*

92. **L. biflorus,** *R. Br. Prod.* 545. An erect shrub, with very spreading
branches, sometimes very straggling, glabrous or minutely pubescent.
Leaves oblong-linear or linear-lanceolate, with a fine and rigid but sometimes
very short point, flat or convex, shining above, mostly under ½ in. long. Peduncles axillary, exceedingly short, with two pendulous flowers or rarely only
one besides the rudiment. Bracteoles not half so long as the calyx. Sepals
about 1½ lines long, acute but not narrow. Corolla-tube as long as the
calyx ; lobes nearly as long as the tube. Anthers attached about the middle,
obtuse, without sterile tips. Hypogynous scales acuminate, distinct or
slightly connate at the base. Ovary 5-celled ; style rather long.—DC. Prod.
vii. 751 ; F. Muell. Fragm. iv. 104 ; *Styphelia biflora*, Spreng. Syst. i. 659 ;
L. sparsus, A. Cunn. ; DC. Prod. vii. 751 ; *L. similis*, Sond. in Pl. Preiss.
i. 321.

N. S. Wales. Port Jackson, *R. Brown ;* Liverpool plains, *A. Cunningham ;* Arbuthnot Range, *Fraser ;* Darling Downs, *F. Law, Mrs. Ford.*
Victoria. Between Fryers Creek and Elphinstone, Haidinger Range, Mount Barkly
and Mount Ligar (from the latter localities seen in bud only), *F. Mueller.*

93. **L. setiger,** *R. Br. Prod.* 545. An erect bushy or straggling shrub,
glabrous or the branches minutely pubescent. Leaves linear-lanceolate,
tapering into a fine rigid point, flat or with recurved margins, shining above,
mostly about ½ in. long. Peduncles axillary, slender, 2 to 4 lines long,
bearing 2 to 4 pendulous flowers or rarely only one besides the rudiment.
Bracts very small ; bracteoles not above ½ line long. Sepals about 2 lines
long, rigid, narrow, acute. Corolla-tube shorter than the calyx, the lobes
as long as the tube. Anthers attached about or above the middle, obtuse
without sterile tips. Hypogynous scales acuminate, free or shortly united.
Ovary 5-celled.—DC. Prod. vii. 751 ; *Styphelia setigera*, Spreng. Syst. i.
659 ; F. Muell. Fragm. vi. 45.

N. S. Wales. Port Jackson to the Blue Mountains, *R. Brown, Sieber, n.* 94, and
several others. F. Mueller unites this with *L. biflorus,* but the long peduncles and narrow
sepals appear to me to be constant, and give it a very different aspect.

94. **L. exolasius,** *F. Muell. Fragm.* vi. 34. Branches pubescent.
Leaves oblong-linear or lanceolate, tapering into a pungent point, with recurved or revolute margins, not exceeding ½ in. Peduncles axillary,
shorter than the leaves, mostly with 2 or 3 flowers. Bracts and bracteoles
about ¾ line long. Sepals rather above 2 lines, rigid, narrow, acute. Corolla villous outside, the tube rather shorter than the calyx, the lobes nearly
as long as the tube. Anthers attached above the middle, obtuse, without
sterile tips. Hypogynous scales lanceolate, acuminate, free or slightly connate at the base. Ovary 5-celled ; style rather long.—*Styphelia exolasia*, F.
Muell. Fragm. vi. 34.

N. S. Wales. Near Camden, *Leichhardt.* This is very closely allied to *L. setiger* differing in the more revolute leaves and in the hairs outside the corolla.

95. **L. Fraseri,** *A. Cunn. in Ann. Nat. Hist.* ii. 47, *not of DC.* A low diffuse or prostrate shrub, with short ascending or erect branches, glabrous or minutely pubescent. Leaves from oval-oblong and scarcely 2 lines long to linear-oblong and nearly ½ in. long, abruptly contracted into a fine rigid point, flat or convex, shining above, striate underneath. Peduncles axillary, very short, bearing only a single flower besides the rudiment. Bracts minute; bracteoles not ½ line long, broad with a minute point. Sepals about 1 line or rather more, acute. Corolla-tube broad, nearly 2 lines long, the lobes shorter. Anthers attached about the middle, obtuse, without sterile tips. Hypogynous disk deeply lobed. Ovary 5-celled, 5-angled; style rather long, usually hairy towards the base.—Hook. f. Fl. Tasm. i. 251, and Handb. N. Zeal. Fl. 178; *L. nesophilus,* DC. Prod. vii. 752; *L. Bellignianus,* Raoul, Choix, Pl. N. Zel. 18. t. 12; *Pentachondra mucronata,* Hook. f. in Hook. Lond. Journ. vi. 270; *L. Stuartii,* F. Muell.; Sond. in Linnæa, xxvi. 249.

N. S. Wales. Mudgee road, Blue Mountains, *Woolls.*
Victoria. Summit of Mount Wellington, dry banks of the Wombayn and Upper Genoa rivers, mountains on the Macalister and Mitta-Mitta, *F. Mueller.*
Tasmania. Hampshire hills, mouth of the Detention river, near Hobarton, Lake Elcho, *J. D. Hooker* and others.

The species is also in New Zealand. F. Muell. Fragm. iv. 105 and vi. 46, suggests its being a variety of *L. juniperinus,* but, besides the habit and foliage, the form of the flowers appears to me to be widely different. A. Cunningham had inadvertently, under the name of *L. Fraseri,* sent both this and *L. multiflorus* to De Candolle, who, not having the means of identifying the latter as Brown's species, selected it to represent Cunningham's name, whilst Cunningham, about the same time, but rather earlier in the precise date, published as *L. Fraseri* the New Zealand plant, for which the name must now be considered as fixed.

96. **L. hirtellus,** *F. Muell. Herb.* Branches and foliage hirsute with short spreading hairs. Leaves oblong-elliptical, tapering into a pungent point, flat or with recurved margins, 2 to 4 lines long. Peduncles axillary, very short, bearing 1 to 3 flowers. Bracts small, mucronate. Sepals 1½ lines long, very acute. Corolla-tube about as long as the calyx, the lobes as long as the tube, mucronate-acute. Anthers attached above the middle, obtuse, without sterile tips. Hypogynous disk short, truncate. Ovary ovoid, 5-celled; style rather long.

S. Australia. Encounter Bay, *Herb. F. Mueller, Whittaker in Herb. Hook.* Near *L. Fraseri,* but appears to be sufficiently distinct in indumentum, calyx, and disk; the specimens are, however, but few and small.

97. **L. ovalifolius,** *Sond. in Pl. Preiss.* i. 324. An erect bushy or straggling shrub of 1 to 2 ft., glabrous or the branches minutely pubescent. Leaves sessile, very spreading or reflexed, obovate-oblong, very shortly mucronate, convex, mostly 2 to 3 lines long. Peduncles axillary, very short and spreading or on luxuriant branches rather longer and recurved, bearing 2 or 3 flowers or only one besides the rudiment. Bracts minute; bracteoles not half so long as the calyx, obtuse or minutely mucronate. Sepals about 1 line long, dry, rather acute. Corolla 2 to 2½ lines long, the lobes longer

than the tube, but erect at the base so as to appear like part of the tube. Anthers attached above the middle, obtuse, without sterile tips. Hypogynous disk truncate or shortly lobed.   Ovary 5-celled; style long and slender.

W. Australia. *Drummond, 1st Coll. n.* 483 ; sandy plain of Quangen, *Preiss, n.* 417 ; Murchison river, *Oldfield.*

98. **L. oxycedrus,** *Sond. in Pl. Preiss.* i. 321.   A shrub of 1 to 2 ft., with erect or more frequently spreading branches, glabrous or minutely pubescent.   Leaves linear oblong or oblanceolate, or when short almost obovate, shortly tapering into a small pungent point, narrowed at the base but usually sessile, flat or slightly convex, finely veined underneath, $\frac{1}{4}$ to $\frac{1}{2}$ in. long.   Peduncles axillary, short, bearing 1, 2 or very rarely 3 flowers, spreading or recurved at the time of flowering, erect when in fruit.   Bracts small ; bracteoles broad, obtuse or minutely mucronate, not half so long as the calyx.   Sepals nearly $1\frac{1}{2}$ lines long, dry, narrow, obtuse or almost acute. Corolla nearly 4 lines long, reddish outside, the tube considerably longer than the calyx, with the lobes erect at the base appearing like a continuation of it.   Anthers attached about the middle or near the top, without sterile tips, but usually emarginate with 2 minute recurved points.   Hypogynous disk truncate or toothed.   Ovary 5-celled ; style filiform.   Fruit very obtuse, scarcely exceeding the calyx.—*L. rubicundus,* F. Muell. Fragm. iv. 102 ; *Styphelia erubescens,* F. Muell. Fragm. vi. 33 ; *L. racemulosus,* var. *pauciflorus,* Sond. in Pl. Preiss. i. 312.

W. Australia. King George's Sound or adjoining districts, *Baxter ;* Lake Leven, Warricup hill, Mount Gairdner, *Maxwell ;* Murchison river, *Oldfield ;* also *Drummond, n.* 123, 482.

Var. *brevifolius.*   Leaves short.—Canning river, *Preiss ;* Gordon plains, *Maxwell ;* also *Drummond, 5th Coll. n.* 309.

In the flowers examined of the short-leaved specimens, I found the anthers attached near the top, the disk truncate and the ovary obtuse, whilst in the long-leaved specimens, the anthers were attached near the middle, the disk toothed and the ovary tapering into the style, but these differences may not prove constant.

99. **L. cuneifolius,** *Stschegl. in Bull. Mosc.* 1859, i. 18.   An erect bushy shrub of several feet, glabrous or the branches scarcely pubescent. Leaves distinctly petiolate, from obovate to oblanceolate, obtuse or with a small callous point, flat or slightly concave, finely veined, mostly about $\frac{1}{4}$ in. long.   Peduncles axillary, very short, bearing 1 or 2 erect flowers.   Bracts very small ; bracteoles very obtuse, scarcely $\frac{1}{3}$ as long as the calyx.   Sepals about 1 line long, dry, obtuse.   Corolla about $2\frac{1}{4}$ lines long, the tube considerably longer than the calyx, the lobes much shorter.   Anthers attached above the middle, obtuse, without sterile tips.   Hypogynous disk obtusely lobed.   Ovary 5-celled, shortly tapering into a style of moderate length.— *L. lissanthoides,* F. Muell. Fragm. iv. 101 ; *Styphelia lissanthoides,* F. Muell. Fragm. vi. 33.

W. Australia, *Drummond, 5th Coll. n.* 324 ; sandy flats on the Phillips river Ranges, *Maxwell.*

100. **L. strictus,** *Benth.*   An erect rigid shrub, glabrous or nearly so. Leaves erect, oblong-lanceolate, tapering into a short rigid point, flat or very

slightly convex, finely veined and often glaucous or whitish underneath, under ½ in. long. Peduncles axillary, exceedingly short, bearing 1 or 2 erect flowers usually longer than the leaf. Bracts very small; bracteoles very obtuse, about half as long as the calyx. Sepals about 1 line long, obtuse, often coloured at the end. Corolla-tube about 3 lines long, the lobes about 2 lines, erect at the base. Anthers attached near the top, very obtuse, without sterile tips. Hypogynous disk short, truncate. Ovary 5-angled, 5-celled; style long and slender.

W. Australia. Between Perth and King George's Sound, *Harvey ;* between Moore and Murchison rivers, *Drummond, 6th Coll. n.* 123. The specimens much resemble those of the eastern *L. Mitchellii,* with similar long flowers, but the calyx is much smaller, besides other minor differences.

101. **L. Mitchellii,** *Benth.* A glabrous and often glaucous shrub of 2 to 3 ft. Leaves sessile, narrow oblong, abruptly contracted into a short, pungent point, flat or slightly concave, rarely ½ in. long. Peduncles axillary, very short, bearing usually only 1 flower besides the rudiment or rarely 2 perfect flowers, which are large for the genus. Bracts minute; bracteoles very broad, truncate, not half so long as the calyx. Sepals nearly 2 lines long, dry, obtuse. Corolla-tube 3 to 3½ lines long; lobes nearly 2 lines. Anthers attached about the middle, obtuse, without sterile tips. Hypogynous disk truncate. Ovary 5-angled, 5-celled; style long, the stigma sometimes very small, sometimes broad and peltate.—*L. cuspidatus,* Mitch. Trop. Austr. 225, 226, not of R. Br.

Queensland. Near Lake Salvator Rosa, *Mitchell ;* in the interior, *Leichhardt ;* Percy Island, *A. Cunningham ;* Mount Hedlow, Rockhampton, *C. E. Porter.*

102. **L. juniperinus,** *R. Br. Prod.* 546. A divaricately branched shrub, with pubescent or hirsute branchlets. Leaves very spreading, linear or oblong-linear, with a fine rigid point and recurved margins, mostly under ½ in. long. Peduncles axillary, very short, bearing usually a single flower, with a single subtending bract or sometimes 2 or 3 empty, very acuminate ones below. Bracteoles about ⅓ line long, broad, obtuse, minutely mucronate. Sepals 1½ to nearly 2 lines long, mucronate-acute. Corolla-tube slender, 3 to 3½ lines long, the lobes very short. Anthers attached near the top, very obtuse, without sterile tips. Hypogynous scales distinct or slightly cohering. Ovary 5-angled, 5-celled; style elongated. Fruit oblong, 2 to 2½ lines long.—DC. Prod. vii. 753; F. Muell. Fragm. iv. 104; Lodd. Bot. Cab. t. 447; *Styphelia juniperina,* Spreng. Syst. i. 658; F. Muell. Fragm. vi. 46; *Lissanthe strigosa,* Sieb. Pl. Exs., not of R. Br.; *Leucopogon Sieberi,* DC. Prod. vii. 751; *Epacris villosa,* Cav. Ic. iv. 27. t. 347; DC. Prod. vii. 763 (from the figure and description).

Queensland. Brisbane river, Moreton Bay, *F. Mueller.*
N. S. Wales. Port Jackson, *R. Brown, Sieber, n.* 96, and several others.
Victoria? Upper Macalister river, *F. Mueller.* The specimens in fruit only, and therefore doubtful.

SERIES 5. CONCAVÆ.—Leaves concave or keeled.

103. **L. rufus,** *Lindl. in Mitch. Three Exped.* ii. 179. An erect shrub

of 2 to 3 or rarely 4 ft., bushy or with divaricate, straggling branches, glabrous or pubescent. Leaves lanceolate to cordate-ovate, tapering into a pungent point, rigid, concave, finely striate, mostly about ½ in. long. Peduncles axillary, exceedingly short, bearing 1, 2 or 3 flowers. Bracts few and small; bracteoles broad, obtuse, ½ to ¾ line long. Sepals about 2 lines long, rigid, obtuse. Corolla-tube nearly as long as the calyx; lobes about as long, recurved. Anthers attached near the top, obtuse, without sterile tips. Hypogynous scales distinct, obovate, mucronate. Ovary oblong, 3-celled in all the specimens examined. Fruit ovoid or oblong, hard, 3 to 4 lines long.—*L. astrolomioides,* F. Muell.; Sond. in Linnæa, xxvi. 249; *Styphelia rufa,* F. Muell. Fragm. vi. 46.

Victoria. Grampian mountains, *Mitchell, Wilhelmi;* Mount Abrupt, dry, rocky hills between Broken and Ovens rivers, Futter's Range, *F. Mueller.*
S. Australia. Marble Range, *Wilhelmi;* Lofty Range, Torrens and Cnkaparinga rivers, Encounter Bay, *F. Mueller;* Kangaroo Island, *E. G. Sealy, Waterhouse.*
Tasmania. Lockwoods, *Bissill;* St. Arnaud, *Stair* (a fragment in Herb. F. Muell. with broader leaves than usual).

104. **L. conostephioides,** *DC. Prod.* vii. 753. An erect, straggling shrub of about 1 ft., glabrous or the branches minutely pubescent. Leaves erect or spreading, lanceolate or linear-lanceolate, rigid, tapering into a pungent point, concave, finely striate, ¼ to ½ in. long. Peduncles axillary, short, usually spreading or recurved, 1- to 3-flowered. Bracts minute; bracteoles about ⅓ as long as the calyx. Sepals fully 1¼ lines long, obtuse or shortly mucronate, dry and smooth. Corolla about 3 lines long, the lobes as long as the tube. Anthers linear, attached about the middle, obtuse, without sterile tips. Hypogynous scales distinct or scarcely adhering, narrow ovate, acuminate. Ovary striate, 2- or 3-celled, usually glabrous; style long and slender, glabrous or slightly hairy.—Sond. in Pl. Preiss. i. 321; *L. rigidus,* A. Cunn.; DC. Prod. vii. 753; *Styphelia conostephioides,* F. Muell. Fragm. vi. 34.

W. Australia. King George's Sound and adjoining districts, *Baxter, Maxwell,* thence to Swan River, *Harvey, Drummond, 1st Coll. n.* 16, 479, *Preiss, n.* 405, and Murchison river, *Oldfield.* Very nearly allied to *L. rufus* and to *L. deformis,* and almost intermediate between the two.

105. **L. deformis,** *R. Br. Prod.* 546. A straggling shrub, with wiry branches like those of *L. virgatus,* and similar foliage, but the inflorescence of *L. rufus.* Leaves erect or spreading, linear or linear-lanceolate, tapering into a short point, rigid, concave, finely veined, 2 to 3 lines long. Peduncles axillary, exceedingly short, bearing usually a single flower besides the rudiment. Bracts very small; bracteoles broad, acute, ciliolate, not half so long as the calyx. Sepals 1¼ lines long, dry, acute, minutely ciliolate. Corolla shortly exceeding the calyx, the lobes as long as the tube. Anthers attached about the middle, obtuse, without sterile tips. Hypogynous disk readily separating into obtuse scales. Ovary densely villous, 3-celled; style long and slender.—DC. Prod. vii. 754; *Styphelia deformis,* Spreng. Syst. i. 658.

N. S. Wales. Moist heaths, Sydney and Botany Bay, *R. Brown.* This species, which I have seen in no other collection, is allied to the two preceding ones, but more slender, with smaller flowers, and a very villous ovary.

106. **L. pogonocalyx,** *F. Muell. Herb.* An erect shrub, with virgate branches, glabrous or minutely pubescent. Leaves erect, ovate or ovate-lanceolate, rigid, with a short pungent point, concave, strongly striate underneath, 2 to 3 lines long. Peduncles axillary, short, bearing 2 or 3 erect flowers. Bracts and bracteoles broad, obtuse, about ⅓ as long as the calyx. Sepals 1 line long, obtuse, fringed at the end with prominent cilia. Corolla about 2 lines long, the lobes as long as the tube, very densely bearded with white hairs as in most species of the section *Perojoa.* Anthers attached at the top, obtuse, without sterile tips. Hypogynous scales distinct or slightly cohering. Ovary 3-angled, 3-celled; style very short.

W. Australia. Mount Manypeak, *Maxwell.*

107. **L. breviflorus,** *F. Muell. Fragm.* iv. 102. A glabrous shrub of about 1 ft. Leaves from oblong-lanceolate to linear-lanceolate, shortly tapering into a fine rigid point, contracted into a short petiole, rigid, concave or nearly flat, smooth, with scarcely prominent veins, 2 to 4 lines long. Peduncles axillary, very short, bearing 1 or 2 erect flowers. Bracts very small; bracteoles very obtuse, about half as long as the calyx. Sepals dry, obtuse, about 1 line long. Corolla 2 lines long, the lobes longer than the tube. Anthers attached near the top, obtuse, without sterile tips. Hypogynous scales nearly free, shortly acuminate or toothed. Ovary 5-celled; style rather long. Fruit pendulous, obovoid, very obtuse, not exceeding the calyx.

W. Australia. Israelite Bay, *Maxwell;* rocky declivities of Stirling Range, *F. Mueller.* Allied to *L. conostephioides* and *L. cuneifolius.* Differs from the former in its less concave leaves and 5-celled ovary; from the latter in the more concave, narrow acute leaves, short corolla-tube, etc.

108. **L. durus,** *Benth.* A stout, rigid, glabrous shrub. Leaves oblong, linear or lanceolate, with a short, rigid, almost pungent point, contracted into a short petiole, thick and concave, smooth, without prominent ribs, 2 to 4 lines long. Peduncles axillary, exceedingly short, bearing 1, 2 or 3 very small flowers. Bracts and bracteoles exceedingly small. Sepals about ½ line long, very obtuse. Corolla 1½ lines long, the lobes about as long as the tube. Anthers attached at the top, obtuse, without sterile tips. Hypogynous disk separating into obtuse scales. Ovary broad, 5-celled; style very short. Fruit depressed-globular, 1 to 1¼ lines diameter.

W. Australia, *Drummond, 5th Coll. n.* 297.

109. **L. multiflorus,** *R. Br. Prod.* 542. A stout, rigid shrub, with pubescent branches. Leaves crowded, erect or scarcely spreading, linear-lanceolate or lanceolate, tapering into a pungent point, concave, finely veined, mostly about ½ in. long. Peduncles axillary, short, bearing usually 3 or 4 or even more flowers. Bracts and bracteoles very obtuse, not half so long as the calyx. Sepals about 1 line long, narrow but obtuse, ciliate with short, almost woolly hairs. Corolla about 2 lines long, the lobes as long as the tube. Anthers attached above the middle, obtuse, without sterile tips. Hypogynous disk lobed, readily separating into acuminate scales. Ovary 5-celled; style rather long.—DC. Prod. vii. 746; *Styphelia multiflora,* Spreng. Syst. i. 658 (not of F. Muell. Fragm. vi. 32, which is *L. Gilbertii*); *L. Fraseri,* A. Cunn. in DC. Prod. vii. 753 (not of Ann. Nat. Hist.).

W. Australia. Lucky Bay, *R. Brown, Baxter.*

Var. *ulicinus.* Leaves narrower, very rigid. Sepals almost acute, not ciliate. Corolla-tube shorter.—*Drummond.*

110. **L. appressus,** *R. Br. Prod.* 546. An erect shrub of 1 to 2 ft., glabrous or the branches minutely pubescent. Leaves lanceolate or ovate-lanceolate, tapering into a fine pungent point, minutely denticulate-ciliate, rigid, very concave, imbricate, mostly about 3 lines long. Peduncles axillary, exceedingly short, bearing 1 to 3 small flowers. Bracts minute; bracteoles broad, mucronate, not half so long as the calyx. Sepals about 1 line long, obtuse. Corolla very shortly exceeding the calyx, the lobes longer than the tube. Anthers attached near the top, without sterile tips. Hypogynous disk lobed. Ovary 5-celled; style short, glabrous.—*Styphelia appressa,* Spreng. Syst. i. 658.

N. S. Wales. Port Jackson, *R. Brown.* Not seen in any other collection. The plant sent by A. Cunningham to De Candolle, and described by him as *L. appressus,* does not differ from *L. esquamatus.*

111. **L. neoanglicus,** *F. Muell. Herb.* An erect, very rigid shrub of 1 to 2 ft. Leaves oblong or lanceolate, acuminate, with a fine pungent point, quite entire, rigid, concave, imbricate or rarely spreading, mostly 3 to 4 lines long. Peduncles axillary, exceedingly short, bearing 1 or rarely 2 or 3 flowers. Bracts minute; bracteoles broad, mucronate, not half so long as the calyx. Sepals 2 lines long, very acute. Corolla-tube as long as or slightly exceeding the calyx; lobes shorter than the tube. Anthers attached about the middle, obtuse, without sterile tips. Hypogynous disk lobed or separating into distinct scales. Ovary 5-angled, 5-celled; style long, usually hairy; stigma peltate.

Queensland. Stradbrooke Island, *Fraser.*
N. S. Wales. New England, *C. Stuart.*

This may possibly prove a variety of *L. appressus,* but the flowers, especially the calyx, are twice the size, the form of both sepals and corolla different, besides the differences in the style, which do not appear, in this instance at least, to be due to dimorphism as they may possibly be in some, but I believe very few, *Epacrideæ.*

112. **L. obtectus,** *Benth.* A shrub of 1 to 2 ft. or perhaps more, with few long, erect branches completely covered by the glaucous foliage. Leaves broadly cordate-ovate or orbicular, mucronate, rigid, concave, erect, imbricate, 4 to 6 lines diameter. Peduncles axillary, very short, bearing 2 to 3 flowers not exceeding the leaves. Bracts small; bracteoles not half so long as the calyx, broad, mucronate. Sepals nearly 2 lines long, lanceolate, acute. Corolla-tube nearly as long as the calyx; lobes rather shorter. Anthers linear, attached above the middle or near the top, obtuse, without sterile tips. Hypogynous disk deeply lobed or separating into distinct scales. Ovary 5-celled.

W. Australia. Between Moore and Murchison rivers, *Drummond, 6th Coll. n.* 125.

113. **L. crassiflorus,** *F. Muell. Fragm.* vi. 40. Apparently erect and not much branched, attaining 1 or 2 ft., glabrous or the branches minutely pubescent. Leaves broadly obovate or orbicular, very obtuse, erect, very

concave and embracing the stem at the base, slightly spreading at the end, mostly about 1 line or the larger ones 1½ lines diameter.   Peduncles in the uppermost axils, very short, bearing 1 or rarely 2 flowers besides the rudiment, forming a short, terminal, leafy corymb or cluster, sometimes reduced to a single flower.   Bracts small, keeled; bracteoles nearly 1 line long, broad, obtuse or very shortly acuminate.   Sepals about 2 lines long, broad, with shortly mucronate or almost obtuse often spreading tips, rigid, minutely striate.   Corolla nearly 3 lines long, the lobes rigid, acutely acuminate, longer than the tube.   Anthers attached above the middle, tipped by 2 short, acuminate lobes, but without the sterile tips of *Perojoa.*   Hypogynous disk lobed.   Ovary short, 10-ribbed, 5-celled; style slender.—*Styphelia crassiflora,* F. Muell. Fragm. vi. 40.

W. Australia.   Between Moore and Murchison rivers, *Drummond, 6th Coll. n.* 120.

114. **L. strongylophyllus,** *F. Muell. Fragm.* iv. 101.   An erect shrub of 2 to 5 ft., glabrous or the branches minutely pubescent.   Leaves rather crowded, erect or spreading, obovate or orbicular, obtuse or with a minute callous point, concave, prominently striate, contracted into a distinct petiole, 1 to 2 lines diameter.   Peduncles axillary, short, erect or at length recurved, bearing 1 or 2 flowers.   Bracts minute; bracteoles broad, obtuse, not half so long as the calyx.   Sepals about 1 line long, dry, almost acute.   Corolla about 2 lines long, the lobes longer than the tube, but erect at the base.   Anthers attached above the middle, obtuse, without sterile tips.   Hypogynous disk minutely toothed.   Ovary 5-celled; style long and slender. —*Styphelia strongylophylla,* F. Muell. Fragm. vi. 33.

W. Australia.   Sandy plains, Murchison river, *Oldfield.*

115. **L. crassifolius,** *Sond. in Pl. Preiss.* i. 316.   An erect, bushy shrub of about 2 ft.   Leaves erect, oblong-linear, obtuse or with a minute callous point, rather thick, concave, with about 3 ribs prominent underneath, contracted into a short petiole, 2 to 4 lines long.   Peduncles axillary, rather short, few-flowered.   Bracts small; bracteoles obtuse, about half as long as the calyx.   Sepals scarcely above ¾ line long, rather thin, ciliolate, obtuse, with the tips often recurved.   Corolla about 1½ lines long, the lobes as long as or longer than the tube.   Anthers attached at the top, obtuse, without sterile tips.   Hypogynous disk truncate, readily separating into distinct scales.   Ovary 5-celled; style rather long.   Fruit obovoid, erect, about 1¼ lines long.—*Styphelia crassifolia,* F. Muell. Fragm. vi. 33.

W. Australia.   Konkoberup hills, Cape Riche, *Preiss, n.* 386 ; Cape le Grand and in the interior from Cape Paisley, *Maxwell.*   In several of the specimens a few flowers may be met with, probably diseased, having a longer tube to the corolla, and an elongated, apparently barren, ovary, with a short style.   This and the following species, without a close examination, might easily be confounded with *L. flavescens.*

116. **L. corynocarpus,** *Sond. in Pl. Preiss.* i. 322.   An erect, slender but rigid shrub, of 2 or 3 ft.   Leaves erect, oblong-linear, obtuse or with a minute callous point, concave, with about 5 slightly prominent ribs underneath, contracted into a short petiole, 2 to 4 lines long.   Peduncles axillary, short, bearing 2 or 3 erect flowers.   Bracts small; bracteoles obtuse, not

half so long as the calyx. Sepals 1½ lines long, rigid, narrow but obtuse. Corolla-tube very short; lobes twice as long as the tube, but only shortly exceeding the calyx. Anthers attached at the top, obtuse, without sterile tips. Hypogynous scales distinct. Ovary short, 5-celled, glabrous; style rather long. Fruit obovoid, smooth, erect, about 2 lines long.

W. Australia, *Drummond, 3rd Coll. n.* 193; near Cape Riche, *Preiss, n.* 379; Kalgan river, *Oldfield;* Stirling Range, *F. Mueller.* Near *L. crassifolius,* but remarkable for the much longer, rigid sepals, and the nerves of the leaves more numerous and less prominent.

117. **L. Woodsii,** *F. Muell. Fragm.* i. 178, *and* vi. 33. A small, bushy shrub, glabrous or the branches minutely pubescent. Leaves erect, ovate or oblong, obtuse, rather thick, prominently veined, contracted into a short petiole, 1 to 2 lines long. Peduncles axillary, shorter than the leaves, but slender, recurved, all 1-flowered besides the minute rudiment. Bracts very small; bracteoles not half so long as the calyx. Sepals less than 1 line long, dry, acute. Corolla about 2 lines long, the lobes longer than the tube, but erect at the base. Anthers attached about the middle, without sterile tips. Hypogynous disk truncate. Ovary oblong, 5-celled; style rather long. Fruit oblong-clavate, about 2 lines long, pendulous.

Victoria. Near the Glenelg, *Robertson.*
S. Australia. Near Penola, *Woods.*
W. Australia, *Drummond ;* limestone bank behind Eyre's Relief, *Maxwell.*

118. **L. leptanthus,** *Benth.* An erect, bushy shrub, of about 1 ft., our specimens glabrous and somewhat glaucous. Leaves erect, oblong or almost ovate, obtuse or with a very small callous point, slightly concave, prominently veined, 2 to 3 lines long. Peduncles axillary, very short, bearing 1 or 2 flowers. Bracts very small; bracteoles obtuse, not half so long as the calyx. Sepals about ¾ line long, dry, narrow but obtuse. Corolla-tube slender, about 2 lines long, the lobes about 1 line. Anthers attached above the middle, oblong-linear, emarginate, without sterile tips, the filaments longer than usual and the ends of the anthers prominent, but the greater portion enclosed within the erect base of the corolla-lobes, as in other species. Hypogynous disk truncate. Ovary glabrous, 5-celled; style rather long.

W. Australia. Between Moore and Murchison rivers, *Drummond, 6th Coll. n.* 124.

13. ACROTRICHE, R. Br.

(Frœbelia, *Regel.*)

Corolla-tube equal to the calyx or longer; lobes valvate in the bud, spreading, with a tuft of long hairs inside at the end, at first inflexed, afterwards erect, and a tuft of hairs or a hairy scale at the base closing the orifice of the tube. Filaments short, terete, inserted at the top of the tube; anthers oblong, usually short, attached above the middle, very obtuse. Hypogynous disk cup-shaped, truncate or obtusely lobed. Ovary 2- to 10-celled, with 1 ovule in each cell; style short, with a small stigma. Drupe globular or depressed, the mesocarp slightly pulpy, the endocarp 2- to 10-celled, rather hard, the pyrenes usually less consolidated than in most *Leucopogons,* but much more so than in *Trochocarpa.*—Low rigid shrubs, usually

very divaricately branched. Leaves rigid. Flowers small, in little sessile
or shortly pedunculate spikes, condensed into heads or clusters, in the axils
of the previous year's leaves or on the stem below the leaves, each flower sessile within the small subtending bract and 2 bracteoles.

The genus is limited to Australia. Although the technical characters by which it is distinguished may appear of little importance, yet the genus is a very natural one. The inflorescence is peculiar, although an approach to it may be observed in *Monotœa scoparia* and
in *Trochocarpa parviflora*. F. Mueller unites *Acrotriche* with *Styphelia*.

Spikes or clusters mostly in the axils of the previous year's leaves.
  Leaves mucronate or pungent-pointed.
    Leaves lanceolate or oblong, mucronate-acute, ½ in. long or
      more.
      Corolla-tube not exceeding the calyx. Ovary usually 5-celled.
        Leaves scarcely paler underneath . . . . 1. *A. divaricata.*
      Corolla-tube twice as long as the calyx. Ovary 6- to 10-
        celled. Leaves pale or glaucous underneath . . . . . 2. *A. aggregata.*
    Leaves linear-lanceolate, tapering into a pungent point. Ovary
      usually 5- or 6-celled . . . . 3. *A. serrulata.*
    Leaves broadly ovate-lanceolate, tapering to a pungent point.
      Ovary usually 5-celled . . . . 4. *A. patula.*
    Leaves obtuse, from broadly ovate to oblong. Ovary usually
      4-celled . . . . 5. *A. ovalifolia.*
Spikes all below the leaves on the old branches or trunk. Leaves
  mucronate or pungent-pointed.
    Leaves ovate-lanceolate, 4 to 6 lines long. Spikes scattered.
      Calyx short . . . . 4. *A. patula.*
    Leaves oblong-linear or linear-lanceolate, 4 to 6 lines long. Spikes
      scattered. Calyx short and broad . . . . 6. *A. ramiflora.*
    Leaves ovate to lanceolate, 1 to 2 lines long. Spikes crowded.
      Calyx narrow, reddish . . . . 7. *A. depressa.*
    Leaves lanceolate or linear-lanceolate, about ⅓ in. long. Spikes
      very densely crowded. Calyx narrow, red, 2 lines long. . . 8. *A. fasciculifolia.*

1. **A. divaricata,** *R. Br. Prod.* 547. A shrub, attaining sometimes
several feet, but of diffuse or spreading habit, the branches usually shortly
hirsute. Leaves spreading, from oblong-elliptical to narrow-lanceolate, mucronate, acute, flat or slightly concave or convex, scarcely paler underneath,
½ in. long or rather more. Flowers very small, green, in very short spikes or
clusters, nearly sessile in the lower axils. Bracts very small; bracteoles very
broad, about half as long as the calyx. Sepals broad, very obtuse, ¾ line
long. Corolla nearly 1½ lines long, the lobes as long as the tube, the throat
closed with dense tufts of hairs. Ovary (always?) 5-celled, not ribbed.
Fruit nearly globular, 1½ lines diameter.—DC. Prod. vii. 756; *Styphelia
divaricata,* Spreng. Syst. i. 658; F. Muell. Fragm. vi. 44.

Queensland. Towards Moreton Bay and Durval, *Leichhardt.*
N. S. Wales. Port Jackson, *R. Brown, Sieber, n.* 93, and others; near Richmond,
Wilhelmi; Arbuthnot Range (with leaves much revolute in drying), *Fraser.*

2. **A. aggregata,** *R. Br. Prod.* 547. A spreading shrub, with the
habit of *A. divaricata,* the branches glabrous pubescent or shortly hirsute.
Leaves spreading, oblong-lanceolate, mucronate-acute, nearly flat or slightly
convex, pale or whitish underneath, ½ to 1 in. long, Flowers in very short
axillary spikes or clusters, sessile or shortly pedunculate, mostly on the pre-

vious year's shoots. Bracts very small; bracteoles scarcely half so long as the calyx. Sepals almost orbicular, ¾ line long. Corolla-tube twice as long as the calyx, the lobes about 1 line, the throat closed with dense tufts of hairs. Ovary 7- to 10-celled, not ribbed. Fruit depressed-globular, at least 2 lines diameter when perfect.—DC. Prod. vii. 757; *Styphelia aggregata,* Spreng. Syst. i. 657; F. Muell. Fragm. vi. 44.

Queensland. Port Bowen, *R. Brown;* Wide Bay (with larger flowers than usual), *Bidwill;* Brisbane river and Moreton Island, *F. Mueller;* towards Durval, *Leichhardt.*
N. S. Wales. Tweed river and Cape Byron, *C. Moore;* New England, *C. Stuart.*

3. **A. serrulata,** *R. Br. Prod.* 547. A low, prostrate or diffuse shrub, with shortly ascending branches, more or less hirsute or rarely glabrous. Leaves very spreading, linear-lanceolate, tapering into a pungent point, usually ciliate with short, rigid hairs, but sometimes without them, nearly flat, the midrib and often 2 lateral ribs prominent underneath, mostly 3 to 4 lines long. Flowers green, in short, dense, axillary spikes or globular clusters, sessile or very shortly pedunculate on the previous year's wood. Bracts very small; bracteoles less than half as long as the calyx, broad and obtuse. Sepals obtuse, nearly 1 line long. Corolla-tube twice as long as the tube, the lobes scarcely 1 line, the throat closed with more or less hairy scales instead of the tufts of hairs of the two preceding species. Ovary 5- or 6-celled, with as many prominent ribs. Drupe globular, nearly 2 lines diameter.—DC. Prod. vii. 757; Hook. f. Fl. Tasm. i. 253; *Styphelia serrulata,* Labill. Pl. Nov. Holl. i. 45. t. 62 (but not the separate figure of the flower); F. Muell. Fragm. vi. 44; *A. patula,* Hook. f. Fl. Tasm. i. 252, not of R. Br.

N. S. Wales. Brushy hills, Argyle county, and N. of Bathurst, *A. Cunningham;* Glendon, *Leichhardt;* Twofold Bay, *F. Mueller;* Gabo Island, *Maplestone.*
Victoria. Port Phillip, *R. Brown;* Wendu vale, *Robertson;* Wilson's Promontory, Cape Otway, Delatite and Macalister rivers, Dandenong ranges, *F. Mueller;* Creswick diggings, *Whan.*
Tasmania. Port Dalrymple and Storm Bay Passage, *R. Brown;* common on dry hills throughout the colony, *J. D. Hooker.*
S. Australia. Lofty Range, near Adelaide, *F. Mueller.*

A. affinis, DC. Prod. vii. 757, is probably the same plant. F. Mueller refers to it also *A. prostrata,* F. Muell. in Trans. Phil. Soc. Vict.; but I have been unable to find the name in that work.

4. **A. patula,** *R. Br. Prod.* 547. A very rigid, stout, divaricate shrub, of 1 to 2 ft., the branches slightly pubescent. Leaves from almost ovate to ovate-lanceolate, very rigid, tapering into a rigid point, more or less concave and sometimes almost cordate at the base, the veins very obscure, under ½ in. long. Flowers in very short spikes or clusters, almost sessile in the axils on the previous year's wood or sometimes below the leaves. Bracts small; bracteoles about half as long as the calyx. Sepals scarcely 1 line long. Corolla-tube nearly twice as long as the calyx; lobes short, the throat closed with long hairs. Ovary scarcely ribbed, 5-celled.—DC. Prod. vii. 757; *Styphelia patula,* Spreng. Syst. i. 657; F. Muell. Fragm. vi. 44.

S. Australia. Petrel Bay, *R. Brown;* Holdfast Bay, *F. Mueller;* towards Spencer's Gulf, *Warburton;* Kangaroo Island, *F. Mueller.*

5. **A. ovalifolia,** *R. Br. Prod.* 548.   An erect and bushy or diffuse shrub, of 6 in. to 1 ft., glabrous or the branches very minutely pubescent. Leaves petiolate, from broadly ovate to oval-oblong, obtuse or with an obscure callous point, thick, flat or nearly so, the veins not prominent, 3 to 4 lines long.   Flowers in very shortly pedicellate axillary clusters, on the previous year's wood.   Bracteoles about half as long as the calyx.   Sepals broad, very obtuse, about ¾ line long.   Corolla-tube not twice as long as the calyx, the lobes short, the throat closed with tufts of hairs.   Ovary, in all the flowers examined, 4-celled.   Fruit small, globular or slightly depressed.— Bot. Mag. t. 3171 ; *Styphelia ovalifolia*, Spreng. Syst. i. 656 ; *A. subcordata*, DC. Prod. vii. 757 ; Sond. in Pl. Preiss. i. 325.

Victoria.   Portland Bay and Cape Nelson, *Allitt*.

S. Australia.   Memory Cove, *R. Brown ;* Port Lincoln and Marble Ranges, *Wilhelmi ;* Lake Alexandrina, *F. Mueller.*

W. Australia.   King George's Sound and adjoining districts, *R. Brown, A. Cunningham, Preiss, n.* 435, and others ; Swan River, *Drummond,* 1*st Coll. and n.* 31, *Oldfield ;* Bank Cliffs, *Maxwell ;* Stirling Range, *F. Mueller.*

Styphelia cordata, Labill. Pl. Nov. Holl. i. 46. t. 63 (*Acrotriche cordata*, R. Br. Prod. 548), has been ascertained by De Candolle and by Sonder to be the same plant; but the specific name, though older, has been rejected as manifestly inapplicable, except to some apparently very exceptional state.   In the very numerous specimens we have from various localities it is only very rarely that a few leaves show a slightly cordate base.   Even the name *subcordata*, which De Candolle substituted, is but rarely applicable, besides that it is more recent than Brown's name *ovalifolia*.

Var. ? *oblongifolia*.   Leaves elliptical-oblong, 4 to 6 lines long.—Mouth of the Glenelg, *Allitt ;* King George's Sound, *Baxter.*

6. **A. ramiflora,** *R. Br. Prod.* 547.   A stout shrub, with divaricate branches, glabrous or slightly pubescent.   Leaves very spreading, oblong-linear, abruptly contracted into a short, rigid point, with revolute margins, obscurely veined underneath or the midrib only prominent.   Flowers not seen.   Fruiting spikes scattered along the old wood below the leaves, ¼ to ½ in. long.   Sepals short and broad as in the preceding species.   Fruit depressed-globular or shortly pear-shaped, not ribbed, above 2 lines diameter. —DC. Prod. vii. 757 ; *A. Manglesii*, Sond. in Pl. Preiss. i. 326 ; *Styphelia ramiflora*, Spreng. Syst. i. 659.

W. Australia.   Lucky Bay, *R. Brown*, and probably in the same neighbourhood, *Baxter, Drummond, n.* 107.   In foliage and calyx this is very near *A. serrulata*, but the inflorescence is more like that of *A. depressa*, except that the spikes are not so crowded.

7. **A. depressa,** *R. Br. Prod.* 548.   A very divaricately branched, rigid shrub of 1 to 2 ft.   Leaves very spreading or reflexed, from ovate and 1 line long to lanceolate and 2 lines, sometimes cordate, mucronate-acute, flat or convex.   Spikes many-flowered, sessile or nearly so, and crowded on the old wood towards the base of the principal branches, each one ½ to ¾ in. long.   Bracteoles about ½ line long, and bracts still smaller.   Sepals rather narrow, reddish, 1 to 1¼ line long.   Corolla-tube very little longer than the calyx ; lobes short, the throat closed with tufts of hairs.   Ovary 2- or rarely 3-celled, tapering into the short style.—DC. Prod. vii. 757, and Pl. Rar. Jard. Gen. 8. Not. t. 1 ; *Styphelia depressa*, Spreng. Syst. i. 655 ; F. Muell. Fragm. vi. 44.

S. Australia. Kangaroo Island, *R. Brown, F. Mueller;* Murray river and near Ta-munda, *F. Mueller.*

W. Australia. Sand ridges, Bald Head, *Baxter.*

8. **A. fasciculiflora,** *Benth.* A rigid shrub of 1 to 2 ft., with stout ascending stems, the branches and foliage usually hirsute. Leaves lanceolate or linear-lanceolate, tapering into a fine point, convex or with recurved margins, pale underneath, rarely exceeding $\frac{1}{2}$ in. Spikes very densely crowded at the base of the stem or sometimes also at the base of the primary branches, forming in some specimens a mass of 2 to 3 in. diameter, the small bracts and bracteoles and very conspicuous calyces of a reddish colour. Sepals $2\frac{1}{2}$ to nearly 3 lines long, thin, and rather narrow. Corolla-tube scarcely exceeding the calyx or half as long again when the sepals are short; lobes about 1 line long, the throat clothed with shortly hairy scales instead of tufts of hairs. Ovary with 3, 4 (or 5 ?) angles and as many cells : style as long as the corolla-tube. Fruit not seen.—*Frœbelia fasciculiflora,* Regel, Gartenfl. i. 164. t. 18; *Acrotriche ramiflora,* Sond. in Linnæa, xxvi. 251, not of R. Br.; *Styphelia ramiflora,* F. Muell. Fragm. vi. 44.

S. Australia. Forest gullies, Lofty Range, *F. Mueller,* who informs us that the tube of the pink corollas is half-full of a clear, sweet-scented honey.

14. MONOTOCA, R. Br.

Corolla-tube small, campanulate or scarcely cylindrical; lobes 5 or rarely 4, valvate in the bud, spreading, glabrous. Filaments inserted at the top of the tube, short, filiform; anthers wholly or partially enclosed in the corolla-tube or spreading with the lobes. Hypogynous disk truncate, lobed or separating into distinct scales. Ovary 1-celled or rarely 2-celled, with 1 ovule in each cell; style short; stigma small. Fruit a small drupe, with a somewhat pulpy mesocarp, and a hard or crustaceous endocarp, with a single seed.— Shrubs or small trees. Leaves with recurved margins or nearly flat. Flowers small, often more or less unisexual, in axillary or terminal spikes or racemes, or sometimes quite solitary, each flower sessile or pedicellate within the subtending bract, with 2 bracteoles close under the calyx.

The genus is limited to Australia. It is closely allied to *Leucopogon,* differing in the small, almost campanulate corolla with glabrous lobes, and in most cases also by the 1-celled ovary, from whence the name is derived; but in one species, *M. tamariscina,* the ovary is 2-celled, whilst in a single species of *Leucopogon, L. pleurandroides,* it is 1-celled only. The first four of the following species sometimes run so closely one into the other as to make it difficult to distinguish them.

Ovary 1-celled. Flowers 5-merous.

Ovary 2-celled.   Flowers 5-merous, mostly solitary, minute.   Leaves
under 1 line, imbricate, obtuse . . . .   6. *M. tamariscina.*

1. **M. elliptica,** *R. Br. Prod.* 546.   A tall shrub or sometimes a tree
of 20 to 30 ft.   Leaves from broadly elliptical-oblong to almost oblong-
linear, mucronate, slightly convex, pale or whitish and finely veined underneath, under ½ in. long in some specimens, from ½ to 1 in. in others.
Flowers pedicellate, few or many together, forming short racemes, either terminal or also axillary, and sometimes exceeding the leaves or growing out
into leafy branches, with a few solitary axillary flowers.   Subtending bracts
membranous, very deciduous ; bracteoles not half so long as the calyx, and
close under it.   Sepals half as long as the corolla, broad, and very obtuse.
Corolla from scarcely 1 line to 1¼ line long, campanulate, the lobes recurved,
shorter than the tube.   Hypogynous disk truncate or sinuate-toothed.   Ovary
1-celled, tapering into a short style.   Fruit ovoid, 1½ to nearly 2 lines long.
—DC. Prod. vii. 755 ; F. Muell. Fragm. vi. 58 ; *Styphelia elliptica*, Sm.
Bot. N. Holl. 49 ; *M. albens*, R. Br. Prod. 547 ; DC. Prod. vii. 755.

Queensland. Moreton Island, *F. Mueller.*
N. S. Wales. Port Jackson and Blue Mountains, *R. Brown, Sieber, n.* 98, 99, and
others ; Twofold Bay, *F. Mueller.*
Victoria. Sealer's Cove, Port Albert, Wilson's Promontory, also in the Baw-Baw
mountains at an elevation of 4500 ft., *F. Mueller.*
Tasmania. Sand banks on the seacoast, *Story.*

In the seacoast specimens the flowers are decidedly diœcious, the males much larger than
the females.   In many of the smaller-flowered mountain specimens the anthers and ovules
appear to be all perfect in the same flower.   In R. Brown's specimens there is considerable
difference in the shape of the leaf between *M. elliptica*, from Port Jackson, and *M. albens,*
from Grose river, but this difference entirely disappears in other specimens.

2. **M. lineata,** *R. Br. Prod.* 547.   A tall shrub or small tree, closely
resembling *M. elliptica* in foliage, but the peduncles are all short, axillary,
and few-flowered, the flowers smaller and sessile or nearly so within the very
small subtending bract, which is usually persistent or sometimes very
minute, or even quite deficient, when the spike is reduced to a single flower.
Corolla more open than in *M. elliptica*, with a very short tube, so as to be
almost rotate.   Fruit ovoid, about 1 line long.—DC. Prod. vii. 755 ; Hook.
f. Fl. Tasm. i. 252 ; *Styphelia glauca*, Labill. Pl. Nov. Holl. i. 45. t. 61.

Tasmania. Kent's Group, Bass's Straits, and Derwent river, *R. Brown ;* abundant on
the skirts of damp forests, etc., *J. D. Hooker.*   The Victorian specimens formerly referred
to this species appear to me rather to belong to *M. elliptica.*

3. **M. scoparia,** *R. Br. Prod.* 547.   An erect, bushy shrub of 2 or
3 ft., glabrous or the branches minutely pubescent.   Leaves oblong-linear,
mucronate, convex or with revolute margins, pale or glaucous and finely
veined underneath, rarely exceeding ½ in.   Flowers in little axillary clusters
of 2 to 4 or sometimes solitary, usually reflexed, the common peduncle exceedingly short.   Bracts very small, broad, membranous, persistent ; bracteoles about half as long as the calyx.   Sepals a little more than ½ line
long, very obtuse.   Corolla about 1 line long, the lobes as long as the tube,
much less spreading than in *M. elliptica*, and thickened at the end.   Hypogynous disk truncate or toothed.   Ovary 1-celled.   Drupe about 1 line long.—

DC. Prod. vii. 756; F. Muell. Fragm. vi. 58; *Styphelia scoparia*, Sm. Bot. N. Holl. 49; *M. patens* and *M. propinqua*, A. Cunn.; DC. Prod. vii. 756.

Queensland. Moreton Island, *F. Mueller.*

N. S. Wales. Port Jackson and Blue Mountains, *R. Brown, Sieber. n.* 97, and others; in the interior, and northward to Cox's River and Mount Lindsay, *A. Cunningham;* sources of the Gwydir, *Leichhardt;* New England, *C. Stuart;* Clarence river, *Beckler;* southward to Twofold Bay, *F. Mueller.*

Victoria. Port Albert, Latrobe river, Mount Disappointment, Grampians, *F. Mueller;* Stringy Bark Forest, *Robertson.*

Tasmania. N.W. coast, *W. V. Bissill, C. Stuart.*

Var. *submutica.* Leaves scarcely mucronate. Upper regions of Mount Useful in Victoria, *F. Mueller;* ascent of Mount Lapeyrouse, in Tasmania, *C. Stuart.*

4. **M. ledifolia,** *A. Cunn.; DC. Prod.* vii. 756. A stiff, bushy shrub, glabrous or the branches scarcely pubescent. Leaves petiolate, oblong, very obtuse, rather thick, flat, the margins slightly thickened, pale or white underneath, the nerves coarser than in the other species, 2 to 4 lines long. Flowers solitary in the axils or rarely 2 together, on very short peduncles. Bracteoles not half so long as the calyx. Sepals about ½ line long, ciliolate, obtuse, but narrower than in *M. scoparia.* Corolla fully 1 line long, the lobes rather longer than the tube, and narrower than in *M. scoparia.* Ovary 1-celled.

N. S. Wales. Bleak open places in the Blue Mountains, *A. Cunningham.* The species requires further investigation from more varied specimens.

5. **M. empetrifolia,** *R. Br. Prod.* 547. A low, diffuse or prostrate, much-branched shrub, glabrous or the branches minutely pubescent. Leaves more petiolate than in any other species, spreading or reflexed, oblong, mucronate, thick, the margins revolute, smooth and shining above, white underneath, 2 to 4 lines long. Spikes distinctly pedunculate, though much shorter than the leaves, nodding, few-flowered. Bracts and bracteoles very small, broad, persistent. Sepals 4, broad, obtuse, about ½ line long. Corolla about 1 line long, broadly campanulate, with 4 broad lobes, fully as long as the tube, and not thickened at the tip. Hypogynous scales distinct. Ovary 1-celled.—DC. Prod. vii. 756; Hook. f. Fl. Tasm. i. 252.

Tasmania. Mount Wellington, *R. Brown;* Mount Wellington, Lake St. Clair, and other mountains at an elevation of 3000 to 5000 ft., *J. D. Hooker.* F. Mueller, Fragm. vi. 59, considers this as a variety of *M. scoparia,* but, besides the foliage, the 4-merous flowers broad corolla and free scales appear to me to be constant in all the specimens examined.

6. **M. tamariscina,** *F. Muell. Fragm.* vi. 79. An erect or diffuse, heath-like shrub of 1 to 2 ft., with the aspect almost of *Oligarrhena*, the branches minutely pubescent. Leaves crowded, imbricate, but very deciduous from the dried specimens, ovate to lanceolate, acute or acuminate, concave, prominently ribbed, mostly about ½ line long and rarely attaining 1 line. Peduncles axillary, much shorter than the leaves, slender, all 1-flowered in our specimens, sometimes 2-flowered according to F. Mueller. Bracts and bracteoles very small. Sepals ovate-lanceolate or acuminate, about ¼ line long. Corolla ¾ line long, slightly pubescent outside, the tube campanulate, the lobes longer than the tube, each with a rather prominent midrib, so as to make the bud 10-angled. Filaments in some flowers nearly as long as the

corolla-lobes, with short, broad anthers, in other flowers short, with semi-abortive anthers. Hypogynous scales small, linear, slightly cohering at the base. Ovary 2-celled; style exceedingly short; stigma minutely 2-lobed.

W. Australia, *Drummond, n.* 86; rocks on the summits of the Stirling Range, *F. Mueller.* Very different from the other species in foliage and in the 2-celled ovary, as well as in geographical station, but appears to be better placed in *Monotoca* than in any other genus.

15. OLIGARRHENA, R. Br.

Corolla-tube campanulate; lobes 4, valvate in the bud, not bearded. Stamens 2, inserted near the top of the corolla-tube, the 2 alternate deficient ones sometimes replaced by minute staminodia; filaments very short, fili-form; anthers included in the tube. Hypogynous disk of 4 distinct scales. Ovary 2-celled; style short. Fruit (according to Sonder) a small, 1-seeded drupe.—A heathlike shrub, with minute leaves. Flowers minute, in small filiform, axillary spikes, forming a leafy, compound raceme.

The single species is limited to West Australia. It is near to some of the very slender *Leucopogons,* but with beardless corollas, the flowers 4-merous as in *Acrotriche empetrifolia,* and differs from all in the deficiency or abortion of two of the stamens.

1. **O. micrantha,** *R. Br. Prod.* 549. An erect, heath-like shrub of 1 to 2 ft., with numerous very slender, almost filiform branches, glabrous or pubescent. Leaves erect, closely appressed, lanceolate or ovate-lanceolate, shortly acuminate or obtuse, concave and embracing the stems, $\frac{1}{2}$ to 1 line long. Spikes axillary, filiform, interrupted, 2 to 3 lines long, solitary in each axil, but very numerous, forming a terminal, compound raceme, crowded with minute white or pale yellow flowers. Bracts and bracteoles about half as long as the calyx. Sepals very thin, ciliate, about $\frac{1}{3}$ line long. Corolla $\frac{3}{4}$ line long, glabrous inside and out, the lobes as long as the tube. Anthers short and broad. Hypogynous scales ovate. Ovary glabrous; style exceedingly short, with a minute stigma. Drupe ovate, striate, scarcely $\frac{1}{2}$ line long (*Sonder*).—DC. Prod. vii. 760; Sond. in Pl. Preiss. i. 326.

W. Australia. Lucky Bay, *R. Brown, Baxter;* between King George's Sound and Cape Riche, *Drummond, n.* 114, *3rd Coll. n.* 195; *Preiss, n.* 2099; Kalgan river, *Old-field;* Stirling Range, *F. Mueller;* ranging along the coast to Cape le Grand, *Maxwell.*

TRIBE II. EPACREÆ.—Ovules several in each cell of the ovary. Style inserted in a central tubular depression, so as to be lateral or basal with re-spect to the carpels. Capsule loculicidally dehiscent.

16. EPACRIS, Cav.

Corolla-tube cylindrical or campanulate; lobes 5, imbricate but not con-torted in the bud, more or less spreading, glabrous. Filaments short, adnate to the corolla-tube (inserted in the throat), but often readily detached almost to the base; anthers attached above the middle, wholly or partially included in the corolla-tube. Hypogynous disk consisting of distinct scales, very rarely cohering in a ring or cup. Ovary 5-celled, with several, usually numerous, ovules or placentas attached to the axis; style long or short, in-serted in a tubular depression, rarely reaching below the middle of the ovary;

stigma small or clavate or dilated.   Capsule loculicidally dehiscent —Shrubs. Leaves sessile or petiolate, articulate on the stem, sometimes embracing it above the base but not sheathing.   Flowers solitary in the upper axils or along the branches, on peduncles usually short.   Bracts numerous, covering the peduncle and imbricate on the calyx, passing gradually into the sepals, and forming an involucre round them.

The genus is limited to Australia and New Zealand, and only one or perhaps two species are common to the two countries.   With all its variations in the foliage and shape of the corolla, it is the most easily recognized in the Order, differing from all except *Lysinema* in foliage and inflorescence, and neatly distinguished from the latter genus by the æstivation of the corolla.   The species, however, are exceedingly difficult to circumscribe by any definite characters, the whole eighteen of the short-flowered ones seeming to pass into each other by small gradations.

F. Mueller observes, Fragm. vi. 71 and 72, that at least *E. impressa* has occasionally several-flowered peduncles.   I looked through the whole of the very numerous specimens in the Kew Herbaria, as well as those of F. Mueller's own collection, and a considerable number of fresh cultivated ones, without being able to find a single peduncle with more than the normal terminal flower even in a rudimentary state ; but F. Mueller has since sent me the specimens on which his observation was founded.   They are evidently branches of a cultivated one, in which a few of the peduncles are in a state of abnormal prolification, and ought, perhaps, to be considered rather as exceptional monstrosities than as evidences against the value of the generic character.

Corolla-tube much longer than the calyx.
  Corolla-tube long and cylindrical, without impressions.   Leaves ovate or ovate-lanceolate, often cordate, mucronate-acute.
    Corolla-limb white, the tube red.   Sepals acuminate, about 2 lines long . . . . 1. *E. longiflora.*
    Corolla all red.   Sepals shortly acute, about 1 line long . . . 2. *E. reclinata.*
  Corolla-tube long or rather short, marked with 5 impressions immediately above the ovary.   Leaves linear-lanceolate to ovate-lanceolate . . . . 3. *E. impressa.*
  Corolla-tube probably long and cylindrical.   Leaves oblong-lanceolate, very shortly mucronate, tapering at the base . . . . . 4. *E. sparsa.*
Corolla-tube shorter than the calyx or very little exceeding it.
  Leaves very obtuse.
    Corolla-tube campanulate.   Style very short.
      Leaves sessile or nearly so, ovate or oblong, under 2 lines long.   Flowers small.
        Leaves mostly imbricate, and about 1 line.   Corolla-tube without prominences inside . . . . 5. *E. petrophila.*
        Leaves mostly spreading, 1 to 2 lines.   Corolla-tube with a raised, transverse line inside . . . . 6. *E. rigida.*
      Leaves distinctly petiolate, very broad, mostly 2 lines long or more . . . . 7. *E. coriacea.*
    Corolla-tube cylindrical.   Style long.
      Stems procumbent or trailing.   Leaves petiolate, obovate to oblong.   Flowers scattered.   Calyx under 2 lines long . . 8. *E. crassifolia.*
      Stems stout, erect.   Leaves petiolate, nearly orbicular.   Flowers in terminal heads.   Calyx above 3 lines.   Corolla-tube included . . . . 9. *E. robusta.*
      Stems erect, virgate.   Leaves narrow, nearly sessile.   Flowers along the branches.   Calyx about 3 lines long.   Corolla-tube exserted . . . . 10. *E. obtusifolia.*
  Leaves rather obtuse or almost acute, petiolate.   Bracts and sepals usually obtuse.   Corolla-tube not longer than the lobes.   Style short.
    Leaves mostly ovate . . . . 11. *E. myrtifolia.*

Leaves mostly elliptical-oblong . . . . 12. *E. exserta.*
Leaves mucronate-acute. Bracts and sepals acute.
Leaves on long petioles, lanceolate. Corolla-lobes much shorter
  than the tube . . . . 13. *E. mucronulata.*
Leaves nearly sessile, linear-lanceolate. Corolla-lobes rather
  shorter than the tube.
  Bracts and sepals ciliate. Style hairy . . . . 14. *E. lanuginosa.*
  Bracts and sepals not ciliate. Style glabrous . . . . . 15. *E. paludosa.*
Leaves mostly ovate, flat or concave. Corolla-lobes rather
  shorter or rather longer than the tube.
  Leaves mostly 2 to 4 lines long . . . . 16. *E. heteronema.*
  Leaves mostly under 2 lines long . . . . 17. *E. serpyllifolia.*
Leaves broad and cordate or very concave at the base, acuminate or
  acute, with spreading points.
  Bracts and sepals obtuse or nearly so.
    Corolla-tube and style very short. Leaves mostly under 2
      lines . . . . 18. *E. microphylla.*
    Corolla-tube longer than the lobes. Style exserted. Leaves
      above 2 lines . . . . 19. *E. acuminata.*
    Corolla-tube shorter than the lobes. Style exserted. Leaves
      above 2 lines. Filaments long and thick . . . . . . 20. *E. apiculata.*
  Bracts and sepals acutely acuminate.
    Leaves mostly under 3 lines long. . . . . 21. *E. pulchella.*
    Leaves mostly about ½ in. long . . . . 22. *E. purpurascens.*

1. **E. longiflora,** *Cav. Ic.* iv. 25. *t.* 344. An erect shrub, with long straggling, usually pubescent branches. Leaves shortly petiolate or almost sessile, ovate or ovate-lanceolate, tapering into a pungent point, rounded or cordate at the base, several nerved, under ½ in. long and in some specimens scarcely ¼ in. Flowers usually drooping, on pedicels of 1 to 2 lines. Bracts acute, often decussate. Sepals acutely acuminate, 2 lines long or rather more. Corolla-tube cylindrical, often slightly curved, ½ to ¾ in. long or even more in garden specimens, crimson-red, except at the end, where it is white as well as the lobes. Hypogynous scales short and very broad, free or cohering in a ring.—DC. Prod. vii. 761; F. Muell. Fragm. vi. 70; *E. grandiflora,* Willd. Spec. i. 834; Sm. Exot. Bot. t. 39; R. Br. Prod. 550; Bot. Mag. t. 982; Lodd. Bot. Cab. t. 21; *E. miniata,* Lindl. Bot. Reg. 1845, t. 5.

N. S. Wales. Port Jackson, *R. Brown, Sieber, n.* 85, and many others; Mount Lindsay, *W. Hill.*

2. **E. reclinata,** *A. Cunn. Herb.* A depressed, straggling, much-branched shrub, usually under 1 ft. high, the branchlets pubescent. Leaves distinctly petiolate, ovate to ovate-lanceolate, acute, rounded or slightly cordate at the base, shining above, smooth or keeled underneath, 2 to 3 lines long. Flowers spreading or drooping, on pedicels of ½ to 1 line. Bracts ovate, acute. Sepals acute, but not acuminate as in *E. longiflora,* not much exceeding 1 line. Corolla-tube cylindrical, 4 to 5 lines long, of a single colour, apparently red, without any trace of the impressions of *E. impressa.* Hypogynous scales short and broad.—*E. ruscifolia,* Sieb. Pl. Exs., not of R. Br.

N. S. Wales. Damp shelving rocks in ravines of the Blue Mountains, *A. Cunningham,* also *Sieber, n.* 82.

3. **E. impressa,** *Labill. Pl. Nov. Holl.* i. 43. *t.* 58. An erect, loosely branched shrub, sometimes flowering when 6 in. high, but attaining several feet, glabrous or shortly pubescent. Leaves sessile, from ovate-lanceolate to lanceolate-linear, tapering into a short and rigid or longer and pungent point, narrowed, rounded or almost cordate at the base, the midrib and often lateral nerves prominent underneath, rarely above ½ in. long and often much smaller. Flowers varying from white to different shades of red, on very short peduncles. Bracts shorter and broader than the sepals, but gradually passing into them. Sepals varying from under 1 line to nearly 1½ lines, more or less acuminate and ciliolate. Corolla-tube varying from scarcely ¼ in. to fully ½ in. long, from almost campanulate to narrow cylindrical, always with 5 impressed cavities outside, alternating with the stamens immediately above the ovary. Anthers very variable in length. Hypogynous scales distinct, truncate. Capsule about 2 lines diameter.—R. Br. Prod. 551; DC. Prod. vii. 762; Hook. f. Fl. Tasm. i. 257; F. Muell. Fragm. vi. 70; Sweet, Fl. Austral. t. 4; Bot. Mag. t. 3407; Bot. Reg. 1839, t. 19; Lodd. Bot. Cab. t. 1691; Maund, Botanist, t. 232; Paxt. Mag. ii. 97, and iv. 126, with figures; *E. variabilis,* Lodd. Bot. Cab. t. 1816; DC. Prod. vii. 761; Paxt. Mag. iv. 125, with a figure; *E. campanulata* (with short red flowers), Lodd. Bot. Cab. t. 1925; DC. Prod. vii. 761; *E. ruscifolia* (with narrow leaves and long flowers), R. Br. Prod. 550; DC. Prod. vii. 761; *E. nivalis* (with white flowers), Lodd. Bot. Cab. t. 1821; Bot. Reg. t. 1831; Bot. Mag. t. 3253; Maund, Botanist, t. 57, altered to *E. nivea,* DC. Prod. vii. 762; *E. ceræflora* (with short white flowers), Grah. in Bot. Mag. t. 3243; DC. Prod. vii. 762.

N. S. Wales. Only at the southern extremity of the colony, Twofold Bay, *F. Mueller.*
Victoria. Port Phillip, *R. Brown;* from the Glenelg, *Robertson, Allitt,* and others, to Gipps' Land, *F. Mueller;* and in the interior to the Grampians, *F. Mueller.*
Tasmania. Derwent river, Port Dalrymple, *R. Brown;* abundant throughout the island, ascending to 2000 ft., *J. D. Hooker.*
S. Australia. Lofty Range, *F. Mueller;* Encounter Bay, *Whittaker, F. Mueller;* Concarara, *Schulzen;* Tattiara country, *Woods.*

Var. *ovata.* Leaves small, ovate or ovate-lanceolate, acute, sometimes almost cordate.— Twofold Bay, *F. Mueller;* Mount Imlay, *L. Morton;* Rocky Cape and Woolncth, *Gunn.*

Var. *grandiflora.* Leaves large, thick, and less acuminate. Flowers long, of a deep purple-red.—Wimmera, *Dallachy* (nearly glabrous); Grampians, *F. Mueller* (glabrous and pubescent); Mount William, *Mitchell* (softly pubescent, *E. tomentosa,* Lindl. in Mitch. Three Exped. ii. 177); near Stowell, *J. Holt* (with very double flowers).

Amidst all the variations in foliage and flowers, this species is always to be recognized by the five cavities on the outside of the tube just above the ovary, which are easily seen in the dry as well as in the fresh state, and of which I have not observed any trace in any other species. The plant referred to as *E. ruscifolia,* Br. in Hook. f. Fl. Tasm. i. 257, is the above var. *ovata.* Brown's own specimens of *E. ruscifolia* belong to the common narrow-leaved form, but with rather long flowers. The N. S. Wales plant, referred doubtfully to *E. impressa* by J. D. Hooker, is the *E. reclinata,* A. Cunn., in which I have never found the cavities characteristic of *E. impressa.*

4. **E. sparsa,** *R. Br. Prod.* 551. A tall shrub, quite glabrous. Leaves shortly petiolate, elliptical-oblong or oblong-lanceolate, minutely mucronulate, flat, smooth and veinless or very obscurely veined, nearly ½ in. long. Flowers shortly pedicellate. Bracts and sepals rather broad, acute, striate,

the sepals about 1½ lines long. Corolla not seen, but the filiform style, ¼ in. long, shows that it has probably a long tube. Ovary entirely of *Epacris.* —DC. Prod. vii. 762.

N. S. Wales. Grose river, *R. Brown.* Specimens without flower of the Sydney Woods collection, Paris Exhibition, 1855, n. 148, *M'Arthur,* said to be from a shrub of 10 to 15 ft., appear from the foliage to be this species, which however, must remain in some measure doubtful till more perfect specimens have been examined.

5. **E. petrophila,** *Hook. f. Fl. Tasm.* i. 261. A low, rigid, depressed or bushy shrub, the branchlets pubescent but concealed by the foliage. Leaves sessile, imbricate and often decussate, ovate, obtuse, thick, concave, prominently keeled, usually about 1 line long, or nearly 2 lines on luxuriant branches, all only ½ line on some slender branchlets. Flowers nearly sessile in the uppermost axils, so as to form little terminal heads. Bracts and sepals obtuse, the latter scarcely above 1 line long. Corolla-tube campanulate, not exceeding the calyx ; lobes rather broad, obtuse, scarcely so long as the tube. Hypogynous scales short, broad, truncate, sometimes slightly cohering. Style very short.—*E. microphylla,* Hook. f. in Hook. Lond. Journ. vi. 272, not of R. Br.

Victoria. Summits of the Munyong, Bogong, Baw-Baw, Mitta-Mitta and other mountains, at an elevation of 4000 to 5000 ft., *F. Mueller.*

Tasmania. Stony places on the summits of the Western mountains, and at Arthur's Lakes, at an elevation of 3700 to 4000 ft., *Gunn.*

This species differs from *E. serpyllifolia* in its thick, very obtuse leaves, bracts, etc., but may not be really distinct from *E. rigida,* differing in the smaller, more imbricate leaves, rather smaller flowers, the filaments more adnate, and the absence of any transverse thickening of the corolla-tube.

6. **E. rigida,** *Sieb. ; Spreng. Syst. Cur. Post.* 64. An erect, rigid, bushy shrub, of 1 or rarely 2 ft., the branchlets scarcely pubescent. Leaves almost sessile, erect or spreading, ovate or ovate-oblong, very obtuse, very thick, the thick keel prominent underneath, mostly 1 to 1½ lines long, but those of the older branches sometimes 2 or even 3 lines when narrow. Peduncles in the uppermost axils usually very short, but sometimes above 1 line long. Sepals 1 to 1¼ line long, obtuse or rarely 1 or 2 of them almost acute. Corolla-tube broad, as long as the calyx, with 5 transverse thickenings inside, forming more or less of a ring just above the ovary ; lobes longer than the tube, broad, obtuse. Hypogynous disk exceedingly short or almost obsolete. Ovary less depressed at the insertion of the style than in most species.—DC. Prod. vii. 762.

N. S. Wales. Frequent in arid situations on the Blue Mountains, *A. Cunningham, Fraser, Sieber, n.* 90, *Fl. Mixt. n.* 488, and others.

Var. *laxa.* Leaves more spreading, the keel sometimes less prominent. Peduncles 1 to 1¼ lines long. Sepals less obtuse.—*E. Muelleri,* Sond. in Linnæa, xxvi. 252.—Blue Mountains, *Siemssen, Clowes.* The specimens examined by Sonder have a very different aspect from the typical form, but others are quite intermediate.

E. alpina of New Zealand is very closely allied to this species and to *E. petrophila,* with the same obtuse sepals and bracts, but with the foliage approaching rather that of *E. serpyllifolia.* The transverse thickenings of the corolla-tube in *E. rigida* were evident in all the flowers examined, but may possibly not be constant in the species.

7. **E. coriacea,** *A. Cunn. ; DC. Prod.* vii. 763. A tall shrub, attain-

ing sometimes 8 to 10 ft. Leaves petiolate, obovate, broadly ovate or almost orbicular, obtuse or with an obscure callous point, thick, concave, smooth, in some specimens scarcely 2 lines, in others 3 to 4 lines long. Flowers axillary along the branches, the peduncles sometimes all under 1 line, sometimes 2 to 3 lines long. Bracts and sepals rigid, obtuse, concave, ciliolate, the sepals scarcely above 1 line long. Corolla-tube campanulate, about as long as the calyx, and the lobes as long as the tube. Hypogynous scales short and broad. Style very short.

N. S. Wales. Blue Mountains and mountain belt near Illawarra, *A. Cunningham;* Sydney Woods, Paris Exhibition, 1855, n. 184, *M'Arthur;* also in *Leichhardt's* collection.

8. **E. crassifolia,** *R. Br. Prod.* 551. Stems procumbent or trailing, short and slender in elevated situations, rather stouter and extending to above a foot near the sea, the branches glabrous or pubescent. Leaves obovate, very obtuse, narrowed into a petiole, the margins usually thickened and sometimes ciliate, the veins few and often obscure, $\frac{1}{4}$ to $\frac{1}{2}$ in. long. Flowers very variable in size, axillary along the branches, the peduncles very short. Bracts and sepals obtuse or rarely almost acute, ciliate, usually decussate, the sepals 2 to nearly 3 lines long. Corolla-tube shortly exceeding the calyx; lobes short and broad. Hypogynous scales short and broad. Stigma rather broad. —DC. Prod. vii. 764.

N. S. Wales. Port Jackson, near the sea (with large leaves and flowers), *R. Brown, Woolls,* and others; summits of the Blue Mountains (with small leaves and flowers), *A. Cunningham, Fraser,* and others.

9. **E. robusta,** *Benth.* An erect, stout, branching shrub of 3 to 4 ft., the branchlets scarcely pubescent. Leaves shortly petiolate, very spreading, obovate or almost orbicular, very obtuse, thick, few-nerved, mostly about $\frac{1}{4}$ in. long. Flowers in the upper axils, rather large, white with a slight yellowish tinge. Bracts and sepals obtuse, scarcely decussate, the sepals fully 3 lines long. Corolla-tube not exceeding the calyx; lobes obtuse, about as long as the calyx.

N. S. Wales or **Victoria.** Granite rocks at the summit of the White Peak mountain, at the head of Genoa river, *F. Mueller.* Allied to *E. obtusifolia* and *E. crassifolia,* differing from the former in foliage, and from the latter in habit.

10. **E. obtusifolia,** *Sm. Exot. Bot.* i. 77. *t.* 40. An erect shrub, with virgate, usually pubescent branches, from under 1 ft. to about 3 ft. high. Leaves oblong-elliptical, obtuse, thick, few-nerved, slightly concave, narrowed into a short petiole, from under $\frac{1}{4}$ in. long when broad to nearly $\frac{1}{2}$ in. when narrow. Flowers white, axillary, usually forming long one-sided leafy racemes, either almost sessile or on peduncles of above 1 line. Bracts and sepals obtuse, ciliate, usually decussate, the sepals about 3 lines long. Corolla sometimes almost campanulate, the tube shortly exceeding the calyx, the lobes broad. Anthers not protruding from the tube. Hypogynous scales distinct, obtuse.—R. Br. Prod. 551; DC. Prod. vii. 762; Hook. f. Fl. Tasm. i. 260; Lodd. Bot. Cab. t. 292.

Queensland. Marshes, Moreton Island, *F. Mueller.*
N. S. Wales. Port Jackson to the Blue Mountains, chiefly in marshy ground, *R. Brown, Sieber, n.* 84, *Fl. Mixt. n.* 487, and many others; New England, *J. Stuart;* Hastings river, *Beckler ;* Clarence river, *Riley.*

Victoria. Near Brighton, *F. Mueller.*
Tasmania. Wet soil, northern and southern shores of the island, *J. D. Hooker.*

11. **E. myrtifolia,** *Labill. Pl. Nov. Holl.* i. 41. *t.* 55. A stout, erect, bushy shrub, of about 1 ft., glabrous or with pubescent branches. Leaves shortly petiolate, crowded, ovate, often shortly acuminate but not acute, concave, several-nerved, mostly 3 to 4 lines long. Flowers crowded in leafy heads or spikes, nearly sessile or on peduncles of about 1 line. Bracts and sepals of a firm consistence, rather obtuse or almost acute, not usually ciliate, the sepals 2½ lines long. Corolla-tube scarcely so long as the calyx, the lobes at least as long as the tube. Anthers partially protruding. Hypogynous scales small.—R. Br. Prod. 551; DC. Prod. vii. 763 (except as to A. Cunningham's specimens, which are *E. serpyllifolia*) ; Hook. f. Fl. Tasm. i. 259.

Tasmania. Recherche Bay, *Labillardière ;* Port Arthur, *J. D. Hooker.*
E. corymbiflora, Hook. f. Fl. Tasm. i. 261. t. 78 A, appears to me to be a slight variety of this species, with smaller, more concave and appressed leaves, but not otherwise different.

12. **E. exserta,** *R. Br. Prod.* 551. An erect shrub, attaining several feet, the branches sometimes short and crowded, sometimes long and virgate. Leaves shortly petiolate, from broadly elliptical-oblong or almost ovate to narrow oblong-lanceolate, obtuse or almost acute, flat or slightly concave, keeled or 3-nerved underneath, 2 to 4 lines long. Flowers either in the uppermost axils forming leafy heads, or more scattered along the branches in leafy spikes. Peduncles very short or rarely 1 line long. Bracts obtuse, sepals rather more acute, varying from 1 to 1½ lines long. Corolla-tube rather shorter than the calyx, the lobes at least as long as the tube. Filaments rather long, so that the anthers protrude rather more than in the allied species. Hypogynous scales broad, but often very small.—DC. Prod. vii. 763 ; Hook. f. Fl. Tasm. i. 260.

Tasmania. Northern parts of the island, *J. D. Hooker ;* South Esk river, Launceston, etc., *Gunn.*
Var. *virgata.* Leaves smaller, approaching those of *E. serpyllifolia.* Flowers rather smaller.—*E. virgata*, Hook. f. in Hook. Lond. Journ. vi. 271, and Fl. Tasm. i. 260. t. 79 A. —Asbestos hills, Yorktown, and between Hobarton and Huon, *Gunn.*

13. **E. mucronulata,** *R. Br. Prod.* 552, *not of Hook. f.* An erect shrub, attaining about 6 ft., glabrous or the branches minutely pubescent. Leaves on rather long petioles, lanceolate, acute, tapering at the base, slightly concave, obscurely keeled or 3-nerved, 3 to 4 lines long. Flowers in the upper axils, on rather long peduncles. Bracts and sepals rather obtuse when old, more acute when young, the sepals about 2 lines long. Corolla-tube cylindrical, slightly enlarged upwards, shortly exceeding the calyx, the lobes much shorter than the tube. Anthers wholly included. Hypogynous scales short and broad.—DC. Prod. vii. 764 ; *E. Franklinii*, Hook. f. Fl. Tasm. i. 261. t. 79 B.

Tasmania. Port Espérance, *R. Brown ;* abundant in annually inundated places on the Franklin river, near Macquarrie harbour, *Gunn.* Brown's specimens are in young bud, Gunn's are past flower, but both appear to belong to one species.

14. **E. lanuginosa,** *Labill. Pl. Nov. Holl.* i. 42. *t.* 57. An erect, rigid shrub, attaining several feet, the branches usually pubescent, sometimes

long and virgate, sometimes short and bushy. Leaves erect or spreading, lanceolate or linear-lanceolate, tapering into a short, rigid point, keeled but otherwise smooth or several-ribbed on the back, $\frac{1}{4}$ to $\frac{1}{2}$ in. long. Flowers nearly sessile, either in the uppermost axils forming dense leafy heads, or forming long leafy spikes along the branches. Bracts and sepals acute or the outer ones rather obtuse, ciliate, scarcely decussate, the sepals 2 to $2\frac{1}{2}$ lines long. Corolla-tube as long as or shortly exceeding the sepals, the lobes rather large. Hypogynous scales lanceolate. Ovary usually pubescent, and the style always more or less hairy.—R. Br. Prod. 551; DC. Prod. vii. 763; Hook. f. Fl. Tasm. i. 258.

Victoria. Portland Road, *Robertson*, Wilson's Promontory, *F. Mueller.*

Tasmania. Port Dalrymple, *R. Brown;* not uncommon in hilly districts in various parts of the island, *J. D. Hooker.*

15. **E. paludosa,** *R. Br. Prod.* 551. An erect shrub, with the habit, acute narrow leaves, and inflorescence of *E. lanuginosa,* and flowers the same or rather larger, except that the rigid, acute bracts and sepals are not ciliate, and the style in all the specimens examined is always perfectly glabrous.— DC. Prod. vii. 762; Lodd. Bot. Cab. t. 1226.

N. S. Wales. Port Jackson to the Blue Mountains, *R. Brown, Sieber, n.* 81, and many others; Illawarra, *A. and R. Cunningham;* in the interior, *M'Arthur;* Berrima, *Woolls.*

Victoria. Munyong mountains, sources of the Mitta-Mitta, Mount Useful, Mount Cobra; snowy regions about Mount Wellington, *F. Mueller.*

Trifling as are the distinctions between this and *E. lanuginosa,* they generally give a different aspect to the plant, and appear to be constant.

16. **E. heteronema,** *Labill. Pl. Nov. Holl.* i. 42. *t.* 56. A rigid shrub, sometimes under 1 ft. high and scrubby, sometimes very tall or even (according to C. Stuart) forming a tree of 20 to 30 ft. Leaves nearly sessile, from broadly ovate to lanceolate, tapering into a short pungent point, not cordate, concave, smooth and almost veinless in the typical form, imbricate or spreading, $\frac{1}{4}$ to nearly $\frac{1}{2}$ in. long. Flowers in the uppermost axils, forming a short, terminal head, rarely lengthening into a short leafy spike. Bracts and sepals acuminate-acute, not ciliate, the sepals usually about 2 lines long. Corolla-tube shorter than the calyx; lobes as long as the tube. Hypogynous scales short and broad; style short.—R. Br. Prod. 551; DC. Prod. vii. 762; Hook. f. Fl. Tasm. i. 259.

Tasmania. Recherche Bay, *Labillardière;* Port Davey, *Milligan;* high heathy places between Franklin and Gordon rivers, *Gunn;* South Port, *C. Stuart.*

Var. ? *planifolia.* Leaves less pungent, flat, the midrib and sometimes the lateral nerves conspicuous underneath. Flowers in a longer, leafy spike. Bracts and sepals acute, but not acuminate, minutely ciliolate. Corolla-tube short, the lobes twice as long.—*E. heteronema,* Hook. Bot. Mag. t. 3257 (incorrect as to the corolla).

N. S. Wales. Swampy lands, Blue Mountains, and N. of Bathurst, *A. Cunningham;* also in *Leichhardt's* collection.

Victoria. Munyong, Baw-Baw, and Mitta-Mitta mountains, Mount Aberdeen, etc., at an elevation of 3000 to 4000 ft., *F. Mueller.*

Tasmania. South Port, *C. Stuart.*

E. dubia, Lindl. Bot. Reg. 1846, t. 38, may be a garden variety of *E. heteronema.* The figure does not, however, correspond with any of our specimens, and the æstivation of the corolla is represented as contorted, but perhaps by an error of the artist.

17. **E. serpyllifolia,** *R. Br. Prod.* 551. A dwarf shrub, either erect and bushy or prostrate and not exceeding 1 ft., glabrous or the branchlets pubescent. Leaves ovate, rather acute or very shortly mucronate, 1 to 2 or rarely when narrow 3 lines long, flat or slightly concave, keeled or obscurely 3-nerved underneath. Flowers in the uppermost axils, forming short leafy heads rarely lengthening out into leafy spikes or becoming lateral by the elongation of the terminal shoot. Bracts and sepals rather broad, acute, often coloured, the sepals nearly 1½ lines long. Corolla-tube scarcely exceeding the calyx, the lobes shorter than the tube. Anthers sometimes partially protruding. Hypogynous scales broad, truncate. Style rather long. —DC. Prod. vii. 763 ; Hook. f. Fl. Tasm. i. 260.

Tasmania. Mount Wellington, *R. Brown* ; summits of all the mountains at an elevation of 3000 to 4000 ft., *J. D. Hooker;* also in Labillardière's collection, but not included in his work.
Victoria. Munyong mountains at an elevation of 5000 to 6000 ft., *F. Mueller.* These specimens have much smaller flowers, but appear to belong to the same species.

Var. *squarrosa.* Leaves usually larger and often recurved. Flowers rather larger.—*E. squarrosa,* Hook. f. Fl. Tasm. i. 259.—Elizabeth river and Oyster Bay, E. coast of Tasmania, *Gunn.*

18. **E. microphylla,** *R. Br. Prod.* 550. An erect shrub with virgate sometimes rather slender branches, often flowering to a considerable length. Leaves cordate, broadly ovate, shortly acuminate or acute in the typical form or sometimes almost obtuse, very concave and broad above the base, erect spreading or reflexed, 1½ to nearly 2 lines long. Flowers numerous along the branches, small, almost sessile or on peduncles of 1 or even 2 lines. Bracts and sepals obtuse or rarely the innermost almost acute, the sepals under 1 line long. Corolla-tube shorter than the calyx, the lobes as long as the tube. Anthers wholly included. Hypogynous scales short. Style very short.—DC. Prod. vii. 760 ; Bot. Mag. t. 3658 ; *E. pulchella,* Sims, Bot. Mag. t. 1170 ; Lodd. Bot. Cab. t. 194, not of Cav. ; *E. rivularis,* Sieb. ; Spreng. Syst. Cur. Post. 64 ; *E. auriculata,* Benth. in Hueg. Enum. 76 ; DC. Prod. vii. 761 ; *E. pedicellata,* DC. Prod. vii. 761.

Queensland. Moreton Bay; *Fitzalan ;* Moreton island, *F. Mueller.*
N. S. Wales. Port Jackson, *R. Brown, Sieber, n.* 88, 89, 91, and *Fl. Mixt. n.* 485, 613, and others ; northward to Hastings river and Mount Mitchell, *Beckler ;* southward to Twofold Bay, *F. Mueller.*

Var. *Gunnii.* Leaves rather larger with longer points, but much less acuminate than in *E. pulchella.* Flowers rather larger than in the typical *E. microphylla,* and the sepals rather more acute.—*E. Gunnii,* Hook. f. in Hook. Lond. Journ. vi. 272, and Fl. Tasm. i. 256. t. 78 B.
N. S. Wales. In the interior ; near Berrima, *Woolls;* upper Murray river, *W. P. Bull.*
Victoria. Cobberas and Dandenong mountains, Genoa Peak, and others of the Australian Alps, *F. Mueller.*
Tasmania. Wet places in mountain districts, chiefly in the interior of the island, at an elevation of 3000 to 4000 ft., *J. D. Hooker.*

19. **E. acuminata,** *Benth.* An erect bushy shrub of 2 to 3 ft., gla-

brous or the branchlets pubescent. Leaves nearly sessile, ovate, acute or tapering into a pungent point, erect concave and clasping the stem at the base, spreading or recurved towards the end, 2 to 3 or rarely 4 lines long. Flowers few, nearly sessile in the uppermost axils. Bracts and sepals rather broad, ciliate, acute but not acuminate, the sepals nearly $1\frac{1}{2}$ lines long. Corolla-tube about as long as the calyx, the lobes obtuse, shorter than the tube. Filaments half as long as the lobes, the anthers wholly exserted. Hypogynous scales ovate-triangular, obtuse. Style exserted.—*E. mucronulata*, Hook. f. Fl. Tasm. i. 258, not of R. Br.

Tasmania. Hobarton, New Norfolk, *Gunn ;* Mount Wellington, *J. D. Hooker ;* N. W. Bay, *C. Stuart.*

20. **E. apiculata,** *A. Cunn. in Field, N. S. Wales,* 340. A straggling shrub, the young branches pubescent or villous but nearly concealed by the foliage. Leaves broadly cordate-ovate, shortly and obtusely acuminate, very concave near the base, 3 to 4 lines long. Flowers along the branches, nearly sessile, mostly shorter than the leaves. Bracts and sepals obtuse or the inner ones almost acute, the sepals about $1\frac{1}{4}$ lines long. Corolla-tube shorter than the calyx, readily splitting into petal-claws, the lobes longer than the tube. Filaments very shortly adnate, the free part thicker than in the other species ; anthers large, included in the corolla-tube. Hypogynous scales very short. Ovary not much depressed round the style, which is rather long and thickened above the base.—DC. Prod. vii. 761.

N. S. Wales. Rare in rocky ravines near King's Table Land, Blue Mountains, *A. Cunningham.* The specimens have much the aspect of *E. purpurascens,* but with smaller flowers, and (if normal in the specimens examined) they show a structure approaching in some respects that of *Lysinema,* although with the æstivation of *Epacris.*

21. **E. pulchella,** *Cav. Ic.* iv. 26. *t.* 345. An erect shrub with virgate branches minutely pubescent. Leaves spreading, nearly sessile, cordate-ovate, acuminate, tapering into a rigid point, broad and very concave near the base, mostly 2 to nearly 3 lines long. Flowers along the branches on very short peduncles. Inner bracts and sepals acutely acuminate, the sepals about 2 lines long. Corolla-tube broad, as long as the calyx, the lobes rather obtuse, scarcely shorter than the tube. Hypogynous scales short and broad. Style rather long.—R. Br. Prod. 550 ; DC. Prod. vii. 760 ; *E. purpurascens,* Sieb. Pl. Exs., not of R. Br.

N. S. Wales. Port Jackson, *R. Brown, Sieber, n.* 87, and others ; Richmond river, *Fawcett.*

E. riparia, R. Br. Prod. 550 ; DC. Prod. vii. 761, from Grose river, appears to be the same species but with larger flatter leaves, mostly 3 to 4 lines long.

22. **E. purpurascens,** *R. Br. Prod.* 550. An erect rigid shrub of several feet, the branches often elongated and completely covered by the leaves. Leaves ovate, acuminate and tapering into a pungent point, broad and rounded or cordate at the base, very concave and embracing the stem in the lower part, spreading or recurved upwards, mostly $\frac{1}{2}$ in. long or rather more. Flowers along the branches sessile or very shortly pedunculate and usually shorter than the leaves, white or more or less tinged with red. Bracts and sepals acutely acuminate, the sepals about 3 lines long. Corolla-tube

not longer than the calyx, the lobes acute.   Anthers included or partially exserted.   Hypogynous scales short and broad.—DC. Prod. vii. 760; Lodd. Bot. Cab. t. 237; *E. pungens,* Sims, Bot. Mag. t. 844, not of Cav.; *E. rubra,* Lodd. Bot. Cab. t. 876; *E. onosmæflora,* A. Cunn. in Field, N. S. Wales, 340; DC. Prod. vii. 763; *Lysinema ruscifolium,* Sieb.; Spreng. Syst. Cur. Post. 64.

N. S. Wales.   Port Jackson to the Blue Mountains, *R. Brown, Sieber, n.* 73, 74, and others.

The species is also in New Zealand.

17. **LYSINEMA**, R. Br.

Corolla-tube cylindrical, entire or separating at the base or altogether into distinct petal-claws; lobes 5, horizontally spreading, contorted in the bud. Filaments free from the base or more or less adnate to the corolla, especially towards the throat; anthers linear, attached at or above the middle, wholly or partially included in the tube.   Hypogynous disk of 5 distinct scales, usually as long as the ovary.   Ovary 5-celled, with several usually numerous ovules in each cell, on a placenta attached to the axis.   Style often thickened above the ovary, inserted in a shallow or deep and tubular depression of the ovary.   Capsule loculicidally dehiscent.—Erect straggling or virgate shrubs. Leaves persistent.   Flowers almost sessile and solitary in the upper axils, forming leafy heads or spikes or rarely more scattered along the branches. Bracts very numerous, imbricate, covering the very short peduncles and calyx, passing gradually into the sepals and forming an involucre round them.

The genus is limited to Australia.   It is very near *Epacris,* but readily distinguished by the more strictly hypocrateriform corolla, with the imbrication of the lobes strictly contorted, not showing one external and one internal lobe as in *Epacris,* besides the frequent tendency to separation of the filaments and petal-claws.   The inflorescence, ovary, and placentation show no constant difference.   The tubular depression of the ovary round the style is deep in *L. pungens,* and rather deep in *L. lasianthum* and *L. ciliatum,* but shallow in the others, at least in the flowers examined, but seems variable according to the stage of growth in which it is observed.   The lower ovules of each placenta are usually reflexed, the upper ones erect and the intermediate ones horizontal as in *Epacris,* but in after growth it is chiefly the upper ones that enlarge, so that the seeds are usually (though not always) all erect from an apparently basal placentæ.

Leaves spreading, acuminate, with pungent points.   Eastern species .   1. *L. pungens.*
Leaves small, erect and acuminate, or spreading and obtuse.   Western
   species.
   Flowers distant in unilateral interrupted spikes.   Corolla-tube to-
     mentose; lobes very small . . . .   2. *L. lasianthum.*
   Flowers close together, few or forming a dense terminal head or
     spike.
     Anthers wholly included.   Petal-claws usually cohering, at least
      at the throat.
      Leaves lanceolate-subulate.   Sepals and bracts scarcely ciliate.
       Corolla-tube under 4 lines; lobes acute . . . . . .   3. *L. conspicuum.*
      Leaves ovate to lanceolate, obtuse.   Sepals and bracts woolly-
       ciliate.   Corolla-tube 5 to 8 lines; lobes obtuse . . .   4. *L. ciliatum.*
      Leaves ovate to lanceolate, obtuse.   Sepals and bracts bor-

dered by crisped teeth or lobes. Corolla-tube ¾ to 1 in.
  long ; lobes obtuse . . . .   **5.** *L. fimbriatum.*
Anthers long, linear, the tips exserted and recurved. Petal-
  claws entirely separate . . . .   **6.** *L. elegans.*

1. **L. pungens,** *R. Br. Prod.* 552.   An erect shrub with long branches,
hoary-pubescent when young but almost concealed by the leaves. Leaves
ovate, acuminate and tapering into a rigid point, broad very concave and
shortly erect immediately above the base, then very spreading, ¼ tc nearly ½
in. long. Flowers white or rarely red, sessile in the upper axils. Calyx and
bracts forming a narrow involucre tapering to the top, 4 to 5 lines long.
Corolla-tube slender, as long as or rather longer than the calyx ; lobes broad,
mucronulate. Filaments free or more or less adnate to the corolla-tube.
Hypogynous scales narrow, acuminate.—DC. Prod. vii. 765 ; *Epacris pungens,* Cav. Ic. iv. 26. t. 346 ; Bot. Mag. t. 1199 ; F. Muell. Fragm. vi. 70 ;
E. attenuata, Lodd. Bot. Cab. t. 38 ; *Lysinema attenuatum,* Link, Enum.
Hort. Berol. i. 211 ; *Epacris rosea,* Lodd. Bot. Cab. t. 863 ; *E. riparia,* Sieb.
Pl. Exs., not of R. Br. ; *Lysinema Sieberi,* Benth. in Hueg. Enum. 76.

Queensland. Moreton Island, *F. Mueller.*
N. S. Wales. Port Jackson, *R. Brown, Sieber, n.* 83, and others.

2. **L. lasianthum,** *R. Br. Prod.* 552.   A slender erect slightly branched
shrub of 1 to 2 ft. Leaves linear, obtuse, erect, concave, mostly 2 to 4 lines
long. Flowers nearly sessile in the upper. axils, but rather distant and all
turned to one side. Sepals and bracts forming a rather broad almost ovoid
involucre, all obtuse, more or less scarious-ciliate, the innermost sepals 3
lines long. Corolla-tube about 4½ lines long, tomentose outside, readily separating into petal-claws ; lobes obtuse, scarcely above 1 line long. Filaments free or slightly cohering to the corolla-tube. Hypogynous scales as
long as the ovary, broad, truncate. Capsule ovoid, as long as the calyx.
Seeds flat.—DC. Prod. vii. 765 ; Sond. in Pl. Preiss. i. 329 ; *Epacris lasiantha,* Poir. Dict. Suppl. ii. 555 ; *Lysinema brevilimbatum,* F. Muell. Fragm.
iii. 142, vi. 70.

W. Australia. Wet places, King George's Sound and adjoining districts, *R. Brown,
Drummond, 5th Coll. n.* 295, *Preiss, n.* 440, *Oldfield, F. Mueller.*

3. **L. conspicuum,** *R. Br. Prod.* 552.   Stems erect, simple or slightly
branched, slender, virgate, 1 to 2 ft. high. Leaves linear or lanceolate-
subulate, erect and appressed, scarcely acute, concave, minutely ciliclate, 2 to
3 lines long. Flowers sweet scented, nearly sessile in the upper axils, forming
a dense cylindrical spike of 1 to 2 in. Sepals and bracts narrow, acuminate,
acute, the innermost sepals 2 to 2½ lines long. Corolla-tube slender, about
4 lines long, the petal-claws and filaments usually cohering but very readily
separating ; lobes acute, nearly 2 lines long. Anthers wholly included,
attached near the top, the lower end usually shortly sterile. Hypogynous
scales linear.—DC. Prod. vii. 765 ; Sond. in Pl. Preiss. i. 329 ; F. Muell.
Fragm. vi. 70 ; *Epacris conspicua,* Poir. Dict. Suppl. ii. 556.

W. Australia. Swamps and peaty places, King George's Sound and the immediate
neighbourhood, *R. Brown, Drummond, n.* 108, *Preiss, n.* 444, and many others.

4. **L. ciliatum,** *R. Br. Prod.* 552.   An erect shrub of from 1 to 3 or

                                                                    R 2

even 4 ft., with long and virgate or short rigid and crowded branches, glabrous or hoary-pubescent. Leaves from ovate crowded spreading decussate and not 2 lines long, to lanceolate erect appressed and above 3 lines long, all obtuse, rather thick and more or less concave. Flowers usually in short dense terminal spikes, rarely more scattered in the upper axils, and very variable in size. Bracts and sepals forming an involucre cylindrical or slightly conical from scarcely 4 to above 6 lines long, all acute or almost obtuse, the margins always fringed with short woolly hairs. Corolla-tube sometimes not longer than the calyx, sometimes exceeding it by 1 or even 2 lines, glabrous or pubescent outside, separating into petal-claws either entirely free or more frequently cohering towards the throat; lobes or petal-laminæ obtuse, 2 to 3 lines long. Stamens entirely free; anthers linear, entirely included in the corolla-tube. Hypogynous scales broadly oblong.—DC. Prod. vii. 765; Sond. in Pl. Preiss. i. 328; F. Muell. Fragm. vi. 70; *L. pentapetalum*, Br. Prod. 552; DC. Prod. vii. 764; Sond. in Pl. Preiss. i. 327; F. Muell. Fragm. vi. 69; *Epacris ciliata* and *E. pentapetala*, Poir. Dict. Suppl. ii. 555; *Lysinema virgatum*, DC. Prod. vii. 765; *L. curvatum*, Lindl. Sw. Riv. App. 25; Sond. in Pl. Preiss. i. 328; *L. spicatum*, Lindl. l. c.; *L. ovatum*, Sond. l. c. 329.

W. Australia. King George's Sound and Lucky Bay, *R. Brown;* apparently very abundant from the south coast to Swan River and Murchison river, *Oldfield,* as it appears in almost all collections, *Drummond, 1st Coll. n.* 78; *Preiss, n.* 441, 442, 443, and many others.

The short crowded ovate leaves and long distant narrow ones, rounded or narrowed at the base, occur on different branches of the same shrub; the length of the corolla-tube, which chiefly distinguished the two supposed species, is exceedingly variable, it always appears longer when the flowering is advanced; the occasionally curved calyx of *L. curvatum* appears to be accidental.

5. **L. fimbriatum,** *F. Muell. Fragm.* iv. 125. vi. 70. An erect shrub of 1 to 2 ft., with long and rather slender virgate branches. Leaves erect and closely appressed, lanceolate or linear, obtuse, keeled, 1 to 2 or rarely 3 lines long. Flowers few, large, in terminal heads. Bracts and sepals forming a cylindrical involucre of 6 to 8 lines, the sepals and especially the bracts obtuse, with their margins split into scarious, crisped teeth, giving them an elegant fringed aspect. Corolla-tube exceeding the calyx by 1 to 2 lines, the petal-claws cohering at the throat, but usually separating lower down; lobes or laminæ 3 to 4 lines long, oblong, obtuse. Filaments free; anthers wholly included in the corolla-tube. Hypogynous scales narrow.

W. Australia. King George's Sound or to the eastward, *Baxter;* moist sandy places, Williams and Tone rivers, *Oldfield;* Upper Kalgan river, base of Stirling Range, *F. Mueller.*

6. **L. elegans,** *Sond. in Pl. Preiss.* i. 327. A shrub of 1 to 2 ft., with erect, rather slender, usually pubescent branches. Leaves erect, appressed, oblong-lanceolate or linear, obtuse, keeled, 1 to 2 lines long on the main branches, often all under 1 line on the side ones, and occasionally opposite. Flowers in compact terminal heads, rarely lengthening into cylindrical spikes of 2 in. Bracts and sepals forming a narrow, cylindrical involucre of 4 to 5 lines, all very narrow, rather obtuse, minutely ciliate or entire. Corolla-tube

nsually 2 to 3 lines longer than the calyx, the petal-claws sometimes slightly cohering at the throat, but usually quite free, especially when the flowering is advanced; lobes or laminæ obtuse, about 2 lines long. Anthers narrow and very long, the lower end often sterile, the upper end more or less protruding from the corolla-tube, and recurved. Hypogynous scales narrow, entire or jagged. Ovary pubescent; style and filaments usually hairy. Capsule narrow.

W. Australia. Swan River, *Drummond, 1st Coll., also n.* 122, *2nd Coll.* 262; Canning river, *Preiss, n.* 439.

18. ARCHERIA, Hook. f.

Corolla-tube rather broad, ventricose-cylindrical or almost campanulate; lobes 5, short, spreading or recurved, imbricate, but not contorted in the bud. Filaments inserted in the throat, very short; anthers short and broad, attached about the middle. Hypogynous disk short, annular or cup-shaped, or obsolete. Ovary 5-celled, with rather numerous ovules, on placentas arising from near the base of the axis; style short or long, inserted in a tubular depression reaching nearly to the base of the ovary; stigma clavate or dilated, more or less distinctly 5-lobed. Capsule loculicidally dehiscent. —Shrubs, often of a straggling habit. Leaves flat. Flowers usually white, pedicellate, in short terminal racemes, which are either leafy, each flower solitary within a floral leaf, or leafless, the subtending bracts small and deciduous, the bracteoles always at the base of the pedicels, and usually deciduous.

The genus is confined to Tasmania and New Zealand, the species of each country all endemic. It is united by F. Mueller with *Epacris*, from which it differs in habit, in the want of the persistent involucrating bracts, and in the almost basal attachment of the style. One species bears a remarkable general resemblance to the South Andine *Prionotes* (*Allodape* or *Lebetanthus* of Endlicher), which, however, has the hypogynous stamens and other characters much nearer to those of *Prionotes*.

Flowers solitary within leaf-like bracts, forming at first a terminal,
    leafy raceme, afterwards lateral. Leaves distichous, acute.
  Sepals obtuse. Ovary pubescent . . . . 1. *A. eriocarpa.*
  Sepals rather acute. Ovary glabrous . . . . 2. *A. hirtella.*
Flowers in a short, terminal, leafless raceme. Leaves irregularly
    crowded, obtuse . . . . 3. *A. serpyllifolia.*
(To this group belong the two New Zealand species.)

1. **A. eriocarpa,** *Hook. f. Fl. Tasm.* i. 263. *t.* 80 B. A straggling, half-climbing shrub, growing to the height of 10 to 12 ft., the branchlets shortly hispid. Leaves nearly sessile, somewhat distichous, ovate to lanceolate, acute and rigidly mucronate, flat, with several nerves slightly prominent on the upper surface, minutely serrulate, 3 to 4 lines long. Flowers on very short pedicels in the upper axils or forming a short terminal raceme. Sepals ovate, obtuse, $1\frac{1}{4}$ lines long, striate in the centre, the margins somewhat scarious, and minutely ciliolate. Corolla-tube ventricose, about 3 lines long, contracted at the throat; lobes small. Hypogynous disk short, thick, sinuate. Ovary hirsute; style filiform; stigma shortly 5-lobed. Capsule about 2 lines long.

Tasmania. Dense forests, S.W. of Lake St. Clair and Macquarrie Harbour, *Gunn,* *Milligan.*

2. **A. hirtella,** *Hook. f. Fl. Tasm.* i. 263. *t.* 81. A large shrub or small tree of 10 to 12 ft., with stout stems and rather spreading branches, hirsute with short spreading hairs. Leaves more or less distichous, ovate, acute, with short pungent points, minutely serrate-ciliate, thick, smooth and shining, obscurely 1-nerved, mostly 3 to 4 lines long. Pedicels in the upper axils, 1 to 2 lines long, forming a short, terminal, leafy raceme. Bracts several, the lower ones small and persistent, the upper ones larger, but very deciduous. Sepals ovate-lanceolate, ciliate, nearly 2 lines long. Corolla-tube ovate-cylindrical, slightly contracted at the throat, fully 3 lines long; lobes short, recurved. Hypogynous disk shortly lobed. Style as long as the corolla.—*Epacris hirtella,* Hook. f. in Hook. Lond. Journ. vi. 271; F. Muell. Fragm. vi. 71.

Tasmania. Dense forests near Macquarrie Harbour and Acheron Valley, S.W. of Lake St. Clair, *Gunn;* foot of Mount Lapeyrouse, *Oldfield.*

3. **A. serpyllifolia,** *Hook. f. Fl. Tasm.* i. 263. *t.* 80 A. A stout, rigid, prostrate or bushy shrub of 6 in. to 1 ft., often clinging closely to the face of the rocks, usually quite glabrous. Leaves crowded, not distichous, ovate-elliptical and obtuse or rather narrower and less obtuse, thick, slightly concave, obscurely 1- or 3-nerved, 3 or rarely 4 lines long. Flowers in short, terminal, dense, sessile racemes, the lower empty bracts small and persistent, the upper subtending ones larger and very deciduous. Sepals 1½ to nearly 2 lines long, narrow-oblong, rather obtuse, ciliate. Corolla-tube broad, almost campanulate, about 2 lines long, hairy inside above the middle; lobes short, somewhat spreading, but not recurved, bearded inside below the middle. Hypogynous disk none in the flowers examined. Ovary glabrous; style very short.—*Epacris micranthera,* F. Muell. Fragm. vi. 72.

Tasmania. Rocks on the summit of Mount Olympus, *Gunn.*

Var. *minor.* Flowers smaller, on shorter pedicels (always?), without bracteoles.—*A. minor,* Hook. f. Fl. Tasm. i. 264. Summit of Mount Sorrell, *Milligan,* and of Mount Lapeyrouse, *Oldfield.*

19. PRIONOTES, R. Br.

Corolla-tube cylindrical or ventricose; lobes 5, imbricate, not contorted in the bud, spreading, glabrous. Stamens hypogynous, free from the corolla; anthers included in the corolla-tube, adnate except at the base, with a prominent dissepiment, though opening in a single slit. Hypogynous disk of minute scales. Ovary 5-celled, with several ovules in each cell, on a small placenta nearly sessile on the axis; style inserted in a tubular depression of the ovary. Capsule loculicidally dehiscent.—Shrubs. Leaves flat, obtusely toothed. Flowers solitary in the upper axils, with several small bracts on the peduncle.

Besides the Australian species, which is endemic, another has been published from extra-tropical South America, which, however, is considered by some botanists as forming a distinct genus, *Allodape* or *Lebetanthus* of Endlicher.

1. **P. cerinthoides,** *R. Br. Prod.* 553. A tall shrub, its slender

branches sometimes prostrate or straggling, sometimes surrounding the trunks of dead or decaying trees, climbing to the height of 20 to 30 ft., and covering them with a dense mass of green foliage and bright crimson flowers. Leaves shortly petiolate, oblong, obtuse, bordered by a few obtuse, callous teeth, rather thick, obscurely veined, mostly 4 to 8 lines long. Flowers in the upper axils, pendulous from long slender peduncles. Bracts very small and distant, acute and ciliate, 1 or 2 of the uppermost sometimes rather larger and close to the calyx. Sepals about 2 lines long, acute, ciliate. Corolla-tube ¾ to 1 in. long, slightly ventricose, and contracted at the throat; lobes very small, broad, obtuse. Capsule twice as long as the calyx. DC. Prod. vii. 766; Hook. f. Fl. Tasm. i. 262; *Epacris cerinthoides,* Labill. Pl. Nov. Holl. i. 43. t. 59.

Tasmania. Recherche Bay. *Labillardière;* sides of Mount Wellington, and dense forests about Macquarrie Harbour, *Milligan, Gunn;* Mount Lapeyrouse, *C. Stuart.*

F. Mueller, Fragm. vi. 69, describes the placentas as being "ex apice columellæ arcuato-pendulæ" as in *Dracophyllum.* In the three flowers from different specimens which I examined, I found the placentation precisely as in *Epacris,* the style inserted in a tubular depression, reaching nearly halfway down the ovary, and the placentas attached immediately below, scarcely stipitate, covered with ovules, the upper ones more or less ascending, the lower ones more or less pendulous.

20. COSMELIA, R. Br.

Corolla-tube cylindrical; lobes 5, spreading towards the end, imbricate but not contorted in the bud. Filaments inserted near the throat, flattened; anthers adnate in the upper half, included in the corolla-tube. Hypogynous disk short, annular. Ovary 5-celled, with numerous ovules in each cell on a placenta attached near the base of the axis; style inserted in a rather deep tubular depression in the ovary.—Shrub, the branches covered with sheathing pungent-pointed leaves. Flowers red, solitary, terminating axillary branchlets or peduncles covered with leaf-like bracts, which pass gradually into the sepals and form an involucre round them.

The genus consists of a single species endemic in S.W. Australia. It is closely allied to *Epacris,* differing chiefly in the leafy nature of the bracts, which give it a peculiar aspect, and in the anthers partially adnate as in *Prionotes.*

1. **C. rubra,** *R. Br. Prod.* 553. An erect, glabrous, not much-branched shrub of 3 to 6 ft., the branches completely covered with the sheathing bases of the leaves until they ultimately fall off, leaving no scars. Leaves broad and concave at the base, tapering into a spreading pungent point, ¼ to ¾ in. (mostly about ½ in.) long, rigid smooth and shining or obscurely veined underneath, the adnate lowest portion which falls off with the rest usually scarious. Peduncles or flowering branchlets very closely covered with similar leaves or bracts, the lower ones small, those about the base of the calyx gradually enlarged, all with spreading tips, but these passing into the sepals which are linear-lanceolate, straight, herbaceous or more or less coloured, acute or acuminate, ¾ to 1 in. long. Corolla-tube scarcely exceeding the calyx; lobes 2 to 2½ lines long.—DC. Prod. vii. 766; Sond. in Pl. Preiss. i. 330; Bot. Reg. t. 1822; Fl. des Serres, t. 1175; *Epacris rubra,* Spreng. Syst. i. 629; *Cosmelia angustifolia,* DC. Prod. vii. 766.

W. Australia. King George's Sound and the immediate neighbourhood, *R. Brown, Drummond, 3rd Coll. n.* 194, *Preiss, n.* 429, and others. I have no hesitation in uniting De Candolle's two species, as the leaves vary in breadth on different branches of the same specimen much more than within the limits assigned to the two (from 2 to 4 lines).

21. SPRENGELIA, Sm.

(Ponceletia, *R. Br.*)

Corolla-tube very short, sometimes separating into distinct petal-claws; lobes 5, very spreading, more or less imbricate in the bud. Stamens hypogynous, shorter than the corolla; anthers connivent or cohering in a ring round the style. Hypogynous disk none. Ovary 5-celled, with several ovules in each cell, on a nearly sessile placenta attached to the axis; style filiform, inserted in a tubular depression of the ovary. Capsule loculicidally dehiscent.—Shrubs. Leaves with a shortly sheathing often membranous base completely covering the branches, very concave and stem-clasping immediately above the base, acute or acuminate with a spreading point, finely veined or almost veinless, the upper ones passing into floral leaves or bracts, the sheathing base of the stem-leaves deciduous with them, leaving the denuded stem without scars. Flowers solitary and terminal, surrounded by numerous leafy bracts. Sepals not of a very different texture from the last bracts. Corolla as long as or scarcely exceeding the calyx.

The genus is limited to Eastern Australia. The two species first known presented differences which seemed to warrant the considering them as distinct genera, but the characters relied upon are invalidated by the third species since discovered, in which they are differently combined.

Petal-claws completely connate into a short tube; laminæ or lobes
  very broad, very much imbricate. Anthers free, glabrous . . 1. *S. Ponceletia.*
Petal-claws as well as the laminæ free and much imbricate. Anthers
  connate, glabrous . . . . 2. *S. ponceletioides.*
Petal-claws scarcely cohering, valvate; laminæ or lobes narrow,
  slightly imbricate. Anthers connate or free, papillose-hirsute or
  rarely glabrous . . . . 3. *S. incarnata.*

1. **S. Ponceletia,** *F. Muell. Fragm.* i. 39, vi. 60. A glabrous erect shrub attaining but few feet. Leaves broad, concave, spreading or incurved, acuminate and pungent-pointed, smooth or finely veined, 2 to 4 lines long, the uppermost floral ones or leafy bracts like the others or with a broader base and crowded round the flower into a kind of involucre. Sepals leaf-like but more lanceolate, almost obtuse, usually about 3 lines long but very variable in breadth. Corolla about as long as the calyx, the very short broad tube not separating into petal-claws, the lobes much longer, very broad and obtuse, almost cordate at the base, very much imbricate. Filaments somewhat flattened. Anthers linear or linear-lanceolate, adnate to below the middle, glabrous, connivent round the style but not cohering. Style rather thick.—*Ponceletia sprengelioides,* R. Br. Prod. 554; DC. Prod. vii. 767.

Queensland. Marshes, Moreton Island, *F. Mueller.*
N. S. Wales. Marshes near Sydney, *R. Brown, Sieber, n.* 71, and others; Port Macquarrie, *Backhouse.*

2. **S. ponceletioides,** *Sond. in Linnæa,* xxvi. 254. A small diffuse

or procumbent shrub, with numerous slender almost filiform branches. Leaves lanceolate-subulate, spreading, concave at the short broad base, tapering to a rigid point, 2 to 3 lines long, the floral ones more lanceolate, shorter than the calyx and forming an involucre round it. Sepals lanceolate, very acute, about 3 lines long. Corolla rather exceeding the calyx, the petals quite free or scarcely cohering above the base, the very short broad claws much imbricate as well as the broadly oblong-lanceolate lobes or laminæ. Anthers glabrous, cohering in a ring round the style, each with a small inflected point at the base.—F. Muell. Fragm. vi. 60 ; *Poncaletia monticola,* A. Cunn. ; DC. Prod. vii. 768.

N. S. Wales. On rocks perpetually wet near Campbell's Cataract, Blue Mountains, *A. Cunningham;* near Sydney (or Blue Mountains ?), *Vernon.*

3. **S. incarnata,** *Sm. Tracts,* 272. *t.* 2. An erect shrub, sometimes low and straggling, more frequently a few feet high and said to attain sometimes 8 to 12 ft. Leaves above the broad concave base tapering into a spreading or recurved long or short rigid or pungent point, varying from under ¼ in. to near ½ in. long, the floral ones similar but smaller, the innermost with scarious margins but similar points, much shorter than the calyx and forming an involucre round it. Sepals from under 3 to fully 4 lines long, linear or linear-lanceolate, coloured. Corolla pink, about equal to the calyx, the petals almost free, the very short claws valvate and slightly cohering, the lobes or laminæ rotately spreading, slightly imbricate in the bud in the upper half, the lower portion valvate or slightly reduplicate. Filaments filiform ; anthers connivent or more frequently cohering in a ring round the style, more or less covered outside with transparent hairs or papillæ or rarely quite glabrous. Style filiform.—R. Br. Prod. 555 ; DC. Prod. vii. 768 ; Hook. f. Fl. Tasm. i. 264 ; F. Muell. Fragm. vi. 59 ; Lodd Bot. Cab. t. 262 ; *Poiretia cucullata,* Cav. Ic. t. 343 (not good).

N. S. Wales. Port Jackson to the Blue Mountains, *R. Brown, Sieber, n.* 72, and others ; in the interior (with very large flowers), *M'Arthur.*
Victoria. Near Portland, *Robertson, Allitt,* and thence to Gipps' Land, *F. Mueller,* and Mount William in the Grampians, at an elevation of 4000 ft., *Wilhelmi.*
Tasmania. King's Island and Port Dalrymple, *R. Brown ;* common throughout the island, *J. D. Hooker.*
S. Australia. Encounter Bay, *F. Mueller.*

S. montana, R. Br. Prod. 555 ; DC. Prod. vii. 768 ; Hook. f. Fl. Tasm. i. 265, is a small mountain form from Tasmania (summit of Mount Wellington, *R. Brown*), distinguished by shorter points to the leaves and free glabrous anthers, but these characters do not always go together, the coherence of the anthers is always slight and they frequently separate as the flowering advances, whether hairy or not. The degree of hairiness is very variable, and the anthers are sometimes quite free in specimens from N. S. Wales as well as in Tasmanian ones. *S. propinqua,* A. Cunn. ; DC. Prod. vii. 768 ; Hook. f. Fl. Tasm. i. 264 (*S. macrantha,* Hook. f. in Hook. Lond. Journ. vi. 273), distinguished by free hairy anthers, must, therefore, with *S. montana,* be united with *S. incarnata,* as suggested by J. D. Hooker.

22. ANDERSONIA, R. Br.

(Atherocephala, *DC.* Homalostoma *and* Sphincterostoma, *Stschegl.*)

Corolla-tube cylindrical or contracted above the ovary ; lobes 5, valvate in the bud, more or less bearded inside, at least at the base, recurved at the end or revolute. Stamens hypogynous, free from the corolla, the filaments filiform or flattened ; anthers attached below the middle or adnate nearly to the base. Hypogynous disk either annular and 5-toothed or 5-lobed, or consisting of 5 distinct scales, sometimes very small (or quite obsolete?). Ovary 5-celled, with several not very numerous ovules, on placentas attached to the base of the axis ; style usually long, inserted in a tubular depression reaching to below the middle or almost to the base of the ovary ; stigma small or slightly dilated. Capsule loculicidally dehiscent.—Shrubs. Leaves with a shortly sheathing often membranous base, rigid, broad concave and stem-clasping immediately above the base, acute acuminate or tapering into an erect or spreading point, finely veined or almost veinless, the upper ones passing into the floral leaves or bracts, the sheathing base of the stem-leaves deciduous with them, leaving the stem smooth without scars. Flowers either solitary and terminal surrounded by numerous bracts, or solitary within each floral leaf, sessile between 2 keeled bracteoles and collected in terminal heads. Sepals of a different texture from the bracts. Corolla shorter than or scarcely exceeding the calyx, white pink or blue.

The genus is limited to Western Australia. F. Mueller proposes to unite it with *Sprengelia*, of which it has the foliage, but the shape of the corolla is very different, independently of the æstivation, which is generally strictly valvate, or, if one or two of the lobes may be very slightly dilated near the end, it appears to me that they are rather induplicate than imbricate, or, at any rate, very obscurely imbricate and much less so than even in *Sprengelia incarnata*.

SECT. I. **Cephalanthesis.**—*Flowers in terminal heads, each one solitary and sessile in the axil of the floral leaf or subtending bract, between 2 keeled and complicated bracteoles.*

Floral leaves longer than the flowers. Corolla-tube contracted
   above the ovary. Tall robust shrubs with large leaves
   (**Sphincterostoma**).
   Leaves erect, lanceolate. Filaments much flattened . . . . 1. *A. colossea.*
   Leaves with long spreading points. Filaments scarcely flattened. 2. *A. patricia.*
Floral leaves as long as the flowers. Corolla-tube not contracted
   above the ovary. Leaves with spreading points.
   Flowers about ½ in. long. Floral leaves with long points. An-
     thers linear, adnate . . . . 3. *A. grandiflora.*
   Flowers about ¼ in. long. Leaves small, linear-subulate, the
     floral ones with long points. Anthers versatile . . . . . 4. *A. setifolia.*
   Flowers under ¼ in. long. Floral leaves with short points.
     Corolla scarcely hairy inside. Filaments flat ; anthers linear
     adnate . . . . 5. *A. involucrata.*
     Corolla densely bearded at the throat. Filaments scarcely
     flattened ; anthers ovate, attached at the middle . . . . 6. *A. homalostoma.*
Floral leaves much shorter than the flowers.
   Leaves with spreading or squarrose points, often twisted.
     Sepals 3 to 4 lines long, convolute when dry . . . . . . 7. *A. sprengelioides.*
     Sepals 5 to 6 lines long, nearly flat when dry . . . . . . 8. *A. latiflora.*
   Leaves with erect or incurved triquetrous points. Floral leaves
     very broad and shortly pointed.
     Leaves with a broad base, rigidly acuminate. Corolla shorter
      than the calyx ; lobes bearded to the middle . . . . . 9. *A. gracilis.*
     Leaves linear or subulate, with a small broad base. Corolla-

lobes exceeding the calyx, narrow and densely bearded nearly
  to the end . . . . 10. *A. aristata.*
  Leaves mostly under 1 line long. Corolla-lobes short, sparingly bearded . . . . 11. *A. parvifolia.*

SECT. II. **Monanthesis.**—*Flowers solitary and terminal, surrounded by numerous bracts.*

Leaves very spreading, acuminate, usually undulate or twisted (very
  rarely erect with a long point). Anthers shorter than the
  filament.
  Leaves lanceolate, acuminate. Corolla-lobes longer than the
  tube.
    Low diffuse or flexuose shrub. Flowering branchlets rather
      long and distant . . . . 12. *A. depressa.*
    Erect shrub. Flowering branchlets very short, usually crowded
      into a dense spike-like panicle . . . . 13. *A. cærulea.*
  Leaves subulate, acuminate. Corolla-lobes shorter than the
    tube . . . . 14. *A. subulata.*
Leaves spreading, with an incurved erect point. Anthers longer
  than the filament . . . . 15. *A. heterophylla.*
Leaves small, erect or closely appressed. Low densely corymbose
  shrubs.
  Leaves about 2 lines. Sepals 3 to 4 lines long. Anthers shortly
    ovate . . . . 16. *A. brachyanthera.*
  Leaves scarcely above 1 line long, closely appressed, with short
    points. Sepals 2 to 2½ lines. Filaments glabrous. Anthers
    linear . . . . 17. *A. brevifolia.*
  Leaves mostly 1 to 1½ lines long, with narrow points. Anthers
    shortly linear.
    Sepals nearly 3 lines long. Filaments usually hairy . . . 18. *A. variegata.*
    Sepals scarcely 1½ lines long. Filaments usually glabrous . 19. *A. micrantha.*

SECT. 1. CEPHALANTHESIS. Flowers in terminal heads, each one solitary and sessile in the axil of the floral leaf or subtending bract, between 2 keeled and complicated bracteoles.

1. **A. colossea,** *F. Muell. Fragm.* vi. 63. A robust probably tall shrub. Leaves crowded, almost imbricate, not recurved, broadly lanceolate, concave, mostly 1½ in. long, the upper floral ones nearly similar but smaller. Flowers sessile, solitary within the almost boat-shaped floral leaves and shorter than them, forming a short terminal leafy spike. Bracteoles linear-lanceolate, keeled, shorter than the calyx. Sepals linear-lanceolate, acute, 5 to 6 lines long. Corolla shorter than the calyx, the tube slightly constricted above the ovary, pubescent inside above the middle; lobes acute, rather longer than the tube, glabrous. Filaments flattened, rigid; anthers very long, adnate nearly to the base. Hypogynous disk lobed. Ovules not numerous.—*Sphincterostoma axilliflorum,* Stschegl. in Bull. Mosc. 1859, i. 22; *Sprengelia colossea,* F. Muell. Fragm. vi. 63.

W. Australia. *Drummond, 5th Coll. n.* 301.

2. **A. patricia,** *F. Muell. Fragm.* vi. 79. A robust shrub attaining 10 ft. Leaves broadly lanceolate, acuminate with a long point, spreading, ¾ to above 1 in. long, the floral ones shorter with a very broad base and more abruptly acuminate. Flowers sessile and solitary within the floral leaves and

shorter than them, forming a terminal leafy head. Bracteoles acutely keeled, complicate, acute, shorter than the calyx. Sepals complicate, very acute, about 4 lines long. Corolla whitish, the tube thickened inside and constricted above the ovary and slightly pubescent above the constriction ; lobes very acute, longer than the tube, glabrous. Filaments slightly flattened ; anthers long, linear, adnate almost to the base. Hypogynous disk toothed or shortly lobed. Ovules not numerous.—*Sphincterostoma echinocephalum,* Stschegl. in Bull. Mosc. 1859, i. 23 ; *Sprengelia patricia,* F. Muell. Fragm. vi. 79.

W. Australia. *Drummond, 4th Coll. n.* 155 ; rocks on the ridges of the Stirling Range, *F. Mueller.*

3. **A. grandiflora,** *Stschegl. in Bull. Mosc.* 1859. i. 21. A densely-branched, glabrous shrub, our specimens not exceeding 6 in. Leaves crowded, broadly lanceolate, tapering into a long, spreading point, more or less undulate or twisted, ½ in. long or rather more. Flowers few, in a dense, terminal, leafy head, each one sessile within a leaf-like bract as long as the flower. Bracteoles keeled and acutely acuminate, about half as long as the calyx. Sepals scarlet, linear-lanceolate, acute, ½ in. long or rather more. Corolla deep red, nearly as long as the calyx ; lobes longer than the tube, erect with shortly revolute tips, slightly hairy inside. Filaments hairy, flattened, especially towards the base ; anthers linear, adnate nearly to the base. Hypogynous disk lobed. Ovules numerous.—*Sprengelia spirophylla* or *Andersonia spirophylla,* F. Muell. Fragm. vi. 62.

W. Australia, *Drummond, 5th Coll. n.* 330.

4. **A. setifolia,** *Benth.* Small and diffuse, the specimens seen not exceeding 4 in. Leaves crowded and very spreading, linear-subulate, with the broad base very small, 2 to 3 lines long ; the floral ones rather longer, subulate-acuminate from a broad lanceolate base. Flowers in a dense terminal head, each one solitary within the floral leaf and much shorter than it. Bracteoles shorter than the calyx, keeled, acuminate. Sepals lanceolate, acute, ciliate, about 1½ lines long. Corolla not exceeding the calyx, the lobes longer than the tube and only slightly bearded inside at the base. Filaments not dilated ; anthers shortly oblong-linear, attached about the middle. Hypogynous scales very small or none. Ovary slightly hairy. Ovules few.

W. Australia. King George's Sound, *Collie ;* also in *Herb. R. Brown,* with unexpanded flowers.

5. **A. involucrata,** *Sond. in Pl. Preiss.* i. 331. Low, diffuse, and very much branched, or rarely more erect and attaining nearly 1 ft. Leaves very squarrose, acuminate and often twisted, mostly 2 to 4 lines long, the floral ones broadly ovate, shortly acuminate, about as long as the flowers. Flowers small, in a terminal globular head, each one sessile within a floral leaf or bract. Bracteoles very prominently keeled, acuminate, as long as the bracts and calyx. Sepals lanceolate, the ends usually spreading and acute, about 2 lines long. . Corolla nearly as long as the calyx, the tube slightly ventricose ; lobes shorter than the tube, sparingly bearded or nearly glabrous.

Filaments broad and flat, glabrous or hairy; anthers narrow-linear, adnate nearly to the base. Hypogynous scales very small. Ovules several, but usually only one in each cell ripening.—*Sprengelia involucrata*, F. Muell. Fragm. vi. 62.

W. Australia, *Drummond, n.* 105, 108, 111, *2nd Coll. n.* 256, 257; Sussex district, *Preiss, n.* 263.

6. **A. homalostoma,** *Benth.* Low and diffuse or more frequently erect, and often above 1 ft. high. Leaves spreading and sometimes squarrose, acuminate and acute, straight or twisted, rarely above ¼ in. long; the floral ones broad, acuminate, about as long as the flowers. Flowers small, pink, in dense terminal globular heads or oblong spikes, sometimes 1 in. long, each one sessile within a floral leaf. Bracteoles keeled, acuminate, shorter than the calyx. Sepals broadly lanceolate, acutely acuminate, the scarious margins entire or ciliate, about 2 lines long. Corolla often slightly exceeding the calyx, the lobes nearly as long as the tube, densely bearded at the base with hairs closing the throat, otherwise glabrous. Filaments slightly flattened; anthers short and broad, attached in the middle. Hypogynous scales narrow.—*Homalostoma simplex*, Stschegl. in Bull. Mosc. 1869, i. 21.

W. Australia, *Drummond, 5th Coll. n.* 332, *Gilbert, n.* 24; King George's Sound and adjoining districts, *A. Cunningham, Baxter, Oldfield, Maxwell;* Hay river, *F. Mueller.* The habit is sometimes that of *A. involucrata*, but the flowers are very different

7. **A. sprengelioides,** *R. Br. Prod.* 554. An erect shrub, either densely branched, corymbose, and under 1 ft., or sometimes taller and loosely branched. Leaves crowded, spreading, broad and concave at the base, contracted into a long point, undulate or twisted, 3 to 5 lines long, the floral ones broader with shorter points. Flowers pink, in short, dense, terminal heads, each one sessile within a floral leaf. Bracteoles leaf-like, keeled, much shorter than the floral leaf. Sepals much longer than the floral leaves, varying from scarcely 3 to above 4 lines, more or less convolute when dry, the margins ciliate or not. Corolla nearly as long as the calyx, the lobes much shorter than the tube, recurved at the end and slightly bearded. Filaments somewhat flattened, glabrous or hairy; anthers oblong-linear, attached below the middle. Hypogynous scales short, broad, and truncate, or united in a truncate or shortly-toothed cup. Style slender, glabrous or minutely pubescent.—DC. Prod. vii. 766; Bot. Mag. t. 1645; Lodd. Bot. Cab. t. 263; *A. sprengelioides, A. Lehmanniana, A. patens,* and *A. Fraseri,* Sond. in Pl. Preiss. i. 330, 331; *Sprengelia Andersoni,* F. Muell. Fragm. vi. 64

W. Australia. King George's Sound, *R. Brown, Fraser,* etc., and thence to Swan River, *Drummond, 1st Coll. n.* 496, *also n.* 104, 109, *3rd Coll. n.* 184, 192; *Preiss, n.* 451, 457, and many others; eastward to Point Malcolm and Cape Arid, *Maxwell.* The locality, N. S. Wales, given by Sonder for *A. Fraseri*, is owing to an error in the label.

Var. *pubescens.* Upper leaves pubescent, the points shorter, less spreading or incurved. —*A. pubescens*, Sond. in Pl. Preiss. i. 331; Swan River, *Drummond, 1st Coll. n.* 495. Sonder describes the filaments as shorter and broader upwards, but that is in the bud only; in the old flowers I find them the same as in the other forms.

8. **A. latiflora,** *F. Muell. Fragm.* vi. 61. An erect shrub, with long branches. Leaves in the typical form squarrose, shortly acuminate and

acute, under ¼ in. long, the floral ones very broadly ovate, acuminate, about
2 lines long. Flowers in dense terminal heads, each one sessile within a
floral leaf. Bracteoles keeled, acuminate, as long as the floral leaves. Sepals linear but rather broad, nearly ½ in. long. Corolla rather shorter
than the calyx, the lobes much shorter than the tube, sparingly bearded. Filaments scarcely flattened ; anthers linear, attached below the middle. Hypogynous scales very obtuse or truncate. Ovary and slender style glabrous or
slightly pubescent.—*Sprengelia latiflora*, F. Muell. Fragm. vi. 61.

W. Australia, *Drummond, n.* 106.
Var. *longifolia.* Leaves much acuminate and twisted, ¾ in. long.—*Drummond, n.* 110,
2nd *Coll. n.* 253.

9. **A. gracilis,** *DC. Prod.* vii. 767. An erect shrub of above 1 ft.,
with slender branches. Leaves mostly narrow, with an erect or incurved,
broad and keeled or triquetrous point, 2 to 3 lines long, the floral ones very
broadly ovate, shortly acuminate. Flowers in dense terminal heads. Bracteoles very prominently keeled, scarcely acuminate, shorter than the floral
leaves. Sepals linear, 4 to 5 lines long. Corolla as long as the calyx,
the lobes about as long as the tube, bearded to about the middle. Filaments
long, not dilated ; anthers linear, attached below the middle. Hypogynous
scales short and broad. Ovules not numerous.—Sond. in Pl. Preiss. i. 331.

W. Australia. Swan River, *Drummond,* 1st *Coll. ;* Darling Range, *Preiss, n.* 456.

10. **A. aristata,** *Lindl. Swan Riv. App.* 25. An erect, much-branched,
heath-like shrub of 1 to 2 ft., the branches slightly pubescent. Leaves from
a short broad base, linear or lanceolate-subulate, erect, mostly mucronulate,
concave and keeled, rarely above 2 lines long, the floral ones often longer,
with a much broader base. Flowers in a dense terminal head, each one sessile within a floral leaf. Bracteoles short, leaf-like. Sepals narrow-linear, 3
to 4 lines long, thin and coloured, scarcely acute, but convolute so as to be
almost awn-like when dry. Corolla white, rather longer than the calyx;
lobes as long as the tube, revolute and densely bearded, except the tips.
Filaments filiform ; anthers linear, attached below the middle, more or less
prominent. Hypogynous scales oblong. Style long. Ovules 8 to 10 in
each cell.—Sond. in Pl. Preiss. i. 334; *Atherocephala Drummondii*, DC.
Prod. vii. 755 ; *Sprengelia aristata*, F. Muell. Fragm. vi. 63.

W. Australia. Swan River, *Drummond,* 1st *Coll., Fraser, Preiss, n.* 449, *Oldfield.*

11. **A. parvifolia,** *R. Br. Prod.* 554. A small, much-branched shrub,
with slender branches. Leaves from a broad base, linear, erect, tapering into
a short, rigid, appressed or incurved, triquetrous point, rarely above 1 line
long, the floral ones similar. Flowers pink, usually few, in terminal heads,
each one sessile within a floral leaf. Bracteoles mucronate-acute, keeled, of
the length of the floral leaves. Sepals coloured, about three lines long.
Corolla about as long as the calyx, the lobes shorter than the tube, sparingly
bearded. Filaments scarcely flattened ; anthers attached a little below the
middle. Hypogynous scales broadly oblong, truncate. Ovules 6 to 8 in
each cell.—DC. Prod. vii. 767 ; *Sprengelia parvifolia*, F. Muell. Fragm. vi.
62 ; *Andersonia brachyota*, F. Muell. Fragm. iv. 125 ; *Sprengelia brachyota*,
F. Muell. Fragm. vi. 62.

W. Australia. King George's Sound and adjoining districts, *R. Brown, Baxter, Drummond,* 3*rd Coll. n.* 183, 5*th Coll. n.* 333 ; heaths, Upper Kalgan and Stirling Range, *F. Mueller;* onward to Middle Mount Barren, Point Malcolm, and towards Eyre's Relief, *Maxwell.*

SECT. 2. MONANTHESIS.—Flowers solitary and terminal, surrounded by numerous floral leaves or bracts without distinct bracteoles.

12. **A. depressa,** *R. Br. Prod.* 554. A low, prostrate or diffuse, much-branched shrub, under 6 in. or rarely more, erect and weak, but not exceeding 1 ft. Leaves with a short, broad base, subulate-acuminate, very spreading, often twisted, ¼ to ½ in. long, glabrous or rarely ciliate and the upper ones pubescent. Flowers solitary at the ends of the branches or branchlets, surrounded by floral leaves or bracts as in *A. cærulea,* but the general inflorescence loose or rarely almost paniculate, and never so crowded as in *A. cærulea,* and the flowers themselves usually larger than in that species. Sepals greenish-white, 4 to 5 lines long or sometimes nearly 6 lines, glabrous or hairy. Corolla blue, shorter than the calyx, the lobes at least as long as the tube. Filaments glabrous or hairy ; anthers linear, attached below the middle. Hypogynous scales broad, obtuse. Style more or less hairy or rarely quite glabrous.—DC. Prod. vii. 767 ; *Sprengelia depressa,* F. Muell. Fragm. vi. 63 ; *Andersonia squarrosa,* R. Br. Prod. 554 ; DC. Prod. vii. 767 ; *A. prostrata,* Sond. in Pl. Preiss. i. 333.

W. Australia. King George's Sound and adjoining districts, *R. Brown, Baxter, Drummond,* 5*th Coll. n.* 331, *Preiss, n.* 458, *Harvey, Oldfield, Maxwell.*

Var. *ciliata.* Leaves ciliate, the upper ones and the calyx hirsute.—*A. lanuginosa,* A. Cunn. ; DC. Prod. vii. 767.—King George's Sound, *A. Cunningham.* To this variety appears to belong the *Melichrus squarrosus,* Sond. in Pl. Preiss. i. 304, described from Preiss's specimens, n. 418, which are young plants, not yet in flower or fruit. The spiral style of *A. prostrata,* Sond., appears to me to be a slightly diseased or monstrous state.

13. **A. cærulea,** *R. Br. Prod.* 554. An erect shrub of 1 to 2 ft. Leaves from a broad base tapering into a long point, usually undulate or twisted, pubescent when young and sometimes retaining their pubescence with marginal cilia when old, above ½ in. long on the main stems, smaller on the branches. Flowers singly terminating short leafy peduncles or branchlets, but usually numerous and crowded into a more or less compound, spike-like, leafy panicle. Leaves on the branchlets or peduncles numerous, lanceolate-subulate, the lower ones scattered, the upper ones forming an involucre round the calyx, sometimes nearly as long as the sepals, but usually shorter. Sepals pink, linear-lanceolate, obtuse or mucronate, marked with fine parallel veins, usually 3 to 4 lines long. Corolla blue, shorter than the calyx, the lobes rather longer than the tube, bearded inside below the middle. Filaments filiform ; anthers linear, attached below the middle. Hypogynous scales broad, obtuse. Ovary glabrous or shortly pubescent ; ovules not numerous ; style usually hairy.—DC. Prod. vii. 767 ; Sond. in Pl. Preiss. i. 333 ; *Sprengelia cærulea,* F. Muell. Fragm. vi. 64.

W. Australia. Dry heaths, King's George's Sound, *R. Brown;* abundant in the adjoining districts and extending far into the interior, and perhaps to Swan River, *Fraser, Drummond, n.* 107, 2*nd Coll. n.* 257, *Preiss, n.* 450, *Oldfield, Maxwell,* and many others (not N. S. Wales, as given by Sonder, from an error in the labels attached to some of Fraser's plants).

Var. *brevifolia.* Leaves rigid, with shorter points; infloreseence not so dense.—Point Possession, *Collie ;* Point Irwin, *Oldfield ;* Phillips river, *Maxwell.*

Var. ? *stricta.* Leaves with long, erect, almost appressed points.—King George's Sound, *Herb. Hook.*

14. **A. subulata,** *Benth.* An erect, slightly-branched shrub of 6 in. to 1 ft., with the habit and foliage nearly of *A. cærulea,* but smaller and more slender. Leaves from a lanceolate base, subulate-acuminate, usually twisted, under ½ in. long. Flowers solitary on leafy peduncles, sometimes forming a compound, spike-like, leafy panicle, nearly as compact as in *A. cærulea,* sometimes more approaching the loose inflorescence of *A. depressa.* Involucrating bracts or floral leaves often squarrose. Sepals about 3 lines long, linear, coloured, the veins scarcely conspicuous. Corolla rather shorter than the calyx, the lobes much shorter than the cylindrical tube, bearded inside. Anthers oblong, not linear, attached about the middle.

W. Australia. Swan River, *Drummond, 1st Coll. n.* 109, *2nd Coll. n.* 255. The inflorescence is between those of *A. depressa* and *A. subulata,* the corolla and anthers different from both.

15. **A. heterophylla,** *Sond. in Pl. Preiss.* i. 333. An erect shrub of ½ to above 1 ft., with rather slender branches. Leaves 2 to 3 lines long, with a rather long, incurved, keeled or triquetrous point, the upper ones with a broader base and shorter point, closely imbricate, and forming an involucre round the solitary terminal flowers, and much shorter than the calyx. Sepals linear or linear-lanceolate, nearly 4 lines long. Corolla rather shorter than the calyx, the lobes longer than the tube, bearded to above the middle. Filaments filiform, nearly as long as the corolla-tube, but shorter than the narrow-linear anthers. Hypogynous scales broad and short.—*A. macranthera,* F. Muell. Fragm. iv. 124 ; *Sprengelia brachynema,* F. Muell. Fragm. vi. 61.

W. Australia. Swan River, *Drummond, n.* 4 ; Darling Range, *Preiss, n.* 460; sand flats, Eyre's Relief, *Maxwell.* I have not seen Preiss's specimen, but Sonder's description agrees well with Drummond's plant, and this is the only species in which the filaments are not longer than the anthers.

16. **A. brachyanthera,** *F. Muell. Fragm.* vi. 61. A shrub, with numerous erect branches, our specimens (with the root attached) not above 1 ft. high. Leaves imbricate, acuminate, with erect, keeled or triquetrous points, about 2 lines long, the floral ones rather more lanceolate, the innermost linear-lanceolate and acute, forming an involucre more than half as long as the calyx. Flowers solitary and terminal, but becoming lateral by the elongation of the branch. Sepals linear-lanceolate, 3 to 4 lines long. Corolla shorter than the calyx, the lobes rather longer than the tube, densely bearded inside. Filaments scarcely flattened, more or less hairy ; anthers ovate, attached near the middle. Hypogynous scales short and broad.— *Sprengelia brachyanthera,* F. Muell. Fragm. vi. 61.

W. Australia, *Drummond.* This species is more robust than the three following ones, and remarkable for the short anthers like those of *A. homalostoma.*

17. **A. brevifolia,** *Sond. in Pl. Preiss.* i. 332. An erect, densely-branched shrub of 1 to 3 ft. Leaves scarcely above 1 line long, imbricate, and closely appressed, ovate, acute, pubescent, the floral ones rather longer,

narrower, and more acuminate, forming an involucre above half as long as the calyx. Flowers solitary and terminal, but sometimes on very short branchlets, forming a terminal corymb. Sepals 2 to $2\frac{1}{2}$ lines long, linear-lanceolate, often pubescent and ciliolate. Corolla about as long as the calyx, the lobes rather shorter than the tube, scarcely bearded, but the tube hairy inside from the middle to the throat. Filaments filiform; anthers linear, attached near the base. Hypogynous disk shortly lobed. Ovules not numerous. Capsule oblong.—*Sprengelia brevifolia,* F. Muell. Fragm. vi. 61.

W. Australia, *Drummond, 5th Coll. Suppl. n.* 70; sandy plains, York District, *Preiss, n.* 462.

18. **A. variegata,** *Sond. in Pl. Preiss.* i. 334. A small, much-branched shrub, with the habit, the narrow acuminate leaves of 1 to 2 lines, and the inflorescence of *A. micrantha,* but with much larger flowers. Sepals pink, nearly or quite 3 lines long. Corolla blue, as long as the calyx, the lobes as long as the tube, bearded to above the middle. Filaments and style more or less hairy. Anthers rather shorter than in *A. micrantha.* Hypogynous scales small. Ovules few.

W. Australia. In the interior, *Preiss;* between King George's Sound and Swan River, *Harvey;* along the Kalgan river, *F. Mueller.* From *A. brevifolia,* this species differs chiefly in the long narrow point and small base to the leaves, and the shorter anthers.

19. **A. micrantha,** *R. Br. Prod.* 554. A small, densely-branched, heath-like shrub, rarely attaining 1 ft., and usually about half that height. Leaves erect, narrow, acuminate, appressed or incurved, minutely ciliolate, 1 or rarely nearly 2 lines long, the floral ones linear, obtuse or mucronulate, forming an involucre shorter than the calyx. Flowers solitary and terminal, on short branchlets. Sepals linear-lanceolate, about $1\frac{1}{2}$ lines long. Corolla about as long as the calyx, the lobes as long as the tube, bearded to above the middle. Filaments slightly flattened, glabrous; anthers linear. Hypogynous scales small, oblong. Style slightly glandular-pubescent. Ovules few.—DC. Prod. vii. 767; Sond. in Pl. Preiss. i. 332 ; *Sprengelia micrantha,* F. Muell. Fragm. vi. 61.

W. Australia. King George's Sound and adjoining districts, *R. Brown, Baxter, Drummond, 3rd Coll. n.* 185, *Preiss, n.* 461, and many others.

23. RICHEA, R. Br.

(Cystanthe, *R. Br.;* Pilitis, *Lindl.*)

Corolla ovoid oblong or conical, circumsciss near the base, the lobes not separating, the upper calyptriform portion falling off entire, leaving a persistent ring or cup. Stamens hypogynous; anthers attached at or below the middle, entire or 2-lobed. Hypogynous disk none or consisting of distinct scales. Ovary 5-celled, with several ovules in each cell, attached to a placenta suspended from an erect or recurved stipes proceeding from the axis. Style inserted in a shallow or deeper and tubular depression of the ovary; stigma usually small, rarely larger and obscurely 5-lobed. Capsule loculicidally dehiscent.—Shrubs or small trees, with a Monocotyledonous aspect, the trunk often not branched. Leaves crowded or imbricate at the ends of or along the branches, very concave above the sheathing bases, which fall off with the leaf, leaving annular scars on the denuded stem and branches. Flowers in dense terminal simple heads, or in compound spike-like or branched panicles. Filaments usually persistent long after the flowering is over.

The genus is limited to Australia and, with one exception, to Tasmania. With the habit of *Dracophyllum*, it differs in the calyptriform corolla, falling off without expanding the lobes. I have followed F. Mueller in reducing the two genera proposed by Brown to sections, for, of the two characters relied upon by Brown, that derived from inflorescence is the same as that which separates the two sections of *Dracophyllum*, and the one derived from the hypogynous scales is invalidated by the subsequent discovery of *Pilitis*, which has the inflorescence of *Cystanthe*, with the hypogynous scales of *Richea*.

SECT. I. **Cystanthe.**—*Spikes simple, contracted into short heads, sessile above the last leaves. Bracts persistent, with 1 flower in each axil, sessile between two or more keeled bracts or bracteoles.*

No hypogynous scales. Leaves broad, very spreading or squarrose.
  Leaves shortly acuminate. Filaments filiform, smooth. Anthers
    entire . . . . 1. *R. sprengelioides.*
  Leaves with long points. Filaments thickened and papillose up-
    wards. Anthers deeply 2-lobed . . . . 2. *R. procera.*
Hypogynous scales conspicuous. Leaves narrow, erect or spread-
  ing, but not recurved (*Pilitis*).
  Leaves under ½ in. long, subulate-acuminate. Stamens scarcely
    exceeding the calyx . . . . 3. *R. acerosa.*
  Leaves above 1 in. long, linear-lanceolate. Stamens 2 or 3
    times as long as the calyx . . . . 4. *R. Milligani.*

SECT. II. **Dracophylloides.**—*Spike or panicle compound, elongated. Bracts and bracteoles very deciduous.*

Panicle narrow and spike-like. Leaves under 2 in. long.
  Panicle or spike pedunculate, interrupted. Corolla 2 lines long.
    Stamens scarcely exceeding the calyx. Leaves very spreading 5. *R. Gunnii.*
  Panicle or spike nearly sessile, dense. Corolla 4 lines long.
    Stamens 2 or 3 times as long as the calyx. Leaves mostly
    erect, imbricate . . . . 6. *R. scoparia.*
Panicle branched, oblong or pyramidal. Leaves above 6 in. long.
  Corolla 4 to 5 lines long. Leaves 6 in. to 1 ft. . . . . . 7. *R. dracophylla.*
  Corolla under 2 lines long. Leaves 3 to 5 ft. . . . . . . 8. *R. pandanifolia.*

SECT. 1. CYSTANTHE.—Spikes simple, contracted into short heads sessile above the last leaves. Bracts persistent, with 1 flower in each axil, sessile between two or more keeled bracts or bracteoles.

1. **R. sprengelioides,** *F. Muell. Fragm.* vi. 68. A bushy shrub, usually low, but attaining sometimes several feet, usually glabrous. Leaves broadly ovate-lanceolate, tapering into a short rigid point, straight or slightly undulate or twisted, ¼ to ½ in. long, the floral ones gradually smaller. Flowers in terminal, globular, leafy heads, each one nearly sessile within a floral leaf, between 2 to 6 distichous, complicate, keeled bracts or bracteoles. Sepals lanceolate, acute, concave, 2 to 2½ lines long. Corolla much longer than the calyx, the deciduous calyptra narrow-conical, the persistent annular base about half as long as the calyx. Filaments filiform, glabrous, much longer than the calyx; anthers linear, attached rather above the middle, not lobed, opening from end to end in 2 valves. No hypogynous scales.—*Cys-*

tanthe sprengelioides, R. Br. Prod. 555; DC. Prod. vii. 769; Hook. f. Fl.
Tasm. i. 265; F. Muell. Fragm. i. 38.

Tasmania. Derwent river and Mount Wellington, *R. Brown, Gunn;* Western mountains, *Lawrence, Archer, C. Stuart;* Mount Sorrell, *Milligan.*

2. **R. procera,** *F. Muell. Fragm.* vi. 68.   Closely resembles *R. sprengelioides,* and included in it as a variety by most authors.   It is usually a
taller shrub, attaining 6 to 8 ft. according to Gunn, or 20 to 30 ft. according to C. Stuart, the leaves longer and narrower, and usually very spreading
or recurved, varying from ½ to 1 in.   Inflorescence, calyx, and corolla of
R. sprengelioides.   Filaments thickened and papillose upwards; anthers
attached below the middle, completely divided above the point of attachment
into 2 lobes, each opening in a longitudinal furrow, and scarcely or very
tardily opening below the point of separation.   No hypogynous scales.—
Cystanthe procera, F. Muell. Fragm. i. 38; Hook. f. Fl. Tasm. ii. 366.

Tasmania. Summit of Mount Wellington, *J. D. Hooker* and others; between Hobarton and Huon river, and summits of the Western mountains, *Gunn.* Originally distributed as a variety of *Cystanthe sprengelioides.*

3. **R. acerosa,** *F. Muell. Fragm.* vi. 69.   An erect bushy shrub, usually
1½ to 2 ft. high.   Leaves erect or spreading, lanceolate-subulate, tapering
into a pungent point, mostly about ½ in. long, the upper ones smaller, passing into the triangular-lanceolate, short, acute floral leaves.   Flowers in a
short, dense, terminal head, each one nearly sessile within a floral leaf, with
3 to 6 distichous, complicate, keeled bracts or bracteoles.   Sepals broad,
acute, about 2 lines long.   Corolla conical.   Filaments glabrous, rather
shorter than the calyx; anthers undivided as in *R. sprengelioides.*   Hypogynous scales usually bifid, with long points projecting above the ovary, but
very irregular.   Style short.—*Pilitis acerosa,* Lindl. Introd. Nat. Syst. ed. 2.
443; DC. Prod. vii. 769; Hook. f. Fl. Tasm. i. 265. t. 82; *Cystanthe
acerosa,* F. Muell. Fragm. i. 38.

Tasmania. Arthur's Lake and summits of Western and other mountains. *Lawrence,
Gunn.*

4. **R. Milligani,** *F. Muell. Fragm.,* vi. 69.   An erect shrub, the
branches denuded of leaves except a short tuft at the end.   Leaves lanceolate, erect, tapering into a rigid point, mostly about 1 in. long, the floral
ones shorter, with a broad base.   Flowers in short terminal heads, each one
nearly sessile within a floral leaf, with 2 to 4 keeled and complicate bracts or
bracteoles, the inner ones and outer sepals about 4 lines long.   Sepals 5,
acute, the outer ones very much like the bracts but less keeled, the inner ones
smaller, narrower, and flatter.   Corolla not seen, except the truncate, persistent base.   Filaments filiform, nearly 1 in. long; anthers narrow-linear,
not lobed.   Hypogynous scales broad, retuse or shortly 2-lobed.—*Pilitis
Milligani,* Hook. f. Fl. Tasm. i. 266. t. 83; *Cystanthe Milligani,* F. Muell.
Fragm. i. 38.

Tasmania. Mount Sorrel, Macquarrie Harbour, *Milligan.*

SECT. 2. DRACOPHYLLOIDES.—Spike or panicle compound, elongated.
Bracts and bracteoles very deciduous.

s 2

5. **R. Gunnii,** *Hook. f. in Hook. Lond. Journ.* vi. 273, *and Fl. Tasm.* i. 267. *t.* 86. An erect shrub of 1 to 3 ft. Leaves lanceolate, tapering into a pungent point, usually spreading or recurved, rarely erect or incurved, the larger ones above 1 in. long, the upper ones with a shorter point, passing into the floral leaves or bracts, which are broadly membranous and coloured at the base, enclosing the clusters of buds, and terminating in a rigid point, but fall off before the flowers expand. Flowers in clusters or short spikes, sessile along the simple rhachis of a spike-like, pedunculate panicle of 2 to 4 in., each flower nearly sessile within a bract and 2 bracteoles like the sepals, but narrower. Sepals ovate, acute, almost petal-like, scarcely 1¼ lines long. Corolla ovate-conical, about 2 lines long, circumsciss very near the base. Filaments filiform, about as long as the calyx; anthers oblong, shortly 2-lobed. Hypogynous scales broad.

Victoria. Mitta-Mitta, Cobberas and Munyong mountains and others of the Australian Alps at an elevation of 4000 to 6000 ft., *F. Mueller.*

Tasmania. Summit of Mount Wellington and of the Western mountains at an elevation of 4000 to 5000 ft., *J. D. Hooker* and others.

6. **R. scoparia,** *Hook. f. in Hook. Lond. Journ.* vi. 273, *and Fl. Tasm.* i. 267. A stout, rigid, erect shrub, usually about 2 ft. high, but attaining in some situations 5 ft. or more. Leaves lanceolate to almost subulate, with a rigid, often pungent point, erect and imbricate or somewhat spreading, 1 to 2 in. long, the ragged remains of the old sheaths often remaining long persistent, but at length leaving annular scars only as in the other species, the floral ones exceedingly deciduous, shortly broad, and coloured at the base with a long rigid point. Flowers white pink or orange-coloured, in sessile clusters or very short spikes, in a simple spike-like panicle more dense than in *R. Gunnii,* and usually sessile, each flower sessile or on a short thick pedicel, with a narrow membranous bract and 2 bracteoles at the base. Sepals broad, petal-like, scarcely ¾ line long. Corolla obovoid-oblong, about 4 lines long, circumsciss near the base. Filaments filiform, about 3 lines long; anthers shortly 2-lobed. Hypogynous scales broad, truncate.—Hook. Ic. Pl. t. 850; *R. dracophylla,* R. Br. Prod. 555 (as to the alpine form); Guillem. Ic. Pl. Austral. t. 3.

Tasmania. Mount Wellington, *R. Brown, J. D. Hooker;* Western mountains and others at an elevation of 3000 to 4000 ft., *J. D. Hooker* and others.

7. **R. dracophylla,** *R. Br. Prod.* 555 (*partly*).—A stout shrub, attaining 10 to 12 ft. or (according to Brown) arborescent in certain situations, but scarcely taller than *R. scoparia* on the summits of mountains. Leaves crowded at the ends of the branches, lanceolate at the base, tapering into long subulate points, 6 in. to 1 ft. or even longer when luxuriant, the very deciduous floral ones or bracts with a very broad base of 1 in. or less, and a long rigid point. Flowers white or pink, in narrow, dense, terminal, sessile panicles of 3 to 6 in. or even more, the secondary branches and peduncles often crowded at the base of the primary ones as in *Dracophyllum verticillatum,* Labill. Bracts and bracteoles very deciduous, narrow, 1 to 3 lines long. Sepals very broad, mostly mucronate, about ½ line long. Corolla obovoid-oblong, 4 to 5 lines long, circumsciss very near the base. Fi-

laments about 3 lines long; anthers shortly lobed. Hypogynous scales broad.—Hook. f. Fl. Tasm. i. 267.

Tasmania. Mount Wellington, *R. Brown, J. D. Hooker,* and others ; Recherche Bay, *C. Stuart.* R. Brown considered this as a luxuriant form of the same species, of which *A. scoparia* is the alpine representative ; but both J. D. Hooker and Gunn observed that the differences were equally prominent where the two grew together. Brown's only specimen of the large form is a remarkably luxuriant one, with a very long, dense panicle, approaching *R. pandanifolia* in foliage, but in fruit only, and therefore somewhat doubtful. Those of the alpine form are rather numerous, and precisely our *R. scoparia.*

8. **R. pandanifolia,** *Hook. f. Fl. Ant.* i. 50, *and Fl. Tasm.* i. 266. *t.* 84. A shrub or tree, attaining sometimes 36 ft. but often not above 20 ft., and stunted in elevated situations, the simple or sparingly branched naked trunk attaining 6 to 9 in. diameter, and crowned by a large tuft of long wavy leaves like those of a *Pandanus,* often 3 to 5 ft. long, tapering into a long point, and bordered by minute, cartilaginous teeth. Panicles ovate, 2 to 3 in. long, looser and broader than in *R. dracophylla,* on peduncles of 4 to 8 in., but usually almost concealed by the leaves. Bracts imbricate, those at the base of the peduncle distichous, keeled, acuminate, 1 to 2 in. long ; those under the panicle longer and narrower, the innermost 3 to 4 in. long ; those within the panicle small and membranous, but almost all as well as the bracteoles fallen off from our specimens. Branches of the panicle racemose, the lower ones often 1 in. long. Flowers small, like those of *R. Gunnii.* Sepals very short and broad. Corolla broadly ovate-conical, 1 to 1½ lines long. Stamens usually exceeding the calyx.

Tasmania. Dense mountain forests in the interior and S.W. parts of the island, *Gunn, Backhouse,* and others.

24. DRACOPHYLLUM, Labill.

(Sphenotoma, *sect. R. Br., gen. Don.*)

Corolla-tube cylindrical, oblong or almost campanulate ; lobes 5, spreading, imbricate in the bud. Stamens hypogynous or the filaments more or less adnate to the corolla-tube ; anthers included in the tube, attached at or above the middle, entire or 2-lobed. Hypogynous disk of 5 distinct scales. Ovary 5-celled, with several ovules in each cell, attached to a placenta suspended from an erect or recurved stipes proceeding from the axis. Style inserted in a shallow or deep and tubular depression of the ovary ; stigma small or larger and shortly 5-lobed.—Shrubs or small trees, having then a monocotyledonous aspect. Leaves crowded at the ends of the branches or imbricate along them, very concave above the sheathing bases, which fall off with the leaf, leaving annular scars on the denuded stem or branches. Flowers in terminal compound racemes or panicles or simple spikes or heads.

The genus is spread over Australia, New Caledonia, New Zealand, and the Antarctic Islands. The Australian species are all endemic, with the exception of one which appears to be the same as a New Zealand one. The annular scars of the leaves and peculiar placentation distinguish it from all except *Richea,* from which it differs in the expanding corolla-lobes.

Flowers in a compound raceme or panicle with very deciduous bracts. Corolla without longitudinal folds at the base of the lobes.

Leaves 6 in. to 3 ft. long. Panicle long, not secund, with very
   numerous flowers . . . .   1. *D. Milligani.*
Leaves 3 to 6 in. long. Panicle secund, raceme-like . . . .   2. *D. secundum.*
Flowers in a simple spike or head, solitary in the axils of persistent
   bracts sessile between 2 bracteoles. Corolla with a narrow
   tube, the throat almost closed by longitudinal folds at the base
   of the lobes (**Sphenotoma**).
Spreading stem-leaves reaching to the base of the spike.
   Leaves very spreading or recurved, mostly about 1 in. long.
   Spikes mostly cylindrical. Corolla-lobes much shorter than
   the tube . . . .   3. *D. squarrosum.*
   Leaves very spreading or recurved, mostly above 2 in. long.
   Spikes mostly ovate. Corolla-lobes nearly as long as the
   tube . . . .   4. *D. Drummondii.*
   Leaves straight, not recurved, ½ to 1 in. long. Spikes mostly
   ovate. Corolla-lobes nearly as long as the tube . . . .   5. *D. phlogiflorum.*
Upper portion of the flowering branch long and peduncle-like
   with appressed leaves, the spreading leaves only at their base
   or on barren branches.
Corolla-lobes nearly as long as the tube.
   Leaves straight, not recurved, ½ to 1 in. long . . . . .   6. *D. capitatum.*
   Leaves recurved, under ¼ in. long . . . .   7. *D. gracile.*
Corolla-lobes much shorter than the tube. Leaves straight,
   not recurved, under ⅓ in. long . . . .   8. *D. parviflorum.*
Flowers solitary. Corolla-lobes with longitudinal folds at their base.
Small tufted plant with densely imbricate small leaves . . . .   9. *D. minimum.*

The extra Australian species belong partly to the first and third of the above groups, and
partly to a fourth group. The *Sphenotomas* are endemic in West Australia.

1. **D. Milligani,** *Hook. f. in Hook. Ic. t.* 845 *and Fl. Tasm.* ii. 367.

A shrub attaining nearly the height of *Richea pandanifolia*, but more
branching, with large terminal crowns of leaves, those under the panicles
from 6 in. to above 3 ft. long, widened at the base, with long recurved or
curled points, the margins entire. Panicles in the original specimens not
above 6 in. long and not much branched, in C. Stuart's specimens 1½ ft.
long, densely crowded with innumerable flowers, the rhachis often pubescent.
Floral leaves broad, membranous, shortly pointed, falling off very early.
Flowers shortly pedicellate along the branches of the panicle. Bracts and
bracteoles narrow-linear, very deciduous. Sepals coloured, lanceolate, concave or keeled, about 2 lines long. Corolla-tube broad, contracted at the
throat, shorter than the calyx ; lobes shortly exceeding the calyx, recurved at
the end. Filaments quite free. Hypogynous scales truncate or 2-lobed.
Style short, with a distinctly 5-lobed stigma.

Tasmania. Mount Sorrel, *Milligan ;* Mount Lapeyrouse, *C. Stuart.*

2. **D. secundum,** *R. Br. Prod.* 556.

An erect shrub, not much
branched, flowering at 1 ft. high, but attaining several feet. Leaves spreading, linear-lanceolate, tapering into long subulate points, 2 to 4 in. long or
more when luxuriant. Panicle narrow and raceme-like, sessile above the last
leaves, often 6 in. long, rather dense but the flowers all pedicellate and turned
to one side. Floral leaves broad and membranous with rigid points, falling
off long before the flowers expand. Bracts and bracteoles exceedingly deciduous. Sepals lanceolate, acutely acuminate, about 3 lines long. Corolla-

tube oblong-cylindrical, slightly contracted at the throat, 4 to 5 lines long ; lobes short, recurved. Filaments exceedingly slender, adnate to the corolla-tube. Anthers ovate or oblong. Hypogynous scales small, truncate.—DC. Prod. vii. 769 ; F. Muell. Fragm. vi. 65 ; Bot. Mag. t. 3264 ; Guillem. Ic. Fl. Austral. t. 1 ; Reichb. Icon. Exot. t. 108 ; *Epacris secunda*, Poir. Dict. Suppl. ii. 556 ; *Prionotes secunda*, Spreng. Syst. i. 631.

N. S. Wales. Port Jackson to the Blue Mountains, *R. Brown, Sieber, n.* 69, and many others ; Illawarra, *A. Cunningham, Backhouse.* The filaments are represented in the Bot. Mag. as free, I have always found them adnate to the corolla-tube.

3. **D. squarrosum,** *R. Br. Prod.* 556. An erect slender shrub of 5 to 6 ft. or rarely taller, not much branched, the young shoots usually and the inflorescence always more or less pubescent. Leaves very spreading or re-curved or rarely more erect, narrow-lanceolate, tapering into a subulate point, the margins often softly ciliate, usually about 1 in. long, but in some speci-mens scarcely above ½ in., in others nearly 2 in. Flowers in dense cylindrical sessile terminal spikes of about 1 in., each one sessile within a floral leaf or bract between 2 bracteoles, the lower bracts sometimes elongated and leaf-like, but usually all ovate to lanceolate, acuminate, ciliate, persistent, as long as or longer than the calyx ; bracteoles lanceolate, acuminate, keeled, shorter than the calyx. Sepals lanceolate, acutely acuminate, ciliate, 3 to 4 lines long, often pubescent. Corolla-tube cylindrical, pubescent, nearly as long as the calyx ; lobes broadly obovate, horizontally spreading, about half as long as the tube, with longitudinal ridges at the base almost closing the orifice of the tube. Filaments adnate to above the middle of the tube ; anthers 2-lobed at the base. Hypogynous disk very short, truncate. Ovary pubescent. —*Sphenotoma squarrosum*, G. Don, Gen. Syst. iii. 785 ; DC. Prod. vii. 771 ; Sond. in Pl. Preiss. i. 336 ; F. Muell. Fragm. vi. 66 ; *Epacris squarrosa*, Poir. Dict. Suppl. ii. 556.

W. Australia. Wet places, King George's Sound and adjoining districts, *R. Brown, Baxter, Drummond, 2nd Coll. n.* 242, *Preiss, n.* 445, *Oldfield, Maxwell, F. Mueller.*

4. **D. Drummondii,** *Benth.* A stout species, probably tall. Leaves spreading, lanceolate, tapering to a pungent point, the margins softly ciliate, at least when young, 1½ to 3 in. long, the upper ones small and appressed, giving the last 1 or 2 in. of the flowering branch the appearance of a pe-duncle. Flowers rather large, in an ovate-oblong dense simple spike of 1 in. or rather more, each one solitary within the bract and sessile. Bracts broadly lanceolate, very acute, ciliate, the upper ones ovate and mucronate ; bracteoles much shorter, complicate. Sepals 4 lines long, rather broad, acute. Corolla-tube glabrous, as long as the calyx ; lobes very broad, nearly as long as the tube, with very prominent longitudinal ridges at the base closing the orifice. Filaments less adnate than in *D. squarrosum* ; anthers 2-lobed at the base. Hypogynous scales small. Ovary glabrous.

W. Australia, *Drummond, 3rd Coll. n.* 191.

5. **D. phlogiflorum,** *F. Muell. Fragm.* vi. 65 (*partly*). An erect branching shrub of 3 to 4 ft., closely allied to *D. capitatum*, and perhaps a variety. Leaves linear-lanceolate, crowded, erect or spreading, not recurved, ½ to 1 in. long up to the base of the spike, without the intervention of any

peduncle with closely appressed leaves.   Spikes short and almost globular, and flowers of *D. capitatum*, but usually larger, the sepals above 4 lines long and the corolla-lobes about 3 lines.   Filaments more or less adnate.   Hypogynous scales ovate.—*Sphenotoma dracophylloides*, Sond. in Pl. Preiss. i. 335.

W. Australia, *Drummond;* Kojonerup hills near Cape Riche, *Preiss, n.* 446; thence to E. Mount Barren, *Maxwell;* Stirling Range, *F. Mueller.*

6. **D. capitatum,** *R. Br. Prod.* 556.   A slender shrub, with few long virgate branches, attaining 3 or 4 ft., but sometimes almost simple and not half so high, usually glabrous.   Lower leaves and those of the barren shoots erect or spreading but not recurved, linear-lanceolate, very acute, $\frac{1}{2}$ to 1 in. long, those of the long peduncle-like extremities of the flowering branches smaller and closely appressed.   Flowers in short and globular or rarely ovate-oblong heads or spikes, each one sessile within a lanceolate or almost ovate acute or acuminate bract about as long as the calyx; the bracteoles much shorter.   Sepals about 3 lines long.   Corolla-tube glabrous, as long as the sepals, contracted upwards; lobes very broad, nearly as long as the tube, with longitudinal ridges at the base almost closing the orifice.   Filaments slightly adnate to the tube in the middle.   Hypogynous scales broad, often united in a truncate disk.   Ovary glabrous.—Bot. Mag. t. 3624; Lodd. Bot. Cab. t. 1846; *Sphenotoma capitatum*, Lindl. Bot. Reg. t. 1515; DC. Prod. vii. 771; Sond. in Pl. Preiss. i. 335; *Epacris capitata*, Poir. Dict. Suppl. ii. 556.

W. Australia.   Sandy places, King George's Sound and adjoining districts, *R. Brown, Drummond, 2nd Coll. n.* 258, *Preiss, n.* 447, *Oldfield, F. Mueller;* Cape Riche, *Maxwell.*

This species, with *D. phlogiflorum* and *D. parviflorum*, might almost be considered as varieties of *D. gracile*, and are all very closely connected with *D. squarrosum* and *D. Drummondii.*

7. **D. gracile,** *R. Br. Prod.* 556.   A very slender shrub, attaining 3 or 4 ft., with long virgate flowering branches, the inflorescence usually pubescent, but otherwise glabrous or nearly so.   Lower leaves and those of the barren branches lanceolate-subulate, acutely acuminate, very spreading or recurved, under $\frac{1}{2}$ in. long, those of the long peduncle-like extremities of the flowering branches small and closely appressed.   Flowers in dense ovate or oblong spikes of $\frac{1}{2}$ to 1 in., each one sessile and solitary within a striate ciliate acuminate bract as long as the calyx; bracteoles much shorter, acuminate with ciliate keels.   Sepals $2\frac{1}{4}$ to nearly 3 lines long, lanceolate, striate but thin, mucronate-acute.   Corolla-tube slender, cylindrical, as long as the calyx, not so much contracted upwards as in *D. capitatum;* lobes broad, very nearly as long as the tube, with longitudinal folds at the base.   Filaments adnate to above the middle of the tube.   Hypogynous disk short and annular or obsolete.   Ovary glabrous or minutely pubescent.—Bot. Mag. t. 2678; Lodd. Bot. Cab. t. 1346; *Epacris gracilis*, Poir. Dict. Suppl. ii. 556; *Sphenotoma gracile*, Sweet, Fl. Austral. t. 44; DC. Prod. vii. 771; Sond. in Pl. Preiss. i. 336; F. Muell. Fragm. vi. 66.

W. Australia.   Common in wet places, King George's Sound and adjoining districts, *R. Brown, Baxter, Drummond, n.* 102, *2nd Coll. n.* 261; *Preiss, n.* 448, and others.

8. **D. parviflorum,** *F. Muell. Herb.* Stem branching at the base, with erect virgate simple branches of above 1 ft.   Leaves lanceolate-subulate, rarely ½ in. long, mostly ciliate; the lower ones more or less spreading but not recurved, all the rest closely appressed.   Flowers in dense ovate spikes of about ½ in., each one sessile within an ovate or ovate-lanceolate mucronate-acute bract, at least as long as the calyx ; bracteoles much shorter, keeled, mucronate.   Sepals about 2 lines long, mucronate-acute.   Corolla-tube as long as the calyx ; lobes obovate, scarcely half as long as the tube. Filaments slightly adnate.   Hypogynous scales broad, obtuse.   Ovary glabrous.

W. Australia.   Thomas river and Cape le Grand, *Maxwell.*   F. Mueller, Fragm. vi. 66, mentions this as a variety of *D. gracile,* but besides the foliage, the small corolla-lobes are remarkable, and the characters by which this and the five preceding species are distinguished, are none of them of more importance.

9. **D. minimum,** *F. Muell. Fragm.* i. 39, vi. 65.   A small tufted moss-like plant, forming dense masses of a few inches in diameter.   Leaves imbricate, with a short broad base, acuminate and rather acute, 2 to 3 lines long.   Flowers solitary terminal and sessile amongst the last leaves, which are as long as the calyx and form a kind of involucre round it.   Sepals lanceolate, acute, about 2 lines long.   Corolla-tube rather broad, cylindrical, as long as the calyx ; lobes rather broad, very spreading, shorter than the tube, with longitudinal folds at the base, not quite so prominent as in *D. gracile* and its allies.   Filaments adnate nearly to the top of the corolla-tube ; anthers ovate-oblong.   Hypogynous scales broad, truncate.   Style rather short.—Hook. f. Fl. Tasm. ii. 367.

Tasmania.   Summit of Mount Lapeyrouse, in tufts, *Oldfield.*   *D. muscoides,* Hook. f., from New Zealand, appears to be the same species; the leaves are usually smaller, but I can see no other difference.

ORDER LXVIII. **PLUMBAGINEÆ.**

Calyx tubular, often enlarged and scarious or petal-like at the top, with 5 prominent ribs usually ending in as many teeth.   Corolla regular, of 5 petals, free or more or less united, contorted-imbricate in the bud.   Stamens 5, inserted at the base of the corolla or petals, opposite to them, and often more or less adnate to them ; anthers versatile, 2-celled, the cells opening in longitudinal slits.   Ovary 1-celled, with 1 ovule suspended from a filiform placenta erect from the base.   Styles 5, distinct or united at the base.   Capsule 1-seeded, indehiscent or opening irregularly.   Seed solitary ; testa thin ; albumen rarely abundant, usually scanty or none; embryo straight, radicle superior.—Herbs or rarely undershrubs or shrubs.   Leaves radical or alternate, entire or (in species not Australian) lobed.   Flowers in terminal heads spikes or panicles.

The Order, although a small one, is widely dispersed over most parts of the globe, chiefly in maritime districts.   Of the three Australian genera, one has nearly the range of the Order ; the second extends over the New as well as the Old World, but only in the warmer regions ; the third is further limited to the tropical seacoasts of the Old World.   The Order is allied to *Primulaceæ* in the position of the stamens and the 1-celled ovary, but differs in the compound pistil with a solitary ovule, and in habit.

Calyx tubular, smooth. Styles free or nearly so. Stigmas terminal, capitate. Fruit narrow, elongated, exserted. Shrub, with petiolate leaves . . . .   1. ÆGIALITIS.
Calyx dilated at the top. Styles free, stigmatic in the upper part. Fruit enclosed in the calyx. Herb with radical leaves . . . .   2. STATICE.
Calyx tubular, glandular-muricate. Styles united up to the base of the stigmatic branches. Fruit enclosed in the calyx. Erect or half-climbing, leafy shrub . . . .   3. PLUMBAGO.

1. ÆGIALITIS, R. Br.

(Ægialinites, *Presl.*)

Calyx tubular, with 5 prominent ribs ending in short teeth with induplicate margins. Petals slightly cohering at the base. Stamens slightly adhering to the base of the petals. Styles free or scarcely cohering at the base; stigmas small, terminal, capitate. Fruit narrow, elongated. Seed without albumen.—Shrubs. Leaves broad, petiolate. Flowers nearly sessile along the branches of a dichotomous panicle.

The genus is limited to a single species, extending to the seacoasts of some parts of tropical Asia.

1. **Æ. annulata,** *R. Br. Prod.* 426. A low glabrous shrub, the branches marked with the annular scars of fallen leaves. Leaves on long, winged, sheathing petioles, broadly ovate or almost orbicular, obtuse, entire, coriaceous, with numerous fine parallel veins diverging from the midrib, 2 to 4 in. long. Panicle with few rigid branches, not much exceeding the leaves. Flowers very shortly pedicellate, solitary within erect, concave bracts, which enclose the calyx and are nearly as long, with two much smaller bracteoles at the base of the calyx. Calyx 3 to $3\frac{1}{2}$ lines long, the ribs smooth. Petals white, narrow, shortly exceeding the calyx. Stamens not exceeding the corolla, the short adnate base of the filaments dilated and thickened. Fruit linear, incurved, 5-angled, $1\frac{1}{2}$ in. long, scarcely above 1 line broad, the pericarp thinly coriaceous. Seed filling the cavity; testa membranous; embryo divided nearly to the middle into linear cotyledons, enclosing a linear-conical plumula.—Gaud. in Freyc. Voy. t. 51; Boiss. in DC. Prod. xii. 621; *Ægialinites annulata,* Presl, Bot. Bem. 103.

Queensland. Seacoast, amongst the *Rhizophoras,* Prince of Wales Islands, *R. Brown;* Howick's group, *F. Mueller;* Cape York, *Daemel;* Port Denison, *Fitzalan;* Fitzroy river, *Thozet.*

It is also found on the coasts of Timor, and of the Malayan Peninsula. The name was altered by Presl, as being preoccupied in zoology,—a reason now generally acknowledged to be insufficient.

2. STATICE, Linn.

(Taxanthema, *R. Br.*)

Calyx more or less expanded at the top into a dry, membranous, coloured and slightly 5-lobed limb, each lobe traversed by a green or dark nerve. Petals slightly united at the base. Styles free, ending in linear-terete stigmas. Fruit included in the calyx. Seed more or less albuminous.—Herbs or rarely undershrubs. Leaves usually radical. Flowers solitary or 2 or 3 together in little spikelets, forming one-sided spikes, arranged in diehotomous or trichotomous panicles, or rarely in simple spikes.

The largest genus of the Order, ranging chiefly over maritime districts in the northern hemisphere, with very few southern species. The only Australian one extends northward as far as Japan.

1. **S. australis,** *Spreng. Syst.* i. 959. Stock short and thick. Leaves all radical, obovate-oblong, 1½ to 3 in. long, quite entire, narrowed into a petiole of very variable length. Scape angular, 9 to 18 in. high, repeatedly forked so as to form a broad corymbose panicle, with a small green bract under each branch, and in some specimens there are a few entire or forked barren branches, ½ to 1¼ in. long, at the base of the panicle. Flowers numerous, in short, dense, unilateral spikes, formed of little clusters or spikelets of 2 or 3 flowers each, or of single flowers. Calyx-lobes pale pink, broad and undulate, the ribs usually hairy outside at the base, produced into short points or almost obtuse. Petals yellow, rather longer than the calyx when first flowering. Seed oblong; albumen very scanty on one side only of the embryo; cotyledons oblong; radicle superior, very short.—Boiss. in DC. Prod. xii. 642; Hook. f. Fl. Tasm. i. 301; *Taxanthema australis,* R. Br. Prod. 426; *Statice taxanthema,* Rœm. et Sch. Syst. vi. 798.

Queensland. Port Curtis, *M'Gillivray;* Port Denison, *Fitzalan;* Fizroy river, *Thozet.*

Victoria. Port Albert, *F. Mueller;* Port Phillip, *R. Brown, Gunn.*

Tasmania. Port Dalrymple, *R. Brown;* common along the coast in mud at high-water mark, *J. D. Hooker.*

The species is also in New Caledonia and in Japan, and probably also on the coasts of China, for I cannot find any difference in the *S. japonica,* Sieb. et Zucc.; and *S. sinensis,* Gir. (Bot. Reg. 1845. t. 63), can scarcely be distinguished by the calyx-lobes usually rather more obtuse. The yellow flowers readily separate the whole group from the common European *S. Limonium.*

3. PLUMBAGO, Linn.

Calyx tubular, with 5 prominent ribs, more or less glandular-muricate, ending in short teeth. Corolla with a cylindrical tube and spreading lobes. Stamens hypogynous. Style filiform, with 5 filiform branches stigmatic from the base. Fruit included in the calyx.—Perennials or shrubs, with leafy branches. Flowers sessile, in simple terminal spikes.

The genus, although comprising but very few species, ranges over the temperate and tropical regions of the New as well as the Old World. The only Australian species is a common Asiatic one. Another Indian species, distinguished chiefly by its red flowers and more herbaceous stem, *P. rosea,* Linn., is said to have established itself near Rockhampton, in Queensland, as an escape from gardens.

1. **P. zeylanica,** *Linn.; Boiss. in DC. Prod.* xii. 692. A shrub attaining several feet, the long weak branches sometimes half-climbing, glabrous except the short, glandular, viscid bristles on the inflorescence, and especially on the calyx. Leaves petiolate, ovate, obtuse acute or acuminate, the larger ones 2 to 3 in. long but mostly smaller, the petiole dilated at the base into a stem-clasping ring or sometimes forming prominent auricles. Flowers white, each one sessile within a small broadly ovate acuminate bract and 2 much smaller bracteoles. Calyx 4 to 5 lines long, the ribs strongly glandular-muricate, neither contracted nor expanded at the top. Corolla-

tube slender, much longer than the calyx ; lobes obovate, about as long as
the exserted part of the tube.   Stamens included in the tube.   Nut much
shorter than the calyx, contracted at the top and at the base, 5-angled.—
Bot. Reg. 1846. t. 23 ; Wight, Illustr. t. 179.

N. Australia. Port Essington, *Armstrong ;* Victoria river, *F. Mueller.*

Queensland. Keppel and Shoalwater Bays and Thirsty Sound, *R. Brown ;* Barnard
Isles, *M'Gillivray ;* Port Denison, *Fitzalan ;* Broad Sound, *Bowman* (said to have blue
flowers) ; Rockhampton, *Dallachy ;* Moreton Bay, *F. Mueller* and others.

N. S. Wales. Hunter's River, *R. Brown ;* Richmond river, *Fawcett ;* Hastings river,
Beckler ; New England, *C. Stuart.*

The species extends over tropical Africa and Asia and the Pacific Islands, to the Sandwich Islands.

Order LXIX. PRIMULACEÆ.

Calyx usually of 5, sometimes 4, 6 or 7 divisions or teeth, free or rarely
the tube shortly adnate to the ovary.   Corolla regular, more or less divided
into as many lobes or teeth as divisions of the calyx, imbricate and often contorted in the bud, rarely wanting.   Stamens as many as lobes of the corolla,
inserted in the tube or at the base, opposite the lobes.   Ovary 1-celled, with
1 or more ovules attached to or immersed in a central placenta, usually quite
free, thick and globular, rarely ovoid and connected with the top of the cavity.   Style single, with a capitate stigma.   Fruit a capsule, usually dehiscent.   Seeds albuminous.—Herbs or very rarely undershrubs.   Leaves
opposite or alternate, undivided except when growing under water, without
stipules.   Flowers axillary or terminal.

A widely-spread Order, inhabiting chiefly the northern hemisphere, and often rising in
high mountains to great elevations, with a few southern species, and but very few within the
tropics, except in mountain districts.   The three Australian genera are all widely spread
over the area of the Order.   *Myrsineæ*, which generally replace *Primulaceæ* in the warmer
regions of the globe, scarcely differ from them, except in their shrubby or arboreous habit.

Calyx entirely free.   No staminodia.
Capsule opening in 5 or 10 valves or irregularly dehiscent.   Corolla
rotate or campanulate.   Leaves whorled opposite or alternate . . 1. Lysimachia.
Capsule circumsciss.   Corolla rotate or campanulate.   Leaves opposite or alternate . . . . 2. Anagallis.
Calyx partially adnate.   Staminodia or scales alternating with the lobes
of the corolla . . . . 3. Samolus.

1. LYSIMACHIA, Linn.

Calyx free, deeply 5-cleft.   Corolla rotate or campanulate, deeply 5-lobed.
Stamens 5.   Capsule opening in 5 or 10 valves or rarely bursting irregularly.
Placenta globular.—Perennials, with erect or procumbent stems.   Leaves opposite or rarely whorled or scattered, or in species not Australian, alternate or
tufted.   Flowers yellow or (in a few species not Australian) white, solitary,
on axillary pedicels or in terminal racemes.

A considerable genus, having a wide range in the northern hemisphere, but limited in the
southern hemisphere to the two Australian species, one of which is endemic, the other the
same as a widely dispersed Asiatic one.

Stem tall, erect. Leaves narrow, scattered or in whorls of 3. Flowers
　in a short, terminal raceme or oblong panicle . . . . 1. *L. salicifolia.*
Stem weak or trailing. Leaves petiolate, ovate. Flowers axillary and
　solitary . . . . 2. *L. japonica.*

1. **L. salicifolia,** *F. Muell. Herb.* Stems tall, erect, simple, glabrous
or minutely glandular-pubescent in the upper part. Leaves scattered or almost collected into whorls of 3, lanceolate or linear-lanceolate, acute, entire,
narrowed into a short petiole, 3 to 4 in. long. Flowers yellow, in a short,
terminal, compound raceme or oblong panicle. Pedicels slender, 3 to 4 lines
long, almost verticillate or in clusters of 3 to 5 on a short common peduncle,
with small linear bracts at their base. Calyx-segments narrow, nearly 2 lines
long. Corolla divided nearly to the base into rather broad, obtuse lobes
nearly 3 lines long. Stamens much shorter than the corolla, the filaments
dilated at the base, but scarcely connate. Style elongated, but very deciduous.

　N. S. Wales. New England, twenty miles from Timbarra, *C. Stuart.*
　Victoria. Snowy River, *F. Mueller.* The aspect of the plant is very nearly that of
the northern Asiatic *L. dahurica,* but the flowers appear smaller, in a single terminal
raceme. The plant is, however, very rare, and the few specimens seen are not in very good
flower.

2. **L. japonica,** *Thunb. Fl. Jap.* 83 ; *Ic. Fl. Jap. t.* 16. Stems weak,
usually creeping and rooting at the base, with shortly ascending simple
branches of 6 in. to 1 ft., more or less pubescent as well as the foliage.
Leaves opposite or rarely in whorls of 3, or the uppermost alternate, petiolate,
ovate, sometimes cordate, $\frac{1}{2}$ to 1 in. long. Flowers yellow, solitary in the
axils, on recurved peduncles shorter than the leaves. Calyx-segments narrow, acute, nearly as long as the corolla. Corolla deeply divided into broad
obtuse lobes. Stamens shorter than the corolla, the filaments dilated at the
base and shortly united. Capsule 5-valved.—Duby in DC. Prod. viii. 67 ;
Klatt, Monogr. Lysim. 34. t. 19 ; *L. maculata,* R. Br. Prod. 428 ; Duby in
DC. Prod. viii. 66 ; *L. debilis,* Wall., Duby, l. c.

　N. S. Wales. Hunter's River, *R. Brown ;* near Timbarra, New England, *C. Stuart.*
　The species is widely dispersed over the hilly parts of tropical and eastern Asia, from Ceylon
and the Neilgherries to the Himalaya, Formosa, and Japan.

2. ANAGALLIS, Linn.

(Euparea, *Gærtn. ;* Micropyxis, *Duby.*)

Calyx free, deeply 5-cleft. Corolla rotate or campanulate, deeply 5-lobed.
Stamens 5. Capsule opening transversely by a circular fissure across the
middle (circumsciss). Placenta globular.—Annuals or perennials, with
creeping procumbent or diffuse stems. Leaves opposite or alternate. Flowers
pink red or blue, axillary and solitary.

　A small genus, widely dispersed over the temperate and warmer regions of the globe, although in some countries only as introduced weeds. Of the two Australian species, one
belongs to the latter class, the other may be indigenous, although common to tropical Asia,
Africa, and America.

Leaves opposite. Corolla quite rotate. . . . . 1. *A. arvensis.*
Leaves alternate. Corolla broadly campanulate, almost rotate. Flowers
　very small. . . . . 2. *A. pumila.*

Euparea amœna, Gærtn.; Duby in DC. Prod. viii. 68, said to be from New Holland from a mistake of Gærtner's, is *Anagallis alternifolia*, Cav., from South America; the specimens in the Banksian Herbarium described by Gærtner are from Terra del Fuego.

*1. **A. arvensis,** *Linn.; Duby in DC. Prod.* viii. 69. A much-branched, procumbent, glabrous annual, extending from 6 in. to nearly 1 ft. Leaves opposite, sessile, broadly ovate, obtuse, ¼ to ½ in. long. Flowers red or blue, on pedicels considerably longer than the leaves, and rolled back as the capsule ripens. Calyx-segments acute. Corolla rotate, spreading to about 3 lines diameter.

A common weed of cultivation in Europe, temperate Asia, and the whole Mediterranean region, now spread over a great part of the world, and more or less established in **N. S. Wales, Victoria, Tasmania, S. Australia,** and **W. Australia.**

2. **A. pumila,** *Swartz, Fl. Ind. Occid.* i. 345. A slender, glabrous, diffuse, much-branched annual, rarely exceeding 6 in. Leaves alternate, nearly sessile, orbicular or broadly ovate, obtuse, rarely above ¼ in. long. Flowers very small, on pedicels nearly as long as the leaves. Calyx-segments 5 or sometimes 4, lanceolate, acute. Corolla broadly campanulate, deeply divided into acute lobes, rather longer than the calyx, spreading to about 1½ lines diameter. Stamens shorter than the corolla, the filaments slightly dilated at the base, usually contracted at the throat as it withers, and persisting over the capsule. Style filiform, sometimes very short. Capsule small, circumsciss. Seeds numerous, slightly compressed.—*Centunculus pentandrus*, R. Br. Prod. 427; *Micropyxis pumila*, Duby in DC. Prod. viii. 72; *Micropyxis tenella*, Wight, Ic. t. 1585; *Centunculus tenellus*, Duby, l. c.; Wight, Ic. t. 2000.

N. Australia. Gulf of Carpentaria, *F. Mueller.*
Queensland. Shoalwater Bay, *R. Brown.*

The species is widely distributed over S. America, E. India, and W. tropical Africa. Although usually 5-merous, some specimens occur in which all or nearly all the flowers are 4-merous.

3. SAMOLUS, Linn.

Calyx campanulate, the tube partially adhering to the ovary; lobes 5. Corolla-tube short or broad; lobes 5, spreading, with 5 small or filiform scales or staminodia alternating with the lobes. Stamens 5, opposite the lobes. Capsule half-inferior, the free part opening in 5 valves opposite the calyx-lobes. Placenta ovoid, the axis usually produced to the apex of the cavity.—Perennials or annuals. Leaves alternate. Flowers white, in loose terminal racemes.

A small genus, confined to the southern hemisphere with the exception of one species generally diffused over most parts of the world. The Australian species include the two which have a most general area, and a third endemic one, which is, however, very nearly allied to some of the endemic Australian forms of *S. repens.*

Leaves obovate, chiefly radical. Flowers not 2 lines diameter, with a
   very short tube, in a loose terminal raceme . . . . 1. *S. Valerandi.*
Leaves rather thick, the radical ones obovate or oblong, the stem ones
   lanceolate or oblong. Flowers above ½ in. diameter, the tube as long
   as the calyx. Flowers axillary or in a short terminal raceme . . . 2. *S. repens.*
Leaves few, oblong, radical. Stems leafless. Flowers of *S. repens* . . 3. *S. junceus.*

S. Valerandi, *Linn. ; Duby in DC. Prod.* viii. 73. A glabrous, bright green annual or perennial, with a tuft of obovate, spreading, radical leaves. Flowering stems 3 or 4 in. to above 1 ft. high, simple or branched, bearing a few obovate or oblong leaves, and loose terminal racemes of small white flowers. Pedicels filiform, with a small bract about the middle. Corolla not 2 lines diameter and sometimes much smaller, the tube very short. Staminodia linear, small. Capsule small, globular, crowned by the small calyx-lobes.

N. S. Wales, Blue Mountains, *R. Brown ;* Tweed river, *C. Moore ;* Richmond river, *Fawcett ;* New England, *C. Stuart.*

Victoria. Clifton morass, Snowy River, *F. Mueller.*

This species is common in most of the temperate and warmer regions of the globe, especially in maritime districts. The Australian specimens belong to the smallest flowered variety, which has been considered by some as a distinct species, and is a common American but comparatively rare European form.

2. **S. repens,** *Pers. Syn.* i. 171. A perennial, with a more or less tufted stock emitting creeping stolons, but very variable in habit and stature, glabrous or hoary with minute, almost scale-like glands. Stems simple or branched, sometimes prostrate, with most of the pedicels axillary, sometimes erect or ascending, ½ to 1 ft. high, leafy, with the flowers chiefly collected in a short terminal raceme. Radical leaves petiolate, obovate or oblong ; stem-leaves usually small, either linear or lanceolate and acute, or oblong obtuse and petiolate. Pedicels long or short, without bracts except the subtending one, which is often adnate to the pedicel, so as to appear inserted on it. Flowers variable in size, but usually spreading to 4 or 5 lines diameter. Calyx-tube adnate to about the middle of the ovary ; lobes acute, longer than or of the length of the tube. Corolla-tube broad, usually about as long as the calyx-lobes ; lobes ovate, obtuse, as long as the tube. Staminodia filiform. Capsule half-inferior ; placenta stipitate, ovoid, ending in a long horn-like point connected with the apex of the cavity. Seeds numerous, globular or angular.—*S. littoralis,* R. Br. Prod. 428; Duby in DC. Prod. viii. 73 ; Hook. f. Fl. Tasm. i. 301 ; Nees in Pl. Preiss. i. 337 ; Lodd. Bot. Cab. t. 435 ; *Sheffieldia incana,* Labill. Pl. Nov. Holl. i. 40. t. 54.

N. S. Wales. Near the sea, Port Jackson, *R. Brown* and others.

Victoria. Along the coast from the Glenelg to Gipps' Land, *F. Mueller* and others.

Tasmania. Abundant along the coast, especially on saline mud, *J. D. Hooker.*

S. Australia. Port Lincoln, *R. Brown ;* common in salt marshes and on rocks along the coast, *F. Mueller* and others.

W. Australia. King George's Sound and Goose Island Bay, *R. Brown ;* along the coast from King George's Sound round to Swan and Murchison rivers, *Drummond, n.* 135, 248, 656, *Preiss, n.* 1241, *Oldfield,* and others, and eastward towards the Great Bight, *Maxwell.*

The species extends to New Zealand and New Caledonia. It is exceedingly variable in stature, foliage, etc., and, besides the diversity indicated in the above description, the following forms, all from W. Australia, might almost be regarded as species.

Var. *floribundus.* Erect and paniculately branched, with numerous flowers, in dense corymbose racemes. Leaves few. Calyx-lobes very acute.—Murchison river, *Oldfield ;* Swan River, *Preiss, n.* 1237.

Var. ? *paucifolius.* Erect, rigidly branched, 1 to 1½ ft. high. Leaves few. Flowers few and distant, on long peduncles.—Swan and Murchison rivers.

Var. ? *ambiguus.* Erect and branched, with scarcely any leaves besides the radical ones, and a smaller plant with shorter peduncles than the var. *paucifolius.—S. ambiguus,* R. Br. Prod. 429; Duby in DC. Prod. viii. 73.—King George's Sound and Murchison river.— The last two varieties connect the species with *S. junceus,* and might be almost equally well regarded as a distinct intermediate species, as proposed by Brown. There are some specimens, however, which might be equally well ranked with some of the common forms of *S. repens. S. parviflorus,* Nees in Pl. Preiss. i. 337, from the Victoria district, W. Australia, Preiss, n. 1239, may perhaps be reckoned among these ambiguous forms, but the specimens are very imperfect, and may belong to the var. *floribundus.*

3. **S. junceus,** *R. Br. Prod.* 429. A perennial, with a shortly creeping stock. Radical leaves few, petiolate, oblong, obtuse. Stems erect, rushlike, simple or slightly branched, 1 to 2 ft. high, leafless except a few minute linear scales. Flowers few, distant, on short pedicels, in a terminal raceme, otherwise entirely as in *S. repens,* of which this may possibly prove to be an extreme form.—Duby in DC. Prod. viii. 73; Nees in Pl. Preiss. i. 338.

W. Australia. King George's Sound and adjoining coasts, *R. Brown, Harvey, Drummond, n.* 134, *Oldfield;* Cape le Grand and Esperance Bay, *Maxwell;* mouth of Swan River, *Preiss, n.* 1238.

Order LXX. MYRSINEÆ.

Flowers regular. Calyx of 5 or rarely 4 divisions or teeth, free or rarely the tube adnate to the ovary. Corolla regular, more or less divided into as many lobes or teeth or rarely distinct petals as the divisions of the calyx, imbricate and often contorted in the bud. Stamens as many as divisions of the corolla and opposite to them, inserted in the tube or at the base. Ovary 1-celled, with several ovules, usually peltate, attached to or immersed in a central placenta, usually quite free, thick, and globular. Style single, with a terminal, capitate or small stigma. Fruit an indehiscent berry or drupe, or very rarely splitting lengthwise on one side. Seeds albuminous (except in *Ægiceras*).—Trees or shrubs, the foliage and inflorescence usually marked with resinous dots. Leaves alternate, simple, entire or toothed, without stipules. Flowers small, in axillary clusters, racemes or panicles, or rarely in terminal panicles.

A considerable Order, widely distributed over the tropical and subtropical regions of the globe. Of the five Australian genera, three are common to the New and the Old World, the other two limited to the Old World. The Order only differs from Primulaceæ in the woody stem and succulent indehiscent fruit. From the Sapotaceæ, it is readily distinguished by the 1-celled ovary.

Ovary wholly or partially inferior . . . . 1. Mæsa.
Ovary inferior. Petals free to the base . . . . 2. Samara.
Ovary superior. Petals united in a short tube with a deeply lobed limb.
    Flowers in axillary clusters . . . . 3. Myrsine.
    Flowers in umbels usually pedunculate and sometimes forming panicles.
        Ovary and fruit obtuse, globular or ovoid. Anther cells not divided
            transversely . . . . 4. Ardisia.
        Ovary and fruit acuminate, becoming long and curved. Anther cells
            transversely divided into several pits . . . . 5. Ægiceras.

1. MÆSA, Forsk.

(Bæobotrys, *Forst.*)

Calyx-tube adnate to the ovary, the limb 5-lobed. Corolla 5-lobed. Stamens 5, with slender filaments and short anthers. Ovary inferior or half-inferior; style short. Berry crowned by the calyx-lobes or teeth.—Trees or shrubs. Flowers small, in simple or compound racemes, either axillary or rarely terminal. Bracts at the base of the pedicels, and 2 bracteoles under the flower usually very small.

The genus is limited to the tropical regions of the Old World. The two Australian species are endemic, both readily distinguished from the Asiatic ones by their long simple racemes.

Woody climber. Leaves cordate or rounded at the base. Ovary almost entirely inferior . . . . 1. *M. dependens.*
Erect tree. Leaves tapering at the base. Ovary half-superior . . 2. *M. haplobotrys.*

1. **M. dependens,** *F. Muell. Fragm.* v. 107. A tall woody climber, the typical form quite glabrous. Leaves shortly petiolate, ovate, very shortly acuminate, irregularly sinuate-toothed or almost entire, cordate or rounded at the base, distantly penniveined, without any or very few intermediate veinlets, usually 3 to 4 in. long, but on barren branches often twice that size. Flowers rather distant, in slender racemes often twice as long as the leaves, these racemes sometimes 2 or 3 together, but usually if not always simple. Pedicels about as long as the calyx or at length rather longer. Calyx-tube ovoid; lobes broad, rounded, entire or minutely ciliolate, about ½ line long. Corolla-tube broad, rather longer than the calyx; lobes spreading, shorter than the tube, imbricate in the bud. Stamens included in the corolla-tube; anthers about as long as the filament. Ovary inferior, except the very short conical summit, which tapers into a very short style; stigma capitate. Fruit ovoid, crowned by the closed calyx-lobes. Seeds immersed in the placenta, but not seen ripe.

Queensland. Rockingham Bay, *Dallachy.*

Var. *pubescens,* F. Muell. Branches, foliage, and inflorescence softly pubescent or shortly villous. Pedicels rather shorter, and young fruit more globular than in the glabrous form. —Rockingham Bay, *Dallachy.*

2. **M. haplobotrys,** *F. Muell. Fragm.* v. 161. A small erect tree, quite glabrous. Leaves narrow-ovate to elliptical-oblong, acuminate, entire or irregularly sinuate-toothed, narrowed into a petiole, which is usually short, distantly penniveined with but few cross veinlets, 3 to 6 in. long. Flowers in simple racemes, longer or rather shorter than the leaves. Pedicels very short. Calyx-tube very short and adnate to the base of the ovary, but enlarging after flowering; lobes broad, almost acute, about ½ line long. Corolla-tube broad, nearly as long as the calyx, the lobes broad, much imbricate, at least as long as the tube. Anthers ovate, as long as the flattened filaments. Ovary at the time of flowering more than half-superior, tapering into a very short style. Young fruit nearly globular, crowned by the closed calyx-lobes. Seeds not seen perfect.

Queensland. Rockingham Bay, *Dallachy.*

2. SAMARA, Linn.

(Choripetalum, *A. DC.*)

Calyx free, deeply 4-lobed. Petals 4, distinct, spreading. Stamens in-

274   LXX. MYRSINEÆ.   [*Samara.*

serted at the base of the petals and longer than them, with filiform filaments
and short anthers. Ovary superior; style short.—Shrubs, often half-trail-
ing or climbing. Flowers small, in short axillary racemes. Bracts and
bracteoles very small.

A small genus dispersed over tropical Asia and Africa. The only Australian species is
endemic.

1. **S. australiana,** *F. Muell. Fragm.* vi. 164. A tall woody climber,
quite glabrous. Leaves ovate-elliptical or obovate, obtuse or shortly acuminate, narrowed into a petiole, thinly coriaceous, shining above, prominently
penniveined, and with intermediate veinlets, 1½ to 3 in. long. Flowers very
small, in short, loose, axillary racemes, the rhachis rarely above ½ in. long
and the pedicels ¼ in. Calyx-lobes 4, about ½ line long. Petals 4, about 1
line long, imbricate in the bud. Filaments exceedingly short; anthers
ovate. Ovary tapering into a style, sometimes nearly as long as the petals,
but usually much shorter; stigma broad. Drupe hard, globular, about 3
lines diameter. Seeds apparently embedded in the globular placenta, but
destroyed by grubs in all the fruits opened.—*Choripetalum australianum,*
F. Muell. Fragm. iii. 36.

N. S. Wales. Macleay and Clarence rivers, *Beckler.* Very different in foliage from
any extra-Australian species known to me. Leichhardt's specimens from the Araucaria
woods, Burnett river, mentioned by F. Mueller, seem to me to belong to the same species.
They are, however, very imperfect, in fruit only.

3. MYRSINE, Linn.

Calyx free, 4- or 5-lobed. Corolla deeply 4- or 5-lobed. Stamens inserted at the base of the corolla-lobes, with very short filaments; the anthers
much longer, erect and lanceolate. Ovary superior; style short, with a
capitate or fringed stigma.—Shrubs or small trees. Leaves coriaceous,
entire or rarely toothed. Flowers small, often polygamous, in umbels or
clusters, sessile in the axils or at the nodes, usually on the old wood. Bracts
minute, deciduous.

A considerable genus, spread over the tropical and subtropical regions of both the New
and the Old World. The four Australian species appear to be endemic, although two of
them are very nearly allied to a common south Asiatic one.

Corolla urceolate, the lobes shorter than the tube . . . . . . 1. *M. urceolata.*
Corolla-lobes much longer than the short open tube.
  Leaves nearly sessile or the petiole not 2 lines long . . . . . 2. *M. crassifolia.*
  Leaves on petioles of ¼ to 1 in. long.
    Calyx not ⅓ as long as the corolla. Corolla-lobes narrow . . 3. *M. variabilis.*
    Calyx campanulate, more than half as long as the corolla. Co-
    rolla-lobes broad and thick . . . . 4. *M. achradifolia.*

1. **M. urceolata,** *R. Br. Prod.* 534. A small tree, quite glabrous.
Leaves elliptical-oblong or broadly lanceolate, obtuse, entire, narrowed into
a rather long petiole, 2 to 3 in. long. Pedicels slender, 2 to 3 lines long.
Calyx very small, with short broad lobes. Corolla narrow-urceolate, about
1¼ lines long, the lobes (usually 4) scarcely half as long as tube. Anthers
nearly sessile, near the base of the corolla-tube and included in it.—A. DC.
Prod. viii. 96.

Queensland. Endeavour river, *Banks and Solander, A. Cunningham.* Resembles the narrow-leaved forms of *M. variabilis,* except in the corolla, which is less divided than in almost any species.

2. **M. crassifolia,** *R. Br. Prod.* 534. A small tree, quite glabrous. Leaves very shortly petiolate or almost sessile, oblong, lanceolate, elliptical or almost obovate, obtuse, quite entire, 3 to 4 in. long or sometimes more, usually but not always more coriaceous than in *M. variabilis,* and much less so than in *M. achradifolia.* Pedicels almost always shorter than the flowers and sometimes not above ½ line long. Calyx small, with acute lobes often longer than in *M. variabilis.* Corolla divided nearly to the base into 4 or 5 narrow lobes a little more than 1 line long when fully out. Anthers obtuse, almost sessile at the base of the lobes and not exceeding them. Fruit small, globular.—A. DC. Prod. viii. 96 (as to the Australian plant) ; *M. subsessilis,* F. Muell. Fragm. iv. 81 ; *M. porosa,* F. Muell. Fragm. v. 108.

Queensland. Port Bowen, *R. Brown ;* Endeavour river, *A. Cunningham ;* Moreton Bay, *Leichhardt ;* Rockingham Bay, *Dallachy.*
N. S. Wales. Clarence river, *Beckler ;* Richmond river, *C. Moore ;* Mount Lindsay, *W. Hill.*

The two forms distinguished by F. Mueller as species differ chiefly in the longer or shorter pointed calyx-lobes. The Norfolk Island plant, referred to *M. crassifolia* by Endlicher and by A. Cunningham (A. DC. Prod. viii. 96), has more coriaceous leaves on much longer petioles. All are nearly allied to the smaller-leaved forms of the E. Indian *M. capitellata.*

3. **M. variabilis,** *R. Br. Prod.* 534. A glabrous tree, attaining 30 to 40 ft. Leaves mostly obovate-oblong, obtuse or rarely shortly acuminate, entire or irregularly bordered by acute teeth, coriaceous and 1½ to 2½ in. long or in other specimens nearly twice as large and thinner, narrowed into a petiole varying from scarcely ¼ in. to above ½ in. Pedicels usually about 2 lines long. Calyx very small, with broad short lobes. Corolla when perfect 1½ lines long or rather more ; divided nearly to the base into 4 or 5 acute narrow lobes, valvate in the bud or nearly so. Anthers nearly sessile at the base of the lobes, rather large but not exceeding the lobes. Drupe globular, 2 to 3 lines diameter.—A. DC. Prod. viii. 94.

Queensland. Keppel Bay, Broadsound, Shoalwater Bay, *R. Brown ;* Rockhampton, *Thozet ;* Nerkool Creek, *Bowman ;* Brisbane, *F. Mueller ;* Ipswich, *Nernst.*
N. S. Wales. Port Jackson to the Blue Mountains, *R. Brown, Sieber, n. 262,* and many others ; northward to Hastings and Clarence rivers, *Beckler* and others ; George and Tweed rivers, *C. Moore ;* New England, *C. Stuart ;* southward to Illawarra, *A. Cunningham ;* and Twofold Bay, *F. Mueller.*
Victoria. All over the forest creeks of Gipps' Land, *F. Mueller ;* near Melbourne, *Adamson.*

4. **M. achradifolia,** *F. Muell. Fragm.* vi. 164. A moderate sized tree. Leaves on rather long petioles (½ to 1 in.) oblong-elliptical, obtuse, thickly coriaceous, smooth and shining above, 4 to 8 in. long. Pedicels 1 to 2 lines long. Flowers larger and more globular than in most species. Calyx broadly campanulate, about 1¼ lines long, the lobes broad, not longer than the tube. Corolla not twice as long as the calyx, deeply divided into broad very thick valvate lobes. Anthers nearly sessile at the base of the lobes, shorter than them, ovoid, obtuse.

Queensland. Rockingham Bay, *Dallachy.* The foliage is that of the largest-leaved forms of *M. capitellata,* Wall., but the flowers are different.

4. ARDISIA, Linn.

Calyx free, 5-(or 4- ?)lobed. Corolla deeply 5-(or 4- ?)lobed, the lobes spreading, contorted-imbricate in the bud. Stamens inserted at the base of the corolla-lobes ; filaments short ; anthers erect, lanceolate, the slits of the cells sometimes not reaching the base. Ovary superior. Style subulate, usually long and persistent, the stigma not enlarged. Berry or drupe globular.—Trees or shrubs or, in species not Australian, almost herbaceous. Flowers not so small as in most genera, usually in short umbel-like racemes, axillary or terminal, solitary or several together in branching panicles. Corolla white or pink, frequently spotted.

A large genus, widely dispersed over the tropical and subtropical regions of the New as well as the Old World. The Australian species appear both to be endemic.

Umbels several, in a slender but not large terminal panicle. Corolla induplicate-valvate . . . . 1. *A. pseudojambosa.*
Umbels solitary on short terminal (or axillary ?) peduncles. Corolla contorted-imbricate . . . . 2. *A. brevipedata.*

1. **A. pseudojambosa,** *F. Muell. Fragm.* iv. 81. A tree, attaining about 30 ft., quite glabrous. Leaves obovate-oblong, acuminate, narrowed into a short petiole, entire or obscurely sinuate, rather thin, the pinnate veins numerous and usually conspicuous, 3 to 5 in. long. Umbels several in a terminal panicle, which is sometimes sessile and shorter than the last leaves, sometimes pedunculate, longer and loose, with slender divaricate branches. Pedicels 3 to 4 lines long. Flowers small. Calyx deeply divided into segments of about ¾ line. Corolla almost rotate, the tube short, the lobes about 1 line long, ovate, acute, with membranous margins, induplicate-valvate in the bud. Anthers nearly sessile, lanceolate, acuminate, the points exceeding the corolla-lobes. Style long and slender. Drupe globular, about 4 lines diameter.

Queensland. Rockhampton and Rockingham Bay, *Dallachy ;* Rockhampton and Thozet's Creek, *Thozet ;* Fitzroy river, *Bowman.*
N. S. Wales. Richmond river, *C. Moore,* apparently the same, but in fruit only.

2. **A. brevipedata,** *F. Muell. Fragm.* vi. 163. A tall glabrous shrub. Leaves oblong-elliptical or broadly lanceolate, acuminate, entire, narrowed into a very short petiole, finely penniveined, 3 to 5 in. long. Peduncles terminal, very short, bearing usually a single umbel of rather numerous small flowers. Pedicels filiform, ¼ to ½ in. long. Calyx open, about 1 line diameter, broadly and shortly 5-lobed. Corolla-tube short and broad ; lobes acutely acuminate, about 1 line long, contorted in the bud. Anthers almost sessile, lanceolate, acuminate but shorter than the corolla-lobes. Style slender, about as long as the corolla. Fruit globular, 3 to 4 lines diameter, ripening usually a single seed. Albumen mealy.

Queensland. Rockingham Bay, *Dallachy.*
A. repandula, F. Muell. Fragm. iv. 82, from "the Brush going to Burun," near Moreton Bay, *Leichhardt,* appears to be this species, with a few but not all of the leaves minutely

and obscurely crenulate. The specimen mentioned by F. Mueller from Richmond river, *C. Moore*, appears to be rather *A. pseudojambosa.* Both are very imperfect and in fruit only.

5. ÆGICERAS, Gærtn.

Calyx free, 5-cleft. Corolla with 5 spreading lobes. Stamens with subulate filaments; anthers lanceolate, the cells divided transversely into several pits. Ovary superior; style subulate, acute; stigma terminal, very small. Fruit cylindrical, incurved, opening as the seed grows in one or two longitudinal slits. Seed without albumen; cotyledons thick and fleshy.—Maritime trees or shrubs, with the habit of *Rhizophoræ,* and, as in those trees and in *Ægialitis,* the seed is said to germinate before the fruit falls off. Flowers white, in umbels or in very short umbel-like racemes, axillary or terminal.

The genus is probably limited to a single species spread over the seacoasts of tropical Asia. It has by some been separated from the Order on account of the peculiar anthers and exalbuminous seeds, but the other characters are quite those of *Myrsineæ.*

1. **Æ. majus,** *Gærtn.; A. DC. Prod.* viii. 142. A glabrous shrub or small tree. Leaves obovate, very obtuse, 2 to 3 in. long, quite entire, narrowed into a petiole at the base, coriaceous and evergreen. Umbels axillary and terminal, nearly sessile. Pedicels stiff, 3 to 5 lines long. Calyx nearly 3 lines long, with very obtuse stiff much-imbricate segments, closely covering the tube of the corolla, which is about their length. Corolla-lobes about the same length, spreading or reflexed, stiff and very acute. Stamens shortly exserted. Ovary very acute, growing out into a curved horn-like fruit.— Wight, Illustr. t. 146; *Æ. fragrans,* Kœn., R. Br. Prod. 534.

N. Australia. Albert river, *Henne;* Port Essington, *Armstrong.*

Queensland. Marshes on the seacoast, Shoalwater Bay, *R. Brown;* from Cape York to Moreton Bay, *F. Mueller* and many others.

N. S. Wales. Port Jackson, *R. Brown* and others; Hunter's River, *A. Cunningham;* Hastings river, *Beckler;* Clarence river, *Wilcox.*

The species extends from Ceylon and the Indian peninsula to the Archipelago and South Pacific Islands. The several species distinguished by Blume, A. Richard, and Presl, are probably not really different from the common one.

Order LXXI. SAPOTACEÆ.

Flowers regular. Calyx free, of 4 to 8 imbricate segments. Corolla more or less divided into as many or rarely twice as many lobes. Perfect stamens either as many as corolla-lobes (or as the inner ones when the lobes are in several series) and opposite to them, or rarely twice as many, besides which there are frequently small scales (or staminodia?) alternating with the lobes of the corolla, or staminodia alternating with the perfect stamens. Ovary superior, 2- or more-celled, with 1 ovule in each cell, erect pendulous or laterally attached. Style simple, with an entire or very slightly lobed stigma. Fruit a berry or drupe, usually indehiscent. Seeds either with a fleshy albumen and flat cotyledons or without any albumen but with thick fleshy cotyledons. Radicle short, inferior.—Trees or shrubs, with the juice very frequently milky. Leaves alternate, entire, without any (or with small very deciduous ?) stipules. Flowers axillary, solitary or clustered. Bracts small or none.

An Order widely distributed over both the New and the Old World within the tropics, and not spreading far beyond them either northward or southward. Of the five Australian genera, three are dispersed over the greater part of the area of the Order, the two other small ones are, as far as known, endemic.

Calyx-segments, stamens and corolla-lobes 5, without scales or sta-
minodia. Albumen scanty or none . . . . 1. CHRYSOPHYLLUM.
Calyx-segments, stamens and corolla-lobes 5 or 6, with small scales
in the throat of the corolla alternating with the lobes.
No hypogynous disk.
Seeds without albumen ; cotyledons thick and fleshy . . . 2. SERSALISIA.
Seeds albuminous ; cotyledons thin and broad . . . . . 3. ACHRAS.
Ovary surrounded by an annular hirsute disk . . . . . . 4. HORMOGYNE.
Calyx-segments 6 or 8. Corolla-lobes twice or three times as many.
Stamens as many as calyx-lobes, with petal-like staminodia be-
tween them. Seeds albuminous . . . . 5. MIMUSOPS.

(*Bassia* has 6, 8, or rarely more calyx-segments and corolla-lobes, twice as many stamens, and seeds without albumen, with thick fleshy cotyledons. See *Sersalisia ? galactoxylon*, p. 279.)

1. CHRYSOPHYLLUM, Linn.

Calyx-segments, corolla-lobes, stamens and cells of the ovary 5 each or rarely in species not Australian 6 to 8 each. No staminodia, nor any scales to the corolla. Seeds usually one or few ; testa hard and smooth ; hilum lateral reaching at least halfway up ; albumen scanty or none ; cotyledons thick and fleshy.—Trees or shrubs. Leaves usually rusty or silvery-tomentose underneath. Flowers small, clustered.

The genus is chiefly tropical-American, with a few African and Asiatic species. The only Australian one is endemic, and very different from any other in several respects.

1. **C. pruniferum,** *F. Muell. Fragm.* vi. 26. A small tree, the branches and underside of the leaves tomentose-villous with rust-coloured stellate hairs. Leaves shortly petiolate, ovate-elliptical or obovate-oblong, shortly and abruptly acuminate, penniveined, reticulate with transverse veinlets, glabrous above, mostly 3 to 4 in. long but sometimes larger. Flowers closely sessile in axillary clusters. Calyx hirsute with rusty hairs ; lobes 5, oblong, about 1 line long, the inner ones with glabrous membranous margins. Corolla-tube shorter than the calyx ; lobes 5, longer than the tube, glabrous, reflexed, much imbricate in the bud. Stamens 5 ; filaments filiform, longer than the corolla-lobes ; anthers ovate-triangular, acute, with the point turned downwards, and the whole anther turned outwards in the bud, inwards when open. Style as long as the corolla. Ovary very villous, 5-celled, with 1 pendulous ovule in each cell. Fruit about 1 in. diameter ("the size of an Orleans plum"), not seen in a good state, the exocarp apparently thin and succulent, the endocarp crustaceous and elegantly veined. Seeds 1 or 2, large; albumen none (or thin ?) ; cotyledons large thick and fleshy; radicle not prominent.

Queensland. Rockingham Bay, *Dallachy.*
N. S. Wales. Bellinger river, *C. Moore.*

The species differs from the genus in the long exserted stamens and small anthers and in the pendulous ovules.

2. SERSALISIA, R. Br.

Calyx-segments, corolla-lobes, stamens, and ovary-cells 5. Scales (or staminodia?) 5, in the throat of the corolla, alternating with the lobes. Ovules laterally attached. Seeds solitary or few; testa hard, shining; hilum lateral, above half as long as the seed; albumen none; cotyledons thick and fleshy.—Trees or shrubs, glabrous or tomentose. Flowers sessile or pedicellate, clustered.

The genus is limited to two Australian species, one of them as yet doubtful. Except the seed, there is no character to separate it from *Achras*.

Leaves more or less tomentose or villous underneath . . . . 1. *S. sericea.*
Leaves quite glabrous . . . . 2. *S. galactoxylon*

1. S. sericea, *R. Br. Prod.* 530. A tree of stunted growth the young branches and underside of the leaves silky-pubescent tomentose or villous, rust-coloured or hoary. Leaves shortly petiolate, mostly ovate or obovate, but varying from nearly orbicular and under 2 in. diameter to broadly-elliptical and 2 to 3 in. long, coriaceous, glabrous above when full-grown. Flowers almost sessile, in axillary clusters, the pedicels rarely 1 line long. Calyx-segments 1½ lines long, rather thick, broadly ovate, pubescent. Corolla-tube rather longer than the calyx, the lobes much shorter, the scales in the throat narrow and acute, but petal-like. Anthers on short filaments, turned outwards but opening laterally, shorter than the corolla. Ovary villous, 5-celled, with 1 laterally-attached ovule in each cell. Fruit ovoid, under 1 in. long, succulent. Seeds 1 or 2, with a lateral (broad?) hilum, no albumen, and thick, fleshy cotyledons.—A. DC. Prod. viii. 177; *Sideroxylon sericeum,* Ait. Hort. Kew. i. 262.

N. Australia. Barren hills, Roe's river, York Sound, *A. Cunningham;* Upper Victoria and Gilbert rivers, *F. Mueller;* Gulf of Carpentaria, *R. Brown.*

Queensland. Bay of Islets and Endeavour river, *Banks and Solander;* Keppel Bay, Thirsty Sound, Shoalwater Bay, Broadsound, *R. Brown;* Port Denison, *Fitzalan* and others; Edgecombe Bay, *Dallachy;* Rockhampton, *Thozet;* Cleveland Bay, Crocodile Creek, etc., *Bowman.*

N. S. Wales? A specimen in leaf only, from Tweed river, *C. Moore,* may be this species.

There is considerable diversity in the form and indumentum of the leaf in different specimens, and the fruits are known in very few cases only. Those sent by Dallachy agree with Brown's, as above described. With one of Thozet's specimens are a couple of large globular fruits, like those of *Mimusops parvifolia,* but twice as large. Either, therefore, some of the fruits are not correctly matched, or there are more than one species confounded under this one.

2. S.? galactoxylon, *F. Muell.* A tall tree, copiously exuding a milky juice, glabrous except the rusty-pubescent young shoots. Leaves shortly petiolate, narrowly obovate-oblong or oblong-cuneate, very obtuse, thinly coriaceous, penniveined, but the veins scarcely conspicuous on the upper side, 3 to 5 in. long. Flowers unknown. Fruits on pedicels of about ½ in., ovoid, 1 to 1½ in. long, contracted at both ends, the pericarp apparently thin and succulent. Seeds 1 or 2, with a smooth shining testa, and large broad lateral hilum; no albumen; cotyledons very thick and hard; radicle scarcely prominent.—*Bassia galactoxyla,* F. Muell. Fragm. vi. 27.

Queensland. Rockingham Bay, *Dallachy.* The genus of this plant cannot be ascertained until the flowers shall have been seen. It is, however, so totally unlike any species of *Bassia* known to me, and so very near to some species of *Achras,* with the fruit almost of *Sersalisia sericea,* that it appears to me to be a much more probable congener of the latter than of *Bassia.*

3. ACHRAS, Linn.

(Sapota, *A. DC.*)

Calyx-lobes, corolla-lobes, stamens, and ovary-cells 5, or rarely 6 or more, or the ovary-cells in species not Australian twice as many. Scales (or staminodia?) in the throat of the corolla alternating with the lobes. Ovules laterally attached. Seeds solitary or few, rarely all perfect; testa hard, shining; hilum lateral, linear or broad, above half as long as the seed; albumen copious, fleshy; cotyledons broad, flat, usually thin.—Trees or shrubs, glabrous or tomentose. Flowers sessile or pedicellate, clustered.

The genus is distributed over the tropical regions of the New as well as the Old World. Of the nine Australian species, one may perhaps extend to Norfolk Island, the others are all endemic, and all belong to the section *Oligotheca,* A. DC., with the ovary-cells of the same number as the calyx- and corolla-lobes. The West Indian *A. Sapota,* cultivated in tropical Asia, alone constitutes the typical section, with the ovary-cells twice as many as calyx- and corolla-lobes. In the following key I have included *Sersalisia,* which can only be distinguished by the seed.

Flowers sessile or on pedicels shorter than the flowers.
  Leaves broad, on short petioles, silky or tomentose underneath.
    (No albumen.) . . . . *Sersalisia sericea.*
  Leaves broad, on long petioles, softly pubescent on both sides.
    (Seeds unknown.) . . . . 1. *A. Arnhemica.*
  Leaves narrow, glabrous . . . . 2. *A. Pohlmaniana.*
Pedicels longer than the flowers.
  Leaves perfectly glabrous on both sides.
    Leaves distinctly acuminate. Hilum linear.
      Petioles rather long; minor veinlets scarcely conspicuous. 3. *A. xerocarpa.*
      Petioles short. Leaves thin, the minor veinlets conspicuous. 4. *A. chartacea.*
    Leaves scarcely acuminate, reticulate; petioles rather long.
      Hilum broad . . . . 5. *A. laurifolia.*
    Leaves scarcely acuminate, coriaceous, much reticulate; petioles rather short. Hilum linear . . . . 6. *A. australis.*
    Leaves narrow, very obtuse. (No albumen.) . . . . . *Sersalisia galactoxylon.*
  Leaves more or less tomentose-silky or pubescent underneath or on both sides, usually very obtuse.
    Leaves 3 to 5 in. long, obovate, minutely silky underneath.
      Calyx-segments orbicular . . . . 7. *A. obovata.*
    Leaves mostly 1 to 2 in. long, ovate or obovate, pubescent.
      Calyx-segments ovate . . . . 8. *A. myrsinoides.*
    Leaves small, obovate or orbicular. Flowers mostly solitary.
      Ovary surrounded by an annular disk . . . . . . *Hormogyne cotinifolia.*

There are in A. Cunningham's collection, from York Sound, N.W. coast, specimens of apparently another species of *Achras,* but with globular fruits much harder than in any species known to me. In the absence of flowers I do not feel sufficiently certain as to their genus to describe them.

1. **A. Arnhemica,** *F. Muell. Herb.* Branches softly pubescent. Leaves

ovate or oval-elliptical, obtuse, contracted at the base, softly silky-pubescent
on both sides, 2 to 4 in. long, on petioles of above 1 in. Flowers in dense
clusters in the lower axils or at the old nodes; softly pubescent, the pedicels
shorter than the flower. Calyx-segments usually 6, very broad, orbicular,
obtuse, concave, the 2 outer ones villous, the others less so and ciliolate, all
rather above 1 line diameter. Corolla-lobes 6, truncate; filaments in the
flowers examined filiform, with abortive anthers. Scales of the corolla-throat
small, linear. Ovary surrounded by a very dense ring of hairs (6-celled?);
ovules laterally attached. Style short, thick, glabrous. Fruit not seen, 4-
seeded according to F. Mueller.

N. Australia. Sea Range, *F. Mueller.*

2. **A. Pohlmaniana,** *F. Muell. Fragm.* v. 184. A tall shrub, the
branchlets thick, clothed as well as the petioles with soft appressed hairs.
Leaves crowded at the ends of the branches, narrowly obovate-oblong, obtuse, contracted into a rather long petiole, rather thin, green on both sides,
the principal veins more or less silky-hairy, otherwise glabrous, 3 to 4 in.
long. Flowers in clusters on the old wood below the leaves, almost sessile
or the pedicels shorter than the calyx. Calyx-segments 5, very much imbricate, orbicular, obtuse, 1½ lines diameter, silky-hairy outside. Corolla about
as long as the calyx, the lobes broad, very obtuse; scales of the throat small
and entire. Stamens inserted in the broad tube, the anthers imperfect in
the flowers examined. Ovary surrounded by a dense ring of long hairs, tapering into a short thick glabrous style, 5-celled; ovules laterally attached.
Fruit globular, above 1 in. diameter, the flesh hard. Seeds 5 in the fruit
opened, about ½ in. long, ovate somewhat compressed; hilum linear, lateral;
albumen not very thick; cotyledons large, flat, thin; radicle very short.—
Sapota Pohlmaniana, F. Muell. l. c.

Queensland. Brisbane river, Moreton Bay, *A. Cunningham, F. Mueller;* Queensland woods, London Exhibition, 1862, n. 98, *W. Hill;* Rockhampton, *Thozet;* Rockingham and Edgecombe Bays, *Dallachy;* Port Denison, *Fitzalan.*

3. **A. xerocarpa,** *F. Muell. Herb.* A tree of 20 to 30 ft., the young
shoots minutely tomentose. Leaves when full grown quite glabrous, on long
petioles, oval-elliptical or oblong, acuminate, penniveined but smooth and
shining on both sides, 3 to 5 in. long. Flowers clustered in the axils, on
pedicels of 4 to 6 lines. Calyx-segments ovate, almost acute, not so broad
as in *A. laurifolia,* and scarcely 2 lines long. Corolla shortly exceeding the
calyx, the lobes broad; scales of the throat oblong or linear. Ovary 5-celled;
style rather long but thick. Fruit nearly globular, very succulent. Seeds
usually 2 or more, compressed; hilum linear; albumen copious; cotyledons
very broad and flat.

Queensland. Rockingham Bay, *Dallachy.*

4. **A. chartacea,** *F. Muell. Herb.* Very near *A. xerocarpa,* but perfectly glabrous. Leaves obovate-oblong or elliptical, shortly and obtusely
acuminate, much thinner and more acuminate than in *A. xerocarpa,* narrowed
into a short petiole, with an intra-marginal vein on each side usually decurrent almost into the petiole. Flowers not seen. Fruit and seeds of *A. xero-*

carpa, but the peduncle rather shorter, and the calyx-segments apparently rounder and more obtuse.

Queensland. Tamoshanter Point, *Dallachy.*

5. **A. laurifolia,** *F. Muell. Herb.* Glabrous or the young shoots slightly pubescent. Leaves oval-oblong or broadly oblong-elliptical, obtuse or scarcely acuminate, narrowed into a short petiole, thinly coriaceous, shining above, 3 to 5 in. long. Flowers clustered in the axils on pedicels of 4 to 6 lines. Calyx-segments 5, broadly ovate, 2 lines long, usually subtended by a bract like them, but rather shorter and broader. Corolla only seen in bud, but apparently exceeding the calyx, the lobes broad and almost truncate ; scales of the throat subulate. Anthers large. Ovary very villous, tapering into a thick glabrous style ; ovules laterally attached near their base. Fruit black, obovoid, ¾ in. long, the pericarp not thick. Seeds 1 or 2, the hilum very broad, and more than half as long as the seed ; albumen copious ; cotyledons very broad and flat.—*Sersalisia laurifolia*, A. Rich. Sert. Austral. 84. t. 31 ; A. DC. Prod. viii. 177 ; *S. glabra*, A. Gray in Proc. Amer. Acad. Sc. v. 327.

Queensland. Moreton Bay, *Backhouse, Fraser ;* Rockingham Bay, *Dallachy.*
N. S. Wales. Woolongong, *American Exploring Expedition.*

In foliage, this differs but little from *A. australis* and *A. xerocarpa ;* the leaves are, however, less acuminate than in the former ; the flowers are rather larger than in *A. xerocarpa,* the sepals not so broad as in *A. australis,* and the pedicels much longer, but the chief difference is in the seed, if I have correctly matched it. The specimens I have seen are: a typical one in flower of A. Gray's species ; one in flower also, which I believe to have been from the collection which supplied A. Richard's species ; and a few in fruit from Dallachy.

6. **A. australis,** *R. Br. Prod.* 530. A tree attaining sometimes a great elevation, quite ·glabrous except a slight appressed pubescence on the very young shoots. Leaves shortly petiolate, from elliptical-oblong and shortly and obtusely acuminate to broadly obovate-oblong and very obtuse, mostly 3 to 4 in. long but sometimes larger, usually much reticulate. Flowers in axillary clusters or almost solitary on pedicels of 2 to 3 lines, more globular than in *A. xerocarpa* and *A. laurifolia.* Calyx-segments 5, broadly orbicular, about 2 lines diameter. Corolla scarcely exceeding the calyx, the lobes short and spreading ; scales of the throat slightly dilated upwards. Anthers on very short filaments near the base of the corolla-tube. Ovary densely villous, tapering into a short glabrous style, 5-celled ; ovules laterally attached near their base. Fruit 1 in. diameter. Seeds few, large, compressed, the hilum on the inner edge more than half as long as the seed, much broader than in *A. xerocarpa,* narrower than in *A. laurifolia.*—*Sapota australis,* A. DC. Prod. viii. 175.

Queensland. Rockhampton and Rockingham Bay, *Dallachy ;* Brisbane river, woods of the Paris Exhibition, 1855, *Macarthur, n.* 44.
N. S. Wales. Hunter's River, *R. Brown ;* Hastings, Clarence, and Richmond rivers, *Beckler, Wilcox, C. Moore ;* Tweed river, and woods of the London Exhibition, 1862, n. 28, *C. Moore ;* Illawarra, *Macarthur ;* Kiama, *Harvey ;* and apparently the same plant, but in leaf only, Lord Howe's Island, *Milne.*

It is possible that when better known *A. xerocarpa, A. australis,* and *A. laurifolia* may prove to be varieties only of one species ; but, as far as our specimens go, there seem to be constant, although but slight, differences in the flowers as well as in the seeds.

7. **A. obovata,** *F. Muell. Herb.* A small tree attaining about 20 ft., the branches and underside of the leaves minutely and often sparingly silky-pubescent. Leaves obovate or broadly elliptical, obtuse, narrowed into a short petiole, coriaceous, glabrous above, 3 to 5 in. long. Flowers in axillary clusters on pedicels about as long as the flowers, but attaining 3 lines when in fruit. Calyx-segments a little above 1 line diameter, orbicular, ciliolate, membranous, the outer ones rather thicker and pubescent. Corolla small, spreading sometimes to nearly 3 lines diameter, the lobes broad and obtuse; scales of the throat ovate-acuminate or linear, sometimes very small, or one or two of them quite deficient. Anthers often shortly exceeding the corolla. Ovary densely hairy, at least round the circumference, tapering into a short thick glabrous style, 5-celled; ovules laterally attached. Fruit obovoid or oblong, ½ in. long or more, usually 1-seeded. Seed obovate, compressed; hilum linear, lateral; albumen not very thick; cotyledons broad, thin, and flat; radicle very short.—*Sersalisia obovata*, R. Br. Prod. 530; A. DC. Prod. viii. 177; *Sideroxylon argenteum*, Spreng. Syst. i. 666 (partly).

Queensland. Endeavour river, *Banks and Solander;* Albany island, *W. Hill;* Howick's group, *F. Mueller.* In establishing the genus *Sersalisia,* Brown had seen the seed only of *S. sericea,* and was not aware that in the present species it was albuminous.

Chrysophyllum myrsinodendron, F. Muell. Fragm. vi. ined., from Herbert river, *Dallachy,* appears to me to be a form or state of this species; the flowers are much smaller, but apparently imperfect, the scales in the throat of the corolla are small and irregular, and sometimes 1 or 2 of them abortive, but in the flowers I examined I never found them wholly deficient. There appears to me to be some kind of dimorphism or partial unisexuality in several species of *Achras.*

8. **A. myrsinoides,** *A. Cunn. mss.* A slender twiggy shrub or small tree, the young branches and leaves more or less pubescent or villous, the full-grown leaves glabrous above or sometimes on both sides. Leaves shortly petiolate, mostly ovate or broadly elliptical, but sometimes obovate, obtuse, narrowed at the base, 1 to 2 or rarely 3 in. long, the veins usually prominent. Flowers not numerous, in axillary clusters or almost solitary, on recurved pedicels longer than the flower (usually about 3, sometimes 4 lines), tomentose-pubescent. Calyx-segments 5, of which the 2 outer tomentose and about 2 lines long. Corolla-tube as long as the calyx; lobes 5, shorter than the tube, broad, truncate; scales of the throat linear. Ovary very villous; ovules laterally attached. Fruit not seen.

N. Australia. Sea Range, *F. Mueller.*
Queensland. Rodd's Bay, *A. Cunningham;* Brisbane river, Moreton Bay, *Fraser, W. Hill;* Breakfast Creek, *Leichhardt;* Queensland woods, London Exhibition, 1862, n. 49, *W. Hill.*
N. S. Wales. Sydney woods, Paris Exhibition, 1855, n. 27 and 40, *C. Moore.*

4. HORMOGYNE, A. DC.

Calyx-segments, corolla-lobes, stamens, and ovary-cells 5. Scales (or staminodia?) 5, in the throat of the corolla alternating with the corolla-lobes. Ovary surrounded by a hirsute disk; ovules solitary in each cell, laterally attached. Fruit unknown.—Shrub. Leaves small. Flowers axillary, mostly solitary.

The genus is limited to the single species endemic in Australia. The flowers differ from those of *Achras, Sersalisia, Sideroxylon,* and allied genera chiefly in the hypogynous disk surrounding the ovary, which may not in itself be sufficient to constitute a distinct genus, but in the absence of the fruit and seed it is impossible to say to which of them it ought to be referred.

1. **H. cotinifolia,** *A. DC. Prod.* viii. 176. A straggling or diffuse shrub with slender branches, the young shoots sprinkled with a few hairs, otherwise glabrous. Leaves shortly petiolate, obovate or almost orbicular, very obtuse, under 1 in. long. Flowers solitary in the axils, on recurved pedicels of 1 to 2 lines. Calyx-segments 5, ovate, slightly pubescent or glabrous, about 1 line long. Corolla-tube exceeding the calyx; lobes short, truncate. Scales in the corolla-throat slightly dilated towards the top Ovary within the disk glabrous as well as the rather long style.—Deless. Ic. Sel. v. t. 37; *Sersalisia cotinifolia,* F. Muell. Fragm. v. 161.

Queensland. Moreton Bay, *W. Hill;* Rockhampton, *Dallachy, Thozet;* Nerkool Creek and Fitzroy river, *Bowman;* Port Denison, *Fitzalan.*

N. S. Wales. Shaded woods, Dividing Range, north of Liverpool Plains, *A. Cunningham;* Sydney woods, Paris Exhibition, 1855, n. 20, *C. Moore.*

5. MIMUSOPS, Linn.

Calyx-segments 6 to 8. Corolla-lobes 3 times or rarely twice as many as calyx-segments, in 3 rarely 2 rows, without scales in the throat. Stamens 6 to 8, inserted in the throat of the corolla opposite the inner lobes, without intervening staminodia. Filaments short; anthers lanceolate, turned outwards, but opening laterally. Ovary 6- to 8-celled; ovules laterally attached near the base or almost erect. Seeds more or less compressed; testa hard and shining; hilum either very small ovate and basal, or more or less elongated and lateral. Albumen copious; cotyledons broad and flat.—Trees or shrubs. Leaves usually with fine parallel veins. Flowers axillary, on recurved pedicels, usually larger than in *Achras.*

The genus is dispersed over the tropical regions of the New as well as the Old World, extending into extratropical South Africa. Of the two Australian species, one extends to the Indian Archipelago, the other is not quite identical with, but closely allied to a common E. Indian one.

Calyx-segments, stamens, and inner corolla-lobes 8. Leaves ovate, not
white underneath . . . . 1. *M. parvifolia.*
Calyx-segments, stamens, and inner corolla-lobes 6. Leaves broad,
whitish underneath . . . . 2. *M. Browniana.*

1. **M. parvifolia,** *R. Br. Prod.* 531. A stout bushy tree, often very low on the seacoast, attaining in other situations 30 to 40 ft., the branchlets, young foliage, and inflorescence clothed with a loose rusty pubescence, which more or less disappears from the full-grown leaves. Leaves on rather long petioles, ovate or elliptical, obtuse or acuminate, finely veined, green on both sides or pale underneath, 2 or rarely 3 in. long. Flowers solitary or 2 together in the upper axils, on pedicels of $\frac{1}{2}$ in. or more. Calyx-segments 8 (rarely 7), lanceolate, acuminate, about 3 lines long, the 4 outer ones pubescent, the inner ones thinner and more glabrous. Corolla-tube very short; lobes lanceolate, 16 outer ones in 2 or 3 rows, 8 inner ones broadly lanceolate, very acute and shortly stipitate. Perfect stamens 8. Filaments very

short ; anthers sagittate, acuminate ; staminodia 8, alternating with the stamens, linear-lanceolate, with long awn-like points and a few long cilia near the base.   Ovary densely hirsute, 8-celled ; ovules erect from the inner angle.   Fruit nearly globular.   Seeds erect, oblong, more or less compressed ; hilum small, orbicular, almost basal ; albumen copious.—A. DC. Prod. viii. 203 ; F. Muell. Fragm. v. 162.

N. Australia. Careening Bay, N.W. Coast, *A. Cunningham.*
Queensland. Cape York, *W. Hill;* islands of Cape Flinders, *A. Cunningham;* estuary of the Burdekin, *Fitzalan ;* Port Denison, Edgecombe and Rockingham Bays, *Dallachy.*

Very nearly allied to the common Indian *M. Elengi,* differing chiefly in the looser, usually rusty pubescence, in the longer peduncles, smaller flowers, and much narrower calycine segments.

2. **M. Browniana,** *Benth.*   A tree of irregular growth.   Leaves on rather long petioles, obovate, broadly ovate or almost orbicular, very obtuse, thickly coriaceous, finely and almost parallel veined, whitish or almost rusty underneath.   Flowers in the upper axils (clustered at the ends of the branches) on recurved pedicels shorter than the petioles.   Calyx-segments 6, in two rows, ovate-lanceolate, acute, about 3 lines long, reflexed under the fruit but not enlarged.   Corolla (in Malayan specimens) scarcely longer than the calyx, with 12 outer and 6 inner lobes, all narrow, acute, the inner ones almost stipitate.   Stamens 6, opposite the inner lobes ; anthers acuminate ; staminodia 6, alternating with them, petal-like, as long as the corolla, jagged at the end.   Ovary 6-celled ; ovules erect from the inner angle.   Fruit ovoid or almost globular, at least 1 in. long.   Seeds 1 or 2, large, more or less compressed ; hilum broadly linear, more than half the length of the seed. —*M. Kauki,* R. Br. Prod. 531, not of Linn.; *M. Kauki,* var. *Browniana,* A. DC. Prod. viii. 203.

Queensland. Islands off Cape Fear, *R. Brown ;* Cape Flinders, Rodd's Bay, and Endeavour river, *A. Cunningham ;* islands off Cape Bedford, Howick's Group, *F. Mueller ;* also in the Indian Archipelago (*Java, Horsfield*) and the Malayan Peninsula (*Griffith*).

The Australian specimens I have seen are in fruit only, with more or less of the persistent calyx, but they agree in every respect with Griffith's and Horsfield's specimens, from which I have described the flower.   Hermann's Cingalese specimen of Linnæus's *M. Kauki,* which Brown thought might be the same, appears to me to be the *M. indica,* A. DC., differing from our plant in several respects.   *M. Bojeri,* A. DC., described from specimens cultivated in the botanical garden of Mauritius, and referred (with *M. Browniana*) to *M. Kauki* by Miquel, cannot well, from the character given, be the same as *M. Browniana.*

ORDER LXXII. **EBENACEÆ.**

Flowers regular, usually diœcious.   Calyx free, 3- to 5-lobed or rarely with 6 or 7 lobes.   Corolla-lobes as many as those of the calyx, imbricate in the bud, usually contorted.   Stamens inserted in the base of the corolla or on the torus within it, indefinite, usually from 10 to 20 in the males, fewer and sterile in the females ; anthers erect, linear or lanceolate, the cells opening in longitudinal slits.   Ovary free, 3- or more-celled, with 1 or 2 pendulous ovules in each cell.   Styles as many or half as many as cells, distinct or united into a simple or 2-cleft style, with small terminal stigmas.   Fruit a

berry, usually indehiscent. Seeds few, albuminous; radicle superior; cotyledons foliaceous.—Trees or shrubs, the juice not milky. Leaves alternate,
entire, without stipules. Flowers axillary, the females often solitary, the
males usually clustered or in small cymes.

The Order, a small one, is dispersed over the tropical and subtropical regions of the New
as well as the Old World, extending northward to the United States and to the Caucasus,
and southward to the Cape of Good Hope. Of the three Australian genera, one, the principal one of the Order, extends over its whole area, another is limited to the Old World,
the third is either endemic or extends only to the Indian Archipelago. Many of the species
here described may require future revision, for the specimens seen, even when very numerous, are usually either males in flower or females in fruit. The female flowers, usually
solitary in the axils, and therefore much less conspicuous than the males, are rarely gathered
by collectors.

Calyx- and corolla-lobes 4 to 6 each. Ovary-cells usually twice as many,
  with 1 ovule in each . . . . 1. Diospyros.
Calyx- and corolla-lobes 4 or 5 each. Ovary-cells the same number, with
  2 ovules in each . . . . 2. Cargillia.
Calyx-lobes, corolla-lobes, and ovary-cells 3 each, with 2 ovules in each cell 3. Maba.

1. DIOSPYROS, Linn.

Calyx-lobes usually 4, sometimes 5 or 6. Corolla-lobes as many. Ovary
with usually twice as many cells as calyx- and corolla-lobes, with 1 ovule in
each cell (2 ovules to each carpel separated by spurious but complete dissepiments). Styles usually 4, more or less connate at the base and bifid at the
summit. Fruit globular or ovoid.—Trees or shrubs, with the habit and inflorescence of the Order.

Leaves pubescent, at least underneath. Fruit glabrous . . . . . . 1. D. cordifolia.
Leaves glabrous. Fruit more or less hairy . . . . 2. D. hebecarpa.

1. **D. cordifolia,** *Roxb. Pl. Corom.* i. 38. *t.* 50. A tree (the trunk
and older branches bearing prickles), the young branches and foliage softly
pubescent. Leaves petiolate, ovate oval-oblong or ovate-lanceolate, obtuse
or acuminate, cordate round or narrowed at the base, at first thin and soft, at
length more coriaceous, with the veins impressed on the upper side, 1 to
2 in. long. Male flowers usually 3 together on short recurved pedicels, the
females solitary. Calyx-lobes 4, ovate, obtuse, ciliate, about 1 line long.
Corolla urceolate, nearly 3 lines long, glabrous; lobes 4, shorter than the
tube. Stamens in the males about 16, inserted in pairs at the bottom of the
corolla, the filaments very short, the outer anthers longer than the inner ones.
Fruit globular, glabrous, about ½ in. diameter or rather more, resting on the
calyx, which is flat, with enlarged recurved lobes.—A. DC. Prod. viii. 230;
Wight, Ill. t. 148 ; *D. rugosula,* R. Br. Prod. 526 ; *D. rugulosa,* A. DC. Prod.
viii. 229 ; *D. punctata,* Dcne. Herb. Tim. 79 ; A. DC. Prod. viii. 230.

N. Australia. Victoria river, *F. Mueller.* Also in Timon and E. India. Our specimens agree perfectly well with Roxburgh's as well as with those we have from Timor, but
not quite so well with Roxburgh's figure. The leaves are rarely truly cordate, notwithstanding the name.

2. **D. hebecarpa,** *A. Cunn.* A tree of 25 ft., the adult foliage and
branches quite glabrous. Leaves from broadly ovate to oval-oblong, very
obtuse or shortly and obtusely acuminate, reticulate, not very coriaceous,

shortly contracted into a short petiole, 2 to 3 in. long. Flowers not seen. Fruit on a very short pedicel, the calyx forming a thick flat disk of ½ in. diameter, with sharp edges, and 4 short broad reflexed lobes. Berry ¾ to 1 in. diameter, covered with short hairs which sometimes wear off at the base. Seeds 8, in pairs, but with perfect dissepiments between them, compressed; albumen cartilaginous; embryo about ⅔ the length of the seed; cotyledons flat, nearly as long as the radicle.

Queensland. Cape York, *W. Hill;* N.E. coast, *A. Cunningham.* The foliage is nearly that of *D. Ebenum.*

2. CARGILLIA, R. Br.

Calyx-lobes usually 4 or 5. Corolla-lobes as many. Ovary 4- or 5-celled with 2 ovules in each cell, not separated by any spurious dissepiment; styles more or less connate. Fruit globular.—Trees or shrubs, with the habit and inflorescence of the Order.

As far as hitherto known there are no species besides the Australian ones, of which three are endemic, the remaining one (*C. laxa*), which is also in the Indian Archipelago, may prove to be a *Diospyros.*

Flowers mostly 4-merous.
Leaves glabrous, 4 to 6 in. long or more, obtuse. Fruit nearly 1 in.
    diameter . . . . 1. *C. laxa.*
Leaves slightly hairy underneath, 3 to 4 in. long . . . . 2. *C. mabacea.*
Leaves glabrous, under 3 in. long . . . . 3. *C. australis.*
Flowers mostly 5-merous. Leaves glabrous, under 2 in. long . . . 4. *C. pentamera.*

1. **C. laxa,** *R. Br. Prod.* 526. A handsome tree attaining 50 ft. (*W. Hill*), glabrous except the flowers. Leaves petiolate, from oval-oblong to oblong-elliptical, usually broad, obtuse, coriaceous, shining above, opaque underneath, 4 to 8 in. long or even more, almost always drying black. Flowers sessile, the males clustered. Calyx pubescent, 4-lobed, about 1½ lines long. Corolla about twice as long as the calyx, silky-pubescent, 4-lobed. Stamens 16 to 18. Female flowers not seen. Fruiting calyx broadly cup-shaped or opening flat, closely appressed to the fruit. Berry globular, attaining nearly 1 in. diameter. Seeds compressed, with a brown shining testa. Radicle longer than the ovate cotyledons. Albumen cartilaginous.—A. DC. Prod. viii. 243; *Diospyros maritima,* Blume, Bijdr. 669; A. DC. Prod. viii. 234; *Cargillia maritima,* Hassk.; Miq. Fl. Ind. Bat. ii. 1049; *C. megalocarpa,* F. Muell. Fragm. v. 163; *Maba megalocarpa,* F. Muell. Fragm. v. 163.

N. Australia. Gulf of Carpentaria, opposite Groote Island, *R. Brown;* Escape Cliffs, *Hulls.*
Queensland. Cape York, *W. Hill.*

The species is also on the coast of Timor, South Java, and Samoa Islands. Until the female flowers shall have been examined, it is not certain whether this may not be a species of *Diospyros,* but there does not appear to have been any septum between the 2 ovules of each carpel.

2. **C. mabacea,** *F. Muell. Fragm.* v. 162. A tree of about 20 ft., the branchlets and veins on the underside of the leaves strigose-pubescent.

Leaves elliptical-oblong, shortly acuminate, narrowed into a short petiole, rather thin, drying black, the veins prominent underneath, 3 to 4 in. long. Male flowers 4-merous, in small dense sessile cymes or clusters. Calyx 1½ lines long, the lobes shorter than the tube, nearly glabrous. Corolla silky-pubescent, nearly twice as long as the calyx. Stamens about 16. Female flowers and fruit not seen, but the fruit, according to C. Moore, a scarlet berry.—*Maba quadridentata*, F. Muell. Fragm. v. 162.

N. S. Wales. Tweed river, *C. Moore.* The affinities of this plant must remain doubtful until the female flowers and fruit shall have been examined.

3. **C. australis,** *R. Br. Prod.* 526. A small tree, glabrous or the young parts very minutely mealy-pubescent. Leaves from oblong to oval-elliptical, obtuse, narrowed into a short petiole, coriaceous, penniveined and reticulate, but the veins less conspicuous on the underside, 1½ to nearly 3 in. long. Male flowers several together, in little axillary clusters or dense cymes. Calyx about 1 line long, 4-lobed. Corolla about 2½ lines long, the tube as long as the calyx; lobes 4, contorted in the bud. Stamens about 12, slightly cohering to the base of the corolla; anthers acuminate. Female flowers solitary or 2 or 3 together, larger than the males; stamens usually fewer and imperfect. Ovary 4-celled, with 2 ovules in each cell without any trace of dissepiment between them. Fruiting calyx enlarged to nearly ½ in. diameter. Berry globular.—A. DC. Prod. viii. 243; Bot. Mag. t. 3274; *Maba Cargillia*, F. Muell. Fragm. v. 162.

Queensland. Brisbane river, Moreton Bay, *F. Mueller* and others; Rockhampton, *Dallachy;* Crocodile Creek, *Bowman.*
N. S. Wales. Port Jackson to the Blue Mountains, *R. Brown* and others; Berrima and Richmond river, *C. Moore;* Hastings and Macleay rivers, *Beckler;* Illawarra, *A. Cunningham* and others.

4. **C. pentamera,** *F. Muell. Fragm.* iv. 82. A tree attaining a considerable height, glabrous or the young shoots slightly silky. Leaves oblong-lanceolate or elliptical, slightly acuminate but not acute, contracted into a very short petiole, coriaceous, shining and reticulate above, opaque and less veined underneath, 1½ to 2½ in. long. Male flowers in very shortly pedunculate clusters of 3 to 5. Calyx 1 line long, 5-lobed. Corolla twice as long as the calyx, 5-lobed, the tube short. Stamens 15 to 20. Female flowers not seen. Fruits solitary, sessile, the subtending calyx enlarged, appressed or the lobes slightly spreading. Berry globular, about ½ in. diameter.—*Maba pentamera*, F. Muell. Fragm. v. 163.

Queensland. Brisbane river, *Fraser;* Queensland woods, London Exhibition, 1862, n. 51, *W. Hill.*
N. S. Wales. Clarence and Richmond rivers, *C. Moore;* Sydney woods, Paris Exhibition, 1855, n. 49, *Macarthur*, n. 30, *C. Moore;* London Exhibition, 1862, n. 44, *C. Moore.*

3. MABA, Forst.

Calyx-lobes usually 3. Corolla-lobes as many. Ovary 3-celled, with 2 ovules in each cell, without any or rarely with an imperfect spurious dissepiment between them. Styles connate. Fruit ovoid or rarely globular.—

Trees or shrubs, with the habit and inflorescence of the Order, the leaves and fruits often but not always smaller than in *Diospyros*.

The genus extends over the tropical regions of the Old World, and more especially over the Indian Archipelago and the islands of the South Pacific. The Australian species appear to be endemic, but may require some further comparison with the New Caledonian ones when these shall be better known. F. Mueller, in his 'Report of the Intercolonial Exhibition of 1867,' proposes to unite *Cargillia* and *Maba* with *Diospyros*, but, as far as known, the want of the septum between the two ovules of each carpel appears to be constant, and the union would entail the reducing the whole Order to a single genus, of which the present genera would be sections,—a nominal rather than any real change, involving great inconvenience without any corresponding advantage.

Leaves (mostly 3 to 4 in. long) shortly and obtusely acuminate.
  Calyx, corolla, and fruit silky-hairy.
    Leaves quite glabrous . . . . 1. *M. laurina.*
    Leaves with appressed silky hairs, at least underneath . . . . 2. *M. sericocarpa.*
Leaves (mostly 3 to 4 in. long) shortly acuminate, glabrous. Fruit glabrous.
  Calyx not above 1 line long. Corolla small.
    Female flowers solitary, almost sessile . . . . 3. *M. hemicycloides.*
    Male flowers in loose dichotomous cymes . . . . 4. *M. laxiflora.*
  Calyx at the time of flowering 2 lines long, with broad obtuse lobes. Flowers in dense sessile clusters . . . . 5. *M. fasciculosa.*
Leaves (rarely 3 in.) obovate, ovate or oblong, very obtuse.
  Leaves (mostly 2 to 3 in.) strongly reticulate. Berry globular.
    Fruiting calyx very flat and open . . . . 6. *M. compacta.*
    Fruiting calyx cup-shaped . . . . 7. *M. reticulata.*
  Leaves not much reticulate. Berry ovoid. Fruiting calyx cup-shaped.
    Leaves mostly about 2 in., obovate, oval or oblong . . . . 8. *M. geminata.*
    Leaves mostly about 1 in. or less, orbicular, obovate or oblong-cuneate . . . . 9. *M. humilis.*

1. **M. laurina,** *R. Br. Prod.* 527. A small tree, the young shoots sprinkled with a few appressed hairs. Leaves shortly petiolate, ovate-oblong, very obtuse or shortly and obtusely acuminate, rigid, coriaceous, shining above, reticulate, 3 to 4 in. long. Flowers nearly sessile, the males few together, the females solitary, larger than in any of the following species. Calyx nearly globular, about 3 lines long, silky-villous with rusty hairs; lobes short, rounded. Corolla yellowish-white (*R. Brown*), the tube shortly exserted, the lobes half as long as the tube, all silky-villous outside. Stamens in the males 9, the filaments alternately free and united in pairs, none in the females. Ovary rudimentary but villous in the males, very villous, 3-celled or rarely 2-celled in the females, with 2 ovules in each cell, without any spurious dissepiment between them. Fruit not seen.

Queensland. Cumberland islands, *R. Brown* (*Herb. R. Brown*).

2. **M. sericocarpa,** *F. Muell. Fragm.* v. 164. A tree with slender branchlets, silky-pubescent with rust-coloured hairs. Leaves on very short petioles, oval-elliptical or oblong, shortly and obtusely acuminate, coriaceous, covered or sprinkled with appressed hairs, which are more abundant on the under side, becoming nearly glabrous above when old, the reticulate veinlets scarcely conspicuous, 3 to 4 in. long. Male flowers only seen in very young bud, they are then 3 to 5 together on an exceedingly short peduncle, densely

silky-villous, the calyx tubular-conical, about 3 lines long, the corolla as yet enclosed in it, and the stamens not numerous. Female flowers not seen. Fruits solitary, nearly sessile. Calyx cup-shaped, appressed, with broad triangular lobes nearly as long as the berry, which is globular, silky-hairy, 4 to 5 lines diameter, but as yet unripe.

Queensland. Rockingham Bay, *Dallachy.*

M. cupulosa, F. Muell. Fragm. v. 164, from the same locality, appears to me to be either the same plant in a more advanced state or a slight variety, less hairy, with the fruit more ovoid, and the tips of the calyx-lobes shortly spreading. I do not find the calyx concrete with the fruit, but separated from it by the hairs which line the calyx-tube and cover the berry. Both *M. sericocarpa* and *M. cupulosa* may, when known in all their stages, prove to be a variety of *M. laurina.*

3. **M. hemicycloides,** *F. Muell. Herb.* A small tree, with slender branches, quite glabrous except the fruit. Leaves shortly petiolate, ovate or oval-oblong, shortly and obtusely acuminate, thinly coriaceous, finely veined, 3 to 4 in. long. Flowers not seen. Fruit nearly sessile and solitary. Calyx very small, glabrous, 3-lobed, spreading. Berry ovoid or globular, minutely silky-hairy, about $\frac{1}{2}$ in. long in the specimens, but not quite ripe.

Queensland. Rockingham Bay, *Dallachy.* The foliage and small calyx are nearly those of *M. laxiflora,* of which it may possibly prove to be the female.

4. **M. laxiflora,** *Benth.* A tall shrub, quite glabrous. Leaves shortly petiolate, oval-oblong or elliptical, acuminate, coriaceous, shining but reticulate above, obscurely veined underneath, mostly 3 to 4 in. long. Flowers small, rather numerous, in loose sessile dichotomous cymes longer than the petioles. Calyx short and broad, slightly 3-lobed. Corolla scarcely 2 lines long, with 3 broad very obtuse lobes. Stamens about 20. Ovary rudimentary, quite glabrous as well as the whole flower. Female plant not seen.

Queensland. Rockhampton, *O'Shanesy.* The foliage is nearly that of *M. fasciculosa,* but the flowers are much smaller, differently shaped, and the inflorescence much looser.

5. **M. fasciculosa,** *F. Muell. Fragm.* v. 163. A tall tree, quite glabrous. Leaves petiolate, oval-oblong or elliptical, obtuse or obtusely acuminate, shining above and scarcely opaque underneath, not much reticulate, 3 to 5 in. long. Female flowers often numerous, in axillary clusters or short cymes. Calyx about 2 lines long, the lobes deeper and more spreading at the time of flowering than in other species. Corolla-tube short, the lobes broad. Staminodia none in the female flowers. Ovary ovoid, 3-celled, tapering into a short thick style. Ovules 2 in each cell, but separated by an incomplete spurious dissepiment. Fruit globular, about $\frac{1}{2}$ in. diameter, the fruiting calyx enlarged, cup-shaped, with short broad recurved lobes.

Queensland. Brisbane river, *F. Mueller;* Queensland woods, London Exhibition, 1862, n. 100, *W. Hill;* Rockhampton, *Dallachy.*

6. **M. compacta,** *R. Br. Prod.* 528. An erect shrub of 4 or 5 ft., quite glabrous when in fruit. Leaves on rather longer petioles than in the other species, oval or oval-oblong, very obtuse, coriaceous, much reticulate, 2 to 3 in. long. Flowers unknown. Fruits closely sessile and solitary, the calyx expanded quite flat under it, with short broad recurved lobes. Berry

globular, about ½ in. diameter. "Seeds with a simple testa; enbryo fully as long as the albumen ; radicle terete; cotyledons shorter than the radicle, ovate, flat, the tips inflected" (*R. Brown, ms.*).—A. DC. Prod. viii. 242.

N. Australia. Islands of the N. coast of Arnhem's Land, *R. Brown.* Some imperfect specimens from Low Island, *Henne,* and from Rockingham Bay in Queensland, *Dallachy,* may belong to the same species.

7. **M. reticulata,** *R. Br. Prod.* 528. A tree attaining 20 to 30 ft., glabrous except the flowers. Leaves shortly petiolate, obovate to oblong, obtuse, much reticulate on the upper side, 2 to 3 in. long. Male flowers several, in short racemes or clusters. Calyx glabrous or nearly so, 1 to 1¼ lines long, shortly and obtusely 3-lobed. Corolla silky-pubescent, the tube scarcely so long as the calyx, the lobes rather longer. Stamens 9 to 12. Female flowers solitary, without stamens or staminodia. Ovary 3-celled, with 2 ovules in each cell, without any spurious dissepiment between them. Berry globular, in a somewhat enlarged cup-shaped calyx.—A. DC. Prod. viii. 241 ; *M. interstans,* F. Muell. Fragm. v. 163.

Queensland. Prince of Wales and Cumberland Islands, *R. Brown ;* Cape York, *M'Gillivray, Daemel ;* Rockingham Bay, *Dallachy.* F. Mueller observed some of the flowers to be 4-merous. I have not met with any such in the specimens examined.

8. **M. geminata,** *R. Br. Prod.* 527. A small tree, with an irregular dense or spreading head, very nearly allied to *M. humilis,* but with larger leaves, usually more ovate than obovate, more coriaceous and shining, but always obtuse, 1½ to 2 in. long, the reticulate veinlets scarcely conspicuous. Male flowers densely clustered, rather small. Fruits solitary or 2 together. Berries ovoid, in a cup-shaped appressed calyx, rather larger than in *M. humilis.*—A. DC. Prod. viii. 242.

Queensland. Keppel Bay, *R. Brown ;* Brisbane river, Moreton Bay, *A. Cunningham, Fraser, F. Mueller ;* Queensland woods, London Exhibition, 1862, n. 50, *W. Hill ;* Rockhampton and Edgecombe Bay, *Dallachy ;* Port Denison, *Fitzalan ;* Dawson river, *Bowman.*

M. littorea, R. Br. Prod. 527, from among *Rhizophoras* on the N. coast, seems to me to be the same species.

9. **M. humilis,** *R. Br. Prod.* 527. A bushy shrub or small tree, glabrous except the flowers. Leaves obovate or obovate-oblong, very obtuse, narrowed into a short petiole, not very coriaceous, more or less reticulate, mostly ½ to 1 in. but sometimes fully 1½ in. long. Male flowers in small clusters, females usually solitary. Calyx about 1½ lines long, nearly glabrous, 3-lobed to about the middle. Corolla very silky, about 2 lines long. Stamens in the males few, in the females none in the flowers examined. Ovary 3-celled, with 2 ovules in each cell, without any spurious dissepiment between them. Berries ovoid, under ½ in. long, in a closely appressed cup-shaped calyx.—A. DC. Prod. viii. 242 ; *M. obovata,* R. Br. Prod. 527 ; A. DC. Prod. viii. 241.

N. Australia. Islands of the Gulf of Carpentaria, *R. Brown ;* Victoria river, *F. Mueller ;* Sweers Island, *Henne .*

Queensland. Broad Sound, *R. Brown ;* Dawson, Burnett, and Gilbert rivers, *F. Mueller ;* Burdekin river *Fitzalan ;* Rockhampton, *Dallachy, O'Shanesy ;* Nerkool Creek, *Bowman.*

Brown's specimens of *M. humilis* are females in fruit from Broad Sound, those of *M. obovata* are males in flower from the Carpentaria islands, but appear to me to belong to one and the same species.

ORDER LXXIII. STYRACACEÆ.

Flowers regular, hermaphrodite. Calyx-tube usually more or less adnate to the ovary, the limb 5 or rarely 4-lobed. Corolla regular, deeply divided into as many lobes as the calyx or rarely (in species not Australian) twice as many, imbricate or valvate in the bud. Stamens usually indefinite, sometimes only twice as many or equal in number to the corolla-lobes, attached in one or more series to the base or within the tube of the corolla, those of the outer series usually alternating with the corolla-lobes. Ovary more or less inferior or rarely quite superior, 2- to 5-celled, with 2 or more ovules in each cell, either all pendulous or the upper ones erect. Style undivided; stigma capitate, entire or lobed. Fruit more or less succulent and indehiscent or rarely opening in valves. Seed usually solitary, the embryo in the axis of a fleshy albumen.—Trees or shrubs. Leaves alternate, entire or toothed, without stipules. Flowers axillary, solitary or in simple or branched racemes.

A small Order, dispersed over the tropical and subtropical regions of Asia and America, with very few African species and only one extending into Europe. The Australian genus is the principal one in Asia and America.

1. SYMPLOCOS, Linn.

Calyx 5-lobed. Corolla-lobes imbricate in the bud, and not contorted, the petals sometimes almost free. Stamens more than twice as many as corolla-lobes. Fruit a berry, crowned by the calyx-lobes. Cotyledons much shorter than the radicle.—Trees or shrubs, the foliage often turning yellowish in drying. Flowers in axillary, simple or branched, spikes or racemes.

The genus ranges over tropical and subtropical Asia and America, but appears to be deficient in Africa. Of the two Australian species, one, extending to the islands of the South Pacific, is a slight variety of a common Asiatic one, the other appears to be endemic, both belonging to a series in which the limitation of species is not very definite.

Flowers sessile or nearly so. Petals about 1½ lines diameter . . . 1. *S. spicata*.
Flowers distinctly pedicellate. Petals about 3 lines diameter . . . 2. *S. Thwaitesii*.

1. **S. spicata,** *Roxb.*; *A. DC. Prod.* viii. 254, var. *australis*. A moderate sized tree, quite glabrous. Leaves usually oval-elliptical or oblong-elliptical, but varying from obovate to lanceolate, obtuse or shortly acuminate, entire or irregularly toothed, contracted into a petiole, mostly about 4 in. long but sometimes much larger, smooth and often shining but scarcely so much so as in *J. Thwaitesii*. Flowers small, sessile or nearly so and often numerous, in axillary spikes sometimes simple but more frequently branched into a panicle of 1 to 2 in. Bracts and bracteoles small and very deciduous. Calyx-lobes exceedingly short, broad. Petals about 1½ lines long, cohering in a ring with the stamens, which are sometimes obscurely 5-adelphous. Fruit in the Australian form ovoid, contracted at the top.—Seem. Fl. Vit. 153 ; *S. Stawellii*, F. Muell. Fragm. v. 60.

Queensland. Rockingham Bay, *Dallachy.*
N. S. Wales. Richmond river, *C. Moore.*

The species is widely spread over E. India and the Indian Archipelago, where the berry is usually quite globular. The southern form, with the longer berry contracted at the top, is precisely the same as that found in the Fiji islands and the New Hebrides. There is, as far as I can perceive, no other difference between the two forms.

2. **S. Thwaitesii,** *F. Muell. Fragm.* iii. 22. v. 211. A shrub or tree, attaining sometimes a considerable size, quite glabrous, closely resembling *S. spicata* in habit and characters, but the leaves are usually firmer and more shining, and the flowers considerably larger and more distinctly pedicellate, forming simple or branched racemes, the pedicels however rarely exceed 1 line. Calyx-lobes broad and obtuse, the innermost much larger than the outer ones. Petals nearly 3 lines long. Fruit when young oblong, contracted at the top, but not seen ripe.

Queensland. Rockingham Bay, *Dallachy.*
N. S. Wales, *Backhouse.* Woods of the Paris Exhibition, 1855, n. 51, *Macarthur;* Hastings river, *Beckler;* Richmond, Bellinger, and Macleay rivers, *C. Moore.*

The species is very closely allied to *S. grandiflora,* Wall., from Silhet, but the leaves are more rigid, less acuminate, and the pedicels shorter.

ORDER LXXIV. JASMINEÆ.

Flowers regular. Calyx-free, usually small, the limb of 4 or 5 or rarely more teeth or lobes or rarely truncate and entire. Corolla with a long or short tube, and 4 or 5 or rarely more lobes, or divided to the base into 4 petals or rarely, in genera not Australian, 2-petaled or wanting. Stamens 2, adhering to the base of the corolla, on opposite sides of the ovary, the filaments usually short; anthers 2-celled, the cells opening in longitudinal slits. Ovary 2-celled, with 2 or rarely 1 or 3 ovules in each cell, laterally attached in the young state but becoming pendulous or ascending according to the growth of the ovary. Fruit succulent or capsular, entire or 2-lobed, 2-celled or reduced to a single cell and seed. Seeds with or without albumen. Embryo straight.—Trees or shrubs, sometimes climbing, very rarely reduced to herbs. Leaves opposite or very rarely alternate, entire or pinnate. Flowers in axillary or terminal panicles, sometimes reduced to short simple racemes or sessile clusters.

A small Order dispersed over the greater part of the warmer or temperate regions of the globe. Of the five Australian genera, four have a very wide range, two in the Old World only, two in both the New and the Old, the fifth is endemic. The Order is often limited to the genera *Jasminum* and *Menodora* with more than 4 corolla-lobes and the seed ascending; the others forming the Order *Oleineæ*, with a more constantly 4-merous corolla and pendulous seeds; but the two Orders, or suborders, are intimately connected into a well-defined group by the position of the stamens, exceptional among *Gamopetalæ.*

SUBORDER I. **Jasmineæ.**—*Corolla-lobes* 5 *or more. Ovules (often solitary) and seeds ascending or erect. No albumen.*

Fruit succulent, indehiscent . . . . 1. JASMINUM.

SUBORDER II. **Oleineæ.**—*Corolla-lobes or petals* 4. *Ovules and seeds pendulous.*

Corolla with a short tube and 4 lobes. Seed albuminous.
Fruit a drupe. Panicles axillary or rarely terminal . . . . . 2. OLEA.

Fruit a berry.   Panicles all terminal . . . .   3. LIGUSTRUM.
Petals 4, quite distinct or connected in pairs by the stameus.   Fruit a
   drupe.
Seed albuminous.   Racemes axillary, simple . . . .   4. NOTELÆA.
Seed without albumen.   Panicles axillary, rarely reduced to a single
   sessile cluster . . . .   5. CHIONANTHUS.

1. JASMINUM, Linn.

Corolla-tube cylindrical, the limb spreading, 5- to 8-lobed, the lobes imbricate, often contorted in the bud.   Stamens included in the tube.   Ovary
(at the time of flowering) entire or notched, 2-celled, with 1 ovule (or in
species not Australian sometimes 2 or even 3 ovules) in each cell, laterally
attached, but becoming erect as the ovary enlarges ; style inserted in the
notch, minutely 2-lobed at the tip.   Berry 2-lobed almost to the base, or entire by the failure of 1 carpel.   Seed usually solitary in each lobe, erect,
without albumen ; cotyledons thick and fleshy, radicle scarcely prominent.—
Shrubs or climbers.   Leaves opposite or rarely alternate, either pinnate with
3 (or more in species not Australian) leaflets or apparently simple, being reduced to 1 leaflet, the petiole being then articulate.   Flowers white or
yellow, in axillary or terminal trichotomous panicles or rarely almost solitary.
Bracts very small in all the Australian species.

A considerable genus, dispersed over the warmer regions of the Old World, with one or
two S. American species.   Of the seven Australian species, two extend over the islands of
the S. Pacific and perhaps of the Eastern Archipelago, another is closely allied to, if not
identical with, a common S. Asiatic one ; the others are endemic, but not presenting any very
marked distinctive characters.

Leaves all or almost all 3-foliolate.   Calyx truncate or very shortly
   and obtusely toothed.
   Leaflets mostly ovate, 2 to 3 in.   Panicle usually broad   . . .   1. J. didymum.
   Leaflets ovate or oblong, ½ to 1½ iu.   Panicles narrowed or re-
      duced to simple racemes . . . .   2. J. racemosum.
   Leaflets mostly linear or lanceolate.   Panicles short   . . . .   3. J. lineare.
Leaves simple (unifoliolate), the petiole articulate below the middle.
   Leaves mostly ovate, penniveined.
      Calyx-teeth much shorter than the tube or obsolete . . . . .   4. J. simplicifolium.
      Calyx-teeth subulate, much longer than the tube . . . . . .   5. J. æmulum.
   Leaves mostly oblong or lanceolate, 3- or 5-nerved.   Calyx-teeth
      as long as or much longer than the tube . . . .   6. J. calcareum.
   Leaves mostly linear, penniveined.   Calyx-teeth subulate, longer
      than the tube . . . .   7. J. suavissimum.

1. **J. didymum,** *Forst. Prod.* 3.   A tall woody climber, usually glabrous or the inflorescence minutely pubescent, but sometimes the foliage and
young branches pubescent or villous.   Leaves 3-foliolate, with rather long
petioles and petiolules or very rarely a few of the lower leaves 1-foliolate ;
leaflets usually orbicular or broadly ovate and obtuse, but sometimes ovate
and acuminate or ovate-lanceolate and acute, mostly 2 to 3 in. long, penniveined and more or less distinctly 3- or 5-nerved at the base.   Flowers small
for the genus and often numerous, in loose trichotomous cymes or panicles,
axillary or terminating short axillary branchlets, and often exceeding the
leaves.   Pedicels short.   Calyx under 1 line long, truncate or very shortly

toothed. Corolla-tube 3 to 4 lines long ; lobes 5 or 6 or rarely 4, less than half as long as the tube. Berry usually globular and 1-seeded, about 5 lines diameter, rarely didymous and 2-seeded. Seed globular, the inner integument formed of an undulated network in the Australian as well as in the Taitian specimens.—DC. Prod. viii. 311 ; *J. divaricatum,* Br. Prod. 521 ; DC. Prod. viii. 311 ; Labill. Sert. Austr. Caled. t. 27 ; *J. parviflorum,* Dcne. Herb. Tim. 77 ; DC. Prod. viii. 310.

N. Australia. Victoria river, *F. Mueller;* islands of the Gulf of Carpentaria, *R. Brown, Henne ;* Port Essington, *Armstrong ;* Quail Island, *Flood ;* Escape Cliffs, *Hulls.* **Queensland.** Keppel Bay, *R. Brown ;* Rodd's Bay, *A. Cunningham ;* Rockhampton, *O'Shanesy ;* Moreton Bay, *F. Mueller.*
The species is also in the S. Pacific and Society Islands.

Var. *pubescens.* Branches and foliage pubescent or villous with spreading hairs.—*J. Dallachii,* F. Muell. Fragm. iv. 150.—Seaview Range, Rockingham Bay, *Dallachy.* Among the Taitian specimens are some almost if not quite as villous as Dallachy's ; but in both countries the more frequent state appears to be quite or very nearly glabrous.

2. **J. racemosum,** *F. Muell. Fragm.* i. 19. A slender glabrous shrub, either erect and bushy or the branches elongated and somewhat twining. Leaves opposite or alternate, 3-foliolate with short petiolules ; leaflets from broadly ovate to narrow-oblong or the lateral ones orbicular, all very obtuse, finely and often obscurely penniveined, shining above, the terminal one often above 1 in. long, the lateral ones shorter or rarely all similar. Flowers small, in axillary and terminal panicles, sometimes trichotomous, sometimes more simple and almost reduced to slender racemes. Calyx truncate or with very short teeth. Corolla-tube nearly 3 lines long ; lobes 5 to 8, at least half as long as the tube. Berry usually globular and 1-seeded, rarely didymous.

Queensland. Thirsty Sound, *R. Brown ;* Araucaria Ranges, sources of the Brisbane, *F. Mueller ;* Broad Sound, Suttor river, Nerkool Creek, etc., *Bowman ;* Rockhampton, *Dallachy* and others ; Port Denison, *Fitzalan.* Very near *J. didymum* (*J. divaricatum,* Br., in which Brown's specimens were included in his herbarium) on the one hand, and, on the other hand, sometimes scarcely to be distinguished from the broad-leaved specimens of *J. lineare.*

3. **J. lineare,** *R. Br. Prod.* 521. An erect shrub, either quite dwarf or bushy and attaining 6 to 8 ft., or with elongated, somewhat twining branches, minutely hoary-pubescent or rarely quite glabrous. Leaves opposite or the upper ones alternate, 3-foliolate, the common petiole short ; leaflets usually lanceolate or linear, obtuse or acute, the terminal one 1 to 4 in. long, the lateral ones shorter, or rarely all, especially the lateral ones, oblong or elliptical. Flowers in axillary trichotomous panicles, often numerous, but rarely exceeding the leaves. Calyx-teeth exceedingly short. Corolla-tube under 3 lines long ; lobes 5 or 6, scarcely shorter than the tube. Ovules solitary in each cell of the ovary. Berry simple and ovoid or rarely didymous.—DC. Prod. viii. 311 ; Hook. Ic. Pl. t. 831 ; *J. Mitchellii,* Lindl. in Mitch. Trop. Austr. 365 ; *J. Bidwillii,* Vis. Pl. Hort. Patav. 1858, 6.

N. Australia. Nichol Bay, *Gregory's Expedition ;* Gregory river, *Landsborough.* **Queensland.** Dawson and Burdekin rivers, *F. Mueller ;* Narran river and near Mount Kennedy, *Mitchell ;* head of the Suttor river, *Sutherland.* **N. S. Wales.** Liverpool plains and bushy country on the Lachlan, *A. Cunningham ;*

Namoi river, *C. Moore;* Murray river, *Mitchell, F. Mueller,* and thence to the Darling and to the Barrier Range, *Goodwin and Dallachy, Victorian and other Expeditions.*

S. Australia. Head of Spencer's Gulf, *R. Brown;* thence to the Murray, *F. Mueller* and others; in the interior, *M'Douall Stuart's Expedition.*

W. Australia. Sharks' Bay, *Milne.*

4. **J. simplicifolium,** *Forst. Prod.* 3. A woody climber or sometimes a tree, glabrous or softly pubescent. Leaves opposite, simple, mostly ovate, shortly acuminate, and 1½ to 2 in. long, but varying from broadly ovate-cordate to ovate-lanceolate or oblong-elliptical, very obtuse or acutely acuminate, and from 1 to nearly 3 in. long; the petioles rather long or sometimes short, articulate below the middle. Flowers white, in terminal trichotomous cymes usually loose and many-flowered, but sometimes compact and few-flowered, but the flowers always pedicellate. Calyx small, the teeth shorter than the tube, and often almost obsolete. Corolla-tube 4 to 5 lines long; lobes acute, rather shorter than the tube.—Bot. Mag. t. 980; *J. gracile,* Andr. Bot. Rep. t. 127; R. Br. Prod. 521; DC. Prod. viii. 309; Bot. Reg. t. 606; *J. geniculatum,* Vent. Choix, t. 8; *J. australe,* Pers. Syn. i. 8; DC. Prod. viii. 306; *J. acuminatum,* R. Br. Prod. 521; DC. Prod. viii. 307; *J. confusum,* DC. Prod. viii. 309; and probably also *J. funale,* Dcne. Herb. Tim. 77; DC. Prod. viii. 308.

N. Australia. Arnhem's Land, *R. Brown.*

Queensland. Keppel and Shoalwater Bays, *R. Brown;* Wide Bay, *Bidwill;* Port Curtis, *M'Gillivray;* Port Denison, *Fitzalan;* Curtis island, *Henne, Thozet;* Rockingham Bay, *Dallachy;* Rockhampton, *O'Shanesy;* Ipswich, *Nernst;* Peak Downs, *F. Mueller.*

N. S. Wales. Hunter's River, *R. Brown;* Richmond, Hastings, and Clarence rivers (where it is said to be arborescent), *Beckler, Wilcox,* and others.

Var. *molle.* Branches, foliage, and inflorescence softly pubescent.—*J. molle,* R. Br. Prod. 521; DC. Prod. viii. 307.—Victoria river, *F. Mueller,* Arnhem's Land and islands of the Gulf of Carpentaria, *R. Brown, Landsborough.*

The species is also in the islands of the South Pacific.

5. **J. æmulum,** *R. Br. Prod.* 521. A woody climber, glabrous or softly pubescent. Leaves opposite, simple, ovate, acute or acutely acuminate or rarely obtuse, penniveined, mostly 1½ to 2 in. but sometimes 3 in. long, not cordate, on petioles of 2 to 4 lines articulate below the middle. Flowers in compact terminal cymes, but each one on a rigid pedicel of 1 to 3 lines. Calyx-tube nearly 1 line long; lobes subulate, more than twice as long as the tube. Corolla-tube rarely under ½ in. long, and sometimes 7 or 8 lines; lobes 6 to 9, acute, shorter than the tube. Berries obovoid or almost globular, about 4 lines diameter, rarely didymous.—DC. Prod. viii. 302.

N. Australia. Islands of the Gulf of Carpentaria and Arnhem's Land, *R. Brown* (one of the specimens marked by mistake as *J. molle* in Herb. R. Br.); Adams Bay, *Hulls* (all belonging to the softly pubescent variety); Port Essington, *Armstrong* (entirely glabrous or slightly pubescent).

Queensland. Cape York, *M'Gillivray, Daemel;* Rockingham Bay, *Dallachy* (all glabrous).

F. Mueller, Fragm. vi. 86, describes this as *J. Forstenii,* Blume. If it be really the same species, Brown's name has the priority by fifteen years. The specimen we have from Blume of his *J. Forstenii* (a Celebes plant) is much nearer the common E. Indian *J. undulatum,* Willd., from which it may indeed sometimes be difficult to distinguish *J. æmulum.* The true *J. undulatum* has, however, differently shaped leaves and a more compact inflorescence, the flowers being very nearly or even quite sessile ; it is spread over the Archipelago, and was gathered by A. Cunningham in Timor, but I have seen no Australian specimen.

6. J. calcareum, *F. Muell. Fragm.* i. 212. Stems woody, short and erect or elongated and twining, quite glabrous as well as the foliage in all the specimens seen. Leaves opposite, simple, oblong-lanceolate or rarely ovate-lanceolate, obtuse acute or acuminate, thick, 3-nerved or 5-nerved when broad, narrowed into a petiole articulate near the base, 1½ to 2½ in. long. Flowers in terminal, rather dense, trichotomous cymes. Calyx-lobes linear, usually thick, in some specimens not longer than the tube, in others twice or even three times as long. Corolla-tube 5 to 7 lines long ; lobes 7 to 10, rather broad. Berry rather large, globular, rarely didymous.

W. Australia. Between Moore and Murchison rivers, *Drummond, 6th Coll. n.* 136 ; Greenough flats, Champion Bay, and Murchison river, *Oldfield.*

A specimen in Herb. F. Mueller from Central Australia (collector not mentioned) is referred to this species by F. Mueller ; but it is too imperfect to determine accurately.

7. J. suavissimum, *Lindl. in Mitch. Trop. Austr.* 355. Stems from a woody branching base erect, about 1 ft. high, but occasionally elongated and twining, glabrous as well as the foliage in all our specimens. Leaves on very short petioles, opposite or rarely alternate, simple, linear or linear-lanceolate, acute, 1 to 2 in. long. Flowers in terminal cymes of 3 to 5 or solitary in the upper axils and then on long peduncles. Calyx-teeth subulate ; longer than the tube. Corolla-tube 3 to 4 lines long ; lobes 5 to 7, very acute, and usually as long as the tube. Fruit small, sometimes didymous. —F. Muell. Fragm. i. 183 ; *J. dianthifolium*, Vis. Pl. Hort. Patav. 1858. 7.

Queensland. Burnett river, *F. Mueller ;* Maranoa river, *Mitchell ;* Brisbane river, Moreton Bay, *Bidwill, C. Stuart ;* Warwick, *Beckler.*

N. S. Wales. Clarence river, *Beckler ;* New England, *C. Stuart ;* Darling Downs, *Law.*

2. OLEA, Linn.

Calyx short, 4-toothed. Corolla with a short tube and 4 lobes, slightly imbricate or valvate in the bud, rarely (in a species not Australian) wanting. Ovules 2 in each cell of the ovary, pendulous. Style short. Fruit a drupe, the endocarp usually hard. Seed solitary or rarely 2 ; albumen copious, fleshy.—Trees or rarely shrubs. Leaves opposite, entire. Flowers small, in axillary panicles or clusters, rarely also terminal.

The genus is widely dispersed over the warmer regions of the Old World, with one North American species. The only Australian species, with the inflorescence less exclusively axillary than in most others, extends only to New Caledonia.

1. O. paniculata, *R. Br. Prod.* 523. A tree of moderate size, quite glabrous. Leaves on rather long petioles, ovate-lanceolate or elliptical, acuminate, 2 to 3 or rarely 4 in. long, penniveined underneath but not conspicuously reticulate. Flowers small, all pedicellate in loose trichotomous panicles, terminal or more frequently in the upper axils, and sometimes exceeding the leaves. Calyx scarcely 1 line diameter. Corolla-tube short and broad, the lobes rather above 1 line diameter. Anthers as long as the corolla, on very short broad filaments. Style very short ; stigma either broad, thick, and shortly 2-lobed, or clavate and unilateral.   Drupe ovoid, resembling that of the common olive.—DC. Prod. viii. 287.

Queensland. Keppel Bay, *R. Brown ;* Brisbane river, Moreton Bay, *Fraser, F. Mueller ;* Rockhampton, *Thozet ;* Rockingham Bay, *Dallachy.*
N. S. Wales. Ash Island and near Newcastle, *R. Brown ;* Clarence river, *Beckler, C. Moore ;* Northern woods, London Exhibition, 1862, n. 84.

The species is also in New Caledonia (*Deplanche, n.* 70). The panicles are often apparently terminal, but even then there are generally two peduncles, with the terminal shoot ultimately growing out between them.

3. LIGUSTRUM, Linn.

Calyx small, 4-toothed.   Corolla with a short tube and 4 lobes, valvate or slightly imbricate in the bud.   Ovules 2 in each cell of the ovary, pendulous. Style short.   Fruit a berry.   Seeds 4 or fewer ; albumen copious, fleshy or almost cartilaginous.—Shrubs or rarely trees. Leaves opposite, entire. Flowers white, rather small, in terminal trichotomous panicles.

The genus is spread over the temperate and mountain-tropical regions of Asia and Europe. The only Australian species is endemic, but very closely allied to two of the East Asiatic ones. As a genus, *Ligustrum* is only to be distinguished from *Olea* by the endocarp, scarcely or not at all hardened, and generally by the inflorescence.

1. **L. australianum,** *F. Muell. Fragm.* v. 20.   A shrub of several feet, glabrous except the minutely pubescent inflorescence.   Leaves shortly petiolate, ovate or ovate-lanceolate, obtusely acuminate, evergreen and smooth on both sides, without prominent veins, $1\frac{1}{2}$ to 3 in. long.   Flowers small and very numerous, in a broad pyramidal terminal panicle.   Pedicels short. Calyx truncate.   Corolla-tube exceedingly short ; lobes about 1 line long. Filaments about half as long as the corolla.   Style short, cylindrical, with a clavate stigma.   Fruit unknown.

Queensland. Dalrymple Gap, Rockingham Bay, *Dallachy.* Very nearly allied to *L. lucidum,* a Chinese species much cultivated in European gardens, but the flowers appear to be considerably smaller. The several evergreen species of *Ligustrum* require, however, a careful revision, and may prove to be nearly all varieties of one species having a wide range in Eastern Asia.

4. NOTELÆA, Vent.

Calyx small, 4-toothed.   Corolla small, of 4 petals, quite distinct or connected in pairs by means of the stamens, induplicate-valvate in the bud. Ovules 2 in each cell of the ovary, pendulous.   Style short.   Fruit a drupe. Seed solitary ; albumen copious, fleshy or almost cartilaginous, more or less ruminate.—Trees or shrubs. Leaves opposite, entire. Flowers small, in short simple axillary racemes, sometimes reduced to sessile clusters.

The genus is limited to Australia. The species are all very closely allied to each other, scarcely differing except in the uncertain character derived from the venation of the leaves and the size of the fruit, which is very difficult to judge of from dried specimens.

Leaves with prominent anastomosing or reticulate veins, at least on the
    upper side.
Leaves ovate or ovate-lanceolate, rounded or cordate at the base or
    very shortly tapering into a short petiole . . . . 1. *N. ovata.*
Leaves narrowed into a petiole, usually rather long.

Leaves mostly broadly lanceolate.
  Leaves reticulate on both sides. Fruit above ½ in. diameter . 2. *N. longifolia.*
  Leaves scarcely reticulate underneath, but copiously dotted.
    Fruit small . . . . 3. *N. punctata.*
Leaves narrow-lanceolate; veins very oblique and prominent
  above, scarcely conspicuous underneath.
    Fruit about ¼ in. diameter when ripe . . . . 4. *N. microcarpa.*
Leaves thick, smooth, very obscurely or not at all veined.
  Leaves mostly lanceolate, without thickened margins . . . . . 5. *N. ligustrina.*
  Leaves linear or scarcely linear-lanceolate, with thickened nerve-like
    margins . . . . 6. *N. linearis.*

1. N. ovata, *R. Br. Prod.* 524. Glabrous or pubescent. Leaves very
shortly petiolate, ovate or broadly ovate-lanceolate, obtuse or acute, cordate
broadly rounded or very shortly contracted at the base, coriaceous, much reticulate and more regularly so than in *N. longifolia*, 1½ to 2 in. long. Racemes short and few-flowered, but usually pedunculate and most frequently
inserted rather above the axils. Pedicels short. Flowers and fruit of *N.
longifolia.*

N. S. Wales. Port Jackson, *R. Brown ;* Grose river, *Miss Atkinson.*

2. N. longifolia, *Vent. Choix, t.* 25. A tall shrub or small slender
tree, the branches and under side of the leaves in the typical form or rarely
both sides of the leaves pubescent with short hairs, sometimes soft and
dense, sometimes scarcely visible without a lens or the whole plant quite
glabrous. Leaves ovate-lanceolate or lanceolate, acute or acuminate, tapering
into a petiole often rather long, coriaceous but sometimes thin, prominently
and irregularly reticulate on both sides, 2 to 6 in. long. Flowers small, in
axillary racemes rarely 1 in. long, and flowering usually from near the base.
Pedicels at first nearly as long as the flowers, much longer when in fruit.
Calyx exceedingly small, obscurely toothed, opening flat. Petals scarcely 1
line long, broad, concave. Anthers large, almost sessile. Fruit ovoid or
globular, said to be about ½ in. diameter when fresh, of a dark bluish colour
with a very succulent mesocarp and hard endocarp, often appearing much
smaller and drier in dried specimens but evidently unripe.—R. Br. Prod.
523 ; DC. Prod. viii. 291 ; *N. reticulata*, DC. l. c. ; *Olea apetala*, Andr.
Bot. Rep. t. 316, not of Vahl ; *N. ovata*, Endl. Iconogr. t. 55, not of R. Br. ;
N. venosa, F. Muell. in Trans. Vict. Inst. 1855, 131 ; and in Hook. Kew
Journ. viii. 163 ; *N. rigida*, Sieb. Pl. Exs.

Queensland. Brisbane river, Moreton Bay, *A. Cunningham, Fraser, F. Mueller ;*
Rockingham Bay, *Dallachy.*
N. S. Wales. Port Jackson to the Blue Mountains, *R. Brown, Sieber, n.* 274, and
many others (Sydney woods, Paris Exhibition, 1855, *Moore, n.* 34, *Macarthur, n.* 45, 67,
187) ; northward to Hastings, Macleay, and Clarence rivers, *A. Cunningham, Beckler,*
and others ; southward to Illawarra, *A. Cunningham* and others.
Victoria. Woods of the eastern part of Gipps' Land, *F. Mueller.*

The northern and southern specimens belong almost entirely to the glabrous form, the
pubescent one is chiefly about Port Jackson and in the Blue Mountains to New England,
some of C. Stuart's specimens from the latter station being densely and softly pubescent all
over. Sieber's specimens, n. 274, have remarkably long and narrow leaves, but much
nearer in shape to those figured by Ventenat than those of the common glabrous specimens.

3. **N. punctata,** *R. Br. Prod.* 524.   Very near the glabrous form of
N. longifolia.   Leaves oblong-lanceolate, 2 to 3 in. long, tapering into a
rather long petiole, reticulate above as in *N. longifolia,* but the under surface
scarcely showing any veins and densely covered with small raised dots.
Flowers of *N. longifolia.*   Fruit ovoid, much smaller than in that species,
and scarcely larger than in *N. microcarpa.*— DC. Prod. viii. 291.

Queensland. Shoalwater Bay, *R. Brown;* Rockingham Bay, *Dallachy.*

4. **N. microcarpa,** *R. Br. Prod.* 524.   A small tree with slender
branches, glabrous, but the young parts often whitish.   Leaves narrow-
lanceolate, acuminate, tapering into a short petiole, 2 to 4 in long, the veins
prominent on the upper side, the primary ones very oblique and some of them
nearly parallel to the margin, anastomosing with the irregular netted veinlets,
all scarcely conspicuous on the under side.   Raceme short and often dense,
sessile in the axils.   Flowers nearly of *N. ligustrina,* but the fruit scarcely
half as large, usually globular and under ¼ in. diameter.—DC. Prod. viii.
291.

Queensland. Broad Sound, *R. Brown;* Wide Bay, woods of the Paris Exhibition,
1855, n. 82, *C. Moore;* Castle Creek, *Bowman;* Rockhampton, *Thozet;* also in *Leichhardt's* and in *Mitchell's* collections.
N. S. Wales. Liverpool plains, *A. Cunningham;* St. Aubyns, *Backhouse;* New
England, *C. Stuart.*

5. **N. ligustrina,** *Vent. Choix, under n.* 25.   A tall shrub or small tree,
said to attain sometimes 30 ft., glabrous, but the young shoots often hoary
or whitish.   Leaves lanceolate, usually narrow, obtuse or rarely acute, contracted into a short petiole, rather thick, smooth, without thickened margins,
quite veinless or with a few scarcely conspicuous and never prominent veins.
Racemes short and sessile, but the pedicels often rather long.   Calyx-lobes
from ⅓ to ½ as long as the petals, which are rather smaller and less induplicate than in *N. longifolia.*   Style as in that species exceedingly short, with
an entire or 2-lobed stigma.   Fruit nearly globular, varying from white to
purple and every shade of pink or red (*Gunn*), nearly ½ in. diameter when
fully ripe.—R. Br. Prod. 524; DC. Prod. viii. 291; Hook. f. Fl. Tasm. i.
268.

Victoria. On the Upper Yarra and in Gipps' Land, *F. Mueller.*
Tasmania. Port Dalrymple and Derwent river, *R. Brown;* common by banks of
rivers and skirts of woods, *J. D. Hooker.*

6. **N. linearis,** *Benth.*   An erect bushy shrub of 3 to 5 ft., quite glabrous, the young branches not at all white.   Leaves linear or very narrowly
linear-lanceolate, acute, tapering into a very short petiole, 1 to 2 in. long,
thick and smooth, with thickened nerve-like margins, the veins quite inconspicuous on both sides.   Flowers small in short sessile racemes.   Fruit
apparently as small as in *N. microcarpa.*

N. S. Wales. Mount Mitchell, Ann river, *Beckler;* mountain gullies, New England,
C. Stuart.

5. CHIONANTHUS, Linn.

(Linociera, *Swartz*.)

Calyx small, 4-lobed. Corolla of 4 narrow, small or elongated, petals, quite distinct or slightly connected at the base, induplicate-valvate in the bud. Ovules 2 in each cell of the ovary, pendulous. Style short. Fruit a drupe. Seed usually solitary, without albumen; cotyledons thick and fleshy, sometimes slightly ruminate.—Trees or shrubs. Leaves opposite, entire. Flowers in axillary panicles rarely reduced to sessile clusters.

The genus is dispersed over the warmer regions of both the New and the Old World. Of the three Australian species, one appears to be the same as a common one in the Indian Archipelago, the two others, closely allied to it, are endemic.

Flowers more or less pedicellate in loose trichotomous panicles . . . 1. *C. ramiflora*.
Flowers sessile or nearly so in clusters arranged in thyrsoid or trichoto-
mous panicles . . . . 2. *C. picrophloia*.
Flowers sessile in a single sessile axillary cluster . . . . 3. *C. axillaris*.

1. **C. ramiflora,** *Roxb. Fl. Ind.* i. 107. A glabrous tree, attaining a considerable size, but flowering when still a shrub. Leaves broadly elliptical-oblong, shortly and obtusely acuminate, narrowed into the petiole, penni-veined and studded over with minute dots, 6 to 9 in. long or even more on barren shoots. Panicles axillary or below the leaves, trichotomous, much and divaricately branched but shorter than the leaves, the ultimate pedicels 1 to 2 lines long and often clustered. Calyx about ½ line long. Petals oblong-linear, 1½ lines long, connected in pairs by the stamens. Fruit ovoid, often ½ in. long.—Wight, Ic. t. 734; *Linociera ramiflora*, DC. Prod. viii. 297; *Chionanthus effusiflora*, or *Linociera effusiflora*, F. Muell. Fragm. iv. 83.

Queensland. Cape York, *M'Gillivray, W. Hill;* Rockingham Bay and ranges near Rockhampton, *Dallachy;* Crocodile Creek, *Bowman.* The species was first described from the Molluccas; we have specimens from Tavoy, Tenasserim, and the Philippines.

2. **C. picrophloia,** *F. Muell. Fragm.* iii. 139. *t.* 24. A tree of con-siderable height, quite glabrous. Leaves obovate-oblong elliptical-oblong or broadly lanceolate, obtuse or acuminate, narrowed into a rather long petiole, 3 to 6 in. long. Flowers sessile in almost globular clusters, forming thyrsoid panicles shorter than the leaves, and the individual flowers rather larger than in *C. ramiflora.* Fruit ovoid-oblong, attaining 1 in. in length.

Queensland. Rockhampton, *Thozet.*

3. **C. axillaris,** *R. Br. Prod.* 523. A glabrous tree, the branches whitish. Leaves shortly petiolate, oblong-elliptical, shortly acuminate, 3 to 4 in. long, distantly penniveined. Flowers small in dense sessile simple axillary spikes not above ½ in. long, the rhachis pubescent as well as the minute broad ciliolate bracts. Calyx-lobes minute. Petals glabrous, very narrow, about 1 line long. Fruit small, but not seen ripe.—DC. Prod. viii. 295.

Queensland. Endeavour river, *Banks and Solander.*

ORDER LXXV. APOCYNEÆ.

Flowers regular. Calyx free, divided nearly or quite to the base into 5 segments or sepals imbricate in the bud, bearing occasionally small glands or scales inside at the base. Corolla regular, with 5 spreading lobes, contorted-imbricate or rarely valvate in the bud, the throat sometimes closed with a corona of scales, and frequently hairy. Stamens 5, inserted in the tube, alternating with the corolla-lobes; anthers erect, turned inwards, 2-celled, the cells opening in longitudinal slits, either free and enclosed in a part of the tube usually swollen, or sometimes exserted and connate or connivent in a cone or ring round the style; the pollen not collected in masses but the auricles at the base of the anthers or the tips occasionally without pollen. Ovary of 2 carpels usually distinct, sometimes united in a 2-celled ovary with axile placentas or in a 1-celled ovary with 2 parietal placentas; ovules several, few or numerous in 2 or more rows. Style single or styles 2, distinct at the base but united upwards; stigma usually thickened, mitriform membranous or bulbous at the base, terminating in a short entire or bifid point. Fruit either a single drupe or berry, or more frequently each carpel forms a follicle opening along the inner edge or a drupe or berry. Seeds pendulous or rarely ascending or peltately attached, often bearing a coma or tuft of long hairs, usually albuminous; embryo straight, with flat or rarely convolute cotyledons.— Trees shrubs or twiners or very rarely perennial herbs, the juice frequently milky. Leaves opposite or whorled, very rarely alternate, entire, without any or with small almost gland-like intrapetiolar or interpetiolar stipules. Flowers usually cymose, on simple or compound and paniculate peduncles, axillary or terminal. Bracts at the base of the branches of the inflorescence and of the pedicels usually very small, rarely larger, coloured and deciduous; bracteoles on the pedicels none or very rare and small.

The Order is abundantly represented in the tropical and subtropical regions of the New and the Old World, with a very few species in the more temperate districts of the northern and southern hemispheres, but does not extend to arctic or high alpine regions. Of the twelve Australian genera, none are perhaps absolutely endemic, one (*Lyonsia*), the most numerous in endemic species, extends probably to New Caledonia, another (*Ochrosia*) from the Mauritius to the islands of the Pacific, the others are all Asiatic, two of them extending to Africa, and one of them to tropical America. The Order is closely allied to *Asclepiadeæ*, differing chiefly in the indefinite free pollen-granules.

Anthers wholly included in the corolla-tube, and usually free.
　Ovary single, the carpels completely united from the base.
　　Ovary 1-celled with 2 parietal placentas. Tall woody
　　　climbers. Fruit a berry. Seeds albuminous . . . . 1. CHILOCARPUS.
　　Ovary 2-celled with axile placentas. Fruit a berry or drupe.
　　　Throat of the corolla with a corona of scales sometimes
　　　　united in a ring. Tall woody climbers . . . . . 2. MELODINUS.
　　　No corona. Erect trees or shrubs, often spinous . . . 3. CARISSA.
　Ovary of 2 distinct carpels united only by the whole or the
　　summit of the style.
　　Ovules few (4 to 6 in each carpel). Fruit of 1 or 2 inde-
　　　hiscent drupes or berries. Mostly trees or shrubs.
　　　Placentas almost dividing each carpel into 2 cells. Leaves
　　　　long alternate or crowded. Bracts large, deciduous.
　　　　Flowers large . . . . 4. CERBERA.
　　　Placentas scarcely prominent. Leaves mostly whorled.
　　　　Bracts small.
　　　　Drupes small, 1-seeded or of 2 or more 1-seeded arti-
　　　　　cles, the endocarp not thick . . . . 5. ALYXIA.

Drupes rather large, the endocarp of 2 bony longitudinal parallel lobes hollow inside with 1 or 2 flat seeds
between them . . . . 6. OCHROSIA.
Ovules numerous. Fruit follicular, dehiscent or without any
hardened endocarp.
Follicles short, usually pulpy and scarcely dehiscent.
Seeds without hairs. Erect trees or shrubs. Leaves
opposite . . . . 7. TABERNÆMONTANA.
Follicles long and linear, dehiscent. Seeds bordered by
hairs, very long at one or both ends. Erect trees or
shrubs. Leaves whorled or opposite . . . . . . 8. ALSTONIA.
Follicles long and linear, dehiscent. Seeds with a coma
at the hilum. Tall woody climber. Leaves opposite. 9. ICHNOCARPUS.
Anthers exserted and cohering or connivent in a cone or ring
round the stigma. Stems twining. Seeds with a coma.
Throat of the corolla with a corona of scales sometimes united
in a ring. Carpels distinct or nearly so. Coma at the end
furthest from the hilum . . . . 10. WRIGHTIA.
No corona. Carpels united in a 2-celled ovary. Coma at the
hilum.
Corolla-lobes imbricate in the bud . . . . 11. PARSONSIA.
Corolla-lobes valvate in the bud . . . . 12. LYONSIA.

1. CHILOCARPUS, Blume.

Calyx without glands. Corolla-tube cylindrical, slightly swollen round
the anthers; lobes spreading, without scales at the throat, contorted in the
bud. Anthers lanceolate, included in the tube. Ovary single, 1-celled, with
2 parietal placentas and numerous ovules. Style single with a thickened
mitriform stigma. Berry ovoid, indehiscent. Seeds numerous, very rugose,
without hairs; albumen copious; cotyledons cordate-ovate, flat.—Tall woody
climber. Leaves opposite. Flowers small in nearly sessile axillary cymes.

Besides the Australian species, which is endemic, there are two or three from the Indian
Archipelago and Malayan Peninsula.

1. **C. australis,** *F. Muell. Fragm.* ii. 90. A tall woody climber, quite
glabrous. Leaves elliptical-oblong, shortly acuminate, sessile and rounded
at the base or narrowed into a short petiole, penniveined, the veins more distant and less regular than in the Javanese species, 2 to 4 in. long. Peduncles
axillary, very short, with about 5 flowers, or the flowers more numerous in
almost sessile cymes. Bracts minute. Calyx-segments about ½ line long,
orbicular, very obtuse, slightly fringed. Corolla yellow, the tube slender,
about 2 lines long, the lobes very broad, about 1 line long, oblique or slightly
auriculate at the base on the inner angle, the left-hand edge overlapping in
the bud. Anthers in the middle of the tube. Fruit about 2 in. long.
Seeds enveloped in the pulp, ovate, compressed, about 3 lines long, very rugose; albumen much ruminate; embryo excentrical in the seeds examined.
—*Melodinus chilocarpoides,* F. Muell. Fragm. vi. 118.

Queensland. Rockingham Bay, *Dallachy;* Rockhampton, *Thozet;* Crocodile Creek,
Bowman.

N. S. Wales. Clarence river, *Beckler;* Richmond river, *Fawcett.*

It does not appear for what reason F. Mueller now removes this plant to *Melodinus.* I
do not find either the scales in the corolla throat, or the 2-celled ovary of the latter genus.

2. MELODINUS, Forst.

Calyx without any or with very few small glands. Corolla-tube cylindrical, slightly swollen round the anthers; lobes spreading, contorted in the bud; throat with 5 or 10 small erect scales, either free or united in a ring or cup. Anthers oblong or lanceolate, included in the corolla-tube. Ovary single, 2-celled; ovules numerous; style filiform, with a thickened conical stigma. Fruit ovoid or globular, succulent, indehiscent. Seeds (where known) without hairs, albuminous.—Tall woody climbers. Leaves opposite. Flowers in terminal or axillary cymes, rarely reduced to very few or single flowers. Bracts very small.

A small genus extending over tropical Asia and the islands of the South Pacific. The Australian species are both endemic, and differ from the rest of the genus in their narrow acute corolla-lobes, and from all, except the Norfolk Island *M. Baueri*, in their axillary inflorescence.

Pubescent. Peduncles solitary, 3-flowered. Corolla-lobes shorter than
    the tube . . . . 1. *M. acutiflorus.*
Glabrous. Peduncles usually 2 in each axil, 1-flowered. Corolla-lobes
    as long as the tube . . . . 2. *M. Guilfoylei.*

1. **M. acutiflorus,** *F. Muell. in Trans. Phil. Soc. Vict.* ii. 71. Branches and under side of the leaves softly pubescent. Leaves shortly petiolate, broadly lanceolate or the lower ones almost ovate, obtusely acuminate, glabrous above, penniveined with the principal veins rather distant, 3 to 4 in. long. Peduncles axillary, not much longer than the petioles, 3-flowered, pubescent. Calyx-segments acuminate, rather above 1 line long, with a few small glands inside at the base. Corolla-tube nearly 3 lines long; lobes narrow, acute, about 2 lines long, the upper part of the tube villous inside, the throat-scales irregularly united in an undulate or lobed ring. Anthers in the middle of the tube. Ovary glabrous. Fruit not seen.

Queensland. Brisbane river, Moreton Bay, *W. Hill.*

2. **M. Guilfoylei,** *F. Muell. Fragm.* vi. 118. A tall woody climber, quite glabrous. Leaves very shortly petiolate or almost sessile, oblong-lanceolate, acuminate but obtuse, rounded at the base, coriaceous, penniveined, somewhat shining, 2 to 3 in. long. Peduncles usually 2 in each axil, slender, 3 to 4 lines long, 1-flowered, but with 2 or 3 pairs of minute bracts. Calyx-segments very acute, about 1 line long, without glands. Corolla-tube 2½ lines long, the lobes at least as long, narrow, acute; throat-scales oblong, distinct. Anthers in the middle of the tube, lanceolate. Ovary glabrous. Fruit not seen.

Queensland. Rockingham Bay, *Dallachy.*

3. CARISSA, Linn.

Calyx without glands. Corolla-tube cylindrical, slightly swollen round the anthers; lobes spreading, contorted in the bud, the throat without scales. Anthers oblong or lanceolate, included in the corolla-tube. Ovary single, 2-celled; ovules several in each cell, in 2 rows; style filiform, with a more or less thickened or conical stigma. Fruit ovoid or globular, succulent, indehiscent. Seeds usually 1 or 2, without hairs, albuminous.—Shrubs or trees, often armed with opposite axillary spines. Leaves opposite Flowers in terminal or axillary cymes. Bracts very small.

The genus is dispersed over tropical and southern Africa, East India, and the Indian Archipelago. The Australian species are endemic.

Leaves 2 to 3 in. long, acuminate. Pedicels longer than the calyx . . 1. *C. laxiflora.*
Leaves on the flowering branches under 2 in. Flowers sessile or the
    pedicels shorter than the calyx.
  Leaves ovate rhomboidal or orbicular, those of the smaller branches
      broadly elliptical.
    Glabrous or very slightly pubescent . . . . 2. *C. ovata.*
    Young leaves densely pubescent, adult ones very scabrous . . . 3. *C. scabra.*
  Leaves lanceolate or narrow-elliptical . . . . 4. *C. lanceolata.*

1. **C. laxiflora,** *Benth.* A shrub? the branches looser than in the other species, quite glabrous, our specimens without spines. Leaves on very short petioles, broadly ovate to ovate-lanceolate, acute, smooth and shining above, with few distant arcuate primary veins, 2 to 3 in. long. Flowers in rather loose terminal cymes on 2 simple or 1 forked or 3-fid peduncle, the pedicels 2 to 3 lines long. Calyx-segments narrow, acute, not 1 line long but unequal. Corolla-tube about 4 lines long, the lobes very acute, nearly 2 lines long, the right-hand edge overlapping in the bud. Anthers below the top of the tube, oblong, not apiculate. Ovules several in each cell of the ovary. Fruit not seen.

Queensland. Cape York, *M'Gillivray.*

2. **C. ovata,** *R. Br. Prod.* 468. An erect, much-branched shrub of 3 or 4 ft., quite glabrous or rarely the young shoots minutely pubescent, more or less armed with opposite horizontally divaricate simple or rarely forked spines, which appear to be abortive peduncles. Leaves ovate rhomboidal or almost orbicular, obtuse or shortly acute, coriaceous, penniveined but the lower veins sometimes very near the base, usually $\frac{1}{2}$ to $\frac{3}{4}$ in. long on the flowering branches, but sometimes twice that size, especially on sterile branches. Flowers in small compact sessile or shortly pedunculate axillary cymes. Calyx-segments lanceolate-subulate, about 1 line long or the inner ones smaller. Corolla-tube nearly 4 lines long, the lobes scarcely 1$\frac{1}{2}$ lines, obliquely ovate or oblong, obtuse or scarcely acute, the right-hand edges overlapping in the bud. Anthers above the middle of the tube, minutely apiculate. Fruit ovoid, usually 1-seeded, $\frac{1}{2}$ to $\frac{3}{4}$ in. long.—A. DC. Prod. viii. 334.

N. Australia ? Victoria river, *Bynoe* (in Herb. Hook., but possibly some error).

Queensland. Thirsty Sound, *R. Brown;* Brisbane river, Moreton Bay, *A. Cunningham, F. Mueller,* and many others; Rockhampton, *Dallachy* and others; Port Nelson, *A. Cunningham;* Fitzroy and Bowen rivers, *Bowman;* Port Denison, *Fitzalan, Dallachy;* in the interior, Mooni and Maranoa rivers, *Mitchell;* Armadillo, *Barton.*

N. S. Wales. Clarence river, *Beckler.*

F. Mueller, Fragm. iv. 45, unites this and the two following species under the name of *C. Brownii.*

3. **C. scabra,** *R. Br. Prod.* 468. A spinous shrub, with the habit of *C. ovata,* but the branches and young leaves densely pubescent. Leaves

ovate, mucronate, with revolute margins, shining but very scabrous on both
sides with minute tubercles the bases of the old hairs, ½ to ¾ in. long.
Flowers in terminal or rarely axillary sessile clusters. Calyx-segments more
subulate and corolla rather longer than in *C. ovata*, but otherwise the same.
—A. DC. Prod. viii. 334.

Queensland. Prince of Wales Island, *R. Brown.*

4. **C. lanceolata,** *R. Br. Prod.* 468. An erect, divaricately-branched
glabrous shrub. Leaves lanceolate or elliptical, acute or rarely obtuse, narrowed into a short petiole, very obliquely penniveined, from under 1 in. to
about 1½ in. long. Flowers in small compact sessile or very shortly pedunculate cymes, mostly terminating short leafy branchlets. Calyx-segments
more subulate, and corolla-lobes narrower, longer, and more acute than in
C. ovata, but otherwise the same. Berry ovoid, about 3 lines long, with 1
or 2 smooth or scarcely wrinkled seeds.

N. Australia. Victoria river, *F. Mueller;* islands of the Gulf of Carpentaria, *R.
Brown;* Strangways river, *Waterhouse.*
Queensland. Suttor river, *Bowman;* Flinders river, *Sutherland.*

F. Mueller unites this with *C. ovata.* If it be the same species, it is a very marked and
apparently constant variety.

4. **CERBERA,** Linn.

Calyx without glands. Corolla-tube cylindrical, slightly swollen round
the anthers, and the throat usually somewhat dilated, without scales; lobes
ovate, spreading, contorted in the bud. Anthers included in the corolla-
tube. Ovary of 2 distinct carpels, united by a single style, each carpel incompletely divided by a very prominent placenta bearing 2 superposed ovules
on each side; stigma conical, often 2-lobed at the tip. Fruit (from the
abortion of one carpel) a single nearly globular drupe, flattened on one side,
with a woody endocarp, usually 1-seeded. Seeds without hairs.—Shrubs
or small trees. Leaves alternate, crowded on the young branches. Flowers
rather large, in terminal cymes or panicles. Bracts large, deciduous.

A genus of very few species, extending over tropical Asia and the islands of the South
Pacific. The Australian species is one of the most widely spread.

1. **C. Odollam,** *Gærtn.; A. DC. Prod.* viii. 353. A glabrous, erect,
tall, bushy shrub or tree, with thick herbaceous branches. Leaves oblong or
lanceolate, acuminate, in some specimens 4 to 6 in. long, in others attaining
1 ft., narrowed into a rather long petiole, the primary veins transverse and
parallel. Flowers white with a yellowish throat, sweet-scented, in a dense
terminal pedunculate cyme. Bracts coloured, ½ to 1 in. long, very deciduous. Calyx-segments oblong or lanceolate, obtuse or acute, about ½ in.
long, deciduous with the bracts. Corolla-tube usually above 1 in. long, but
variable in length; lobes obtuse or shortly acuminate, from less than half as
long to nearly as long as the tube.—Wight, Ic. t. 441 (with a short corolla-
tube); *C. Manghas,* Bot. Mag. t. 1845 (with a long corolla-tube).

Queensland. Cape York, *M'Gillivray, W. Hill;* between the Dawson and Mackenzie
rivers, *F. Mueller;* Rockingham Bay, *Dallachy.* Widely distributed over the maritime

districts of E. India, the Archipelago, and the Pacific Islands.   This appears to be the plant described by Linnæus after Osbeck as *C. Manghas*, although Burman's plant quoted by Linnæus is correctly referred by Thwaites to *Tabernæmontana dichotoma*.

5. ALYXIA, R. Br.

Calyx without glands.   Corolla-tube cylindrical, slightly swollen round the anthers, the lobes spreading, contorted in the bud, the throat without scales.   Anthers enclosed in the tube.   Ovary of 2 distinct carpels, united by a single style, with a capitate or oblong stigma; ovules few (about 4 to 6) in each carpel, in 2 rows.   Fruit of 1 or 2 distinct drupes or berries, either ovoid or oblong and 1-seeded or each consisting of 2 or rarely more ovoid or oblong 1-seeded articles placed end to end.   Seeds furrowed or concave on the inner face, albuminous, without hairs.—Shrubs usually glabrous.   Leaves in whorls of 3 or 4, or rarely opposite.   Flowers in small heads or clusters, or in short spikes or spike-like panicles, terminal or axillary, sessile or shortly pedunculate.   Bracts very small.

A small genus, extending over the islands of the Pacific and into southern Asia.   The Australian species are all endemic.

Flowers in terminal sessile heads or clusters.
  Leaves obtuse, smooth above . . . . 1. *A. buxifolia.*
  Leaves acute, mostly pungent, marked above with parallel veins.
    Veins of the leaves oblique and very prominent.   Corolla-tube
      under 3 lines long . . . . 2. *A. ruscifolia.*
    Veins of the leaves transverse, numerous, and fine.   Corolla-tube
      fully 4 lines long . . . . 3. *A. ilicifolia.*
Inflorescence axillary, shortly pedunculate.
  Peduncles bearing each a simple head or cyme of 3 to 7 flowers . . 4. *A. obtusifolia.*
  Peduncles simple or branching at the base, the flowers singly sessile
    along the rhachis.   Corolla-tube scarcely exceeding the calyx . . 5. *A. spicata.*
  Peduncles bearing clusters of flowers sessile or pedunculate along the
    rhachis.   Corolla-tube twice or three times as long as the calyx   6. *A. thyrsiflora.*

1. **A. buxifolia,** *R. Br. Prod.* 470.   A low rigid spreading or bushy shrub, quite glabrous, the branches often dichotomous.   Leaves opposite or in whorls of 3, shortly petiolate, oval-oblong obovate or almost orbicular, very obtuse, thick and rigid, the margins recurved, mostly ½ to 1 in. long. Flowers few (3 to 7) together in little terminal sessile cymes or clusters. Pedicels very short, rarely exceeding 1 line.   Calyx-segments ovate, acute, scarcely above ¼ line long.   Corolla-tube 3 to 4 lines long, the throat partially closed by a thickish ring; lobes not half so long as the tube.   Anthers mucronulate, near the top of the corolla-tube.   Ovary surrounded by a dense ring of almost scale-like hairs.   Fruits orange-coloured, usually 1-seeded or of 2 joints rarely 3 or 4, each one ovoid, 3 to 4 lines long.   Seeds very broad, but folded over the intruded placenta so as to appear ovoid, the testa wrinkled and folded; albumen ruminated.—A. DC. Prod. viii. 348; Hook. f. Fl. Tasm. i. 269; F. Muell. Rep. Burdek. Exp. 16; *A. capitellata*, Benth. in Hueg. Enum. 81; Lehm. Pl. Preiss. i. 366.

N. S. Wales.   Twofold Bay, *A. Cunningham.*
Victoria.   Port Phillip, *R. Brown ;* along the coast from the Glenelg to Gipps' Land, *F. Mueller* and others.

Tasmania. Kent's Group, Bass's Straits, *R. Brown*; rocky places, north coast, *J. D. Hooker.*

S. Australia. Along the coast, St. Vincent's and Spencer's Gulfs, etc., *F. Mueller* and others.

W. Australia. Goose Island Bay, *R. Brown*; along the south and west coasts, *Drummond, n.* 88, *Suppl. n.* 40, 52, *Preiss, n.* 1299, and others, up to Murchison river, *Oldfield*, and Sharks' Bay, *Milne.*

2. **A. ruscifolia,** *R. Br. Prod.* 470. A tall, handsome, glabrous shrub. Leaves in whorls of 4 or rarely of 3 only, from broadly ovate-elliptical to narrow-lanceolate, acute with a short pungent point, contracted into a very short petiole, the margins recurved or revolute, coriaceous and shining, obliquely and prominently veined on the upper side, $\frac{3}{4}$ to $1\frac{1}{2}$ in. long. Flowers small, in sessile terminal heads, and the flowers themselves sessile. Calyx-segments lanceolate, ciliolate, nearly 1 line long. Corolla-tube under 3 lines long, the lobes about half as long, the throat scarcely thickened inside. Ovary surrounded by hairs at the base. Fruits orange-coloured, consisting of 1 or 2 carpels, each with 1 or 2 ovoid-globular articles about 4 lines long. Seed with a wrinkled testa and ruminate albumen.—A. DC. Prod. viii. 347; Bot. Mag. t. 3312; Lodd. Bot. Cab. t. 1811; *A. Richardsonii*, Sweet in Loud. Hort. Brit. 67 (according to A. DC.).

Queensland. Sandy Cape, Harvey's Bay, *R. Brown*; Cape Cleveland, Endeavour river and Brisbane river, *A. Cunningham*; Burdekin river, *F. Mueller*; Curtis island, *Henne*; Port Denison, *Fitzalan*; Rockhampton, *Dallachy*; Ipswich, *Nernst.*

N. S. Wales. Macleay and Clarence rivers, *Beckler*; Richmond river, *Fawcett*; Lord Howe's Island, *Milne.*

3. **A. ilicifolia,** *F. Muell. Fragm.* iv. 149, v. 212. Very near the broad-leaved form of *A. ruscifolia*, but the leaves are larger ($1\frac{1}{2}$ to 3 in. long), the veins more numerous, finer, and transverse, and the recurved margins have occasionally a few small teeth or prickles. Flowers rather larger. Corolla-tube slender, fully 4 lines long, the lobes narrow, nearly 3 lines. Articles of the fruit ovoid, about 4 lines long.

Queensland. Rockingham Bay, *Dallachy.*

4. **A. obtusifolia,** *R. Br. Prod.* 470. A glabrous shrub. Leaves usually in whorls of 3, obovate or oblong, very obtuse, narrowed into a very short petiole, smooth and shining above, the margins slightly recurved, the veins parallel and fine, but very obscure and not so numerous as in *A. spicata*, mostly 1 to 2 in. long. Flowers 3 to 7 together, in little axillary heads or cymes, the common peduncle slender, 2 to 4 lines long, with 3 small bracts under the head, the flowers sessile or on short thick pedicels. Calyx-segments ovate, obtuse, about $\frac{1}{2}$ line long. Corolla-tube twice or three times as long as the calyx, the lobes very broad, fully as long as the tube. Anthers mucronate. Articles of the fruit nearly globular, 3 to 4 lines diameter.—A. DC. Prod. viii. 346.

Queensland. Shoalwater Bay and Broad Sound, *R. Brown*; Cape Cleveland and Endeavour river, *A. Cunningham*; Curtis Island, *Henne.*

5. **A. spicata,** *R. Br. Prod.* 470. A shrub, glabrous except the inflorescence, the branchlets angular. Leaves in whorls of 3 or rarely 4, oval or

elliptical, obtuse, narrowed into a short petiole, smooth and shining above, the margins slightly recurved, the veins numerous, almost parallel, visible especially on the under side, $1\frac{1}{2}$ to $2\frac{1}{2}$ in. long. Flowers small and singly sessile along the simple or rarely branched rhachis of axillary shortly pedunculate spikes, scarcely above $\frac{1}{2}$ in. long, and usually minutely pubescent. Calyx about 1 line long, divided to about the middle. Corolla-tube very shortly exceeding the calyx, the lobes narrow, acute, nearly as long as the tube. Anthers almost acute. Ovary pubescent. Fruit larger than in *A. buxifolia*, the articles usually solitary, varying from globular to ovoid.— A. DC. Prod. viii. 346.

N. Australia. Islands of the Gulf of Carpentaria, *R. Brown, Henne.*
Queensland. Prince of Wales Islands, *R. Brown;* Albany Island, Cape York, *M'Gillivray;* Rockingham Bay, *Dallachy.*

A. tetragona, R. Br. Prod. 470 ; A. DC. Prod. viii. 346, from Endeavour river, *R. Brown,* appears to be the same species with the leaves in whorls of 4, which is sometimes the case on the main branches of the common 3-leaved form.

6. **A. thyrsiflora,** *Benth.* A glabrous shrub. Leaves usually in whorls of 3, obovate or oblong, very obtuse, narrowed into a petiole of 2 or 3 lines, the margins slightly recurved, smooth and shining above, the veins parallel but very obscure and not so numerous as in *A. spicata.* Flowers 3 to 7 together in little heads or clusters, pedicellate or nearly sessile along the rhachis of small axillary thyrsoid or spike-like panicles, the upper ones sometimes singly sessile, all rather smaller than in *A. obtusifolia.* Calyx-segments narrow and acute. Corolla-tube slender, 2 or 3 times as long as the calyx, the lobes ovate, much shorter than the tube. Ovary glabrous. Articles of the fruit 4 to 5 lines long.

Queensland. Endeavour river, *R. Brown, A. Cunningham;* Albany Island, *W. Hill;* Burdekin river and Port Denison, *Fitzalan, Dallachy;* Lizard Island, *A. Cunningham.*

6. **OCHROSIA,** Juss.

(Bleekeria, *afterwards* Lactaria, *Hassk.*)

Calyx without glands. Corolla-tube cylindrical, slightly swollen round the anthers, the lobes spreading, contorted in the bud, the throat without scales. Anthers enclosed in the tube. Hypogynous disk of minute glands or none. Ovary of 2 distinct carpels united by the style; stigma conical; ovules few (4 to 6) in each carpel, in 2 rows. Fruit a drupe, usually single by the abortion of the other carpel, somewhat compressed from front to back, the endocarp very thick and bony, divided longitudinally along the inner face into 2 thick, cylindrical, usually hollow portions, the real cell between them containing 1 or rarely 2 broad flat seeds. Albumen fleshy ; cotyledons broad and flat ; radicle superior.—Trees, with abundant milky juice. Leaves opposite or whorled, the primary veins transverse. Flowers in pedunculate cymes, terminal or in the uppermost axils. Bracts very small.

A small genus, dispersed over the Mascarene and South Sea Islands. Of the two Australian species, one is endemic, the other extends to New Caledonia and the Fiji islands. The floral characters are nearly those of *Alyxia,* but the fruit is very different.

Leaves mostly in whorls of 3 or 4, obovate-oblong and obtuse. Drupes
acuminate . . . . 1. *O. elliptica.*
Leaves mostly opposite, oblong-lanceolate and acuminate.   Drupes obtuse .   2. *O. Moorei.*

1. O. elliptica, *Labill. Sert. Austr. Caled.* 25. *t.* 30.   A tree with a
milky juice, quite glabrous.   Leaves in whorls of 3 or 4, rarely here and
there opposite, from obovate-oblong to broadly elliptical, very obtuse or
shortly acuminate, contracted into a petiole, coriaceous, the transverse veins
numerous and parallel, 3 to 6 in. long or even more, on luxuriant barren
shoots.   Flowers sessile, in small dense corymbose cymes shortly pedunculate
in the uppermost axils.   Calyx-segments ovate-oblong, obtuse, very concave, thickened in the middle, about 1 line long.   Corolla-tube slender,
about 4 lines long, the lobes linear, about 3 lines.   Anthers lanceolate,
acute.   Hypogynous glands scarcely conspicuous.   Drupes scarlet, acuminate, 1 to 2 in. long, each of the parallel portions of the endocarp about 4 to
5 lines diameter, with a large cavity more or less filled with pith, the real
cell very much compressed.   Seeds nearly orbicular, bordered by a narrow
wing-like margin.—A. DC. Prod. viii. 357; *O. parviflora,* Hensl. in Ann.
Nat. Hist. (not Ann. Bot.) i. 345; A. DC. l. c.; Seem. Fl. Vit. 158;
Bleekeria kalocarpa, Hassk. Retzia, i. 40; *Lactaria calocarpa,* Hassk. in
Ned. Kruidk. Arch. iv. 9; F. Muell. Rep. Burdek. Exped. 15; Fragm. iii.
110.

Queensland.   Keppel Bay, *Thozet;* Edgecombe Bay, *Fitzalan, Dallachy;* Brook
Island, *Dallachy.*   Also in New Caledonia and in the Fiji and other islands of the S. Pacific.
The Society Island plant in the Hookerian herbarium, referred here by Seemann, has very
narrow long leaves and much smaller flowers, and may be *O. mariannensis,* A. DC.   The Su-
matra plant, sent by Miquel as *Lactaria calocarpa,* Hassk., is totally different, and seems to
be a *Tabernæmontana.*   Labillardière's specimens agree quite with the Australian ones; but
in the figure, by a mistake of the artist, the seeds are placed in the pithy cavities of the
endocarp, instead of in the real cell.   The species is nearly allied to the original *O. borbo-*
nica, but has a much smaller and more dense inflorescence, and longer acuminate fruits.

2. O. Moorei, *F. Muell.*   A slender tree, quite glabrous.   Leaves
mostly opposite, narrow-oblong or oblong-lanceolate, acuminate, tapering
into a short petiole, the transverse veins not near so close and numerous as
in *O. elliptica* and the texture thinner, 3 to 6 in. long.   Flowers sessile in
dichotomous cymes in the uppermost axils, the common peduncle longer than
the petioles, but the whole inflorescence much shorter than the leaves.
Calyx-segments thick and concave, like those of *O. elliptica,* but rather
longer.   Corolla-tube about 4 lines long, scarcely swollen round the anthers;
lobes about 3 lines long.   Anthers small, close under the throat.   Drupes
scarlet, 1½ to 2 in. long, obtuse and more flattened than in *O. elliptica.*—
Lactaria Moorei, F. Muell. Fragm. iii. 110, vi. 118.

N. S. Wales.   Richmond and Clarence rivers, *C. Moore.*

7. TABERNÆMONTANA, Linn.

Calyx with a ring of small linear glands inside at the base.   Corolla-tube
cylindrical, slightly swollen round the anthers, the lobes spreading, contorted
in the bud, the throat without scales.   Anthers enclosed in the tube.   No
hypogynous glands or scales.   Ovary of 2 distinct carpels, united by the

style ; stigma thickened, usually with a membranous ring round the base ; ovules numerous in each cell, in 3 or 4 rows. Fruit of 2 (or 1 by abortion) obliquely oblong or nearly globular carpels, distinct or rarely united at the base, more or less fleshy or pulpy, either indehiscent or tardily opening along the inner face. Seeds without hairs ; albumen fleshy.—Trees or shrubs, the branches often dichotomous. Leaves opposite, one of each pair often smaller than the other. Flowers in axillary cymes, usually 2 at the ends of the branches (in the axils of the terminal pair of leaves) or only in the axil of the smaller leaf. Bracts usually very small.

A considerable genus, spread over the tropical regions of the New as well as the Old World. Of the two Australian species, one extends to Timor and to the S. Pacific islands, the other, if really distinct, appears to be endemic.

Quite glabrous. Calyx-segments scarcely obtuse . . . . 1. *T. orientalis.*
Foliage pubescent. Calyx-segments very obtuse . . . . 2. *T. pubescens.*

1. **T. orientalis,** *R. Br. Prod.* 468. A dichotomously-branched shrub or small tree, quite glabrous. Leaves elliptical-oblong, obtusely acuminate, narrowed into a petiole sometimes very short, more frequently ¼ to ½ in. long, distantly penniveined, 2 to 4 in. long or sometimes longer. Cymes pedunculate, 2 together at the ends or in the forks of the branches, or becoming lateral by the development of only one fork, loose and several-flowered but shorter than the leaves. Pedicels as long as or longer than the calyx. Calyx-segments acuminate or almost obtuse in the northern specimens, about 1 line long, with 2 to 4 minute glands inside at the base of each. Corolla-tube usually 5 to 6 lines long, the lobes at least half as long as the tube, obliquely oval-oblong. Ovary glabrous, on a thick torus. Carpels of the fruit ovoid-falcate, usually about ½ in. long, but variable in size, and more or less prominently 3-angled. Seeds 3 or 4 in each carpel, ovate, deeply furrowed or concave on the inner face.—A. DC. Prod. viii. 371.

N. Australia. Arnhem's Land, *R. Brown ;* Sims Island and common along the coast, *A. Cunningham ;* Port Essington, *Armstrong.*
Queensland. Cape York, *M'Gillivray, Daemel ;* Brisbane river, Moreton Bay, *A. Cunningham* and others.
N. S. Wales. Hastings river, *Fraser, A. Cunningham,* and others ; Clarence river, *Beckler ;* Richmond river, *Fawcett ;* Tweed river, *C. Moore.*

T. parviflora, Dcne. Herb. Tim. Descr. 51, or *T. Decaisnii,* A. DC. Prod. viii. 369, from Timor, of which we have flowering specimens communicated by Decaisne, appears to be the same form as the common North Coast one ; so also are the specimens of *T. vitiensis,* Seem., referred to *T. orientalis* by Seemann, Fl. Vit. 159. A Timor specimen from A. Cunningham (var. *petiolata,* A. DC. ?), and some from Fiji, *Milne,* have much smaller flowers, and small acute calycine segments. *T. Cumingiana,* A. DC., from the Philippine Islands, also referred to this species by Seemann, appears to me to differ considerably, in its very much longer flowers, with the anthers considerably below the summit of the tube, and in the very prominently angled, almost winged fruits. The Australian stations given above supply the following rather marked varieties :—

Var. *angustisepala.* Calyx-segments lanceolate, acuminate ; leaves as in the northern form. To this belong the N. S. Wales specimens.

Var. *angustifolia.* Leaves oblong-lanceolate. Calyx-segments very narrow, acuminate. Fruits acuminate, about 1 in. long. This comprises nearly all the Moreton Bay specimens.

2. **T. pubescens,** *R. Br. Prod.* 468. A shrub or tree very near to *T.*

orientalis, and perhaps a variety only, the branches and both sides of the young leaves softly pubescent, the older leaves becoming glabrous or nearly so on the upper side, the leaves otherwise as in *T. orientalis* or rather larger. Inflorescence of *T. orientalis,* but more or less hairy. Flowers larger. Calyx-lobes about 1 line long, very obtuse, hirsute. Corolla-tube fully ½ in. long, and the lobes more than half the length of the tube. Fruit like that of *T. orientalis* and as variable in size, from ½ to 1 in. long.—A. DC. Prod. viii. 376.

N. Australia. North coast, *R. Brown ;* Goulburn Islands, *A. Cunningham.*
Queensland. Cape York, *M'Gillivray, Daemel ;* Port Denison, *Fitzalan ;* Rockingham and Edgecombe Bays, *Dallachy.*

T. ebracteata, R: Br. Prod. 468, A. DC. Prod. viii. 376, from Groote Island, Gulf of Carpentaria, appears to be a less luxuriant form of the same species, with the leaves more obtuse and from 2 to 4 in. long.

8. ALSTONIA, R. Br.

(Blaberopus, *A. DC.*)

Calyx without any or with very minute glands. Corolla-tube cylindrical, more or less swollen round the anthers, the lobes spreading, contorted in the bud, the throat without scales. Anthers enclosed in the tube. No hypogynous scales. Ovary of 2 distinct carpels united by the style ; stigma ovoid or conical ; ovules numerous, in about 4 rows in each carpel. Fruit of 2 long linear follicles. Seeds oblong, compressed, peltately attached, bordered with hairs of which those at each end usually very long ; albumen scanty.— Trees or tall shrubs with a milky juice. Leaves in whorls of 3 or more, or in a few species opposite. Flowers in terminal corymbose cymes, usually 1 in the axil of each leaf of the terminal whorl. Bracts small.

The species are not numerous, spread over tropical Asia and the Pacific islands. Of the seven Australian species, one is the commonest of the Asiatic ones, another, if correctly identified, is also in Java, the other five are endemic.

Corolla-lobes with the left-hand margin overlapping. Leaves
 whorled, with transverse parallel veins.
 Flowers sessile or nearly so in dense cymes, pubescent. Ovary
  villous . . . . 1. *A. scholaris.*
 Flowers pedicellate in loose cymes, glabrous as well as the ovary. 2. *A. verticillosa.*
Corolla-lobes with the right-hand margin overlapping.
 Leaves in whorls of 3 or 4.
  Veins numerous, nearly transverse and parallel . . . . . 3. *A. ophioxyloides.*
  Veins distant, more or less arcuate and anastomosing.
   Leaves oval elliptical or oblong, softly villous underneath . 4. *A. villosa.*
   Leaves linear, glabrous . . . . 5. *A. linearis.*
 Leaves opposite. Veins distant.
  Quite glabrous . . . . 6. *A. constricta.*
  Softly pubescent . . . . 7. *A. mollis.*

1. **A. scholaris,** *R. Br.*; *A. DC. Prod.* viii. 408. A tree attaining 80 to 90 ft., usually glabrous except the minutely pubescent inflorescence. Leaves in whorls of 5 to 7, broadly petiolate, obovate-oblong, very obtuse, shortly contracted at the base, coriaceous, with transverse parallel veins, smooth and shining above, opaque and pale or whitish underneath, 4 to 6 in.

long. Peduncles in the axils of the terminal whorls shorter than the leaves, each bearing 1 or 2 whorls of secondary peduncles, and each of these a dense cyme of nearly sessile flowers. Calyx-segments ovate, pubescent, rather above 1 line long. Corolla-tube 3 to 4 lines long; lobes pubescent outside, much shorter than the tube, the left-hand edges overlapping in the bud, the throat closed by a dense ring of hairs. Ovary hirsute at the top. Follicles 1 ft. long or even more. Seeds about 3 lines long, the hairs at each end longer than the seed itself.—Wight, Ic. t. 422.

Queensland. Port Denison, *Fitzalan ;* Edgecombe and Rockingham Bays, *Dallachy ;* Broad Sound, *Bowman.* The species is widely spread over tropical Asia and Africa.

2. **A. verticillosa,** *F. Muell. Fragm.* vi. 116. A tree of strong growth, attaining sometimes a considerable height, with abundance of milky juice. Leaves in whorls of 4 to 7, long-lanceolate, rather obtuse, narrowed into a petiole of $\frac{1}{4}$ to $\frac{1}{2}$ in. or sometimes very short, pale underneath, the transverse veins parallel and numerous, mostly 3 to 4 in. long but sometimes only 2 in. Peduncles in the axils of the terminal whorls, each bearing 1 or rarely 2 whorls of secondary peduncles and each of these a loose cyme of flowers on pedicels of 1 to 2 lines, all quite glabrous. Calyx-segments broadly ovate, ciliolate, about $\frac{1}{2}$ line long. Corolla-tube above 1 line long, the lobes about half as long as the tube, the left-hand edges overlapping in the bud, the throat hairy inside. Ovary glabrous ; ovules rather numerous in each carpel, but less so than in *A. scholaris.* Fruit unknown.—*Alyxia actinophylla*, A. Cunn. in Bot. Mag. under n. 3313 ; A. DC. Prod. viii. 346.

N. Australia. Rocks, Montague Sound, *A. Cunningham ;* Port Essington, *Armstrong ;* rocky gullies, sources of the Roper river, *F. Mueller.*

Queensland. Cape York, *Daemel ;* Albany Island, *W. Hill ;* grassy forest land, south shore of Endeavour river, *A. Cunningham.*

The fruit being unknown, the true genus of this plant is somewhat uncertain, but the habit, inflorescence, and ovules are much more those of *Alstonia* than of *Alyxia.*

3. **A. ophioxyloides,** *F. Muell. Fragm.* i. 57. A tree, the branchlets, inflorescence, and under side of the leaves softly tomentose-pubescent. Leaves in whorls of 3, on rather long petioles, elliptical-oblong, shortly and obtusely acuminate, narrowed at the base, glabrous above or nearly so, the primary veins nearly transverse, much more numerous than in *A. constricta,* rather less so than in *A. scholaris,* 2 to 4 in. long. Flowers numerous, in pedunculate cymes of which 5 or 6 are arranged in a terminal sessile umbel, all much shorter than the leaves. Calyx-segments broadly ovate, very obtuse, minutely ciliolate, about 1 line long. Corolla-tube about 2 lines long, bearded inside towards the throat ; lobes broad, obtuse, shorter than the tube, the right-hand edges overlapping in the bud. Ovary glabrous. Follicles 8 to 10 in. long or sometimes more Seeds broadly oblong, silky hairy, the marginal hairs at each end about as long as the seed itself.

N. Australia. Upper Victoria and Fitzmaurice rivers, *F. Mueller.*

4. **A. villosa,** *Blume? ; F. Muell. Fragm.* vi. 117. A tall shrub or tree attaining 30 ft., the branches and under side of the leaves softly velvety-pubescent. Leaves in whorls of 3, oval-oblong or elliptical, obtuse or ob-

tusely acuminate, the upper surface at length glabrous or minutely pubescent, the primary veins curved and rather distant, prominent underneath, rounded or acute at the base, 4 to 6 in. long, on petioles of ¼ to ½ in. Peduncles usually solitary and terminal, bearing an umbel of 4 to 8 rays, each with a loose corymbose cyme, the whole inflorescence shorter or longer than the leaves, minutely hoary-pubescent. Calyx-segments about 1 line long, obtuse, the 2 outer ones ovate, the 3 inner usually narrower. Corolla-tube about 2 lines long, the lobes rather shorter, papillose inside as well as the throat but not hairy, the right-hand edges overlapping in the bud. Ovary glabrous. Follicles ½ to 1 ft. long. Seeds about 3 lines long, with a short beak at the upper end, the hairs at the two ends about as long as the seed itself.

Queensland. Rockingham Bay, *Dallachy.* *A. villosa,* Blume Bijdr. 1088 ; A. DC. Prod. viii. 410, or *Blaberopus villosus,* Miq. Fl. Ind. Bat. ii. 440, from Java, is said to have subsessile leaves, and in the only specimen I have seen, belonging to Miquel's var. *petiolata,* the petioles are shorter than in most Australian ones ; the leaves are also in whorls of 4, and smoother than in ours. All this, however, is not incompatible with the identity of the species, supposing the flowers to be the same. Those of the Javanese plant are, I believe, as yet unknown.

5. **A. linearis,** *Benth.* A small tree of robust growth, quite glabrous. Leaves in whorls of 3 or 4, linear, almost acute, contracted into a short slender petiole, the primary veins few distant and oblique, mostly 2 to 3 in. long. Cymes terminal. Flowers not seen. Follicles 8 to 10 in. long. Seeds pubescent, the hairs of each end about as long as the seed itself, as in *A. ophioxyloides.*

N. Australia. Cliffs of Brunswick Bay and Regent's River, N. W. coast, *A. Cunningham.*

6. **A. constricta,** *F. Muell. Fragm.* i. 57. iv. 170. A tall shrub or tree attaining sometimes 40 ft., quite glabrous. Leaves opposite, on long petioles, mostly oblong-lanceolate, but varying from almost ovate to narrow-lanceolate, acute or acuminate, the primary veins distant, oblique, and not very prominent, 3 to 5 in. long. Flowers numerous, in corymbose cymes, either solitary and terminal or 2 together in the forks of the branches and shorter than the leaves. Calyx-segments ovate, almost acute, about ½ line long, with a few minute and irregular glands at the base inside. Corolla-tube about 1 line long, the lobes about 2 lines long, glabrous or slightly bearded inside at the base, the right-hand edges overlapping in the bud. Follicles from 3 to 4 in. to twice that length. Seeds linear, flat or concave, pubescent, 4 to 6 lines long, ciliate with long hairs at the upper end and shorter ones at the lower end.

Queensland. Between the Burnett and Burdekin rivers, *F. Mueller ;* Rockhampton, *Dallachy, Thozet ;* Nerkool and Castle Creeks, *Bowman ;* Natal Downs, *Fitzalan ;* near Mount Pluto, *Mitchell.*

N. S. Wales. Hastings river, *Fraser ;* Clarence river, *Beckler ;* northern woods, London Exhibition, 1862, *C. Moore ;* Darling river, *Goodwin ;* " Bitterbark " of the colonists (*F. Mueller*).

This species differs from the rest of the genus, and approaches *Tabernæmontana* in habit and foliage, and usually in the presence of small glands at the base of the calyx inside. The fruit and seeds are, however, those of *Alstonia.*

7. **A. mollis,** *Benth.* A shrub of about 10 ft., closely resembling *A. constricta,* and perhaps a variety, but the branchlets foliage and inflorescence softly tomentose-pubescent, the leaves scarcely becoming glabrous on the upper surface with age. Leaves opposite, oblong-lanceolate, acute or acuminate, with the venation of *A. constricta,* but larger, mostly 4 to 6 in. long. Flowers numerous, in shortly pedunculate terminal corymbose cymes, the primary branches sometimes umbellate, of the size and structure of those of *A. constricta.* Fruits not seen.

Queensland. Barcoo river, near Mount Northampton, *Mitchell.*

9. ICHNOCARPUS, R. Br.

Calyx without any or with very minute glands inside at the base. Corolla-tube cylindrical, swollen round the anthers, the lobes spreading, contorted in the bud, the throat without scales. Anthers enclosed in the tube. Hypogynous scales 5, linear-clavate, as long as the ovary. Ovary of 2 distinct carpels united by the style; stigma thickened, acuminate; ovules numerous, in 3 or 4 rows in each carpel. Fruit of 2 long linear follicles. Seeds linear, with a long tuft of hairs at the hilum, albuminous.—Tall woody climbers. Leaves opposite. Flowers small, in compact cymes arranged in axillary and terminal thyrsoid panicles. Bracts very small.

A small genus, dispersed over tropical Asia and extending probably into Africa, the only Australian species being the commonest on the whole range. The delimitation of this and other genera separated from *Echites* requires, however, much revision.

1. **I. frutescens,** *R. Br. ; A. DC. Prod.* viii. 435. A tall climber with a milky juice, the young parts and inflorescence minutely rusty-pubescent, the adult leaves glabrous or more or less pubescent underneath, especially the principal nerves. Leaves petiolate, ovate, acuminate, rather thin, shining above, the primary veins prominent underneath, the veinlets elegantly reticulate but not prominent, 2 to 3 in. long. Cymes small and compact, often numerous, forming loose leafy panicles at the ends of the branches. Calyx-segments scarcely 1 line long, pubescent, narrow. Corolla-tube 1½ lines long, the lobes narrow, tapering into a long point, much contorted in the bud, the right-hand edges overlapping, glabrous inside or with a few hairs in the throat. Ovaries hirsute at the top. Follicles slender in Asiatic specimens, not seen in the Australian ones.—*I. leptodictyus,* F. Muell. Fragm. vi. 118.

Queensland. Rockingham Bay, *Dallachy.* Common in E. India and in the Archipelago. The Indian specimens have usually rather narrower leaves, more hirsute underneath, but some are quite similar to the Australian ones.

10. WRIGHTIA, R. Br.

(Balfouria, *R. Br.*)

Calyx with 5 to 10 broad scales inside at the base. Corolla-tube cylindrical, usually short; lobes spreading, contorted in the bud, the throat with a corona of 5 or 10 erect scales, either distinct or united in a ring. Stamens inserted in the throat; filaments very short and broad; anthers sagittate, exserted, united or connivent in a cone round the stigma. No hypogynous disk or scales. Ovary of 2 carpels, distinct or connate, but readily separable; ovules numerous. Fruit long, terete, at length separating into 2 follicles. Seeds oblong, with a tuft of hairs at the end furthest from the hilum.—Trees or shrubs, the branches often slender but not twining. Leaves opposite. Flowers in terminal corymbose cymes, appearing sometimes axillary after the development of one branch only of the fork. Bracts small.

A small genus dispersed over tropical Asia. Of the three Australian species, one is also in the Indian Archipelago, and closely allied to a common E. Indian one; the two others are endemic, and very distinct.

Corolla-lobes with the left-hand edges overlapping, and longer than
    the tube.
    Pubescent. Leaves ovate or elliptical. Corona of 10 distinct
      scales . . . . 1. *W. pubescens.*
    Glabrous. Leaves linear or lanceolate. Corona a truncate or
      toothed ring . . . . 2. *W. saligna.*
Corolla-lobes with the right-hand edges overlapping and shorter
    than the tube. Leaves linear or lanceolate. Corona of 5 or 10
    distinct scales . . . . 3. *W. Cunninghamii.*

1. **W. pubescens,** *R. Br. Prod.* 467. A tall shrub or small tree, the foliage and inflorescence more or less pubescent or velvety-tomentose. Leaves shortly petiolate, ovate to elliptical-oblong, acuminate, the primary veins rather distant and prominent underneath, 2 to 4 in. long. Flowers white, sessile or shortly pedicellate in terminal trichotomous corymbose cymes not exceeding the leaves. Calyx-segments broadly ovate, almost orbicular, nearly 2 lines long, with 1 or 2 ovate scales inside at the base of each. Corolla-tube broad, scarcely exceeding the calyx, contracted at the throat; lobes twice as long as the tube, the left-hand edges overlapping in the bud; corona of 10 erect scales, 5 larger ones alternating with the stamens irregularly several-toothed, 5 smaller behind the stamens 2-toothed. Anthers in a cone of above 3 lines. Carpels of the ovary connate or coherent at the base, distinct at the top; stigma dilated and membranous at the base, with 2 linear lobes. Fruit hard, about 6 in. long, the follicles not separating until maturity.—A. DC. Prod. viii. 405.

N. Australia. Islands off the N. coast of Arnhem's Land, *R. Brown;* Port Essington, *Armstrong;* Fitzmaurice river, *F. Mueller;* also in the Indian Archipelago. Cuming's n. 1293, from the Philippine Islands, appears to be the same; n. 1453 referred here by A. De Candolle has larger, rather differently shaped flowers. *W. tomentosa* of East India has the same foliage, but the corolla-tube is generally rather longer in proportion to the calyx, and the corona is shortly lobed only, not divided to the base. All may, however, prove to be varieties of one species.

2. **W. saligna,** *F. Muell. Herb.* A tall shrub or small tree, quite glabrous. Leaves linear or linear-lanceolate, acutely acuminate, contracted into a short petiole, almost veinless except the midrib, 3 to 5 in. long. Flowers yellow, in short cymes either terminal or apparently axillary from the development of a side branch. Calyx-segments nearly orbicular, ¾ line diameter, with 1 or 2 broad scales at the base of each more than half as long as the calyx. Corolla-tube broad, not twice as long as the calyx, contracted at the throat; lobes narrow, fully twice as long as the tube, the left-hand

edges overlapping in the bud; corona-scales united in a truncate or sinuate-toothed ring round the anthers. Anthers in a cone of about 2 lines. Carpels of the ovary coherent but separable. Stigma dilated at the base, 2-lobed. Fruit hard, acuminate, nearly terete, 6 to 8 in. long, the follicles not separating till maturity.—*Balfouria saligna*, R. Br. Prod. 467; Endl. Iconogr. t. 75; A. DC. Prod. viii. 403.

N. Australia. Barren ridges, Sea Range, Victoria river, and M'Adam Range, *F. Mueller;* islands of the Gulf of Carpentaria, *R. Brown;* Port Essington, *Armstrong;* Albert river, *Henne.*

Queensland. Port Denison, *Fitzalan, Dallachy;* Elliott and Suttor river, *Bowman.*

3. **W. Cunninghamii,** *Benth.* A slender glabrous shrub with the habit and nearly the foliage of *W. saligna.* Leaves linear-lanceolate or lanceolate, acutely acuminate, narrowed into a long slender petiole, the veins irregular and scarcely visible above, impressed as well as the reticulate veinlets underneath, 3 to 5 in. long. Cymes shortly pedunculate, terminal or axillary from the development of a side branch. Calyx-segments ovate, almost acute, with short broad obtuse or acuminate scales inside at the base, about 5 to the whole calyx. Corolla-tube about 3 lines long; lobes rather broad, ovate-falcate, the right-hand edges overlapping in the bud; corona of 5 bifid or 10 simple scales shortly acuminate, all behind the anthers. Anthers shorter than in the preceding species and more ovate with recurved points, but exserted and connivent in a cone or ring round the stigma. Ovary of 2 distinct carpels. Fruit not seen.

W. Australia. Enderby island, Dampier's Archipelago, *A. Cunningham.* Although this plant differs from the other species in several particulars, they do not seem sufficient in the absence of the fruit to constitute a distinct genus.

11. PARSONSIA, R. Br.

(Heligme, *Blume.*)

Calyx with an irregular ring of minute glands inside at the base or with few or none. Corolla-tube cylindrical or nearly globular; lobes spreading, contorted in the bud, the edges slightly overlapping, the throat without scales. Stamens inserted at or below the middle of the tube, the filaments often twisted together under the anthers; anthers oblong lanceolate or linear, wholly or partially exserted, cohering in a cone or ring round the stigma, each with 2 rigid basal lobes usually devoid of pollen. Hypogynous scales 5, as long as the ovary, free or united in a plicate ring. Ovary 2-celled, with numerous ovules in each cell. Stigma surrounded by a ring or membranous expansion at the base, usually 2-lobed. Fruit elongated, nearly terete, separating more or less completely into 2 follicles. Seeds with a coma or tuft of long silky hairs at the hilum.—Tall and woody or slender climbers. Leaves opposite. Flowers in terminal or axillary corymbose cymes. Bracts small.

A small genus, extending over E. India, the Archipelago, the South Pacific islands, and New Zealand. The Australian species appear all to be endemic. The overlapping of the corolla-lobes is sometimes slight, but is easily observed on the unexpanded bud.

Corolla-tube cylindrical or ovoid. Anthers without appendages on the back.

Anthers wholly exserted from the corolla-tube.
  Cymes terminal or on short axillary branches. Leaves narrow
    or broad, not cordate, glabrous or tomentose . . . . . . 1. ·*P. lanceolata.*
  Cymes on long axillary peduncles. Leaves broad, usually cor-
    date. Plant usually rusty pubescent . . . . 2. *P. velutina.*
Anthers with the tips only protruding. Cymes terminal or on
  short axillary branches. Plant glabrous or nearly so . . . . . 3. *P. Leichhardtii.*
Corolla-tube nearly globular. Anthers with a prominent longitudinal
  appendage or wing on the back . . . . 4. *P. ventricosa.*

1. **P. lanceolata,** *R. Br. Prod.* 466. A tall woody climber, glabrous
or the young branches under side of the leaves and inflorescence shortly
pubescent or rarely tomentose-pubescent all over, often but not always
glaucous and not rust-coloured. Leaves shortly petiolate, elliptical-oblong or
lanceolate, more rarely oval or almost orbicular, obtuse mucronate or shortly
acuminate, not cordate, the margins usually recurved, the veins scarcely conspicuous on the upper side, pale or light brown and penniveined underneath,
mostly 2 to 3 in. long. Cymes usually compact, terminal or on short axillary
branches. Calyx-segments unequal, from under 1 line to above 1½ lines
long. Corolla-tube shorter than the calyx, slightly constricted at the attachment of the stamens; lobes about 2 lines long. Filaments spirally twisted
under the anthers; anthers wholly exserted, linear, without dorsal appendages. Fruit 3 to 5 in. long, dividing into rather thin follicles. Seeds
attenuate into a short beak at the hilum, with a silky tuft of ½ to ¾ in.—
A. DC. Prod. viii. 402 ; *Echites lanceolata,* Spreng. Syst. i. 634 ; *Parsonsia
glaucescens,* F. Muell. Fragm. vi. 126.

Queensland. E. coast, *R. Brown ;* Araucaria Ranges, Burnett river, *F. Mueller;*
Port Denison, *Fitzalan ;* Edgecombe and Rockingham Bays, and Rockhampton, *Dallachy*
and others ; heads of the Isaac river, *Bowman ;* Peak Downs, *F. Mueller.*
  N. S. Wales. Hunter's River and Liverpool Plains, *A.Cunningham ;* New England,
C. Stuart ; between the Lachlan and Darling, *Neilson.*

P. mollis, R. Br. Prod. 466; A. DC. Prod. viii. 401 ; *Echites mollis,* Spreng. Syst. i.
634, from Keppel Bay, is a softly tomentose-pubescent form of the same species. Similarly
or even more tomentose specimens are also included among those above quoted, but are evidently much more rare than the glabrous ones. F. Mueller gathered a very small-flowered
glabrous form on the Brisbane river.

2. **P. velutina,** *R. Br. Prod.* 466. A tall woody climber, softly pubescent or villous, the hairs usually rusty or rarely nearly glabrous. Leaves on
rather long petioles, from broadly ovate to ovate-oblong, truncate or cordate
at the base, the primary veins distant and prominent underneath, 2 to 5 in.
long. Cymes rather small and dense, solitary or few on opposite axillary
peduncles. Calyx-segments lanceolate, nearly 1 line long, the tips usually
spreading. Corolla scarcely 2 lines long, the tube shorter than the calyx,
the lobes broad, rather longer than the tube, bearded inside below the middle.
Hypogynous scales truncate, usually convolute. Stamens inserted near the
base of the corolla ; filaments flattened, not twisted, hirsute under the
anthers ; anthers wholly exserted, the basal lobes rather short. Ovary
crowned by 4 small glands. Fruit hard, 3 to 6 in. long, tomentose or glabrous, tardily separating more or less completely into 2 follicles.—A. DC.
Prod. viii. 401 ; *Echites velutina,* Spreng. Syst. i. 634.

N. Australia. Victoria river, *F. Mueller* (in fruit only).
Queensland. Endeavour river, *Banks and Solander;* Keppel Bay, *R. Brown;* Port Denison and Edgecombe Bay, *Dallachy;* Rockhampton, *O'Shanesy;* Broad Sound, *Bowman.*
N. S. Wales. Hastings and Clarence rivers, *Beckler.*

Var.? *glabrescens.* Foliage at length nearly glabrous.—Victoria river, *F. Mueller;* South Goulburn Island, *A. Cunningham;* Cape York, *Daemel.* The specimens of this form are none of them perfect, and require further comparison with the New Caledonian *Lyonsia scabra,* A. DC.

3. **P. Leichhardtii,** *F. Muell. Fragm.* vi. 128. A rather slender twiner, glabrous or the branches and inflorescence minutely pubescent. Leaves petiolate, ovate, acuminate, rounded or cordate at the base, 2 to 4 in. long. Flowers in rather small loose dichotomous cymes, terminal or on short axillary branchlets, the pedicels short. Calyx-segments about 1 line long, lanceolate, with spreading tips. Corolla-tube rather broad, about 1 line long, the lobes acute, as long as the tube, densely bearded inside at the base, the edges slightly overlapping in the bud. Filaments very short, inserted in the base of the corolla-tube; anthers united in a cone, of which the point only protrudes from the tube. Hypogynous scales broad, emarginate. Fruit unknown.—*Lyonsia Leichhardtii,* F. Muell. l. c.

Queensland. Wide Bay, *Leichhardt.* The specimens are few and small. The æstivation of the corolla-lobes is described by F. Mueller as valvate, but their edges certainly overlapped in the buds I examined.

4. **P. ventricosa,** *F. Muell. in Trans. Phil. Inst. Vict.* ii. 71, *Fragm.* vi. 130. A glabrous twiner with slender branches. Leaves petiolate, ovate-lanceolate, acuminate, with the point often long and acute, truncate or cordate at the base, almost membranous, penniveined, 1½ to 2½ in. long. Cymes small, almost umbellate, on slender opposite axillary peduncles much shorter than the leaves. Calyx-segments rather broad, acute, about 1 line long, very spreading, without basal glands. Corolla-tube inflated, nearly globular, contracted at the throat, nearly 2 lines diameter; lobes acute, about as long as the tube, bearded inside at the throat, the edges narrowly overlapping in the bud. Stamens inserted in the middle of the tube; filaments short, not twisted; anthers exserted, connivent in a cone, the basal lobes short broad and spreading, each anther with a longitudinal prominent rather thick wing-like appendage on the back, giving the staminal body the appearance of the winged gynostegia of some *Asclepiadeæ.* Hypogynous scales united in an undulate plicate cup as long as the ovary. Ovary 2-celled. Fruit unknown.

Queensland. Brisbane river, Moreton Bay, *W. Hill, F. Mueller.*
N. S. Wales. Clarence river, *Beckler;* New England, *C. Stuart.*

The habit, inflated corolla-tube, and peculiar anther-appendages distinguish this from all others, and have induced F. Mueller to suggest that it might form a distinct genus under the name of *Gastranthus ventricosus.* A. Gray, Proc. Amer. Acad. v. 333, refers it to *Lyonsia,* under the belief that the corolla-lobes are valvate, which, however, I do not find t be the case in the bud.

12. **LYONSIA,** R. Br.

Calyx with an irregular ring of minute glands inside at the base or with few or none. Corolla-tube cylindrical, often very short; lobes linear or lanceolate, valvate in the bud, bearded inside at the base or along the surface. Stamens inserted at or below the middle of the tube, the filaments short, not twisted or very rarely slightly so ; anthers oblong or linear, exserted, cohering in a cone or ring round the stigma, each with 2 rigid basal lobes usually devoid of pollen. Hypogynous scales 5, as long as the ovary, free or more or less connate. Ovary 2-celled, with numerous ovules in each cell; stigma surrounded by a ring or membranous expansion at the base, usually 2-lobed. Fruit elongated, nearly terete, separating more or less completely into 2 follicles, leaving the placentas free, either 1 in each follicle or the two united into one between the open follicles. Seeds with a coma or tuft of long silky hairs at the hilum.—Tall woody or slender climbers. Leaves opposite. Flowers in terminal or axillary corymbose cymes. Bracts small.

The genus is limited to Australia, unless the New Caledonian *Echites scabra*, Labill., prove to be a true *Lyonsia*, as conjectured by De Candolle. It is, as observed by A. Brown, closely allied to *Parsonsia*, and usually distinguished by the fruit described as capsular and 2-valved, leaving a free central placenta in *Lyonsia straminea*, and separating into 2 follicles in *Parsonsia;* but the real difference is not nearly so much marked as these words would imply. The carpels in all the species cohere until their maturity and the dehiscence is the same in all, leaving the placentas free or adhering to the valves at the end only; but in *L. straminea* the 2 placentas remain closely connate back to back, and fall off in one piece between the valves, whilst in the other species they separate, one remaining in each follicle till it falls with the seeds. A more marked character is that of the valvate corolla-lobes of *Lyonsia*, more especially pointed out by A. Gray, and almost, if not quite exceptional in Apocyneæ. But even this character is deceptive, and must be observed in the bud, for the overlapping in some species of *Parsonsia* is so slight that, from the appearance of the expanded flower, the æstivation has been described as valvate. F. Mueller now therefore proposes to reunite *Lyonsia* with *Parsonsia*.

Corolla-lobes shorter than or scarcely longer than the tube. Slender nearly glabrous plants. Cymes loose, few-flowered, mostly terminal.
　Corolla-lobes flat . . . . 1. *L. lilacina.*
　Corolla-lobes very concave, hood-shaped . . . . 2. *L. induplicata.*
Corolla-lobes lanceolate, not above twice as long as the tube.
　Glabrous or minutely pubescent. Cymes mostly in terminal panicles.
　　Leaves lanceolate elliptical or almost ovate. Eastern species.
　　　Leaves smooth above. Corolla-lobes bearded at the base only. Follicles thin, 2 to 3 in. long, placentas connate . 3. *L. straminea.*
　　　Leaves reticulate above. Corolla-lobes bearded above the middle. Follicles hard, 6 to 8 in. long, placentas usually separate . . . . 4. *L. reticulata.*
　　Leaves long, lanceolate or linear. Western species . . . . 5. *L. diaphanophlebia.*
　Rusty-pubescent or villous. Cymes mostly axillary and opposite.
　　Cymes divaricately branched, the flowers not crowded. Calyx-segments broad and short . . . . 6. *L. Langiana.*
　　Flowers crowded in small clusters on the loosely-branched cymes. Calyx-segments narrow-acute . . . . 7. *L. largiflorens.*
Corolla-lobes linear, 4 or 5 times as long as the tube. Cymes corymbose with numerous flowers.
　Inflorescence all terminal. Leaves ovate, membranous . . . 8. *L. latifolia.*
　Inflorescence axillary and terminal. Leaves coriaceous, on long petioles, broadly oblong, obtuse . . . . 9. *L. oblongifolia.*
　Inflorescence in one axil. Leaves long-lanceolate, acuminate, on short petioles . . . . 10. *L. eucalyptifolia.*

1. **L. lilacina,** *F. Muell.* A slender twiner, the branches almost filiform, pubescent as well as the inflorescence and often the nerves of the leaves
underneath, the foliage otherwise glabrous. Leaves on very short petioles,
oval-oblong or lanceolate, usually mucronate, cordate at the base membranous, the margins often undulate or crisped, 1 to 1½ in. long. Flowers few,
in loose terminal almost filiform cymes. Pedicels mostly longer than the
calyx. Calyx-segments lanceolate, acute, about ¾ line long. Corolla glabrous inside, the tube above 1 line long, the lobes acute, as long as the tube,
valvate in the bud. Anthers united in a half-exserted cone. Hypogynous
scales lanceolate, as long as the ovary. Fruit separating into membranous
follicles of 3 or 4 in., with a placenta in each.—*Parsonsia lilacina,* F. Muell.
Fragm. vi. 127.

Queensland. Rockhampton, *Dallachy* ; Scrubby Creek, *Leichhardt.*
N. S. Wales. Paramatta, *A. Cunningham ;* Clarence river, *Beckler ;* Richmond
river, *C. Moore.*

2. **L. induplicata,** *F. Muell.* A rather slender twiner, glabrous or
the inflorescence minutely pubescent. Leaves petiolate, from ovate to oblong-
elliptical or lanceolate, acuminate or acute, quite entire, penniveined, narrowed at the base, 1½ to 3 in. long. Flowers in loose terminal cymes, sometimes axillary from the development of the side branches. Pedicels varying from very short to ½ in. long. Calyx-segments narrow, fully ½ line long,
with spreading tips. Corolla-tube rather broad, under 1 line long, bearded
inside towards the throat ; lobes scarcely so long as the tube, very concave,
almost hood-shaped, mucronate, induplicate-valvate in the bud. Hypogynous
scales narrow-oblong. Carpels of the ovary united but separable. Fruit unknown.—*Parsonsia induplicata,* F. Muell. Fragm. vi. 129.

N. S. Wales. Hastings river, *Beckler ;* Lansdowne river, *C. Moore.*

3. **L. straminea,** *R. Br. Prod.* 466. A woody climber, scaling trees
to a great height, the pendulous branches rooting when they reach the ground.
Leaves shortly petiolate, ovate-lanceolate or lanceolate, often acuminate, glabrous, smooth and shining above, paler underneath, the primary veins oblique
and usually prominent underneath, 2 to 3 or sometimes 4 in. long, in loose pedunculate trichotomous cymes or panicles, terminal on the young shoots or
axillary on the older ones, minutely pubescent. Calyx-segments narrow, acute,
about ¾ line long. Corolla-tube about as long as the calyx ; lobes lanceolate,
acute, about twice as long as the tube, bearded inside at the base with reflexed
hairs. Stamens inserted at the base of the corolla-tube ; anthers linear, connivent in an exserted cone or cylinder, the basal lobes long and linear.
Hypogynous scales large, obtuse. Ovary pubescent. Fruit 2 to 3 in. long,
tardily dividing into rather thin follicles, the 2 placentas closely connate by
the backs and separating from the follicles in a single plate.—A. DC. Prod.
viii. 401 (partly) ; Hook. f. Fl. Tasm. i. 270 ; F. Muell. Pl. Vict. t. 58.

N. S. Wales, *Backhouse ;* Blue Mountains, *Miss Atkinson.*
Victoria. Mount Disappointment and Mount Juliette, *F. Mueller.*
Tasmania. Port Dalrymple, *R. Brown ;* northern parts of the island, in deep shady
ravines, etc., *J. D. Hooker.*

4. **L. reticulata,** *F. Muell. Rep. Burdek. Exped.* 16. An immense

woody climber. Leaves petiolate, oblong-elliptical or broadly lanceolate, obtuse acute or acuminate, shining above and pale underneath as in *L. straminea*, but much more prominently veined and reticulate, and usually 3 to 4 in. long. Flowers in terminal corymbose pedunculate cymes or panicles, minutely pubescent as in *L. straminea*. Calyx-segments nearly 1 line long. Corolla about 3 lines long, the lobes scarcely longer than the tube, bearded inside almost to the end. Stamens inserted higher up the tube than in *L. straminea;* anthers united in an exserted cone or cylinder. Hypogynous scales united in a 5-lobed cup. Fruit hard, 6 to 8 in. long, tardily separating into 2 follicles, each one (in some fruits at least) carrying off its own placenta.

Queensland. Keppel Bay, *R. Brown;* Moreton Bay and Port Denison, *Fitzalan;* Rockhampton, *Dallachy.*

N. S. Wales. Hunter's and Williams rivers, *R. Brown;* Hastings river, *Fraser* and others ; Clarence river, *Beckler* and others ; Paramatta, *Woolls;* Illawarra, *A. Cunningham.*

This species was confounded with *L. straminea* by R. Brown and others, but well distinguished by F. Mueller. A. De Candolle's description of *L. straminea* in the Prodromus is taken from A. Cunningham's specimens of *L. reticulata.*

5. **L. diaphanophlebia,** *F. Muell.* A tall woody climber, glabrous or minutely pubescent, very closely allied to *L. straminea*. Leaves on short petioles, narrow-lanceolate or almost linear, acute and acuminate, glabrous and smooth above, paler and sometimes pubescent underneath, the primary veins oblique, usually prominent, 4 to 8 in. long. Cymes terminal or on short axillary branchlets, usually pubescent. Calyx-segments lanceolate, about 1 line long. Corolla pubescent outside, the tube about as long as the calyx, the lobes not twice as long, lanceolate, bearded inside to above the middle. Filaments very short ; anthers mucronate-acute, the basal lobes short. Hypogynous scales broad and very obtuse. Ovary pubescent. Fruit unknown.—*Parsonsia diaphanophlebia,* F. Muell. Fragm. ii. 158.

W. Australia, *Drummond,* (*3rd Coll. ?*) *n.* 225, *Clarke;* Murray river, *Oldfield.*

6. **L. Langiana,** *F. Muell. Fragm.* vi. 128. A very tall woody climber, the branches, under side of the leaves, and inflorescence rusty-pubescent. Leaves shortly petiolate, ovate or ovate-lanceolate, acuminate, rounded or somewhat cordate at the base, shining and sprinkled with a few hairs on the upper side, penniveined with transverse reticulations usually conspicuous on the upper surface, mostly 3 to 4 in. long. Flowers small, " yellow," pubescent, in very loose divaricately-branched cymes, on axillary peduncles often as long as the leaves. Pedicels rather thick, 1 to 2 lines long. Calyx-segments broad, about ½ line long. Corolla-tube broad, about 1 line long ; lobes about as long as the tube, with a ring of reflexed hairs inside at the base. Anthers nearly as long as the lobes, united in a cone, with short spreading basal lobes. Hypogynous scales truncate. Ovary glabrous. Fruit unknown. —*Parsonsia Langiana,* F. Muell. Fragm. vi. 128.

Queensland. Rockingham Bay, *Dallachy.*

7. **L. largiflorens,** *F. Muell. Herb.* A tall woody climber, the young shoots and inflorescence minutely rufous-pubescent, the adult leaves nearly

glabrous. Leaves petiolate, broadly ovate, acuminate, truncate or cordate at the base, thinly coriaceous, 3 to 5 in. long. Flowers in compact cymes on the branches of loose cymes or umbels, on axillary peduncles, the whole inflorescence sometimes longer than the leaves. Pedicels mostly longer than the calyx. Calyx-segments pubescent, narrow, acuminate, about 1 line long. Corolla pubescent, about 2½ lines long, the lobes lanceolate, not twice as long as the tube, bearded inside at the base. Anthers wholly exserted. Hypogynous scales slightly united at the base. Ovary glabrous. Fruit unknown.

Queensland. Rockingham Bay, *Dallachy.*
N. S. Wales. Tweed river, *C. Moore.*

8. **L. latifolia,** *Benth.* A tall climber, glabrous except the minutely pubescent inflorescence. Leaves on long petioles, broadly ovate, acuminate, truncate or cordate at the base, membranous, 2½ to 3½ in. long. Flowers numerous, in dense terminal pedunculate cymes, often several inches broad. Calyx-segments short. Corolla-tube exceedingly short; lobes narrow-linear, densely bearded at the base, about 2 lines long. Stamens, hypogynous scales and ovary of *L. eucalyptifolia.* Fruit unknown.

Queensland. Wide Bay, *Bidwill;* Rockingham Bay, *Dallachy.* The flowers are those of *L. eucalyptifolia* or rather smaller, but the inflorescence appears to be constantly terminal, and the leaves very different in shape and consistence.

9. **L. oblongifolia,** *Benth.* This is considered by F. Mueller as a variety of *L. eucalyptifolia,* of which it has the numerous flowers in broad cymes, with the same very short corolla-tube and long linear lobes, but the inflorescences are generally terminal as well as axillary, and the leaves are very differently shaped, being broadly oblong, very obtuse, 3 to 4 in. long and 1 to 1½ in. broad, on long petioles, and much thicker and more coriaceous than in *L. latifolia.* Fruits (not yet ripe) hard, rather thick, and 2 to 3 in. long, probably twice that size when full grown.

Queensland. Rockhampton, *O'Shanesy;* Scrubby Creek, *Bowman.*

10. **L. eucalyptifolia,** *F. Muell.* A tall woody climber, minutely pubescent or glabrous. Leaves petiolate, lanceolate or linear-lanceolate, acuminate, often rather thick, the veins scarcely prominent or impressed above, the under surface pale, with rather distant prominent primary veins, 4 to 8 in. long or even more. Cymes axillary, but only in one axil of the pair of leaves, often several inches broad but shorter than the leaves, the flowers numerous. Calyx-segments lanceolate, acute, above 1 line long, but often cohering or united to more than half their length, the tips spreading. Corolla-tube exceedingly short; lobes linear, 2½ to 3 lines long, shortly bearded inside at the base, revolute when open. Filaments slender, pubescent and slightly twisted under the anthers; anthers mucronate-acute, forming a cylinder nearly as long as the corolla, the basal lobes very short. Hypogynous scales narrow. Ovary glabrous. Fruit unknown.—*Parsonsia eucalyptifolia,* F. Muell. Fragm. ii. 159.

Queensland. Brigalow scrub on the Suttor and between the Dawson and Mackenzie rivers, *F. Mueller;* Nerkool Creek and Flinders river, *Bowman;* Warwick, *Beckler.*

N. S. Wales. Peel's Range, *Fraser;* St. George's Range, *A. Cunningham;* between the Darling and Lachlan rivers, *Burkitt;* Mutanie and Barrier Ranges, *Victorian Expedition;* Castlereagh river, *C. Moore.*

Order LXXVI. ASCLEPIADEÆ.

Flowers regular. Calyx free, divided nearly or quite to the base into 5 segments or sepals, imbricate in the bud, bearing frequently 5 or more small glands at the base inside. Corolla regular, with 5 teeth or lobes, contorted or valvate in the bud, with or without scales or appendages in the throat alternating with the lobes. Stamens 5, inserted at the base or near the base of the corolla, the filaments short, connate or rarely free, the anthers always connate in a tube (called *gynostegium*) enclosing the style; anthers 2-celled, or by the subdivision of the cells more or less completely 4-celled, the cells opening inwards, the connectivum produced into a short, truncate or rarely acute appendage, or more frequently terminating in an inflexed membrane; corona consisting of variously shaped glandular membranous or fleshy appendages attached to the back of the filaments or anthers, sometimes united in a cup or ring, quite deficient in a few genera; pollen consolidated into 1 or 2 masses in each cell of the ovary, attached (when the anther opens) in pairs or in fours (1 or 2 from each of the adjoining anthers) to small processes of the stigma placed between the anthers, and ultimately detached from the stigma and carrying off the pollen-masses. Ovary of 2 distinct carpels, with several usually numerous ovules attached to the inner angle; styles united immediately above the ovary, and thickened within the anthers into an angular body, usually called the stigma, although not wholly stigmatic; the summit in the centre either truncate or more or less protruding in a conical or elongated, beak-like, entire or 2-lobed process. Fruit of 2 follicles or frequently 1 only from the abortion of the other carpel. Seeds usually pendulous, with a long silky tuft of hairs or coma at the hilum, compressed, often bordered; testa usually brown, smooth or rough; albumen thin; embryo straight; cotyledons foliaceous; radicle short, superior.—Herbs, with a perennial, sometimes tuberous rootstock, or more or less woody stock, or shrubs or very rarely trees. Stems or branches frequently twining; juice usually milky. Leaves almost always opposite, entire; stipules none or very obscure. Flowers often small, in racemes or cymes often reduced to umbels, axillary or more frequently on one side of the branch between the petioles. Bracts small, at the base of the branches and pedicels; bracteoles on the pedicels none or very rare and small.

Like *Apocyneæ*, the Order is abundantly dispersed over the tropical regions of both the New and the Old World, and represented by a few extratropical species in the southern as well as the northern hemisphere, but does not extend to arctic or high alpine regions. Of fourteen Australian genera, two extend nearly over the whole range of the Order; another is in S. America as well as in Asia and Africa; five more are in the warmer regions of Asia and Africa; two are limited to Eastern Asia; one is only represented by a slightly different section in extratropical or hilly regions of the Old World; and the remaining three, comprising only four species, are endemic. The Order is nearly allied to *Apocyneæ*, but, with a somewhat different habit, it is neatly distinguished by the definite pollen-masses, and their peculiar adherence to bodies detached from the style. In determining the species of this Order, it is absolutely necessary that the number and position (pendulous horizontal or erect) of these pollen-masses should be carefully studied, and secondly that the configuration of the corona be attended to, for whilst there is great general resemblance in the majority of species belonging to very different genera, the genera themselves are better defined than might have been expected from characters apparently so artificial.

TRIBE I. **Periploceæ.**—*Pollen-masses* (1, 2 or 4 to each anther) granular, evidently *consisting of* 4 or more agglomerated granules. Filaments more or less free (connate in all the other tribes).

Corolla-tube cylindrical ; limb spreading. Corona of 5 scales, inserted in the tube behind the filaments . . . . 1. GYMNANTHERA.

TRIBE II. **Secamoneæ.**—*Pollen-masses smooth,* 4 to each anther.

Corolla rotate ; lobes contorted-imbricate. Corona of 5 laterally-compressed segments, vertically adnate to the gynostegium . . 2. SECAMONE.

TRIBE III. **Euasclepiadeæ.**—*Pollen-masses smooth,* 2 to each anther, pendulous. *Corolla-lobes imbricate in the bud, usually contorted (very slightly so in* Gomphocarpus and Asclepias).

Corona of 5 distinct hood-shaped segments attached to the corolla-tube, which is broad but distinct . . . . *ARAUJA.
Corona of 5 distinct saccate or inflated segments at the back of the anthers, with an exterior corona below them, consisting of 5 protuberances or a slightly prominent ring. Corolla nearly rotate.
  Outer corona very prominent. Plant with succulent leafless branches (in the Australian species) . . . . 3. SARCOSTEMMA.
  Outer corona scarcely prominent. Leaves (in the Australian species) linear . . . . 4. PENTATROPIS.
Corona of 5 distinct hood-shaped or concave segments, without any external ring or protuberances. Stems erect. Corolla rotate.
  Corona-segments without internal appendages . . . . . *GOMPHOCARPUS.
  Corona-segments each with an internal appendage . . . . *ASCLEPIAS.
Corona membranous, the segments either distinct but not saccate or united in a lobed cup round the gynostegium. Corolla nearly rotate.
  Corona of 5 segments, quite distinct in the Australian species . 5. VINCETOXICUM.
  Corona with 5 or 10 marginal lobes and 5 or 10 additional protuberances or lobes inside the margin . . . . 6. CYNANCHUM.

TRIBE IV. **Marsdenieæ.**—*Pollen-masses smooth,* 2 to each anther, horizontal or erect. Corolla-lobes contorted-imbricate (very slightly so in some Tylophoræ).

Pollen-masses small, globular, laterally attached. Corolla rotate. Corona of 5 segments, more or less fleshy ; adnate to the gynostegium and sometimes forming a ring round it . . . . . 7. TYLOPHORA.
Pollen-masses oblong, erect. Corolla urceolate or rotate. Corona of 5 segments usually fleshy or saccate and adnate at the base with a short erect free or adnate point . . . . 8. MARSDENIA.
Pollen-masses erect. Corolla (small) with a short broad tube and spreading lobes. Corona none or of 5 scarcely prominent glands at the base of the gynostegium . . . . 9. GYMNEMA.

TRIBE V. **Stapelieæ.**—*Pollen-masses smooth,* 2 to each anther, erect. Corolla-lobes valvate (often induplicate).

Corolla rotate. Corona a scarcely conspicuous undulate ring. (Leaves linear or none) . . . . 10. MICROSTEMMA.
Corolla urceolate. Corona of 5 linear 2-cleft segments. (Leaves small orbicular fleshy) . . . . 11. DISCHIDIA.

Corolla rotate. Corona of 5 very prominent segments.
Corona-segments expanded into horizontally spreading laminæ
or disks. (Leaves fleshy) . . . . 12. Hoya.
Corona-segments erect, membranous, peltately attached to the
back of the anthers. (Leaves membranous) . . . . . . 13. Thozetia.
Corolla-tube elongated ; lobes connivent at the tips. Corona membranous, cup-shaped, with 5, 10, or 15 lobes. (Leaves cordate.) 14. Ceropegia.

The introduced plants belonging to the genera marked above with the asterisk*, are the following .—

Arauja albens, G. Don ; Dcne. in DC. Prod. viii. 534 (*Physianthus albens*, Mart. Nov. Gen. et Sp. t. 32, Bot. Mag. t. 3201, Bot. Reg. t. 1759). A twiner, with ovate or lanceo-late leaves, truncate or cordate at the base, white underneath as well as the young shoots. Flowers large for the Order. Calyx-segments broad, leafy. Corolla with a broad tube and a campanulate 5-lobed limb. Corona of 5 fleshy segments attached to the corolla-tube at the base, the upper portion hood-shaped or convex with revolute margins. Pollen-masses pendulous.—A native of S. Brazil, said to have spread from gardens into the neighbourhood of Moreton Bay.

Gomphocarpus fruticosus, Br. ; Dcne. in DC. Prod. viii. 557 ; Bot. Mag. t. 1628. A tall erect plant more or less woody at the base, the young parts hoary-pubescent. Leaves linear or linear-lanceolate, several inches long. Flowers white, in pedunculate umbels. Co-rolla rotate. Corona attached to the gynostegium, of 5 laterally compressed broad erect segments as long as the anthers, truncate at the top. Follicles inflated, membranous, covered with long soft prickles.—A native of Africa, introduced with the early colonists, and long since established near Port Jackson and other parts of N. S. Wales, *R. Brown* and others.

Asclepias curassavica, Linn. ; Dcne. in DC. Prod. viii. 566 ; Bot. Reg. t. 81. An erect perennial of 2 or 3 ft., usually glabrous. Leaves lanceolate, rather long. Umbels peduncu-late, many-flowered. Corolla rotate, orange-red with a yellow gynostegium. Corona attached to the gynostegium, of 5 ovate erect hood-shaped segments, each with an inner curved horn longer than the outer segment. Follicles glabrous and smooth.—A native of the West Indies, now spread in great abundance over most inhabited tropical regions, and already common in several parts of Queensland.

TRIBE 1. PERIPLOCEÆ.—Pollen-masses (1, 2 or 4 to each anther) gra-nular, evidently consisting of 4 or more agglomerated granules. Filaments more or less free.

1. GYMNANTHERA, R. Br.

Corolla with a cylindrical tube and spreading lobes, contorted in the bud, the right-hand edges overlapping. Filaments distinct, inserted in the throat. Corona of 5 scales, inserted in the corolla-throat behind the filaments. Pollen-masses 4 to each anther, granular; attached in fours (2 from each ad-joining anther) to filiform appendages of the style.—Twiner with milky juice. Leaves herbaceous. Cymes loose, on interpetiolar or almost axillary peduncles.

The genus is limited to a single species, endemic in Australia. It is the only Australian *Asclepiadea* with granular pollen-masses.

1. **G. nitida,** *R. Br. Prod.* 464. A tall glabrous twiner, woody at the base. Leaves opposite, on rather long petioles, from ovate to oblong-ellipti-cal, obtuse acute or mucronate, 2 to 3 or rarely 4 in: long, of a rather firm consistence. Flowers of a greenish-white, the cymes shortly pedunculate, with rather elongated branches and numerous small bracts. Pedicels 3 to 4 lines long. Calyx-segments ovate, ¾ line long, with an irregular broken ring

of small glands inside at the base, often united into small lobed scales alternating with the segments.   Corolla-tube about 3 lines long, the lobes nearly as long, broadly ovate.   Corona-scales broad, mucronate or denticulate. Filaments broad and flat, connivent but not connate.   Anthers acuminate, glabrous, the cells subdivided by vertical partitions between the pollen-masses. Stigma very shortly 2-lobed.   Follicles divaricate, rather slender, 2 to 3 in. long.—Dcne. in DC. Prod. viii. 493.

N. Australia.   Islands of the Gulf of Carpentaria, *R. Brown ;* Sandy Island, Victoria river, *F. Mueller ;* Sweers Island, *Henne ;* Forster's Range and Daly waters, *Waterhouse.*
Queensland.   Port Curtis, *M'Gillivray ;* Nerkool Creek, Sattor river, and Mount Wyatt, *Bowman.*

TRIBE II. SECAMONEÆ.—Pollen-masses smooth, 4 to each anther.

2. SECAMONE, R. Br.

Corolla rotate, deeply 5-lobed, the lobes contorted in the bud, the right-hand edges overlapping.   Corona of 5 laterally compressed segments, vertically adnate to the gynostegium at the base, with free falcate or ligulate points, usually shorter than the gynostegium.   Filaments connate.   Pollen-masses 4 to each anther, nearly smooth, erect, attached in fours, 2 from each adjoining anther, to small appendages of the styles (as in the genera with 2 masses to each anther).   Stigma very short and obtuse.—Stems from a woody base, straggling trailing or twining.   Leaves herbaceous, often pellucid-dotted.   Flowers very small in loose axillary or interpetiolar solitary cymes.

The genus extends over tropical Asia, southern Africa, and the Mascarene Islands, the Australian species both endemic, unless one proves really to be the same as a Philippine Island one.

Leaves ovate-acuminate or lanceolate . . . . 1. *S. elliptica.*
Leaves ovate, obtuse or scarcely acute . . . . 2. *S. ovata.*

1. **S. elliptica,** *R. Br. Prod.* 464.   A rather slender twiner or low straggling shrub, quite glabrous or the inflorescence slightly pubescent. Leaves petiolate, ovate to lanceolate, acuminate, rounded or acute at the base, the veins scarcely conspicuous, 1 to 1½ or rarely 2 in. long.   Cymes shortly pedunculate, several-flowered, but shorter than the leaves, the peduncles and pedicels glabrous or scarcely pubescent.   Calyx-segments membranous, orbicular, ciliolate, about ½ line long.   Corolla " yellow " with a very short tube, nearly rotate, deeply lobed, spreading to about 2 lines diameter, with small introrse auricles or appendages at the base of the lobes.   Corona of 5 small laterally compressed segments, adnate to the gynostegium at the base, with free incurved falcate points considerably shorter than the gynostegium. Follicles divaricate, 2 to 2½ in. long, tapering into a long point.—Dcne. in DC. Prod. viii. 504 ; *S. emetica,* F. Muell. Fragm. v. 161, not of R. Br.

N. Australia.   Islands of the Gulf of Carpentaria, *R. Brown ;* Port Essington, *Armstrong.*
Queensland.   Port Curtis, *M'Gillivray ;* Port Denison, *Fitzalan ;* Fitzroy river, *Bowman ;* Rockingham Bay, *Dallachy ;* Rockhampton, *O'Shanesy ;* Moreton Bay, *F. Mueller, C. Stuart ;* Peak Downs, *F. Mueller ;* Mackenzie river, *Leichhardt.*
N. S. Wales.   Clarence river, *C. Moore.*

The species extends perhaps to the Philippine Islands, for the *S. attenuata*, Dcne., does not appear to me to differ from it. The common Indian *S. emetica* is, however, much more distinct, in its purple flowers, less imbricate corolla-lobes, shorter corona and longer style.

2 ?. **S. ovata,** *R. Br. Prod.* 464. Of this I have only seen a single specimen in R. Brown's herbarium, of which the flowers are gone. It appears to be distinct in its broader more ovate and obtuse or scarcely acute leaves, which are shortly petiolate, 1 to 1½ in. long.—Dcne. in DC. Prod. viii. 504.

N. Australia. Islands of the Gulf of Carpentaria, *R. Brown.* The species must remain very doubtful until it shall be again observed.

TRIBE III. EUASCLEPIADEÆ.—Pollen-masses smooth, 2 to each anther, pendulous. Corolla-lobes imbricate, usually contorted (very slightly so in *Gomphocarpus* and *Asclepias*).

3. SARCOSTEMMA, R. Br.

Corolla rotate or nearly so, the lobes contorted in the bud, the right-hand edge overlapping. Corona double, the outer one annular or cup-shaped, usually at the base of the gynostegium, the inner one of 5 segments at the back of the anthers, fleshy or saccate at the base, with a free erect point. Anthers terminating in a membrane. Pollen-masses 2 to each anther, oblong or clavate, pendulous. Stigma short, obscurely notched.—Stems in the Australian typical section leafless and somewhat fleshy. Flowers in sessile lateral umbels.

The genus in its typical form is spread over tropical Asia and southern and eastern Africa, the Australian species is endemic, but nearly allied to a widely spread Asiatic one. The other sections, in which the stems are leafy, are limited to South America.

1. **S. australe,** *R. Br. Prod* 463. A glabrous leafless somewhat fleshy twiner, woody at the base, the branches terete, often articulate at the nodes, the leaves replaced by minute opposite scales. Umbels sessile on one side of the nodes between the scales. Pedicels about ¼ in. long. Calyx-segments ovate, obtuse, scarcely ½ line long. Corolla deeply divided into ovate obtuse lobes of about 2 lines. Outer corona adnate to the base of the gynostegium and about half its length, much undulate and sinuate but not lobed; segments of the inner corona saccate, nearly as long as the anthers. Follicles rather narrow, 2 to 3 in. long.—Endl. Iconogr. t. 64.

N. Australia. Intercourse Island, Dampier's Archipelago, *A. Cunningham.*
Queensland. Albany Island, *W. Hill;* E. coast, *R. Brown;* Curtis Island, *Henne;* Howick's group, *F. Mueller;* Port Denison, *Fitzalan;* dry ridges near Rockhampton, *Dallachy, O'Shanesy.*
N. S. Wales. New England, *C. Stuart;* Hastings river, *Beckler;* Mount Goningbery, *Victorian Expedition.*
S. Australia. Bird Island, *R. Brown;* Lake Torrens, *F. Mueller;* Lake Gillies, *Burkitt;* Mount Searl, *Warburton.*
W. Australia. Murchison river, *Oldfield, Drummond, 6th Coll. n.* 144.

4. PENTATROPIS, R. Br.

(Rhyncharrhena, *F. Muell.*)

Corolla nearly rotate, the lobes contorted in the bud (the right-hand edge

Pentatropis.] LXXVI. ASCLEPIADEÆ. 329

usually overlapping). Corona double, the outer one of a slightly prominent ring or 5 scarcely conspicuous obtuse lobes at the base of the gynostegium or near the top of the filaments, the inner (or upper) one of 5 adnate segments more or less saccate or inflated at the base, the erect points adnate and sometimes produced beyond the anthers. Anthers terminating in a small membrane. Pollen-masses 2 to each anther, oblong or clavate, pendulous. Stigma short.—Twiners. Leaves herbaceous, linear in the Australian species. Flowers in simple umbels on solitary interpetiolar peduncles.

The genus is spread over tropical Asia and Africa. The Australian species, all endemic, closely resemble each other, and differ from all the extra-colonial ones in their linear revolute leaves.

Corolla divided to about the middle.
  Outer annular corona at the base of the gynostegium ; inner segments distant from it, saccate and almost spurred at the base . 1. *P. linearis.*
  Outer annular corona close under the inner inflated but scarcely saccate segments . . . . 2. *P. atropurpurea.*
Corolla divided nearly to the base . . . . 3. *P. quinquepartita.*

1. **P. linearis,** *Dene. in DC. Prod.* viii. 536. A slender twiner, minutely hoary-pubescent or glabrous. Leaves shortly petiolate, linear, with revolute margins, 1 to 2 in. long. Umbels shortly pedunculate, the pedicels slender, 4 to 6 lines long. Calyx-segments narrow, spreading, about 1 line long. Corolla very broadly campanulate or almost rotate, the lobes ovate, spreading to nearly 3 lines diameter. Outer corona a very slightly prominent undulate and sinuate ring at the base of the gynostegium ; inner corona of 5 segments laterally compressed, adnate to the back of the anthers, the lower end free and saccate or almost spurred, the upper end produced into a short adnate point. Pollen-masses small, pendulous or almost horizontal.

W. Australia, *Drummond, 1st Coll. n.* 667.

2. **P. atropurpurea,** *Benth.* A slender glabrous twiner. Leaves linear with revolute margins, tapering into a short petiole, 1 to 2 in. long. Flowers dark purple, in simple or irregularly compound umbels on short peduncles, the pedicels about ¼ in. long. Calyx-segments narrow, about 1 line long. Corolla almost rotate, divided to about the middle into broad acute lobes, spreading to about 4 lines diameter. Outer corona a scarcely prominent undulate ring at the top of the very short filaments ; inner corona of 5 segments adnate to the back of the anthers, inflated but scarcely saccate at the lower end, the upper end an adnate point shortly produced beyond the anthers. Follicles 2 to 3 in. long, not acuminate.—*Rhyncharrhena atropurpurea,* F. Muell. Fragm. i. 128.

Queensland. Suttor desert, *F. Mueller.*

3. **P. quinquepartita,** *Benth.* A slender glabrous twiner, closely resembling *P. atropurpurea,* with the same linear revolute leaves. Umbels in our specimens simple and few-flowered. Flowers of *P. atropurpurea,* except that the corolla is divided nearly to the base into acutely acuminate lanceolate lobes of 3 to 3½ lines, slightly bearded inside. Corona of *P. atropurpurea,* but with the points of the segments scarcely protruding beyond the anthers.

Follicles 5 to 6 in. long, acuminate.—*Rhyncharrhena quinquepartita,* F. Muell. Fragm. i. 128.

N. S. Wales. Desert of the Murray and Darling rivers, *Dallachy ;* tributaries of the Upper Darling river, *Bowman.*

5. VINCETOXICUM, Mœnch.

Corolla nearly rotate, deeply divided into 5 lobes contorted in the bud, the right-hand edges usually overlapping.   Corona simple, of 5 membranous erect segments, either distinct from the base or (in species not Australian) united in a lobed cup encircling the gynostegium.   Anthers terminating in a membrane.   Pollen-masses 2 to each anther, pendulous, but sometimes laterally attached below the top.   Stigma short, obtuse, truncate or minutely 2-lobed.—Herbs, with twining trailing or rarely erect stems.   Leaves herbaceous or rarely somewhat fleshy.   Peduncles interpetiolar, bearing 2 or more umbels or an irregular cyme, or rarely a single umbel.

The genus is widely dispersed over the temperate as well as the warmer regions of the globe, but the Australian species are all endemic.   They belong to a section which has been proposed as a genus, under the names of *Blyttia* and *Haplostemma,* distinguished by the corona being divided quite to the base instead of being lobed only, but the separation is neither natural nor yet has any relation to geographical distribution, for the section would include *V. fruticulosum* and *V. sibiricum.*   The whole genus is very closely allied to *Cynoctonum* and *Cynanchum,* being distinguished from the latter by the want of any inner lobes to the corona, and from both by the more deeply lobed or 5-partite corona.

Corona-segments nearly as long as the gynostegium, truncate, obtuse or
 scarcely acuminate.   Leaves ovate or ovate-lanceolate, sometimes cordate, membranous . . . . 1. *V. ovatum.*
Corona-segments abruptly acuminate, the point short, curved over the
 gynostegium.   Corolla-lobes ovate-lanceolate.   Leaves ovate, slightly
 fleshy . . . . 2. *V. elegans.*
Corona-segments tapering into erect points shortly exceeding the gynostegium.   Corolla-lobes narrow, acuminate.   Leaves oblong or linear,
 slightly fleshy . . . . 3. *V. carnosum.*
Corona-segments with slender points more than twice as long as the
 gynostegium.   Leaves cordate-ovate, acuminate, membranous . . . 4. *V. leptolepis.*

1. **V. ovatum,** *Benth.*   A rather slender twiner, glabrous or the inflorescence minutely pubescent.   Leaves on slender petioles, ovate, obtuse, acute or shortly acuminate, rounded or cordate at the base, 1 to 2 or rarely 3 in. long.   Umbels usually 2 at some distance from each other on a slender peduncle, the flowers often exceeding the leaves.   Pedicels 3 to 6 lines long. Calyx-segments about 1 line long, ovate, thin.   Corolla spreading to about 3 lines diameter, deeply divided into narrow ovate lobes.   Corona-segments very broad, almost orbicular, obtuse or very shortly acuminate, rather shorter than the anthers, shortly adnate at the base along the centre.   Pollen-masses small, obovoid.

Queensland.   Brisbane river, *F. Mueller ;* Rockhampton, *O'Shanesy, Thozet ;* Table Mountain, *Bowman ;* Mogill scrub, *C. Stuart.*

2. **V. elegans,** *Benth.*   A tall glabrous twiner.   Leaves on slender petioles, ovate, acute or mucronate, not cordate, rather fleshy (*Woolls*) penni-veined, 1 to 2 in. long, the upper ones sometimes narrow.   Flowers few, in

clusters or umbels on forked or dichotomous peduncles often longer than the leaves. Pedicels ½ in. long or more. Calyx-segments ovate, very small. Corolla pure white (*Woolls*), spreading to 5 or 6 lines diameter, very deeply divided into oblong-lanceolate lobes, slightly pubescent inside. Corona-segments broad, nearly the length of or rather longer than the anthers, abruptly acuminate, the point incurved over the gynostegium. Pollen-masses oblong, pendulous but attached below the top.

N. S. Wales. Cabramatta, *Woolls;* Ash Island, *Miss Scott* (fragmentary specimens, but apparently the same).

3. **V. carnosum,** *Benth.* Quite glabrous, trailing or twining and rather slender. Leaves oblong oblanceolate lanceolate or linear, obtuse mucronate or acute, rather thick and almost veinless, 1 to 2 in. long, narrowed into a short petiole. Flowers "yellowish-green," in simple umbels or very short racemes, on peduncles usually under ½ in. long, the pedicels rather longer. Calyx-segments narrow, acute, about 1½ lines long. Corolla deeply divided into narrow acuminate lobes of 3 to 3½ lines. Corona-segments lanceolate, acuminate, longer than the anthers. Pollen-masses oblong, small. Follicles rather broad, acuminate, 2 to 3 in. long.—*Oxystelma carnosum*, R. Br. Prod. 462; Dcne. in DC. Prod. viii. 543.

N. Australia. Cygnet Bay, N.W. coast, *A. Cunningham;* Victoria river, *F. Mueller;* islands of the Gulf of Carpentaria, *R. Brown.*
Queensland. Port Curtis and Moreton Island, *M'Gillivray;* Rockingham Bay, *Dallachy.*
N. S. Wales. Port Macquarrie, *Backhouse.*

Brown, in referring this with doubt to *Oxystelma,* suggested that it was not a congener with the Indian *O. esculentum,* which is so widely different in aspect. Decaisne had no authentic specimen to examine; and, not recognizing a narrow-leaved one he afterwards saw in the Hookerian herbarium, suggested that it might be a new *Pentatropis:* there is, however, no trace of the outer corona of that genus, and the inner segments are quite different. It appears to me to agree well with *Vincetoxicum,* and certainly to be a congener of the other Australian species I have referred to the latter genus.

4. **V. leptolepis,** *Benth.* A twiner, quite glabrous or the inflorescence minutely pubescent. Leaves on long petioles, broadly cordate-ovate, acutely acuminate, 1½ to 2½ in. long. Flowers small, in irregular loose cymes or peduncles usually exceeding at least the upper leaves. Calyx-segments about 1 line long, oblong, ciliolate. Corolla spreading to rather above 2 lines diameter, divided to below the middle. Corona-segments lanceolate, tapering into a subulate-acuminate ligula, more than twice as long as the anther.

Queensland. Mount Elliott, *Fitzalan;* Elliott river, *Bowman.* The foliage is much like that of several *Cynancha.*

6. CYNANCHUM, Linn.

Corolla nearly rotate, deeply divided into 5 lobes contorted in the bud, the right-hand edges usually overlapping. Corona membranous, forming a loose cup or tube round the anthers, the margin 10-lobed with the addition of 5 or 10 inner lobes (reduced in *C. erubescens* to prominent ribs). Anthers terminating in a membrane. Pollen-masses 2 to each anther, pendulous. Stigma

short, obtuse, truncate or minutely 2-lobed.—Herbs with twining, trailing or
rarely erect stems. Leaves herbaceous, cordate, usually several-nerved at the
base. Flowers in the Australian species in loose irregular cymes rarely contracted into umbels (in other species usually umbellate), on solitary interpetiolar peduncles.

The genus, as limited by Decaisne, is confined to the Old World, extending as far as the
Mediterranean region of Europe. The Australian species are all endemic. Decaisne, who
had no opportunity of examining any specimens, thought, from Brown's short characters,
that they should be rather referred to *Cynoctonum,* but the *C. floribundum* and *C. pedunculatum* have the inner lobes of the corona characteristic of *Cynanchum* very prominent. In
C. erubescens they are considerably reduced, thus almost connecting the two genera, but yet
sufficiently conspicuous to retain the species in *Cynanchum.*

Corona-lobes 10, truncate or crenulate, not exceeding the anthers,
    and 10 inner scarcely prominent keels . . . .   1. *C. erubescens.*
Corona-lobes 20, subulate-acuminate, much exceeding the anthers.
    No stipule-like leaves . . . .   2. *C. floribundum.*
Corona-lobes 20, 10 short obtuse denticulate or partially acuminate,
    and 10 inner rather longer acuminate keels. A pair of stipule-
    like leaves usually at the base of the petioles.
Glabrous or the inflorescence minutely pubescent . . . . .   3. *C. pedunculatum.*
Whole plant softly pubescent . . . .   4. *C. puberulum.*

1. **C. erubescens,** *R. Br. Prod.* 463. A twiner, slightly pubescent,
especially the young shoots and inflorescence, the adult foliage usually glabrous. Leaves petiolate, ovate, deeply cordate with broad rounded auricles,
acute, 1 to 3 in. long. Flowers having sometimes, but not always, a slight
reddish tint, in rather loose cymes, the peduncles not much exceeding the
petioles. Calyx pubescent, the segments ovate, almost acute, scarcely 1 line
long. Corolla pubescent outside, glabrous inside, shortly exceeding the
calyx, very deeply divided into ovate lobes. Corona loosely plicate, not exceeding the anthers, the margins shortly 10-lobed, 5 lobes alternating with
the anthers rather longer than the others, thin, truncate or crenulate, each
lined with 2 slightly prominent inner keels, 5 opposite the anthers thickened
inside, but scarcely or shortly produced into distinct laminæ. Follicles
rather broad, acuminate, with prominent almost winged angles, about ? in.
long.—*Cynoctonum erubescens,* Dcne. in DC. Prod. viii. 529.

Queensland. Endeavour river, *Banks and Solander, R. Brown;* Rockingham Bay,
Dallachy.

2. **C. floribundum,** *R. Br. Prod.* 463. Erect or the branches sometimes slightly flexuose or twining, the inflorescence and sometimes the foliage
hoary-pubescent, the older leaves usually glabrous. Leaves on rather long petioles, cordate, ovate or lanceolate, acuminate, 1 to 2 in. long. Cymes rather
dense, rarely contracted to umbels. Calyx-segments ovate-lanceolate, acute,
ciliate, about 1 line long. Corolla deeply divided into lobes of 2 to 3 lines.
Corona deeply divided into 20 subulate-acuminate lobes exceeding the anthers, of which 10 outer ones parallel to the gynostegium, and 10 inner at
right angles to them forming prominent keels. Pollen-masses oblong. Follicles fusiform, acuminate, more or less winged, 1 to 2 in. long.—*Cynoctonum
floribundum,* Dcne. in DC. Prod. viii. 529.

N. Australia. Nichol Bay, N.W. coast, *Gregory's Expedition.*

Queensland. Wentworth and Gilbert rivers, *F. Mueller;* Suttor river and Broad Sound, *Bowman.*

S. Australia. Head of Spencer's Gulf, *R. Brown;* low hills, Elder's Range, Lake Torrens, etc., *F. Mueller;* Mount Searle, *Warburton;* Cooper's Creek, *Bowman.*

The species is very variable in the breadth of the leaves, the size of the flowers, and the pubescence of the inflorescence, but readily recognized by the prominent subulate points of the corona-lobes.

3. **C. pedunculatum,** *R. Br. Prod.* 463. A twiner, either glabrous or the inflorescence minutely pubescent. Leaves on rather long petioles, deeply cordate, ovate, acuminate, $1\frac{1}{2}$ to $2\frac{1}{2}$ in long, usually with two small accessory stipule-like leaves at their base on the side opposite to the peduncle. Cymes several-flowered, on peduncles often as long as the leaves or even longer. Calyx-segments narrow, acute, about half as long as the corolla. Corolla deeply divided into narrow rather acute lobes of 2 to 3 lines. Corona shortly 20-lobed, 10 outer lobes short obtuse or denticulate or some of them shortly acuminate, the 10 inner keels in pairs one on each side of the lobes opposed to those of the corolla, more or less acuminate and usually longer than the outer ones. Fruit not seen.—*Cynoctonum pedunculatum,* Dcne. in DC. Prod. viii. 529.

N. Australia. Montague Sound, N.W. coast, *A. Cunningham;* Victoria river, *F. Mueller;* islands of the Gulf of Carpentaria, *R. Brown.*

4. **C. puberulum,** *F. Muell. Herb.* Very near *C. pedunculatum,* and perhaps a variety, but it is all over softly hoary-pubescent, the flowers are larger, and several of the outer lobes of the corona as well as the inner keels are produced into long subulate points as in *C. floribundum.* Follicles fusiform, acuminate, hard, about 4 in. long and 1 in. broad below the middle.

N. Australia. Rocks, Upper Victoria river, *F. Mueller.*

TRIBE IV. MARSDENIEÆ.—Pollen-masses smooth, 2 to each anther, horizontal or erect. Corolla-lobes imbricate, contorted.

7. TYLOPHORA, R. Br.

Corolla rotate or nearly so, deeply divided into 5 lobes contorted in the bud, the right-hand edge usually overlapping. Corona of 5 segments. adnate to the back of the anthers, usually prominent and thick or slightly inflated at the base, with short erect adnate or recurved tips. Anthers terminating in a small membrane. Pollen-masses 2 to each anther, small, nearly globular, horizontal or slightly ascending. Stigma short, obtuse or minutely 2-lobed. — Stems erect or more frequently twining. Leaves herbaceous. Flowers in one or more clusters or umbels, along a small or branched interpetiolar peduncle.

The genus is spread over tropical Asia and Africa and southern Africa, but the Australian species appear to be all endemic. It comes very near to those species of *Marsdenia* which have rotate corollas, and is most readily distinguished by the small globular pollen-masses.

Stems erect, pubescent.
Leaves narrow-lanceolate . . . . 1. *T. erecta.*
Leaves ovate-lanceolate . . . . 2. *T. macrophylla.*

Twiners.
Flowers (nearly 1 in. diameter) solitary or very few in a simple
umbel. Corona-segments horizontal . . . . 3. *T. grandiflora.*
Flowers (4 to 5 lines diameter) in 1, 2 or several umbels. Leaves
broad, cordate. Corona-segments vertical.
Calyx hirsute. Corona-segments slightly saccate, not spurred . 4. *T. floribunda.*
Calyx glabrous. Corona-segments spurred at the base . . . 5. *T. calcarata.*
Flowers (about 3 lines diameter) in 1 or 2 umbels. Leaves ovate
to lanceolate, not cordate . . . . 6. *T. barbata.*
Flowers (2 to 3 lines diameter) in several umbels or clusters in a
divaricate almost filiform cyme.
Corolla-lobes oval-oblong, obtuse . . . . 7. *T. Woollsii.*
Corolla-lobes tapering into a long linear point . . . . . . 8. *T. paniculata.*
Flowers (scarcely 1½ lines diameter) in several umbels, sessile along
slender simple peduncles. . . . . 9. *T. flexuosa.*

1. **T. erecta,** *F. Muell. Herb.* Stems erect, simple, 1 to 2 ft. high, the
whole plant softly tomentose-pubescent or the upper surface of the leaves at
length nearly glabrous. Leaves shortly petiolate, lanceolate, rather acute, 3
to 6 in. long. Flowers in few clusters or sessile umbels forming a rather
compact cyme or in a single umbel on interpetiolar peduncles shorter than
the leaves. Pedicels 2 to 6 lines long. Sepals narrow, acute, above 1 line
long. Corolla spreading to about 3 lines diameter. Corona-segments thick
fleshy and very prominent, slightly compressed laterally and tapering upwards
into an adnate point. Fruit not seen.

Queensland. Burdekin river, *F. Mueller;* Sellheim and Bowen rivers, *Bowman.*

2. **T. macrophylla,** *Benth.* Apparently tall and erect, softly tomentose-pubescent. Leaves on short petioles, ovate-lanceolate, acuminate,
rounded or cordate at the base, 4 to 6 in. long. Umbels in our specimens 2
on each peduncle, the peduncles exceeding the leaves. Bracts linear. Pedicels 2 to 4 lines long. Calyx-segments narrow, at least 1½ lines long.
Corolla spreading to nearly 4 lines diameter, dark coloured, pubescent outside, glabrous inside, the lobes rather broad. Corona-segments somewhat
fleshy, very prominent and horizontally spreading at the base, tapering upwards into a short adnate point. Follicles rather narrow, acuminate, fully 3
in. long.

N. Australia. Port Essington, *Armstrong;* Adam's Bay, *Hulls.*

3. **T. grandiflora,** *R. Br. Prod.* 460. A rather slender twiner, the
stems and foliage pubescent. Leaves on slender petioles, ovate to ovate-
lanceolate, shortly and acutely acuminate, more or less cordate, 1 to 2 in.
long. Flowers large for the genus, purple, on filiform pedicels of ½ to 1 in.,
usually only 1 or 2 rarely 3 on a short interpetiolar peduncle. Calyx-segments ovate, about 1 line long. Corolla spreading to about 1 in. diameter,
the lobes ovate-oblong obtuse. Gynostegium small. Corona-segments
thick, rounded, horizontally spreading and united in a ring at the base.
Follicles acuminate, about 3 in. long. Seeds broad.—Dcne. in DC. Prod.
viii. 612 ; *Hoya grandiflora,* Spreng. Syst. i. 843.

N. S. Wales. Hunter's River, *R. Brown,* also *Herb. F. Mueller,* but the collector
not indicated.

4. **T. floribunda,** *Benth.* A slender twiner, slightly pubescent. Leaves petiolate, broadly ovate-cordate, shortly acuminate, 3 to 4 in. long. Flowers of a deep purple, rather large and numerous, in several umbels, forming corymbose cymes on interpetiolar peduncles shorter than the leaves. Calyx-segments narrow, hirsute, about 1 line long. Corolla spreading to about 5 lines diameter; lobes ovate-lanceolate, almost acute. Corona-segments very prominent, almost saccate, slightly compressed laterally, tapering into a short adnate point, forming 5 thick radiating wings to the gynostegium.

N. S. Wales. Mount Warning, Tweed river, *C. Moore* (*Herb. F. Mæell.*, a single specimen).

5. **T. calcarata,** *Benth.* A glabrous twiner. Leaves on rather short petioles, broadly cordate-ovate, shortly acuminate, membranous, about 4 in. long in the specimens seen, with 3 or 4 small glands at the base of the limb on the upper side as in the section *Gongronema* of *Gymnema.* Flowers apparently dark green, in 1 or 2 umbels on a short peduncle, the pedicels ½ in. long or more. Calyx-segments lanceolate, glabrous, scarcely above ½ line long. Corolla spreading to fully 4 lines diameter, the lobes almost acute and only very slightly overlapping in the bud so as to appear almost valvate. Corona-segments very prominent, laterally compressed, forming 5 vertical radiating wings to the gynostegium, the lower outer angle produced into a short spur, the upper one shortly turned up into an obtuse point.

Queensland. Rockhampton, *Thozet.* Evidently very near *T. floribunda,* differing, however, in the colour of the flowers, in the corona-segments, etc.

6. **T. barbata,** *R. Br. Prod.* 460. A slender glabrous twiner. Leaves on slender petioles, ovate or ovate-lanceolate, acute, not cordate, membranous, 1 to 2 in. long. Flowers of a dingy purple, not numerous, in 1 or rarely 2 umbels on interpetiolar peduncles not exceeding the leaves, and filiform as well as the pedicels. Calyx-segments acute, about 1 line long. Corolla spreading to nearly 3 lines diameter, the lobes ovate, obtuse, slightly bearded inside. Gynostegium short. Corona-segments fleshy, forming globular protuberances on the backs of the anthers. Follicles acuminate, 2 to 3 in. long—Dcne. in DC. Prod. viii. 609 ; *Hoya barbata,* Spreng. Syst. i. 843.

N. S. Wales. Port Jackson to the Blue Mountains, *R. Brown, A. and R. Cunningham,* and others; Illawarra, *A. Cunningham, Ralston.*
Victoria. Mallee Cotta Inlet, Lake King, Lake Tyers, Snowy River, Boggy Creek, etc., *F. Mueller.*

7. **T. Woollsii,** *Benth.* A glabrous rather slender twiner. Leaves on rather long petioles, ovate or broadly lanceolate, more or less cordate, mucronate or shortly acuminate, 2 to 3 in. long. Flowers small, in loose irregular paniculate cymes not exceeding the leaves. Calyx-segments about ¾ line long. Corolla (purple?) spreading to about 2½ lines diameter, the lobes oval-oblong, obtuse. Gynostegium very short. Corona-segments broad, slightly prominent, forming a ring round the base of the gynostegium. Follicles slender, 2 to 3 in. long.

N. S. Wales. Paramatta, *Woolls.*

8. **T. paniculata,** *R. Br. Prod.* 460. A tall slender twiner, glabrous or slightly pubescent. Leaves on slender petioles, ovate or ovate-lanceolate, acutely acuminate, 1 to 2 in. long on the flowering branches. Umbels few-flowered, in loose slender cymes exceeding the leaves. Pedicels filiform, 4 to 6 lines long. Calyx-segments lanceolate, thin, nearly ½ line long. Corolla-lobes from a short broad base tapering into long linear almost filiform points, spreading to nearly 5 lines diameter. Corona-segments adnate at the base, broadly ovate, obtuse, the tips prominent and somewhat spreading, shorter than the gynostegium.—Dcne. in DC. Prod. viii. 612; *Hoya paniculata,* Spreng. Syst. i. 843.

N. S. Wales. Hunter's river, *R. Brown;* Clarence river, *Beckler.*

9. **T. flexuosa,** *R. Br. Prod.* 460. A rather slender twiner, glabrous or the inflorescence pubescent. Leaves ovate-lanceolate to linear-lanceolate, acutely acuminate, cordate when broad, the larger ones 2 to 3 in. long or even more, the upper ones small and narrow. Flowers very small, in little clusters or sessile umbels along a slender flexuose peduncle, often longer than the leaves. Pedicels filiform, 1 to 2 lines long. Calyx-segments about ¾ line long. Corolla scarcely 1½ lines diameter. Corona-segments rather thick, but not very prominent, acuminate, the adnate points reaching to about half the length of the anthers.—Dcne. in DC. Prod. viii. 612; *Hoya flexuosa,* Spreng. Syst. i. 843.

N. Australia. Upper Victoria river, *F. Mueller;* islands of the Gulf of Carpentaria, *R. Brown.*

8. MARSDENIA, R. Br.

(Leichhardtia, *R. Br.*)

Corolla with a short broad tube or nearly rotate, the limb spreading, divided into 5 lobes contorted in the bud, the right-hand edge usually over-lapping. Corona of 5 segments adnate to the gynostegium at the base, sometimes with free basal auricles or almost peltate, the upper end erect and free, shorter than the anthers or scarcely exceeding them. Anthers terminating in a membrane. Pollen-masses 2 to each anther, oblong or rarely obovoid, erect. Stigma (or summit of the style) obtuse or terminating in a cone or in a long beak.—Stems twining or rarely erect. Leaves herbaceous. Flowers in an irregular cyme or panicle or more frequently in a simple umbel on an interpetiolar peduncle.

The genus, as at present constituted, is dispersed over the tropical regions of the New as well as the Old World, with one species extending to the East Mediterranean region. The Australian species are, however, as far as hitherto ascertained, all endemic.

Flowers small in compound cymes. Corolla rotate. Leaves broad or narrow, not cordate.
 Stigma short and very obtuse.
  Erect or scarcely twining. Leaves usually broad. Corolla-lobes shortly bearded . . . . 1. *M. cinerascens.*
  Twiner. Leaves usually narrow. Corolla-lobes glabrous . 2. *M. flavescens.*
 Stigma narrow-conical or rostrate. Leaves narrow. Corolla-lobes bearded at the base . . . . 3. *M. cymulosa.*

Flowers often rather large, in cymes or in simple umbels. Corolla
　more or less urceolate (except *M. Hullsii*). Leaves large,
　cordate.
　　Flowers in cymes. Stigma short . . . .　4. *M. velztina.*
　　Flowers in simple umbels.
　　　Stigma very short and obtuse. Corolla ¼ in. long . . . .　5. *M. Hullsii.*
　　　Stigma narrow-conical or rostrate. Corolla ½ in. long . .　6. *M. araujacea.*
Flowers in simple umbels. Corolla more or less urceolate (except
　in *M. longiloba*). Leaves broad or narrow, not cordate.
　Stigma long and rostrate.
　　Leaves ovate, shortly and obtusely acuminate. Corolla-lobes
　　　shortly bearded below the middle . . . .　7. *M. rostrata.*
　　Leaves lanceolate, acute. Corolla-throat densely bearded　.　8. *M. Fraseri.*
　Stigma narrow-conical, not much exceeding the anthers.
　　Leaves oval oblong or lanceolate.
　　　Corolla glabrous inside, the lobes twice as long as the tube.　9. *M. longiloba.*
　　　Corolla bearded in the throat, the lobes scarcely longer than
　　　　the tube . . . .　10. *M. suaveolens.*
　　Leaves linear. Corolla-lobes shorter than the tube or not
　　　longer.
　　　Corolla-lobes not 1 line long; tube pubescent inside . .　11. *M. leptophylla.*
　　　Corolla glabrous inside, the lobes nearly 2 lines long . .　12. *M. Leichhardtiana.*
　Stigma very short and obtuse.
　　Glabrous or nearly so. Leaves mostly lanceolate, acute.
　　　Corolla-throat with a prominent ring inside and more or
　　　less bearded . . . .　13. *M. viridiflora.*
　　Glabrous or nearly so. Leaves lanceolate acute. Corolla
　　　glabrous inside with an inflexed scale at the throat at each
　　　sinus . . . .　14. *M. coronata.*
　　Pubescent. Leaves ovate to broadly lanceolate. Corolla gla-
　　　brous inside . . . .　15. *M. microlepis.*

1. **M. cinerascens,** *R. Br. Prod.* 461. Shrubby, diffuse, suberect or
shortly twining, the young shoots and inflorescence usually hoary-pubescent,
the adult foliage glabrous. Leaves on rather long petioles, in some specimens broadly ovate or almost orbicular, in others oval-oblong or oblong-lanceolate, obtuse or shortly acuminate, 1½ to 2 in. long or rarely longer.
Flowers small, in compound cymes shorter than the leaves, on axillary or interpetiolar peduncles, or in much longer irregular terminal panicles, consisting of several umbels as in *M. flavescens.* Calyx-segments broad, obtuse,
hirsute, about ½ line long. Corolla nearly rotate, spreading to about 1½
lines diameter, the lobes ovate-oblong, obtuse, shortly bearded inside. Corona-segments short, broad, obtuse, spreading at the tips. Pollen-masses
obovoid, smaller than in most species. Stigma very obtuse, truncate. Follicles tomentose, 2 in. long, acuminate from a rather broad base.—Dcne. in
DC. Prod. viii. 614.

N. Australia. N.W. coast, *Bynoe;* Cygnet Bay and S. Goulburn Island, *A. Cunningham;* islands of the Gulf of Carpentaria, *R. Brown;* Port Essington, *Armstrong.*
Allied to *M. flavescens,* and approaching also *Tylophora flexuosa;* the pollen-masses almost
intermediate between those of *Tylophora* and *Marsdenia.*

2. **M. flavescens,** *A. Cunn. in Bot. Mag. t.* 3289. A rather tall
twiner, more or less tomentose-pubescent all over, except the upper surface
of the leaves. Leaves shortly petiolate, oval-oblong to oblong-lanceolate,
obtuse acute or acuminate, rounded truncate or rarely contracted at the base,

the larger ones 2 in. long when broad or 3 in. when narrow. Flowers small, of a pale greenish-yellow, rather numerous, in pedunculate corymbose cymes, shorter than the leaves, opposite and axillary or solitary and more or less lateral, consisting of several umbels, the pedicels very short. Calyx-segments oval-oblong, obtuse, scarcely above ½ line long. Corolla nearly rotate, spreading to a diameter of 1½ lines, the lobes oblong, obtuse, glabrous inside. Corona-segments broad, with 2 obtusely prominent auricles at the base, the free summit obtuse and much shorter than the anthers. Pollen-masses oblong. Stigma short, obtuse. Follicles narrow, acuminate, about 2 in. long.—Dcne. in DC. Prod. viii. 614.

N. S. Wales. Port Jackson, *Bynoe, McArthur, Woolls;* northward to Hastings and Clarence rivers, *Beckler;* New England, *C. Stuart;* also in *Leichhardt's* collection; southward to Illawarra, *A. Cunningham;* the roots edible (*Woolls*).

3. **M. cymulosa,** *Benth.* Softly but shortly tomentose-pubescent. Leaves oblong-lanceolate, acute or rather obtuse, rounded or narrowed at the base, 2 to 4 in. long, on petioles of ¼ to ½ in. Flowers small, in little umbels or clusters arranged in dichotomous or trichotomous cymes, pedunculate in the upper axils, often forming leafy panicles but scarcely exceeding the leaves. Pedicels short. Calyx-segments obtuse, about ½ line long. Corolla nearly rotate, deeply divided into narrow obtuse lobes of about 1 line, bearded inside at the base as well as the very short tube. Corona-segments with the adnate base not prominent, the erect part broad, obtuse, membranous, nearly as long as the anthers. Stigma narrow-conical, tapering into a short beak. Fruit not seen.

Queensland. Chin-Chin Creek, *Bowman.*

4. **M. velutina,** *R. Br. Prod.* 461. A tall climber, shortly but softly pubescent. Leaves on long petioles, broadly ovate-cordate, shortly acuminate, 3 to 4 in. long. Flowers in compact cymes, either almost sessile or on dichotomous interpetiolar peduncles nearly as long as the petioles. Pedicels 1 to 3 lines long. Calyx-segments broad, obtuse, the margins more or less scarious, about 1½ lines long. Corolla pubescent outside, the broad tube as long as the calyx, thickened inside at the throat into a prominent glabrous ring; lobes spreading, obtuse, rather longer than the tube. Corona-segments vertically adnate, laterally compressed, with 2 prominent angles or keels on the back, more or less confluent at the base with those of the adjoining segments, tapering at the top into free incurved points as long as or rather longer than the anthers. Pollen-masses linear-oblong. Stigma thick, obtuse or shortly 2-lobed, not exceeding the anthers. Fruit not seen. —Dcne. in DC. Prod. viii. 614.

N. Australia. Islands of the Gulf of Carpentaria, *R. Brown;* Port Essington, *Armstrong;* Adams Bay, *Hulls.* In foliage this plant resembles *Parsonsia velutina,* but the flowers are very different.

5. **M. Hullsii,** *F. Muell. Herb.* A glabrous twiner. Leaves petiolate, broadly ovate and deeply cordate, abruptly acuminate, with a long or short acute point, 2 to 3 in. long. Flowers numerous, in simple umbels on very short lateral peduncles, the pedicels slender, at length ½ in. long or more. Calyx-segments about 1 line long, minutely ciliate. Corolla with a very

short broad tube, almost rotate, the broad lobes spreading to 4 or 5 lines
diameter, quite glabrous, with a prominent annular membrane half closing
the throat. Gynostegium very short and broad. Corona-segments fleshy
but not very prominent, almost united in a ring, the erect portion exceedingly
short. Pollen-masses oblong-clavate, on a long slender stipes. Stigma very
obtuse. Follicles rather thick, acuminate, 3 to 4 in. long.

N. Australia. Adams Bay, *Hulls.*

6. **M. araujacea,** *F. Muell. Fragm.* vi. 135. A large twiner, the
branches and foliage sprinkled with appressed hairs. Leaves on rather long
petioles, cordate-ovate, shortly acuminate, membranous, 4 to 5 in. long.
Flowers large, in dense simple umbels, on lateral peduncles shorter than the
petioles, the pedicels, bracts, and calyx shortly pubescent. Calyx-segments
nearly orbicular, very obtuse, 3 lines long. Corolla about ½ in. long, the
tube very broad, slightly contracted at the throat, the lobes rather longer,
bearded inside with long hairs up the centre and round the base. Corona-
segments broadly oblong, peltately adnate, almost cartilaginous, the margins
free, the upper end exceeding the anthers and abruptly contracted into a
short point. Pollen-masses linear-oblong. Stigma with a narrow conical
beak not exceeding the corona.

Queensland. Rockingham Bay, *Dallachy.*

7. **M. rostrata,** *R. Br. Prod.* 461. A rather stout twiner, either glabrous or the young shoots and inflorescence or the whole plant tomentose-
pubescent. Leaves on rather long petioles, ovate to almost orbicular, shortly
and obtusely but usually abruptly acuminate, not cordate or very slightly so
when very broad. Flowers sweet-scented, of a greenish-yellow or nearly
white, numerous, in simple dense globular umbels, on interpetiolar peduncles
shorter than the petioles. Pedicels 2 to 3 lines long. Calyx-segments
broad, obtuse, about 1 line long. Corolla glabrous outside, the broad tube
not contracted at the throat, nearly as long as the calyx, glabrous inside;
lobes rather longer, spreading, obtuse, very shortly bearded below the middle.
Corona-segments with an adnate, scarcely prominent base, the free part erect,
incurved, nearly as long as or rarely longer than the anthers. Stigma produced into a flexuose beak, already as long as the corolla before it opens.
Pollen-masses oblong. Follicles broad, acuminate, not above 2 in. long.—
Dcne. in DC. Prod. viii. 616.

Queensland. Keppel Bay, *R. Brown;* Brisbane river, *Fraser ?* (too young to determine accurately).
N. S. Wales. Hunter's River, *R. Brown;* Hastings and Clarence rivers, *Beckler*
and others; Tweed river, *C. Moore;* Newcastle, Durval, etc., *Leichhardt;* Blue Mountains, *Miss Atkinson;* Lord Howe's Island (with larger flowers), *Milne;* woods and shady
thickets, Illawarra, *A. Cunningham;* Twofold Bay, *F. Mueller.*
Victoria. Snowy and Broadribb rivers, *F. Mueller.*

8. **M. Fraseri,** *Benth.* A rather slender twiner, our specimens quite
glabrous. Leaves shortly petiolate, lanceolate, acute, rounded at the base, 2
to 3 in. long. Umbels simple and solitary, rather dense, on interpetiolar peduncles of ½ to 1 in. Pedicels 2 to 3 lines long. Calyx-segments very
broadly ovate, very obtuse, about 1 line long. Corolla-tube broad, as long

z 2

as the calyx, slightly contracted at the throat, which is closed by a dense ring of long inflexed hairs, the tube also slightly hairy inside ; lobes ovate, obtuse, shorter than the tube. Corona-segments with the adnate part scarcely conspicuous, the free erect part lanceolate, acute, about as long as the anthers. Stigma rostrate as in *M. rostrata.* Fruit not seen.

Queensland. Cliffs, Moreton Bay, *Fraser, A. Cunningham.* A specimen of F. Mueller's from Moreton Island, in very young flower, may also be the same.

9. **M. longiloba,** *Benth.* A glabrous twiner. Leaves petiolate, ovate-lanceolate or lanceolate, acuminate, rounded or truncate at the base, 1½ to 2½ in. long. Umbels simple and solitary, on peduncles rarely exceeding the petioles. Pedicels 3 to 4 lines long. Calyx-segments ovate-lanceolate, rather obtuse, about 1 line long. Corolla with a very broad and short tube, glabrous inside, the limb spreading to 4 or 5 lines diameter, the lobes twice as long as the tube, rather narrow, obtuse or almost acute, with 5 prominent scales in the throat alternating with the lobes. Corona-segments each with 2 very prominent vertical auricles on the adnate base, the free part erect, narrow, short or nearly as long as the anthers. Pollen-masses narrow-oblong. Stigma narrow-conical, about as long as the anther-membranes.

N. S. Wales. Hastings river, *Beckler ;* Tweed river, *C. Moore.*

10. **M. suaveolens,** *R. Br. Prod.* 461. Rather stout, either nearly erect or with long twining branches, glabrous or minutely pubescent. Leaves on short petioles, oval-oblong or lanceolate or the lower ones ovate, obtuse, rounded at the base, rather thick, the veins scarcely conspicuous, 1½ to 3 in. long. Umbels simple and solitary, rather dense, on short interpetiolar peduncles. Calyx-segments ovate, obtuse, ciliolate, scarcely 1 line long. Corolla-tube broad, about as long as the calyx, the lobes rather longer than the tube, oblong, obtuse, bearded inside from the base to the middle or higher up, as well as the upper part of the tube. Corona-segments with an adnate base, sometimes scarcely prominent, sometimes horizontally protruding and forming almost an outer corona, the erect free portion oblong, obtuse, as long as or longer than the anthers. Pollen-masses small, but oblong and erect. Stigma conical. Follicles not broad, acuminate, 2 to 4 in. long.—Dcne. in DC. Prod. viii. 614 ; Rudge in Trans. Linn. Soc. x. t. 21 ; Bot. Reg. t. 489 ; Guillem. Ic. Pl. Austral. t. 9.

N. S. Wales. Port Jackson to the Blue Mountains, abundant, *R. Brown* and many others

11. **M. leptophylla,** *F. Muell. in Herb. Hook.* A slender, slightly pubescent twiner. Leaves narrow-linear, contracted into a short petiole, 1 to 2 in. long. Umbels simple and solitary, rather dense, on short interpetiolar peduncles. Pedicels short. Calyx-segments oblong, rather obtuse, nearly 1 line long. Corolla-tube nearly 1½ lines long, contracted upwards, the upper part pubescent inside, with raised lines decurrent from the sinus of the lobes to about halfway ; lobes much shorter than the tube, broad, obtuse. Corona-segments small, peltate, shortly auriculate at the base, the free erect part very short. Pollen-masses short but erect. Stigma narrow-conical, almost rostrate.

Queensland. Burdekin river, *F. Mueller.* A single specimen in Herb. Hook.

12. **M. Leichhardtiana,** *F. Muell. Fragm.* v. 160.   Stems woody at
the base, with long twining branches, the young parts glaucous or white-
tomentose.   Leaves linear, obtuse, contracted into a very short petiole, rather
thick, 1½ to 3 or even 4 in. long.   Flowers large for the genus, in simple and
solitary dense umbels, on short interpetiolar peduncles.   Pedicels short.
Calyx-segments ovate, obtuse, about 2 lines long.   Corolla-tube 2 lines long,
contracted at the throat with a thickened ring prominent inside, but less so
than in *M. viridiflora* and quite glabrous ; lobes rather shorter than the
tube, obtuse.   Corona-segments small, the adnate part broad and slightly
auriculate at the base, the free erect part very short.   Stigma conical, obtuse,
shorter than the anthers.   Pollen-masses oblong.   Follicles thick, ovoid or
ovoid-oblong, scarcely acuminate, 1½ to 3 in. long.—*Leichhardtia australis,*
R. Br. in App. Sturt. Voy. 18.

N. Australia. Hooker's and Sturt's Creek, *F. Mueller.*
Queensland. Narran river, the pods eaten by the natives under the name of "Dou-
bah," *Mitchell.*
N. S. Wales. Lachlan, Murray, and Darling rivers, *Victorian and other Expeditions.*
S. Australia. Daly Waters, *Waterhouse ;* between Stokes Range and Cooper's Creek,
Wheeler.
W. Australia. Murchison river, *Oldfield.*

13. **M. viridiflora,** *R. Br. Prod.* 461.   A twiner, quite glabrous or
with a rusty pubescence on the young shoots and inflorescence.   Leaves
either long-lanceolate with a broad sometimes almost hastate base with
rounded auricles, or all linear-lanceolate, rarely almost ovate, acute or rarely
obtuse, 2 to 4 in. long.   Flowers green, in simple umbels, solitary or rarely
2 together on short interpetiolar peduncles.   Pedicels slender, 3 to 4 lines
long.   Calyx-lobes ovate, very obtuse, about 1 line long.   Corolla-tube
broad, rather shorter than the calyx, half closed at the throat by a prominent
ring and bearded inside, the hairs either in tufts opposite each sinus or forming a complete ring ; lobes ovate, obtuse, spreading to nearly 3 lines diameter.
Corona-segments auriculate at the base, the free erect part obtuse, much
shorter than the anthers.   Stigma very short, obtuse.   Follicles about 3 in.
long, above 1 in. broad, scarcely acuminate.—Dcne. in DC. Prod. viii. 615.

N. Australia. Victoria river, *F. Mueller* (a single specimen in Herb. Hook.) ;
Adams Bay, *Hulls.*
Queensland. Keppel Bay and Broad Sound, *R. Brown ;* Brigalow scrub, from the
Dawson and Burdekin to the Burnett rivers, *F. Mueller ;* Rockhampton, *Dallachy* and
others ; Suttor river and Nerkool Creek, *Bowman.*
N. S. Wales. Cumberland and Camden, "Native Potato," *Woolls.*

14. **M. coronata,** *Benth.*   A rather slender twiner, glabrous or slightly
pubescent.   Leaves petiolate, lanceolate, acute, narrowed or rounded at the
base, 1 to 2 in. long.   Umbels few-flowered, simple and solitary on short
interpetiolar peduncles.   Pedicels 4 to 6 lines long.   Calyx-lobes ovate-
oblong, obtuse, rather exceeding 1 line.   Corolla-tube broad, as long as the
calyx, slightly contracted at the throat with a small inflexed scale opposite
each sinus ; lobes as long as or rather shorter than the tube, broad and very
obtuse, all quite glabrous inside.   Corona-segments broad, thick, and hori-

zontally prominent at the base, the erect free part acute, nearly as long as the anthers.   Pollen-masses oblong-clavate.   Stigma very short and obtuse.

Queensland.   Brisbane river, Moreton Bay, *F. Mueller.*

15. **M. microlepis,** *Benth.*   A twiner, softly pubescent or the foliage at length glabrous.   Leaves petiolate, ovate oblong or broadly lanceolate, obtuse or mucronate, truncate or almost cordate at the base, 1 to 1½ in. long.   Umbels many-flowered, simple, solitary and nearly sessile.   Calyx-segments pubescent, nearly 1 line long.   Corolla-tube nearly as long as the calyx, slightly contracted and thickened inside at the throat, but quite glabrous ; lobes shorter than the tube, and apparently less spreading than in *M. viridiflora,* sometimes slightly pubescent outside.   Corona-segments peltate, the free erect part obtuse and very short.   Stigma short and obtuse.

Queensland.   Port Curtis, *M'Gillivray ;* Burdekin river, *F. Mueller ;* head of Boyd river, *Leichhardt ;* Nerkool Creek, *Bowman* (all pubescent) ; Rockhampton, *Dallachy* (nearly glabrous).

9. GYMNEMA, R. Br.

(Bidaria *and* Gongronema, *Endl.*)

Corolla with a short broad tube, the limb spreading, divided into 5 lobes, contorted in the bud.   Corona none or reduced to 5 scarcely prominent protuberances at the base of the gynostegium.   Anthers terminating in a membrane.   Pollen-masses 2 to each anther, obovoid or oblong, erect.   Stigma short and obtuse, or conical, or rarely elongated.—Stems erect or twining. Leaves herbaceous.   Flowers small, umbellate, the umbels either solitary or 2 together on short interpetiolar peduncles or axillary and opposite.

The genus is dispersed over the tropical and subtropical regions of the Old World. Of the six Australian species, one is a common East Indian one, another extends to the islands of the South Pacific, the remaining four are endemic.

SECT. I. **Gymnema.**—*Corolla with small scales in the throat alternating with the lobes. No corona.*

Leaves ovate to ovate-lanceolate.   Umbels mostly in pairs .   .   .   .   1. *G. sylvestre.*
Leaves narrow-lanceolate.   Umbels mostly solitary .   .   .   .   .   .   2. *G. Muelleri.*

SECT. II. **Bidaria.**—*Corolla without scales in the throat.   Corona none or of very obscure glands at the base of the gynostegium.*

Leaves ovate, ½ to 1 in. long .   .   .   .   3. *G. brevifolium.*
Leaves oval-oblong or lanceolate, 1 to 2 in. long, the veins very
    oblique .   .   .   .   4. *G. trinerve.*
Leaves narrow-linear   .   .   .   .   5. *G. stenophyllum.*

SECT. III. **Gongronema.**—*Corolla without scales in the throat.   Corona consisting of small glands or protuberances at the base of the gynostegium.   Small glands at the base of the upper surface of the leaves.*

Leaves obovate-oblong or elliptical .   .   .   .   6. *G. micradenia.*

SECT. I. GYMNEMA.—Corolla with small scales in the throat alternating with the lobes.   No corona.

1. **G. sylvestre,** *R. Br. in Trans. Wern. Soc.* i. 33.   A twiner, either softly-pubescent all over or the upper side of the leaves nearly or quite gla-

brous. Leaves petiolate, from broadly ovate to ovate-lanceolate, shortly acuminate, rounded or rarely contracted, or almost cordate at the base, obliquely penniveined, 1 to 2 in. long. Umbels usually 2 together on very short peduncles. Pedicels 1 to 2 lines long. Calyx-segments broad, very obtuse, under 1 line long. Corolla very shortly exceeding the calyx, divided to about the middle into ovate lobes, with scales in the throat alternating with the lobes and decurrent halfway down the tube. No corona. Stigma broadly and obtusely conical or almost globular, shortly exceeding the anthers. Follicles narrow, acuminate, 2 to 3 in. long.—Dcne. in DC. Prod. viii. 621; Wight, Ic t. 349 ; *G. geminatum*, R. Br. Prod. 462 ; Dcne. in DC. Prod. viii. 623.

N. Australia. Islands of the Gulf of Carpentaria, *R. Brown.* The species is also in Ceylon and the Indian Peninsula, and has probably a much wider range in tropical Asia. R. Brown indicated the close resemblance of the Australian to the Indian plant; and, after a careful analysis of the flowers of both, I can detect no character to separate them unless the scales of the corolla may be slightly more decurrent in the former than in the latter.

2. **G. Muelleri,** *Benth.* A more or less pubescent twiner, woody at the base. Leaves lanceolate, acutely acuminate, rounded or narrowed at the base, narrow and 2 to 6 in. long on luxuriant branches, ovate-lanceolate and 1 to 2 in. on shorter less twining branches. Umbels dense, on very short peduncles, solitary and interpetiolar or rarely axillary and opposite. Calyx-segments ovate, obtuse, about 1 line long. Corolla-tube rather shorter than the calyx, bearded inside with reflexed hairs, the throat with very few prominent hirsute scales alternating with the lobes and decurrent halfway down ; lobes ovate, obtuse, shorter than the tube. No corona. Stigma short, obtuse.

N. Australia. Upper Victoria river, *F. Mueller.*

SECT. II. BIDARIA.—Corolla without scales in the throat. No corona.

3. **G. brevifolium,** *Benth.* Apparently erect, pubescent. Leaves on very short petioles, ovate, obtuse, with recurved margins, becoming at length glabrous above, ½ to 1 in. long. Umbels usually opposite, on very short peduncles. Pedicels 1 to 2 lines long. Sepals herbaceous, pubescent, obtuse, about ¾ line long. Corolla-tube broad, about as long as the calyx, slightly contracted at the throat ; lobes very obtuse, shorter than the tube, all as well as the throat and tube glabrous inside, without scales. No corona. Follicles pubescent, not thick, acuminate, about 2 in. long.

Queensland. Rockhampton, *Thozet ;* Princhester Creek, *Bowman.*

4. **G. trinerve,** *R. Br. Prod.* 462. Nearly erect or shortly twining, pubescent all over but the hairs few on the upper side of the leaves. Leaves shortly petiolate, oval-oblong to oblong-lanceolate, obtuse, very obliquely penniveined or 3-nerved at the base, 1 to 2 in. long. Umbels solitary, sessile. Pedicels 1 to 2 lines long. Calyx-segments herbaceous, pubescent, obtuse but not broad, above 1 line long. Corolla shortly exceeding the calyx, the tube short and broad ; lobes obtuse, longer than the tube, the throat without scales, but half-closed with tufts of hairs alternating with the lobes and sometimes forming a complete ring. Corona none except **very** obscure glands or protuberances at the base of the gynostegium. Stigma not prominent.—*Bidaria trinervis*, Dcne. in DC. Prod. viii. 624.

N. Australia. Islands of the Gulf of Carpentaria, *R. Brown;* also in *Leichhardt's* collection without the locality.

5. **G. stenophyllum,** *A. Gray in Proceed. Amer. Acad. Sc.* v. 335. Stems shrubby at the base, erect and slender (or the branches rarely elongated and twining ?), minutely pubescent as well as the inflorescence and young foliage.　Leaves very narrow-linear, flat or with recurved margins, contracted into a very short petiole, 3 to 5 in. long.　Flowers not numerous, in shortly pedunculate interpetiolar umbels.　Sepals almost acute, about ¾ line long.　Corolla-tube rather shorter than the calyx, broadly campanulate; lobes ovate-lanceolate, rather longer than the tube, bearded inside below the middle as well as the tube, and the throat almost closed with tufts of hairs alternating with the lobes, but no scales.　Corona none except obscure protuberances at the base of the gynostegium.　Stigma narrow-conical, longer than the anthers.—Seem. Fl. Vit. t. 31.

N. Australia. Victoria river, *F. Mueller.*

Also in the Fiji Islands, if my identification is correct.　Seemann's specimens have the leaves not so long, and the peduncles shorter, or the umbels almost sessile.　The corollas are also described as more rotate and less hairy inside, but the Australian and Fiji plants are otherwise exceedingly similar.　Our specimens from Fiji have no perfect corollas.

SECT. III. GONGRONEMA.—Corolla without scales in the throat.　Corona consisting of 5 small glands or protuberances at the base of the gynostegium. Leaves with small glands at the base of the lamina on the upper side.

6. **G. micradenia,** *Benth.*　A rather slender but tall twiner, minutely but softly pubescent.　Leaves obovate-oblong or elliptical, abruptly and shortly acuminate, mostly rounded at the base, becoming glabrous above when full grown, with 2 or 3 minute glands on the upper surface immediately above the petiole, 1 to 1½ in. long.　Umbels small, sometimes shortly developed into cymes, on short interpetiolar peduncles, solitary or rarely opposite and axillary.　Calyx-segments ovate, obtuse, not ½ line long. Corolla-tube broad, as long as the calyx, slightly contracted at the throat but without scales or hairs inside; lobes short, obtuse.　Corona of 5 scarcely prominent small glands or protuberances at the base of the gynostegium. Stigma shortly and obtusely conical.　Follicles acuminate, about 1 line long.

Queensland. Brisbane river, *F. Mueller;* Ugly Creek, *C. Stuart;* Rockhampton, *O'Shanesy.*

10. MICROSTEMMA, R. Br.

Corolla rotate, deeply divided into 5 acuminate lobes, valvate in the bud. Corona an undulate ring round the gynostegium below the anthers, and more prominent between the anthers than opposite to them.　Anthers without terminal membranes.　Pollen-masses 2 to each anther, ovoid, erect, laterally attached below the middle.　Stigma short, obtuse.—Herbs, with tuberous rhizomes and erect stems.　Leaves linear or none.　Flowers in sessile umbels.

The genus is limited to Australia, but is nearly allied to the Asiatic *Pentasacme.*　It was accidentally omitted in De Candolle's ' Prodromus.'

Corolla bearded inside . . . . 1. *M. tuberosum.*
Corolla quite glabrous . . . . 2. *M. glabriflorum.*

1. **M. tuberosum,** *R. Br. Prod.* 459. Stems from a tuberous rhizome, erect, slender, simple or slightly branched, glabrous. Leaves either all replaced by minute scales or a few in the upper part of the plant long and narrow-linear. Umbels sessile or very shortly pedunculate at the upper nodes, consisting of few dark-purple flowers, on filiform pedicels of 3 to 4 lines. Calyx-segments small. Corolla spreading to about 4 lines diameter, bearded inside with long purple hairs, the lobes acuminate. Gynostegium very short, the anther-cells prominent. Corona not very conspicuous. Follicles narrow-linear, acuminate, fully 3 in. long.—Endl. Iconogr. t. 60 ; F. Muell. Fragm. i. 58.

N. Australia. Islands of the Gulf of Carpentaria, *R. Brown.*
Queensland. Burdekin river, *F. Mueller.*

2. **M. glabriflorum,** *F. Muell. Fragm.* i. 58. Very near *M. tuberosum,* with a similar habit and inflorescence, but the flowers smaller, the corolla not bearded inside, and the coronal ring rather more prominent.

N. Australia. Sea range, Victoria river, *F. Mueller.* Only two small specimens were found, and those had very few flowers. It may prove to be a variety of *M. tuberosum.*

11. DISCHIDIA, R. Br.

Corolla urceolate ; lobes 5, spreading, valvate in the bud. Corona of 5 segments attached to the base of the gynostegium, erect, linear, and bifid at the end. Anthers terminated by a membrane. Pollen-masses 2 to each anther, oblong, erect. Stigma obtuse.—Herbs usually creeping over the stems of trees and rooting at the nodes. Leaves fleshy, some of them (in species not Australian) often converted into pitchers. Flowers small, in axillary or interpetiolar umbels or clusters.

The genus is spread over East India, and more especially the Indian Archipelago, where the only Australian species is also found.

1. **D. nummularia,** *R. Br. Prod.* 461. A succulent milky-juiced epiphyte more or less mealy-white, the slender stems creeping over the trunks and branches of trees, rooting at the nodes, and apparently attaching themselves by means of disk-like expansions of the fibres, the upper branches loose and hanging. Leaves on very short petioles, nearly orbicular, thick and fleshy, not exceeding ½ in. diameter. Flowers very small, in little sessile axillary or interpetiolar clusters, the pedicels very short. Calyx-segments minute. Corolla under 1½ lines long, the tube inflated, the lobes narrow, longer than the tube. Corona-segments as long as the gynostegium, at first erect incurved, with short subulate inflected lobes, at length spreading with recurved lobes. Follicles membranous, about 1 in. long, acuminate, recurved at the end. Seeds very small, with a very copious silky coma.—Dcne. in DC. Prod. viii. 632.

Queensland. Cape York, *M'Gillivray, Daemel ;* Endeavour river, *Banks and Solander ;* Rockingham Bay, *Dallachy.* This species is also in the Indian Archipelago and the Malayan Peninsula.

12. HOYA, R. Br.

Corolla rotate, the lobes flat or with reflexed margins, valvate in the bud. Corona of 5 rather thick fleshy segments attached to the gynostegium, horizontally spreading or expanded into variously-shaped disks. Anthers terminated by a membrane; pollen-masses 2 to each anther, erect. Stigma obtuse or scarcely prominent.—Stems twining or trailing. Leaves thick and fleshy or in a few species not Australian membranous. Flowers often fleshy or waxy, in pedunculate interpetiolar simple umbels.

The genus is widely spread over tropical Asia, and more sparingly represented in Africa on the one hand, and the South Pacific Islands on the other. Of the three Australian species, one extends over the Indian Archipelago to South China, another is in the islands of the South Pacific, the third may be endemic, although closely allied to a Pacific Island one.

Flowers white or pink. Corona-segments very convex, and spreading into a horizontal star . . . . 1. *H. carnosa.*
Flowers white or pink. Corona-segments expanded into a concave disk, the outer margin very obtuse, the inner acuminate . . . . 2. *H. australis.*
Flowers yellow. Corona-segments expanded into slightly concave ovate almost acute disks, the inner margin very short and obtuse . 3. *H. Nicholsoniæ.*

1. H. carnosa, *R. Br.* (as to the Linnean plant); *Dcne. in DC. Prod.* viii. 636. A succulent glabrous twiner. Leaves from broadly ovate-cordate to ovate-oblong, obtuse or shortly acuminate, thick and fleshy, 2 to 3 in. long, on short petioles. Flowers white, mixed with pink in the centre, succulent, in rather large simple umbels on short interpetiolar peduncles, the pedicels ¾ to 1½ in. long, pubescent. Corolla spreading to fully ½ in. diameter, broadly 5-lobed, densely papillose on the upper side, the margins recurved, the gynostegium spreading like a closely appressed star in the centre. Corona-segments forming the horizontal rays of the star, alternating with the corolla-lobes, ovate-lanceolate, convex, shining above (wrinkled in drying), the margins revolute.—*Asclepias carnosa,* Linn.; Bot. Mag. t. 788; Sm. Exot. Bot. ii. t. 70.

Queensland. Cape York, *Jardine* (*Herb. F. Mueller*). The species is also in South China, and probably in the Indian Archipelago. The Australian plant at first doubtfully referred to this species by Brown, is the following one.

2. H. australis, *R. Br.; Traill in Trans. Hort. Soc.* vii. 28. A succulent glabrous twiner or epiphyte. Leaves on short petioles, ovate obovate or nearly orbicular, obtuse or shortly acuminate, rounded or rarely almost cordate at the base, thick and fleshy, 2 to 3 in. long or rarely more. Flowers white tinged with pink in the centre, in simple umbels on interpetiolar peduncles rarely exceeding the petioles, the pedicels slender, ½ to 1 in. long or even more. Calyx-segments about 1 line long. Corolla spreading to ⅓ in. diameter, broadly 5-lobed, the upper surface nearly smooth and glabrous except towards the edges which are slightly papillose and not reflexed. Corona-segments expanded into concave (at first almost cup-shaped) horizontally spreading laminæ, very obtuse on the outer margin, the inner margin acuminate and incurved, the back prominently 2-keeled.—*H. carnosa ?,* R. Br. Prod. 460 (as to the Australian plant); *H. bicarinata,* A. Gray in Proc. Amer. Acad. Sc. v. 335; *H. Dalrympliana,* F. Muell. Rep. Burdek. Exped. 16.

Queensland. Endeavour river, *Banks and Solander*, *R. Brown;* Port Denison, *Fitzalan;* Rockhampton, Rockingham and Edgecombe Bays, *Dallachy;* Brisbane river, Moreton Bay, *Backhouse, F. Mueller;* Ugly Creek, *C. Stuart.*
N. S. Wales. Clarence river, *Herb. F. Mueller.*

The species is also in the Fiji and Samoa Islands.

3. **H. Nicholsoniæ,** *F. Muell. Fragm.* v. 159. A glabrous succulent epiphyte clinging to the trunks of trees, the branches often twining and emitting fibres not confined to the nodes. Leaves ovate or elliptical, acuminate, contracted into a rather short petiole, thick and fleshy, very obliquely penniveined and 3-nerved at the base or almost quintuplinerved, 2 to 3 in. long or when luxuriant above 4 in. Flowers yellow rather numerous in the umbel, the pedicels ½ to ¾ in. long. Corolla if spread open nearly ½ in. diameter, deeply lobed, quite glabrous, the lobes acute and curved over the gynostegium in the dried specimen. Corona-segments expanded into ovate slightly concave disks, the outer margin almost acute, the inner margin very broad short and obtuse, the back with 2 broadly-prominent involute keels.

Queensland. Rockingham Bay, *Dallachy.* The foliage is that of *H. diptera*, Seem., from the Fiji Islands, which is also said to have yellow flowers, but the very imperfect flowers on our specimens appear to have the corolla pubescent, with broader lobes.

13. THOZETIA, F. Muell.

Corolla nearly rotate, deeply divided into 5 lobes induplicate-valvate in the bud. Corona of 5 erect membranous segments, peltately attached to the back of the anthers, the upper end free. Anthers terminating in a membrane. Pollen-masses 2 to each anther, obovate-oblong, erect. Stigma conical.— Tall twiner. Leaves herbaceous. Flowers in reflexed interpetiolar racemes, the persistent rhachis much lengthened and covered with the scars of fallen pedicels.

The genus is limited to a single Australian species, allied to *Hoya* in its flowers, with the corona of *Marsdenia.*

1. **T. racemosa,** *F. Muell. Herb.* Apparently a tall woody twiner, the branches and petioles hirsute with spreading hairs. Leaves cordate-ovate, rather narrow, 4 to 5 in. long, glabrous above, more or less hirsute underneath. Flowers rather large, dark coloured, in reflexed interpetiolar nearly sessile racemes, the rhachis lengthening to 2 in. or more and densely covered with the prominent scars of the fallen pedicels, interspersed with small persistent bracts. Pedicels 3 to 5 lines long. Calyx slightly hirsute, the segments lanceolate, about 2 lines long. Corolla-lobes narrow-lanceolate, about 5 lines long. Corona-segments peltate, the lower free portion broad, the upper free portion oblong, truncate or sinuate, rather longer than the anthers. Stigma narrow-conical. Follicles glabrous, acuminate..

Queensland. Near Rockhampton, *Thozet.* The corollas I have seen are already expanded, but they have every appearance of having been induplicate-valvate in the bud, as described in F. Mueller's notes.

14. CEROPEGIA, Linn.

Corolla with a distinct often elongated tube usually swollen at the base;

lobes acute or acuminate, incurved and connivent or cohering at the tips, valvate in the bud. Corona inserted on the gynostegium, campanulate or rotate at the base, with 10 or 15 lobes in 2 rows (or rarely only 5), the inner ones usually longer acuminate and connivent over the gynostegium. Anthers without any terminal membrane. Pollen-masses 2 to each anther, erect or incurved. Stigma obtuse.—Stems usually twining from a tuberous rhizome. Leaves membranous or fleshy. Flowers often few and rather large, in axillary or interpetiolar cymes or umbels.

A considerable genus, widely spread over tropical Asia and Africa, although rare in the Indian Archipelago. The only Australian species is one of the few from the latter region.

1. **C. Cumingiana,** *Dcne. in DC. Prod.* viii. 643. A glabrous twiner. Leaves ovate or ovate-oblong, always cordate at the base, shortly and acutely acuminate, thin and membranous when dry, penniveined, 3 to 4 in. long, on petioles of ½ to 1½ in. Flowers rather numerous in the typical specimen, few in the Australian one, in a shortly-branched cyme sometimes contracted into an umbel, the common peduncle usually longer than the petiole, the pedicels from ½ to 1 in. long. Calyx-segments subulate-acuminate, 1 to 1¼ lines long. Corolla-tube 1 in. long, including the campanulate throat, which spreads to ½ in. diameter; lobes shorter than that diameter, broad, acute, erect, arcuate and cohering at the tips. Corona loosely campanulate at the base, with 10 short lanceolate or oblong outer lobes in pairs usually sprinkled with a few long hairs, and 5 inner linear ones, twice as long as the outer ones and connivent over the gynostegium or cohering at the tips. Follicles (in the Philippine Island plant) very long and linear.—F. Muell. Fragm. v. 159.

Queensland. Near Somerset, Cape York, *Jardine,* a single specimen in Herb. F. Mueller. Also in the Philippine Islands. The glabrous stems, cordate leaves, and small calyx-segments, readily distinguish this species from all others in the Kew collections. On comparing the gynostegium and corona of the Australian and Philippine specimens, I find no difference; the outer corona-lobes are very slightly hairy in both, although the hairs are sometimes reduced to 1 or 2 to each lobe. The Javanese plant figured as *C. Cumingiana* in Bot. Mag. t. 4349, differs in the corolla-tube densely hairy inside, the lobes much longer and acuminate, and in the much more clavate inner corona-segments, and appears to be the *C. curviflora,* Hassk., altered to *C. Horsfieldiana,* by Miquel, Fl. Ind. Bat. ii. 528, the only species hitherto found in Java.

Order LXXVII. LOGANIACEÆ.

Flowers regular. Calyx free, with 4 or 5 teeth lobes or segments, very rarely reduced to 2 or in a very few species 6 or 7. Corolla with 4, 5, or, in genera not Australian, more than 5 lobes, valvate contorted or otherwise imbricate in the bud. Stamens as many as corolla-lobes, alternate with them, inserted in the tube or very rarely, in a genus not Australian, reduced to 1; anthers 2-celled, the cells opening longitudinally. Ovary free, 2-celled or very rarely 3- to 5-celled or imperfectly divided; style single or separating into 2 at the base, with an entire or 2-lobed stigmatic summit; ovules 1 or more in each cell of the ovary or to each placenta. Fruit a berry or capsule. Seeds albuminous. Embryo straight, often oblique, with leafy or small cotyledons.—Trees shrubs climbers or rarely herbs. Leaves opposite, usually connected by interpetiolar stipules or at any rate by a raised line.

A small Order, ranging over the tropical regions of the New as well as the Old World, with a few extratropical species in the southern hemisphere and in North America. Of the five Australian genera, one is sparingly represented in New Zealand, another has a very few South Asiatic species, a third belongs chiefly to the tropical Asiatic and Polynesian region, a fourth is in tropical and subtropical Asia and America, the fifth is spread over tropical Asia, Africa, and America. The Order is a somewhat heterogeneous group of genera, differing from *Rubiaceæ* in the free ovary and less developed stipules, and closely connecting that Order with *Apocyneæ*, *Gentianeæ*, and *Scrophularineæ*.

Herbs. Corolla-lobes valvate. Capsules 2-lobed or truncate.
Flowers 5-merous, in dichotomous cymes . . . . 1. MITREOLA.
Flowers 4-merous, on simple solitary or clustered pedicels . . . . 2. MITRASACME.
Herbs undershrubs or shrubs. Corolla-lobes imbricate, not contorted.
Flowers usually 5-merous. Capsule septicidally dehiscent . . . . 3. LOGANIA.
Shrubs or trees. Corolla-lobes imbricate-contorted.
Fruit a capsule. Flowers small in axillary clusters or cymes . . . 4. GENIOSTOMA.
Fruit a berry. Flowers usually large in terminal cymes or panicles . 5. FAGRÆA.
Shrubs trees or woody climbers. Corolla-lobes valvate. Fruit a berry.
Leaves 3- or 5-nerved . . . . 6. STRYCHNOS.

1. MITREOLA, Linn.

Calyx 5-cleft. Corolla-tube cylindrical, hairy inside at the throat; lobes 5, valvate in the bud. Stamens 5, included in the corolla-tube. Ovary 2-celled, with several ovules in each cell, the summit broad with 2 styles distinct at the base but united in a small capitate stigma, very rarely separating to the end at the time of flowering. Capsule broad at the top, truncate or 2-lobed, the carpels more or less diverging or separating when ripe and opening along their inner margin. Seeds numerous, small; embryo linear, with small cotyledons.—Herbs. Flowers small, in dichotomous cymes. Bracts small.

The genus is dispersed over tropical Asia and tropical and northern America, and consists of but very few species. The Australian one is not uncommon in tropical Asia.

1. **M. oldenlandioides,** *Wall.; A. DC. Prod.* ix. 9. An erect annual, of ½ to 1½ ft., glabrous or nearly so. Leaves ovate or oval-oblong, acuminate, narrowed into a rather long petiole, 1 to 2 in. long; stipules minute. Cymes dichotomous, terminal or in the upper axils; the flowers scarcely 1 line long, nearly sessile along the slender branches or more pedicellate in the forks. Corolla not twice as long as the calyx; styles very short, united at the stigma. Capsule 1½ to nearly 2 lines broad at the top, the lobes forming 2 horns very divaricate at the base with the ends shortly incurved. Seeds very small, ovoid or oblong, smooth.—Benth. in Journ. Linn. Soc. i. 91; Hook. Ic. Pl. t. 827.

N. Australia. Arnhem's Land, *F. Mueller.* We have the same species from the East Indian Peninsula, from Burmah and from Java.

2. MITRASACME, Labill.

Calyx campanulate, 4-lobed or rarely 2-lobed. Corolla-tube short and broad or elongated and cylindrical; lobes 4, spreading, valvate and sometimes reduplicate in the bud. Stamens 4, inserted in the tube; anthers included or rarely exserted. Ovary 2-celled, with several ovules in each cell; style usually simple at first but splitting at the base as the flowering advances; stigma capitate or minutely 2-lobed. Capsule globular or ovoid or compressed, usually truncate or 2-lobed at the top, opening along the inner margins of the lobes or carpels. Seeds small, ovoid or globular.—Herbs usually small or slender. Leaves opposite, without stipules. Flowers small, usually white, often yellowish at the throat, rarely all yellow, either solitary in the upper axils or the upper pedicels crowded into a terminal cluster or irregular umbel without bracteoles.

The genus is chiefly Australian, represented in tropical Asia by three species, of which two, or perhaps all three, are identical with Australian ones; the other Australian species are all endemic.

SECT. I. **Mitragyne.**—*Calyx* 4-*lobed.*

Perennials (sometimes flowering the first year?). Stems leafy, much-branched, prostrate diffuse or tufted. Flowers axillary or the upper ones in a sessile umbel.
Calyx divided almost to the base. Small alpine plants.
Leaves oblong, coriaceous. Calyx-lobes unequal. Corolla-
lobes as long as the tube. Capsule with 2 incurved horns. 1. *M. Archeri.*
Leaves obovate or orbicular. Calyx-lobes equal. Corolla-
lobes shorter than the tube. Capsule broad, truncate . . 2. *M. montana.*
Calyx divided to the middle.
Corolla-lobes much longer than the tube. Anthers more or less exserted.
Leaves ovate to oblong-lanceolate . . . . 3. *M. serpyllifolia.*
Leaves linear-lanceolate or oblong-linear . . . . . . 4. *M. paludosa.*
Corolla-lobes shorter than the tube. Anthers included . . 5. *M. pilosa.*
Perennials (sometimes flowering the first year?). Stems leafy, erect or ascending. Corolla-lobes as long as the tube.
Flowers (small) in the upper axils or the uppermost in an almost sessile umbel.
Nearly glabrous. Leaves small ovate or ovate-lanceolate . 6. *M. alsinoides.*
Pubescent or hirsute. Leaves narrow . . . . 4. *M. paludosa.*
Flowers on long filiform pedicels, mostly forming loose umbels on long terminal peduncles . . . . 7. *M. polymorpha.*
Annuals. Leaves all radical, rosulate, usually withering as the flowering advances. Corolla-tube cylindrical.
Corolla-tube about ½ in. long. Anthers included.
Calyx-lobes short and broad. Capsule nearly globular . . 8. *M. longiflora.*
Calyx-lobes long narrow and acute. Capsule twice as long as broad . . . . 9. *M. elata.*
Corolla-tube about ¼ in. long. Anthers exserted . . . . . 10. *M. exserta.*
Corolla-tube under ¼ in. long. Anthers included.
Corolla-tube 2 or 3 times as long as the calyx.
Calyx-lobes short but acute . . . . 11. *M. tenuiflora.*
Calyx-lobes very short, obtuse, broad and thick . . . . 12. *M. ambigua.*
Corolla-tube shortly exceeding the calyx, rather broad, the lobes very short . . . . 13. *M. nudicaulis.*
Annuals. Stems erect, leafy at least at the base. Pedicels long and slender, the upper ones usually forming an umbel.
Corolla-tube cylindrical, 4 to 5 lines long, lobes rather shorter . 14. *M. connata.*
Corolla-tube broad, scarcely exceeding the calyx; lobes as long as or longer than the tube.
Very glabrous. Stems often slightly twining. Corolla-tube 1½ to 2 lines long . . . . 15. *M. lævis.*

Glabrous or pubescent, not twining. Corolla-tube about 1
　　line long . . . . 16. *M. indica.*
Corolla-tube ovoid or cylindrical, the lobes much shorter than
　　the tube.
　　Umbels dense and many-flowered.
　　　Flowers not 2 lines long. Anthers included . . . . . 17. *M. stellata.*
　　　Flowers 3 lines long. Anthers exserted . . . . . . 18. *M. Cunninghamii.*
　　Umbels loose and few-flowered, or pedicels long and solitary.
　　　Leaves obtuse or scarcely acute.
　　　　Leaves ovate. Corolla (white) under 2 lines long . . 19. *M. pygmæa.*
　　　　Leaves linear. Corolla (yellow) about 3 lines long . . 20. *M. lutea.*
　　　Leaves and calyx-lobes subulate-acute. Plants under 3 in.
　　　　high.
　　　　Stem branching. Capsule ovoid . . . . 21. *M. multicaulis.*
　　　　Stem usually simple. Capsule globular . . . . . 22. *M. laricifolia.*
Annuals. Stems leafy, densely branched, under 2 in. high, the
　　flowers shorter than or scarcely exceeding the leaves.
　　Calyx very angular, with short divaricate very acute lobes . . 23. *M. prolifera.*
　　Calyx with long linear very hispid lobes and a very short tube.
　　　Plant of 1 to 2 in. . . . . 24. *M. gentianea.*
　　　Plant not exceeding ¼ in. . . . . 25. *M. phascoides.*

SECT. II. **Plecocalyx.**—*Calyx 2-lobed. Small erect branching annuals.*

Styles connivent and cohering at the top at the time of flowering.
　　Plant usually of 2 to 4 in. . . . . 26. *M. paradoxa.*
Styles distinct and parallel (except in the young bud). Plant
　　usually under 1 in. . . . . 27. *M. distylis.*

SECTION I. MITRAGYNE, *Endl.*—Calyx 4-lobed.

1. **M. Archeri,** *Hook. f. Fl. Tasm.* ii. 368. A small glabrous perennial, forming dense tufts of about ½ in. diameter or the prostrate stems extending to 1 or 2 in. Leaves crowded, often imbricate, oblong, obtuse,
coriaceous, shining, contracted and sometimes minutely ciliate at the base,
about 1 line long. Flowers nearly sessile in the uppermost axils. Calyx-
tube exceedingly short, encircling or almost adnate to the base of the ovary,
the lobes leaf-like, the 2 outer ones above 1 line long and obtuse, the 2 inner
ones narrower, more acute and rather shorter. Corolla-tube broad, scarcely
1 line long, the lobes about as long as the tube, glabrous inside. Anthers
nearly sessile in the throat, broadly ovate, with minute tips. Ovary tapering
into 2 short styles, connivent but not connate. Capsule with 2 erect
conical points, not truncate.

Tasmania. Western Mountains, *Archer.*

2. **M. montana,** *Hook. f. Fl. Tasm.* i. 274. *t.* 88 C. A dwarf prostrate
perennial forming dense patches, much smaller than *M. pilosa,* and less
slender than *M. serpyllifolia,* glabrous or slightly hirsute. Leaves obovate
or orbicular, very obtuse, rather thick, sometimes slightly ciliate, 1 to 2 lines
long. Flowers in the typical form either nearly sessile in the uppermost
axils or on pedicels of 1 to 2 lines long, but in one specimen the pedicels
attain ½ in. Calyx-tube exceedingly short, adnate to the base of the ovary;
lobes lanceolate, about 1 line long. Corolla-tube broad, about 1 line long,
glabrous inside; lobes not half so long as the tube. Anthers nearly sessile

in the throat, ovate with recurved tips.   Ovary tapering into 2 short connivent but not connate styles.   Capsule much compressed, broadly truncate, the
small erect styles forming short points to the very divaricate lobes or angles.

Victoria.   Baw-Baw Mountains at an elevation of 4500 ft., *F. Mueller.*
Tasmania.   Mount Wellington and Black Bluff Mountains, *Gunn;* Recherche Bay,
Oldfield.

3. **M. serpyllifolia,** *R. Br. Prod.* 454.   A perennial, with creeping
prostrate intricately branched filiform stems extending to a few inches, glabrous or sprinkled with a few hairs.   Leaves ovate to oblong-lanceolate,
obtuse, 1 to 2 lines long, glabrous or ciliate.   Flowers axillary, almost
sessile.   Calyx divided to about the middle or rather deeper, about 1 line
long.   Corolla-tube exceedingly short; lobes spreading to about 2 lines
diameter, the throat glabrous.   Stamens inserted in the very short tube,
nearly half as long as the lobes; anthers small, ovate.   Styles diverging at
the base, connate at the top.   Capsule obovoid-globular, slightly compressed.
—A. DC. Prod. ix. 11; Hook. f. Fl. Tasm. i. 274; *M. perpusilla,* Hook. f.
in Hook. Lond. Journ. vi. 275.

N. S. Wales.   Grose river, *R. Brown;* Mount Tomah, *A. Cunningham.*
Victoria.   Haidinger range at an elevation of 5000 ft., *F. Mueller.*
Tasmania.   Sandy soil, Western Mountains, Hampshire Hills, etc., *Gunn.*

This plant sometimes resembles *M. pilosa,* but is at once known by the deeply lobed
almost rotate corolla and exserted stamens.

4. **M. paludosa,** *R. Br. Prod.* 453.   A perennial with creeping prostrate
or ascending branched stems of 3 to 6 in., more or less pubescent or hirsute.
Leaves oblong-linear or linear-lanceolate, acute or obtuse, 3 to 4 lines long.
Flowers in the upper axils on short pedicels or the uppermost pedicels
longer in a small terminal leafy umbel.   Calyx about 1 line long, divided to
about the middle.   Corolla-tube broadly campanulate, scarcely above ½ line
long; lobes spreading to nearly 2 lines diameter.   Anthers shortly exserted,
but less so than in *M. serpyllifolia.*   Capsule nearly globular, shorter than
the calyx.—A. DC. Prod. ix. 11; *M. diffusa,* Benth. in Journ. Linn. Soc. i.
93; *M. pilosula,* F. Muell. Fragm. i. 134.

Queensland.   Burnett river, *F. Mueller;* Rockingham Bay, *Dallachy;* Stradbrooke
Island, *Fraser.*
N. S. Wales.   Port Jackson, *R. Brown, Woolls;* near Durval, *Leichhardt.*

This species approaches *M. montana* on the one hand and *M. alsinoides* on the other, but
differs from both in flowers as well as in foliage.

5. **M. pilosa,** *Labill. Pl. Nov. Holl.* i. 36. *t.* 49.   A perennial, with
diffuse or prostrate stems, sometimes rooting at the base, and forming dense
patches of 1 ft. or more, sometimes short and ascending to a few inches,
more or less hirsute with rigid hairs.   Leaves ovate orbicular or broadly
oblong, obtuse, ciliate with long hairs, 2 to 3 lines long.   Pedicels axillary,
not exceeding the leaves in the typical form.   Calyx broad, above 1 line
long, hispid, the lobes ovate.   Corolla-tube broad, as long as the calyx, the
throat slightly contracted and densely bearded inside; lobes broad, obtuse,
much shorter than the tube.   Anthers ovate, with recurved, afterwards erect
tips, not exserted.   Style split at the base.   Capsule globular, with 2 con-

nivent beaks. Seeds granular.—R. Br. Prod. 454; A. DC. Prod. ix. 11;
Hook. f. Fl. Tasm. i. 274.

N. S. Wales. Blue Mountains, *A. Cunningham.*
Victoria. Mount William, *F. Mueller;* Victoria Ranges, *Wilhelmi;* Portland, *Allitt.*
Tasmania. Port Dalrymple, *R. Brown;* common about Circular Head and Hobarton, *J. D. Hooker* and others.

Var. *Stuartii*, Hook. f. Pedicels elongated, sometimes above 1 in. long.—Near Port Sorrell, *C. Stuart;* Macquarrie Harbour, *A. Cunningham.*

6. **M. alsinoides,** *R. Br. Prod.* 453. Apparently perennial, erect, 3 to 4 in. high, glabrous or very slightly scabrous-pubescent. Leaves almost sessile, ovate or ovate-lanceolate acute, or obtuse, mostly 1 to 2 lines long. Flowers very small, on slender pedicels of 3 to 4 lines, in the upper axils. Calyx broad, very open, about 1 line long, divided to the middle. Corolla-tube very broad, shorter than the calyx; lobes longer than the tube, the throat not bearded inside. Anthers broad, with a small inflexed tip, not exserted. Style split at the base. Capsule small globular.—A. DC. Prod. ix. 11.

Queensland. Moreton Bay, *Backhouse.*
N. S. Wales. Port Jackson, *Bauer* (*R. Brown*).

I have not seen the typical specimens. The species is not in Brown's herbarium. I have described it from Backhouse's specimens, which answer well to Brown's character.

7. **M. polymorpha,** *R. Br. Prod.* 452. A perennial but flowering sometimes the first year so as to appear annual, glabrous scabrous-pubescent or hirsute, usually much branched near the base, 6 in. to 1 ft. high. Leaves rather crowded on the lower part, linear-lanceolate or oblong, mostly obtuse, with recurved margins, about ¼ in. long or rarely attaining 5 lines. Peduncles terminal, slender, simple or sparingly branched, terminating in a loose irregular umbel of 3 to 5 flowers, the pedicels long and filiform. Calyx broadly campanulate, from scarcely more than 1 line to nearly 2 lines long, the lobes acute, as long as the tube. Corolla-tube broad, about as long as the calyx; lobes spreading, at least as long as the tube, the throat bearded inside. Anthers ovate, with minute tips. Style splitting at the base. Capsule small, globular.—A. DC. Prod. ix. 10; *M. squarrosa, M. cinerascens,* and *M. canescens,* R. Br. Prod. 452, 453; A. DC. Prod. ix. 10; *M. prolifera,* Sieb. Pl. Ex., not of R. Br.; *M. Sieberi,* A. DC. Prod. ix. 10; *M. hirsuta,* Presl, Bot. Bem. 104.

Queensland. Endeavour river, *Banks and Solander.*
N. S. Wales. Port Jackson to the Blue Mountains, *R. Brown, Sieber, n.* 170, and many others; northward to Hastings river, *Beckler;* southward to Illawarra, *A. Cunningham;* Twofold Bay, *F. Mueller.*

M. polymorpha, Br., is nearly glabrous with narrow leaves; *M. squarrosa* is more pubescent, with shorter broader often recurved leaves; *M. cinerascens,* with the leaves of *M. polymorpha,* has the stems hirsute; and *M. canescens* is hirsute all over. But the degree of hairiness is not always in correspondence with the breadth of the leaves, and both characters are very variable. The tuft of a few hairs at the tip of the calyx-lobes occurs occasionally in the glabrous as well as in the hirsute forms.

Var. *calycina.* Calyx-lobes very broad and obtuse.—N. S. Wales, *Vicary.*

8. **M. longiflora,** *F. Muell. in Herb. Hook.* A slender erect annual,

resembling *M. elata* in the small rosulate radical leaves, in the elongated stems, leafless except small scales, in the irregularly umbellate inflorescence, and in the long flowers, but the calyx is much shorter (1 to 1¼ line long), with very short obtuse or scarcely acuminate lobes, and the capsule is globular. Stigmatic lobes rather long, linear, somewhat dilated and flattened.— *M. elata*, F. Muell. Fragm. i. 132, not of R. Br.

N. Australia. Grassy rocks, Wickham river and Depôt Creek, Victoria river, *F. Mueller.*

9. **M. elata,** *R. Br. Prod.* 453. A slender erect glabrous annual of 1 to 1½ ft., simple or slightly branched. Leaves radical, rosulate, obovate-oblong, mostly 3-nerved, ½ to 1 in. long, those of the stem reduced to few distant pairs of minute scales. Flowers large, few, in simple umbels at the ends of the stem or branches. Calyx narrow, about 2 lines long, the lobes as long as the tube. Corolla-tube 6 to 8 lines long; lobes rather more than half that length. Stamens inserted near the base; anthers linear, included. Stigmatic lobes linear. Capsule ovoid-oblong, acuminate, shortly exceeding the calyx.—A. DC. Prod. ix. 11.

N. Australia. Islands of the Gulf of Carpentaria, *R. Brown* ; Port Essington, *Armstrong.*
Queensland. Burdekin river, *Bowman;* Table land of South Alligator river, *Leichhardt.*

10. **M. exserta,** *F. Muell. Fragm.* i. 131. An erect branching annual of 1 to 1½ ft., glabrous except the scabrous-hirsute base. Leaves radical, apparently obovate-oblong or oblanceolate, but almost withered away from the specimens seen, those of the stem reduced to minute scales. Flowers in terminal umbels, usually compact as in *M. stellata,* but sometimes loose and often proliferous. Calyx strongly ribbed, scarcely 1⅓ lines long, the lobes rather shorter than the tube. Corolla-tube 3½ to 4 lines long, slightly dilated at the throat; lobes more than half as long as the tube. Stamens inserted about the middle of the tube; anthers narrow-oblong, shortly exserted. Style shortly exserted, the stigmatic lobes short, almost orbicular. Capsule nearly globular.

N. Australia. Sandy, rocky, often inundated places on the Victoria river, *F. Mueller.*

11. **M. tenuiflora,** *Benth.* A slender annual of ½ to 1 ft., with the habit of *M. elata* and *M. exserta.* Leaves radical, rosulate, oblong or lanceolate, those of the stem reduced to minute almost microscopic scales. Umbels terminal, several-flowered, the filiform pedicels ½ to 1 in. long. Calyx not ¾ line long, without the prominent ribs of *M. exserta,* the lobes much shorter than the tube, but more or less acute. Corolla-tube slender, nearly 3 lines long, the lobes more than half that length. Anthers and style included. Capsule globular, scarcely 1 line diameter.

N. Australia. Upper Victoria river, *F. Mueller.* Our specimens have only a single perfect corolla, so that I have not been able to ascertain the shape of the anthers and stigma. The plant may possibly be a variety of *M. longiflora,* but with the flowers and fruit not above half as large.

12. **M. ambigua,** *R. Br. Prod.* 454. A slender annual. Leaves

radical, rosulate, oblong, 2 to 3 lines long. Stems filiform, erect, usually forked, leafless except minute broad obtuse bracts at the branching and sometimes a pair lower down. Flowers very small, in simple or double loose umbels. Calyx nearly ¾ line long, the lobes very short, broad, obtuse, thick and somewhat concave. Corolla-tube cylindrical, twice as long as the calyx, the lobes very short and obtuse.—A. DC. Prod. ix. 12.

Queensland. Endeavour river, *Banks aud Solander.*

13. **M. nudicaulis,** *Reinw. in Blume Bijdr.* 849 ? A slender annual, the filiform erect stems usually 3 to 4 in., rarely 6 in. high. Leaves radical, rosulate, oblong or lanceolate, obtuse, scabrous or hirsute, 2 to 3 lines long, those of the stem reduced to minute scales. Flowers very small, in loose irregular terminal umbels, sometimes reduced to 1 or 2 flowers on filiform pedicels. Calyx about ½ line long, the lobes very acute, rather longer than the tube. Corolla-tube cylindrical, scarcely more than 1 line long, the throat slightly constricted, the lobes rather shorter than the tube. Stamens inserted in the middle of the tube; anthers orbicular or reniform, included in the tube. Stigmatic lobes oblong-linear. Capsule very small, globular.—A. DC. Prod. ix. 12 ; *M. chinensis,* Griseb. in Pl. Meyen. 51.

N. Australia. Providence Hill and, near M'Adam Range, *F. Mueller.* The species is in Khasia and in South China, and, if the identification with Reinwardt's plant be correct, also in the Moluccas.

14. **M. connata,** *R. Br. Prod.* 454. An erect annual, usually 1 to 2 ft. high, simple or scarcely branched, the foliage and base of the stem more or less hirsute, the inflorescence glabrous. Leaves in few pairs near the base of the stem, oblong-lanceolate or linear-lanceolate, obtuse or mucronate-acute, more or less prominently 3-nerved, often above 1 in. long, the upper ones few, very remote, not 2 lines long. Flowers few, on long unequal pedicels, forming 1 or 2 irregular umbels. Calyx tubular-campanulate, about 2 lines long, the lobes shorter than the tube. Corolla-tube 4 to 6 lines long ; lobes oblong or oblong-linear, shorter than the tube. Stamens inserted below the middle of the tube, which is there bearded inside; anthers narrow, included in the tube. Stigma included, clavate. Capsule nearly globular, scarcely exceeding the calyx.—A. DC. Prod. ix. 11 ; *M. constricta,* F. Muell. Fragm. i. 131.

N. Australia. Rocky hills and along streams, Victoria river, Hooker's and Sturt's Creeks, *F. Mueller ;* islands of the N. coast of Arnhem's Land, *R. Brown ;* barren elevated cliffs, Goulburn islands, *A. Cunningham ;* Port Essington, *Armstrong.*
Queensland. Cape York, *Dæmel ;* Endeavour river, *Banks and Solander ;* Cape river, *Bowman.*

The length of the corolla-lobes appears to be variable, but always, when fully opened out, more than half that of the tube. F. Mueller describes the stigmatic lobes as a length divergent. I have always found the stigmatic end of the style clavate and undivided in the open flower as well as in the bud ; there may, therefore, be perhaps some degree of unisexuality or dimorphism.

15. **M. lævis,** *Benth. in Journ. Linn. Soc.* i. 93. A glabrous annual, with slender elongated stems, erect but weak or often trailing or twining. Leaves in distant pairs, linear or lanceolate, acute or obtuse. Pedicels long

2 A 2

and filiform, often exceeding 2 in., axillary or in irregular terminal clusters or umbels. Flowers white, like those of *M. indica,* but larger. Calyx 1½ lines or after flowering 2 lines long; lobes very acute, about as long as the tube. Corolla-tube broad, fully 2 lines long, slightly constricted at the throat, which is copiously bearded inside; lobes broad, obtuse, longer than the tube. Stamens inserted near the base; anthers narrow, acuminate, included. Stigmatic lobes clavate, truncate. Capsule globular, not exceeding the calyx, the styles remaining connate or shortly splitting or at length broadly separating at the base.—*M. subvolubilis,* F. Muell. Fragm. i. 133.

N. Australia. Grassy plains near Steep Head, Victoria river, *F. Mueller;* Port Essington, *Armstrong.* The latter specimens are not good, but one of them is partially twining; the flowers are smaller than in F. Mueller's, but too much shrivelled to ascertain their precise size and proportions. The species is altogether very near *M. indica.*

16. **M. indica,** *Wight, Ic. t.* 1601. A weak, slender, glabrous or scabrous-pubescent annual, branching from the base, often only 2 or 3 in. high, but sometimes 6 to 8 in. or even more, and almost trailing. Leaves ovate ovate-lanceolate or rarely oblong, 2 to 4 lines long. Pedicels filiform, variable in length but usually about ½ in., all in the upper axils or sometimes the 2 uppermost pairs forming a terminal umbel. Calyx 1 to 1¼ lines long, the lobes acute, at least as long as the tube, with the points often very divergent. Corolla-tube broad, scarcely exceeding the calyx, the throat shortly bearded, the lobes broad, very obtuse or retuse, at least as long as the tube. Stamens inserted below the middle of the tube; anthers ovate, included. Styles separating at the base very early; stigmatic lobes short and broad. Capsule small, globular, the persistent styles parallel or connivent at the tips.—Benth. in Journ. Linn. Soc. i. 92, with the synonyms there adduced.

N. Australia. Swamps at the foot of Providence Hill and Victoria river, *F. Mueller;* Port Essington, *Armstrong.*
Queensland. Brisbane river, Moreton Bay, *F. Mueller;* Rockhampton, *O'Shanesy.*
N. S. Wales. Hastings river, *Beckler;* New England, *C. Stuart.*
Var. *latifolia.* Leaves ovate.—York Sound and Hunter's River, N.W. coast, *A. Cunningham.*
The species is also in the East Indian Peninsula.

17. **M. stellata,** *R. Br. Prod.* 454. An annual (sometimes lasting a second year?) scarcely branched, often exceeding 1 ft. but sometimes not above 2 or 3 in. high, glabrous or the leafy portion scabrous-pubescent. Leaves lanceolate or linear-lanceolate, with rather long sheathing bases, the midrib very prominent underneath, ½ to 1 in. long, usually occupying about half the stem, the other half a long peduncle bearing 1 or several compound compact umbels of small flowers. Calyx-lobes acute, about as long as the tube. Corolla almost campanulate, not 2 lines long, the throat not contracted, bearded inside; lobes very short. Stamens inserted below the middle of the tube, with tufts of hairs at their insertion; anthers small, ovate, included. Capsule nearly globular, but with a rather long point.—A. DC. Prod. ix. 11.

Queensland. Endeavour river, *Banks and Solander, A. Cunningham;* Moonlight Creek (dwarf specimens), *Bowman.*
Var. *latifolia.* Leaves broadly lanceolate or oval-oblong.

N. Australia. Port Essington, *Armstrong;* Depot Creek, Victoria river (fragmentary specimens, apparently this species), *F. Mueller.*

18. **M. Cunninghamii,** *Benth.* An annual with ascending or erect stems, the leafy portion pubescent or hirsute. Leaves narrow-lanceolate, acute, 1-nerved, $\frac{1}{4}$ to 1 in. long the sheathing base short. Peduncles long, glabrous, bearing a compact umbel of several flowers. Calyx broad above 1 line long, the lobes acute, as long as the tube. Corolla-tube broad, slightly contracted above the ovary, about 3 lines long, the throat slightly bearded inside; lobes less than half as long as the tube. Stamens inserted at the contraction of the tube, with tufts of hairs at their insertion; anthers ovate, shortly exserted. Stigmatic lobes rather long, the style not split at the base. Young capsule nearly globose, acuminate.

Queensland. Endeavour river, *A. Cunningham.* Allied to *M. stellata,* but the flowers much larger, with exserted anthers.

19. **M. pygmæa,** *R. Br. Prod.* 453. A slender annual, usually pubescent, the leafy portion of the stem very short. Leaves few, not rosulate at the base but often stellate a little above the base, obovate ovate or oblong, mostly under $\frac{1}{2}$ in. long. Flowering stems or peduncles above the stellate leaves, 2 to 6 in. high, leafless except small scale-like bracts under the umbel and quite glabrous, bearing a terminal umbel of few small flowers on filiform pedicels. Calyx scarcely $\frac{3}{4}$ line long, with acute lobes not so long as the tube. Corolla almost campanulate, the tube $1\frac{1}{2}$ lines long, the lobes short and broad. Stamens inserted at or below the middle of the tube, with tufts of hairs at their insertion; anthers oblong, included in the tube. Style not split at the base, the stigmatic lobes shortly oblong. Capsule globular, very small.—A. DC. Prod. ix. 11.

Queensland. Bay of Inlets and Bustard Bay, *Banks and Solander;* Pine Port, *R. Brown;* Wide Bay, *Bidwill;* Cape river, *Bowman;* Rockhampton, *O'Shanesy;* Brisbane river, Moreton Bay, *F. Mueller.*

M. capillaris, Wall.; A. DC. Prod. ix. 11, from E. India, scarcely differs from this species in its usually narrower leaves and larger flowers.

20. **M. lutea,** *F. Muell. Fragm.* i. 133. A slender erect annual of about 6 in., glabrous or with a few short rigid hairs on the margins of the leaves and on the stem below them. Leaves few in the lower part of the stem, the 2 uppermost pairs often approximate so as to appear whorled, linear, obtuse or scarcely acute, rarely exceeding $\frac{1}{2}$ in. Umbels few-flowered, irregularly simple or double, on a slender peduncle, with filiform pedicels. Calyx nearly 1 line long, the lobes acute. Corolla yellow (*F. Mueller*), the tube rather broad, about 2 lines long, the throat sparingly bearded, the lobes not half so long as the tube. Stamens inserted at the base of the tube; anthers included. Style not split at the base at the time of flowering, the stigmatic summit clavate-capitate. Capsule nearly globular scarcely exceeding the calyx.

N. Australia. Moist grassy places, head of Sturt's Creek, *F. Mueller.*

21. **M. multicaulis,** *R. Br. Prod.* 453. An erect annual of 2 or 3 in., nearly simple or branching from the base, pubescent at least at the base.

Leaves linear-subulate, acute, 2 to 4 lines long. Pedicels slender, often above ½ in. long, scarcely umbellate. Flowers small. Calyx narrow, the lobes subulate or very acute, as long as the tube. Corolla-tube cylindrical, longer than the calyx, the throat not contracted; lobes much shorter than the tube. Anthers small, oblong, included in the tube. Capsule ovoid, shorter thàn the calyx.—A. DC. Prod. ix. 10; *M. ramosa,* R. Br. Prod. 453; A. DC. l. c.

N. Australia. Islands of the Gulf of Carpentaria, *R. Brown ;* Main Camp, Victoria river, *F. Mueller.*

Queensland. Endeavour river, *Banks and Solander.*

The specimens I have seen of this plant are few and not very satisfactory. It appears to be near *M. laricifolia,* but usually pubescent, more branching, with a much narrower calyx and capsule.

22. **M. laricifolia,** *R. Br. Prod.* 453. A slender but rigid, glabrous or slightly pubescent annual, usually 1 to 2 in. high. Leaves linear-subulate, very acute, 2 to 4 lines long, the upper ones usually crowded. Pedicels filiform in the upper axils, often longer than the rest of the plant. Calyx about ¾ line long, the lobes very acute. Corolla about 2 lines long, the tube cylindrical, the lobes very obtuse, about half as long as the tube. Anthers linear-sagittate, included. Stigmatic lobes oblong. Capsule small, globular. —A. DC. Prod. ix. 10.

N. Australia. Victoria river, *F. Mueller.*

Queensland. Endeavour river, *Banks and Solander.*

23. **M. prolifera,** *R. Br. Prod.* 453. A little densely branched erect or diffuse annual of ½ to 1 in., minutely scabrous-pubescent. Leaves lanceolate or oblong-lanceolate, 1½ to 3 lines long. Pedicels clustered, scarcely so long as the leaves. Calyx very prominently 4-angled, about 1 line long, the lobes divaricate, mucronate or almost aristate, shorter than the tube. Corolla-tube about as long as the calyx, the throat bearded inside; lobes rather shorter than the tube. Stamens inserted below the middle of the tube; anthers ovate, included, Style split at the base from the time of flowering; stigmatic lobes oblong. Capsule nearly globular, shorter than the calyx. Seeds very small.—A. DC. Prod. ix. 10.

N. Australia. Victoria river, near the Main Camp, *F. Mueller.*

Queensland. Shoalwater Bay, *R. Brown.*

Var. ? *diffusa.* More diffuse, with rather broader leaves, forming tufts of about 1 in. diameter.—Macadam Range, *F. Mueller.*

Var. ? *major.* More erect, less branched, attaining 2 in.; leaves also rather larger.— Endeavour river, *Banks and Solander.*

Some varieties of *M. polymorpha* occur frequently in herbaria under the name of *M. prolifera.*

24. **M. gentianea,** *F. Muell. Fragm.* i. 130. An erect, dichotomously branched, corymbose annual of 1 to 2 in. or rarely higher, more or less hirsute with rigid hairs. Leaves linear oŕ linear-lanceolate, the larger ones ½ in. long or rather more. Peduncles in the upper axils, slender, shorter than the leaves. Calyx hispid, about 3 lines long, the tube very short, the lobes linear and acute. Corolla-tube about as long as the calyx, the throat dilated and bearded inside; lobes ovate, about half as long as the tube. Stamens

inserted below the middle of the tube; anthers linear-sagittate, included. Style not split at the time of flowering, the stigmatic lobes very short. Capsule about half as long as the calyx, tapering into 2 beaks. Seeds very small, black, smooth and shining.

N. Australia. Inundated banks of Victoria river, *F. Mueller.*

25. **M. phascoides,** *R. Br. Prod.* 454. A minute plant, forming sessile stemless tufts of leaves and flowers, not above ¼ in. high. Leaves linear, hispid with rather long hairs. Flowers nearly sessile in the axils. Calyx-tube very short, the lobes linear and hispid like the leaves. Corolla not seen. Capsule globular, but the summit widely gaping, the 2 short styles connivent at the tips.—A. DC. Prod. ix. 11.

Queensland. Endeavour river, *Banks and Solander.*

SECT. II. PLECOCALYX, *G. Don.*—Calyx 2-lobed.

26. **M. paradoxa,** *R. Br. Prod.* 454. A slender erect branching annual, usually 3 to 4 in., rarely 6 in. high, glabrous or with a few hairs about the lower leaves. Leaves oblong oblanceolate or linear, connate and shortly sheathing at the base, 2 to 3 lines long. Pedicels filiform, ½ to 1½ in. long, the upper ones sometimes forming a terminal umbel. Calyx at the time of flowering a little above 1 line long, with 2 broad herbaceous lobes scarcely so long as the tube, enlarged after flowering to nearly 2 lines, with the lobes diverging and acute. Corolla-tube contracted above the ovary, the upper part broad, almost campanulate, scarcely exceeding the calyx, the lobes shorter than the tube. Stamens inserted below the middle of the tube; anthers included, orbicular-reniform. Style split at the base from the time of flowering; stigmatic lobes dilated upwards. Capsule included in the calyx. —A. DC. Prod. ix. 11; Hook. f. Fl. Tasm. i. 274. t. 88 A (incorrect as to the style); Nees in Pl. Preiss. i. 365; *M. divergens,* Hook. f. in Hook. Lond. Journ. vi. 276; *M. nuda,* Nees in Pl. Preiss. ii. 239 (with few leaves close together at the base of the stem).

Victoria. Dry sandy places, Port Phillip, and about Melbourne, *F. Mueller, Harvey,* and others; Mount Abrupt and the Grampians, *F.·Mueller;* Portland, *Allitt;* Wendu Vale, *Robertson.*

Tasmania. Not uncommon in poor land near the sea, *J. D. Hooker.*

S. Australia. Crystal Brook, Mount Burr, *F. Mueller.*

W. Australia. Dry sandy places from King George's Sound, *R. Brown, Preiss, n.* 2377, and others, to Swan River, *Drummond,* 1st *Coll. n.* 700, *Preiss, n.* 2240, and others; moist mossy rocks, Stirling Range, *F. Mueller.*

27. **M. distylis,** *F. Muell. in Trans. Phil. Soc. Vict.* i. 20, *and in Hook. Kew Journ.* viii. 164. A little erect or ascending annual, simple or slightly branched, from ½ to 1 in. high. Leaves oblong-linear, rather thick, 1 to 2 lines long. Pedicels filiform, 2 to 4 lines long. Calyx broad, very shortly and obtusely 2-lobed, about 1 line long when in flower, slightly enlarged when in fruit. Corolla as long as the calyx, campanulate, the lobes very short. Stamens included in the tube; anthers short and broad. Styles sometimes cohering in the bud, but quite distinct and parallel at the time of flowering, the stigmatic ends slightly dilated and not cohering. Capsule shorter than the calyx. Seeds with a loose reticulate testa.—Hook. f. Fl. Tasm. i. 274. t. 88 B.

Victoria. Around swamps near Mount William, in the Grampians, *F. Mueller.*
Tasmania. Near Georgetown, *Gunn.*

3. LOGANIA, R. Br.

(Euosma, *Andr. not of Willd.*)

Calyx 5-cleft, rarely 4-cleft. Corolla campanulate or with a cylindrical tube; lobes 5, rarely 4, imbricate in the bud. Stamens 5, rarely 4, inserted in the tube; anthers linear or ovate, included or exserted but shorter than the corolla. Ovary 2-celled, with several ovules or rarely only a single one in each cell; style simple, with a capitate or oblong undivided stigma. Capsule ovoid oblong or globular, septicidally dehiscent, the carpels almost separating, opening inwards by a longitudinal slit, leaving the placentæ at length free without any pulp. Seeds ovoid or more or less peltate.—Herbs undershrubs or shrubs. Leaves opposite, connected by a raised stipular line or short sheath or rarely with small setaceous stipules. Flowers white or rarely flesh-coloured, usually small, in terminal or rarely axillary cymes or panicles, or sometimes solitary, often more or less unisexual. Bracts small.

The genus is limited to Australia, with the exception of a single species found in New Zealand.

SECT. I. **Eulogania.**—*Calyx-segments obtuse. Stamens inserted in the middle of the tube; anthers included. Stems shrubby, at least at the base. Flowers more or less diœcious.*

Leaves from lanceolate to orbicular, flat or with recurved margins.
  Corolla bearded in the throat.
    Flowers in pedunculate cymes, forming a terminal thyrsoid panicle, leafy at the base.
      Leaves lanceolate, acuminate . . . . 1. *L. longifolia.*
      Leaves shortly ovate or elliptical . . . . 4. *L. ovata.*
    Flowers in sessile or shortly pedunculate compact terminal cymes.
      Leaves obovate or orbicular, obtuse or shortly and obtusely acuminate, $1\frac{1}{2}$ to $2\frac{1}{2}$ in. long . . . . 2. *L. latifolia.*
      Leaves obovate or orbicular, very thick, smooth and shining, $\frac{3}{4}$ to $1\frac{1}{2}$ in. long . . . . 3. *L. crassifolia.*
      Leaves from broadly cordate-ovate to oval-elliptical, $\frac{1}{2}$ to 1 in. long . . . . 4. *L. ovata.*
      Leaves from obovate to oval, very obtuse, under $\frac{1}{2}$ in. long . . 5. *L. buxifolia.*
Leaves linear. Flowers small, in small loose terminal cymes.
  Leaves with revolute margins.
    Flowers 5-merous. Corolla bearded at the throat . . . . . 6. *L. stenophylla.*
    Flowers 4-merous. Corolla glabrous inside. Ovules solitary 7. *L. micrantha.*
  Leaves flat or concave.
    Corolla glabrous inside or nearly so. Cymes mostly pedunculate . . . . 8. *L. linifolia.*
    Corolla bearded in the throat. Cymes mostly sessile, and shorter than the leaves . . . . 9. *L. fasciculata.*
Leaves linear or lanceolate. Cymes axillary . . . . 10. *L. floribunda.*
Leaves sessile, deeply cordate. Cymes axillary and terminal, in a long narrow thyrsoid leafy panicle . . . . 11. *L. cordifolia.*

SECT. II. **Stomandra.**—*Calyx-segments acute or rarely rather obtuse. Stamens inserted in the throat; anthers exserted (but not exceeding the corolla-lobes). Herbs or undershrubs. Flowers hermaphrodite.*

Stems erect, rigid or virgate. Leaves narrow or none.
  Stems leafless. Flowers at the nodes. Calyx-segments not 1 line long, rather obtuse . . . . 12. *L. nuda.*

Stems more or less leafy. Calyx-segments long, linear, acute.
  Leaves linear, with revolute margins, mostly reduced to small
    scales. Flowers at the nodes or solitary and terminal . . . 13. *L. spermacocea.*
  Leaves linear, with revolute margins. Flowers solitary and ter-
    minal. Corolla densely bearded at the throat . . . . . 14. *L. callosa.*
  Leaves linear or lanceolate. Flowers terminal, solitary or few
    in a broad cyme. Corolla not bearded . . . . 15. *L. campanulata.*
Low, branching, slender, erect or diffuse, leafy herbs. Leaves small,
  ovate-oblong or lanceolate.
Flowers few, sessile in little terminal heads . . . . 16. *L. serpyllifolia.*
Flowers all solitary in the axils . . . . 17. *L. pusilla.*

Sect. I. Eulogania, *DC.*—Calyx-segments obtuse. Stamens inserted
in the middle of the tube; anthers included. Stems shrubby, at least at
the base. Flowers more or less dioecious, the males with an abortive ovary,
the style apparently developed but with a narrow stigma ; the females with
abortive or imperfect anthers.

1. **L. longifolia,** *R. Br. Prod.* 456. An erect glabrous shrub or
undershrub, attaining 5 or 6 ft. or even more. Leaves narrow-elliptical or
lanceolate, acuminate acute or almost obtuse, narrowed into a rather long
petiole or rarely almost sessile, distinctly penniveined, 1½ to 3 in. long or
sometimes even longer. Flowers smaller than in *L. latifolia* and more numerous, in pedunculate cymes, terminal or in the upper axils, usually forming
a loose terminal leafy panicle. Calyx-segments about ½ in. long. Corolla-
tube rather longer than the calyx, the throat bearded inside. Anthers in the
males ovate-oblong. Stigma in the females ovate-capitate, in the males
cylindrical or clavate. Ovules numerous in each cell on a small almost
stipitate placenta. Capsule ovoid-oblong, about 3 lines long.—DC. Prod.
ix. 25 ; Nees in Pl. Preiss. i. 367.

S. Australia. Sturt river, Mount Lofty, Mount Remarkable, *F. Mueller ;* Onkaparinga river, *Whittaker.*

W. Australia. King George's Sound and adjoining districts, *R. Brown, A. Cunningham,* and others ; Swan river, *Fraser, Drummond, 1st Coll.; Preiss, n.* 1245, 1246, and
others.

Var. *brevifolia.* Leaves shorter broader and less acuminate, almost intermediate between
this species and *L. latifolia.—*Bremer Bay, *Maxwell ;* Crystal Brook, *F. Mueller.*

Var. *subsessilis.* Leaves nearly sessile, with recurved margins.—Onkaparinga river, *F.
Mueller.*

This and the four following species seem to pass gradually one into the other, and might
well be considered as varieties of one species. F. Mueller, Fragm. vi. 132, proposes to join
three of them under the name of *L. vaginalis,* retaining *L. crassifolia* and *L. buxifolia,*
both of which appear to me to be much nearer to *L. ovata* than this is to *L. longifolia.*

2. **L. latifolia,** *R. Br. Prod.* 455. An erect glabrous and often glaucous
shrub or undershrub of about 3 ft. Leaves from broadly obovate to oblong-
elliptical, obtuse or very shortly acuminate, narrowed into a short petiole,
opaque or shining above but not so thick as in *L. crassifolia,* mostly 1½ to
2½ in. long. Flowers in dense terminal sessile or shortly pedunculate cymes.
Calyx-segments ovate, ciliolate, about 1 line long. Corolla-tube rather
longer, broad, bearded inside at the throat. Stamens, ovary, and style of *L.
longifolia.* Capsule ovoid-oblong, about 4 lines long. — Spreng. Neue

Entdeck. i. t. 2 (a narrow form); DC. Prod. ix. 25 ; Nees in Pl. Preiss. i. 366 ; *Exacum vaginale,* Labill. Pl. Nov. Holl. i. 37. t. 51.

W. Australia. King George's Sound, *Labillardière, R. Brown, Drummond,* 5*th Coll. n.* 344, *Preiss, n.* 1244, and others; eastward to Cape Arid and Espérance Bay, *Maxwell.*

3. **L. crassifolia,** *R. Br. Prod.* 455. A rigid diffuse shrub, the branches and inflorescence scabrous, otherwise glabrous. Leaves on very short petioles, broadly ovate or orbicular, very obtuse, thick smooth and shining, with recurved nerve-like thickened margins, $\frac{3}{4}$ to $1\frac{1}{2}$ in. long. Flowers numerous, in compact terminal almost sessile cymes, the pedicels very short. Calyx, corolla, stamens, ovary, style, and capsule, entirely as in *L. latifolia.*—DC. Prod. ix. 25.

S. Australia. Memory Cove and Port Lincoln, *R. Brown, Wilhelmi ;* Guichen Bay and Lake Alexandrina, *F. Mueller.*

W. Australia. King George's Sound or to the eastward (Lucky Bay ?), *Baxter.* Probably a strictly maritime variety of *L. latifolia.*

4. **L. ovata,** *R. Br. Prod.* 455. An erect glabrous shrub. Leaves sessile or very shortly petiolate, orbicular ovate obovate or broadly elliptical, obtuse or acute, rounded slightly cordate or shortly contracted at the base, rather thick and often shining above, but more distinctly veined than in *L. crassifolia,* and the margins not so thick, $\frac{1}{2}$ to 1 in. long. Flowers in small compact cymes, on simple or branched peduncles, terminal or in the upper axils. Calyx-segments ovate, very obtuse, $\frac{1}{2}$ to $\frac{3}{4}$ line long. Corolla-tube about as long as the calyx, bearded inside at the throat, the lobes broad, obtuse, about as long as the tube. Anthers rather large in the males, abortive in the females. Ovary rudimentary in the males, but the style developed with a rather large clavate stigma ; stigma in the females almost globular. Capsule ovate, acuminate, about 2 lines long.—DC. Prod. ix. 25 ; *L. elliptica,* R. Br. Prod. 455 ; DC. Prod. ix. 25.

Victoria. Mouth of the Glenelg, *Robertson, F. Mueller ;* Grampians ?, *Mitchell.*

S. Australia. S. coast, *R. Brown ;* Port Lincoln and Encounter Bay, *Whittaker ;* Guichen Bay, *Schulzen ;* Kangaroo Island, *Waterhouse.*

W. Australia. King George's Sound, *R. Brown, Baxter ;* Flinders Bay, *Collie ;* Champion Bay, *Oldfield.*

Besides the foliage, the flowers are smaller and the inflorescence looser than in *L. latifolia* and *L. crassifolia.* Some specimens seem almost to pass into the short-leaved forms of *L. longifolia,* but with smaller much more compact cymes. The leaves are sometimes very nearly those of *L. crassifolia,* sometimes again almost as small as in *L. buxifolia.*

5. **L. buxifolia,** *F. Muell. Fragm.* vi. 132. An erect glabrous shrub of 1 to 2 ft. Leaves from obovate to oval, very obtuse, contracted into a short petiole, thick and almost veinless, 3 to 4 lines long. Flowers in small compact cymes, on short terminal peduncles, their structure entirely as in *L. ovata,* or rather smaller. Capsule, as in that species, ovoid, about 2 lines long.

W. Australia, *Drummond,* 5*th Coll. n.* 345 ; South-west Bay and West Mount Barren, *Maxwell.*

6. **L. stenophylla,** *F. Muell. Fragm.* i. 128. A small shrub, quite glabrous. Leaves linear, obtuse, rather thick, with revolute margins, mostly

about ½ in. long. Flowers small, in small rather loose terminal cymes, usually pedunculate, the pedicels very short. Calyx-segments obtuse, under 1 line long. Corolla-tube shorter than the calyx, bearded inside at the throat; lobes about as long as the tube; the males with rather long anthers, a flat abortive ovary and a cylindrical clavate stigma; the females with abortive anthers, a very obtuse or emarginate ovary and broadly capitate stigma, the ovules numerous. Capsule ovoid obtuse, about 2 lines long.

W. Australia. Phillips river, *Maxwell.*

7. **L. micrantha,** *Benth. in Journ. Linn. Soc.* i. 94. A small divaricately-branched shrub, glabrous or the young shoots minutely pubescent. Leaves very shortly petiolate, linear, obtuse, rather thick, with recurved or thickened margins, under ½ in. long. Flowers minute, 4-merous, in little cymes or clusters terminal or in the upper axils, on peduncles of 1 to 2 lines, sometimes almost solitary. Calyx-segments very obtuse, scarcely as long as the corolla-tube. Corolla altogether about ¾ line long, the lobes very obtuse, about as long as the tube, the throat not bearded. Anthers in the males rather large, ovate; ovary abortive; stigma small oblong. Ovary in the females large, with a broad sessile stigma, and a single ovule in each cell peltately attached to the prominent placenta. Capsule ovoid, as long as the calyx. Seeds solitary in each cell, flattened; testa minutely pitted; albumen firm; embryo with a rather long radicle.

W. Australia. Towards Cape Riche, *Drummond, 5th Coll. n.* 252, 253.

8. **L. linifolia,** *Schlecht. Linnæa,* xx. 605. An erect branching shrub, quite glabrous or rarely pubescent. Leaves linear, obtuse, rather thick, flat or concave, narrowed into a short petiole, mostly under ½ in. long except on the main branches where they attain sometimes ¾ in. Flowers rather small, in small pedunculate terminal cymes. Calyx-segments obtuse; corolla and stamens of *L. ovata,* except that the corolla is either quite glabrous inside or only very slightly bearded at the base of the lobes as in *L. floribunda,* from which this is at once distinguished by the terminal inflorescence.

N. S. Wales. Between the Murrumbidgee and the Darling rivers, *F. Mueller.*
Victoria. Wimmera and Murray desert, *Dallachy.*
S. Australia. From the Murray to St. Vincent's Gulf, *F. Mueller;* Encounter Bay, *Whittaker.*

9. **L. fasciculata,** *R. Br. Prod.* 456. A small diffuse or procumbent much-branched shrub, the branches and young shoots pubescent or glabrous. Leaves linear oblong or somewhat spathulate, obtuse, thick, flat or slightly concave, the margins never recurved, about ¼ in. long. Flowers small, in little compact terminal sessile cymes, often not exceeding the leaves, and sometimes reduced to 1 or 2 flowers, the pedicels very short. Calyx-segments oval-oblong, very obtuse, ciliolate, ¾ line long. Corolla-tube nearly as long as the calyx, bearded inside at the throat, the lobes as long as the tube. Anthers in the males rather large, and the stigma thick and large, but the ovary abortive. Female flowers rather smaller. Capsule ovoid, contracted at the top, about 2 lines long.—DC. Prod. ix. 26; *L. bracteolata,* Nees in Pl. Preiss. i. 367.

W. Australia. King George's Sound, *R. Brown.* Dry limestone ridges, Bald Head,

Preiss, n. 1249, and thence eastward to Espérance Bay, *Maxwell.* The shortness of the lateral flowering branches often make the cymes appear to be axillary.

10. **L. floribunda,** *R. Br. Prod.* 456. An erect shrub, attaining several feet, glabrous or slightly pubescent, the branches more or less angular. Leaves lanceolate or linear, acute acuminate or rather obtuse, narrowed into a short petiole, flat or with revolute margins, pale underneath, with a prominent midrib, otherwise almost veinless, $1\frac{1}{2}$ to 3 in. long. Flowers small, usually 5-merous but occasionally 4-merous, in axillary trichotomous cymes or panicles much shorter than the leaves, rarely reduced to almost simple racemes. Calyx-segments broad and obtuse, shorter than the corolla-tube, minutely ciliolate. Corolla 1 to $1\frac{1}{2}$ lines long, the lobes very broad and obtuse, shorter than the tube, the throat often slightly thickened inside and glabrous or slightly pubescent or bearded. Anthers included in the tube, ovate. Female flowers usually rather smaller than the males. Capsule rather narrow, about 2 lines long.—DC. Prod. ix. 25 ; Endl. Iconogr. t. 57 ; Lodd. Bot. Cab. t. 1118 ; *Euosma albiflora,* Andr. Bot. Rep. t. 520 ; *Logania angustifolia,* Sieb. in Spreng. Syst. Cur. Post. 59 ; DC. Prod. ix. 25.

N. S. Wales. Port Jackson and Blue Mountains, very common, *R. Brown, Sieber, n.* 288, 290, and *Fl. Mixt. n.* 509, 511, and many others ; northward to Clarence river, *Beckler ;* New England, *C. Stuart ;* southward to Illawarra, *Shepherd.*
Victoria. Delatite river, Buffalo Range, *F. Mueller.*

Var. *brevifolia.* Leaves lanceolate-elliptical, acute at both ends, under 1 in. long.— Paramatta, *A. Cunningham.*

L. revoluta, R. Br. Prod, 456, DC. Prod. ix. 25, from Grose river, *R. Brown,* Upper Genoa river, *F. Mueller,* is a form with very narrow, much revolute, and rather shorter leaves, which, in other localities, passes gradually into the commoner form. The corolla appears to be quite glabrous inside, but so it is in many specimens of the larger-leaved varieties.

This species is usually distinguished from most others by " lateral setaceous distinct stipules." The short, sometimes scarcely prominent, sheath or raised line connecting the leaves, appears to me to be nearly the same in all the species, although in *L. floribunda* it is more decidedly pubescent-ciliate. The truly axillary inflorescence appears, however, to be a constant distinctive character. The flowers are frequently, but perhaps not quite so constantly, diœcious as in the other species.

11. **L. cordifolia,** *Hook. in Mitch. Trop. Austr.* 341. A tall glabrous and glaucous erect perennial or undershrub, the branches smooth and terete. Leaves sessile, deeply cordate with rounded auricles, ovate or lanceolate, acute, 3- or 5-nerved at the base, 1 to 2 in. long. Flowers small, in little compact trichotomous cymes forming numerous thyrsoid panicles, terminal or in the upper axils, sessile and compact or looser on long peduncles. Calyx-segments broad and obtuse, scarcely ciliolate, not $\frac{1}{2}$ line long. Corolla about 1 line long, the lobes very broad, obtuse, shorter than the broad tube, quite glabrous inside. Anthers in the males ovate, included ; stigma ovoid ; ovary abortive. Female flowers usually rather smaller than the males, in more compact inflorescences, the calyx-segments less obtuse ; anthers quite abortive and the filaments minute ; ovary prominent, with a thick stigma on a very short style. Capsule about 2 lines long, somewhat compressed. Seeds oblong, much flattened, the testa minutely reticulate.

Queensland. Among rocks near Mount Pluto, *Mitchell.*

SECT. 2. STOMANDRA, *R. Br.*—Calyx-segments acute or rarely rather obtuse. Stamens inserted in the throat of the corolla ; anthers usually linear, exserted but not exceeding the corolla-lobes.   Herbs   or   undershrubs. Flowers hermaphrodite.

12. **L. nuda,** *F. Muell. Fragm.* i. 129.   An undershrub, with erect rush-like leafless stems, the branches opposite or sometimes clustered, usually terete and sometimes ending in a weak spine.   Leaves replaced by minute scales.   Flowers opposite or in opposite clusters of 3 or 4 or even more at the nodes, the pedicels very short.   Calyx-segments ovate, rather obtuse, scarcely exceeding 1 line.   Corolla campanulate, pubescent, about 4 lines long, the lobes longer than the tube, very spreading, bearded inside at the base as well as the throat.   Stamens inserted in the sinus of the lobes. Anthers narrow oblong.   Ovary pubescent, tapering into the style, the stigma less thickened than in the allied species.   Ovules about 10 in each cell, on a broad peltate placenta.   Capsule 2 to 2½ lines long.

N. S. Wales. Desert of the Murray and Darling, *Victorian Expedition.*
W. Australia, *Drummond.* Specimens very bad, with here and there a solitary bud or capsule and no expanded flowers, and therefore somewhat doubtful.

13. **L. spermacocea,** *F. Muell. Fragm,* vi. 134.   A perennial with numerous erect rigid Ephedra-like striate stems, 6 to 9 in. high, glabrous pubescent or hirsute.   Leaves linear, erect with revolute margins, sometimes ¼ to ½ in. long, sometimes nearly all reduced to small scales.   Flowers solitary in the upper axils on very short peduncles, bearing 1 or sometimes 2 pairs of small leaves or bracts.   Calyx-segments linear, acute, minutely ciliate, prominently nerved, longer than the corolla-tube.   Corolla campanulate, 4 to 5 lines long, the lobes as long as the tube, the throat slightly pubescent inside.   Stamens inserted in the sinus of the lobes ; anthers oblong. Ovules very numerous ; stigma ovoid-oblong.

W. Australia. Swan River, *Drummond, 1st Coll. n.* 651 ; Murchison river and Champion Bay, *Oldfield.* In some specimens both the anthers and the stigma are longer than in others, but in both cases the ovules appear to be quite perfect.

14. **L. callosa,** *F. Muell. Fragm.* vi. 134.   The single specimen seen has a hard base with several erect simple stems under 6 in. high.   Leaves erect, linear, obtuse, with revolute margins, ciliate below the middle, otherwise glabrous, 3 to 4 lines long.   Flowers pubescent, solitary and terminal, of the size, shape, and structure of those of *L. campanulata,* except that the corolla-lobes are longer than the tube, and densely bearded inside near the base.

W. Australia, *Drummond (Herb. F. Mueller).* The species requires further investigation from better specimens.

15. **L. campanulata,** *R. Br. Prod.* 456.   A perennial, with erect virgate simple or slightly branched stems of 1 to 2 ft., glabrous except the inflorescence which is more or less pubescent.   Leaves linear or linear-lanceolate, erect, obtuse or acute, with revolute margins, ½ to 1 in. long. Flowers solitary and terminal or few in a very loose broad terminal cyme, pedicellate in the forks or terminating the branches.   Calyx-segments lanceo-

late-subulate, longer than the corolla-tube. Corolla narrow-campanulate, $\frac{1}{2}$ to $\frac{3}{4}$ in. long, the lobes about as long as the tube, the left-hand edges usually overlapping in the bud excepting the fifth, which is entirely inside, pubescent inside and out but without the longer beard of *L. callosa.* Stamens inserted under the sinus of the lobes ; anthers linear. Ovules very numerous in each cell of the ovary ; stigma long and cylindrical. Capsule not seen.—DC. Prod. ix. 26 ; De Vr. in Pl. Preiss. ii. 240 ; F. Muell. Fragm. vi. 133 ; Hook. Ic. Pl. t. 832.

W. Australia. King George's Sound, *R. Brown* and others ; Swan River, *Drummond, 1st Coll. n.* 650 ; near Seven-mile Bridge, *Preiss, n.* 1885 ; Tone river, Gardiner Range, Bremer Bay, E. Mount Barren, Moir's Inlet, *Maxwell.*

16. **L. serpyllifolia,** *R. Br. Prod.* 456. A slender perennial or undershrub, usually much branched, erect or diffuse, glabrous or shortly hispid, 6 in. to 1 ft. high. Leaves scarcely petiolate, ovate or lanceolate, obtuse or almost acute, the margins recurved, glabrous or ciliate, 2 to 4 lines or very rarely nearly $\frac{1}{2}$ in. long, the connecting stipular line usually shortly ciliate. Flowers usually 3 to 5 together almost sessile in little terminal leafy heads. Calyx-segments linear-lanceolate, subulate-acuminate, ciliate, often nearly as long as the corolla. Corolla white or flesh-coloured, campanulate, $2\frac{1}{2}$ to 3 lines long, the lobes longer than the tube, conspicuously veined and bearded inside as well as the throat. Stamens inserted under the sinus of the lobes ; anthers linear. Capsule ovate, not exceeding the calyx.—DC. Prod. ix. 26 ; F. Muell. Fragm. vi. 133 ; *L. hispidula,* Nees in Pl. Preiss. i. 368, and *L. centaurium,* Nees, l. c. ii. 240 (from the descriptions given).

W. Australia. King George's Sound, *R. Brown* and many others, to Swan River and Champion Bay, *Oldfield* and others, *Drummond, n,* 66, 116, 144, 145, 222 ; eastward to Cape le Grand, *Maxwell.*

Var. *angustifolia.* Leaves lanceolate or linear-lanceolate, about $\frac{1}{2}$ in. long, contracted into a short petiole ; flowers sometimes but not always rather larger.—*L. hyssopoides,* Nees in Pl. Preiss. i. 368.—*Drummond, n.* 146, 221, *Preiss, n.* 1242 ; Cape Leschenault, *Oldfield.*

17. **L. pusilla,** *R. Br. Prod.* 456. A small glabrous or pubescent, procumbent or diffuse herb or undershrub, the branches ascending to a few inches. Leaves from obovate to elliptical or oblong, obtuse, with recurved or revolute margins, narrowed into a short petiole, $\frac{1}{4}$ to nearly $\frac{1}{2}$ in. long. Flowers solitary in one axil of the pair of leaves, sessile or shortly pedicellate. Calyx-segments lanceolate-subulate or linear, acute, with almost scarious edges, longer than the corolla-tube. Corolla about 3 lines long, veined as in *L. serpyllifolia,* slightly hairy inside, the lobes about as long as the tube. Stamens inserted under the sinus of the lobes ; anthers oblong. Ovary tapering into the style ; stigma ovoid. Capsule rather acute, about as long as the calyx.—DC. Prod. ix. 26 ; Endl. Iconogr. t. 58.

N. S. Wales. Port Jackson, *R. Brown, A. and R. Cunningham, Woolls.*

4. GENIOSTOMA, Forst.

Calyx 5-lobed. Corolla-tube shortly cylindrical, hairy inside at the throat ; lobes 5, spreading, imbricate, usually contorted in the bud. Stamens 5 ;

anthers exserted. Ovary 2-celled; ovules several in each cell; style simple,
with a thick capitate or oblong stigma. Capsule opening in 2 broad spreading valves, leaving the erect placentas consolidated with the axis in a pulpy
mass enclosing the seeds. Embryo cylindrical, nearly as long as the albumen.—Shrubs. Flowers small, in opposite axillary sessile cymes or clusters.
Bracts small.

The genus extends from the Mascarene Islands to the Indian Archipelago, the islands of
the South Pacific, and New Zealand. The only Australian species is perhaps endemic, but
closely resembles the commonest South Pacific form. In some Polynesian species the imbricate corolla-lobes are, according to A. Gray, not strictly contorted.

1. **G. australianum,** *F. Muell. Fragm.* v. 19. A tall shrub or small
tree, quite glabrous. Leaves shortly petiolate, oblong-lanceolate, acuminate,
contracted at the base, under 2 in. long in some specimens, 3 to 4 in. long in
others; stipules a short truncate sheath. Flowers white, in sessile dichotomous cymes shortly exceeding the petioles, the pedicels scarcely 1 line long,
each with a pair of minute bracteoles at or below the middle. Calyx under
1 line long, the lobes acuminate. Corolla-tube broad, about as long as the
calyx, the lobes rather longer, more or less bearded inside, at least at the
throat, contorted in the bud, the left-hand edges overlapping. Filaments
very short and flat; anthers opening rather broad, nearly as long as the
corolla-lobes. Capsule ovoid or almost globular, but rather acute, about
¼ in. diameter, the valves when open rather thick, concave, recurved, the placentas remaining long persistent, enveloping the seeds in a thin pellicle,
which when soaked swells into a pulpy mass.

Queensland. Rockingham Bay, *Dallachy*. Very closely allied to the narrow-leaved
forms of *G. rupestre*, Forst., and probably a variety only, with rather larger flowers, the inflorescence usually more compact, and the fruits less obtuse.

5. FAGRÆA, Thunb.

Calyx 5-cleft. Corolla-tube usually expanded at the top into a campanulate throat, the limb spreading, often oblique; lobes 5, rather unequal, imbricate in the bud. Stamens 5, usually shortly exserted. Ovary more or
less completely 2- or rarely 3-celled, the placentas often not meeting in the
centre, at least at an early stage; ovules several to each cell or placenta;
style single, with a peltate stigma. Fruit succulent, indehiscent. Seeds
immersed in pulp; albumen copious; embryo very small.—Trees or shrubs.
Leaves coriaceous. Flowers usually rather large, in terminal raceme-like or
corymbose panicles, rarely reduced to a single flower. Bracts small, with 2
bracteoles under the calyx.

The genus extends over East India, Ceylon, and the Indian Archipelago. Of the two
Australian species, one appears to be common in the eastern portion of that area, the other,
as far as known, is endemic.

Leaves broad, rounded at the base. Flowers in clusters or cymes along a
  simple elongated terminal rhachis . . . . 1. *F. racemosa.*
Leaves narrow, contracted at the base. Flowers in a short sessile terminal cyme . . . . 2. *F. Muelleri.*

1. **F. racemosa,** *Jack in Roxb. Fl. Ind. ed. Wall.* ii. 35. A tall shrub

or small tree, quite glabrous. Leaves oval-oblong or rarely ovate, shortly
acuminate, rounded at the base, mostly 8 in. to 1 ft. long, on petioles of from
¼ to 1 in., the stipules forming a short interpetiolar sheath or ring. Flowers
of a dirty yellowish-white, in clusters or cymes along the simple rhachis of a
raceme-like terminal nodding panicle, which is pedunculate between the last
pair of leaves, and varies from 2 or 3 in. to above 1 ft. in length. Pedicels
thick, 2 to 3 lines long. Calyx-lobes broad, 1 to 1½ lines long. Corolla-
tube about ¾ in. long, the throat broadly campanulate, the lobes broad, 3 to 4
lines long. Ovary with 2 or rarely 3 parietal placentas, often short in the
bud, but meeting in the centre and completely dividing the ovary into 2 or 3
cells after the flowering is over.—A. DC. Prod. ix. 29 ; *F. volubilis,* Jack in
Roxb. Fl. Ind. ed. Wall. ii. 36 (A. DC. Prod. ix. 30), according to Jack in
Mal. Misc. ii. 82 ; *F. morindifolia,* Blume in Rumphia ii. 32. t. 79, and
(analysis only) t. 73. f. 2 ; Mus. Bot. i. 169; A. DC. Prod. ix. 29, and probably also several of those described as allied species by Blume, Mus. Bot. i.
169, 170 ; *F. Thwaitesii,* F. Muell. Fragm. ii. 137.

N. Australia. Providence Hill, *F. Mueller.* Extends over the Indian Archipelago
to the Malayan Peninsula and the Philippine Islands. It is probably from some accidentally
weak flexuose branch that *F. volubilis* came to be described as a twiner. All authors, including Jack himself, speak of it as a shrub growing sometimes into a small tree.

2. **F. Muelleri,** *Benth.* A glabrous tree or shrub. Leaves opposite,
crowded at the ends of the branches, elliptical-oblong or lanceolate, shortly
acuminate, contracted into a petiole, thick, obscurely veined except the prominent midrib, 3 to 6 in. long. Peduncles terminal, short and apparently
few-flowered. Calyx-lobes orbicular, thick, rather above 1 line diameter.
Corolla not seen. Fruit red, under ½ in. diameter, ripening 1 or 2 seeds but
with 6 to 8 unenlarged ovules to each cell or placenta.—*Gardneria fagræacea,*
F. Muell. Fragm. vi. 130.

Queensland. Rockhampton Bay, *Dallachy.* Although the corolla is unknown, I have
ventured to remove this to *Fagræa,* of which it has the foliage, inflorescence, calyx, and
fruit ; for the number of ovules, as well as the inflorescence, show that it cannot be a
Gardneria.

6. STRYCHNOS, Linn.

Calyx 4- or 5-lobed. Corolla with a short or cylindrical tube and 4 or 5
spreading lobes, valvate in the bud. Stamens 5, inserted in the tube, the
anthers usually exserted. Ovary 2-celled, with several ovules in each cell.
Style simple, with a capitate or obscurely 2-lobed stigma. Fruit a globular
indehiscent berry, with the rind usually hard. Seeds imbedded in pulp,
more or less compressed, and often reduced to one or two in each fruit.—
Shrubs trees or woody climbers. Leaves opposite, 3-nerved or 5-nerved at
the base, with transverse reticulate veinlets, often smooth and shining. In
the climbing species there are usually spirally recurved hooks in one of the
axils (not observed in any Australian specimens), in which case the subtending leaf is usually reduced to a small bract, whilst the opposite leaf remains
normal. Flowers in axillary or terminal cymes clusters or panicles.

The genus is dispersed over the tropical regions of the New and the Old World. The
Australian species are both endemic, unless one of them proves to be really a variety only of
a widely spread South Asiatic one.

Flowers in corymbose cymes. Corolla-tube narrow, twice as long as
the lobes . . . . 1. *S. lucida.*
Flowers in thyrsoid panicles. Corolla-tube exceedingly short, the lobes
longer. . . . . 2. *S. psilosperma.*

1. **S. lucida,** *R. Br. Prod.* 469. An erect, divaricately-branched shrub,
often tall, glabrous except the minutely pubescent inflorescence. Leaves
ovate, obtuse, obtusely acuminate or almost acute, 3- or 5-nerved, thinly
coriaceous, shining above, more or less glaucous underneath, 1½ to 3 in. long,
contracted into a very short petiole. Flowers 5-merous, in corymbose trichotomous cymes, shortly pedunculate above the last pair of leaves. Calyx-
lobes scarcely above ½ line long, ciliolate. Corolla-tube cylindrical, about 3
lines long, very slightly hairy inside; lobes narrow, about 1½ lines long.
Anthers almost sessile in the throat. Ovary glabrous, with numerous ovules
in each cell; style either much shorter than the corolla-tube with a peltate
stigma or nearly as long as the whole corolla with a smaller stigma. Berry
globular, orange-brown, 1 to 1½ in. diameter. Seeds few, flat, orbicular,
about 5 lines diameter; testa membranous, densely silky-hairy; albumen
cartilaginous, splitting into 2 halves; cotyledons broadly ovate, about 1 line
long; radicle short, at one edge of the seed.—DC. Prod. ix. 16.

N. Australia. Islands of the Gulf of Carpentaria, *R. Brown;* Cambridge Gulf,
Regent's River, and Goulburn Island, *A. Cunningham;* Treachery Bay and Victoria river,
F. Mueller; N.W. coast, *Marten.*

The species is also in Timor, if *S. ligustrina,* Blume, Rumphia, i. 68. t. 25 from that
island, be really the same. F. Mueller, Fragm. iv. 44, refers the whole to *S. nux-vomica,*
Linn., a widely-spread E. Indian species, in which he is probably right, although, as far as I
can judge from the specimens I have seen, the Indian tree has usually broader leaves on a
much longer petiole and not glaucous underneath, the corolla-tube not so slender, and the
fruit is said to be much larger.

2. **S. psilosperma,** *F. Muell. Fragm.* iv. 44; vi. 131. A glabrous
shrub, with weak but scarcely climbing branches (occasionally spinescent,
F. Mueller). Leaves broadly ovate, shortly acuminate, smooth and shining
when old, 3- or 5-nerved, contracted into a very short petiole, 1½ to 2 in.
long. Flowers 5-merous, in small thyrsoid or short panicles, axillary and
terminal, rarely exceeding the leaves. Calyx minute Corolla not 2 lines
long, the tube very short; lobes rather longer than the tube, broad and thick,
bearded inside at the base. Anthers nearly sessile in the throat. Berry
globular, about ½ in. diameter. Seeds usually solitary, orbicular, glabrous,
not shining.

Queensland. Percy Island, *A. Cunningham;* Edgecombe Bay, Mount Archer and
Mount Elliott, *Dallachy.*

Order LXXVIII. GENTIANEÆ.

Calyx of 4 or 5, rarely more, lobes or segments. Corolla usually regular,
with 4 or 5, rarely more, lobes, contorted or otherwise imbricate or induplicate in the bud. Stamens as many as corolla-lobes, and alternate with
them, inserted in the tube. Anthers versatile, with 2 parallel cells opening
longitudinally or in terminal pores. Ovary 1-celled, but with 2 parietal placentas often projecting into the cavity so as partially to divide it into 2 or 4

cells, or rarely completely 2-celled; ovules numerous; style single, entire or with 2 short stigmatic lobes. Fruit a capsule, opening septicidally in 2 valves or rarely indehiscent or succulent. Seeds small, with a fleshy albumen. Embryo small, straight, with short cotyledons.—Herbs, very rarely in species not Australian shrubs, usually glabrous and bitter. Leaves opposite and entire in the principal tribe, alternate or clustered in the *Menyantheæ*. Stipules none. Flowers usually in cymes or corymbose panicles, rarely clustered or solitary.

The Order is chiefly abundant in the temperate or mountainous regions of the northern hemisphere, with a few tropical or southern species. Of the seven Australian genera, four have a wide range over the New and the Old World; one extends from South Africa on the one side to New Zealand on the other; another belongs to the tropical Asiatic flora; the seventh is endemic but monotypic, and may not be definitively maintained.

TRIBE I. **Eugentianeæ.**—*Terrestrial plants. Leaves opposite. Corolla-lobes contorted or otherwise imbricate in the bud. Testa of the seeds membranous.*

Corolla-tube cylindrical or short; lobes spreading, contorted in the bud. Style deciduous.

Calyx divided nearly to the base. Anthers at length recurved at the tips. Ovary completely 2-celled	1. SEBÆA.
Calyx shortly lobed. Anthers straight or at length twisted. Ovary 1-celled with parietal placentas	2. ERYTHRÆA.

Corolla-tube cylindrical; lobes spreading, imbricate in the bud.

Calyx narrow tubular, shortly 4-toothed	3. CANSCORA.
Corolla campanulate or rotate. Style persistent	4. GENTIANA.

TRIBE II. **Menyantheæ.**—*Aquatic or marsh plants. Leaves radical or alternate, sometimes floating. Corolla-lobes with broad margins, induplicate in the bud. Testa of the seeds crustaceous.*

Capsule opening at the top in 4 valves. Marsh plants. Flowers in loose cymose panicles	5. VILLARSIA.
Capsule indehiscent, usually ovoid. Plants usually aquatic. Leaves with a broad lamina. Pedicels clustered or rarely solitary . .	6. LIMNANTHEMUM.
Fruit globular, indehiscent, pulpy. Creeping plant with linear tufted leaves. Pedicels axillary, solitary	7. LIPAROPHYLLUM.

TRIBE. I. EUGENTIANEÆ.—Terrestrial plants. Leaves opposite. Corolla-lobes contorted or otherwise imbricate in the bud (the margins induplicate at the base in some species not Australian). Testa of the seeds membranous.

1. SEBÆA, R. Br.

Calyx deeply 5- rarely 4-cleft. Corolla-tube cylindrical; lobes 5, rarely 4, spreading, contorted in the bud. Stamens inserted in the throat; anthers opening in longitudinal slits, at length recurved at the tips. Ovary completely 2-celled; style deciduous; stigma clavate or capitate, often shortly 2-lobed. Capsule septicidally 2-valved, the margins introflexed, separating from the central placenta, which remains entire or splits. Seeds small and numerous.—Annuals. Leaves opposite. Flowers yellow or white, in terminal dichotomous cymes. Bracts usually small.

The genus, as now limited, extends over Southern Africa, and is also in New Zealand, but ought perhaps to include the South American *Schuebleria*. Of the two Australian species, one is the same as the New Zealand one, the other appears to be endemic. Like *Exacum*, the genus belongs to the very few *Gentianeæ* with a completely 2-celled ovary, and is readily distinguished from *Exacum* by the anthers.

Flowers pale yellow, 5-merous, in loose cymes . . . . 1. *S. ovata.*
Flowers nearly white, 4-merous, in compact cymes . . . . 2. *S. albidiflora.*

1. **S. ovata,** *R. Br. Prod.* 452. A glabrous, erect, simple or slightly branched annual, rarely exceeding 6 to 8 in. and sometimes not above half that height. Leaves in distant pairs, sessile, ovate or orbicular-cordate, rarely above ½ in. long. Flowers small, pale yellow, in a rather loose terminal dichotomous cyme, those in the forks very shortly pedicellate. Bracts narrow, acute. Calyx-segments about 2 lines long, lanceolate, acute, prominently keeled or almost winged. Corolla as long as the calyx, the lobes much shorter than the tube. Anthers linear, tipped with a gland, recurved after fading. Style usually short. Capsule oblong.—Griseb. in DC. Prod. ix. 53; Hook. f. Fl. Tasm. i. 270; *Exacum ovatum,* Labill. Pl. Nov. Holl. i. 38. t. 52; *Erythræa chloræfolia,* Lehm. Pl. Preiss. ii. 239.

N. S. Wales. Port Jackson, *R. Brown, Woolls;* between the Upper Bogan and Lachlan rivers, *L. Morton.*
Victoria. Common about Melbourne, *Adamson, Harvey,* and others; Skipton, *Whan;* Wendu Vale, *Robertson.*
Tasmania. Port Dalrymple, *R. Brown;* common in pasture lands at Circular Head and Launceston, *J. D. Hooker* and others.
S. Australia. Bugle Range, Gawler Town, *F. Mueller;* Kangaroo Island, *Heazenræder.*
W. Australia. Swan River, *Drummond, 1st Coll.;* Goderich district, *Preiss, n.* 1962; grassy places, Blackwood river, *Oldfield;* basaltic ridges, Stirling Range, *F. Mueller.*

2. **S. albidiflora,** *F. Muell. in Trans. Phil. Soc. Vict.* i. 46, *and in Hook. Kew Journ.* viii. 164. A small, glabrous, erect, nearly simple annual, with the foliage and habit of the smaller specimens of *S. ovata,* except that the cyme is usually more simple and compact. Floral leaves and bracts small, ovate, obtuse. Calyx-segments 4, ovate or broadly oblong, obtuse, scarcely keeled except at the base. Corolla whitish, nearly as long as the calyx, with 4 short, broadly ovate, obtuse lobes.—Hook. f. Fl. Tasm. ii. 367.

Victoria. Saline pastures, Port Phillip to Port Fairy, Mount Emu Creek, *F. Mueller.*
Tasmania. Near George Town, *C. Stuart.*

2. ERYTHRÆA, Pers.

Calyx more or less 5- or 4-lobed. Corolla-tube cylindrical; lobes 5 or 4, spreading, contorted in the bud. Stamens inserted in the tube; anthers opening in longitudinal slits, at length spirally twisted. Ovary 1-celled, with 2 parietal placentas; style deciduous; stigma capitate or 2-lobed. Capsule 2-valved, the margins of the valves involute and bearing the placentas. Seeds numerous, small.—Annuals. Leaves opposite. Flowers pink white or yellow, in dichotomous terminal cymes, either corymbose with small bracts, or with few elongated leafy branches.

The genus is widely spread over the temperate regions of the globe, and occurs also within the tropics. The only Australian species is also in some of the Pacific islands, and is scarcely distinct from a common Mediterranean one.

1. **E. australis,** *R. Br. Prod.* 451. An erect glabrous annual, from under 6 in. to 1½ ft. high, the branches few and not very spreading. Leaves sessile, ovate-oblong elliptical or lanceolate, mostly obtuse, the lower ones stem-clasping, rarely 1 in. long. Flowers nearly sessile along the more or

2 B 2

less elongated branches of the once-forked or dichotomous cyme, with a leafy bract under each flower, thus forming one-sided interrupted leafy spikes. Calyx narrow, 3 to 4 lines long, with 4 rarely 5 angles and acute teeth or lobes. Corolla-tube usually exceeding the calyx, but sometimes shorter; lobes ovate or oblong, much shorter than the tube. Capsule oblong, shorter than the calyx. Seeds small, reticulate-striate. — Griseb. in DC. Prod. ix. 60 ; Hook. f. Fl. Tasm. i. 271.

N. Australia. Islands of the Gulf of Carpentaria, *R. Brown*; Victoria river, *F. Mueller ;* Nichol Bay, N.W. coast, *Ridley's Expedition.*

Queensland. Common on the coast, *R. Brown ;* Rockingham Bay, *Dallachy ;* Rockhampton, *Thozet ;* on the Maranoa, *Mitchell.*

N. S. Wales. Port Jackson or Blue Mountains, *Sieber, n.* 493, *Backhouse, Woolls,* and others ; northward to Hastings, Macleay, and Richmond rivers, *Beckler* and others; New England, *C. Stuart ;* southward to Illawarra, *A. Cunningham ;* in the interior on the Darling, *Victorian Expedition.*

Victoria. Wendu Vale, *Robertson ;* Skipton, *Whan ;* Yarra and Macalister rivers, *F. Mueller.*

Tasmania. Wet saline marshes on the banks of the Tamar, *Gunn.*

S. Australia. South coast, *R. Brown ;* round St. Vincent's Gulf and Torrens river, *F. Mueller ;* Spencer's Gulf, *Warburton.*

W. Australia. King George's Sound, *R. Brown,* and thence round to Swan and Murchison rivers, *Drummond, n.* 702, *Preiss, n.* 1959, *Oldfield,* and others, and eastward towards the Great Bight, *Maxwell.*

The species is also in New Caledonia and in the Loochoo Islands. As far as I can ascertain, it only differs from *E. spicata* (a species common in the Mediterranean region, and eastward at least as far as Affghanistan) in the flowers, usually but not constantly 4-merous. I find, however, occasionally 5-merous flowers in Australian specimens (*e. g.* in Mitchell's and in some of Gunn's), and, on the other hand, 4-merous flowers have been observed in some East Mediterranean specimens. Indeed, *E. babylonica,* Griseb., distinguished by that character, is probably a variety only of *E. spicata,* to which *E. australis* may be eventually referred.

3. CANSCORA, Lam.

(Orthostemon, *R. Br.*)

Calyx narrow, shortly 4-toothed. Corolla-tube cylindrical ; lobes 4, slightly unequal, imbricate in the bud. Stamens inserted in the tube, usually unequal ; anthers opening in longitudinal slits. Ovary 1-celled, with 2 parietal placentas ; style deciduous ; stigmatic lobes broad. Capsule 2-valved, the margins of the valves involute and bearing the placentas. Seeds numerous, small.—Annuals. Leaves opposite. Flowers pink, on filiform pedicels, terminal or in the upper parts of the branches.

The genus consists of very few tropical Asiatic species, the commonest of which is the same as the Australian one.

1. **C. diffusa,** *R. Br.; Griseb. in DC. Prod.* ix. 64. A glabrous, erect, dichotomous annual of 6 in. to 1 ft., with very numerous divaricate almost filiform branches, slightly 4-angled. Leaves ovate, acute, 3-nerved, the lower ones sometimes contracted into a petiole and $\frac{1}{2}$ to 1 in. long, the upper ones sessile with a broad base and under $\frac{1}{2}$ in. Flowers small, pink, on filiform pedicels, terminal or in the upper forks. Calyx 2 to nearly 3 lines long. Corolla-tube shorter than the calyx ; lobes small, the 2 inner ones rather smaller and less deeply separated than the outer ones, with the filament of

the intervening stamen rather shorter than the three others. Capsule rather shorter than the calyx.—*Pladera virgata,* Roxb.; Hook. Bot. Misc. iii. t. suppl. 25, and other synonyms given by Griseb. l. c.; *Canscora tenella,* Wight, Ic. t. 1327; *Orthostemon erectus,* R. Br. Prod. 451; Griseb. in DC. Prod. ix. 63.

N. Australia. Between Victoria and Fitzmaurice rivers, *F. Mueller.*
Queensland. Endeavour river, *Banks and Solander, A. Cunningham.*

The species is common in tropical Asia, extending on the one hand to tropical Africa, on the other to the Archipelago and New Caledonia. The irregularity of the corolla is very slight, but quite as much in the Australian specimens as in many of the Indian ones.

4. GENTIANA, Linn.

Calyx tubular, 5- rarely 4-lobed. Corolla with a cylindrical or short and broad or campanulate tube and a spreading 5-lobed rarely 4-lobed limb. Anthers opening in longitudinal slits. Ovary 1-celled, with 2 parietal placentas; style persistent after the flower fades, often reduced to sessile stigmatic lobes. Capsule 2-valved, the margins of the valves involute and bearing the placentas. Seeds numerous, small.—Herbs, annual or perennial. Leaves opposite. Flowers in the Australian species blue purple or white, terminal or in the upper axils.

A large and beautiful genus, spread over the northern hemisphere, especially in mountainous districts both of the New and the Old World, penetrating into the tropics, but only represented in Australia by a single species, which is also in New Zealand. In some northern species the margins of the corolla-lobes are produced and induplicate in the bud, forming accessory lobes in the sinus.

1. **G. montana,** *Forst. Prod.* 21. An erect glabrous plant, usually annual, sometimes with a single slender stem from a tuft of rosulate radical leaves, sometimes branching at the base into a dense tuft of robust stems, and in a few specimens appearing to form a perennial stock, varying from under ½ ft. to above 1½ ft. in height. Lower leaves petiolate obovate or spathulate, upper ones more sessile, oblong lanceolate or rarely almost ovate. Flowers terminal or in the upper axils, on long peduncles, forming either a very loose trichotomous panicle or sometimes, especially when the flowers are large, a compact corymb. Calyx deeply divided into linear or lanceolate herbaceous lobes, more than half as long as the corolla. Corolla broadly campanulate, white or blue, or more or less striped or passing into pink or purplish and yellowish at the base, varying from scarcely ¾ in. to above 1½ in. diameter when open, very deeply divided into obovate lobes. Anthers small. Stigmas 2, sessile.—R. Br. Prod. 450; Griseb. in DC. Proc. ix. 99; Hook. f. Fl. Tasm. i. 271; *G. Grisebachii,* Hook. f. in Hook. Ic t. 636; *G. pleurogynoides,* Griseb. in DC. Prod. ix. 99.

N. S. Wales. Campbell and Argyle counties, *A. Cunningham.*
Victoria. Frequent in the Australian Alps, at an elevation of 4000 to 5000 ft., *F. Mueller;* Port Phillip, *Gunn.*
Tasmania. Derwent river, *R. Brown;* abundant throughout the island, growing both in wet and dry places, ascending to 4000 ft., *J. D. Hooker.*
S. Australia. Crest of Mount Gambier, Tilly's Swamp, *F. Mueller.*

Var. *saxosa.* Stock perennial (but of very short duration ?). Flowers large, in compact corymbs.—*G. saxosa,* Forst. Prod. 21; Griseb. in DC. Prod. ix. 89.—Mount Buller, in

Victoria, *F. Mueller. G. Diemensis,* Griseb. Gent. 224, and in DC. Prod. ix. 90, described from a specimen of Gunn's in Herb. Hook., which I cannot now find, may be this form ; but the numerous specimens sent at different times by Gunn, as far as I have seen, all appear to be annual.

All the above forms occur in New Zealand, including the typical *G. montana,* which is a small weak plant, with a single one or very few flowers, on long solitary peduncles.

TRIBE II. MENYANTHEÆ.—Aquatic or marsh plants. Leaves radical or alternate, sometimes floating. Corolla-lobes with broad margins, induplicate in the bud. Testa of the seeds crustaceous.

5. VILLARSIA, Vent.

Calyx 5-cleft nearly to the base. Corolla broadly campanulate, almost rotate, the tube short and broad ; lobes 5, with broad entire or fringed margins, induplicate in the bud. Ovary surrounded by 5 minute hypogynous glands, 1-celled with 2 parietal placentas; stigmatic lobes usually broad. Capsule 1-celled, opening at the top in 4 valves. Seeds few or numerous, with a crustaceous testa.—Herbs, usually growing in marshes. Radical leaves on long petioles. Flowering stems erect, paniculate and leafless besides small bracts, or branching with few alternate leaves (rarely, in a species not Australian and perhaps not a congener, reduced to a 1-flowered scape). Flowers yellow or white.

Besides the Australian species, which are endemic, there is one in S. Africa, and one or two in N.W. America. The corollas in this and the following genus are so rarely preserved in dried specimens in a state fit for examination, that the characters derived from them require verification on the living plant, and those founded on the seeds may perhaps not prove so constant as they are assumed to be.

Stems leafy, branched.
Calyx-segments broad, margins flat, much imbricate. Seeds ovoid,
  granular. Stems tall and stout . . . . 1. *V. calthifolia.*
Calyx-segments broad, margins undulate, recurved. Seeds compressed, smooth . . . . 2. *V. congestiflora.*
Calyx-segments lanceolate, acute. Seeds compressed, smooth.
  Flowers capitate. Calyx woolly-hairy . . . . 3. *V. capitata.*
  Flowers loosely paniculate. Calyx glabrous . . . . . . 4. *V. latifolia.*
Stems paniculate, leafless or with a single leaf under the primary
  branch.
Corolla yellow. Calyx with a distinct obconical base.
  Corolla-lobes very broad, much longer than the calyx. Eastern
    species . . . . 5. *V. reniformis.*
  Corolla-lobes narrow, shortly exceeding the calyx. Western
    species.
    Stems above 6 in. (1 to 2 ft.), many-flowered . . . . . 6. *V. parnassifolia.*
    Stems under 6 in., very few- or almost 1-flowered . . . . 7. *V. violifolia.*
Corolla white. Calyx almost obtuse at the base.
  Corolla-lobes with a longitudinal keel or wing inside. Seeds
    hispid . . . . 8. *V. lasiosperma.*
  Corolla-lobes without any longitudinal appendage inside. Seeds
    smooth or tuberculate . . . . 9. *V. albiflora.*

1. **V. calthifolia,** *F. Muell. Fragm.* vi. 140. A stout erect plant, attaining 7 or 8 ft. (*Drummond*), the stems more or less leafy. Radical leaves on long thick petioles, orbicular or almost reniform, coarsely crenate, deeply cordate, attaining 6 to 8 in. diameter ; stem-leaves several but not numerous,

smaller and on shorter petioles. Flowers yellow, in rather dense terminal corymbose panicles, somewhat larger than in *V. latifolia.* Calyx-lobes broadly ovate, the margins not undulate, much imbricate in the bud. Corolla apparently not much longer than the calyx, the lobes bearded or fringed inside at the base, but the margins entire. Capsule ovoid-oblong, 4 to 5 lines long, opening at the top in 4 valves, of which the short broad tips sometimes cohere and fall off together, leaving the valves truncate; but in some capsules the tips also separate and remain on the recurved valves. Seeds ovoid, minutely granular-tuberculate; funicles short.

W. Australia, *Drummond,* 3rd *Coll. Suppl.;* granite summits of Mount Perongerup, *F. Mueller.*

2. **V. congestiflora,** *F. Muell. Fragm.* vi. 141. Stems sometimes tall, stout, and erect, sometimes weak and diffuse, more or less leafy. Radical leaves on long petioles, mostly orbicular or almost reniform, entire or coarsely crenate, often 2 in. diameter, the stem-leaves and the radical ones of small weak specimens much smaller, ovate or orbicular. Flowers yellow, in broad dichotomous leafy cymes, sessile or nearly so, solitary in the forks, often clustered at the ends of the branchlets, but less so than in *V. capitata* and quite glabrous. Calyx-lobes ovate-lanceolate, the margins usually undulate and recurved. Corolla-lobes shortly exceeding the calyx, bearded inside at the base. Anthers linear. Ovules few to each placenta, on filiform funicles. Capsule rather shorter than the calyx, opening in 4 short recurved valves. Seeds few, orbicular, slightly compressed, smooth and shining or opaque but not granular, with a slightly prominent small strophiole.

W. Australia. Swan and Murchison rivers, *Drummond, Oldfield.*

3. **V. capitata,** *Nees in Pl. Preiss.* i. 365. Stems erect, slightly branched, leafy, scarcely above 6 in. high. Leaves on long petioles, broadly ovate orbicular or reniform, coarsely sinuate-toothed or entire, under 1 in. long and often under ½ in. Flowers (yellow?) sessile in compact globular or depressed heads of about ½ in. diameter, on long peduncles, with 2 broadly-ovate outer bracts forming an involucre to the head, the inner bracts under each flower much narrower, membranous and ciliate. Calyx membranous, covered with long hairs which give the head a woolly aspect; segments lanceolate, subulate-acuminate. Corolla scarcely exceeding the calyx, the lobes slightly bearded inside at the base, the margins entire. Stigmatic lobes broad, almost petal-like. Capsule shorter than the calyx, acuminate, opening in (4?) valves. Seeds small, smooth and shining, black or pale-coloured.— *V. involucrata,* Hook. Ic. Pl. t. 725.

W. Australia. Swan River, *Drummond, n.* 7, *Preiss, n.* 1956, *Clarke* · Serpentine river, *Oldfield.*

4. **V. latifolia,** *Benth. in Hueg. Enum.* 81. An erect plant, often tall, with much of the habit of *V. parnassifolia,* but always with 2 or more fully developed leaves on the stem. Radical leaves on long petioles, from ovate to broadly orbicular, not at all or scarcely cordate, entire or sinuate-toothed, 1 to 3 in. diameter, those of the stem rather smaller on shorter petioles. Flowers yellow, in a loose terminal panicle as in *V. parnassifolia,* but rather

smaller. Calyx-segments lanceolate, acute. Corolla-lobes shortly exceeding the calyx, with entire margins, bearded inside at the base. Ovary attached by a very broad base ; ovules rather few ; style with oblong stigmatic lobes. Capsule shorter than the calyx, opening in 4 valves. Seeds few, smooth and shining, with a prominent tubercle or strophiole at the hilum.—Nees in Pl. Preiss. i. 364.

W. Australia. Swan River, *Drummond, 1st Coll. n. 3, also n.* 163, *Huegel, Preiss, n.* 1955 ; Blackwood river, *Oldfield ;* Cape Leeuwin, *Collie.* Drummond's n. 4 appears to be a slender form of the same species, with ovate, often narrow, rather small leaves.

5. **V. reniformis,** *R. Br. Prod.* 457. An erect plant, varying in stature from 6 in. to 3 ft. Leaves usually all radical, in a dense tuft, on long petioles from ovate to orbicular or reniform, more or less cordate at the base, entire or slightly sinuate-toothed, rather thick, obscurely several-nerved, mostly 1 to 2 in. long. Flowering stems paniculately branched, leafless except small linear or lanceolate bracts, or rarely bearing a single petiolate leaf at the first branching. Calyx-lobes lanceolate, usually acute, 3 to 4 lines long. Corolla yellow, spreading to from ¾ to 1 in. diameter ; lobes broad with entire or slightly denticulate margins, copiously bearded or fringed inside at the base. Anthers linear. Stigmatic lobes ovate, rather thick. Capsule opening at the top in 4 valves. Seeds smooth or granular-tuberculate.—F. Muell. Fragm. vi. 139 ; *V. parnassifolia,* R. Br. ; Griseb. in DC. Prod. ix. 136 ; Hook. f. Fl. Tasm. i. 272 (all as to the eastern plant only); *Menyanthes exaltata,* Sims, Bot. Mag. t. 1029 ; *V. exaltata,* F. Muell. Fragm. vi. 140.

Queensland. Brisbane river, Moreton Bay, *F. Mueller.*
N. S. Wales. Port Jackson, *R. Brown, Woolls,* and others ; Hastings river, *Beckler.*
Victoria. In marshes and stagnant waters from Gipps' Land, *F. Mueller,* to the Glenelg, *Robertson.*
Tasmania. Port Dalrymple, *R. Brown ;* abundant in marshes and lakes, ascending to 3000 ft., *J. D. Hooker.*
S. Australia. Around St. Vincent's Gulf, *F. Mueller, Blandowski.*

Menyanthes sarmentosa, Sims, Bot. Mag. t. 1328, referred by Grisebach, through some mistake, to *Limnanthemum geminatum,* appears to me to be a stoloniferous specimen of *V. reniformis. V. reniformis,* Lindl. Bot. Reg. t. 1533, seems rather to represent the true *V. parnassifolia.*

F. Mueller proposes to take up Sims's specific name of *exaltata* because the leaves are rarely truly reniform, but neither is the stem always tall.

6. **V. parnassifolia,** *R. Br. Prod.* 457 (*partly*). A tall erect plant with the habit of some of the more slender specimens of *V. reniformis,* but with very much smaller corollas. Radical leaves on long petioles, ovate or almost orbicular, entire or sinuate-crenate, slightly cordate or rounded at the base, mostly under 1 in. long. Flowering stems 1 to 2 ft. high, loosely paniculate, leafless except small lanceolate bracts or rarely with a single petiolate ovate-lanceolate leaf at the first branching. Calyx more distinctly obconical at the base than in most species ; segments lanceolate, acute, about 3 lines long. Corolla yellow, the lobes shortly exceeding the calyx, with entire margins, but fringed or bearded inside at the base. Anthers linear-lanceolate. Ovules numerous. Style in the perfect flowers short and thick,

with 2 ovate stigmatic lobes; in other flowers the style is long linear-subulate and undivided, but in these I do not find any perfect ovules. Capsule opening at the top in 4 valves. Seeds small, somewhat compressed, smooth or minutely granular.—Nees in Pl. Preiss. i. 364 ; F. Muell. Fragm. vi. 140 ; *Swertia parnassifolia,* Labill. Pl. Nov. Holl. i. 72. t. 97.

W. Australia. Swamps about King George's Sound and adjoining districts, *R. Brown, Preiss, n.* 1957, *Harvey,* and others ; extending eastward to Goose Island Bay, *R. Brown,* and Cape le Grand, *Maxwell.*

Brown had originally confounded this with some of the slender specimens of the eastern *V. reniformis,* but his herbarium shows that he subsequently recognized the distinction.

7. **V. violifolia,** *F. Muell. Fragm.* vi. 138. A dwarf plant which may possibly prove to be a very much reduced form of *V. parnassifolia* Leaves all radical, on almost filiform petioles, ovate orbicular or reniform, from under ½ in. to nearly 1 in. diameter. Flowering stems leafless or with a single leaf, bearing but very few flowers or only a single one. Flowers small, yellow, with very acute calyx-lobes and entire corolla-lobes as in *V. parnassifolia,* and the base of the calyx similarly obconical. Capsule acute, 4-valved. Seeds few, smooth.

W. Australia. Tweed river, *Oldfield;* Don river, *Maxwell.*

8. **V. lasiosperma,** *F. Muell. Fragm.* vi. 137. A tall erect plant resembling *V. parnassifolia* and *V. albiflora.* Leaves all radical, on long petioles, ovate, entire or sinuate-toothed, usually narrowed at the base, 1 to 1½ in. long, usually much thicker than in *V. parnassifolia,* and the thick petiole more dilated at the base. Flowering-stems attaining 2 to 3 ft., with the loose panicle and white flowers of *V. albiflora,* but the corolla larger and the lobes, besides the beard or fringe at the base, have on their inner surface a longitudinal raised keel or wing. Style short or long, the stigmatic lobes usually smaller than in *V. albiflora.* Capsule ovoid-globular, opening in 4 short valves. Seeds compressed and concave when dry with a somewhat prominent border, but swelling out when soaked, hispid with short rigid hairs.

W. Australia. Swamps near King George's Sound, *F. Mueller.*

9. **V. albiflora,** *F. Muell. Fragm.* ii. 21, vi. 138. A tall erect plant, closely resembling *V. parnassifolia,* but usually stouter, the leaves broader and the flowers white. Radical leaves on long petioles, orbicular or almost reniform, entire or sinuate-toothed, broadly cordate or truncate at the base, mostly 1 to 2 in. diameter. Stems attaining 2 ft. or more, leafless besides the small bracts, or with a single petiolate leaf at the first branching. Flowers in a large loose panicle as in *V. parnassifolia,* and of the same size or rather smaller. Calyx obtuse or scarcely obconical at the base, the segments acute, from 2 to nearly 3 lines long. Corolla-lobes bearded inside at the base, but without the longitudinal central wing of *V. lasiosperma.* Capsule as long as or longer than the calyx, opening at the top in 4 valves. Style rather short, with ovate acute stigmatic lobes. Seeds ovate, minutely granular or muricate, the tubercle at the hilum very small.—*V. reniformis,* Nees in Pl. Preiss. i. 364, not of R. Br.

W. Australia. Swan River, *Drummond, 1st Coll. Suppl. n.* 9 (4 in Herb. F. Muell.), *Preiss, n.* 1958, *Oldfield ;* Blackwood river, *Oldfield.*

6. **LIMNANTHEMUM,** Gmel.

Calyx 5-cleft nearly to the base.´ Corolla almost rotate, the tube short and broad ; lobes 5, with broad entire or fringed margins induplicate in the bud. Ovary surrounded by 5 minute hypogynous glands, 1-celled with 2 parietal placentas ; stigmatic lobes broad, sometimes petal-like. Capsule ovoid or oblong, indehiscent or bursting irregularly. Seeds few or numerous, with a crustaceous testa.—Herbs either aquatic and floating or creeping in swamps. Leaves ovate or broad. Pedicels 1-flowered, either in the tufts of leaves or 2 together or clustered at the nodes of the weak stems or close to an almost terminal sessile leaf, so as to appear to be inserted on the petiole. Flowers yellow or white.

The genus, although with very few extra-Australian species, is widely dispersed over the temperate and warmer regions of the globe. Of the seven Australian species, one is common in most tropical countries, the six others are endemic. F. Mueller proposes to re-unite this genus with *Villarsia.* That would, however, entail the re-uniting the whole tribe of *Menyantheæ* into the Linnæan genus *Menyanthes,* which would then be divided into sections corresponding with the present genera, a course which does not appear to be of much practical advantage, whilst it would add above twenty names to the synonymy.

Leaves almost sessile under the cluster of pedicels which appear as
    if inserted on the petiole. Flowers white.
  Leaves 2 to 8 in. diameter. Calyx 3 to 4 lines long . . . .   1. *L. indicum.*
  Leaves under 1 in. diameter. Calyx 1 line long. Stems filiform.   2. *L. minimum.*
Leaves all on long petioles. Pedicels usually 2 together at the
    nodes of almost leafless stems. Flowers yellow.
  Leaves mostly crenate. Corolla-lobes with a longitudinal central
    wing inside . . . .   3. *L. crenatum.*
  Leaves entire. Corolla-lobes without any longitudinal wing.
    Calyx-segments above 2 lines long. Corolla much larger.
      Leaves usually with a deep narrow sinus. Seeds smooth
        or tuberculate, glabrous. . . . .   4. *L. geminatum.*
      Leaves usually truncate or with a broad open sinus at the
        base. Seeds densely velvety-pubescent . . . . . .   5. *L. hydrocharoides.*
    Calyx-segments under 1½ lines long. Corolla small. Stems
      filiform. Leaves ovate orbicular or with a broad open
      sinus.
      Flowering stems elongated, many-flowered . . . . .   6. *L. exiliflorum.*
      Flowering stems short, few-flowered . . . .   7. *L. exiguum.*

1. **L. indicum,** *Thw. Enum. Pl. Ceyl.* 205, *not of Griseb.* Stems from a submerged tuft formed the preceding year, simple, resembling a petiole, bearing a single terminal floating leaf, nearly sessile, and just below the real exceedingly short petiole an unilateral tuft of pedicels and young stems, which tuft ultimately sinks, the old tuft and stem dying away. Leaves orbicular or broadly oval, deeply cordate, usually entire, with a close or open sinus, of a thick consistence, palmately veined and reticulate, but the veins not prominent and often very obscure, not above 2 in. diameter in some specimens, above 8 in. in others. Pedicels usually numerous in the clusters, 1 to 2 in. long. Calyx-segments lanceolate, membranous, 3 to 4 lines long. Corolla white, varying from 6 to 10 lines diameter, the margins fringed, the

inner surface more or less bearded at the base with pedicellate g ands, but without any longitudinal wing. Anthers linear-sagittate, on short flat filaments. Style short or long, with 2 broad lobed and spreading stigmatic laminæ. Ovules on 2 broad parietal placentas. Seeds smooth tubercular or muricate.—*Menyanthes indica*, Linn. Spec. 207 ; Bot. Mag. t. 658 ; *Villarsia indica*, Vent. Choix, t. 9 ; Wight in Hook. Bot. Misc. iii. t. Suppl. 30 (*V. macrophylla* in the text) ; *Limnanthemum Kleinianum*, Griseb. Gent 344, and in DC. Prod. ix. 139 ; Seem. Fl. Vit. t. 33 ; *L. Wightianum*, Grisɔb. ll. cc. ; *Villarsia nympheæfolia*, Fras. in Hook. Bot. Misc. i. 257 ; *Limnanthemum Fraserianum*, Griseb. Gent. 346, and in DC. Prod. ix. 140, and probably also *L. Forbesianum, L. Thunbergianum, L. Ecklonianum, L. orbiculatum*, and *L. Humboldtianum*, Griseb. Gent. 345 to 348, and in DC. Prod. ix. 139, 140 ; *Villarsia trachysperma*, F. Muell. Fragm. vi. 136.

N. Australia. Ponds near S. Alligator river towards Macadam Range, *F. Mueller.*

Queensland. Logan river (not Swan River, as quoted by mistake by Grisebach), *Fraser ;* Wide Bay, *Bidwill ;* Moreton Bay, *F. Mueller ;* Rockhampton, *O'Shanesy ;* Rockingham Bay, *Dallachy ;* Broad Sound and Fitzroy river, *Bowman ;* also in *Leichhardt's* collection.

N. S. Wales. Clarence river, *Beckler.*

The species appears to inhabit the pools and lakes of all tropical countries, at least I am unable to appreciate the characters by which it has been attempted to distinguish the Asiatic, African, and American specimens, and even to subdivide each of these. In Australia, as in tropical Asia (*Thwaites*) and America (*A. de St. Hilaire*), the seeds vary from perfectly smooth or sprinkled towards the edge with a few tubercles, to being densely muricate all over, the style is long or short, and the cilia, or pedicellate glands at the base of the lobes, are certainly very variable, although in most cases it is scarcely possible to ascertain even their presence or absence from dried specimens.

As suspected by Thwaites, the specimen of Wight's, from which Grisebach drew up his character of *L. indicum*, proves to be *L. cristatum*, whilst there is every reason to refer to the present species the *Menyanthes indica* of Linnæus, as well as the two figures quoted by him, viz. Rumph. Herb. Amb. vi. t. 72, f. 3, and Rheede, Hort. Malab. xi. t. 28 (referred by Grisebach by mistake to *L. cristatum*).

2. **L. minimum,** *F. Muell. Fragm.* i. 40. A small plant, differing from *L. indicum* chiefly in size. Stems filiform. Leaves mostly broadly reniform or deeply cordate, but passing here and there into the orbicular or obovate leaves of *L. exiguum*, rather thick, quite entire, scarcely above ½ in. diameter. Pedicels clustered at the base of the leaves as in *L. indicum*. Calyx-segments scarcely 1 line long. Corolla white, about 4 lines diameter, the lobes denticulate on the margin and fringed inside at the base (*F. Mueller*). Capsule about as long as the calyx, with few small tuberculate seeds.—*Villarsia minima*, F. Muell. Fragm. iv. 128, vi. 137.

N. Australia. York Sound, N.W. coast, *A. Cunningham ;* lagoons on the Fitzmaurice river, *F. Mueller.*

3. **L. crenatum,** *F. Muell. in Trans. Phil. Soc. Vict.* i. 17, and *in Hook. Kew Journ.* viii. 164. Stems floating or creeping, emitting at the nodes tufts of leaves, or a single leaf and a cluster of pedicels, or a short flowering-branch, as in *L. geminatum*, which this species closely resembles. Leaves broadly orbicular-cordate or reniform, irregularly crenate, 1 to 3 in. diameter, usually thick. Pedicels 1 to 2 in. long. Calyx-lobes narrower and more acute than in *L. geminatum* and more united at the base. Corolla yellow,

above 1 in. diameter, the lobes fringed on the margin, bearded inside at the base with a few long cilia, and bearing a prominent longitudinal wing along their centre. Style conical, the stigmas (in the flowers examined) erect, much dilated, the margins reduplicate and deeply fringed, assuming the appearance of 4 longitudinal wings to a conical stigma. Capsule membranous, 4 to 5 lines long. Seeds smooth.— *Villarsia crenata,* F. Muell. Fragm. iv. 127.

N. Australia. Albert river and Bentinck Island, *Henne.*
Queensland. Rockhampton, *O'Shanesy;* Burdekin and Condamine rivers, *Bowman.*
N. S. Wales. Paramatta, *Woolls.*
Victoria. Murray river, *F. Mueller;* Wimmera, *Dallachy.*

4. **L. geminatum,** *Griseb. Gent.* 346, *and in DC. Prod.* ix. 140 (*as to Brown's plant*). Stems floating in water or creeping in mud, emitting at the nodes either a tuft of leaves without flowers or a single leaf with a more or less elongated flowering branch, along which the pedicels are in pairs or rarely 3 together with a pair of lanceolate bracts at the base, but not paniculately branched as in *Villarsia.* Leaves broadly ovate-orbicular or almost reniform, deeply cordate with a close or open sinus, quite entire or very obscurely sinuate, digitately veined, rather thick, from under 1 to about 4 in. diameter, but usually rather small. Pedicels 1 to 2 in. long. Calyx-segments obtuse, much shorter than the corolla. Corolla yellow ($\frac{1}{2}$ to $\frac{3}{4}$ in. diameter?), the lobes scarcely fringed on the margin, sparingly bearded at the base, and without the longitudinal wing of *L. crenatum.* Ovules spreading over the greater part of the inner surface of the ovary. Lobes large, petal-like, spreading, crenulate or fringed. Capsule membranous. Seeds numerous, slightly compressed, smooth tubercular or muricate, $\frac{1}{4}$ to fully $\frac{1}{2}$ line diameter.—*Villarsia geminata,* R. Br. Prod. 457; F. Muell. Fragm. vi. 137.

N. Australia. Albert river and Bentinck Island, *Henne;* stagnant waters near Macadam Range, *F. Mueller* (unless these specimens belong to *L. hydrocharoides*).
Queensland. Shoalwater Bay, *R. Brown;* Brisbane river, *F. Mueller;* Moreton Bay, *O'Shanesy.*
N. S. Wales. Nepean river, *R. Brown;* Sydney, *Clowes;* Paramatta, *Woolls;* New England, *C. Stuart.*
Victoria. Avon river, Mitta-Mitta, Omeo, etc., *F. Mueller.*

The seeds seem to be as variable in this as in other species as to their size, and the smooth or muricate surface. I have not seen the seeds, however, of Brown's or other tropical specimens, which may possibly belong to the following species or variety.

5. **L. hydrocharoides,** *F. Muell. Herb.* A creeping plant, closely resembling the smaller and less developed specimens of *L. geminatum.* Leaves in dense tufts, on long petioles, broader and less deeply cordate than in *L. geminatum,* and often only truncate at the base. Corolla rather smaller than in that species and the seeds larger, densely and minutely velvety-pubescent.—*Villarsia hydrocharoides,* F. Muell. Fragm. vi. 139.

Queensland. Rockingham Bay, *Dallachy.* The indumentum of the seed which characterizes this species is very different both from the granular tubercles occasionally observed in so many species of *Limnanthemum* and *Villarsia,* and from the rigid hairs of the seeds of *Villarsia lasiosperma.*

6. **L. exiliflorum,** *F. Muell. Fragm.* v. 46. A slender plant with creeping stolons. Leaves tufted, on long petioles, ovate-orbicular or reniform, truncate or cordate at the base, from under ½ in. to about ¾ in. long. Flowering stems very slender, decumbent or ascending, simple and leafless or once or rarely twice branched with a small petiolate leaf at the branch, ½ to 1 ft. long. Pedicels filiform, usually 2 together or clustered at the nodes. Flowers very small, yellow. Calyx-segments ovate, membranous, scarcely 1½ lines long. Corolla shortly exceeding the calyx, the lobes slightly fringed on the margin (or entire?), bearded with a minute cilia inside towards the base. Capsule ovoid, not exceeding the calyx. Seeds globular, with minute reticulations or glands, giving them a pubescent aspect when dry.—*Villarsia exiliflora*, F. Muell. Fragm. v. 46; vi. 137.

Queensland. Stagnant waters, Rockingham Bay, *Dallachy.*

7. **L. exiguum,** *F. Muell. Fragm.* i. 40. A small species, the filiform floating or creeping stems emitting at the nodes tufts of leaves with or without pedicels or a short branch with the pedicels in pairs as in *L. geminatum,* thus bearing the same relation to that species which *L. minimum* does to *L. indicum.* Leaves on filiform petioles, ovate obovate or almost orbicular, not cordate in any of our specimens, under ½ in. long in the Tasmanian ones. Pedicels filiform, often solitary. Flowers yellow, very small, the calyx scarcely above 1 line long. Corolla-lobes shortly exceeding the calyx, not fringed (but slightly bearded inside at the base?). Style short, with short broad stigmatic lobes not petal-like.—*Villarsia exigua,* F. Muell. Fragm. vi. 137; Hook. f. Fl. Tasm. ii. 368.

Tasmania. South Port, *C. Stuart.*

W. Australia? Drummond's specimens, n. 17, with longer filiform (floating?) stems and rather larger ovate leaves, appear to belong to this species, but they are very imperfect.

7. LIPAROPHYLLUM, Hook. f.

Calyx deeply 5-cleft. Corolla almost rotate, deeply divided into 5 lobes with broad margins, induplicate in the bud. No hypogynous glands. Ovary 1-celled, with 2 parietal placentas; stigmatic lobes short and broad. Fruit globular, indehiscent, pulpy or succulent. Seeds with a crustaceous testa.— Small creeping plant. Leaves tufted, linear. Pedicels short, 1-flowered.

The genus is confined to a single species, endemic in Tasmania, and might perhaps, as suggested by J. D. Hooker and by F. Mueller, be united with *Limnanthemum ;* but, as far as can be judged from the small number of good specimens seen, the inflorescence and fruit appear different as well as the foliage. The absence of hypogynous glands in *Liparophyllum,* and their supposed constant presence in *Limnanthemum,* require further verification.

1. **L. Gunnii,** *Hook. f. in Hook. Lond. Journ.* vi. 472, *and Fl. Tasm.* i. 273. *t.* 87. A small glabrous perennial, with a creeping rhizome emitting long thick fibres. Leaves tufted, narrow-linear, obtuse, entire, thick and fleshy, with a sheathing membranous base, 1 to 1½ in. long. Pedicels usually solitary in the tufts of leaves, and much shorter than them. Calyx-segments linear, somewhat fleshy, about 2 lines long. Corolla-lobes ovate, with the margins slightly undulate and sprinkled with a few hairs inside. Stamens inserted in the sinus of the lobes, the filaments short. Ovary

tapering into a short style, with 2 short broad stigmatic lobes.   Fruit about ¼ in. diameter.   Seeds orbicular, somewhat compressed.—*Limnanthemum* or *Villarsia Gunnii*, Hook. f. Fl. Tasm. ii. 368.

Tasmania.   Wet sandy soil on the margins of alpine lakes, *Gunn*.

ORDER LXXIX. HYDROPHYLLACEÆ.

(Hydrophyllaceæ *and* Hydroleaceæ, *DC*.)

Flowers regular.   Calyx free, of 5 divisions.   Corolla with a short or rarely elongated tube, and 5 spreading lobes, imbricate and sometimes contorted in the bud.   Stamens 5, inserted at the base of the corolla-tube and alternating with its lobes; anthers 2-celled, the cells opening in longitudinal slits.   Ovary superior, entire, either 1-celled with 2 parietal or free placentas or rarely 2-celled with the placentas on the dissepiment; style terminal, bifid or divided to the base into 2 distinct styles; stigmas obtuse or capitate; ovules numerous or rarely reduced to 2 to each placenta and then laterally attached.   Fruit a capsule, opening in 2 valves, the margins alternating with the placentas or rarely opposite the dissepiment.   Seeds with a thin usually reticulate testa, and copious fleshy albumen.   Embryo straight, usually small and distant from the hilum.—Herbs or rarely undershrubs, often hispid. Leaves alternate or rarely the lower ones opposite, entire lobed or divided. Flowers usually blue or white, in one-sided spikes or racemes, often rolled back when young and sometimes branching into dichotomous cymes as in Boragineæ or forming small and compact cymes or clusters.   Bracts usually present under the pedicels and often leaf-like; bracteoles rarely present.

A small Order, chiefly American.   The only Australian genus is also American, but extends into tropical Asia and Africa, and differs from the remainder of the Order in its completely 2-celled ovary and capsule, and in the septifragal dehiscence of the latter.

1. **HYDROLEA,** Linn.

Calyx divided nearly to the base.   Corolla almost rotate, with a very short tube.   Ovary 2-celled, with very numerous ovules in each cell on a broad spongy placenta attached along the central line to the dissepiment.   Styles 2.   Capsule opening in 2 valves parallel to the dissepiment.—Herbs or undershrubs. Leaves entire.   Flowers blue, usually in short and compact racemes or cymes.

A genus of few species, all apparently American, including the two Australian ones, of which one and perhaps both are also in the tropical regions of the Old World.

Annual, glabrous or nearly so, without spines . . . . 1. *H. zeylanica*.
Perennial, glandular-pubescent and hirsute, armed with axillary spines   2. *H. spinosa*.

1. **H. zeylanica,** *Vahl; Chois. in DC. Prod*. x. 180.   Said to be always annual, but the stems creep and root at the base often to a considerable length, ascending to about 1 ft., the whole plant quite glabrous or the inflorescence and calyxes hispid with a few hairs.   Leaves lanceolate, entire, narrowed at the base and sometimes shortly petiolate, mostly 1½ to 2½ in. long.   Flowers in rather compact simple or branched racemes or cymes in the upper axils, usually shorter than the leaves or the upper ones forming a

terminal panicle. Bracts small and narrow. Calyx-segments lanceolate, striate, about 3 lines long. Corolla scarcely exceeding the calyx, divided to about the middle into broad lobes. Stamens and styles shorter than the corolla; anthers sagittate. Capsule membranous, shorter than the calyx. Seeds very small and numerous.—Wight, Ill. t. 167; Ic. Pl. t. 601.

N. Australia. Gulf of Carpentaria, *F. Mueller.* Abundant in tropical Asia and Africa, less so in tropical America.

2. **H. spinosa,** *Linn.; Chois. in DC. Prod.* x. 181. An erect perennial or undershrub of 1 to 2 ft., glandular-pubescent and hispid with spreading hairs, more or less viscid and armed with axillary spreading spines of ¼ to ½ in. Leaves lanceolate or ovate-lanceolate, acute, contracted into a short petiole, from under 1 in. to about 2 in. long. Flowers larger than in *H. zeylanica,* in compact leafy cymes terminating the stems and upper branchlets. Calyx-segments narrow-lanceolate, acuminate, above 3 lines long. Corolla exceeding the calyx, divided to about the middle into broad lobes. Stamens and styles rather long; anthers lobed at both ends. Ovary, capsule, and seeds of *H. zeylanica.*—Bot. Reg. t. 566.

Queensland. Cape York, *Daemel* (*Herb. F. Mueller*). Very abundant in tropical America; not as yet known from the Old World, excepting the Timor specimen in the Banksian Herbarium referred to by Choisy.

Order LXXX. BORAGINEÆ.

Flowers regular or nearly so. Calyx free, of 5 rarely 4 or 6 or more divisions or teeth or rarely irregularly split. Corolla with a long or short tube, and 5 rarely 4 or 6 or more lobes, imbricate or induplicate in the bud. Stamens as many as corolla-lobes or very rarely fewer, inserted in the corolla-tube and alternate with its lobes; anthers 2-celled, the cells opening in longitudinal slits or rarely in terminal pores. Ovary superior, entire or 4-lobed rarely 2-lobed, either 4- or 2-celled with 1 ovule in each cell or 2-celled with 2 ovules in each cell (in all cases formed of 2 carpels); style terminal or inserted between the lobes; ovules laterally attached, ascending or pendulous. Fruit either a drupe with the endocarp entire or separating into 2 carpels or 4 pyrenes, or dry and separating into 4 rarely 2 nuts. Seed with a thin testa; albumen none or scanty; embryo straight; cotyledons flat and rather thick or rarely folded; radicle short.—Herbs, usually rough with coarse hairs, or in the drupaceous genera sometimes trees or shrubs with a softer indumentum or glabrous. Leaves alternate or very rarely opposite, usually undivided, entire or toothed, very rarely deeply lobed. Flowers in one-sided spikes or racemes, rolled back when young and often forked or dichotomous or rarely in irregularly-branched panicles or solitary. Bracts often not immediately subtending the pedicels and sometimes entirely wanting; bracteoles very rarely present.

A considerable Order, the herbaceous genera chiefly spread over the northern hemisphere with a very few tropical or southern species; the frutescent drupaceous genera chiefly tropical in the New as well as the Old World. Of the twelve indigenous Australian genera, five belong to the tropical Flora of the New as well as the Old World; a sixth is also tropical, but limited to the Old World; five are southern representatives of genera chiefly inhabiting the temperate regions of the northern hemisphere ; and one only (*Halgania*) is endemic in Australia. The Order differs from Verbenaceæ and Labiatæ in the more regular and isomerous flowers ; from Solaneæ in the ovules, 1 or 2 only to each carpel ; from Convolvulaceæ, besides the habit, in the flat or rarely folded, not contortuplicate, cotyledons. The two great tribes into which it is divided are separated by the same character which distinguishes Verbenaceæ and Labiatæ; but in the case of Boragineæ the passage is too gradual to admit of the adoption of two distinct Orders; *Trichodesma*, for instance, has the entire ovary and terminal style of *Ehretia* and *Halgania*, with the 4 nuts and much of the habit of Borageæ ; and *Heliotropium*, so closely connected with *Tournefortia* by the style, has the ovary sometimes almost as much lobed as in Borageæ.

TRIBE I. **Cordieæ.**—*Ovary entire (sometimes lobed in* Heliotropium), *with a terminal style. Trees shrubs or herbs. Fruit drupaceous or dry.*

Fruit a drupe, indivisible or separable into pyrenes. Trees shrubs or woody climbers (rarely herbs in species not Australian).
Style twice forked . . . . 1. CORDIA.
Style 2-lobed, without any thick ring . . . . 2. EHRETIA.
Style simple, with a fleshy ring round the entire or slightly 2-lobed summit (the ring deficient in *T. sarmentosa*) . . . 3. TOURNEFORTIA.
Fruit dry (except in *Halgania*), separable into 4 nuts or into 2 2-seeded carpels. Herbs undershrubs or low shrubs.
Style 2-lobed. Prostrate herb with lobed leaves . . . . . 4. COLDENIA.
Style simple with a fleshy ring round the summit or at the base of a terminal cone or point . . . . 5. HELIOTROPIUM.
Style simple, filiform, with a minute stigma. Anthers exserted, connivent in a long cone or cylinder.
Fruit separable into 2 2-celled carpels. Seeds albuminous. Anther-points straight . . . . 6. HALGANIA.
Fruit 4-lobed, the lobes forming separable nuts, the centre undivided and persistent. No albumen. Anther-points long and twisted . . . . 7. TRICHODESMA.

TRIBE II. **Borageæ.**—*Ovary 4- or 2-lobed, the style inserted between the lobes, and more or less lateral with reference to the carpels. Herbs, very rarely (in species not Australian) shrubby. Fruit separable into 4 or 2 small nuts.*

Corolla rotate. Anthers exserted and connivent in a long cone or cylinder . . . . * BORAGO.
Corolla oblique. Anthers exserted, the stamens unequal . . . * ECHIUM.
Corolla with a cylindrical (sometimes very short) tube and spreading limb. Anthers included or, if exserted, equally so.
Nuts attached by a broad concave base to the slightly convex receptacle. Corolla-throat closed by very prominent scales . * ANCHUSA.
Nuts erect, attached by a small base to a slightly convex receptacle.
Corolla-throat without any scales . . . . * LITHOSPERMUM.
Corolla-throat with small scales. Nuts small, very smooth and shining . . . . 8. MYOSOTIS.
Nuts erect, obliquely attached to the shortly pyramidal torus. Corolla without scales. Nuts reticulate . . . . 9. ERITRICHIUM.
Nuts erect, laterally attached to the narrow conical receptacle. Corolla-scales small or none.
Nuts 4 . . . . 10. ECHINOSPERMUM.
Nuts 2 . . . . 11. ROCHELIA.
Nuts depressed, spreading, obliquely attached to the convex receptacle . . . . 12. CYNOGLOSSUM.

The introduced plants belonging to the genera marked above with the asterisk * are the following :—

Borago officinalis, Linn. ; DC. Prod. x. 35. A rough hispid erect annual or biennial with spreading branches. Lower leaves petiolate, obovate or oblong, upper ones narrow. Flowers blue, pedicellate, in loose forked cymes. Calyx deeply cleft. Corolla rotate, with a very short tube and short broad erect scales in the throat opposite the lobes. Anthers exserted, connivent in a dark-coloured erect cone in the centre of the flower, each filament with an erect appendage or scale behind the anther. Nuts 4, attached to the convex receptacle by a broad concave base.—About Adelaide in S. Australia.

Echium violaceum, Linn.; DC. Prod. x. 22. A coarse biennial, hispid with stiff almost prickly hairs, with large spreading radical leaves. Stems erect but very spreading, with narrow leaves. Flowers at first purplish, becoming blue, in long one-sided spikes. Calyx deeply divided. Corolla often 1 in. long, the tube very open, the limb oblique, with short almost erect teeth or lobes, the stamens protruding and unequal in length. Nuts 4, wrinkled.—Victoria and Tasmania.

Anchusa officinalis, Linn.; DC. Prod. x. 42. A coarse, hirsute, erect biennial or perennial of about 2 ft. Radical leaves petiolate and often long; stem-leaves lanceolate. Flowers blue, in dense, one-sided forked spikes, which often lengthen considerably, each flower with a small leafy bract at the base. Calyx deeply divided. Corolla-tube closed by very prominent scales. Stamens included in the tube. Nuts 4, rather large, broad and obtuse, wrinkled, attached to the hemispherical receptacle by their broad concave base.—S. Australia.

Lithospermum arvense, Linn.; DC. Prod. x. 74. An erect hard annual of 1 ft. or more, more or less hoary with appressed hairs. Leaves narrow lanceolate. Flowers small and white in the upper axils, the upper ones forming terminal leafy cymes. Calyx deeply cleft. Corolla with a cylindrical tube, not closed by scales, the lobes spreading. Stamens included in the tube. Nuts 4, erect, hard, wrinkled, attached to the slightly convex receptacle by a small base.—Established in several localities in Queensland, Victoria, Tasmania, and S. Australia, probably also in N. S. Wales, although I have not seen specimens from thence.

TRIBE I. CORDIEÆ.—Ovary entire, with a terminal style. Trees shrubs or herbs. Fruit drupaceous or dry.

1. CORDIA, Linn.

Calyx tubular or campanulate, 5-toothed or irregularly toothed or lobed. Corolla-tube cylindrical or funnel-shaped, the limb 5- or sometimes 6- or more-lobed. Stamens inserted in the tube; anthers included or exserted. Ovary entire, 4-celled, with 1 pendulous ovule in each cell; style terminal, twice forked. Fruit a drupe, the endocarp hard, with 4 cells or fewer by abortion. Seeds without albumen; testa thin; cotyledons longitudinally folded; radicle superior.—Trees or shrubs, glabrous scabrous-pubescent or villous. Leaves entire or toothed. Flowers in cymes, sometimes contracted into heads, at first terminal, but often becoming lateral by the growth of the branch. Bracts small or none.

A considerable genus, spread over the tropical regions of both the New and the Old World. Of the three Australian species, one is common in tropical Asia; another perhaps limited to the Indian Archipelago, the islands of the South Pacific, and the S. E. coast of Africa, and cultivated only in India; the third still further confined to the South Pacific islands.

Flowers large. Corolla-tube much longer than the calyx. Drupe enveloped in the enlarged calyx . . . . 1. *C. subcordata.*
Flowers small. Corolla-tube not exceeding the calyx. Drupe seated on the open calyx.
    Usually villous. Calyx cylindrical, strongly ribbed . . . . . . 2. *C. aspera.*
    Glabrous or scabrous. Calyx campanulate, smooth . . . . . . 3. *C. Myxa.*

1. **C. subcordata,** *Lam.; DC. Prod.* ix. 477. A moderate-sized

spreading tree, the young shoots pubescent. Leaves on rather long petioles, very broadly ovate, acute or acuminate, often slightly cordate at the base, the upper surface scabrous with small scattered hairs, bearded underneath with short hairs in the axils of the principal veins or along their whole length, 4 to 6 in. long. Flowers not numerous, large, white, in shortly pedunculate loose cymes. Calyx tubular, about ½ in. long, without prominent ribs, shortly and irregularly lobed, enlarged after flowering and closing over the fruit. Corolla broadly funnel-shaped, 2 or 3 times as long as the calyx, with 5 to 7 broad lobes, much shorter than the tube, spreading to from 1 to 1½ in. diameter. Anthers included in the tube. Style forked, each branch with 2 spathulate stigmatic lobes. Drupe nearly globular but contracted at the top, about 1 in. diameter, completely enveloped in the enlarged calyx.—Seem. Fl. Vit. t. 34 ; *C. orientalis*, R. Br. Prod. 498.

N. Australia. Islands of the Gulf of Carpentaria, *R. Brown, Henne ;* Port Essington, *Armstrong ;* Escape Cliffs, *Hulls.*
Queensland. Hope Islets, *M'Gillivray.*

The species is also on the Mozambique coast and Comoro Islands, and in the Indian Archipelago extending to the Philippines and Pacific Islands ; in India perhaps only where cultivated.

2. **C. aspera,** *Forst. Prod.* 18. A tree of 20 to 30 ft., the young shoots rusty-pubescent or villous. Leaves petiolate, ovate, acuminate with the point sometimes much elongated, irregularly toothed or rarely entire, membranous, sprinkled above with short scattered hairs, the veins underneath scabrous pubescent or hirsute, 3 to 6 in. long in some specimens, twice as large in others. Flowers small, in shortly pedunculate, rather dense cymes. Calyx scarcely above 2 lines long, tubular, hirsute, with 10 or 12 prominent ribs and 5 or 6 small linear teeth. Corolla-tube cylindrical, scarcely so long as the calyx ; lobes ovate, undulate and crisped, much shorter than the tube. Stamens scarcely exserted ; anthers small. Style forked, with spathulate shortly 2-lobed branches. Drupe ovoid-pyramidal, not exceeding ¼ in. and sometimes much smaller, resting on the broad open calyx ; putamen hard, very rugose, ripening usually only a single seed.—DC. Prod. ix. 499 ; Seem. Fl. Vit. 169. t. 35 ; F. Muell. Fragm. vi. 114.

Queensland. Rockingham Bay, *Dallachy.*

Var. *inciso-dentata.* Leaves (all young in the specimens seen) irregularly and deeply toothed, almost lobed, but the teeth perhaps not so prominent in the adult state.—*C. lacerata,* F. Muell. Fragm. v. 193.—Cape York, *Daemel.*

The species is also in the islands of the South Pacific.

3. **C. Myxa,** *Linn. ; DC. Prod.* ix. 479. A handsome tree, with a dense coma, glabrous or the foliage scabrous-pubescent. Leaves on rather long petioles, from ovate to orbicular, very obtuse or shortly acuminate, entire or irregularly sinuate, 3- or 5-nerved at the base, usually 2 to 3 in. long. Flowers not large, polygamous, in loose pedunculate cymes or panicles. Calyx membranous, about 3 lines long, entire and closed over the corolla in the bud, opening irregularly into short lobes without prominent ribs when the flower expands, hardened, broadly cup-shaped, and irregularly and broadly toothed or lobed under the fruit. Corolla-tube oblong-cylindrical, slightly contracted at the throat, nearly as long as the calyx, glabrous inside and out ;

lobes narrow, recurved, as long as the tube. Stamens exserted, but not exceeding the corolla-lobes ; anthers oblong-linear. Style short, with 4 long filiform branches stigmatic along the inner side. Drupe ovoid or nearly globular, " pale yellow or slightly pink," the pulp very viscid, the putamen very hard, usually 1- or 2-celled, with 1 seed in each cell.—Wight, Illustr. t. 169 ; *C. dichotoma*, Forst. ; R. Br. Prod. 498 ; *C. Brownii*, DC. Prod. ix. 499 ; *C. latifolia*, Roxb. ; DC. Prod. ix. 478 ; *C. ixiocarpa*, F. Muell. Fragm. i. 59.

Queensland. Broad Sound, *R. Brown;* E. coast, *A. Cunningham ;* Gilbert river, *F. Mueller ;* Rockingham Bay, *Dallachy ;* Port Denison, *Fitzalan ;* Nerkool Creek, *Bowman ;* Rockhampton, *Thozet, Dallachy.*

The specimens with all the flowers male have usually a looser, more slender inflorescence than those in which all or nearly all the flowers are perfect. The species is dispersed over tropical Asia, from Ceylon to the Philippines, and is also sent from tropical Africa, but in many places it is cultivated only.

2. EHRETIA, Linn.

Calyx deeply divided into 5 segments. Corolla with a short or cylindrical tube ; limb of 5 spreading lobes, imbricate in the bud. Stamens inserted in the tube ; anthers exserted or rarely almost included. Ovary 2-celled with 2 ovules in each cell, or 4 celled with 1 ovule in each cell ; style terminal, more or less 2-lobed or forked, the lobes entire, without any prominent ring. Fruit a drupe, the endocarp forming 2 2-seeded or 4 1-seeded pyrenes. Seeds with a membranous testa and usually scanty albumen ; cotyledons ovate, not folded.—Trees or shrubs, often glabrous. Leaves entire or toothed. Flowers rather small, usually white, in panicles or cymes, either terminal in the upper axils or rarely all axillary. Bracts small. Fruits usually much smaller than in *Cordia.*

The genus is widely distributed over the tropical regions of the New as well as the Old World. Of the Australian species, one or perhaps two are common E. Indian ones, the four or three others are endemic.

SECT. I. **Euehretia.**—*Flowers in irregularly-branched panicles. Ovary 4-celled, with 1 ovule in each cell.*

Leaves glabrous underneath, rather narrow . . . . 1. *E. acuminata.*
Leaves pubescent underneath, usually broad . . . . 2. *E. pilosula.*

SECT. II. **Bourreria.**—*Flowers in divaricately dichotomous cymes. Ovary 2-celled, with 2 ovules in each cell.*

Corolla-tube longer than the calyx ; lobes shorter. Cymes mostly
    terminal. Leaves oblong or lanceolate.
Leaves long-lanceolate, with very oblique veins. Anthers included
    in the tube or scarcely exserted . . . . 3. *E. saligna.*
Leaves oblong or oblong-lanceolate, the veins rather oblique.
    Anthers quite exserted . . . . 4. *E. membranifolia.*
Corolla-tube not longer than the calyx ; lobes twice as long.
    Cymes small, lateral. Leaves mostly ovate . . . . . . 5. *E. lævis.*

SECT. I. EUEHRETIA, *DC.*—Flowers in irregularly-branched panicles. Ovary 4-celled, with 1 ovule in each cell. Fruit of 2 2-seeded pyrenes.

1. **E. acuminata,** *R. Br. Prod.* 497. A tall shrub or tree of 20 to
                                                                    2 c 2

30 ft., quite glabrous, except the inflorescence, which is slightly pubescent. Leaves petiolate, oval or elliptical-oblong, shortly and obtusely acuminate, usually narrowed at the base, serrate with callous teeth, 3 to 6 in. long. Flowers sessile and crowded on the branchlets of dense thyrsoid panicles, terminal and in the upper axils, scarcely exceeding the leaves. Calyx-segments about ¾ line long, nearly orbicular, ciliate. Corolla-tube exceedingly short; lobes spreading to about 3 lines diameter. Anthers exserted. Ovary 4-celled, with 1 laterally attached ovule in each cell; style-lobes in the typical form very short, clavate, truncate and usually connivent. Fruit globular, 2 to 3 lines diameter, the endocarp separating into 2 hard 2-celled 2-seeded pyrenes.—DC. Prod. ix. 503.

Queensland. Moreton Bay, *F. Mueller.*

N. S. Wales. Port Jackson and Blue Mountains, *R. Brown* and others ; Emu plains, *A. Cunningham ;* Clarence, Hastings, and Richmond rivers, *Beckler, C. Moore,* and others ; Illawarra, *A. Cunningham, M'Arthur.*

Var. *laxiflora.* Leaves less toothed, sometimes almost entire. Style divided nearly to the base. To this belong all the specimens both northern and southern, except those from Port Jackson and the Blue Mountains. The differences in the style may, however, depend on a certain degree of unisexuality. The common East Indian *E. serrata,* Roxb., may not be specifically distinct from *E. acuminata.*

2. **E. pilosula,** *F. Muell. Fragm.* v. 20. A handsome tree of 20 to 30 ft., closely allied to *E. acuminata,* and probably a variety, with the leaves broader, rounded or almost cordate at the base, and more or less pubescent or villous underneath. Inflorescence, flowers, and fruit the same as in *E. acuminata.*

Queensland. Rockingham Bay and Herbert river, *Dallachy.*

SECT. II. BOURRERIA, *DC.*—Flowers in divaricately dichotomous cymes. Ovary 2-celled, with 2 ovules in each cell. Fruit of 4 1-seeded pyrenes.

3. **E. saligna,** *R. Br. Prod.* 497. A shrub or tree, quite glabrous, the foliage apparently glaucous. Leaves long-lanceolate or linear, tapering to a fine point, quite entire, contracted into a rather long petiole, rather thick, very obliquely veined, 2 to 5 in. long. Flowers in divaricately dichotomous shortly pedunculate cymes. Calyx-segments narrow-ovate, almost acute, ¾ line long. Corolla-tube cylindrical, fully 1 line long; lobes very spreading, nearly as long as the tube. Anthers included in the tube or very shortly protruding. Ovary 2-celled, with 2 ovules in each cell. Style deeply forked, with obtuse stigmas. Fruit about 2 lines diameter, containing 4 distinct, not very hard pyrenes.—DC. Prod. ix. 504.

N. Australia. Islands of the Gulf of Carpentaria, *R. Brown ;* Victoria river, *F. Mueller.*

Queensland? N.E. Australia, precise station not given, *Fitzalan.*

4. **E. membranifolia,** *R. Br. Prod.* 497. A shrub or tree, quite glabrous, with slender branchlets. Leaves oblong or oblong-lanceolate, obtuse, membranous, obliquely veined but not near so much so as in *E. saligna,* 1½ to 2½ in. long, on slender petioles. Flowers small, in loosely divaricate, shortly pedunculate, dichotomous cymes. Calyx-segments broad, rounded, ciliate, about ½ line long. Corolla-tube ¾ to nearly 1 line long, the lobes

rather shorter than the tube. Anthers wholly exserted. Ovary 2-celled, with 2 ovules in each cell; style shortly or deeply forked.—DC. Prod. ix. 504.

Queensland. Broad Sound, *R. Brown, Bowman;* Araucaria range, *F. Mueller;* Rockhampton, *Dallachy, O'Shanesy;* Suttor river and Crocodile Creek, *Bowman.*

5. **E. lævis,** *Roxb. Pl. Corom.* i. 42. *t.* 56. A tall shrub or tree, quite glabrous. Leaves ovate oval or elliptical, acutely acuminate or rarely obtuse, quite entire, rounded or tapering at the base, shining above, mostly 3 to 5 in. long. Flowers small, in axillary, shortly pedunculate, dichotomous cymes, much shorter than the leaves. Calyx-segments narrow, under 1 line long. Corolla-tube about as long as the calyx, the lobes twice as long, spreading. Anthers exserted. Ovary 4-celled, with 1 ovule in each cell. Style more or less forked. Drupe small, containing 4 1-seeded pyrenes.— DC. Prod. ix. 505; Wight, Ic. t. 1382.

Queensland. Cape York, *W. Hill.* The species is common in East India and Ceylon. Wight's figure represents well the Australian form, Roxburgh's an equally common East Indian one with more obtuse leaves.

3. TOURNEFORTIA, Linn.

Calyx deeply divided into 5 segments. Corolla-tube cylindrical; lobes 5, spreading, imbricate or induplicate in the bud. Stamens inserted in the tube: anthers included. Ovary entire, 4-celled, with 1 pendulous ovule in each cell; style terminal, entire, the stigmatic summit entire or notched, surrounded by a prominent fleshy ring (except in *T. sarmentosa*). Fruit a drupe, with 4 1-seeded pyrenes, either quite distinct or more or less cohering in pairs or quite united in a 4-celled putamen. Seeds with or without albumen, the cotyledons not folded.—Trees shrubs or woody climbers, rarely (in species not Australian) almost herbaceous, tomentose villous or rarely almost glabrous. Leaves undivided and entire. Flowers usually white or nearly so, in unilateral spikes, arranged in dichotomous cymes, usually without bracts. Fruits small.

The genus is widely distributed over the tropical regions of the New and the Old World. Of the three Australian species, two are in the Indian Archipelago, and extend from the islands of the African coast to those of the South Pacific, the third is endemic. The genus is a natural one, distinguished from *Heliotropium* by the drupaceous fruit, from *Ehretia* chiefly by the style. The fleshy annulus round the summit, which *Tournefortia* has in common with *Heliotropium*, and which Fresenius (in Mart. Fl. Bras.) considers as an absolute character of the proposed Order of *Heliotropieæ*, is, however, wanting in *T. sarmentosa.*

Corolla-lobes longer than the tube, imbricate in the bud. Foliage and
　inflorescence very densely silky-tomentose . . . . 1. *T. argentea.*
Corolla-lobes shorter than the tube, plicate and induplicate in the bud.
　Foliage tomentose hirsute or nearly glabrous.
Corolla-lobes rather acute. Stigma surrounded by a fleshy ring.
　Erect tree or shrub . . . . 2. *T. mollis.*
Corolla-lobes obtuse or retuse. Stigma not thickened. Branches
　long weak or climbing . . . . 3. *T. sarmentosa.*

1. **T. argentea,** *Linn. f.; R. Br. Prod.* 497. A tall erect shrub, the branches, foliage, and inflorescence densely tomentose or villous with soft

silky hairs. Leaves crowded at the ends of the thick branchlets, obovate ovate or obovate-oblong, soft and thick, 4 to 6 in. long, narrowed into a short petiole. Flowers small, sessile, and numerous, in large, dense, terminal dichotomous cymes or panicles. Calyx-segments orbicular, ciliate, about 1 line diameter. Corolla shortly exceeding the calyx, the lobes broad, obtuse, longer than the tube, imbricate in the bud, glabrous inside, sprinkled with a few hairs outside. Anthers ovate, mucronulate. Hypogynous disk prominent or obscure. Style exceedingly short, the thick fleshy annulus surrounding the notched summit appearing almost sessile on the ovary. Fruit glabrous, nearly globular, with 2 opposite furrows, but containing 4 distinct pyrenes. Seeds without any (or with a thin?) albumen.—DC. Prod. ix. 514.

N. Australia. Islands of the Gulf of Carpentaria, *R. Brown, Henne;* Sims Island, *A. Cunningham.*

Queensland. Albany Island, *W. Hill;* Barnard Isles, *M'Gillivray;* Howick's group, *F. Mueller;* estuary of the Burdekin, *Fitzalan;* Rockingham Bay, *Dallachy.*

The species extends along the seacoasts of Eastern tropical Africa, the Mauritius, Ceylon, the Eastern Archipelago, New Caledonia, and the islands of the Pacific.

2. **T. mollis,** *F. Muell. Fragm.* i. 59. An erect not much branched shrub of several feet, the branches and foliage velvety-pubescent. Leaves from broadly ovate to ovate-lanceolate, obtuse or obtusely acuminate, often rugose, 2 to 4 in. long or longer on barren shoots. Cymes dichotomous, pedunculate, the pedicels exceedingly short, without bracts. Calyx-segments lanceolate, pubescent, about 1 line long. Corolla pubescent outside, the tube 1½ to 2 lines long, somewhat contracted upwards; lobes broad, much plicate, rather acute, about one-third as long as the tube, induplicate in the bud. Ovary tapering into a short style, the stigma with a thick broad fleshy ring, round a slightly-depressed obscurely 2-lobed centre. Fruit nearly globular, above 2 lines diameter when full grown, containing either 4 pyrenes or 2 pyrenes each one readily divisible into 2. Seeds with a rather thick albumen.

N. Australia. Montague Sound, *A. Cunningham.*

Queensland. Burdekin river, *F. Mueller;* Edgecombe Bay, Port Denison, and Herbert river, *Dallachy.*

3. **T. sarmentosa,** *Lam. Illustr.* i. 416. A tall shrub with weak branches, or sometimes climbing to a great height, glabrous or sprinkled with short rigid hairs. Leaves petiolate, ovate, acuminate, entire, 2 to 4 in. long, either glabrous or sprinkled with short hairs above, more or less pubescent or hirsute underneath. Flowers usually white, sessile along the divaricate branches of terminal dichotomous cymes, without bracts. Calyx-segments lanceolate, shortly pubescent, under 1 line long. Corolla-tube angular, varying from 1½ line to above 2½ lines in length; lobes broad, obtuse or retuse, undulate-plicate, with a thick midrib, induplicate in the bud. Ovary tapering into a short style, which is very shortly lobed at the top, but without the thickened ring of the other species. Fruit ovoid-globular, slightly compressed, the endocarp thick, of a loosely cellular texture although hard, with 4 very small real cells, and sometimes separating or separable into 2 2-celled

pyrenes.—DC. Prod. ix. 516; *T. orientalis*, R. Br. Prod. 497; DC. Prod. ix. 516; *T. acclinis*, F. Muell. Fragm. iv. 95.

Queensland. Cape York, *Daemel;* Endeavour river, *Banks and Solander*, *R. Brown;* Port Denison, *Fitzalan;* Edgecombe and Rockingham Bays, *Dallachy*, Rockhampton, *Thozet* and several others; Broad Sound and Amity Creek, *Bowman;* Port Mackay, *Nernst.* The species is also in the Mauritius, Timor, the Philippines, and probably in other islands of the Indian Archipelago.

The Timor specimens, from the collection described by Decaisne, have the flowers rather smaller than usual; the Philippine Island specimens referred to by De Candolle are rather more hairy; some of R. Brown's have remarkably long flowers, the slender corolla-tube above 2½ or almost 3 lines long; but in all these respects the Queensland specimens from other collections are variable, and all appear to belong to one species, remarkable for its climbing habit and exceptional style. According to Thozet, the flowers are blue; other collectors describe them sometimes as whitish, sometimes as pure white. If they really pass into a bluish tint, it would seem probable that the common East Indian *T. viridiflora*, Wall., which is not to be distinguished in the dried state, but is said always to have green flowers, ought to be considered as a variety only of *T. sarmentosa*.

4. COLDENIA, Linn.

(Lobophyllum, *F. Muell.*)

Calyx deeply divided into 4 or 5 segments. Corolla with a short cylindrical tube; lobes 4 or 5, spreading, imbricate in the bud. Stamens inserted in the tube; anthers included. Ovary entire, 4-celled, with 1 pendulous ovule in each cell; style terminal, bifid or divided to the base into 2 styles, with a capitate or clavate stigma on each branch or style. Fruit more or less 4-lobed, dry or scarcely succulent, separating into 2 hard 2-celled carpels or finally into 4 1-seeded nuts. Seeds with a very thin (or without any?) albumen, the cotyledons not folded.—Hispid herbs. Leaves toothed or lobed. Flowers small, solitary in the axils, the upper ones often forming one-sided leafy spikes.

Besides the Australian species, which is a common one in tropical Asia and Africa, the genus has been extended by A. Gray so as to comprise several from N.W. and W. tropical America.

1. **C. procumbens,** *Linn.; DC. Prod.* ix. 558. A hard prostrate, hirsute annual, the branches radiating from the crown of the root to a considerable length, but not rooting. Leaves petiolate, obovate or oblong, much undulate, wrinkled and crenate or obtusely lobed, the largest ones scarcely exceeding 1 in., the floral ones usually very small. Flowers nearly sessile in their axils, often forming leafy spikes on the branchlets, usually 4-merous, at least in the Australian specimens. Calyx-segments herbaceous, ovate-lanceolate, not 1 line long. Corolla scarcely exceeding the calyx, the lobes broad, shorter than the tube, glabrous inside. Style bifid. Fruit depressed-globular, glandular-hispid, about 1 line long, more or less distinctly 4-lobed and often with prominent ribs between the lobes.—*Lobophyllum tetrandrum*, F. Muell. in Hook. Kew Journ. ix. 21.

N. Australia. Victoria river, Sturt's Creek, and Upper Roper river, *F. Mueller.*
S. Australia. Cooper's Creek, *Howitt's Expedition.*

The species is common in a great part of tropical Asia and Africa.

5. HELIOTROPIUM, Linn.

(Schleidenia, *Endl.*)

Calyx deeply divided into 5 segments. Corolla with a cylindrical tube; lobes 5, spreading, plicate and imbricate in the bud. Stamens inserted in the tube; anthers often mucronate or acuminate and sometimes cohering by their tips, included or the tips slightly protruding. Ovary entire, 4-celled, with 1 laterally attached or pendulous ovule in each cell; style terminal, short or long, the stigma or stigmatic summit broadly umbrella-shaped or with a fleshy ring surrounding the base of a more or less distinct central cone or point. Fruit more or less 2- or 4-lobed or furrowed, separating into 4 1-seeded nuts, or in species not Australian into 2 hard 2-seeded carpels. Seeds with a scanty or rarely with a rather thick albumen.—Herbs undershrubs or rarely shrubs, with appressed and strigose or with rigid and spreading hairs, very rarely glabrous. Flowers usually small, sessile or pedicellate in one-sided simple or once- or twice-forked spikes, with or without bracts, which when present are often not immediately under the pedicels.

The genus is widely dispersed over the tropical and subtropical regions of the globe, a few species extending beyond the tropics both in the northern and the southern hemispheres. Of the twenty-one Australian species, three belong to the warmer regions of the Old World, chiefly abundant in Africa and western Asia; one is widely spread over the temperate and subtropical regions of the Old World; another extends over the seacoasts of almost all warm countries; the remaining sixteen, as far as known, are all endemic. All the Australian species belong to the genus or section of *Heliotropium* proper as limited by De Candolle, or to *Heliotropium* and *Schleidenia* as defined by Fresenius (in Mart. Fl. Bras.). The section *Heliophytum*, with the fruit separating into 2 2-seeded carpels, established by De Candolle as a genus, comprises the *H. indicum*, Linn., a very common S. Asiatic weed, but which does not appear to have been yet found in Australia.

SECT. I. **Platygyne.**—*Stigma nearly sessile, umbrella-shaped, without any distinct central cone.*

Glabrous and glaucous prostrate perennial . . . . 1. *H. curassavicum.*

SECT. II. **Euheliotropium.**—*Stigma a thick ring at the base of or round a central cone or point. Anthers obtuse or shortly acuminate, not cohering (except in* H. brachygyne ?). *Throat of the corolla not bearded.*

Flowers in scorpioid forked or rarely simple spikes, without bracts.
    Leaves petiolate or contracted at the base.
    Style shorter than the stigma.
        Leaves ovate, flat. Anthers obtuse . . . . 2. *H. europæum.*
        Leaves oblong or lanceolate, rugose, undulate. Anthers very
          shortly acuminate . . . . 3. *H. undulatum.*
    Style longer than the stigma.
        Leaves oblong or lanceolate, contracted into a very short petiole 4. *H. asperrimum.*
        Leaves ovate, undulate, on long petioles . . . . 5. *H. crispatum.*
Flowers few, sessile within the uppermost leaves or forming leafy
    spikes. Leaves sessile, small.
    Leaves linear-lanceolate, rather crowded. Anthers scarcely
        pointed. Style longer than the stigma . . . . 6. *H. fasciculatum.*
    Leaves lanceolate, almost imbricate. Anther-points long. Style
        longer than the stigma . . . . 7. *H. vestitum.*
    Leaves lanceolate or linear, distant. Anther-points long. Style
        very short . . . . 8. *H. brachygyne.*
    Leaves ovate or ovate-lanceolate, crowded. Anther-points short.
        Stigma all but sessile . . . . 9. *H. epacrideum.*

Sect. III. **Schleidenia.**—*Stigma a thick ring round a central cone or point. Anthers acuminate, cohering by the minutely hairy tips. Corolla-throat bearded or pubescent inside.*

Corolla-tube scarcely swollen. Style shorter than the stigma. Calyx-segments very unequal.

Leaves obovate-oblong or lanceolate, flat. Spikes without bracts 10. *H. ovalifolium.*
Leaves linear with revolute margins. Spikes bracteate . . . 11. *H. strigonum.*
Leaves lanceolate. Flowers axillary, scarcely forming leafy spikes 8. *H. brachygyne.*
Corolla-tube swollen round the anthers at or above the middle. Style as long as or longer than the stigma.

Leaves lanceolate or linear-lanceolate, flat or with recurved margins, mostly acute and above ¼ in. long. Bracts leaf-like. Plant hirsute or pubescent.

Stems long and prostrate. Corolla-limb longer than the tube. Bracts small . . . . 12. *H. prostratum.*
Stems diffuse. Bracts longer than the calyx, petiolate or contracted at the base . . . . 13. *H. bracteatum.*
Stems ascending or erect. Bracts longer than the calyx, sessile.

Leaves rather crowded, narrow. Nuts usually 4 . . . . 14. *H. pauciflorum.*
Leaves lanceolate. Fruit of a single conical nut, the others abortive . . . . 15. *H. conocarpum.*
Leaves oblong-lanceolate, obtuse, crowded, under 2 lines long. Bracts leaf-like, imbricate. Plant cottony-white . . . . . 16. *H. filaginoides.*
Leaves all narrow-linear, with revolute margins. Bracts rarely exceeding the calyx, usually few and small. Nuts usually scabrous-pubescent.

Stigmatic cone very short.

Erect slightly-branched annual. Hairs spreading. Corolla-tube very slender . . . . 17. *H. ventricosum.*
Stem paniculately branched. Hairs appressed. Calyx-segments acuminate.

Calyx 2 lines long. Leaves linear . . . . 18. *H. tenuifolium.*
Calyx 1 line long. Leaves filiform . . . . 19. *H. paniculatum.*
Stigmatic cone long and narrow. Calyx-segments 1 line long. Leaves linear . . . . 20. *H. Cunninghamii.*
Lower leaves ovate, petiolate, crowded. Flowering branches elongated, with few narrow small bract-like leaves. Nuts very hispid . . . . 21. *H. diversifolium.*

Sect. I. Platygyne.—Stigma nearly sessile, umbrella-shaped, without any distinct central cone.

1. **H. curassavicum,** *Linn.; DC. Prod.* ix. 538. A much-branched prostrate glabrous and glaucous perennial, often somewhat succulent, spreading sometimes to 2 or 3 ft. Leaves linear oblanceolate or oblong, usually obtuse and narrowed into a short petiole, rarely obovate, rather thick, veinless except the midrib, ¼ to 1 in. long. Spikes once-forked or rarely simple, terminal or lateral. Flowers sessile, without bracts. Calyx-segments obtuse, nearly 1 line long. Corolla white or with a yellow eye, the tube about 1 line long, the throat not bearded; lobes broad, as long as the tube. Anthers nearly sessile at the base of the tube. Ovary depressed-globular, capped by an umbrella-shaped almost sessile stigma, often broader than the ovary itself. —Lehm. Pl. Preiss. i. 348; Bot. Mag. t. 2669.

N. S. Wales. Darling river, *Neilson, Mrs. Ford.*
Victoria. Wimmera, *Dallachy.*

S. Australia. Murray river, *F. Mueller.*
W. Australia. Swan River, *Drummond*, 1st *Coll. n.* 35 ; Avon river, *Preiss, n.*
1933; Murchison river, *Oldfield.*

The species is frequent in sandy places, chiefly on the seacoast, in North and South America, South Africa, and the Pacific Islands. The ovary and fruit are occasionally but rarely 3-merous (with 6 ovary-cells and nuts).

SECT. II. EUHELIOTROPIUM.—Stigma a thick ring at the base of or round a central cone or point. Anthers obtuse or shortly acuminate, not usually cohering. Throat of the corolla not bearded inside.

2. **H. europæum,** *Linn.; DC. Prod.* ix. 534. An erect divaricately-branched or spreading rather hard annual, scabrous-pubescent or hirsute, rarely above 1 ft. high. Leaves on rather long petioles, oval, obtuse, entire but undulate when young, green on both sides, the principal veins very prominent underneath, ¾ to 1½ in. long. Spikes once-forked or rarely simple, hirsute, without bracts. Calyx-segments lanceolate, hirsute, about 1 line long. Corolla white, the tube about 1 line long, the throat glabrous inside, the lobes shorter than the tube. Anthers obtuse. Style shorter than the stigmatic cone, which is usually pubescent. Nuts pubescent.—*H. glandulosum*, R. Br. Prod. 493 ; DC. Prod. ix. 542 ; *H. lacunarium*, F. Muell. in Hook. Kew Journ. viii. 167, and in Trans. Phil. Soc. Vict. i. 20.

N. S. Wales. Lagoons on the Murray, Murrumbidgee, and Darling rivers, *F. Mueller.*
S. Australia. Head of Spencer's Gulf, *R. Brown ;* in the interior, *M‘Douall Stuart.*

The species is very abundant in the Mediterranean region of the northern hemisphere, extending eastward to Affghanistan and westward to the Canary Islands. I can find no character to distinguish the Australian specimens from the northern ones. Brown's specimens have a somewhat different aspect, but they are evidently stunted, gathered late in the season, in fruit, and retaining only the small leaves of the smaller branches, and might be easily matched amongst old autumn specimens from the dry wastes of southern Europe.

3. **H. undulatum,** *Vahl; DC. Prod.* ix. 536. Stems much branched, diffuse or ascending, hard and almost woody at the base, scabrous-pubescent and hispid with spreading hairs. Leaves shortly petiolate, oval-oblong, lanceolate or almost linear, obtuse, with undulate recurved margins, very rugose. Spikes at first dense, at length elongated, once- or twice-forked or rarely simple, without bracts. Calyx-segments nearly equal, about ¾ line long. Corolla-tube as long as the calyx, the throat glabrous inside ; lobes shorter than the tube. Style shorter than the long stigmatic cone. Nuts rugose, above 1 line long.—*H. arenarium*, F. Muell. Fragm. vi. 116.

N. Australia. Nicholson and Albert rivers, *F. Mueller.*
S. Australia. Lake Gregory, *Babbage's Expedition.*
W. Australia. Murchison river, *Oldfield, Drummond, 6th Coll. n.* 134.

The species is common in northern Africa and W. Asia, extending from the Cape de Verd Islands to Affghanistan. I can find no difference whatever in the Australian specimens except that the nuts are rather larger, but their size is variable in the African specimens.

4. **H. asperrimum,** *R. Br. Prod.* 493. A perennial with erect or ascending stems of 1 to 1½ ft., the branches and foliage very scabrous and sprinkled with rigid spreading hairs. Leaves shortly petiolate, oblong or lanceolate, obtuse, with undulate revolute margins but not very rugose, 1 to

2 in. long. Spikes short and dense, once- or sometimes twice-forked, rarely simple, without bracts. Calyx-segments linear or lanceolate, hispid, 2 lines long. Corolla-tube shortly exceeding the calyx, the throat glabrous inside; lobes much plicate, very broad, expanding to $\frac{1}{4}$ in. diameter. Anthers oblong, very shortly acuminate. Style filiform; stigmatic cone long and narrow, but shorter than the style.—DC. Prod. ix. 542 ; *H. foliatum*, Lehm. Pl. Preiss. ii. 238 (from the description given) not of R. Br.

S. Australia. Head of Spencer's Gulf, *R. Brown;* sterile rocky and sandy places, Flinders' Range, Cudnaka, etc., *F. Mueller ;* Mount Searle, *Warburton.*

W. Australia, *Drummond,* 5th *Coll. n.* 337, *Maxwell.*

5. **H. crispatum,** *F. Muell. Herb.* Much branched and very hispid. Leaves on rather long petioles, ovate, acute or obtuse, rugose, with undulate margins, $\frac{3}{4}$ to $1\frac{1}{4}$ in. long. Spikes dense, forked or simple, very hispid, without bracts. Flowers large as in *H. asperrimum.* Calyx-segments linear, 2 lines long at the time of flowering, 3 lines when in fruit, one usually rather larger than the others. Corolla-tube hispid, scarcely exceeding the calyx, the throat glabrous inside; lobes nearly as long as the tube. Anthers oblong, very shortly acuminate. Style filiform; stigmatic cone long and narrow, but shorter than the style. Nuts much shorter than the calyx, the seed-bearing part tuberculate, but one side, almost one half, of each nut smooth and empty.

N. Australia. N.W. coast, *Bynoe;* Hammersley Range, *Maitland Brown.* The leaves are nearly those of *H. europæum,* but much more rugose; the flowers more like those of *H. asperrimum.*

6. **H. fasciculatum,** *R. Br. Prod.* 494. An erect hard annual, looking almost woody at the base, with rigid divaricate branches and hispid with rigid, half-spreading hairs. Leaves rather crowded, sessile, linear-lanceolate or lanceolate, acute, with thickened nerve-like rigidly ciliate margins, all except a few on the main stem under $\frac{1}{4}$ in. long. Flowers few, scattered, sessile amongst the upper floral leaves, scarcely forming leafy spikes. Calyx-segments lanceolate, rather acute, hispid, above 1 line long. Corolla hirsute, the tube rather longer than the calyx, the throat glabrous inside. Anthers ovate-triangular, scarcely acuminate. Style terete, more than twice as long as the ovary; stigma very broad, the thick cone rather shorter than the style.—DC. Prod. ix. 547.

N. Australia. Rocky sandstone ranges, Victoria river, *F. Mueller*; islands of the Gulf of Carpentaria, *R. Brown.*

7. **H. vestitum,** *Benth.* Erect, branching, densely hirsute with spreading hairs. Leaves crowded, almost imbricate, sessile, lanceolate, with slightly recurved margins, about $\frac{1}{4}$ in. long. Flowers few, sessile within the last floral leaves, which are as long as the calyx, forming dense leafy spikes. Calyx-segments linear-lanceolate, rather obtuse, villous, about 2 lines long. Corolla villous, rather longer than the calyx, the tube very short, the throat glabrous inside. Anthers with rather long connivent points. Ovary tapering into a style twice as long as itself, the stigmatic cone very short.

N. S. Wales, *Clowes* in Herb. Hook., the precise station not given. A very distinct species, with the stamens almost of the section *Schleidenia,* but the corolla-tube is not at all ventricose and quite glabrous inside.

8. **H. brachygyne,** *Benth.* Diffuse or much branched and hirsute, probably annual, some specimens almost erect and scarcely 6 in. high, others very diffuse with branches of above 1 ft. Leaves lanceolate or oblong-linear, obtuse, with revolute margins, narrowed at the base but sessile, under ½ in. long, not nearly so crowded as in *H. vestitum.* Flowers few in the uppermost axils, forming very short leafy spikes, the floral leaves or bracts as long as or longer than the calyx. Calyx-segments about 1¼ lines long, rather obtuse and somewhat unequal. Corolla-tube shorter than the calyx, not ventricose nor bearded in the throat; lobes shorter than the tube. Anthers with rather long connivent points. Style very short, with a broad thick stigma and a very small central cone. Fruit depressed-globular, the nuts scabrous.

Queensland, *Mitchell;* Burnett river, *F. Mueller.* Near *H. vestitum,* but the leaves are not imbricate, the flowers much smaller, the style much shorter, etc.

9. **H. epacrideum,** *F. Muell. in Herb. Hook.* An erect, hard, much-branched annual of 6 to 8 in., looking almost woody at the base, the branches rather softly pubescent or villous. Leaves crowded, sessile, ovate or lanceolate, mostly obtuse, with nerve-like somewhat recurved margins, ciliate with rigid hairs, 2 to 3 or rarely 4 lines long. Flowers few, scattered, sessile amongst the upper leaves, scarcely forming leafy spikes. Calyx-segments ovate, obtuse, ciliate, about 1¼ lines long. Corolla very villous, the tube very short, the throat glabrous inside; lobes longer than the calyx, convolute towards the end. Anthers very shortly acuminate. Stigma very broad and thick, almost sessile on the ovary, with an exceedingly short truncate and pubescent central cone.

N. Australia. Sturt's Creek, *F. Mueller.*

SECT. III. SCHLEIDENIA.— Stigma a thick ring round a central cone or point as in *Euheliotropium.* Corolla-tube usually swollen round the anthers, the throat bearded or pubescent inside. Anthers acuminate, cohering by the minutely hairy tips.

The characters are precisely those of the genus *Schleidenia* as defined by Fresenius (in Mart. Fl. Bras.) on S. American species.

10. **H. ovalifolium,** *Forsk.; R. Br. Prod.* 493. A procumbent or diffuse annual, hoary or white all over with appressed rigid or silky hairs. Leaves petiolate, oval obovate or oblong, very obtuse in the common form, entire, the veins not very prominent, mostly ¾ to 1 in. long. Spikes rather slender, simple or once-forked, without bracts. Calyx-segments hirsute, not 1 line long, one much broader than the other four. Corolla-tube hirsute, not 1 line long, the throat bearded inside with a ring of reflexed hairs sometimes reduced to very few; lobes rather shorter than the tube. Anthers inserted below the middle of the tube, ovate, acuminate, the points nearly as long as the cells and usually cohering at the tips by minute terminal tufts of hairs. Stigma with a linear-conical point, longer than the very short style. Nuts small, hispid.—*H. coromandelianum,* Retz; DC. Prod. ix. 541; Wight, lc. t. 1388.

N. Australia. Victoria river, *F. Mueller;* islands of the Gulf of Carpentaria and opposite mainland, *R. Brown, Henne, Landsborough.*

Queensland. Port Denison, *Fitzalan;* Rockingham and Edgecombe Bays, *Dallachy;* Rockhampton, *O'Shanesy;* Broad Sound, Nerkool Creek, Bowen river, *Bowman.*

S. Australia. Cooper's Creek, *Howitt's Expedition;* all over central Australia, *M'Douall Stuart.*

Var. *oblongifolium,* DC. Erect slender and not much branched, the leaves narrower and less obtuse.—*H. gracile,* R. Br. Prod. 493.—Islands of the Gulf of Carpentaria, *R. Brown;* Keppel Bay, *Thozet.*

The species is common in tropical and northern Africa and East India.

11. **H. strigosum,** *Willd.; DC. Prod.* ix. 546. An erect or diffuse much-branched annual, more or less hoary or sprinkled with appressed rigid hairs. Leaves linear, obtuse or scarcely acute, with revolute margins, under 1 in. long. Flowers small, distant, forming slender interrupted scarcely scorpioid spikes, interspersed with small bracts, at least in the lower part. Calyx-segments about 1 line long, strigose-hispid, rather obtuse, the outer ones broader than the inner. Corolla-tube not exceeding the calyx, slightly swollen round the anthers, the throat bearded inside; lobes rather shorter than the tube. Anthers acuminate, cohering by the tips of the points. Style short, the stigmatic cone broad, about as long as the style. Nuts nearly globular, shortly pubescent.

N. Australia. Cygnet Bay, N.W. coast, *A. Cunningham;* Hooker's Creek, *F. Mueller.*

Queensland. Bowen river, *Bowman.*

The species is widely spread over the warmer regions of northern Africa and western Asia.

12. **H. prostratum,** *R. Br. Prod.* 494. Stems prostrate, 1 to 2 ft. long in the specimens seen, strigose as well as the foliage with short hairs. Leaves distant, linear-lanceolate or lanceolate, acute, contracted at the base but scarcely petiolate, flat or the margins scarcely recurved, ½ to 1 in. long. Flowers distant, in a long interrupted terminal spike, the bracts oblong, leafy, but scarcely exceeding the calyx. Calyx-segments lanceolate, obtuse, strigose, 1¼ line long. Corolla-tube as long as the calyx, ventricose at or above the middle, bearded in the throat; limb longer than the tube, broadly plicate, shortly lobed. Anthers acuminate and cohering at the tips. Stigmatic cone about half as long as the filiform style.—DC. Prod. ix. 548.

N. Australia. Islands of the Gulf of Carpentaria, *R. Brown.* Apparently a very distinct species, but the specimens seen not numerous.

13. **H. bracteatum,** *R. Br. Prod.* 493. A hispid annual, either erect and branching or with several stems ascending from the base, under 1 ft. high. Leaves more or less petiolate, lanceolate, rather acute, narrowed at the base, flat or the margins slightly recurved, from under ½ to about ¾ in. long. Spikes terminal, leafy, 1-sided but scarcely scorpioid, the leafy bracts more or less petiolate or very much contracted at the base, as long as or longer than the calyx. Calyx-segments lanceolate, about 1¼ lines long. Corolla-tube rather shorter than the calyx, ventricose at or below the middle, the throat bearded inside, the lobes shorter than the tube. Anthers acumi-

nate, cohering at the tips. Stigmatic cone slender, but shorter than the fili-form style. Nuts minutely pubescent, the segments of the fruiting calyx usually elongated and spreading.—DC. Prod. ix. 547 ; *H. foliatum*, R. Br. Prod. 493 ; DC. Prod. ix. 548.

N. Australia. Islands of the Gulf of Carpentaria, *R. Brown.* The specimens of *H. foliatum* seem to me to represent a rather luxurious form of *H. bracteatum*, with larger, more distinctly petiolate bracts.

Var. *leptostachyum.* Diffuse and much branched, 1 ft. long or more. Leaves shortly petiolate, acute. Style rather shorter.

Queensland. Cape York, *Daemel.*

14. **H. pauciflorum,** *R. Br. Prod.* 493. An erect, much-branched, very hispid, leafy annual, under 6 in. high. Leaves very shortly petiolate, linear-lanceolate, with revolute margins, under $\frac{1}{2}$ in. long. Flowers shortly pedicellate, rather distant, with leafy bracts between them longer than the calyx, forming single terminal leafy spikes or racemes scarcely scorpioid. Calyx-segments about $1\frac{1}{2}$ lines long, lanceolate, hispid, the outer ones rather larger than the inner. Corolla-tube shorter than the calyx, bearded inside at the throat, the lobes about as long as the tube. Anthers acuminate, the points cohering by their minutely hairy tips. Stigmatic cone rather long, but shorter than the filiform style. Nuts small, scabrous with short hairs.— DC. Prod. ix. 547.

N. Australia. Islands of the Gulf of Carpentaria, *R. Brown.*
Queensland. Suttor River, *Thozet ;* Nerkool Creek, *Bowman.*

15. **H. conocarpum,** *F. Muell. in Herb. Hook.* Much branched, with a hard base but perhaps annual, hoary with appressed hairs, and the leaves, bracts, and calyx-segments copiously ciliate with long spreading hairs. Leaves sessile, lanceolate or almost linear, the margins slightly recurved, the lower ones above $\frac{1}{2}$ in. long, the upper ones small. Spikes simple or once-forked, the leafy bracts lanceolate, at least as long as the calyx. Calyx-segments about 2 lines long, the outer ones lanceolate, the inner narrow, all densely ciliate, especially at the base. Corolla-tube slender, nearly as long as the calyx, ventricose above the middle, the throat pubescent inside. Anthers acuminate, the points cohering at the tips. Style long and slender, the stigmatic cone small. Nuts only one ripening in each calyx in all those examined, rather large with an acutely conical point, slightly scabrous. Al-bumen scanty as in all the allied species, but the embryo very much curved, not straight as in the species that ripen the four nuts.

N. Australia. Sturt's Creek, *F. Mueller.*

16. **H. filaginoides,** *Benth.* A dwarf erect much-branched corymbose plant, apparently shrubby, clothed in every part with a white cottony-wool, the specimen seen complete with the root, but scarcely 4 in. high. Leaves crowded and almost imbricate towards the end of each year's shoot, oblong-lanceolate, obtuse, with recurved margins, not 2 lines long. Flowers small, in dense terminal cymes almost contracted into heads, sessile within the imbri-cate bracts, which are as long as the calyx. Calyx-segments oblong-lanceo-late, obtuse, rather above 1 line long. Corolla not 2 lines long, the lobes

about as long as the tube, with a ring of hairs in the throat. Anthers acuminate with long exserted points, connected by the minute hairs of their tips. Style very short, the stigmatic ring broad and cup-shaped, with a small central cone. Fruit not seen.

S. Australia. Cooper's Creek, *Howitt's Expedition.*

17. **H. ventricosum,** *R. Br. Prod.* 494. An erect corymbosely-branched annual of ¼ to 1 ft., becoming diffuse when old, hirsute with rigid strigose or spreading hairs. Leaves narrow-linear, acute, with revolute margins, rarely above ½ in. long. Flowers in short compact spikes, interspersed with long leaf like bracts, all as well as the calyxes very hispid. Calyx-segments narrow, shorter than the corolla-tube. Corolla-tube about 2 lines long, slender to above the middle, ventricose below the throat, which is pubescent inside. Anthers small, acuminate, the points connected by the minutely hairy tips. Style long and filiform, with a short stigmatic cone. Fruit tipped by the persistent base of the style, the nuts small, minutely strigose or nearly glabrous.—DC. Prod. ix. 547.

N. Australia. Depot Creek, Victoria river, *F. Mueller;* Gulf of Carpentaria, *R. Brown;* Port Essington, *Armstrong, Leichhardt.*
Queensland. Port Denison, *Fitzalan.*

18. **H. tenuifolium,** *R. Br. Prod.* 494. Stems erect, branching, hard or almost woody at the base, 1 ft. high or more, more or less hoary as well as the foliage with short appressed hairs. Leaves linear, acute or almost obtuse, with revolute margins, ¾ to 1½ in. long. Flowers distant, forming interrupted once- or twice-forked spikes, interspersed especially at the base with a few bracts, which are rarely entirely wanting. Calyx glabrous or strigose, the segments very acute, about 2 lines long, the 2 outer ones often larger than the others. Corolla-tube about as long as the calyx, ventricose above the middle, shortly bearded in the contracted throat, the limb nearly as long as the tube. Anthers acuminate, the points cohering by their minutely hairy tips. Style slender, the stigma broadly 4-lobed with a small central cone. Fruit tipped by the persistent base of the style, the nuts slightly scabrous-pubescent. Embryo usually curved.—DC. Prod. ix. 547.

N. Australia. Victoria river, *F. Mueller;* islands of the Gulf of Carpentaria, *R. Brown,* and on the main land, *Landsborough.*
Queensland. Suttor river, *F. Mueller, Bowman;* Bowen river, *Bowman.*

19. **H. paniculatum,** *R. Br. Prod.* 494. Erect and paniculately branched with the habit of *H. tenuifolium,* but much more slender, more or less hirsute or nearly glabrous. Leaves linear-filiform, with revolute margins. Panicles at length very divaricate, with dichotomous filiform branches. Flowers very much smaller than in *H. tenuifolium,* but otherwise similar. Calyx 1 line long, strigose, with acutely acuminate segments. Corolla-tube rather longer than the calyx, ventricose above the middle. Stamens and style of *H. tenuifolium.* Nuts small, ovoid, acute, very shortly scabrous-pubescent.—DC. Prod. ix. 547 ; *H. glabellum,* R. Br. Prod. 494 ; DC. Prod. ix. 548.

N. Australia. York Sound, N.W. coast, *A. Cunningham;* Cygnet Bay, *Bynoe;*

Victoria river and Sturt's Creek, *F. Mueller;* islands of the Gulf of Carpentaria, *R. Brown.*

20. **H. Cunninghamii,** *Benth.* An erect slender paniculately-branched annual, hoary with short strigose hairs, evidently near *H. paniculatum,* but the inflorescence more dense and the style different, at least as far as observed. Leaves narrow-linear, with revolute margins, under 1 in. long. Flowers much closer in the spikes than in *H. paniculatum,* the bracts very small. Calyx-segments about 1 line long, rather acute. Corolla-tube longer than the calyx, strigose outside, ventricose above the middle, bearded in the throat. Anthers of *H. paniculatum.* Style shortly filiform, about as long as the long narrow stigmatic cone. Nuts very small, minutely scabrous-pubescent.

N. Australia. Dampier's Archipelago, N.W. coast, *A. Cunningham.* Possibly a variety of *H. paniculatum,* connecting it in some measure with *H. strigosum.*

21. **H. diversifolium,** *F. Muell. in Herb. Hook.* Stems much branched and decumbent at the base, in some specimens apparently annua in others evidently perennial and almost woody at the base, the slender flowering branches extending to from 3 or 4 in. to above twice that heigh strigose with appressed hairs as well as the foliage. Leaves crowded in the lower part of the stem and on the short sterile branches, petiolate, ovate or ovate-lanceolate, ¼ to ½ in. long, those of the flowering branches small and distant, linear or linear-lanceolate. Flowers small, distant, forming long slender interrupted spikes, with a few small bracts, usually much below each flower. Calyx-segments lanceolate, acute, strigose, above 1 line long. Corolla-tube 1½ lines long, ventricose above the middle, shortly bearded in the throat, the lobes shorter than the tube. Anthers acuminate, the points cohering by the minutely-hairy tips. Style long and filiform, the stigmatic ring undulate or lobed, with a short central cone. Nuts very hispid, with a deep cavity on each side of the central angle of the inner face.

N. Australia. Cygnet Bay, N.W. coast, *A. Cunningham, Bynoe;* Sturt's Creek, *F. Mueller.*

6. HALGANIA, Gaudich.

Calyx deeply divided into 5 segments. Corolla with a very short tube, rotate with 5 broad lobes in the bud. Stamens 5, inserted in the throat, the filaments very short and flat; anthers erect, connate in a short cylinder contracted into a long straight beak, formed of the linear terminal appendages of the anthers, enclosing the style. Ovary entire, 4-celled, with 1 pendulous ovule in each cell; style terminal, filiform, with a minute stigma. Fruit dry or the exocarp very slightly succulent, separating into 2 carpels, each with a crustaceous 2-celled endocarp. Seeds 1 in each cell, more or less albuminous; embryo terete, straight.—Undershrubs or small much-branched shrubs, rarely entirely herbaceous, more or less tomentose or hirsute, or rarely glabrous. Leaves entire or toothed. Flowers blue purple or white, often rather large and showy, in terminal or at length lateral cymes, sometimes reduced to short simple spikes or almost to single flowers. Bracts few or none.

The genus is limited to Australia. The flowers, including the ovary and style, are very nearly those of *Trichodesma*, but the fruit is rather that of the section *Heliophytum* of *Heliotropium*. The species, though usually very different in aspect, are very difficult to characterize, as most of them appear connected by occasional intermediate specimers.

Leaves more or less toothed or rarely entire, flat concave or complicated.
  Scabrous-pubescent or nearly glabrous shrubs or undershrubs.
    Leaves entire or obtusely toothed. Corolla-lobes broad, usually obtuse.
    Leaves mostly entire, obovate oblong or cuneate. Calyx-lobes obtuse . . . . 1. *H. solanacea.*
    Leaves mostly obtusely toothed, at least at the end . . . . 2. *H. littoralis.*
    Undershrub. Stems erect, corymbose at the top, hispid as well as the foliage with long spreading hairs. Corolla-lobes acute . 3. *H. corymbosa.*
    Undershrubs. Corolla-lobes rather acute. Inflorescence white, with appressed silky hairs . . . . 4. *H. sericiflora.*
  Glandular-glutinous, small branching shrubs, nearly glabrous or with a very short appressed pubescence.
    Leaves mostly complicate, the margins undulate-plicate, with very prominent teeth . . . . 5. *H. Preissiana.*
    Leaves narrow, mostly shortly 3-toothed at the end only . . 6. *H. strigosa.*
Leaves quite entire, thick, convex or with recurved margins.
  Leaves mostly obtuse, $\frac{1}{2}$ to $\frac{3}{4}$ in. long. Outer calyx-lobes broad. Flowers in cymes . . . . 7. *H. lavandulacea.*
  Leaves under $\frac{1}{4}$ in. long, few and distant. Flowers almost solitary. Calyx-lobes linear . . . . 8. *H. integerrima.*

H. Lehmanniana, Sond. in Pl. Preiss. ii. 238, from York district, *Preiss, n.* 2336, is entirely unknown to me, but, from the character given, it would appear to be one of the numerous forms of *H. Preissiana.*

1. **H. solanacea,** *F. Muell. in Hook. Kew Journ.* ix. 21. An erect branching undershrub, the stems and foliage covered with a close pubescence, with more or less of longer appressed or loose hairs, almost silky on the inflorescence and calyx. Leaves obovate or cuneate-oblong, very obtuse, entire (or sparingly toothed?), flat, penniveined with the midrib prominent underneath, narrowed into a short petiole, $\frac{3}{4}$ to $1\frac{1}{2}$ in. long. Cymes loose, with 1 or 2 leafy bracts at the base, or without any bracts. Calyx-segments lanceolate, obtuse, 3 to 4 lines long. Corolla-lobes shortly exceeding the calyx, apparently obtuse, but not seen perfect. Anthers pubescent. Fruit obtuse, 2 to 2$\frac{1}{2}$ lines long.

N. Australia. Upper Victoria river and Sturt's Creek, *F. Mueller;* between the Bonney river and Mount Morpeth, *M'Douall Stuart.*

2. **H. littoralis,** *Gaudich. in Freyc. Voy. Bot.* 449. *t.* 59. A shrub or undershrub, with ascending or erect rigid branching stems, more or less hirsute with short appressed and rigid or longer and loose hairs, or the foliage nearly glabrous. Leaves cuneate-oblong, obtuse, bordered especially at the end by a few coarse broad mostly obtuse teeth, entire and narrowed at the base, thick and firm, $\frac{1}{2}$ to $1\frac{1}{2}$ in. long. Cymes without bracts, compact and corymbose when many-flowered, but often reduced to very few flowers, the inflorescence usually villous with rather long soft hairs. Calyx-segments linear or lanceolate, rather obtuse, varying at the time of flowering from scarcely 3 to above 4 lines in length, and sometimes still longer after flowering. Corolla-lobes broad, rather obtuse, spreading to a diameter of $\frac{3}{4}$ to $1\frac{1}{4}$

in. Anthers pubescent, the beak longer than the cells. Fruit oblong, shorter than the calyx, transversely rugose when dry.—DC. Prod. x. 177.

W. Australia. Sharks' Bay, *Gaudichaud, Denham;* Dirk Hartog's Island, *A. Cunningham;* Swan and Murchison rivers, *Drummond, n.* 52, 653, *and 6th Coll. n.* 132.

Var. *glabrifolia.* Leaves rather broad and quite glabrous.—*H. Bebrana,* F. Muell. Fragm. i. 209.—South Hutt river and Port Gregory, *Oldfield.*

Drummond, n. 122, appears to be a small-leaved form of the same species, which is a very variable one, and several of our specimens are very indifferent.

3. **H. corymbosa,** *Lindl. Swan Riv. App.* 40. Erect and not much branched, apparently herbaceous but with a hard base, 1 to 1½ ft. high, hirsute all over with long spreading hairs. Leaves from linear-cuneate to oblong-lanceolate, coarsely and acutely toothed, contracted at the base. Cymes corymbose, with few elongated branches, sometimes almost reduced to simple 1-sided racemes. Calyx-segments linear, acute, 3 to 4 lines long. Corolla of a deep blue, as large as in *H. littoralis,* but the lobes much more acute. Fruit of *H. littoralis,* but the carpels appear sometimes to open longitudinally at the edge.

W. Australia. Swan River, *Drummond, 1st Coll. n.* 32.

4. **H. sericiflora,** *Benth.* Apparently an undershrub, with ascending or diffuse branching stems, hirsute as well as the foliage with appressed or scarcely spreading hairs, the upper leaves as well as the inflorescence and calyx more or less densely silky with white hairs. Leaves narrow-oblong or linear, all but the uppermost bordered with a few distant teeth, not so acute as in *H. corymbosa,* mostly ½ to 1 in. long or rarely longer. Cymes often many-flowered. Calyx-lobes narrow and acute, and corolla of a deep blue with acute lobes as in *H. corymbosa,* but the flowers altogether usually smaller. Fruit not seen.

W. Australia. Murchison river, *Oldfield.*

5. **H. Preissiana,** *Lehm. Pl. Preiss.* i. 347. An undershrub, with a thick woody stock and numerous slightly branched stems, or a more branching shrub, the stem and foliage sprinkled with short appressed rigid hairs, as in *H. strigosa,* but with less of the scabrous glandular indumentum. Leaves from obovate to oblong-cuneate, more or less conduplicate, the end and margins undulate, with very prominent and acute teeth, much contracted at the base or shortly petiolate, ½ to 1 in. long. Cymes loose and few-flowered. Flowers usually but not always larger than in *H. strigosa,* with the same acute calyx-segments and bright blue corolla.

N. S. Wales. Wellington Valley and Croker's Range, *A. Cunningham;* New England, *C. Stuart;* Clarence river, *Beckler.*
W. Australia. Port Gregory and S. Hutt river, *Oldfield;* Plantagenet and Stirling Ranges, *Maxwell.*

H. anagalloides, Endl. in Ann. Wien. Mus. ii. 204 ; Lehm. Pl. Preiss. i. 348, from the neighbourhood of King George's Sound, *Roe, Preiss,* which I have not seen, must be nearly allied to *H. Preissiana,* with the essential characters the same, but with the leaves only 1½ lines long, and the flowers very small.

6. **H. strigosa,** *Schlecht. Linnæa,* xx. 614. An erect branching shrub,

the stems and foliage scabrous with a minute glandular tomentum, and more
or less sprinkled or covered with short rigid appressed hairs. Leaves narrow-
oblong or linear-cuneate, mostly 3-toothed at the end, and sometimes with 1
or 2 small teeth on each side lower down, the midrib very prominent underneath, othewise flat, complicate or the margins very slightly recurved, rarely
above $\frac{1}{2}$ in. long except in some very luxuriant Western specimens. Flowers
smaller than in *H. corymbosa*, but with the same acute calyx-segments and
deep blue corollas, the lobes often acute or shortly acuminate but sometimes
quite obtuse. Fruit, where known, transversely rugose and rather acute.—
H. tuberculosa, Schlecht. Linnæa, xx. 615; *H. cyanea*, Lindl. Swan Riv.
App. 40.

Queensland. Armadillo, *Barton.*
N. S. Wales. Bogan river, *Mitchell;* Lachlan, Murray, and Darling rivers, *Victorian
and other Expeditions.*
Victoria. Murray river, *Dallachy.*
S. Australia. Sand scrub, *Behr;* rocky ridges, Cudnaka, *F. Mueller;* Port Lincoln,
Wilhelmi; Venus and Streaky Bays, *Warburton;* Lake Gillies, *Burkitt;* in the interior,
M'Douall Stuart.
W. Australia, *Drummond,* n. 121, 402, 654; Oldfield river, S.W. Bay, Phillips and
Gardner flats, *Maxwell.*

The eastern specimens have usually the calyx scarcely above 2 lines long, and the corolla
also small; some of the western specimens have the flowers quite as small, in others they are
considerably larger, but the plant has never the long spreading hairs of *H. corymbosa*, nor
the dense silky hairs of *H. sericiflora.*

7. **H. lavandulacea,** *Endl. in Ann. Wien. Mus.* ii. 205. An erect
branching shrub, attaining sometimes 2 ft. or more, the branches hoary
with very short appressed rigid hairs, and often somewhat viscid. Leaves
oblong or lanceolate, obtuse, rather thick, with recurved or revolute quite entire margins, rounded at the base and sometimes shortly petiolate, sprinkled
above with very short appressed hairs, hoary or white and tomentose under-
neath, mostly $\frac{1}{2}$ to $\frac{3}{4}$ in. long or under $\frac{1}{2}$ in. on the smaller branches.
Flowers not numerous, in short dense cymes interspersed with a few small
leafy bracts. Calyx-segments very unequal, the outer ones broad, above 2
lines long, with recurved herbaceous tips, the inner ones shorter and linear.
Corolla spreading to about $\frac{1}{2}$ in. diameter, the lobes rather obtuse. Anthers
glabrous outside, the cells villous inside. Fruit obtuse, not seen ripe.—*H.
andromedæfolia*, F. Muell. Fragm. i. 209.

N. S. Wales. Between the Murray and Darling rivers, *F. Mueller.*
W. Australia, *Roe, Drummond,* n. 336.

8. **H. integerrima,** *Endl. in Ann. Wien. Mus.* ii. 205'. An erect,
much-branched shrub, more slender than the other species, the branchlets
white with a close tomentum, becoming sometimes nearly glabrous when old.
Leaves small and scattered, linear or oblong, very obtuse, quite entire, convex, pubescent when young or at length glabrous above, more tomentose underneath, the margins thickened mostly under $\frac{1}{4}$ in. long. Flowers few in
the cymes and often solitary, the size of those of *H. strigosa*. Calyx-segments linear, minutely but rigidly glandular-pubescent. Anthers very slightly
pubescent. Fruit shortly acuminate.

W. Australia, *Roe, Drummond, 5th Coll. n.* 96. Although I have not seen the typical specimens of this and the preceding species, Endlicher's descriptions leave no doubt as to their identity.

7. TRICHODESMA, R. Br.

Calyx deeply divided into 5 segments. Corolla with a very short tube, almost rotate, with 5 acuminate lobes contorted in the bud. Stamens 5, inserted in the throat, the filaments very short and flat; anthers erect, linear, ciliate, cohering by the hairs in a cylinder contracted into a long spirally-twisted beak formed of the terminal appendages of the anthers. Ovary entire, 4-celled, with 1 pendulous ovule in each cell; style terminal, filiform, with a minute stigma. Fruit of 4 1-seeded nuts, attached by their whole inner face, which when detached leave 4 cavities in the thick persistent prominently 4-angled axis. Seeds without albumen; embryo straight, with a very short radicle.—Coarse hispid hoary or silky herbs. Leaves opposite or alternate, usually entire. Flowers in terminal one-sided simple or rarely forked racemes, usually accompanied by bracts.

The genus comprises very few species dispersed over the warmer regions of Asia and Africa. The only Australian species extends over nearly the whole range of the genus. Formerly included in the genus *Borago*, and still doubtfully referred to the tribe of Borageæ, *Trichodesma* differs in the entire ovary with a terminal style, and is in fact very nearly allied to *Halgania*. The fruit, however, does not, as in that genus, separate into distinct carpels, but the endocarp, hardening round each seed, forms 4 pyrenes or nuts, which detach themselves from the persistent remainder of the pericarp.

1. **T. zeylanicum,** *R. Br. Prod.* 496. A coarse hard annual, usually erect, not much branched, and often attaining several feet, the indumentum very various, sometimes close and hoary or longer and silky, more frequently consisting of short rigid appressed hairs or long loose scattered ones, or the various hairs intermixed, the longer ones usually arising from prominent tubercles. Leaves in the Australian specimens mostly alternate or the lower ones opposite, more rarely nearly all (as is usually the case in Indian specimens) opposite, linear, linear-lanceolate or rarely broadly oblong-lanceolate, obtuse, often 3 to 4 in. long, the margins usually recurved. Flowers pale blue, in simple racemes, with a leafy bract under each always shorter than the pedicel. Calyx-segments lanceolate, acuminate, $\frac{1}{4}$ to $\frac{1}{2}$ in. long at the time of flowering, narrow or broad, valvate or reduplicate, often cohering at the base, sometimes much enlarged round the fruit, but without the reflexed auricles of *T. indicum*. Corolla-lobes broad, longer than the calyx, the points narrow, spirally-twisted in the bud as well as the long anther-points. Nuts smooth and shining.—A. DC. Prod. x. 172, with the synonyms adduced; Bot. Mag. t. 4820.

N. Australia. N.W. coast, *A. Cunningham* and others; Victoria river, *F. Mueller* islands of the Gulf of Carpentaria and adjoining mainland, *R. Brown* and others.

Queensland. Keppel and Shoalwater Bays, *R. Brown;* common from Cape York to Moreton Bay and in the interior, *A. Cunningham, F. Mueller,* and many others.

N. S. Wales. Between Stokes' Range and Cooper's Creek, *Wheeler.*

S. Australia. Head of Spencer's Gulf, *R. Brown;* Elder's and Flinders' Ranges, *F. Mueller;* Cooper's Creek, *Howitt's Expedition;* Mount Searle, *Warburton.*

W. Australia. Sharks' Bay, *Milne;* Murchison river, *Oldfield, Drummond, 6th Coll. n.* 133; Flinders' Bay, *Collie.*

Var. *latisepalum,* F. Muell. Calyx-segments short and broad, almost cordate but not auriculate.—Hooker's and Sturt's Creeks and Burdekin river, *F. Mueller.*

Var. *sericeum.* Stem and foliage very hoary with a close soft tomentum.—*T. sericeum,* Lindl. in Mitch. Trop. Austr. 258.—Victoria river, *F. Mueller ;* Belyando river, *Mitchell.*

The species is common in the E. Indian peninsula, in Ceylon, and in tropical Africa.

TRIBE II. BORAGEÆ.—Ovary 4- or 2-lobed, the style inserted between the lobes and more or less lateral or basal with reference to the carpels. Herbs or very rarely, in species not Australian, shrubs. Fruit separable into 4 or 2 small nuts, leaving a persistent flat convex conical or rarely elongated receptacle or axis.

8. MYOSOTIS, Linn.

(Exarrhena, *R. Br.*)

Calyx deeply divided into 5 segments or, in species not Australian, 5-toothed. Corolla with a cylindrical tube, with 5 small scales in the throat, the limb spreading, 5-lobed. Stamens inserted in the tube ; anthers included or exserted. Ovary 4-lobed ; style filiform, inserted between the lobes ; stigma small, usually capitate. Nuts 4, smooth and shining, erect, attached by a small basal area. Seeds without albumen ; radicle short.—Herbs usually hispid. Leaves entire. Flowers blue or white, in simple or forked one-sided spikes or racemes, without bracts.

The genus is chiefly abundant in the temperate regions of the northern hemisphere, especially in the Old World, more rare in North America, tropical Asia, and in the extratropical regions of the southern hemisphere. Of the two Australian species, one extends to New Zealand, the other is endemic.

Corolla-lobes shorter than the tube. Anthers included or the tips
 scarcely protruding. (Flowers very small) . . . . . . 1. *M. australis.*
Corolla-lobes as long as the tube. Anthers wholly exserted . . . . 2. *M. suaveolens.*

1. **M. australis,** *R. Br. Prod.* 495. An erect or diffuse hispid annual (or perennial ?), the stems usually branching from the base, sometimes slender and under 6 in. high, sometimes long and weak, extending to 1 or 2 ft. Lower leaves on long petioles, from obovate-oblong to oblanceolate or linear-spathulate, the stem ones more sessile and smaller, the uppermost sometimes very small sessile and cordate-ovate. Flowers small, white or yellowish (or rarely blue ?), in scorpioid spikes at first dense but at length often long and interrupted. Calyx-segments narrow-lanceolate, hispid with hooked hairs, $\frac{3}{4}$ to $1\frac{1}{4}$ line long. Corolla-tube rather longer than the calyx, the scales of the throat obtuse and notched, the lobes short, broad, obtuse or retuse. Anthers and style included in the tube or the tips slightly protruding. Nuts shorter than the calyx.—DC. Prod. x. 110 ; Hook. f. Fl. Tasm. i. 279 ; *M. staminea,* Lehm. Pl. Preiss. i. 348.

N. S. Wales. Paterson's River, *R. Brown ;* Blue Mountains, *A. and R. Canningham* and many others ; Illawarra, *A. Cunningham ;* Nangas, *M'Arthur.*

Victoria. Wendu Vale, *Robertson ;* common in the Australian Alps, ascending to 4000 to 5000 ft., *F. Mueller.*

Tasmania. Port Dalrymple and Derwent river, *R. Brown ;* common everywhere, ascending to 4000 ft., *J. D. Hooker.*

S. Australia. Guichen Bay, Lofty Range, *F. Mueller.*

W. Australia, *Drummond, n.* 196, 217; Rottenest Island, *Preiss, n.* 1934; Mount Manypeak river, *Maxwell;* Perongerup Range, *F. Mueller.*

The species is also in New Zealand.

2. **M. suaveolens,** *Poir. Dict. Suppl.* iv. 44. An erect but sometimes weak perennial, forming a thick hard stock, the stems simple or branched, 1 to 2 ft. high, the hairs long and spreading on the stem and often on the margins and midribs of the leaves, appressed on their surface. Leaves oblong linear or lanceolate, mostly acute, sessile and broad at the base or contracted into a short petiole, often decurrent, the lower ones sometimes 3 or 4 in. long, the upper ones small. Flowers white or bluish, in simple or branched racemes, at first dense, at length several inches long, the pedicels short. Calyx-segments narrow, 3-nerved, usually about 2 lines long but variable, hispid with hooked hairs. Corolla-tube as long as the calyx; scales of the throat short and broad; lobes broad, as long as the tube. Stamens inserted in the throat; filaments filiform; anthers narrow-oblong, wholly exserted as well as the style. Nuts shorter than the calyx.—DC. Prod. x. 111; Hook. f. Fl. Tasm. i. 279; *Exarrhena suaveolens,* R. Br. Prod. 495; A. Rich. Sert. Astrol. t. 29.

N. S. Wales. Frequent on rocky margins of creeks in the Blue Mountains, *A. Cunningham.*

Victoria. Frequent in the Australian Alps, *F. Mueller;* Ballarat, *Glendinning.*

Tasmania. Port Dalrymple, *R. Brown;* abundant in light rich soil, ascending to 2000 ft., *J. D. Hooker.*

9. ERITRICHIUM, Schrad.

Calyx deeply divided into 5 segments. Corolla with a cylindrical tube, the throat with 5 minute gibbosities or scales or quite naked, the limb spreading, 5-lobed. Stamens inserted in the tube, the anthers included. Ovary 4-lobed; style filiform, inserted between the lobes, with a small usually capitate stigma. Nuts 4, rugose or reticulate, erect, attached to the shortly pyramidal or convex receptacle by an oblique areole, the inner angle prominent. Seeds without albumen; radicle short.—Herbs with the habit foliage and flowers nearly of *Myosotis,* but the inflorescence usually with bracts.

There is a considerable number of species dispersed over the temperate and mountainous regions of Europe and Asia, and in America descending from the United States along the line of the Andes to Chile. The only Australian one is endemic. The genus is closely allied to *Myosotis* and *Echinospermum,* having the same habit and flowers, with the fruit intermediate, as it were, between the two, the receptacle more prominent than in *Myosotis,* much less so than in *Echinospermum,* the nuts neither smooth and shining as in the former nor muricate as in the latter.

1. **E. australasicum,** *A. DC. Prod.* x. 134. Stems usually numerous, tufted diffuse or ascending, rarely nearly simple and erect, mostly under 6 in. long, the whole plant hispid, the hairs often yellowish on the young shoots. Leaves linear, obtuse, the lower ones sometimes almost opposite, rarely exceeding ½ in., the upper ones smaller. Flowers very small (white?), nearly sessile in the axils of the bracts, forming simple one-sided leafy spikes. Calyx-segments very hispid, linear, scarcely 1 line long. Corolla scarcely exceeding the calyx, the lobes shorter than the tube. Anthers small. Style

short. Nuts shorter than the calyx, rugose, with much-raised reticulations. —*Heliotropium elachanthum,* F. Muell. in Linnæa, xxv. 424.

Victoria. Wimmera, *Dallachy;* Skipton, *Whan* (the latter specimens very young and somewhat doubtful).

S. Australia. Pastures, Rocky Creek, *F. Mueller.*

W. Australia, *Drummond, n.* 505.

10. ECHINOSPERMUM, Swartz.

Calyx deeply divided into 5 segments. Corolla with a cylindrical tube, the throat with 5 small scales inside; lobes 5, spreading. Stamens inserted in the tube; anthers included. Ovary 4-lobed; style inserted between the lobes, with a small usually capitate stigma. Nuts 4, usually more or less muricate with hooked prickles, erect, laterally attached to a narrow-conical receptacle. Seeds without albumen, radicle short.—Herbs with the habit foliage and flowers of *Myosotis,* but the flowers usually interspersed with bracts.

A considerable genus dispersed over the temperate and mountainous regions of Europe and Asia, but scarcely tropical. The only Australian species appears to be endemic.

1. **E. concavum,** *F. Muell. Fragm.* ii. 139; vi. 116. An annual, strigose or hoary with appressed hairs, looser on the main stems; stems either erect nearly simple and under 6 in. high or longer, diffuse and almost woody at the base. Leaves linear or oblanceolate, the larger ones above 1 in. long, the upper ones passing into the small floral leaves or bracts. Flowers in one-sided leafy racemes, the pedicels at first very short, but lengthening to ¼ in. or more when in fruit. Calyx-segments about ¾ line long, lanceolate, enlarging after flowering. Corolla about 1 line long, the tube with a ring of obtuse scales inside above the anthers, the lobes obovate-oblong, much shorter than the tube. Nuts about 2 lines long, very concave on the back, with thick raised almost involute margins bordered by stout conical glochidiate prickles, the enclosed area tuberculate.

N. S. Wales. Darling river, *Victorian Expedition;* between Stokes Range and Cooper's Creek, *Wheeler.*

Victoria. Wimmera, *Dallachy.*

S. Australia. Near Gawler Town, *F. Mueller.*

W. Australia, *Drummond, n.* 165.

11. ROCHELIA, Reichenb.

(Maccoya, *F. Muell.*)

Calyx deeply divided into 5 or more segments. Corolla with a cylindrical sometimes incurved tube, the throat with or without scales inside; limb spreading, 5-lobed or rarely 4- or 6-lobed. Stamens 5 or rarely fewer, included in the tube. Ovary 2-lobed, 2-celled, with 1 ovule in each cell; style inserted between the lobes, with a small usually capitate stigma. Nuts 2, erect, rugose, laterally attached to the narrow-conical receptacle. Seeds without albumen; radicle short.—Herbs with the habit of *Myosotis,* the inflorescence usually interspersed with bracts, the fruiting calyx often more or less hardened at the base round the nuts.

The genus comprises very few species from the Mediterranean region of the northern hemisphere and from western Asia. The only Australian species is endemic, but is very nearly allied to one of the east Mediterranean ones. The genus is nearly allied to *Echinospermum,* but has only 1 cell and ovule to each carpel.

1. **R. Maccoya,** *F. Muell. ms.* A hispid annual with several procumbent or ascending stems, a few inches in length. Leaves linear, the radical ones 1 in. long or even more, those of the stem few and small. Flowers on very short pedicels, usually below the floral leaves or bracts, the upper ones forming an irregular one-sided leafy raceme. Calyx about 2 lines long, oblique, incurved, divided into from 7 to 9 rather unequal linear-segments. Corolla shorter than the calyx, with 4 to 6 very short obtuse lobes, without scales in the throat. Stamens usually 4. Nuts enclosed in the hardened base of the calyx-segments.—*Maccoya plurisepala,* F. Muell. Fragm. i. 127.

N. S. Wales. Murray Desert towards the Darling river, *F. Mueller.* In the few flowers I opened, I found 7 calyx-segments and only 4 stamens, the corolla-lobes sometimes 5, sometimes 6; F. Mueller has observed as many as 9 calyx-segments. The species is nearly allied to *R. cancellata,* Boiss., which has a similar multiplication of calyx-segments but with the normal number of 5 corolla-lobes and stamens.

12. CYNOGLOSSUM, Linn.

(Omphalodes, *Mœnch.*)

Calyx deeply divided into 5 segments. Corolla with a short broad tube, the throat closed with scales opposite the lobes, the limb spreading, almost rotate, 5-lobed. Anthers enclosed in the tube. Ovary 4-lobed; style shortly filiform, inserted between the lobes, with a small usually capitate stigma. Nuts 4, depressed, attached by the inner end of the under surface or by the inner edge to the convex or hemispherical receptacle, the upper surface usually more or less covered with short hooked prickles or bordered by a raised often toothed margin. Seeds without albumen; radicle short.— Herbs clothed with stiff hairs, either appressed and hoary or spreading, sometimes reduced to scattered tubercles. Leaves entire. Flowers blue purplish or rarely white, in one-sided simple or forked racemes, with or without bracts.

The genus is widely dispersed over the temperate and warmer regions of the Old World. The Australian species appear to be all endemic.

Diffuse or straggling. Leaves ovate, petiolate. Floral leaves or
  bracts at or near all the pedicels . . . .   1. *C. latifolium.*
Erect. Leaves lanceolate or oblong.
  Nuts glochidiate all over.
    Pedicels longer than the calyx, with bracts at or near them all,
      or at least the lower ones . . . .   2. *C. suaveolens.*
    Pedicels shorter than calyx. Bracts none . . . .   3. *C. australe.*
  Nuts glochidiate only on the raised margin or rarely along a cen-
    tral raised ridge. Bracts none . . . .   4. *C. Drummondii.*

1. **C. latifolium,** *R. Br. Prod.* 495. A perennial, with diffuse or straggling branching stems, extending sometimes to several feet, scabrous with scattered tubercles, which rarely lengthen into short hairs or prickles. Leaves petiolate, ovate, acute, quintuplinerved, the lower ones often at least 2 in. long, the upper ones gradually smaller and more sessile, ultimately reduced to small bracts. Flowers small, on slender recurved pedicels, usually

by the side of or rather below the floral leaves or bracts. Calyx-segments rather broad, obtuse or shortly acuminate, about 1 line long but somewhat enlarged after flowering. Nuts obovate, spreading, convex, glochidiate all over, attached by the inner end of the under surface.—DC. Prod. x. 156; Hook. f. Fl. Tasm. i. 280.

N. S. Wales. Paterson's River, *R. Brown ;* Bent's Basin, *Woolls ;* New England *C. Stuart ;* Clarence river, *Beckler.*
Victoria. Shady places, Dandenong Mountains and Latrobe river, *F. Mueller ;* Wannon river, *Robertson.*
Tasmania. Very damp shady situations, Circular Head, *Gunn.*

2. **C. suaveolens,** *R. Br. Prod.* 495. An erect stout coarsely-hirsute plant with a perennial stock, the stems slightly branched, 1 to 2 ft. high. Radical and lower leaves on long petioles, lanceolate or oblong, sometimes several inches long; stem-leaves few, on shorter petioles, the upper ones small, sessile, lanceolate. Racemes loose, more or less forked, with small leafy bracts below most of the pedicels. Pedicels longer than the calyx and sometimes $\frac{1}{2}$ in. long, recurved after flowering. Calyx-segments very open, narrow, $1\frac{1}{2}$ lines long. Nuts ovoid, spreading, flat or slightly convex, densely glochidiate outside, obliquely attached by their inner smooth face to the broad very prominent almost hemispherical receptacle. Seed flat or slightly concave. Embryo concave.—DC. Prod. x. 156; Hook. f. Fl. Tasm. ii. 368.

N. S. Wales. Port Jackson, *R. Brown ;* between the Upper Bogan and Lachlan rivers, *L. Morton ;* Wellington Valley, *A. Cunningham.*
Victoria. Port Phillip, *R. Brown ;* common about Melbourne, *F. Mueller* and others ; Skipton, *Whan ;* Wimmera, *Dallachy ;* mouth of the Glenelg, *Allitt.*
Tasmania. Port Dalrymple, *R. Brown ;* common in dry soil, *J. D. Hooker.*

3. **C. australe,** *R. Br. Prod.* 495. An erect stout hispid perennial, usually taller than *C. suaveolens,* and the hairs of the lower part of the stem long and reflexed. Radical and lower leaves on long petioles, the upper ones nearly sessile, all lanceolate or the lower ones oblong, often several inches long. Flowers sweet-scented, light blue or white, in long slender more or less forked racemes, without bracts, the pedicels rarely exceeding the calyx and mostly shorter. Calyx-segments shortly hispid, obtuse, about 1 line long. Nuts very spreading, depressed, obovate, either nearly flat or with a more or less raised and glochidiate margin, the whole surface also more or less glochidiate, attached to the convex or almost hemispherical receptacle by a small smooth portion at the inner end of their under surface. Seeds flat.—DC. Prod. x. 151; Hook. f. Fl. Tasm. ii. 368.

N. S. Wales. Hastings, Clarence, and Macleay rivers, *Beckler ;* Richmond river, *Fawcett ;* Paramatta and Mudgee, *Woolls ;* Illawarra, *A. Cunningham.*
Victoria. Port Phillip, *R. Brown ;* Forest Creek, Snowy River, Darebin Creek, etc., *F. Mueller ;* Wendu Vale, *Robertson.*
Tasmania. Port Dalrymple, *R. Brown ;* common in dry soil, *J. D. Hooker.*

The fruit of this species sometimes almost passes into that of the genus or section *Omphalodes,* although always glochidiate.

4. **C. Drummondii,** *Benth.* A tall erect hispid perennial like *C.*

australe, and perhaps a variety, but with the peculiar fruit of the genus or section *Omphalodes*. Hairs of the plant usually longer and looser than in *C. australe*, the pedicels often rather longer and occasionally a few bracts developed at the base of the raceme, the foliage, inflorescence, and flowers otherwise the same. Nuts depressed, spreading, almost orbicular, with a much-raised membranous shortly fringed border, the enclosed upper concave surface quite smooth or rarely with a slightly raised midrib bearing a few hooked prickles, the under surface convex and quite smooth, the attachment as in *C. australe*.

S. Australia. Mount Remarkable, *F. Mueller;* Mount Searle, *Warburton* (with large flowers).

W. Australia, *Drummond, n.* 504 (with small flowers).

Order LXXXI. CONVOLVULACEÆ.

Flowers regular. Calyx free, persistent, of 5 distinct much imbricated sepals, rarely united in a 5-toothed or 5-lobed calyx. Corolla campanulate or funnel-shaped or rarely rotate or with a cylindrical-tube, the limb usually spreading, 5-angled or 5-lobed, folded in the bud or very rarely imbricate. Stamens 5, inserted in the tube, alternate with the lobes or angles of the corolla, often of unequal length; anthers versatile or almost erect, with 2 parallel cells opening by longitudinal slits. Ovary free, 2-, 3- or 4-celled, rarely divided into 2 or 4 distinct carpels, with 1 or 2 erect or ascending ovules in each cell or carpel or 1-celled with 2 or 4 ovules; style single or more or less divided into 2 entire or 2-fid branches or styles. Fruit either a capsule opening in 2, 3, or 4 or twice as many valves, leaving the dissepiments attached to the axis, or opening transversely, or bursting irregularly, or succulent and indehiscent. Seeds with a small quantity of mucilaginous albumen or without any; cotyledons usually very much folded, rarely straight or imperceptible.—Herbs often twining or rarely shrubs, woody twiners or even trees, or (in *Cuscuta*) leafless, twining parasites. Leaves alternate. Inflorescence various, usually axillary and more or less cymose or peduncles 1-flowered. Bracts and bracteoles usually small or deciduous, rarely large and persistent. Flowers often large and showy, rarely very small.

A considerable Order, widely spread over almost every part of the globe, but most abundant in warm countries. Of the eleven Australian genera, seven are diffused over the whole area of the Order or at least over the warmer regions of both the New and the Old World, two extend over tropical Asia, and the remaining two appear to be endemic. A large proportion of the species also have a very wide geographical range.

Leafy plants (climbing prostrate or erect). Sepals distinct.
Stigma sessile. Corolla-limb of 5 deeply 2-lobed divisions. Tall woody climbers. Flowers small . . . . 1. ERYCIBE.
Style filiform, undivided to the stigma or stigmatic lobes. Corolla-limb 5-angled or 5-lobed.
Stigma or stigmatic lobes globular or nearly so.
Fruiting sepals scarcely altered, or if enlarged, closing over the capsule . . . . 2. IPOMŒA.
Fruiting sepals much enlarged, very spreading, thin and veined 5. PORANA.
Stigmatic lobes linear oblong or rarely ovate.
Stigmatic lobes 2. Ovary completely or imperfectly 2-celled, with 2 ovules to each cell . . . . 3. CONVOLVULUS.

Stigmatic lobes 4 to 8 or rarely 2. Ovary 2-celled, with 1
  ovule in each cell . . . . 4. POLYMERIA.
Style more or less branched below the stigmas or divided to the
  base.
Style-branches 2, with globular stigmas.
  Corolla-limb 5-angled or shortly 5-lobed, folded in the bud.
    Flowers axillary . . . . 6. BREWERIA.
  Corolla-limb of 5 divisions, imbricate in the bud. Flowers in
    terminal heads or spikes . . . . 7. CRESSA.
  (The Asiatic *Poranas* have also sometimes a branched style.)
Style-branches 4, with linear stigmas . . . . 8. EVOLVULUS.
Ovary itself divided into 2 carpels, each with a basal style and con-
  taining 1 or 2 ovules . . . . 9. DICHONDRA.
Leafy plants (low and diffuse). Calyx 5-toothed. Style branched,
  with globular stigmas . . . . 10. WILSONIA.
Leafless parasites with filiform stems. Style branched or divided to
  the base . . . . 11. CUSCUTA.

1. ERYCIBE, Roxb.

Corolla with a short tube, the limb spreading, of 5 deeply 2-lobed divisions,
the lobes in the bud closely folded over each other, the divisions themselves
contorted-imbricate. Ovary 1-celled, with 4 erect ovules ; stigma sessile,
large and thick, divisible into 2, but marked outside with 5 or 10 angles or
furrows (the result of the impression of the folds of the corolla or of the
stamens in the bud). Fruit an indehiscent berry, containing usually a
single seed.—Tall woody evergreen climbers. Leaves entire. Flowers
small, in short dense racemes, cymes or clusters, either all axillary or the
upper ones in a terminal leafless panicle.

The genus consists of very few species, very nearly allied to each other, spread over
tropical Asia. The only Australian one appears to be the one most common in East India
and the Archipelago.

1. **E. paniculata,** *Roxb. ; Pl. Corom.* ii. 31. *t.* 159. A very tall
woody climber, the young branches under side of the young leaves and inflorescence more or less rusty-tomentose or villous, the adult foliage glabrous
or nearly so. Leaves shortly petiolate, oval-elliptical, more or less acuminate,
entire, coriaceous, mostly 3 to 4 in. long. Flowers yellow, in short dense
racemes or compact panicles, the lower ones often axillary and much shorter
than the leaves, the upper ones forming, in the few Australian specimens
seen, a small narrow, terminal panicle, which, in the Indian ones, is usually
large and much branched. Sepals orbicular, a little more than 1 line long,
hairy outside. Corolla-tube scarcely so long as the sepals ; limb spreading
to 3 or 4 lines diameter, pubescent outside, the divisions deeply and broadly
2-lobed. Filaments attached to the base of the tube ; anthers ovoid, acuminate, with rather long points. Berry in the Indian specimens ovoid,
above ½ in. long, not seen in the Australian ones.—DC. Prod. ix. 464 ;
Wight, Illustr. t. 180 (the stigma incorrectly drawn).

Queensland. Rockingham Bay, *Dallachy.* Widely spread over E. India and the
Archipelago, and including probably some other described species, the differential characters
in the whole genus being as yet very vague and uncertain.

2. IPOMŒA, Linn.

(Pharbitis, Batatas, Calonyction, Quamoclit, Aniseia, *and* Skinneria, *Chois.*)

Corolla campanulate or with a cylindrical tube ; the limb spreading, entire, angular or rarely deeply lobed, folded in the bud. Ovary 2- or 3-celled, with 2 ovules in each cell, or more or less perfectly 4-celled by the addition of a spurious dissepiment between the ovules. Style filiform ; stigma capitate, entire, or with 2 short globular or rarely almost ovate lobes. Fruit a dry capsule.—Twining prostrate creeping or rarely low and erect herbs or woody climbers. Leaves entire lobed or divided into distinct segments or leaflets. Flowers often large and showy, axillary, solitary or in dichotomous cymes or rarely in irregular racemes.

A large genus, dispersed over all warm climates, very few species being found without the tropics, either in the New or the Old World. Of the thirty-eight Australian species here enumerated, six or perhaps seven are dispersed over the tropical regions of the New as well as the Old World, five or perhaps six spread over Africa as well as Asia, six appear to be limited to tropical Asia, two extend from the Mascarene to the Pacific Islands, one only extends to the Pacific Islands, two are probably introduced only in Australia, and the remaining fourteen, fifteen, or sixteen, are, as far as hitherto known, endemic in Australia. The distribution of the numerous species into distinct genera has been frequently attempted, but has been practically unsuccessful. The separation of the species with a hypocrateriform corolla and exserted stamens is perhaps the most definite, but a very unnatural one, as it would associate *I. Bona-nox* with *I. Quamoclit*. *Pharbitis* with a 3-merous pistil, is quite as artificial, as it would include *I. dissecta* with *I. hederacea* and its allies besides that the character is sometimes inconstant in the same species. The spurious disₜsepiments of *Batatas* are often very imperfect or disappear altogether. The ovary oᶠ *Skinneria* is not 1-locular, as had been supposed, although the dissepiment dries up as the fruit enlarges. The inequality of the sepals in some species of *Aniseia* is not greater than in several species retained in *Ipomœa*. The spiral twisting of the anthers after emitting their pollen, so characteristic of some of the large-calyxed species is but slight or uncertain in others. And notwithstanding great differences in the form of the corolla, in the dehiscence of the capsule, and indumentum of the seeds, no good natural groups founded upon any of these characters have as yet been proposed. As a whole, the genus *Ipomœa* itself can scarcely be said to be a very well marked nor a very natural one ; it is distinguished from *Argyreia* by the dry capsular fruit usually, but not always, opening in valves, and from *Convolvulus* only by the globular or orbicular stigma or stigmatic lobes. The series of species here proposed are too artificial, and not always sufficiently distinct, to be given as sections, but they are the best I have been able to frame for the determination of the Australian species.

SERIES I. **Digitatæ.** *Leaves digitately divided into deep lobes or distinct segments or leaflets. Flowers of the* Speciosæ, *or rarely of the* Campanulatæ.

Leaves palmately or almost pedately several-lobed. Flowers large in
loose cymes . . . . 1. *I. paniculata.*
Leaves divided into 3 (or 5 ?) distinct obovate lobed segments.
Flowers rathers large. Plant stellate-tomentose . . . . . . 2. *I. Davenporti.*
Leaves divided into 5 or 7 ovate or lanceolate entire segments.
Flowers large. Plant glabrous or hairy.
Sepals nearly equal. Seeds pubescent or hairy. Leaf-segments
usually confluent at the base . . . . 3. *I. palmata.*
Inner sepals nearly twice as large as the outer ones. Seeds glabrous. Leaf-segments quite distinct . . . . 4. *I. quinata.*
Leaves divided into 3 to 7 linear usually pinnatifid segments.
Corolla nearly 2 in. long. (Ovary 2-celled ?) . . . . . . 5. *I. diversifolia.*
Corolla scarcely ½ in. long. Ovary 3-celled . . . . 6. *I. dissecta.*

SERIES II. **Pharbitides.**—*Leaves entire or 3 lobed.*   *Ovary usually 3-celled.*
Sepals usually long and narrow. *Corolla of the* Speciosæ.

Calyx acuminate, above ½ in. long.
  Leaves mostly 3-lobed. Flowers few on the peduncle. Calyx
    hairs usually spreading . . . . 7. *I. hederacea.*
  Leaves mostly entire. Flowers usually several in a compact cyme.
    Hairs usually appressed . . . . 8. *I. congesta.*
Calyx scarcely acuminate, under ½ in. long. Leaves mostly entire . 9. *I. purpurea.*

SERIES III. **Calycinæ.**—*Leaves entire or lobed at the base.* *Ovary 2-celled.*
Sepals large, obtuse (attaining ¾ to 1 in. after flowering). *Corolla large (1½ to 3 in.*
long).

Glabrous or nearly so. Bracteoles small. Sepals very large. Corolla broadly campanulate, about 2 in. long.
    Leaves large, mostly peltate . . . . 10. *I. peltata.*
    Leaves cordate or hastate, acuminate. Petioles and peduncles
      often winged . . . . 11. *I. alata.*
Pubescent. Bracteoles large, membranous, deciduous. Corolla
  campanulate, 1½ in. long or rather more . . . . 12. *I. Turpethum.*
Glabrous or sparingly pubescent. Bracteoles small. Corolla about
  3 in. long, contracted into a tube at the base.
    Leaves acuminate. Outer sepals rather larger than the others . 13. *I. longiflora.*
    Leaves very obtuse, almost reniform. Outer sepals rather smaller
      than the others . . . . 14. *I. costata.*

SERIES IV. **Speciosæ.**—*Leaves entire toothed or lobed at the base.* *Ovary 2-celled*
or spuriously 4-celled. *Sepals moderate or small (rarely attaining ½ in.).* *Corolla large*
(1½ to 3 in. long), often more or less tubular at the base, usually pink purple or white.

Stems prostrate or creeping and rooting at the lower nodes. Mari-
  time plants. Seeds woolly-hairy.
  Leaves rather thick, very obtuse or emarginate. Ovary more or
    less 4-celled.
    Leaves broad, 2 to 3 in. long, the veins prominent . . . . 15. *I. Pes-capræ.*
    Leaves small or narrow, the veins scarcely prominent . . . 16. *I. carnosa.*
  Leaves rather thin. Ovary 2-celled.
    Leaves acute or acuminate. Seeds woolly-hairy . . . . . 17. *I. reptans.*
    Leaves obtuse or acute. Seeds glabrous . . . . 21. *I. denticulata.*
Stems twining (sometimes creeping in *I. denticulata*).
  Corolla 2½ to above 3 in. long.
    Leaves linear-lanceolate, not cordate. Stems slender . . . 18. *I. graminea.*
    Leaves broadly ovate-cordate. Stems woody at the base.
      Pubescent or villous . . . . 19. *I. velutina.*
      Glabrous . . . . 20. *I. abrupta.*
  Corolla 1½ to 2 in. long or scarcely more. Leaves cordate or
    hastate.
    Peduncles usually much shorter than the rather long pedicels.
      Sepals obtuse or mucronate. Leaves mostly obtusely auricu-
        late . . . . 21. *I. denticulata.*
      Sepals acuminate. Leaves mostly hastate . . . . . . 22. *I. gracilis.*
    Peduncles mostly longer than the short pedicels. Leaves cor-
      date.
      Corolla above 1½ in. long. Seeds glabrous . . . . . . 23. *I. sepiaria.*
      Corolla scarcely 1½ in. long. Seeds villous . . . . . . 24. *I. Muelleri.*

SERIES V. **Campanulatæ.**—*Leaves entire toothed or lobed at the base.* *Ovary 2-*
celled. *Corolla moderate or small (rarely exceeding 1 in.), usually broadly campanu-*
late, yellow white or pink.

Stems twining. Flowers several together in pedunculate cymes or
racemes, the pedicels usually short.
Leaves cordate-ovate or lanceolate.   Cymes many-flowered.
Sepals obtuse, coriaceous.   Corolla 1 in. long, white.   Capsule
acuminate . . . . 25. *I. cymosa.*
Leaves narrow, not cordate.   Racemes or cymes loose, few-
flowered. Sepals acute. Corolla small, yellow. Capsule 1 celled 26. *I. linifolia.*
Leaves cordate.   Cymes dense or few-flowered.   Sepals obtuse.
Corolla yellow.   Capsule globular, at length rugose.
Sepals squarrose. Corolla small. Seeds pubescent . . . . 27. *I. chryseides.*
Sepals not squarrose. Corolla nearly 1 in. long. Seeds gla-
brous . . . . 28. *I. flava.*
Stems twining. Flowers solitary or rarely 2 or 3 together, the pe-
duncles and pedicels mostly longer than the calyx.
Leaves petiolate.
Leaves cordate-ovate, entire. Sepals small, obtuse . . . . 29. *I. obscura.*
Leaves ovate-lanceolate, deeply toothed or lobed below the
middle. Sepals narrow, acute . . . . 30. *I. incisa.*
Leaves oblong or linear, not cordate, entire. Sepals large,
acute, the outer ones broad and decurrent . . . . . . 31. *I. uniflora.*
Leaves sessile, narrow, toothed at the base or hastate . . . . 32. *I. angustifolia.*
Stems twining. Flowers small, solitary or several together, sessile
or with very short peduncles and pedicels. Leaves from cordate
to lanceolate.
Pedicels mostly about as long as the calyx. Capsule glabrous . 33. *I. plebeia.*
Flowers nearly sessile. Capsule pubescent or villous . . . . 34. *I. eriocarpa.*
Stems erect or ascending, not twining. Leaves not cordate, usually
narrow. Flowers small.
Plant hairy. Leaves entire toothed or lobed. Flowers nearly
sessile . . . . 35. *I. heterophylla.*
Plant tomentose or densely villous. Leaves entire. Flowers
solitary or in small cymes on a more or less elongated peduncle 36. *I. erecta.*

SERIES VI. **Urceolatæ.**—*Leaves entire. Ovary 2-celled. Corolla small, urceolate,
the tube broad, contracted at the throat.*

Stems twining, villous. Leaves cordate. Flowers in dense almost
sessile cymes . . . . 37. *I. urceolata.*

SERIES VII. **Hypocrateriformes.**—*Leaves entire or pinnatifid. Ovary 2-celled or
spuriously 4-celled. Corolla with a cylindrical narrow tube, and spreading nearly flat
limb.*

Stems twining. Leaves pinnatifid, with linear-subulate lobes.
Flowers red. Ovary 4-celled . . . . 38. *I. Quamoclit.*

SERIES 1. DIGITATÆ.   Leaves digitately divided into deep lobes or distinct segments or leaflets. Flowers large or small, campanulate or tubular
at the base.

1. **I. paniculata,** *R. Br. Prod.* 486. Stems trailing or twining sometimes to a great length, the whole plant glabrous. Leaves palmately or
almost pedately divided to below the middle into 5 to 9 ovate-lanceolate
obtuse or acuminate lobes, the whole leaf often 6 to 8 in. long and broad.
Peduncles longer than the petiole and sometimes 8 to 10 in. long, bearing a
cyme of several large purple or pink flowers. Sepals broad, very obtuse, 3
to 4 lines long. Corolla campanulate, shortly tubular at the base, 2 in. long

or more. Anthers large, undulate but scarcely twisted. Ovary 2-celled or more or less 4-celled by spurious dissepiments between the seeds especially at the top. Capsule ½ in. diameter or even larger. Seeds densely woolly-villous.—Bot. Reg. t. 62 ; *Batatas paniculata,* Chois. Conv. Or. and in DC. Prod. ix. 339 ; *I. insignis,* Andr. Bot. Rep. t. 636 ; Bot. Reg. t. 75 ; Bot. Mag. t. 1790 (a form with less deeply divided leaves, which occurs also in Australia)

N. Australia. Arnhem Bays, *R. Brown.*
Queensland. Cape York, *Jardine;* Palm Island, *Banks and Solander ;* Port Molle, *M'Gillivray ;* Rockingham Bay, *Dallachy.*
A maritime plant, not uncommon on the coasts of tropical Asia, Africa, and America.

2. **I. Davenporti,** *F. Muell. Fragm.* vi. 97. Stems apparently twining, the whole plant hoary with a stellate almost floccose tomentum. Leaves petiolate, divided to the base into 3 (or more?) petiolulate leaflets mostly obovate and more or less lobed, 1 to 2 in. long, the lower ones probably larger. Peduncles longer than the leaves, bearing 1 or 2 rather large white flowers, with persistent leafy ovate or lanceolate acuminate bracts. Sepals acuminate, ½ in. long, stellate-tomentose. Corolla broadly campanulate, about 1½ in. long. Capsule glabrous, not seen ripe. Young seeds pubescent.

N. Australia. Davenport Range, *M Douall Stuart.* The specimens are little more than fragmentary, but the foliage and indumentum are very different from those of any other species known to me.

3. **I. palmata,** *Forsk.; Chois. in DC. Prod.* ix. 386. A glabrous twiner, the old stems often more or less tuberculate or muricate. Leaves digitately divided nearly or rarely quite to the base into 5 to 7 ovate-lanceo-late lanceolate or oblong lobes, obtuse or rarely acute, 1 to 2 in. long. Peduncles usually several-flowered and as long as or longer than the petioles ; pedicels rather long. Sepals broad, obtuse, 3 to 4 lines long, all nearly equal. Corolla purple pink or white, campanulate but contracted into a tube towards the base, 1½ to 2 in. long, the angles or short broad lobes generally terminating in acute points. Ovary 2-celled. Capsule globular, as long as the calyx. Seeds pubescent and usually bordered by long silky hairs.—*I. pendula,* R. Br. Prod. 486 ; Andr. Bot. Rep. t. 613 ; Bot. Reg. t. 632 ; Chois. in DC. Prod. ix. 387, with most of the synonyms adduced (but not *I. Horsfalliæ,* Hook.) ; *I. pulchella,* Roth, and *I. tuberculata,* Rœm. and Schult., Chois. in DC. Prod. ix. 386, with most but not all of the synonyms adduced.

Queensland. Moreton Bay, *Fraser, F. Mueller ;* Ipswich, *Nernst ;* Rockhampton, *O'Shanesy ;* Edgecombe Bay, *Dallachy ;* Mackenzie Island, *Thozet ;* Curtis Island, *Henne.*
N. S. Wales. Port Jackson, *R. Brown ;* Clarence and Hastings rivers, *Beckler ;* Richmond river, *Henderson ;* Tweed river, *C. Moore ;* Lord Howe's Island, *M'Gillivray.*
The species is widely dispersed over tropical Asia, Africa, and America.

4. **I. quinata,** *R. Br. Prod.* 486. A rather slender twiner, glabrous or the stems and foliage more or less hirsute with long spreading hairs. Leaves digitate, with 5 distinct lanceolate or narrow-oblong obtuse entire segments,

1 to 1½ in. long, contracted at the base.   Peduncles usually 1-flowered and shorter than the leaves.   Sepals ovate, the outermost about 3 lines long, the innermost nearly or quite twice as long.   Corolla white or pale pink, campanulate but contracted into a tube towards the base, nearly 2 in. long, very shortly and broadly lobed or angled.   Ovary 2-celled.   Capsule ½ in. long, somewhat acuminate.   Seeds glabrous.—Chois. in DC. Prod. ix. 385; *I. hirsuta*, R. Br. Prod. 486 ; *I. pentadactylis*, Chois. Conv. Or., and in DC. Prod. ix. 385 ; *Convolvulus quinatus*, Spreng. Syst. i. 590.

N. Australia.   N. Coast, *R. Brown, Henne ;* Arnhem N. Bay (the hairy form), *R. Brown.*

Queensland.   Cape York and Port Molle, *M'Gillivray ;* Rockhampton, *Dallachy.*

The species is also in Burmah and S. China.

5. **I. diversifolia,** *R. Br. Prod.* 487.   Stems very slender, trailing or twining, glabrous as well as the foliage.   Leaves digitate, with very narrow linear segments, entire or more frequently toothed or pinnatifid, the central one 1 to 2 in. long, the others much shorter.   Peduncles rather short and thick, mostly 1-flowered.   Sepals oblong-lanceolate, acute, nearly equal or the inner ones rather longer.   Corolla nearly 2 in. long, contracted into a tube towards the base.   Capsule glabrous, globular, nearly as long as the calyx.—*Convolvulus diversifolius*, Spreng. Syst. i. 592.

N. Australia.   Islands of the Gulf of Carpentaria, *R. Brown.*   The leaves are not unlike those of the slender forms of *I. dissecta,* to which Choisy refers it, but the flowers are totally different.

6. **I. dissecta,** *Willd. Phytogr.* 5. *t.* 2.   Stems annual, slender, trailing or twining, glabrous as well as the foliage.   Leaves digitate, with 3, 5 or rarely 7 linear or linear-cuneate segments, acute and once or even twice pinnatifid and toothed.   Peduncles 1- or rarely 2-flowered, short in the Australian specimens, but often longer than the leaves in Indian ones.   Sepals ovate or lanceolate, obtuse or shortly acuminate, all nearly equal, 2 to 2½ lines long, often muricate on the midrib.   Corolla white, campanulate, about twice as long as the calyx.   Anthers oval-oblong, slightly twisted when fading.   Ovary 3-celled ; stigma capitate, 3-lobed.—R. Br. Prod. 487 ; Chois. in DC. Prod. ix. 363 (partly) ; *I. coptica*, Roth ; Chois. in DC. Prod. ix. 384.

N. Australia.   Islands off Cape Wilberforce, *R. Brown ;*   Victoria river, *F. Mueller* Port Essington, *Armstrong.*

Queensland.   Burdekin river, *Bowman.*

The species extends over tropical Asia and Africa.

SERIES 2. PHARBITIDES.—Leaves entire or 3- or 5-lobed.   Ovary usually 3-celled.   Sepals usually long and narrow.   Corolla large, campanulate, more or less tubular towards the base.—*Pharbitis,* Chois.

The species of this group, many of them much cultivated in warm countries for the beauty of their flowers, are mostly nearly allied to each other and difficult to define, nor are the characters of the series constant, some species having been differently placed in *Pharbitis* or in *Ipomœa,* according as the ovaries examined have been 3- or 2-celled.

7. **I. hederacea,** *Jacq. Collect.* i. 124 ; *Ic.* i. *t.* 36.   A tall herbaceous twiner, more or less hirsute, the hairs of the stem reflexed.   Leaves petiolate, broadly cordate, more or less 3- or 5-lobed, the lobes acuminate, the

middle one broad or narrow, (but not linear), contracted or dilated at the base, the lateral ones shorter and broader, the whole leaf from 1½ to 4 in. long. Peduncles short or rarely longer than the petioles, with 2 or 3 nearly sessile flowers at the end. Bracts linear. Sepals lanceolate, acuminate, broader and hispid with long hairs at the base, from ½ in. to nearly 1 in. long. Corolla blue or purple, often above 2 in. long. Ovary almost always 3-celled. —R. Br. Prod. 486 ; Bot. Reg. t. 85 ; *I. Nil*, Roth, Catal. Bot. i. 36 ; *Pharbitis Nil* and *P. hederacea*, Chois. Conv. Or. and in DC. Prod. ix. 343, 344, with most, if not all, of the synonyms quoted.

N. Australia. Victoria river, *F. Mueller.*

Queensland. Booby Island, *Banks and Solander ;* Suttor and Burdekin rivers, *Leichhardt, Bowman ;* Cape and Flinders rivers, *Bowman ;* Rockingham Bay, *Dallachy ;* Rockhampton, *Thozet ;* Moreton Bay, *Backhouse.*

Var. *limbata*, Hook. f. Bot. Mag. t. 5720. Flowers of a deep blue, with a pale or white margin.—*Pharbitis limbata*, Lindl. in Journ. Hort. Soc. v. 33 ; Henfr. in Gard. Mag. Bot. ii., with a fig. copied into Fl. des Serres, t. 608, and Lem. Jard. Fleur. t. 97.—Raised from N. Australian as well as from Javanese seeds.

The species is common in most tropical and subtropical regions of the New as well as the Old World, in some places, perhaps, escaped from cultivation.

8. **I. congesta,** *R. Br. Prod.* 485. A tall hirsute twiner, nearly allied to *I. hederacea*, but generally larger and the hairs less spreading. Leaves broadly cordate-ovate, acuminate, entire (or obscurely 3-lobed ?), usually 3 to 4 in. long. Peduncles longer than in *I. hederacea*, bearing a dense cyme of 3 or more large blue purple or pink flowers. Sepals lanceolate, acuminate, ¾ in. long. Corolla nearly 3 in. long. Ovary 3-celled.—Chois. in DC. Prod. ix. 369 ; *Convolvulus congestus*, Spreng. Syst. i. 601 ; *Pharbitis insularis*, Chois. Conv. Or. and in DC. Prod. ix. 341.

Queensland. Endeavour river, *Banks and Solander ;* Rockingham Bay, *Dallachy.* Also in Norfolk Island and in the islands of the S. Pacific.

*9. **I. purpurea,** *Roth, Catal.* i. 36. Stems twining, more or less hirsute with reflexed hairs or rarely glabrous. Leaves cordate-ovate, acuminate, entire or very rarely somewhat 3-lobed, glabrous or the petioles and veins pubescent, mostly 2 to 4 in. long. Peduncles longer than the petioles, bearing 1, 2 or 3 pedicellate flowers. Bracts small and narrow. Sepals lanceolate, scarcely acuminate, under ½ in. long, mostly hairy at the base. Corolla often above 2 in. long, purple blue pink or rarely white or variegated, campanulate, more or less tubular towards the base. Ovary 3-celled.—*Convolvulus purpureus*, Linn. ; Bot. Mag. t. 113, 1005, 1682 ; *Pharbitis hispida*, Chois. Conv. Or. and in DC. Prod. ix. 341.

Queensland. Curriwillighi, *Dalton.*
N. S. Wales. Darling Downs, *F. Law.*

The species is of American origin, long since cultivated in tropical as well as in European gardens, and become naturalized in many places. The above Australian stations are therefore probably also escapes from cultivation.

SERIES 3. CALYCINÆ.—Leaves entire or lobed at the base. Ovary 2-celled. Sepals large, obtuse, usually ¾ to 1 in. long at least after flowering. Corolla large, campanulate or tubular at the base, above 1½ in. and sometimes 3 in. long.

10. **I. peltata,** *Chois. Conv. Or. and in DC. Prod.* ix. 359. A tall woody twiner, with a milky juice (*Seemann*), covering whole trees with its dark green foliage (*Dallachy*), quite glabrous or the veins of the leaves hairy underneath. Leaves broadly ovate, shortly acuminate, more or less peltate or the upper ones cordate with a narrow sinus, 6 to 10 in. long. Flowers large, usually white, in loose cymes on a common peduncle usually shorter than the petiole. Sepals broad, obtuse, coriaceous, nearly equal, about ¾ in. long when in flower. Corolla broadly campanulate, at least 2 in. long. Anthers large, glabrous in our specimens, woolly according to Choisy. Fruit not seen.

Queensland. Rockingham Bay, *Dallachy.* Also in the Mascarene Islands, in the Indian Archipelago, and in the islands of the S. Pacific. The flowers are white, according to Dallachy and Seemann, white or purplish according to Desrousseaux (Lam. Dict. iii. 672), sulphur-coloured according to Blume, yellow according to Choisy.

11. **I. alata,** *R. Br. Prod.* 484. A tall twiner, quite glabrous. Leaves petiolate, cordate-ovate or ovate-lanceolate, acuminate with the point usually long and fine, sometimes angular or lobed at the base, 3 to 4 in. long, the petiole often winged. Peduncles rather longer than the petiole, often winged, bearing each a single rather large white flower. Sepals very broad, obtuse, coloured, nearly 1 in. long at the time of flowering. Corolla campanulate, about 2 in. long. Fruit not seen.—Chois. in DC. Prod. ix. 369; *Convolvulus alatus,* Spreng. Syst. i. 596.

N. Australia. Islands off Cape Wilberforce, *R. Brown.*

The S. American *I. altissima,* Mart., and the Central American *I. codonantha,* Benth., cannot in the dried state, when in flower, be distinguished from *I. alata.* There may, however, be differences in the seeds, and we have no similar species either from Asia or Africa.

12. **I. Turpethum,** *R. Br. Prod.* 485. A tall twiner, the young parts, foliage, and inflorescence softly pubescent, the old stems often bordered by narrow longitudinal wings. Leaves petiolate, mostly broadly cordate-ovate and acuminate, but sometimes obtuse or angular at the base, 2 to 4 in. long, or when luxuriant twice that size. Peduncles usually shorter than the leaves, bearing a short raceme of few rather large white flowers, but sometimes 1-flowered. Bracts ovate, thin, coloured, ½ to 1 in. long, very deciduous. Pedicels at first short and thick, but lengthening to 1 in. Outer sepals broadly ovate, often ¾ in. at the time of flowering and lengthening to 1 in., the inner ones rather smaller. Anthers large, much twisted when fading. Ovary 2-celled. Capsule much shorter than the calyx, globular, membranous. Seeds glabrous.—Chois. in DC. Prod. ix. 360; Bot. Reg. t. 279; *Convolvulus Turpethum,* Linn.; Bot. Mag. t. 2093; Wight in Hook. Bot. Misc. iii. t. Suppl. 38; *I. anceps* and *I. triquetra,* Rœm. and Schult.; Chois. in DC. Prod. ix. 360; *Argyreia alulata,* Miq. Fl. Ind. Bat. ii. 587.

Queensland. Broad Sound, *R. Brown;* Lizard Island, *M'Gillivray;* Sir Charles Hardy's Island, *Henne;* Suttor river, *Bowman;* Flinders river, *Sutherland;* Rockhampton, *O'Shanesy.* The species extends from the Mauritius and Ceylon over the peninsula of India, the Himalayas, and the Eastern Archipelago, as far north as Formosa, and is also said to be in the West Indies, probably introduced from Asia.

13. **I. longiflora,** *R. Br. Prod.* 484. A tall twiner, glabrous or the

leaves sparingly pubescent underneath. Leaves petiolate, broadly cordate-
ovate, acuminate, entire or somewhat 3-lobed, with rounded auricles, mostly
2 to 4 in. long. Peduncles shorter or rather longer than the petioles, bearing 1 or rarely 2 or 3 large (pale purple or pink?) flowers. Bracts very
small; pedicels short. Sepals obtuse, ½ to ¾ in. long, all nearly equal in
length, but the outer ones broad and almost cordate at the base, of a much
firmer consistence than in *I. Turpethum*, and when in fruit above 1 in. long.
Corolla 2½ to 3 in. long, the tube cylindrical at the base but dilated upwards
and not nearly so slender as in *I. Bona-nox*. Anthers included in the tube.
Capsule ovoid or globular, ¾ to 1 in. diameter or even larger. Seeds large,
minutely silky-pubescent, and usually, but not always, either bordered or
covered all over with long woolly hairs.—*I. macrantha*, Rœm. and Schult.
Syst. iv. 251; *Convolvulus longiflorus*, Spreng. Syst. i. 595.

N. Australia. Islands in the Gulf of Carpentaria, *R. Brown;* Victoria river, *F.
Mueller;* Escape Cliffs, *Hulls.*
Queensland. Burdekin river, the root eaten by the natives, *Bowman;* Rockingham
Bay, *Dallachy*; Rockhampton, *O'Shanesy.*

Choisy in DC. Prod. ix. 345, unites this with *I. Bona-nox*, a species differing widely in
its acuminate sepals, hypocrateriform corolla, exserted stamens, and glabrous seeds. I cannot, however, as far as our specimens go, distinguish *I. jucunda*, Thw. Enum. Pl. Zeyl. 211
and 426, and *Calonyction comosperma*, Boj.; Chois. in DC. Prod. ix. 346, from *I. longiflora;* and if these be really the same, the species has a wide range from E. tropical Africa
to Ceylon, the Indian Archipelago, and the S. Pacific islands.

14. **I. costata,** *F. Muell. in Herb. Hook.* Stems apparently woody and
probably twining, our specimens quite glabrous. Leaves petiolate, cordate,
orbicular or reniform, very obtuse and sometimes emarginate, 1 to 3 in.
broad. Peduncles very short, bearing 1 to 3 large flowers, on pedicels much
longer than the peduncles, the bracts very small or none. Sepals ½ to ⅗ in. long,
lengthening to nearly 1 in. when in fruit, almost acute, the outer ones usually
rugose and rather shorter. Corolla nearly 3 in. long. Ovary 2-celled, with
2 ovules in each cell. Capsule globular, apparently indehiscent or circumsciss about the middle. Seeds pubescent.

N. Australia. Sturt's Creek, *F. Mueller;* Attack Creek, *M'Douall Stuart.* The
species requires further investigation. The fruit shows some approach to that of *Argyreia*,
but the plant has not at all the aspect of that genus.

SERIES 4. SPECIOSÆ.—Leaves entire toothed or lobed at the base. Ovary
2-celled or spuriously 4-celled. Sepals moderate or small, rarely attaining
½ in. Corolla large (1½ to 3 inches long), often more or less tubular at the
base, usually pink purple or white.

15. **I. Pes-capræ,** *Roth, Nov. Sp. Pl.* 109. A glabrous perennial,
with long prostrate creeping or trailing stems. Leaves on long petioles,
oval obovate or orbicular, broadly emarginate or very obtusely 2-lobed, rather
thick, with nearly parallel oblique veins, the lower ones converging at the base
of the leaf, mostly 2 to 3 in. long. Peduncles often as long as the leaves,
bearing 1 or 2 rather large pink flowers on rather long pedicels. Sepals obtuse, about 3 lines long or the inner ones rather longer. Corolla broadly
campanulate, somewhat tubular at the base, about 1½ in. long. Ovary more
or less perfectly 4-celled, at least at the time of flowering. Capsule 2-celled,

ovoid or nearly globular, coriaceous, ½ to ¾ in. long.　Seeds hairy.—*Convolvulus Pes-capræ,* Linn. Spec. Pl. 226 ; *C. maritimus,* Desr. in Lam. Dict. iii. 550 ; *Ipomœa maritima,* R. Br. Prod. 486 ; Bot. Reg. t. 319, and probably all the synonyms adduced by Choisy in DC. Prod. ix. 349, under *I. Pes-capræ* except *I. carnosa.*

N. Australia. Glenelg river and Brecknock Harbour, N.W. coast, *Marten ;* Nichol Bay, *Ridley's Expedition ;* N. coast, *R. Brown ;* Sweers and other islands, and Albert river, *Henne ;* Escape Cliffs, *Hulls.*
Queensland. Torres Straits, *F. Mueller ;* Harvey Bay, Sandy Cape, *R. Brown ;* Port Denison, *Fitzalan ;* Edgecombe Bay, *Dallachy ;* Mackenzie Island, *Sutherland.*
N. S. Wales. Richmond river, *Fawcett, Henderson.*

The species is common on the seacoasts of most tropical countries in the New as well as the Old World. Although placed by Choisy in *Ipomœa,* there is generally a more or less developed, spurious, transverse dissepiment between the ovules and young seeds, subdividing each cell into two.

16. **I. carnosa,** *R. Br. Prod.* 485.　A prostrate or creeping glabrous perennial.　Leaves petiolate, mostly ovate or oblong, very obtuse or emarginate, cordate at the base, thick and somewhat fleshy, penniveined, and ½ to 1 in. long, but sometimes with 1 or 2 very prominent basal lobes on each side, and in some specimens (not Australian) long and narrow with a hastate base.　Peduncles short, bearing 1 or rarely 2 or 3 rather large white flowers.　Sepals rather narrow, 4 to 5 lines long at the time of flowering, subsequently enlarged, the outer ones mucronate-acute, the inner ones obtuse and often rather longer.　Corolla campanulate, about 1½ in. long.　Capsule nearly globular, more or less perfectly 4-celled, about ½ in. diameter, glabrous. Seeds densely woolly-hairy.—*Convolvulus carnosus,* Spreng. Syst. i. 609 ; *Batatas littoralis* and *B. acetosæfolia,* Chois. in DC. Prod. ix. 337, 338, with most, if not all, of the synonyms adduced ; *Convolvulus stoloniferus,* Cyr. Pl. Rar. 14. t. 5 (very good).

N. Australia. Islands of the Gulf of Carpentaria, *R. Brown.* The species is dispersed along the coasts of the warmer regions of Asia, Africa, and America, extending beyond the tropics to the shores of the Mediterranean. Although placed by Choisy in a different genus from *I. Pes-capræ,* it is very nearly allied to it, differing chiefly in the narrower, more fleshy, and less prominently veined leaf, and the spurious dissepiments usually but not always, more perfect and more permanent in the fruit.　Grisebach (Fl. Brit. W. Ind. 471) places it in a section with glabrous seeds, but I have always found them very woolly in American as well as in African and European specimens.

17. **I. reptans,** *Poir. ; Chois. in DC. Prod.* ix. 349.　A glabrous perennial, with long, prostrate, trailing or floating and hollow stems, often rooting at the nodes and sometimes bearing short ascending branches. Leaves on long petioles, from ovate to linear-lanceolate, acuminate, always cordate or hastate at the base, the angles rounded or produced into broad or narrow acute auricles, the leaf usually 2 to 4 in. long, but rarely on the smaller branches a few small ovate-cordate and obtuse ones.　Peduncles 1- or few-flowered.　Sepals rather obtuse, 3 to 4 lines long.　Corolla not so broad as in the two preceding species, more tubular at the base, pink purplish or white, about 1½ in. long.　Filaments hairy at the base.　Ovary 2-celled. Capsule globular, coriaceous, apparently indehiscent or bursting irregularly. Seeds large, woolly, often reduced to 2.

N. Australia. Upper Victoria river, *F. Mueller* ; Adams Bay, *Hulls.*
Queensland. Flinders river, *Sutherland ;* Cape river, *Bowman.*

The species is found in marshy or wet sandy places, or floating in water, in many parts of tropical Asia and Africa.

18. **I. graminea,** *R. Br. Prod.* 485. A slender glabrous twiner. Leaves on short petioles, linear-lanceolate or linear, entire, 4 to 8 in. long. Peduncles 1-flowered, ½ to 1½ in. long (including the pedicel), with very small distant bracts. Sepals oblong, obtuse, ½ in. long or rather more, the outermost one shorter. Corolla 2½ to 3 in. long, contracted into a slender tube. —Chois. in DC. Prod. ix. 367 ; *Convolvulus gramineus,* Spreng. Syst. i. 607.

N. Australia. Islands off Cape Wilberforce, *R. Brown (Herb. R. Br.).* This species is unlike any other one known to me. The corolla is nearly that of *I. longiflora*, but more slender.

19. **I. velutina,** *R. Br. Prod.* 485. A tall twiner, apparently woody at the base, softly velvety-pubescent or villous all over, the hairs usually reflexed on the branches, often silky on the leaves. Leaves petiolate, broadly ovate-cordate, obtuse or shortly acuminate, entire, 2 to 4 in. long. Peduncles about as long as the petioles or sometimes longer, bearing a dichotomous cyme of several large flowers, rarely reduced to a single flower on the side branches. Bracts very deciduous or none. Sepals broad, very obtuse, glabrous or nearly so, 3 to 4 or even 5 lines long. Corolla fully 3 in. long, contracted into a tube at the base.—Chois. in DC. Prod. ix. 369 ; *Convolvulus velutinus,* Spreng. Syst. i. 601.

N. Australia. Islands of the Gulf of Carpentaria, *R. Brown.* A specimen in Herb. Hook., from Clowes's collection, appears to be a uniflorous state of the same species. Brown's specimens have the inflorescence of *I. paniculata*, with a very different foliage, and the corolla of *I. longiflora*, but with a much smaller calyx.

20. **I. abrupta,** *R. Br. Prod.* 485. A tall woody twiner, glabrous or nearly so. Leaves petiolate, cordate-ovate, obtusely acuminate or almost acute, entire, from 2 or 3 in. long, to twice that size. Flowers large, in pedunculate cymes, rarely reduced on lateral branches to single flowers. Sepals obtuse, coriaceous, 3 to 4 or rarely 5 lines long. Corolla fully 3 in. long, contracted into a tube at the base.—Chois. in DC. Prod. ix. 370 ; *Convolvulus abruptus,* Spreng. Syst. i. 596.

N. Australia. N. coast, *Bauer (Herb. Banks)* ; Escape Cliff, *Hulls ;* Gloster Island, *Henne (both in Herb. F. Muell.).*
Queensland. Burdekin river, *Herb. F. Mueller.*

This appears to differ from *I. velutina* chiefly in its glabrous stems and foliage, and may be a variety only of that species. It has the corolla of *I. longiflora*, but a much smaller calyx and the flowers usually cymose ; but the specimens I have seen are all single and fragmentary. The one named by Brown in the Banksian herbarium appears to be a side branch, with 1-flowered peduncles ; the three others have cymose flowers. It is wanting in Brown's own herbarium.

21. **I. denticulata,** *Chois. in DC. Prod.* ix. 379, *not of R. Br.* Glabrous or nearly so ; stems rather slender, prostrate and trailing or twining. Leaves petiolate, deeply cordate, ovate, obtuse or acute, broad or narrow, the basal auricles rounded, with occasionally an acute tooth or angle on the outer

side, the whole leaf 1 to 2 in. long. Pedicels rather long, often above 1 in , solitary or few together on a very short common peduncle, with minute bracts. Sepals obtuse or mucronate, about 4 lines long, coriaceous but often with membranous margins. Corolla 1½ to nearly 2 in. long. Capsule depressed-globular, shorter than the calyx, not wrinkled. Seeds glabrous.—*I. carnea*, Forst. Prod. 15, not of Jacq.; *I. lævigata*, Soland. (not Steud.) in Herb. Banks; *I. littoralis*, Thw. Enum. Pl. Zeyl. 211, not of Blume.

Queensland. Cape York, *Daemel ;* Rockingham Bay, *Dallachy.* Also in Ceylon, the Eastern Archipelago, and in the Society and Sandwich Islands.

I refer this to *Convolvulus denticulatus*, Desr., transferred to *Ipomœa* by Choisy, on the authority of a Society Island specimen, evidently the one described by Forster, determined by Choisy, in the Banksian herbarium, but the specific name is scarcely applicable. Blume's *I. littoralis*, from his short character and from Miquel's description, is evidently a very different plant, probably closely allied to, if not identical with, *I. reniformis*. Some of Dallachy's specimens of *I. denticulata* have the leaves more acute and acutely auriculate, almost like those of *I. gracilis*, but with the calyx of *I. denticulata*.

22. **I. gracilis,** *R. Br. Prod.* 484. A rather slender twiner, glabrous or scabrous-pubescent. Leaves on long petioles, lanceolate-hastate or triangular-cordate, acute, mostly 1 to 2 in. long, the basal lobes or auricles usually acute long and divergent or curved inwards towards the end, but varying in breadth and sometimes but rarely almost obtuse. Peduncles, including the pedicel, as long as the petioles, bearing a single rather large white flower, the bracts very small. Sepals broadly lanceolate, acute or acuminate or rarely obtuse, 4 to 5 lines long or 6 lines when in fruit. Corolla about 2 in. long. Capsule globular, smooth, 5 to 6 lines diameter. Seeds shortly pubescent, with a tuft of longer hairs at the hilum.—Chois. in DC. Prod. ix. 370 ; *Convolvulus gracilis*, Spreng. Syst. i. 604.

N. Australia. Islands of the Gulf of Carpentaria and off Cape Wilberforce, *R. Brown.*

Queensland. Bowen river and Brawl Creek, *Bowman ;* Rockingham Bay, *Dallachy.*

Although, generally speaking, this species is readily distinguished by its aspect from *I. denticulata*, it is difficult to assign any positive limits between the two. The more twining habit, acutely hastate leaves, and acute sepals of *I. gracilis*, are none of them without exceptions, and the seeds have been observed in too few specimens to judge of the constancy of the character derived from their hairs.

23. **I. sepiaria,** *Kœn.; Chois. in DC. Prod.* ix. 370. A twiner, either quite glabrous or the stems hirsute with long spreading or reflexed hairs. Leaves petiolate, broadly cordate-ovate, obtuse or shortly and obtusely acuminate, the basal auricles rounded or angular, mostly 1 to 3 in. long. Peduncles rather rigid, longer than the petioles, bearing a dense cyme or cluster of 3 to 5 pink or white flowers on short pedicels. Bracts small, narrow, often persistent. Sepals ovate-lanceolate, acute or mucronate, varying from 2½ to 4 lines in length, the outer ones often rugose. Corolla about 1½ in. long. Capsule globular, somewhat depressed, smooth. Seeds glabrous.— Wight, Ic. t. 838.

Queensland. Cape river, *Bowman (Herb. F. Muell.).*

N. S. Wales. Between Darling River and Cooper's Creek, *Neilson (Herb. F. Muell.).*

The species is dispersed over E. India and the Eastern Archipelago. I describe it chiefly from Indian specimens, and refer to it the two from Australia above quoted with much hesitation. They are mere fragments, and may belong to *I. Muelleri,* but have much more the aspect of the true *I. sepiaria.* The flowers are rather larger than they are usually in India, but there also it occasionally occurs with similar large flowers.

24. **I. Muelleri,** *Benth.* A glabrous rather slender twiner. Leaves on rather long petioles, very broadly cordate-ovate, obtuse, with rounded basal auricles, entire, 1 to 2 in. long. Peduncles shorter or at length longer than the petioles, bearing 1 to 3 flowers on very short pedicels. Bracts very small. Sepals broad, obtuse or scarcely acuminate, 4 to 5 lines long. Corolla apparently pink, rather above 1 in. long. Capsule globular, smooth, as long as the calyx. Seeds villous.

N. Australia. Nichol Bay, *Walcott ;* Sturt's Creek, *F. Mueller ;* in the interior, lat. 18° 30', *M'Douall Stuart's Expedition.* Evidently nearly allied to *I. sepiaria,* but the flowers are smaller and the seeds villous.

SERIES 5. CAMPANULATÆ.—Leaves entire toothed or lobed at the base. Ovary 2-celled. Corolla moderate or small, rarely exceeding 1 in. in length, usually broadly campanulate, yellow white or pink.

25. **I. cymosa,** *Rœm. and Schult. ; Chois. in DC. Prod.* ix. 371. A rather coarse twiner, glabrous or softly pubescent, usually turning dark brown in drying. Leaves petiolate, from ovate to oblong or lanceolate, shortly acuminate or obtuse, the larger ones broadly cordate or almost sagittate, the narrow ones rounded at the base. Flowers of a pure white or with a yellow eye, in cymes of 6 to 12 or even more, rarely solitary or nearly so, on a short rather thick common peduncle. Sepals 2 to 3 or in some Indian specimens nearly 4 lines long, glabrous, coriaceous, nearly equal in length. Corolla campanulate, 1 in. long or even larger, often hairy outside at the top. Ovary 2-celled. Capsule shortly acuminate, longer than the calyx. Seeds clothed with long soft loose hairs.—Bot. Reg. 1843, t. 24.

Queensland. Rockingham Bay, *Dallachy,* sometimes covering the trees with a sheet of white flowers. The species is common in the greater part of India and the Archipelago, and is scarcely to be distinguished from the *I. umbellata* of tropical America and Africa, except by the colour of the flowers, which, in the latter species, is yellow.

26. **I. linifolia,** *Blume ; Chois. in DC. Prod.* ix. 369. A slender twiner, glabrous or clothed or sprinkled with long silky hairs. Leaves petiolate, narrow-lanceolate, entire, rounded or truncate at the base, 1 to 2 in. long. Peduncles slender, mostly about as long as the leaves, bearing sometimes a forked cyme, more frequently a simple loose 1-sided raceme of 3 or more yellow flowers, the bracts small but persistent. Pedicels nearly as long as the calyx. Sepals ovate or lanceolate, acute, nearly equal, 2½ to nearly 3 lines long. Corolla campanulate, about ¾ in. long. Ovary 2-celled, but the dissepiment usually drying up after flowering. Capsule small, smooth, globular, usually 1-celled. Seeds 4, usually glabrous.—*Skinneria cæspitosa,* Chois. Conv. Or. t. 6, and in DC. Prod. ix. 435.

Queensland. Rockingham Bay, *Dallachy.* Common in India and the Eastern Archipelago. I refer this plant to Blume's *I. linifolia,* on the authority of a specimen received from Miquel under that name, which agrees well with Blume's short diagnosis.

27. **I. chryseides,** *Ker, Bot. Reg. t.* 270. A slender glabrous twiner.

Leaves on long petioles, broadly ovate-cordate or almost hastate, acuminate, 1 to 2 in. long, entire or with 2 broad rounded and sometimes toothed basal lobes. Peduncles as long as the leaves or nearly so, with 2 spreading branches, each bearing 2 to 4 small yellow flowers, with 1 in the fork. Sepals 2 to 2½ lines long, obovate or broadly oblong, truncate or retuse and herbaceous at the end, with a small recurved point in the centre, giving the cyme a squarrose aspect. Corolla broadly campanulate, not twice as long as the calyx. Stamens and style often as long as the corolla or nearly so. Capsule 3 to 4 lines diameter, nearly globular, with 4 raised longitudinal ribs and usually transversely wrinkled when quite ripe. Seeds pubescent.—Chois. in DC. Prod. ix. 382 ; Wight, Ic. t. 157.

Queensland. Rockingham Bay, *Dallachy.* Common in E. India and the Archipelago.

28. **I. flava,** *F. Muell. Herb.* A rather slender glabrous twiner. Leaves petiolate, cordate-ovate, mostly acuminate, with large rounded or angular basal auricles, or the upper ones lanceolate-sagittate, 1 to 2 in. long. Peduncles rather long, bearing an irregular dense cyme of very few yellow or nearly white flowers on short pedicels or sometimes the flower solitary. Bracts very small. Sepals ovate or oblong, obtuse, rigid but smooth and almost membranous, 3 to 4 lines long. Corolla campanulate, apparently about 1 in. long. Anthers rather large. Capsule globular, coriaceous, at length wrinkled, about 4 lines diameter. Seeds glabrous.

N. Australia. Albert river and its tributaries, *F. Mueller.* Abundant on flats subject to inundations on Alligator river, *A. Cunningham.* Raised also in Kew Gardens from a seed gathered on De Grey River in Ridley's Expedition. Allied to *I. chryseides,* but the peduncle longer, the flowers fewer and twice as large, and the calyx different. The single specimen from Kew Gardens had only produced its first flower, but appeared to belong to the same species, although the flower was almost white.

29. **I. obscura,** *Ker, Bot. Reg. t.* 239. A slender glabrous or pubescent twiner. Leaves on slender petioles, broadly and deeply cordate-ovate, acuminate, from under 1 in. to nearly 2 in. long and broad. Peduncles as long as the petioles, bearing 1 or rarely 2 or 3 yellow flowers. Sepals lanceolate, acute, scarcely 2 lines long in the common form. Corolla campanulate, ¾ to 1 in. long. Capsule globular, smooth. Seeds pubescent.—Chois. in DC. Prod. ix. 370 ; *Convolvulus obscurus,* Linn. Spec. Pl. 220 ; *I. luteola,* R. Br. Prod. 485 ; Chois. in DC. Prod. ix. 369 ; *I. Brownii,* Rœm. and Schult. Syst. iv. 252.

Queensland. Keppel Bay, *R. Brown ;* Cape River, *Bowman.* Common in tropical Asia, extending into tropical Africa and eastward to the Archipelago. The Australian specimens seem to have rather smaller flowers than the Indian ones, but they are very imperfect. The species is readily distinguished from its nearest allies by the small calyx.

30. **I. incisa,** *R. Br. Prod.* 486. Prostrate trailing or scarcely twining, pubescent villous or nearly glabrous, the branches rather slender but sometimes very long. Lower leaves broadly ovate-cordate and deeply and irregularly toothed or lobed, especially below the middle ; upper ones oblong or lanceolate, hastate or almost digitate with one long central lobe and several

short lateral ones. Peduncles long and slender, bearing 1 or rarely 2 or 3 pink or purplish flowers, the pedicels as long as the calyx, the bracts minute. Sepals lanceolate or ovate-lanceolate, acute or rather obtuse, about 3 lines long. Corolla campanulate, ¾ to nearly 1 in. long. Ovary 2-celled. Fruiting calyx slightly enlarged, the capsule globular and smooth. Seeds glabrous.—Chois. in DC. Prod. ix. 352 ; *Convolvulus incisus,* Spreng. Syst. i. 609.

N. Australia. Upper Victoria river, *F. Mueller ;* islands of the Gulf of Carpentaria, *R. Brown.*

I. cinerascens, R. Br. Prod. 486 ; Chois. in DC. Prod. ix. 359 (*Convolvulus cinerascens,* Spreng. Syst. i. 609), from the islands off Cape Wilberforce, appears to me to be only a more villous form of *I. incisa,* the flowers perhaps rather smaller.

31. **I. uniflora,** *Rœm. and Schult. Syst.* iv. 247. A glabrous or some-what silky-pubescent rather slender twiner. Leaves oblong to linear, obtuse or mucronate, entire, narrowed into a short petiole, mostly 1 to 3 in. long. Peduncles shorter than the leaves, bearing usually a single white flower, the pedicel as long as or longer than the calyx, the bracts very minute. Sepals leafy, acute, the outer ones broad and shortly decurrent on the pedicel, ½ to ¾ in. long, the inner ones smaller and narrower. Corolla campanulate, longer than the calyx, but rarely exceeding 1 in., more or less hairy outside. Ovary 2-celled. Capsule globular. Seeds glabrous puberulous or bordered by short hairs.—*Aniseia uniflora,* Chois. Conv. Or., and in DC. Prod. ix. 431 ; Wight, Ic. t. 850 ; *A. martinicensis* and *A. ensifolia,* Chois. Conv. Or., and in DC. Prod. ix. 430 ; *A. cernua,* Moric. Pl. Amer. t. 38 ; Chois. in DC. Prod. ix. 431, and perhaps some other species referred by Choisy to *Aniseia.*

Queensland. Rockingham Bay, *Dallachy.* The species is widely dispersed over tropical Asia, Africa, America, and the Pacific islands.

32. **I. angustifolia,** *Jacq. Collect.* ii. 367, *and Ic. Rar. t.* 317, *not of Choisy.* A glabrous annual, with slender prostrate trailing or twining stems, usually drying of a black or brown colour. Leaves on very short petioles or almost sessile, linear or lanceolate, acute or acuminate, cordate or hastate, and often toothed at the base, 1 to 2 or even 3 in. long when very luxuriant. Peduncles slender, longer than the leaves, bearing 1 or 2 small yellowish-white flowers. Sepals ovate-lanceolate or lanceolate-acuminate, about 3 lines long. Corolla campanulate, ½ to ¾ in. long. Stamens rather long. Ovary 2-celled. Seeds glabrous.—*I. filicaulis,* Blume ; Chois. in DC. Prod. ix. 353 ; Bot. Mag. t. 5426 ; *I. denticulata,* R. Br. Prod. 485 ; Bot. Reg. t. 317.

N. Australia. Montague Sound, N.W. coast, *A. Cunningham ;* Camden and Breck-nock Harbours, *Martin ;* Upper Victoria river, *F. Mueller ;* islands of the Gulf of Carpentaria, *R. Brown, Henne ;* Port Essington, *Armstrong.*
Queensland. Cape York, *Daemel ;* Rockingham Bay, *Dallachy ;* Cape River, *Bowman.*

The species is widely dispersed over tropical Africa and Asia. R. Brown's Australian specimens include a narrow-leaved form corresponding precisely with the Guinea plant originally described by Jacquin and by Vahl, together with the broader-leaved form more prevalent in E. India and the Archipelago.

33. I. plebeia, *R. Br. Prod.* 484.   A slender twiner, softly pubescent and sprinkled with rather long hairs, which are reflexed on the branches, scattered on the leaves or sometimes wanting.   Leaves on slender petioles, cordate-lanceolate or the lower ones broadly ovate-cordate, acuminate, entire or obscurely 3-lobed, with the basal auricles rounded, $1\frac{1}{2}$ to 3 in. long. Peduncles slender, 1-flowered, articulate and minutely bracteate near the base (the peduncle much shorter than the pedicel).   Outer sepals ovate-lanceolate, subulate-acuminate, hispid, about 4 lines long ; the inner ones smaller.   Corolla campanulate, fully $\frac{1}{2}$ in. long, often pubescent at the top. Stigma with 2 globular lobes.   Capsule shorter than the calyx, glabrous, 2-celled.   Seeds pubescent.—*Convolvulus plebeius*, Spreng. Syst. i. 604; Chois. in DC. Prod. ix. 412.

Queensland.   Bay of Inlets, *Banks and Solander ;* islands of Moreton Bay, *F. Mueller ;* Walloon and Comet river, *Bowman.*   It does not appear for what reason Choisy removed this plant to the genus *Convolvulus ;* the stigma is certainly that of *Ipomœa*, where Brown placed it.

34. I. eriocarpa, *R. Br. Prod.* 484.   A twining annual, more or less hirsute with rigid hairs, mostly reflexed on the stem, scattered on the under side of the leaves or confined to the margins and principal veins.   Leaves petiolate, from deeply cordate-ovate to lanceolate or hastate, acuminate, 1 to 3 in. long, the upper surface usually glabrous.   Peduncles exceedingly short, bearing 1, 2 or rarely more small flowers either quite sessile or very shortly pedicellate.   Sepals ovate or ovate-lanceolate, acuminate, hirsute, 2 to 3 lines long.   Corolla scarcely exceeding the calyx or sometimes not so long in Australian specimens, rather larger in some extra-Australian ones, slightly hirsute outside.   Stigma with 2 globular lobes.   Capsule globular, pubescent or hirsute, but becoming nearly glabrous when ripe.   Seeds glabrous.—Chois. in DC. Prod. ix. 369 ; *Convolvulus eriocarpus*, Spreng. Syst. i. 598 ; *Ipomœa sessiliflora*, Roth ; Chois. in DC. Prod. ix. 366 ; Wight, Ic. t. 169 (a remarkably luxuriant large-leaved specimen, apparently with an 8-seeded fruit, probably by a mistake of the artist) ; *I. Horsfieldiana*, Miq. Fl. Ind. Bat. ii. 611.

N. Australia.   Upper Victoria river, *F. Mueller.*
Queensland.   Endeavour river, *Banks and Solander ;* Burdekin river, *Bowman, Fitzalan.*

The species is common in tropical Africa and Asia, and is also (probably introduced) in the West Indies.

35. I. heterophylla, *R. Br. Prod.* 487.   Stems erect or ascending, not twining, rather slender, simple or branched, 1 to 2 ft. high, more or less hirsute, as well as the foliage, with long loose hairs, rarely at length nearly glabrous.   Leaves petiolate, lanceolate or oblong, quite entire or bordered by coarse teeth or lobes, especially below the middle, never cordate, 1 to 3 in. long, the upper ones small and narrow.   Peduncles very short or the flowers almost sessile, between 2 linear bracts or bracteoles almost as long as the calyx.   Sepals lanceolate, subulate-acuminate, ciliate and hispid with long hairs, 3 to 4 lines long.   Corolla campanulate, rather longer than the calyx. Stigmas large and broad, usually distinct.   Capsule and seeds glabrous.— Chois. in DC. Prod. ix. 354 ; *I. polymorpha*, Rœm. and Schult. Syst. iv. 254 ;

Convolvulus Brownii, Spreng. Syst. i. 612, altered in the Index to *C. Robertianus.*

N. Australia. Victoria river, *F. Mueller;* islands of the Gulf of Carpentaria, *R. Brown;* Port Essington, *Armstrong.*
Queensland. Cape York, *Daemel;* Flinders river, *Sutherland;* Rockhampton, *O'Shanesy;* Curriwillighi and Armadillo, *Dalton.*

36. **I. erecta,** *R. Br. Prod.* 487. Stems from a perennial base, erect or ascending, simple or slightly branched, softly tomentose or villous as well as the foliage and inflorescence, the hairs intricate on the branches, more appressed on the leaves, and often rust-coloured. Leaves very shortly petiolate, oblong or lanceolate, obtuse or acute, not cordate, 1 to 2 in. long, the upper ones smaller and narrower. Peduncles mostly shorter than the leaves, bearing 1, 2 or 3 pink flowers, the pedicels short, the bracts very small. Sepals ovate-lanceolate, rather acute, softly villous or nearly glabrous, of a somewhat firmer consistence at the base as in many species of *Convolvulus*, the outer ones 3 to 4 lines long, the inner ones smaller. Corolla campanulate, about ¾ in. long. Stigmatic lobes very broadly ovate, recurved. Capsule globular, readily splitting into 6 to 8 valves, as in *Convolvulus parviflorus* and its allies. Seeds glabrous.—Chois. in DC. Prod. ix. 354; *Convolvulus erectus,* Spreng. Syst. i. 612.

N. Australia. Victoria river, *F. Mueller;* islands of the Gulf of Carpentaria, *R. Brown, Henne;* near Caledon Bay, *B. Geell;* in the interior, Attack Creek, Newcastle Water, Strangeways river, etc., *M'Douall Stuart.*

I. pannosa, R. Br. Prod. 487; Chois. in DC. Prod. ix. 356 (*Convolvulus pannosus,* Spreng. Syst. i. 612), from the mainland, Carpentaria, appears to me to be only a densely villous form of the same plant, and *I. biflora,* R. Br. l. c.; Chois. l. c. 367 (*I. diantha,* Rœm. and Schult. Syst. iv. 254; *Convolvulus flexuosus,* Spreng. Syst. i. 612), a slight variety with more slender branches, sometimes almost twining at the extremity. The species, both in habit and in character, shows a slight approach to the genus *Convolvulus*.

SERIES 6. URCEOLATÆ.—Leaves entire. Ovary 2-celled. Corolla small, urceolate, the short broad tube contracted at the throat.

37. **I. urceolata,** *R. Br. Prod.* 485. A tall twiner, softly villous, the hairs of the branches reflexed, those of the foliage almost silky. Leaves petiolate, cordate-ovate, acuminate, entire, 3 to 4 in. long. Flowers numerous, in dense almost sessile cymes or clusters, the pedicels much longer than the calyx. Sepals orbicular or broadly ovate, obtuse, hirsute, scarcely above 1 line long. Corolla-tube ovoid, inflated, contracted towards the throat, about ½ in. long, the limb spreading, scarcely half as long as the tube. Capsule globular, glabrous about 3 lines diameter, 2-celled. Seeds 4, glabrous.—Chois. in DC. Prod. ix. 369; *Convolvulus urceolatus,* Spreng. Syst. i. 601.

Queensland. Endeavour river, *Banks and Solander.* The aspect, inflorescence, and flowers, are almost those of *Lepistemon flavescens,* Blume (which includes *L. Wallichii,* Chois.), but the scales at the base of the stamens surrounding the ovary of that species are wanting in the flower I dissected of *I. urceolata,* and are not mentioned in Brown's notes.

SERIES. 7. HYPOCRATERIFORMES.—Leaves entire or pinnatifid. Ovary 2-celled or 4-celled by spurious dissepiments between the 2 ovules of each cell. Corolla with a cylindrical narrow tube and spreading nearly flat limb. Stamens and style usually exserted from the tube.

*38. **I. Quamoclit,** *Linn. Sp. Pl.* 227. A slender glabrous twiner. Leaves sessile, deeply pinnatifid, with linear-subulate entire segments. Peduncles longer than the leaves, bearing 1 to 3 scarlet flowers on long pedicels thickened upwards. Sepals obtuse, 2 to 3 lines long. Corolla-tube cylindrical, slender, ¾ to nearly 1 in. long; limb short, spreading, shortly 5-lobed. Stamens and style longer than the tube. Ovary 4-celled, with 1 ovule in each cell. Capsule ovoid-globular, glabrous, rather longer than the calyx, completely 4-celled. Seeds glabrous.—Bot. Mag. t. 244 ; *Quamoclit vulgaris,* Chois. in DC. Prod. ix. 336.

Queensland. Rockhampton, *Sutherland,* said to be wild, but probably escaped from a garden or accidentally introduced. The species, believed to be of East Indian origin, has long been extensively cultivated for ornament in almost all warm civilized regions, and has established itself as a weed in the New as well as in the Old World.

3. CONVOLVULUS, Linn.

(Calystegia, *R. Br.;* Jacquemontia, *Chois.*)

Corolla campanulate, entire, angular or rarely lobed. Ovary 2-celled, with 2 ovules in each cell. Style filiform, with 2 ovate oblong linear or subulate stigmatic lobes. Fruit a dry capsule, completely or sometimes incompletely 2-celled.—Twining prostrate creeping or erect herbs, or in species not Australian undershrubs or low shrubs. Leaves entire or rarely toothed, lobed or deeply divided. Flowers axillary, solitary or in corymbose or umbel-like cymes. Seeds glabrous, at least in the Australian species.

A large genus, distributed over the whole area of the Order, less numerous within the tropics than *Ipomœa,* but extending far into the temperate and cooler regions both of the northern and the southern hemispheres. Of the six Australian species, two are spread over the extratropical regions of both hemispheres ; two belong to the tropical Asiatic flora ; the remaining two extend only to New Zealand.

Convolvulus has no character to distinguish it from *Ipomœa* besides the more or less elongated stigmatic lobes of the style, the habit is usually but not always different. For the subdivision of the genus, the largely developed bracts or bracteoles, and the imperfect development of the septum of the ovary, characters which in *Ipomœa* are scattered and isolated, are so far associated in the group *Calystegia* as to constitute a well-marked section, which however appears to me to be still too artificial to adopt it as a genus after Brown and others. The shortness of the stigmatic lobes, upon which Choisy had founded the genus *Jacquemontia* as intermediate between *Ipomœa* and *Convolvulus,* is not nearly so decided in the typical West Indian *Jacquemontias* as in *Convolvulus marginatus.*

Sect. I. **Convolvulus.**—*Bracts small or none. Dissepiment of the ovary usually perfect.*

Flowers solitary or rarely 2 together. Sepals obtuse. Leaves either
very narrow or toothed or lobed . . . . 1. *C. erubescens.*
Flowers in cymes. Sepals acuminate. Leaves cordate, entire.
Softly tomentose. Pedicels short . . . . 2. *C. multivalvis.*
Glabrous or pubescent. Pedicels rather long . . . . 3. *C. parviflorus.*

Sect. II. **Calystegia.**—*Bracts or bracteoles 2, enlarged and enclosing the calyx. Dissepiment of the ovary usually incomplete.*

Leaves hastate, acuminate. Bracteoles broadly cordate, almost orbicular. Calyx 2 lines ; corolla ¾ in. long . . . . 4. *C. marginatus.*

Leaves cordate or hastate, acute or acuminate. Bracteoles ovate, acute,
  longer than the calyx. Sepals 4 to 5 lines ; corolla above 2 in. long 5. *C. sepium.*
Leaves fleshy, reniform or rounded-cordate, obtuse. Bracteoles ovate,
  very obtuse, shorter than the calyx. Sepals 4 to 6 lines; corolla
  about 1¼ in. long . . . . 6. *C. Soldanella.*

SECT. I. CONVOLVULUS.—Bracts small or none. Dissepiment of the
ovary usually perfect.

1. **C. erubescens,** *Sims, Bot. Mag. t.* 1067. A perennial, either glabrous, pubescent or densely tomentose, rarely villous, with a creeping rootstock and slender prostrate trailing or rarely twining stems. Foliage exceedingly variable, the leaves usually more or less sagittate-cordate, the lower
ones ovate-lanceolate, the upper ones passing into narrow-lanceolate or
linear, with diverging entire or lobed basal auricles and from ¾ to 1½ in. long,
but sometimes nearly all small, cordate-ovate obtuse and slightly crenate,
sometimes nearly all narrow-linear with either very minute or long and linear
basal auricles or lobes. Peduncles often as long as the leaves, 1-flowered,
with minute bracts at a distance from the calyx. Sepals 2 to nearly 3 lines
long, ovate, obtuse or almost acute. Corolla pink or white, usually from ½
to ¾ in. long. Ovary and fruit completely 2-celled. Stigmatic lobes linear.
—R. Br. Prod. 482 ; Chois. in DC. Prod. ix. 412 ; Hook. f. Fl. Tasm. i.
275 ; *C. remotus,* R. Br. Prod. 483 ; Chois. in DC. Prod. ix. 412 (a small-
leaved form) ; *C. angustissimus,* R. Br. Prod. 482 (very narrow-leaved specimens) ; *C. adscendens,* De Vr. in Pl. Preiss. i. 346 ; *C. subpinnatifidus,* De
Vr. l. c. 347.

Queensland. Moreton Bay, *A. Cunningham,* and Nerkool Creek, *Bowman* (both
luxuriant specimens, with large leaves and sometimes 2 flowers on the peduncles) ; Rockhampton, *O'Shanesy ;* Curriwillighi, *Barton* (both the common form).

N. S. Wales. Port Jackson to the Blue Mountains, *R. Brown* and many others ;
northward to New England, *C. Stuart, C. Moore,* and Clarence river, *Beckler ;* in the interior to Bathurst Plains, *A. Cunningham,* and to the Barrier Range, *Victorian and other
Expeditions.*

Victoria. Very common in pastures, etc., from the Glenelg to Gipps' Land *Adamson,
F. Mueller,* and others ; Wimmera, *Dallachy.*

Tasmania. Near Risden Cove, *R. Brown* (very narrow-leaved specimens) ; abundant
in good soil, *J. D. Hooker.*

S. Australia. Spencer's Gulf, *R. Brown* (very small-leaved specimens) from the
Murray to St. Vincent's and Spencer's gulfs, *Behr, F. Mueller,* and others ; Lake Torrens,
F. Mueller ; in the northern interior, *M'Douall Stuart.*

W. Australia. Swan River, *Drummond,* 1st *Coll. n.* 652, 3rd *Coll. n.* 8ᵗ, 4th *Coll.
n.* 164, *Preiss, n.* 1924, 1925 ; Murchison river, *Oldfield.*

Among the more remarkable forms or varieties are one with very small flowers from the
Murray river, *F. Mueller,* and one with the leaves very densely tomentose and much-cut
and crisped and the peduncles very short from Cudnaka, *F. Mueller.* The species is also in
New Zealand, and appears to be the Australasian representative of the South African *C. has-
tatus* as well as of the *C. arvensis* of the northern hemisphere. Besides the synonyms given
above, and those quoted by Choisy, it should also probably include *C. acaulis,* Chois. in
DC. Prod. ix. 406, and *C. Preissii* and *C. Huegelii,* De Vr. in Pl. Preiss. i. 346, all referred here by F. Mueller, but of which I have seen no specimens.

2. **C. multivalvis,** *R. Br. Prod.* 483. A twiner, closely allied to
C. parviflorus, and considered by most authors as a variety, with the same
cordate entire leaves and cymose inflorescence, but densely clothed with a

soft close tomentum or velvety pubescence, the peduncles shorter and the capsule longer and much more distinctly splitting into about 8 valves.

N. Australia. Regent's River, N.W. coast, *A. Cunningham, Bynoe;* Glenelg district, *Martin.*

Queensland. Keppel Bay, *R. Brown;* Howick's group, *F. Mueller;* Nerkool Creek and Suttor river, *Bowman.*

Also on the S. coast of New Guinea.

3. **C. parviflorus,** *Vahl; Chois. in DC. Prod.* ix. 413. A tall twiner, glabrous or slightly pubescent. Leaves on rather long petioles, cordate-ovate, acuminate, entire, membranous, 2 to 3 in. long. Peduncles about as long as the leaves, bearing a dense cyme of numerous small flowers. Bracts minute. Sepals ovate, acuminate, pubescent, 2 to 2½ lines long. Corolla very open, above ½ in. diameter. Stamens rather long. Stigmatic lobes linear-oblong, recurved, much shorter than in *C. erubescens.* Capsule small, completely 2-celled, opening in 4 valves, which are sometimes split but much less so than in *C. multivalvis.—C. multivalvis,* var. β, R. Br. Prod. 483.

N. Australia. Islands of the Gulf of Carpentaria, *R. Brown, Henne;* Escape Cliffs, *Hulls;* Port Essington, *Armstrong.*

Queensland. Rockingham Bay, *Dallachy;* Rockhampton, *Thozet, Dallachy,* and others.

The species is widely dispersed over E. India and the eastern Archipelago. The flowers are usually described as white, but are said to be blue by Choisy, l. c., and pink by F. Mueller (Fragm. vi. 99).

SECT. II. CALYSTEGIA.—Bracts or bracteoles 2, enlarged and enclosing . the calyx. Dissepiment of the ovary usually incomplete.

4. **C. marginatus,** *Spreng. Syst.* i. 603. A glabrous twiner. Leaves on rather long petioles, narrow-lanceolate or broad and triangular, hastate or sagittate, the basal auricles or lobes acute, diverging, and often lobed, the whole leaf usually 1½ to 2 in. long, but when luxuriant twice as large. Peduncles rarely exceeding the petiole. Bracts very broadly cordate-ovate, from scarcely longer than the calyx to twice as long. Sepals rarely above 3 lines long. Corolla about ¾ in. long. Ovary very imperfectly 2-celled. Stigmatic lobes ovate, obtuse. Capsule globular, 1-celled. Seeds 4.—*Calystegia marginata,* R. Br. Prod. 483; Chois. in DC. Prod. ix. 434; Hook. f. Fl. N. Zeal. t. 48.

Queensland. Brisbane river, Moreton Bay, *F. Mueller.*

N. S. Wales. Port Jackson to the Blue Mountains, *R. Brown* and others; New England, *C. Stuart;* Clarence and Macleay rivers, *Beckler.*

Victoria. Snowy and Broadribb rivers, *F. Mueller.*

The species is also in Norfolk Island and in New Zealand. Some specimens have the aspect of some varieties of *C. erubescens,* but are at once distinguished by the large bracts.

5. **C. sepium,** *Linn. Sp. Pl.* 218. A tall rather slender herbaceous twiner, quite glabrous or very slightly pubescent, with a creeping perennial rootstock. Leaves from broadly ovate-triangular to lanceolate-hastate, acutely acuminate, cordate with angular basal auricles or lobes, mostly 2 to 4 in. long but sometimes larger. Peduncles longer than the petioles and often as long as the leaves, bearing a single large flower of a pure white or more or less

tinged with pink. Bracts large, ovate or ovate-lanceolate, leafy, acute or scarcely obtuse, longer than the calyx and enclosing it. Sepals ovate-lanceolate or lanceolate-acuminate, rather unequal, 4 to 5 lines long. Corolla 2 to 3 in. long. Ovary incompletely 2-celled, surrounded by a cup-shaped disk. Stigmatic lobes ovate or oblong,, obtuse. Capsule 1-celled.—*Calystegia sepium*, R. Br. Prod. 483; Chois. in DC. Prod. ix. 433; De Vr. in Il. Preiss. i. 345; Hook. f. Fl. Tasm. i. 276.

N. S. Wales. Port Jackson to the Blue Mountains, *R. Brown.*

Victoria. Wilson's Promontory and Merriman's Creek, *F. Mueller; Emu Creek, Whan; Wendu Vale along rivers and springs, Robertson;* Melbourne, *Adamson.*

Tasmania. Port Dalrymple, *R. Brown;* abundant in several parts of the island, *J. D. Hooker.*

W. Australia. Swan River, *Drummond, n.* 219; Port Leschenault, *Preiss, n.* 1926; Tone river, *Maxwell;* Murchison river, *Oldfield.*

The species is abundant in the temperate and subtropical regions of the northern hemisphere, and is also in New Zealand.. When on the seacoast the lower leaves are sometimes thicker, shorter, and more obtuse, but appear to me to be always very different from those of *C. Soldanella.*

6. **C. Soldanella,** *Linn. Sp. Pl.* 226. A glabrous perennial with a creeping rootstock and prostrate trailing or shortly twining stems. Leaves on rather long petioles, broadly rounded-cordate or kidney-shaped, entire or angular-lobed, rather thick, mostly about 1 in. but sometimes 2 in. diameter. Peduncles 1-flowered, about as long as the leaves. Bracts broadly ovate-cordate, very obtuse, rather shorter than the calyx. Sepals nearly ½ in. long, broad and thin, all very obtuse or the inner ones almost acute. Corolla pink or purplish, rather smaller than in *C. sepium.* Ovary incompletely 2-celled, surrounded by a cup-shaped disk. Stigmatic lobes ovate or oblong, usually narrower than in *C. sepium,* much shorter and broader than in *C. erubescens.* Capsule 1-celled.—*Calystegia Soldanella,* R. Br.; Chois. in DC. Prod. ix. 433; Hook. f Fl. Tasm. i. 276; *C. reniformis,* R. Br. Prod. 484.

N. S. Wales. Seacoast, Port Jackson, *R. Brown, Sieber, Woolls;* Illawarra, *A. Cunningham.*

Victoria. Wilson's Promontory, *F. Mueller* (the specimens not in flower, and therefore somewhat doubtful).

Tasmania. Seashore, Circular Head, *Gunn.*

The species is common on the extratropical seacoasts in both the northern and southern hemispheres both of the New and the Old World, including New Zealand, where it varies much more than in the northern hemisphere. F. Mueller (Fragm. vi. 100) reduces it to *C. sepium,* but those who are familiar with the two species, at least in our northern hemisphere, will scarcely agree to the union of forms so constantly distinct.

4. POLYMERIA, R. Br.

Corolla very broadly campanulate, entire or angular. Ovary 2-celled, with 1 ovule in each cell. Style filiform, with several (4 to 8) or very rarely only 2 linear stigmatic lobes. Fruit a dry capsule with 1 or 2 seeds.—Erect prostrate or trailing herbs, rarely twining. Leaves usually entire. Peduncles axillary, bearing 1 to 3 flowers. Bracts very small.

The genus is limited to Australia. Closely allied to *Convolvulus* in habit as well as in character, it differs in the ovules reduced to 2 (1 only in each cell of the ovary), whilst the stigmatic lobes, in all the species except *P. distigma,* are increased in number, probably by their division. Several of the species here enumerated run much one into the other, and they might all well be reduced to two or three.

Leaves linear or lanceolate.
  Leaves glabrous above, fringed with silky hairs. Flowers nearly 1 in.
    long, the sepals very unequal . . . .   1. *P. marginata.*
  Leaves silky or hoary or glabrous on both sides. Sepals nearly equal.
    Stigmatic lobes 6 to 8.
      Flowers about ¾ in. long. Stems usually erect . . . . .   2. *P. longifolia.*
      Flowers about ¼ in. long. Stems usually diffuse . . . . .   3. *P. angusta.*
    Stigmatic lobes 2. Stems erect. Peduncles 2-flowered . . .   4. *P. distigma.*
Leaves cordate, ovate or oblong.
  Outer sepals orbicular-cordate, inner ones narrow . . . . . .   5. *P. calycina.*
  Sepals nearly equal.
    Stem and leaves usually villous or pubescent. Sepals about 3 lines
    long . . . .   6. *P. ambigua.*
    Stem slender. Leaves small or linear, nearly glabrous. Sepals
    about 2 lines long . . . .   7. *P. pusilla.*

1. **P. marginata,** *Benth.* Stems erect, under 1 ft. high, loosely hirsute. Lower leaves petiolate, oblong, obtuse, deeply cordate, under 2 in. long, upper ones nearly sessile, lanceolate or linear-lanceolate, acute, slightly cordate, 2 to 4 in. long, all glabrous on both sides, except a few hairs on the veins underneath, but the margins elegantly fringed with rather long hairs. Peduncles shorter than the leaves, 1-flowered, with linear bracts above the middle. Outer sepals broadly lanceolate, acute, 5 or even 6 lines long, the 2 innermost smaller and much narrower. Corolla nearly 1 in. long. Stigmatic lobes about 8.

Queensland. In the interior, *Mitchell.* This may possibly prove to be a very marked variety of *P. longifolia,* notwithstanding the differences in the indumentum and calyx and the large flowers.

2. **P. longifolia,** *Lindl. in Mitch. Trop. Austr.* 398. Stems from a perennial stock erect, slightly branched, usually about 1 ft. high or shorter, pubescent or villous as well as the foliage with appressed silky hairs. Leaves almost sessile, linear or linear-lanceolate, mucronate-acute, minutely hastate at the base, often above 2 in. long. Peduncles 1-flowered, shorter than the leaves. Sepals oval-oblong, more or less acuminate or acute, about 3 lines long, all nearly equal. Corolla pink, usually about ¾ in. long, but sometimes smaller. Stigmatic lobes usually 6, but sometimes 7 or 8.

Queensland. Near the Gwydir, *Mitchell;* plains of the Condamine, *Leichhardt;* Suttor, Isaacs, Bowen rivers, etc., *Bowman;* Flinders river, *Sutherland;* Armadillo, *Barton.*

N. S. Wales ? Between Darling river and Cooper's Creek, *Neilson* (referable perhaps to *P. angusta*).

The Queensland specimens include some with remarkably narrow-linear leaves and rather smaller flowers.

3. **P. angusta,** *F. Muell. Fragm.* vi. 100 (*partly*). A perennial apparently diffuse or prostrate, softly and densely silky-hairy, otherwise very near *P. longifolia* and perhaps a variety. Leaves mostly nearly sessile and lanceolate or linear, the lower ones more distinctly petiolate and cordate at the base, all silky on both sides. Flowers much smaller than in *P. longifolia.* Sepals

lanceolate, acute, $2\frac{1}{2}$ to 3 lines long, slightly unequal. Corolla apparently scarcely $\frac{1}{4}$ in. long.   Style-branches 6 to 8.

N. Australia. Sturt's Creek, *F. Mueller;* and possibly also in **N. S. Wales,** between Darling river and Cooper's Creek, *Neilson.* F. Mueller included *P. longifolia* under his *P. angusta*, having accidentally overlooked Lindley's older name.   I have ventured to retain F. Mueller's name for the N. Australian form, which at present appears to me distinct, although it is not unlikely that further specimens may show that it is a variety only. F. Mueller describes the ovary as 1-celled.   In the flowers examined, I have always found a dissepiment between the 2 ovules in this as in all other species of the genus.

4. **P. distigma,** *Benth.* Stems erect, hoary-tomentose, with the stature and general aspect of some specimens of *P. longifolia.*   Leaves linear, entire, narrowed into a short petiole, glabrous or nearly so.   Peduncles slender, shorter than the leaves, mostly 2-flowered.   Sepals ovate-lanceolate, acuminate-acute, nearly equal, about 3 lines long.   Corolla fully $\frac{3}{4}$ in. long. Stigmatic lobes 2, linear-cuneate, obtuse.

N. Australia. Glenelg district, N.W. coast, *Martin.* This is evidently allied to *P. longifolia*, and has the 2-ovulate ovary of the genus; but the style, as observed by F. Mueller as well as by myself, is that of *Convolvulus.*

5. **P. calycina,** *R. Br. Prod.* 488. A glabrous or slightly pubescent annual (or sometimes with a perennial creeping rootstock?).   Stems slender, prostrate or creeping.   Leaves on slender petioles, the lower ones ovate, obtuse or emarginate, deeply cordate, under 1 in. long, the upper ones oblong linear or lanceolate, obtuse, slightly cordate or rarely hastate at the base, often above 1 in. long.   Peduncles slender, shorter than the leaves, 1-flowered, with minute bracts at or below the middle.   Outer sepals very broadly ovate or cordate, about 3 lines long, the inner ones shorter, ovate-lanceolate or lanceolate, acuminate.   Corolla 5 to 6 lines long, broadly campanulate, slightly silky-pubescent outside.   Anthers rather long.   Stigmatic lobes about 6.   Capsule shorter than the calyx.   Seeds pubescent or silky-villous in the specimens seen, glabrous according to R. Brown.—Chois. in DC. Prod. ix. 432; Endl. Iconogr. t. 67.

Queensland. Keppel Bay, *R. Brown, Thozet;* Rockhampton, *O'Shanesy;* Gracemere, *Bowman;* Moreton Bay, *C. Stuart.*
N. S. Wales. Port Jackson to the Blue Mountains, *R. Brown* and others; Richmond river, *Fawcett.*
Var.? *mollis.* The whole plant softly pubescent, the characters otherwise the same as in the Eastern form.
W. Australia. Port Walcott, *C. Harper (Herb. F. Muell.).*
Some of the narrow-leaved Eastern specimens might very well, without close examination, be mistaken for some varieties of *Convolvulus erubescens.* The outer sepals, although variable in breadth, are, however, always broader than in the latter species, independently of the generic character.

6. **P. ambigua,** *R. Br. Prod.* 488. An annual (or sometimes perennial?), with long, slender, creeping or trailing stems, occasionally rooting at the lower nodes and sometimes shortly twining at the extremities.   Leaves petiolate, ovate or oblong, obtuse, often mucronate, cordate at the base, usually rugose, glabrous, sparingly pubescent or rarely villous above, more or less villous or silky-hairy underneath, mostly about 1 in. long, but variable in size.   Pe-

duncles usually longer than the petioles, bearing 1 to 3 flowers with minute bracts at the base of the pedicels, and usually 2 small bracteoles on the pedicels. Sepals nearly equal, acuminate, about 3 lines long. Corolla not twice as long, very open. Stigmatic branches 4 to 6. Capsule nearly as long as the calyx. Seeds glabrous or very minutely hoary-pubescent.—Chois. in DC. Prod. ix. 432.

N. Australia. Islands of the Gulf of Carpentaria, *R. Brown;* Victoria river and Sturt's Creek, *F. Mueller.*

Queensland. Thirsty Sound, *R. Brown.*

P. lanata, R. Br. l. c. and Choisy, l. c., from the same N. Australian localities, appears to be a very densely silky-villous variety, the leaves much smaller, broad in R. Brown's specimens, narrow in F. Mueller's, the flowers rather smaller, mostly solitary.

P. quadrivalvis, R. Br. l. c. and Chois. l. c. (the above-quoted specimens from Thirsty Sound, *R. Brown*), appears to me to be another variety, nearly glabrous, with rather small solitary flowers, connecting *P. ambigua* with *P. pusilla.*

7. **P. pusilla,** *R. Br. Prod.* 488. This may be another small slender form of *P. ambigua,* glabrous or slightly pubescent. Stems almost filiform, prostrate or twining. Leaves on slender petioles, from cordate-ovate obtuse or retuse and under 1 in. long, to linear and then entire or hastate at the base. Peduncles 1-flowered, with minute bracts at a distance from the flower. Sepals nearly equal, lanceolate, acuminate, about 2 lines long. Corolla about twice as long, very open. Stigmatic branches usually 4.—Chois. in DC. Prod. ix. 432.

Queensland. Broad Sound, *R. Brown,* also from *Bowman's* collection, with short ovate leaves as in Brown's specimens; Rockhampton, *O'Shanesy,* and Keppel Bay and Fitzroy river, *Thozet,* with linear leaves. Possibly the small-flowered nearly glabrous variety referred above to *P. ambigua* (*P. quadrivalvis,* R. Br.) may be rather a form of *P. pusilla.*

5. **PORANA,** Linn.

(Duperreya, *Gaudich.*)

Sepals much enlarged after flowering, and horizontally spreading under the fruit. Corolla campanulate or tubular-campanulate, angular or 5-lobed, folded in the bud. Ovary 1-celled, with 2 or 4 ovules; style entire or bifid; stigma globular, single or 1 on each branch. Capsule usually 1-seeded by abortion, indehiscent (or 2-valved?).—Tall twiners, often woody at the base. Leaves entire. Flowers usually small, axillary and solitary in the Australian species, paniculate in the Indian ones. Bracts small.

The genus extends over tropical Africa and Asia to the Indian Archipelago. The only Australian species is endemic and extratropical; and, although possessing the essential characters of the Asiatic ones, is very different in foliage and inflorescence. The remarkable fruiting calyx distinguishes this from all other Convolvulaceous genera.

1. **P. sericea,** *F. Muell. Fragm.* vi. 100. A tall but slender twiner, shrubby at the base, the branches and foliage silky-pubescent. Leaves very shortly petiolate, linear or linear-lanceolate, obtuse or scarcely acute, 1 to 1½ in. long. Peduncles axillary, 1-flowered, shorter than the leaves, with 2 small bracts or bracteoles close under the calyx. Sepals broadly ovate, obtuse, silky-pubescent, 2 to 2½ lines long at the time of flowering. Corolla very open, not twice as long as the calyx, said by some to be blue, by others

pale pink. Ovary 1-celled, with 2 ovules ; style undivided, with a large globular stigma. Fruiting sepals broadly ovate, rigidly scarious, elegantly veined, ¾ in. long or even more. Capsule small, ovoid or oblong, membranous, indehiscent, 1-seeded.—*Duperreya sericea,* Gaudich. in Freyc. Voy. Bot. 452. t. 63 ; Chois. in DC. Prod. ix. 436 ; *Ipomœa modesta,* F. Muell. Fragm. ii. 22.

W. Australia. Between Moore and Murchison rivers, *Drummond, 6th Coll. n.* 223 ; Blackwood and Murchison rivers, *Oldfield.*

6. BREWERIA, R. Br.

(Prevostea, *Chois.;* Seddera, *Hochst. and Steud.;* Stylisma, *Nutt.*)

Corolla campanulate, angular or shortly and broadly 5-lobed, folded in the bud. Ovary 2-celled, with 2 ovules in each cell. Style bifid or divided to the base, with a capitate stigma to each branch. Fruit a dry capsule.—Herbs or undershrubs, with erect prostrate trailing or twining stems, often tomentose or silky. Leaves usually entire. Flowers axillary, solitary or rarely 2 or 3 together in the Australian species, the upper ones often forming a leafy spike, or in extra-Australian species the peduncles often several flowered.

The genus, as at present constituted, includes several species from tropical Asia, Africa, and America, but the Australian ones appear to be all endemic. The habit is often that of some species of *Convolvulus,* but the style is very different. A. Gray (Proc. Amer. Acad. v. 336), relying only upon the characters given by R. Brown and by Choisy, thought that *Breweria* could not be separated from *Bonamia,* Thou. ; the study, however, of Thouars' detailed description and figure (in the absence of authentic specimens) shows that the latter has neither the inflorescence nor probably the æstivation of the corolla nor the fruit of *Convolvulaceæ,* but appears to be more closely allied to *Ehretia* and *Cordia,* with the style of the former and the embryo of the latter. A. Gray appears, however, to be quite right in uniting *Stylisma,* Nutt., with *Breweria;* nor can I distinguish the African *Seddera,* Hochst., nor yet the tropical American *Dufourea,* H. B. and K., or *Prevostea,* Chois., for the sole character given for the latter, the enlarged outer sepals, occurs in a very marked way in *B. pannosa,* and to a slight degree in several other species.

Silky-pubescent or shortly hirsute. Bracteoles minute.
Leaves linear or lanceolate . . . . 1. *B. linearis.*
Lower leaves somewhat cordate, upper ones lanceolate . . . . . 2. *B. media.*
All the leaves regularly heart-shaped . . . . 3. *B. brevifolia.*
Densely rusty-tomentose or villous with long hairs. Bracteoles at least
    as long as the calyx.
Sepals slightly unequal. Corolla pink, about 1 in. long . . . . 4. *B. rosea.*
Outer sepals much larger than the inner. Corolla (blue) under ¾ in.
    long . . . . 5. *B. pannosa.*

1. B. linearis, *R. Br. Prod.* 488. Silky-pubescent or hirsute. Stems prostrate or shortly twining. Leaves very shortly petiolate, oblong-linear or narrow-lanceolate, mostly obtuse and about 1 in. long. Peduncles 1-flowered, about half as long as the leaves or sometimes very short, with minute bracts at the base sometimes scarcely perceptible. Sepals lanceolate, acuminate, about 3 lines long, the innermost rather smaller. Corolla apparently small, but not seen perfect. Style in the specimens examined divided to about the middle.—Chois. in DC. Prod. ix. 439.

N. Australia. Islands of the Gulf of Carpentaria, *R. Brown;* Upper Victoria river, *F. Mueller;* mainland, Carpentaria, *Lansborough.*

2. **B. media,** *R. Br. Prod.* 488.   Pubescent or somewhat silky-hairy. Stems prostrate. Leaves shortly petiolate, the lower ones ovate-oblong or ovate-lanceolate, obtuse or acute and often somewhat cordate at the base, the upper ones lanceolate, acute, rarely above 1 in. long.   Peduncles 1-flowered, short, with small bracts at the base.   Sepals ovate-lanceolate, subulate-acuminate, slightly unequal, 2 to 3 lines long.   Corolla (white ?) under ½ in. long.   Ovary hirsute at the top with long hairs.   Styles cohering to the middle, but readily separable to the base.—Chois. in DC. Prod. ix. 438.

N. Australia.   S. Arnhem Bay, *R. Brown ;* Victoria river, *F. Mueller.*
Queensland.   Bowen river, *Bowman.*
N. S. Wales.   Between Darling river and Cooper's Creek, *Neilson.*

Var. *? parviflora.* Stems very slender. Leaves more cordate. Flowers smaller.—Victoria river, *F. Mueller.*

Var. *? villosa.* Much more villous. Flowers small.—Victoria river, *F. Mueller* (the specimens very imperfect).

The circumscription of this species and of *B. linearis* and *B. brevifolia* may require considerable modification when a more complete series of specimens shall be obtained.

3. **B. brevifolia,** *Benth.*   A perennial, with a thick and hard stock and long, slender, prostrate, pubescent stems.   Leaves on short petioles, ovate-cordate, acute or the lower ones rounded and more obtuse, rarely above ½ in. long, glabrous above, more or less hairy underneath.   Peduncles 1-flowered, shorter than the calyx, with minute bracts usually at the base.   Sepals lanceolate, acutely-acuminate, hirsute, about 2½ lines long, slightly unequal.   Corolla about twice as long as the calyx.   Ovary hirsute with long hairs.   Style divided to about the middle.

N. Australia.   Port Essington, *Armstrong.*

4. **B. rosea,** *F. Muell. Fragm.* i. 233.   An undershrub or shrub of 1 to 2 or even 3 ft., densely tomentose or hirsute with ferruginous hairs, especially on the upper leaves and calyxes.   Leaves nearly or quite sessile, ovate obovate or orbicular, mostly obtuse, thick and soft as in *B. pannosa.*   Flowers pink (*Oldfield*), solitary in the axils, nearly sessile, larger than in *B. pannosa*, the upper one forming a leafy terminal spike with the uppermost floral leaves very small.   Sepals lanceolate, 3 to 4 lines long, the inner ones rather narrower than the outer.   Corolla-tube broad but almost cylindrical, fully ½ in. long, the limb broad, spreading to ¾ in. diameter.   Ovary hirsute with long hairs.   Styles free from the base.

N. Australia.   Hammersley Range, N.W. Coast, *M. Brown.*
W. Australia.   Murchison river, *Oldfield.*

5. **B. pannosa,** *R. Br. Prod.* 488.   Stems from a perennial stock, prostrate or twining, the whole plant densely hirsute with soft ferruginous or silky hairs.   Leaves on very short petioles, ovate and acute or the lower ones orbicular and obtuse, thick and soft, under 1 in. long.   Flowers blue (*R. Brown*), solitary in the axils or rarely 2 or 3 together, on short pedicels, with a pair of linear bracts about the middle.   Sepals very hirsute like the rest of the plant, the outer ones broadly ovate, acuminate, 4 to 5 lines long, the inner ones much smaller, and the innermost one linear-lanceolate.   Co-

rolla above ½ in. long, hairy outside. Ovary hirsute at the top with long hairs. Style divided to about the middle.—Chois. in DC. Prod. ix. 438.

N. Australia. Islands of the Gulf of Carpentaria, *R. Brown;* Victoria river, *F. Mueller* (very densely hirsute); Port Essington, *Armstrong* (the hairs rather shorter).

7. CRESSA, Linn.

Corolla tubular-campanulate; lobes 5, contorted (or otherwise imbricate?) in the bud, not plicate. Ovary 2-celled, with 2 ovules in each cell; styles 2, distinct from the base, each with a capitate stigma. Capsule usually 2-valved and 1-seeded by abortion.—A small branching perennial. Leaves entire. Flowers small, in terminal leafy spikes or heads.

The genus is limited to a single species, common to the warmer regions of the New as well as the Old World.

1. **C. cretica,** *Linn.; Chois. in DC. Prod.* ix. 440. An erect or diffuse, much-branched perennial, sometimes almost woody at the base, rarely exceeding 6 in., hoary silky-pubescent or villous all over. Leaves sessile or the lower ones shortly petiolate, ovate-lanceolate, or in specimens not Australian linear, entire, rarely exceeding ¼ in. Flowers sessile in terminal leafy spikes or heads, rarely reduced to a single flower. Sepals broadly obovate, very obtuse, ciliate, about 2 lines long. Corolla very shortly exceeding the calyx, hairy outside. Anthers large, oblong. Ovary villous. Capsule ovoid, exceeding the calyx, rarely ripening more than one smooth seed. — *C. australis,* R. Br. Prod. 490.

N. Australia. Islands of the Gulf of Carpentaria, *R. Brown;* mouth of the Victoria river, *F. Mueller;* Albert river, *Henne.*
Queensland. Broad Sound, *R. Brown;* sandy flats, Port Denison, *W. Hill.*
N. S. Wales. Murray and Darling deserts, *Victorian Expedition.*
S. Australia. Subsaline pastures, Murray river to St. Vincent's Gulf, *F. Mueller.*
W. Australia, *Drummond, n.* 131.

The species is abundantly spread over sandy maritime or saline districts in the warmer regions of the Old and New World, extending to the Mediterranean region of Europe. The flowers in some of the tropical Australian specimens are larger than usual, but not otherwise different.

8. EVOLVULUS, Linn.

Corolla campanulate or tubular at the base, the limb 5-angled or 5-lobed. Ovary 2-celled, with 2 ovules in each cell. Styles 2, filiform, distinct from the base, each divided into 2 branches; stigmas linear, terminating each branch. Fruit a capsule, with 4 seeds or fewer by abortion.—Herbs, not twining, annual or with a short perennial stock. Leaves entire, usually small. Flowers small, on axillary peduncles or in terminal spikes or racemes.

A considerable tropical American genus, of which one or two species are spread also over the warmer regions of the Old World. The only Australian species is the one most common over the whole area.

1. **E. alsinoides,** *Linn.; Chois. in DC. Prod.* ix. 447. A perennial, with a short almost woody stock, but often flowering the first year so as to appear annual, with numerous slender prostrate or erect stems, 6 in. to 1 ft. long, the whole plant more or less silky-hairy. Leaves usually oblong or lanceolate, ses-

sile or nearly so, 3 to 6 lines long, but varying from ovate to almost linear, obtuse or acute. Flowers small, pale blue or white, 1 to 3 together on slender axillary peduncles mostly longer than the leaves, but the lower ones sometimes shorter and the upper ones often long and filiform, forming a loose terminal leafy raceme or narrow panicle. Bracts small under each pedicel. Sepals narrow, acute. Corolla pale blue and white or entirely of one of these colours, very open or almost rotate, about 3 lines diameter.—R. Br. Prod. 489 ; *E. linifolius,* Linn.; R. Br. Prod. 489 ; Chois. in DC. Prod. ix. 449 ; *E. decumbens,* R. Br. Prod. 489 ; *E. villosus,* R. Br. Prod. 489, but perhaps not of Ruiz and Pav.; *E. heterophyllus,* Labill. Sert. Austr. Caled. t. 29 ; Chois. in DC. Prod. ix. 449, and probably some others enumerated by Choisy; *E. pilosus,* Roxb. Fl. Ind. ii. 106.

N. Australia. Islands of the Gulf of Carpentaria, *R. Brown ;* N.W. coast, *Bynoe, Gregory's Expedition,* etc.; Victoria river and Arnhem's Land, *F. Mueller ;* Port Essington, *Armstrong, A. Cunningham ;* in the interior, *M'Douall Stuart's Expedition.*
Queensland. Abundant along the whole coast, *R. Brown* and many others ; and in the interior, *Mitchell, Bowen,* and others.
N. S. Wales. Clarence river, *Beckler ;* New England, *C. Stuart ;* from Darling river to Cooper's Creek, *Victorian and other Expeditions.*
S. Australia. Cooper's Creek, *Howitt's Expedition.*
W. Australia. Port Walcott, *C. Harper* (the following variety only).

Var. *sericeus.* Leaves thicker and very white, with long silky hairs.—*E. argenteus,* R. Br. Prod. 489, not of Pursh.—Islands of the Gulf of Carpentaria, *R. Brown ;* Port Walcott, *C. Harper.*

Linnæus originally distinguished the broad, obtuse-leaved form as an Asiatic, and the narrow acute-leaved as an American species, and since both have been found to be abundant in both the New and the Old World, the distinction has been kept up by Choisy, Grisebach, and others ; but the two run so much one into the other that it has appeared to me impossible to separate them in any long series of specimens. Many have the lower leaves of the one and the upper ones of the other, and often the difference appears to arise from soil and station. I had accordingly in the ' Niger Flora,' as well as in the ' Flora Hongkongensis,' proposed to unite the two under the name of *E. alsinoides ;* F. Mueller also unites them, but prefers the other Linnæan name of *E. linifolius.* Amongst other supposed Asiatic species, *E. angustifolius,* Roxb. Fl. Ind. ii. 107, and *E. gracillimus,* Miq. Fl. Ind. Bat. ii. 629, appears to be a very narrow-leaved state which also occurs in Australia ; the African synonyms have been already given by Choisy, and to these ought probably to be added several American ones requiring further investigation. *E. villosus,* Ruiz and Pavon, appears however to differ in its flowers much larger than in Brown's specimens.

9. DICHONDRA, Forst.

Corolla campanulate, deeply 5-lobed. Ovary of 2 distinct carpels, each with an almost basal style and 1 or 2 ovules ; stigmas capitate. Fruit of 1 or 2 membranous capsules, each with 1 or rarely 2 seeds.—Prostrate creeping small herbs. Leaves entire. Flowers small, axillary.

Besides the Australian species, which is widely spread over the warmer regions of the New as well as the Old World, there is another closely allied to it from Central America.

1. **D. repens,** *Forst. ; Chois. in DC. Prod. ix. 451.* A slender creeping perennial, rooting at the nodes, usually hoary with a minute pubescence, often silky. Leaves on long petioles, orbicular or reniform, 4 to 8 lines or rarely 1 in. diameter. Flowers solitary, on peduncles shorter than the

petioles. Sepals obovate, scarcely 1 line long. Corolla rather shorter than the calyx, yellow. Carpels also shorter than or rarely as long as the calyx, nearly globular.—R. Br. Prod. 491; Hook. f. Fl. Tasm. i. 278; Sm. Ic. Ined. t. 8.

N. Australia. Port Essington, *Armstrong.*
Queensland. Moreton Bay, *F. Mueller;* Rockhampton, *O'Shanesy.*
N. S. Wales. Port Jackson, *R. Brown;* Blue Mountains, *Miss Atkinson.*
Victoria. Port Phillip, *R. Brown;* Wendu vale, *Robertson;* Yarra river, *F. Mueller, Adamson;* Ballarook forest, *Whan.*
Tasmania. Abundant in many parts of the island, *J. D. Hooker.*
S. Australia. Around St. Vincent's Gulf, *F. Mueller, Behr.*
W. Australia, Cape le Grand, *R. Brown;* Vasse river, *Oldfield;* also in *Drummond's* collections, n. 86 and 163.

The species is generally diffused over the tropical regions of both the New and the Old World, especially near the sea, extending northward to the southern United States and to China, and southward to the Cape of Good Hope, extratropical S. America and New Zealand.

10. WILSONIA, R. Br.

Sepals united in a tubular-campanulate shortly 5-lobed or 5-toothed calyx. Corolla with a slender tube and campanulate 5-lobed limb, the lobes imbricate in the bud but with induplicate margins. Ovary 2-celled or almost 1-celled with 1 erect ovule in each cell (or rarely 2 ?). Style divided into 2 filiform branches with capitate stigmas. Fruit a capsule with 1 or 2 seeds. —Prostrate much-branched perennials or undershrubs. Leaves entire, small. Flowers axillary, sessile or nearly so. Bracts none.

The genus is limited to Australia, and almost to the southern shores or to the saline tracts in the interior.

More or less hairy. Leaves lanceolate ovate or orbicular, rarely
　　above 2 lines long.
Silky-pubescent. Leaves very concave, usually distichously im-
　　bricate on the branchlets　. . . . .　1. *W. humilis.*
Loosely hairy. Leaves flat, not imbricate . . . .　2. *W. rotundifolia.*
Glabrous. Leaves linear, 3 to 6 lines long . . . .　3. *W. Backhousii.*

1. **W. humilis,** *R. Br. Prod.* 490. A prostrate much-branched undershrub or shrub, hoary all over with silky hairs, the stems spreading to from 6 in. to 1 ft. Leaves crowded, imbricate on the smaller branchlets and usually distichous, sessile or lanceolate, thick, very concave, from under 1 line to nearly 2 lines long. Flowers sessile and solitary. Calyx silky-hairy, nearly 2 lines long, the teeth or lobes shorter than the tube. Corolla-tube nearly as long as the calyx; lobes shorter than the tube, spreading. Anthers ovoid or oblong, scarcely exserted. Ovary 1-celled at the base, but often, if not always, 2-celled at the top. Capsule shorter than the calyx, membranous, usually 1-seeded.—Chois. in DC. Prod. ix. 450; Lehm. Pl. Preiss. ii. 237; Hook. f. Fl. Tasm. i. 277; F. Muell. Fragm. vi. 101; *Frankenia cymbifolia* (afterwards corrected to *W. humilis*), Hook. Ic. Pl. t. 265.

Victoria. Port Phillip, *R. Brown;* salt marshes near Melbourne, *Adamson* and others; Queenscliff, Station Peak, and salt plains near Mount Abrupt, *F. Mueller.*
Tasmania. Great Swan Port, *Backhouse.*

S. Australia. Port Adelaide, *Blandowsky* ; Spencer's Gulf, *Warburton.*

W. Australia. King George's Sound, *R. Brown, Harvey ;* Phillips and Fitzgerald Ranges, *Maxwell ;* Arthur's Head and Gordon river, *Preiss, n* 2391, 2392; also *Drummond, n.* 106, 138, 220.

Var. *spinescens,* F. Muell. More shrubby with divaricate branches, the smaller ones spinescent.—W. Australia, *Drummond, n.* 82.

2. **W. rotundifolia,** *Hook. Ic. Pl. t.* 410. Stems from a perennial stock, prostrate or diffuse and much-branched, but shorter than in *W. humilis,* and not woody, the whole plant more or less hirsute with rather long hairs, not silky, and sometimes nearly glabrous. Leaves rather crowded but not imbricate, orbicular or ovate, contracted at the base or almost petiolate, thick but flat, rarely exceeding 2 lines. Flowers of *W. humilis* or rather larger, the corolla-tube slender, sometimes but not always quite as long as the calyx, and the stamens and styles rather more exserted. Ovary either completely 2-celled or the dissepiment incomplete at the base. Ovules 1 or, according to Hooker, sometimes 2 in each cell. Seeds 1 or 2 in the capsule.—Chois. in DC. Prod. ix. 450 ; F. Muell. Fragm. vi. 101.

Victoria. Near Melbourne, *Adamson ;* Port Phillip, Station Peak, Lake Omeo and Murray river, *F. Mueller ;* Skipton and Warangau (salt) lake, *Whan ;* Wimmera, *Dallachy.*

S. Australia. Holdfast Bay and Kaiserstuhl, *F. Mueller.*

W. Australia. *Drummond, n.* 18, 657, *and* 335 (the latter specimens more glabrous).

3. **W. Backhousii,** *Hook f. in Hook. Lond. Journ.* vi. 275, *and Fl. Tasm.* i. 277. Stems from a perennial stock, prostrate or diffuse with short ascending branches, the whole plant quite glabrous and rarely extending to more than 6 in. Leaves linear or rarely narrow-oblong, acute or obtuse, flat but thick or almost terete, ¼ to ½ in. long. Calyx glabrous, from under 3 to nearly 4 lines long, the teeth much shorter than the tube. Corolla-tube more slender than in the two preceding species and usually exceeding the calyx. Stamens and style also more exserted than in either of the others, and the anthers narrower. Ovary 2-celled from the bottom.—F. Muell. Fragm. vi. 101.

Victoria. Near Melbourne, *Adamson ;* Port Phillip, Lake Wellington, and near Brighton, *F. Mueller ;* Wimmera, *Dallachy ;* also in *R. Brown's* collection, without any label, probably from Port Phillip.

Tasmania. Kelvedon, Great Swan Port, *Story, Backhouse.*

W. Australia. Middle Mount Barren, *Maxwell ;* Port Gregory, *Oldfield.*

11. CUSCUTA, Linn.

Sepals distinct or united in a 5-lobed rarely 4-lobed calyx. Corolla campanulate, ovoid or globular, with a short 5-lobed or rarely 4-lobed limb. Anthers usually nearly sessile, with a scale below each in the tube of the corolla. Ovary completely or partially 2-celled, with 2 ovules in each cell. Styles 2, distinct or more or less united ; stigmas capitate or acute. Fruit a dry or scarcely succulent capsule, opening transversely or bursting irregularly. Embryo spiral or curved round a fleshy albumen ; cotyledons inconspicuous.—Herbs, with leafless thread-like parasitical stems, bearing usually sessile clusters of small sessile or pedicellate flowers, white or pink.

A considerable genus, dispersed over all warm and temperate regions of the globe. Of the three Australian species, one has a very wide range both in the New and the Old World; another is limited to tropical Asia; the third appears to be endemic; but, notwithstanding the carefully-elaborated monographs of Engelmann, there is still much doubt as to the characters by which the species are to be distinguished. The Australian ones have all of them the sepals united at the base, and distinct styles with capitate stigmas. They have also the very slender filiform stems of *C. epithymum* and its allies, not the firmer ones of *C. monogyna* and others.

Flowers sessile or very shortly pedicellate in globular clusters.
　Calyx-lobes prominently keeled . . . . 1. *C. chinensis.*
　Calyx-lobes not keeled . . . . 2. *C. australis.*
Flowers on pedicels of 3 to 5 lines. Corolla campanulate . . . . 3. *C. tasmanica.*

1. **C. chinensis,** *Lam.; Engelm. in Trans. Acad. St. Louis,* i. 479. Flowers rather small, nearly globular, very shortly pedicellate in globular clusters, sometimes reduced to 2 or 3 flowers and not usually so dense as in *C. australis.* Calyx shorter than the corolla, divided to the middle or rather lower into obtuse lobes, the keels and sutures of the sepals forming 10 rather prominent ribs to the tube. Corolla 1 to 1½ lines long, the lobes rather obtuse. Scales of the tube deeply fringed or lobed. Styles distinct, unequal, rather slender, with capitate stigmas. Capsule bursting irregularly. —*C. carinata,* R. Br. Prod. 491.

Queensland. Bay of Inlets and Cape Grafton, *Banks and Solander (Herb. Mus. Brit.).* Apparently common in tropical Asia, extending from Madagascar and Ceylon to China.

2. **C. australis,** *R. Br. Prod.* 491. Flowers nearly globular, sessile or very shortly pedicellate, in globular clusters, sometimes reduced to two or three flowers, each about 1½ lines diameter, and all the parts minutely glandular-dotted. Calyx shorter than the corolla, divided to below the middle into obtuse lobes, without prominent ribs. Corolla-lobes very obtuse, at length recurved. Scales of the tube bifid or fringed, sometimes very small but often nearly as long as the tube. Ovary much depressed; styles distinct, rather thick, unequal, with capitate stigmas. Capsule depressed, with a broad rhomboidal area between the styles.—*C. obtusiflora,* H. B and K.; Engelm. in Trans. Acad. St. Louis, i. 491.

Queensland. Broad Sound, *R. Brown.*
N. S. Wales. M'Leay and Clarence rivers, *Beckler;* Bent's Basin, *Woolls.*
Victoria. Snowy, Goulburn, and King rivers, *F. Mueller.*

The species is widely dispersed over the warmer parts of America and Asia, extending northwards to the southern United States and to S. Europe.

3. **C. tasmanica,** *Engelm. in Trans. Acad. St. Louis,* i. 512. Pedicels clustered, much longer than the flowers, usually about 4 or even 5 lines long, with minute bracts at their base. Calyx much shorter than the corolla, deeply divided into obtuse lobes, not prominently ribbed. Corolla campanulate, not contracted at the throat, above 1½ lines long, the lobes obtuse, as long as the tube. Scales of the tube large, deeply fringed. Styles distinct, rather long, nearly equal, with large obscurely-lobed capitate stigmas. Capsule short. Seeds with a small hilum.—*C. australis,* Hook. f. Fl. Tasm. i. 278, not of R. Br.

Victoria. Port Phillip, *F. Mueller,* who believes it to be introduced.

Tasmania. Near Hobarton and at George Town, *Gunn.*

This appears to me as to Engelmann a very distinct species, but the sectional character derived by Engelmann from the deeply-lobed concave stigma seems to have been an exceptionally abnormal state in the flower examined. In other flowers I find the stigma larger than in most species, but not very distinctly lobed.

Order LXXXII. SOLANEÆ.

Flowers regular or nearly so. Calyx free, usually with 5, rarely with 4, 6 or 10 teeth lobes or segments. Corolla with 5 or rarely with 4 teeth or lobes, induplicate-plicate or rarely imbricate in the bud. Stamens as many as lobes of the corolla and alternate with them; anthers various, usually 2-celled. Ovary superior, 2-celled or rarely spuriously 4-celled or abnormally 3- or more-celled; style simple, terminal, with an entire or lobed stigma. Fruit an indehiscent berry or rarely a capsule, with several seeds. Embryo usually curved or spiral, surrounding a fleshy albumen, rarely straight in the centre of the albumen.—Herbs shrubs or soft-wooded trees. Leaves alternate, without stipules. Flowers solitary or in centrifugal cymes or unilateral racemes, usually at first terminal but becoming lateral by the elongation of the shoot, rarely axillary, the cymes or racemes usually without bracts, and no bracteoles on the pedicels.

A numerous Order in the tropical and warmer regions of the globe, and more especially S. America, with a comparatively few species straying into more temperate districts both in the northern and the southern hemisphere. Of the seven genera here enumerated, four have nearly the range of the Order; one is a tropical weed spread from America; one has a single Australian representative of an otherwise S. American genus; and one only is endemic. The Order is closely connected with Scrophularineæ, being technically separated by the more regular flowers with the stamens and corolla isomerous. On the other hand, it is allied to Hydrophyllaceæ through *Hydrolea,* which differs chiefly in its divided style and small embryo; and yet nearer to Polemoniaceæ, an Order scarcely distinguished from Solaneæ except by the almost constant tricarpellary ovary and contorted æstivation of the corolla; it is unrepresented in Australia, excepting occasionally by a N.W. American *Collomia* or other annual escaped from a garden.

Fruit an indehiscent berry.
  Corolla rotate or campanulate, folded in the bud.
    Calyx, if enlarged after flowering, not inflated. Corolla rotate
      or very open. Anthers opening in terminal pores or slits . 1. Solanum.
    Calyx inflated over the fruit. Corolla campanulate. Anthers
      opening in longitudinal slits.
      Calyx 5-parted, cordate at the base . . . . 2. Nicandra.
      Calyx shortly 5-lobed . . . . 3. Physalis.
    Corolla contracted into a tube at the base, the lobes imbricate in
      the bud . . . . 4. Lycium.
Fruit capsular, opening in valves. Corolla folded in the bud or with
  induplicate lobes.
    Corolla (small) broadly campanulate. Anthers 1-celled . . . 5. Anthotroche.
  Corolla with a cylindrical or funnel-shaped tube.
    Calyx tubular, circumsciss after flowering, leaving a broad per-
      sistent base. Corolla large. Capsule prickly . . . . . 6. Datura.
    Calyx entirely persistent. Capsule smooth . . . . 7. Nicotiana.

1. SOLANUM, Linn.

Calyx with 5, rarely with 4 or more than 5 teeth or lobes. Corolla rotate or very broadly campanulate, with 5 or rarely 4 angles or lobes, folded in the bud. Filaments usually very short, rarely as long as the anthers; anthers oblong or linear, erect and connivent, either parallel or more frequently tapering upwards and forming a cone round the style, opening at the top in pores or transverse slits, rarely continued down the sides of the anthers, without any prominent connectivum between the cells. Fruit a berry, usually 2-celled rarely 4-celled (the cells divided by a spurious dissepiment) or in species or varieties not Australian several-celled. Seeds several, flattened, with a curved or spiral embryo surrounding a fleshy albumen.— Herbs shrubs or rarely low soft-wooded trees, either unarmed or with prickles scattered on the branches, on the principal veins of the leaves, especially on the upper surface and in some species also on the inflorescence and calyxes, straight and slender in most Australian species, stout and recurved in some others. Leaves alternate, but often in pairs, a smaller one being developed in the axil of the larger one, entire or irregularly toothed lobed or divided. Flowers normally in terminal centrifugal cymes; but, owing to the rapid development of the branch, the inflorescence becomes usually lateral and very often, by the abortion of one branch, reduced to a simple unilateral apparently centripetal raceme or to a single flower. Corolla usually blue purplish or white or in species not Australian yellow, always tomentose outside in the species where the tomentum is stellate, but usually only on the pars exposed in the bud, with the induplicate margins glabrous. Style frequently curved to one side, the stigma slightly dilated, entire or 2-lobed.

A very large genus, spread over the warmer and temperate regions of the globe, but most abundant in tropical America. Besides the introduced species, there are forty-eight described below, of which one is a common weed over the whole range of the Order; another is spread over the tropical regions of the Old as well as the New World; one extends only to New Zealand; another to Timor; and a third only to the islands of the South Pacific; the remaining forty-three are endemic, belonging chiefly to groups sparingly or not at all represented in S. America; and *S. indicum* and other species with short stout prickles, so common in tropical Asia, have not as yet been detected in Australia.

The distinction and determination of the numerous species of this genus (most extravagantly multiplied by Dunal in the 'Prodromus') is attended with peculiar difficulties, the chief characters being derived from the very variable ones of foliage, armature and indumentum. The sections proposed by Sendtner, Dunal, and others break down in several instances, and are scarcely applicable to the Australian species. The three first here enumerated have a marked difference in the anthers, but there are extra-Australian intermediates; the differences in the form of the corolla, often very difficult to ascertain from dried specimens, are seldom in relation to other characters; and the form and colour of the fruit varies in a remarkable degree in some individual species. If, therefore, in the following key I have founded the principal groups or series chiefly upon indumentum and armature, it is not that I regard them as good sections, but only because I have as yet found no better way of leading to the determination of the Australian species.

§ 1. *No prickles. Whole plant glabrous or pubescent with simple hairs (not stellate).*

Anthers very obtuse, parallel, the terminal slits continued more
　　or less down the sides.
　　Annual. Leaves ovate on long petioles. Flowers very small,
　　　　in pedunculate umbels . . . . 1. *S. nigrum.*
　　Perennials or shrubs. Leaves lanceolate or linear, entire or
　　　　pinnatifid. Flowers in short lateral loose racemes.
　　Leaves mostly acute, the longer ones with a few long lobes.
　　　　Flowers large. Berries green or yellow . . . . . 2. *S. aviculare.*
　　Leaves mostly obtuse, very rarely and shortly lobed. Flowers
　　　　moderate. Berries purple . . . . 3. *S. simile.*

Anthers tapering upwards, opening only at the end. Glabrous
    shrub. Leaves broadly lanceolate. Corolla deeply lobed . . *4. *S. pseudocapsicum.*
Flowers unknown. Shrub. Leaves ovate, pubescent underneath 5. *S. Shanesii.*

§ 2. *No prickles. Stellate pubescence or tomentum on the whole plant or rarely on the flowers only.*

Flowers in forked pedunculate cymes.
    Leaves quite glabrous. Cymes loose. Corolla deeply lobed . 6. *S. viride.*
    Leaves (large) very soft and densely tomentose. Cymes dense.
        Leaves shortly acuminate, without stipule-like leaves at the
        base . . . . 8. *S. verbascifolium.*
        Leaves long-acuminate, mostly with small semicircular
        stipule-like leaves at the base . . . . *S. *auriculatum.*
Flowers in simple lateral racemes or clusters.
    Leaves sprinkled with scattered stellate hairs. Corolla deeply
    lobed . . . . 7. *S. tetrandrum.*
    Leaves densely or closely tomentose underneath or on both sides.
    (See § 3, of which several species, especially *S. discolor, S. esuriale, S. furfuraceum,* and
    S. dianthophorum, are occasionally unarmed.)

§ 3. *Prickles slender on the branches and leaves (numerous few or very rare), none on the calyxes. Stellate pubescence or tomentum on the whole plant or rarely on the flowers only.*

Leaves glabrous above except along the veins (rarely scabrous-
    pubescent in *S. violaceum*), tomentose underneath (except in
    S. defensum).
Flowers rather small, the corolla deeply lobed.
    Leaves tomentose and white underneath.
        Leaves ovate or elliptical, rarely above 2 in. long. To-
        mentum very close and short . . . . 9. *S. discolor.*
        Leaves lanceolate, large and broad or small and narrow,
        mostly acute. Tomentum close or loose . . . . . 10. *S. stelligerum.*
        Leaves narrow-oblong obtuse, usually small. Tomentum
        close . . . . 11. *S. parvifolium.*
    Leaves green underneath, glabrous or loosely stellate-hairy.
        Leaves linear or linear-lanceolate, entire or hastate.
        Branches slender. Prickles abundant . . . . . 12. *S. ferocissimum.*
        Leaves oblong-lanceolate, pinnatifid, 5 to 6 in. long. . . 13. *S. defensum.*
Flowers large, the corolla-lobes broad and short.
    Leaves broadly lanceolate. Ovary 2-celled . . . . . . 14. *S. violaceum.*
    Leaves oblong-linear or linear-lanceolate. Ovary 2-celled . 15. *S. amblymerum.*
    Leaves oblong or ovate-oblong, obtuse, cordate at the base.
    Ovary 4-celled . . . . 16. *S. tetrathecum.*
Leaves closely whitish tomentose on both sides (the tomentum
    rarely disappearing at length on the upper side). Calyx-
    teeth very small at the time of flowering.
    Leaves small, mostly broad on very short petioles.
        Leaves under ½ in. long, ovate or broadly oblong . . . . 17. *S. elachophyllum.*
        Leaves orbicular, about ½ in. long. Corolla deeply lobed . 18. *S. orbiculatum.*
        Leaves cordate, ½ to ¾ in. long. Corolla-lobes short and
        broad . . . . 19. *S. oligacanthum.*
    Leaves narrow or on long petioles, mostly above ¾ in. long.
        Leaves ovate oblong or lanceolate, entire or sinuate-toothed.
        Corolla deeply lobed (½ to ¾ in. diameter) . . . . . 20. *S. esuriale.*
        Leaves ovate-lanceolate or lanceolate, mostly lobed at the
        base. Corolla-lobes rather short (½ in. diameter) . . . 21. *S. chenopodinum.*
        Leaves oblong or lanceolate, entire. Corolla with short
        broad lobes (¾ to 1 in. diameter) . . . . 22. *S. Sturtianum.*
Leaves densely and softly tomentose or velvety hirsute on both

sides or at least underneath, sometimes greener and shortly
tomentose above.
Racemes short, few-flowered or pedicels solitary or 2 together .
Calyx divided nearly to the base into narrow segments.
  Calyx-segments subulate-acuminate.
    Leaves mostly entire, scabrous above with scattered
      hairs.
      Stems usually prickly. Flowers mostly racemose . 23. *S. furfuraceum.*
      Prickles exceedingly rare. Flowers mostly in pairs . 24. *S. dianthophorum.*
    Leaves densely velvety-tomentose on both sides, the
      larger ones much sinuate . . . . 25. *S. Dallachii.*
  Calyx-segments lanceolate, acute. Leaves densely velvety
    tomentose or hirsute on both sides . . . . . . 26. *S. densevestitum.*
  Calyx-segments oblong-linear obtuse. Leaves (small)
    densely tomentose . . . . 27. *S. nemophilum.*
Calyx campanulate with broad lobes. (Western species) . . 28. *S. Oldfieldii.*
Racemes or cymes many-flowered, on long very prickly pe-
  duncles. Leaves green and closely tomentose above, white
  and softly tomentose underneath, often lobed . . . . . 29. *S. semiarmatum.*

§ 4. *Prickles slender or rarely thickened at the base on the calyxes as well as on the
branches and generally on the leaves. Stellate pubescence, rarely mixed with simple hairs,
on the whole plant or on the corolla only.*

Leaves green and glabrous or sprinkled with stellate hairs or hir-
  sute on the upper or both sides, sinuate-lobed or pinnatifid.
  Leaves glabrous or sprinkled with very few small hairs. Flowers
    in short loose racemes.
    Leaf-lobes very obtuse and rounded at the end . . . . . 30. *S. sodomæum.*
    Leaf-lobes mostly acute.
      Corolla unarmed . . . . 31. *S. armatum.*
      Corolla armed with prickles . . . . 32. *S. hystrix.*
  Leaves sprinkled with stellate hairs or hirsute, without any
    glandular pubescence.
    Male flowers racemose, female solitary. Berry enclosed in
      the very prickly calyx . . . . 33. *S. cataphractum.*
    Flowers hermaphrodite, in pairs, the pedicels slender, not
      racemose. Calyx-lobes narrow, acuminate . . . . . 34. *S. pungetium.*
    Flowers hermaphrodite, racemose. Calyx-lobes broad, acute 35. *S. eremophilum.*
  Leaves hirsute and glandular-pubescent on both sides. Flowers
    racemose.
    Corolla large, broadly campanulate, very shortly lobed . . 36. *S. campanulatum.*
    Corolla rather small, deeply lobed . . . . 37. *S. adenophorum.*
  Leaves green and glabrous or slightly stellate above, white and
    densely tomentose underneath.
    Leaf-lobes rather acute. Calyx-lobes acuminate . . . . 38. *S. cinereum.*
    Leaf-lobes obtuse. Calyx-lobes short and broad, not acumi-
      nate . . . . 39. *S. lacunarium.*
Leaves nearly equally tomentose on both sides, sinuate-lobed or
  pinnatifid.
  Calyx-lobes narrow, acuminate.
    Calyx-lobes with very prominent keels or midribs. Leaf-
      lobes short, very undulate . . . . 40. *S. petrophilum.*
    Calyx-lobe without prominent ribs. Leaf-lobes deep, very
      obtuse or spathulate . . . . 41. *S. diversiflorum.*
  Calyx-lobes broad, obtuse or acute, much enlarged round the
    fruit after flowering.
    Leaves narrow, shortly tomentose with very numerous
      long prickles. . . . . 42. *S. carduiforme.*

Leaves broad, very densely and softly tomentose with few
prickles . . . . 43. *S. melanospermum.*
Leaves nearly equally, densely and softly tomentose on both sides,
entire or slightly sinuate.
Leaves acute or scarcely obtuse, mostly undulate (1 to 2 in.
long), with very numerous long prickles . . . . . . 44. *S. horridum.*
Leaves obtuse (mostly under 2 in. long), entire or sinuate, with
few or no prickles.
Fruiting calyx membranous, globular, very prickly, com-
pletely enclosing the fruit. Leaves ovate or oblong . . 45. *S. echinatum.*
Fruiting calyx globular, thick, nearly enclosing the fruit.
Leaves orbicular or broadly ovate . . . . 46. *S. lasiophyllum.*
Fruiting calyx open, 6 to 8 lines diameter. Ovary 3-celled 47. *S. ellipticum.*
Leaves mostly acuminate or acute (3 in. long or more), entire,
not at all or scarcely prickly.
Flowering calyx under 3 lines long; fruiting calyx 6 to 8
lines diameter. Ovary 4-celled. . . . . 48. *S. quadriloculatum.*
Flowering calyx ½ in. long; fruiting calyx nearly 2 in. dia-
meter. (Ovary not seen). . . . . 49. *S. phlomoides.*
Leaves lanceolate, rather obtuse (1 to 3 in. long), entire.
Calyx with a globular very prickly tube and long linear lobes.
Flowers large . . . . 50. *S. Cunninghamii.*

There are in the Hookerian as well as in the Muellerian herbarium a few specimens of
what appear to be additional species of *Solanum*, but too imperfect for determination.

§ 1. *Unarmed. Pubescence simple or none.*

1. **S. nigrum,** *Linn. Sp. Pl.* 266. An erect annual or biennial, with
very spreading branches, 1 to nearly 2 ft. high, glabrous or pubescent with
simple hairs, without prickles, but the angles of the stem often raised and
smooth or rough with prominent tubercles. Leaves petiolate, ovate, with
coarse irregular angular teeth or nearly entire, 1 to 2 in. long. Flowers
small and white, in little cymes usually contracted into umbels, on a common
peduncle, from very short to nearly 1 in. long. Calyx 5-toothed or lobed to
the middle. Corolla deeply lobed, 3 to nearly 4 lines diameter. Anthers
very obtuse and short, opening in terminal slits, often at length continued
down the sides. Berry small, globular, usually nearly black, but sometimes
green yellow or dingy red.—R. Br. Prod. 445; Hook. f. Fl. Tasm. i. 288; *Solanum "Morellæ veræ,"* Dun. in DC. Prod. xiii. part i. 45 to 59, as to the
greater number of the supposed species included in the group; *S. rubrum,*
Mill.; Nees in Pl. Preiss. i. 345.

N. Australia. Gilbert river, *F. Mueller.*
Queensland. Broad Sound, *R. Brown;* Port Curtis, *M'Gillivray;* Rockingham
Bay, *Dallachy;* Nerkool Creek, *Bowman;* Rockhampton, *O'Shanesy.*
N. S. Wales. Port Jackson, common, *Banks and Solander* and many others; north-
ward to Hastings river, *Beckler;* New England, *C. Stuart;* southward to Gabo Island,
Maplestone; in the interior on the Darling river, *Victorian Expedition.*
Victoria. About Melbourne, *Adamson, F. Mueller;* Murray river, *F. Mueller.*
Tasmania. Throughout the island on waste places, etc., especially near the sea, *J. D.
Hooker.*
S. Australia. Lofty Range, *F. Mueller;* Kangaroo Island, *Waterhouse.*
W. Australia. Bald Island and Mount Manypeak river, *Maxwell;* Swan River,
Preiss.

This species is a common weed in almost all tropical and temperate parts of the world,
but in many places, as probably in some of the Australian localities, introduced with culti-

vation. The berries are said by several Australian collectors to be frequently eaten. They
vary in colour as in Europe, black yellow or red.

2. S. aviculare, *Forst. Prod.* 18.   An erect glabrous unarmed vigorous
undershrub or shrub, attaining 5 or 6 ft. or even more, flowering the first
year so as then to appear herbaceous.   Leaves lanceolate, acute or rarely almost obtuse, mostly entire on the older shrubby individuals, often pinnatifid
with 1, 2 or 3 lanceolate lobes on each side on the younger ones, especially
the first year, the larger leaves 6 to 10 in. long, but in some specimens all
under 4 in., tapering at the base and often shortly petiolate, in some varieties decurrent so as to form raised angles on the stems.   Flowers few, large,
in short loose pedunculate racemes, mostly lateral.   Pedicels rather long.
Calyx-lobes short, broad, very obtuse or mucronate.   Corolla ¾ to 1 in. diameter, very shortly and broadly lobed.   Filaments filiform, as long as or
longer than the anthers, which are oblong, very obtuse, parallel, opening in
terminal transverse slits, which are at length more or less continued down the
sides and often to the base.   Stigma capitate, slightly 2-lobed.   Berries ovoid
or globular, green or yellow, rather large.—Dun. in DC. Prod. xiii. part i. 69 ;
Hook. f. Fl. Tasm. i. 288 ; *S. laciniatum*, Ait. Hort. Kew. ed. 1, i. 247 ; R.
Br. Prod. 445 ; Bot. Mag. t. 349 ; Dun. l. c. 69 ; *S. reclinatum*, L'Hér. ;
Dun. l. c. 68 ; *S. vescum*, F. Muell. in Trans. Vict. Inst. 1855, 69, in Hook.
Kew Journ. viii. 165 and 336, and Pl. Vict. ii. t. 62.

Queensland.   Brisbane river, *Henne.*
N. S. Wales.   Port Jackson to the Blue Mountains, *R. Brown, Sieber, n.* 255, and
several others ; Sydney woods, Paris Exhibition, 1865, *M'Arthur, n.* 208 ; northward to
Hastings river, *Beckler ;* southward to Twofold Bay, *F. Mueller.*
Victoria.   Port Phillip, *Gunn ;* Portland, *Robertson ;* about Melbourne, *Adamson,
F. Mueller ;* sandy plains at the entrance of Snowy River, Lakes King and Wellington,
sources of the Yarra, *F. Mueller.*
Tasmania.   Islands of Bass's Straits, *R. Brown ;* common in damp shady woods, etc.,
J. D. Hooker.
S. Australia.   Mount Gambier, *F. Mueller.*

The species is also in New Zealand.   F. Mueller distinguishes his *S. vescum* by the sessile decurrent leaves, less deeply lobed corollas, longer filaments in proportion to the anthers,
and edible globular greenish berries, known in Gipps' Land under the name of " Gunyang,"
whilst in the true *S. aviculare* the leaves are not decurrent, the filaments shorter, and the
berry ovoid, yellow, and inedible.   There certainly appear from all accounts to be, in this as
in so many other species, marked varieties in the form, colour, and quality of the fruit, but
I cannot trace, from the materials and notes before me, any correspondence between these
and the forms of the foliage.   F. Mueller's Twofold Bay specimens have the most prominently decurrent leaves with the berries not specially described ; those from Gipps' Land
have more or less decurrent leaves, and globular greenish berries ; and the original New
Zealand as well as the Tasmanian plant have leaves not at all or only slightly decurrent, and
yellowish ovoid berries, but which are eaten, at least in New Zealand.   Another variety,
however, apparently the common Port Jackson one, and which is one of those early cultivated
in European botanic gardens, has the leaves not decurrent, but the berries globular, of a yellowish green.   Sieber's and other Blue Mountain specimens have the leaves somewhat decurrent, and some of these are described as having ovoid edible berries.   There would
appear, therefore, to be several distinct varieties or races, of which two, well distinguished
by F. Mueller, are in Victoria, and one, two, three or more in N. S. Wales, which can only
be characterized by observing them in a living state.   *S. reclinatum*, L'Hér., appears to
have been always described from garden specimens, probably of the same N. S. Wales origin as
Aiton's plant, with a mistaken indication of a Peruvian origin.   A specimen dried in 1822

in the Montpellier garden as authentic, is certainly undistinguishable from the N. S. Wales
S. laciniatum.

3. **S. simile,** *F. Muell. Trans. Phil. Soc. Vict.* i. 19, *and Fragm.* vi.
145. A glabrous erect unarmed undershrub or shrub, closely resembling
entire-leaved specimens of *S. aviculare,* usually not so stout, although attaining 4 or 5 ft. Leaves lanceolate or linear, usually obtuse, contracted into a
short petiole, not decurrent, entire or rarely with 1 or 2 short lobes on each
side near the base, mostly only 2 or 3 in. long. Flowers smaller than in *S.
aviculare,* few in lateral racemes, with a very short or sometimes scarcely any
common peduncle. Calyx and corolla otherwise nearly as in *S. aviculare,*
the corolla not much above ½ in. diameter. Anthers obtuse, parallel, opening at length down the sides. Berry globular ovoid or oblong, usually
smaller than in *S. aviculare,* and purple. Seeds rather large.—*S. laciniatum,*
var., R. Br. Prod. 445 ; Benth. in Hueg. Enum. 82 ; Nees in Pl. Preiss. i.
345 ; *S. fasciculatum,* F. Muell. Fragm. i. 123, vi. 144.

N. S. Wales. Darling river, *Dallachy ;* Murray river, *F. Mueller.*
Victoria. Wimmera, *Dallachy.*
S. Australia. Port Lincoln, *Wilhelmi ;* Spencer's Gulf and Kangaroo Island, *R.
Brown, F. Mueller ;* Mount Serle, *Warburton ;* Lake Gillies, *Burkitt.*
W. Australia. Goose Island Bay, *R. Brown ;* King George's Sound, *Oldfield, F.
Mueller ;* Fitzgerald river, *Maxwell ;* Swan River, *Huegel, Drummond, Oldfield ;* Mur-
chison river, *Oldfield ;* Rottenest Island, *Preiss, n.* 1965.

F. Mueller distinguishes *S. fasciculatum* as a Western species with ovoid berries. I can
find no other character, and there appear to be at least three different forms of fruit all included by F. Mueller as varieties of *S. fasciculatum,*—globular, ovoid, and oblong,—the
latter sometimes at least 1 in. long and very narrow. All three are in West Australia, and
the two extremes in South Australia. The narrowest-fruited specimens have also very
narrow leaves, from Phillips river, *Maxwell,* and Lake Gillies, *Burkitt.*

*4. **S. pseudo-capsicum,** *Linn. ; Dun. in DC. Prod.* xiii. *part* i.
152. A glabrous unarmed erect shrub or undershrub, attaining 3 to 4 ft.
Leaves broadly lanceolate, entire, contracted into a rather long petiole. Pedicels lateral, solitary or 2 or 3 together in a cluster or on a very short common peduncle. Calyx deeply divided into ovate-lanceolate herbaceous segments. Corolla white, rather small, divided to about the middle. Filaments
short ; anthers connivent and tapering upwards. Berry globular, bright red
or yellow.

N. S. Wales. Hastings river, *Beckler.* An introduced plant of somewhat uncertain
origin, now widely diffused in tropical countries, chiefly as a weed or escape from cultivation.

5. **S. ? Shanesii,** *F. Muell. Fragm.* vi. 144. An erect shrub, attaining 8 ft., with slender glabrous branches. Leaves solitary or the upper ones
in pairs, ovate, acuminate, membranous, entire, glabrous above, sprinkled underneath with simple not stellate hairs, 1½ to 2 in. long, the lamina decurrent
on a rather long petiole. Flowers unknown. Fruiting pedicels solitary or
2 together, reflexed. Calyx obtuse, obscurely lobed. Berry globular, red,
about ½ in. diameter.

Queensland. More's Creek, Rockhampton, *Dallachy, O'Shanesy.* The genus as well
as the immediate affinities of this species must remain uncertain until the flowers shall have

been seen. The specimens have rather the aspect of some *Capsicum* allied to *C. sinense*, Jacq., than of a *Solanum*.

§ 2. *Unarmed. Pubescence stellate, at least on the corolla.*

6. **S. viride,** *R. Br. Prod.* 445. An erect undershrub or shrub of 6 to 7 ft. or even more, quite glabrous except the stellate pubescence of the flowers, and sometimes a very few small stellate hairs scattered on the upper leaves. Leaves solitary or in pairs, ovate-oblong, obtuse, shortly acuminate or rather acute, membranous, entire or obscurely sinuate, 3 to 5 in. long, on rather long petioles. Flowers in forked pedunculate cymes, terminal or lateral, the branches of the cyme short, the pedicels often above ½ in. long after flowering, the whole inflorescence and calyx glabrous or slightly stellate-tomentose, the corolla always stellate-pubescent outside. Calyx scarcely above 1 line long at the time of flowering, the lobes obtuse, either very short or separating to the middle. Corolla deeply divided into narrow lobes of 3 to 4 lines. Filaments very short ; anthers connivent and tapering upwards. Berries small, globular, red.—Dun. in DC. Prod. xiii. part i. 190 ; *S. viridifolium*, Dun. l. c. 73.

Queensland. Broad Sound, *R. Brown ;* Cape Grafton, *Banks and Solander ;* Cape York, *Daemel ;* islands off the N.E. coast, *A. Cunningham, M'Gillivray, F. Mueller,* and others ; Port Denison, *Fitzalan ;* Rockingham Bay, *Dallachy ;* Port Mackay, *Nernst.*

7. **S. tetrandrum,** *R. Br. Prod.* 445. An erect unarmed undershrub of 2 to 3 ft., sprinkled with a small stellate tomentum, rather dense on the inflorescence, more scattered on the leaves and sometimes disappearing from the upper surface. Leaves mostly in pairs, petiolate, ovate, obtuse or shortly acuminate, entire or obscurely sinuate, membranous, the larger ones 3 to 6 in. long. Flowers small, in short loose lateral racemes, the common peduncle not so long as in *S. viride* and not at all or very rarely forked. Calyx 2 to 2½ lines long, very tomentose, unequally divided to about the middle. Corolla stellate-pubescent outside, under ½ in. long, divided nearly to the base into narrow lobes. Filaments short; anthers connivent and tapering upwards. Berry small, globular.—Dun. in DC. Prod. xiii. part i. 194 ; Seem. Fl. Vit. 176 ; *S. inamœnum,* Benth. in Hook. Lond. Journ. ii. 228 ; Dun. l. c. 269.

N. Australia. Arnhem N. Bay, and islands of the Gulf of Carpentaria, *R. Brown ;* Goulburn Islands, *A. Cunningham ;* Port Essington, *Armstrong.*

The species is also in the South Pacific islands. The flowers, in this as in *S. viride*, are occasionally, but not always, 4-merous; several 5-merous flowers occur indeed in Brown's own specimens.

Var. ? *floribundum.* Corollas larger, very tomentose, and one of the peduncles of the specimen forked.—From *Leichhardt's* collection, a single specimen in Herb. F. Mueller.

8. **S. verbascifolium,** *Ait. ; Dun. in DC. Prod.* xiii. *part* i. 114. A tall stout unarmed shrub, attaining often 10 to 12 ft., thickly covered with a stellate tomentum often very dense and floccose or velvety, sometimes more scattered on the upper side of the leaves. Leaves ovate, acuminate, entire, soft and thick, often 6 to 8 in. long, on long petioles. Flowers often numerous, in dense pedunculate dichotomous cymes, terminal or at length lateral, the pedicels very short. Calyx densely tomentose, the lobes shorter than the

tube, thick and obtuse. Corolla pale blue or white, under ½ in. diameter. Filaments short; anthers scarcely tapering but opening only at the end. Berry globular, yellow, under ¼ in. diameter.—R. Br. Prod. 444.

Queensland. Broad Sound and Shoalwater Bay, *R. Brown;* Brisbane river, Moreton Bay, *Fraser, F. Mueller;* Rockhampton, *Dallachy;* Nerkool Creek, *Bowman;* Port Denison, *Fitzalan, Dallachy;* Rockingham Bay, *Dallachy.*

N. S. Wales. Clarence river, *Beckler.*

The species is widely dispersed over tropical Asia and America.

S. auriculatum, Ait.; Dun. in DC. Prod. xiii. part 1, 115, a tropical American species, closely resembling *S. verbascifolium,* but more densely woolly, the leaves more acuminate, with a pair of stipule-like small semicircular leaves at the base of most of the petioles, and purple flowers, has been sent from the N. shore, Port Jackson, as an introduced species.

§ 3. *Prickles numerous few or occasional on the stem and often on the leaves, none on the calyxes. Pubescence stellate, at least on the corolla.*

9. **S. discolor,** *R. Br. Prod.* 445.   An erect shrub, with weak half-climbing branches, the young ones as well as the under side of the leaves and inflorescence silvery or hoary with a minute, exceedingly close but dense stellate tomentum.   Prickles few, slender on the branches and veins of the leaves or in some specimens none.   Leaves petiolate, irregularly oval elliptical or broadly oblong, rather obtuse, entire or irregularly sinuate, glabrous and smooth on the upper surface, 1 to 2 in. long in flowering specimens, larger in barren shoots.   Flowers rather small, in simple lateral racemes, few or even solitary with a very short common peduncle on the fruit-bearing specimens, numerous along a slender rhachis but very deciduous upon apparently sterile ones, the pedicels short at the time of flowering, 3 to 4 lines long and thickened under the fruit.   Calyx very small and shortly toothed when in flower, somewhat enlarged and more deeply cleft under the fruit.   Corolla white, deeply lobed, about or under ½ in. diameter.   Berries globular, of a greenish-white, about 4 lines diameter.—Dun. in DC. Prod. xiii. part i. 293; *S. corifolium,* F. Muell. Fragm. ii. 166.

N. Australia. Coen river, Gulf of Carpentaria, *R. Brown.*

Queensland. Araucaria Ranges, Moreton Bay, *F. Mueller.*

This and the three following species (*S. stelligerum, parvifolium,* and *ferocissimum*) are closely allied to each other, having nearly the same flowers and fruit, and differing chiefly in foliage and prickles.

10. **S. stelligerum,** *Sm. Exot. Bot.* ii. 57. *t.* 88.   An erect shrub, sometimes small and slender, sometimes attaining 6 ft. or even more, the branches, under side of the leaves, and inflorescence covered with a stellate tomentum, often loose and floccose.   Prickles straight or slightly recurved on the branches and sometimes on the upper side of the leaves, but not numerous.   Leaves petiolate, lanceolate or ovate-lanceolate, acute or acuminate, very rarely on luxuriant shoots broad and obtuse, usually glabrous and smooth on the upper side except minute stellate hairs along the principal veins, mostly 2 to 4 in. long.   Flowers blue, rather small, in lateral racemes, the common peduncle very short, the pedicels lengthening to about ½ in. or even 1 in. under the fruit.   Calyx under 2 lines long when in flower, with narrow acuminate lobes, somewhat lengthened under the fruit and then often divided to near the base.   Corolla usually under ½ in. diameter, deeply divided into

narrow lobes. Anthers connivent and tapering upwards. Berry red, globular, small.—R. Br. 445 ; Dun. in DC. Prod. xiii. part i. 191.

Queensland. Keppel Bay, *R. Brown;* Brisbane river, Moreton Bay, *A Cunningham, F. Mueller,* and others ; Rockhampton, *Dallachy* and others ; Rockingham Bay, *Dallachy;* Araucaria Ranges, Burnett river, *F. Mueller;* Armadillo, *Barton;* Warwick, *Beckler.*

N. S. Wales. Port Jackson to the Blue Mountains, *R. Brown* and others ; New England, *Leichhardt;* Clarence and Hastings rivers, *Beckler.*

Several of the northern specimens have smaller, narrower leaves than usual, approaching those of *S. parvifolium,* but acute ; one from Cape Byron, *C. Moore* (Herb. F. Muell.), has large broad leaves, more prickly than usual ; and some from the Araucaria Ranges, Burnett river, *F. Mueller,* sent by him as a var. *lucorum,* have the leaves sprinkled on the upper surface with a few stellate hairs. Brown's Keppel Bay specimens have a more rufous tomentum, showing some approach to *S. furfuraceum,* but very much more glabrous on the upper surface of the leaves.

Var. ? *magnifolium.* Leaves broadly ovate, 4 to 8 in. long, mostly sinuate, with several prickles on the upper side. Flowers very few in the imperfect specimens seen, but quite those of *S. stelligerum.*—Murray river, Rockingham Bay, *Dallachy;* mountain brush, Moreton Bay, *Leichhardt* (both in Herb. F. Mueller).

11. **S. parvifolium,** *R. Br. Prod.* 446. A bushy slender shrub, closely allied to the small-leaved varieties of *S. stelligera,* but the leaves are narrow-oblong or almost linear, always obtuse, quite entire or with a short broad lobe on each side near the base, glabrous above, stellate-tomentose underneath, in some specimens not above 1 in. long, in others twice as long. Flowers blue, like those of *S. stelligera.* Calyx about 1 line long when in flower and not 2 lines when in fruit, deeply divided into acuminate lobes. Corolla deeply lobed. Berry small, globular.—Dun. in DC. Prod. xiii. part i. 191 ; *S. leptophyllum,* F. Muell. Fragm. ii. 164.

Queensland. Broad Sound, *R. Brown;* Brigalow Scrub on the Mackenzie and Suttor rivers, *F. Mueller;* in the interior, *Mitchell;* Cape river, *Bowman;* Armadillo, *Barton* (with rather larger flowers).

N. S. Wales. Liverpool Plains, *A. Cunningham;* Macnamara hills, *Fraser;* Mount Murchison, *Dallachy.*

12. **S. ferocissimum,** *Lindl. in Mitch. Three Exped.* ii. 58. A low straggling slender shrub, allied to *S. parvifolium,* the branches rather loosely stellate-tomentose. Prickles long and slender, very numerous on the branches and leaves, none on the calyx. Leaves linear or linear-lanceolate, not so obtuse as in *S. parvifolium,* entire or the larger ones hastately lobed at the base, 1 to 2 in. long, glabrous or with loose stellate hairs especially underneath, without the close tomentum of *S. parvifolium.* Flowers small, blue, in loose racemes, with a very short common peduncle and slender pedicels. Calyx 1 to 1½ lines long in flower, somewhat enlarged in fruit, deeply divided into acuminate lobes. Corolla about ½ in. diameter, deeply lobed.—Dun. in DC. Prod. xiii. part i. 373.

N. S. Wales. Lachlan river, *Mitchell;* between that and the Upper Bogan, *L. Morton;* Darling river, *Bowman, Panton;* Peel's Range, *Fraser, A. Cunningham;* Mount Murchison, *Dallachy, E. Giles.*

13. **S. defensum,** *F. Muell. Fragm.* v. 193. The single specimen described is an erect nearly simple shoot from a woody stock (or from the base of a shrub that has been cut down), stout and rigid, above 1½ ft. high, sca-

brous with scattered stellate hairs. Prickles straight, rather numerous on the stem and leaves, none on the calyxes. Leaves very shortly petiolate, oblong-lanceolate, acuminate, pinnatifid with short obtuse lobes or some sinuate only, 3 to 5 in. long, green on both sides, glabrous above, with a few small scattered stellate hairs underneath. Flowers blue, rather small, resembling those of *C. stelligerum*, in loose lateral racemes, the pedicels rather long even when in flower. Calyx-lobes acuminate, split almost to the base, but not exceeding 2 lines with the fruit far advanced. Corolla about ½ in. diameter, deeply lobed. Anthers tapering upwards. Berry small, globular, but not quite ripe in the specimen.

Queensland. Cape York, *Daemel* (*Herb. F. Mueller*). Very remarkable in the foliage, but that may have been in some measure modified by the circumstances of growth of the only specimen known.

14. **S. violaceum,** *R. Br. Prod.* 445. An erect shrub of several feet, the branches under side of the leaves and inflorescence covered with a stellate tomentum, sometimes dense and close, more rarely loose and floccose. Prickles slender, straight, not numerous, on the branches and sometimes on the upper side of the leaves, none on the calyxes. Leaves petiolate, lanceolate or ovate-lanceolate, acute or rather obtuse, entire or rarely sinuate, often oblique at the base but not cordate, mostly 2 to 4 in. long, glabrous on the upper surface or rarely scabrous with small stellate hairs. Flowers (violet?) large, in lateral racemes, the common peduncle at first very short as well as the pedicels, but both sometimes much lengthened in fruit. Calyx above 2 and often 3 lines long, with acuminate teeth sometimes very short sometimes as long as the tube, somewhat enlarged and more deeply lobed when in fruit. Corolla ¾ to above 1 in. diameter, the lobes short and very broad. Berry globular, larger than in *S. stelligerum*.—Dun. in DC. Prod. xiii. part i. 336 ; *S. Brownii,* Dun. Hist. Solan. 201.

N. S. Wales. Paterson's River, *R. Brown ;* Blue Mountains, *A. Cunningham* and others ; Clarence and Hastings rivers, *Beckler ;* Richmond river, *Fawcett ;* Glendon, *Leichhardt.*

Var. ? *scabrum.* Upper side of the leaves very scabrous. Calyx-teeth short.—N. S. Wales, *Vicary, C. Moore.*

The species sometimes resembles some forms of *S. stelligerum* in foliage, but is at once distinguished by the large and differently-shaped calyx and corolla.

15. **S. amblymerum,** *Dun. in DC. Prod.* xiii. *part* i. 294. An erect shrub of several feet, the branches under side of the leaves and inflorescence covered with a dense stellate tomentum usually close. Prickles slender, straight, on the branches and often on the leaves, none on the calyxes. Leaves shortly petiolate, narrow-lanceolate or almost linear, rather obtuse, entire or with short obtuse lobes near the base, 2 to 4 in. long, the upper surface glabrous and smooth or slightly scabrous with minute scattered stellate hairs. Flowers large like those of *S. violaceum*, in lateral racemes, usually more numerous than in that species, with the common peduncle more developed. Calyx about 3 lines long when in flower, with small acuminate teeth, enlarged and more lobed after flowering. Corolla fully ¾ in. diameter with short broad lobes. Anthers tapering upwards. Ovary 2-celled. Berry globular.

Queensland. Warwick, *Beckler.*

N. S. Wales. Macquarrie river, *A. Cunningham ;* New England, *C. Stuart.*

These specimens were all included by F. Mueller in his *S. tetrathecum*, but besides the differences in foliage, calyx, and corolla, I found the ovary 2-celled only in all the flowers I examined. They may possibly, however, prove to be a narrow-leaved variety of *S. viola-ceum.*

16. **S. tetrathecum,** *F. Muell. Fragm.* ii. 165 (*partly*). A straggling shrub, the branches under side of the leaves and inflorescence covered with a close but dense and soft stellate tomentum. Prickles few on the branches, and in some specimens none. Leaves petiolate, ovate oblong or oblong-lanceolate, very obtuse, slightly cordate at the base, 1 to 3 in. long, the upper side glabrous or sprinkled with scattered stellate hairs. Flowers rather large, few together in lateral racemes, the pedicels at first very short, lengthening to about $\frac{1}{2}$ in. Calyx about 2 lines long, with minute teeth, scarcely en-larged but somewhat lobed when in fruit. Corolla deeply lobed, about $\frac{3}{4}$ in. diameter, of a rather firm consistence. Anthers tapering upwards. Ovary 4-celled, as well as the globular berry.

Queensland. Araucaria ranges, Burnett river, *F. Mueller ;* near Morpeth, *Leich-hardt.* The ovary and fruit in this and *S. tetrathecum* are probably, as in other species, dicarpellary, but each carpel divided by a spurious dissepiment, as in some *Convolvulaceæ* and most *Boragineæ.*

17. **S. elachophyllum,** *F. Muell. Fragm.* ii. 164. A slender straggling shrub, the branches and foliage hoary or silvery with a very close stellate to-mentum, less white on the upper side of the leaves. Prickles slender, abundant on the branches, none on the leaves or calyxes. Leaves ovate obovate or broadly oblong, narrowed into a very short petiole, entire, 3 to 5 or rarely 6 lines long. Flowers solitary or few together in short lateral ra-cemes. Calyx when in flower about 1 line long with small teeth, enlarged and more divided when in fruit. Corolla " violet," about $\frac{1}{2}$ in. diameter, deeply lobed. Berries globular, variegated, nearly $\frac{1}{2}$ in. diameter, the fruit-ing pedicels $\frac{1}{2}$ in. long.

Queensland. Between Mackenzie and Dawson rivers, *F. Mueller.* Differs from all other Australian *Solanums* in its small leaves.

18. **S. orbiculatum,** *Dun. in Poir. Dict. Suppl.* iii. 762, *and in DC. Prod.* xiii. *part* i. 292. A scrubby irregularly spreading shrub usually of 2 to 3 ft., but sometimes twice that height, the branches foliage and inflo-rescence covered with a close but dense stellate tomentum. Prickles rather stout, straight or recurved, rather numerous on the branches, none on the leaves or calyxes. Leaves shortly petiolate orbicular or very broadly and obscurely cordate or almost reniform, very obtuse, entire or slightly sinuate, thick and soft, usually about $\frac{1}{2}$ in or rarely $\frac{3}{4}$ in. diameter. Flowers very few together in lateral racemes, the common peduncle exceedingly short, the pedicels also short at first but lengthening to nearly $\frac{1}{2}$ in. when in fruit. Calyx small with short broad very obtuse teeth and densely tomentose. Corolla densely tomentose, deeply lobed, about $\frac{1}{2}$ in. diameter. Stamens tapering upwards. Berry small, globular.

W. Australia. Sharks' Bay and Dirk Hartog's Island, *Milne ;* Murchison river, *Old-field, Drummond, 6th Coll. n.* 130.

19. **S. oligacanthum,** *F. Muell. in Trans. Phil. Soc. Vict.* i. 19, *and in Hook. Kew Journ.* viii. 167. Evidently closely allied to *S. orbiculatum,* with the same tomentum, prickles, small leaves, inflorescence and flowers, except that the petioles are still shorter, the leaves more cordate, and the corolla apparently less deeply divided.

S. Australia. In the interior, *Sturt,* described from a single small specimen in Herb. F. Mueller.

20. **S. esuriale,** *Lindl. in Mitch. Three Exped.* ii. 43. A low shrub, often under 6 in. high and rarely exceeding 1 ft., the branches inflorescence and both sides of the leaves covered with a close but dense and soft stellate tomentum, rarely somewhat looser underneath. Prickles few and slender on the stems or the whole plant unarmed. Leaves petiolate, ovate oblong or lanceolate, obtuse, entire or sinuate-toothed, mostly $\frac{3}{4}$ to 1 in. long, but in luxuriant specimens narrow-lanceolate entire and 2 to 3 in. long. Flowers solitary or 2 to 4 together, on a very short lateral common peduncle, the pedicels lengthening to $\frac{1}{2}$ in. Calyx under 2 lines when in flower with narrow almost acute teeth, enlarged after flowering and dividing into triangular acuminate lobes. Corolla blue, $\frac{1}{2}$ to $\frac{3}{4}$ in. diameter, deeply lobed. Anthers tapering upwards. Berry globular.—Dun. in DC. Prod. xiii. part i. 373 ; *S. pulchellum,* F. Muell. in Trans. Phil. Soc. Vict. i. 18, and in Hook. Kew Journ. viii. 166.

N. Australia. Sturt's Creek, *F. Mueller.*
Queensland. Ranges about Lake Salvator, *Mitchell;* Upper Burdekin river, *F. Mueller ;* Suttor and Bowen rivers, *Bowman ;* Armadillo and Curriwillighi, *Barton.*
N. S. Wales. Peele's Range, *Mitchell, A. Cunningham, Fraser ;* from the Murray, Lachlan, and Darling to the western frontier, *Victorian and other Expeditions.*
Victoria. Wimmera, Avoca, and Murray rivers, *F. Mueller, Dallachy.*
S. Australia. From the Murray to St. Vincent's and Spencer's gulfs, *F. Mueller* and others; Cooper's Creek, *Howitt's Expedition ;* Purdie's Ponds, *Waterhouse.*

21. **S. chenopodinum,** *F. Muell. Fragm.* ii. 165. A slender divaricate shrub of 2 to 3 ft., with the close stellate tomentum of *S. esuriale,* which however sometimes almost disappears from the upper surface of the old leaves. Prickles slender, few or rarely more numerous on the branches, very rare on the leaves and none on the calyxes. Leaves petiolate ovate lanceolate or lanceolate, the broader ones cordate at the base, rather obtuse, sinuate-lobed towards the base and sometimes hastate, mostly 1 to 2 in. long. Flowers few in short lateral racemes. Calyx scarcely $1\frac{1}{2}$ lines long when in flower with very small teeth, more deeply lobed but scarcely above 2 lines long when in fruit. Corolla blue, about $\frac{1}{2}$ in. diameter, the lobes rather broad and short. Berry globular, shining, rather small.

N. Australia. In the interior, between Mount Blight and Mount Fisher, lat. 20° 20', *M'Douall Stuart's Expedition.*
N. S. Wales. From the Darling river to the Barrier Range, *Victorian Expedition ;* Mount Murchison, *Bonney.*
S. Australia. Cooper's Creek, *Howitt's Expedition.*

The species differs from *S. esuriale* in its taller stature, mostly lobed or hastate leaves, and apparently in the form of the corolla.

22. **S. Sturtianum,** *F. Muell. in Trans. Phil. Soc. Vict.* i. 19, *and in*

Hook. Kew Journ. viii. 166. An erect shrub with the close stellate tomentum and rare prickles of *S. esuriale,* but apparently of taller stature.
Leaves petiolate, oblong or lanceolate, obtuse, entire or scarcely sinuate, ¾ to
1½ in. long. Peduncles usually rather longer than in *S. esuriale* bearing a
short raceme of very few rather large flowers, the pedicels very short at the
time of flowering but lengthening afterwards. Calyx about 2 lines long
when in flower, with short acute teeth, much enlarged and irregularly lobed
when in fruit. Corolla ¾ to 1 in. diameter, with short broad lobes. Anthers
tapering upwards. Berry black, above ½ in. diameter.

N. Australia, Glenelg district, N.W. coast, *Marten.*

S. Australia. In the interior, *Sturt;* Flinders Range and Cooper's Creek, *Howitt's
Expedition;* Mount Searl, *Warburton;* Lake Gillies, *Burkitt.*

The species differs from *S. esuriale* chiefly in the large slightly lobed corolla.

23. **S. furfuraceum,** *R. Br. Prod.* 446 *(the char. wrong as to the
leaves by a clerical error).* An erect spreading shrub of 4 to 6 ft., the
branches and inflorescence covered with a rather loose rusty tomentum.
Prickles straight, slender, not numerous on the branches, very rare on the
leaves and none on the calyxes. Leaves petiolate, ovate or ovate-lanceolate,
acute acuminate or almost obtuse, entire or sinuate, rounded or slightly cordate at the base, not above 2 in. long in our specimens, more or less scabrous
above with stellate hairs sometimes very dense, densely tomentose underneath and often woolly or floccose. Flowers blue in rather dense lateral racemes, the pedicels short. Calyx divided almost to the base into narrow
acuminate lobes, above 2 lines long at the time of flowering, 4 to 6 lines
when in fruit. Corolla rather large, divided to near the middle into broad
lobes. Berry globular, much larger than in *S. stelligerum,* the enlarged
calyx-segments broadly lanceolate, subulate-acuminate.—*Dun.* in DC. Prod.
xiii. part i. 293.

Queensland. Broad Sound, *R. Brown;* Brisbane river, Moreton Bay, *Fraser, F.
Mueller;* Mogile scrub, *C. Stuart;* Rockhampton, *Dallachy, O'Shanesy;* Table mountain,
Bowman.

This species has been frequently misunderstood and the name applied to *S. parvifolium,*
or to varieties of *S. stelligerum,* owing to a clerical error in Brown's diagnosis. In his
notes, as well as in Dunal's detailed description, the leaves are correctly described as
ovate lanceolate and scabrous-tomentose on the upper side. It appears that Brown had
originally intended to give to the present species the name of *S. parvifolium,* which he
afterwards transferred to another, and in writing out the diagnosis of *S. furfuraceum* for
press, retained by mistake the character as to foliage of *S. parvifolium,* the remainder
appertaining to *S. furfuraceum.* Dunal copies Brown's diagnoses without remark, although
in contradiction to the accurate description which follows.

24. **S. dianthophorum,** *Dun. Hist. Sol.* 183, *and in DC. Prod.* xiii.
part i. 192. Perhaps a variety only of *S. furfuraceum,* with the same indumentum, but a more spreading slender shrub without any prickles at all or
very rarely with a very few small slender prickles on the stem. Leaves as in
S. furfuraceum, ovate or ovate-lanceolate entire or slightly sinuate, rarely
above 1 in. long. Flowers solitary or two together on slender pedicels rarely
above ½ in. long. Calyx of *S. furfuraceum.* Corolla rather smaller. Anthers much acuminate. Berry like that of *S. furfuraceum.*—*S. biflorum,* R.
Br. Prod. 445, not of Lour.

Queensland. Bay of Inlets, *Banks and Solander ;* Port Bowen, *R. Brown ;* Perry Islands, *A. Cunningham.*

25. **S. Dallachii,** *Benth.* An erect stout shrub of 6 to 10 ft., the branches inflorescence and foliage densely villous with loose velvety hairs mostly stellate at the base. Prickles slender, very rare on the branches and leaves, none on the inflorescence. Leaves broadly ovate, acuminate, the larger ones 6 to 8 in. long and 4 to 5 in. broad, and mostly sinuate-toothed, the smaller ones entire and resembling those of *S. densevestitum.* Peduncles axillary, often longer than in the allied species, bearing a short raceme almost contracted into an umbel and sometimes forked. Pedicels under $\frac{1}{2}$ in. long when in flower, nearly 1 in. when in fruit. Calyx at the time of flowering nearly 3 lines long, with narrow acuminate teeth or lobes, somewhat enlarged in fruit, and then deeply divided into lanceolate subulate-acuminate lobes. Corolla blue, deeply lobed, about $\frac{1}{2}$ in. diameter. Anthers tapering upwards. Berry yellow, globular, glabrous.—*S. repandum,* F. Muell. Fragm. vi. 145, not of Forster.

Queensland. Rockingham Bay, *Dallachy.* I cannot agree with F. Mueller in referring this plant to *S. repandum,* Forst., notwithstanding a general resemblance in the larger leaves, for in all our specimens Forster's plant differs in the more sessile and denser inflorescence, in the broadly campanulate and broadly lobed calyx (usually larger than is represented in Seemann's figure, Fl. Vit. t. 38), and in the larger hirsute berry. *S. Dallachii* appears to me to be much nearer allied to *S. stelligerum,* and especially to *S. furfuraceum,* differing chiefly in indumentum and in the larger leaves. Those specimens, indeed, from Rockingham Bay which I have mentioned above as a large-leaved doubtful variety of *S. stelligerum,* were included by F. Mueller under *S. repandum.*

26. **S. densevestitum,** *F. Muell. in Herb. Hook.* An erect shrub of several feet, the branches inflorescence and foliage densely villous with loose velvety-stellate hairs sometimes more tomentose but very soft and almost floccose. Prickles slender, very few or rarely rather numerous on the stems, very rare on the leaves and none on the inflorescence or calyxes. Leaves ovate or ovate-lanceolate, rather obtuse, entire or slightly sinuate, often somewhat cordate, thick and soft, 2 to 3 in. long. Flowers solitary or very few in short almost sessile lateral racemes, the pedicels also short. Calyx hispid, divided to the base into lanceolate acute segments about 3 lines long at the time of flowering, longer when in fruit. Corolla about $\frac{3}{4}$ to 1 in. diameter, rather deeply divided into broad lobes.

Queensland. Araucaria ranges, Upper Burnett river, *F. Mueller ;* Brisbane river, Moreton Bay, *F. Mueller* and others ; also in *Leichhardt's* collection.
N. S. Wales. New England, *C. Stuart ;* Hastings river, *Beckler ;* Mount Lindsay, *C. Moore.*

This may possibly prove to be a remarkable variety of *S. furfuraceum,* but, besides the indumentum, the calyx is certainly different and the flower larger.

27. **S. nemophilum,** *F. Muell. Fragm.* ii. 161. A low spreading shrub, the branches inflorescence and foliage covered with a soft thick stellate tomentum. Prickles none in the specimens seen, probably rare on the branches. Leaves ovate oblong or almost lanceolate, rather obtuse, entire, rounded or slightly cordate at the base, thick and soft, 1 to 2 in. long. Flowers violet, solitary or 2 or 3 together on a very short lateral common

peduncle, the pedicels at length nearly ½ in. long. Calyx divided to the base into narrow obtuse thick and woolly segments, 2 to 3 lines long when in flower and but slightly enlarged when in fruit. Corolla above ½ in. diameter, deeply lobed. Berry red, ovoid (*F. Muell.*).

Queensland. Ironbark forest between the Mackenzie and Dawson rivers, *F. Mueller ;* Burnett river, *Haly ;* Flinders river, *Sutherland.* The unripe berries of some specimens appear globular. The species is very near the last three, but the calyx-segments remarkably obtuse, besides the differences in foliage and indumentum.

28. **S. Oldfieldii,** *F. Muell. Fragm.* ii. 161. An erect shrub of 1 to 3 ft., the branches inflorescence and under side of the leaves covered with a soft dense more or less rusty stellate tomentum, sometimes almost floccose, sometimes closer and more hoary, usually shorter and more scabrous on the upper side of the leaves. Prickles small and slender, not numerous on the branches, none on the leaves or inflorescence and sometimes the whole plant unarmed. Leaves petiolate, ovate or oval-oblong, very obtuse, entire sinuate or undulate, thick and soft, mostly 1 to 2 in. long. Flowers rather large, several in pedunculate racemes very rarely once forked. Calyx rusty-villous, broadly campanulate, about 3 lines long, with broad obtuse lobes shorter than the tube at the time of flowering, enlarged and more deeply divided in fruit. Corolla apparently about 1 in. diameter, with short broad lobes. Anthers rather short and scarcely tapering upwards, the filaments longer than in the other species of the group. Ovary 2-celled. Berry globular, yellow, at least ½ in. diameter.

W. Australia, *Drummond, 2nd Coll. n. 224, and Suppl. n. 7 ;* Murchison river and Champion Bay, *Oldfield.*

29. **S. semiarmatum,** *F. Muell. Fragm.* ii. 163. An erect shrub, the branches and inflorescence covered with a dense hoary or white stellate tomentum sometimes floccose. Prickles slender, straight, very numerous on the branches and on the peduncles, very rare on the leaves and none on the calyxes. Leaves petiolate, ovate or ovate-lanceolate and scarcely lobed when small, the larger ones broad and pinnatifid with triangular or lanceolate lobes, green but softly tomentose on the upper side, very white-tomentose underneath, 2 to 4 in. long. Flowers numerous in loose pedunculate lateral simple racemes or more frequently branched cymes, usually as long as the leaves. Pedicels slender, ¼ to ½ in. long at the time of flowering. Calyx turbinate, about 2 lines long, the lobes almost obtuse to subulate-acuminate, longer than the tube, enlarged after flowering and sometimes separating to the base. Corolla ½ to ¾ in. diameter, divided to below the middle. Anthers tapering upwards. Berry globular when young, not seen ripe.

Queensland. In the interior, *Mitchell ;* Connor's River, *Bowman.* The leaves usually scarcely lobed, although a few are larger and more lobed, showing the connection with the typical specimens.

N. S. Wales. Clarence river, *Beckler ;* Richmond river, *C. Moore* (small but evidently luxuriant specimens, with large deeply lobed leaves and ample inflorescence) ; Darling Downs, *Law* (like the Queensland specimens).

The species forms a passage from the third to the fourth group, the prickles being abundant on the peduncles, but the calyxes entirely unarmed.

§ 4. *Prickles on the calyxes as well as on the rest of the plant. Pubescence stellate, rarely almost simple or none.*

*30. **S. sodomæum,** Linn.; Dun. in DC. Prod. xiii. part i. 366.* A spreading or diffuse shrub or herb of 2 to 3 ft., the foliage green but sprinkled as well as the branches with a few small stellate hairs. Prickles stout, often thickened downwards on the stem and leaves, more slender on the calyxes. Leaves deeply pinnatifid, with very obtuse rounded obovate or spathulate lobes, often sinuate, the whole leaf 3 to 6 in. long. Racemes pedunculate, few-flowered, short and simple or rarely once-forked. Calyx divided to the middle into obtuse lobes. Corolla rather large, divided to near the middle into broad lobes. Berries globular, rather large, variegated green and white or at length yellow.—Sibth. Fl. Græc. t. 235.

N. S. Wales and **Victoria.** A native of the Mediterranean region and of S. Africa, early introduced into the neighbourhood of Port Jackson, *R. Brown,* and now naturalized there as well as at Plenty Creek in Victoria, and probably some other places, *F. Mueller* and others.

31. **S. armatum,** *R. Br. Prod.* 446. A diffuse herb or undershrub of 2 to 3 ft., quite glabrous except the corolla, or with a very few small stellate hairs scattered on the young shoots. Prickles slender, numerous on the stems, leaves, inflorescence, and calyxes. Leaves ovate or broadly oblong-lanceolate, acute, sinuate-lobed or pinnatifid, with acute broad or rarely narrow often sinuate lobes, the larger leaves 3 to 4 in. long. Flowers usually 2 or 3 but sometimes more numerous in loose lateral racemes, the common peduncle more or less elongated above the lowest pedicel, the pedicels rather long. Calyx 4 to 5 lines long at the time of flowering, with lanceolate acuminate lobes, and scarcely enlarged afterwards. Corolla $\frac{3}{4}$ to 1 in. diameter, stellate-pubescent outside, the lobes not very deep, acute or sometimes much dilated and obtuse. Filaments short; anthers scarcely tapering upwards. Berry globular, variegated, above $\frac{1}{2}$ in. diameter.—Dun. in DC. Prod. xiii. part i. 295 ; *S. hystrix,* Dun. l. c. 296 and some others, but not of R. Br. ; *S. pungetium,* Sieb. Pl. Exs. not of R. Br.

Queensland. Near Warwick, *Beckler.*
N. S. Wales. Port Jackson to the Blue Mountains, *R. Brown, Sieber, n.* 254, and many others ; Hastings river, *Beckler.*
Victoria. Lake King and shaded valleys Dandenong Ranges, *F. Mueller.*

S. prinophyllum, Dun. in DC. Prod. xiii. part i. 296, from Port Jackson, is probably, from his description, the same as *S. armatum.*

32. **S. hystrix,** *R. Br. Prod.* 446. A diffuse but rigid herb, quite glabrous, every part densely covered with long rather stout straight prickles. Leaves narrow oblong, sinuate or pinnatifid, with a narrow rhachis and short lobes, intensely prickly, about 2 lines long. Flowers few, on branch-like prickly peduncles. Calyx exceedingly prickly, with lanceolate lobes. Corolla pale blue, divided to about the middle, armed outside with a few prickles like those of the calyx. (R. Brown.)

S. Australia. Petrel Bay, *R. Brown.* Brown's herbarium contains only a single specimen without flowers. It resembles *S. cataphractum,* but the prickles on the corolla described by Brown are wanting in that species, as indeed in every other Australian *Solanum* known to me. The above description of the flowers is taken from Brown's notes.

33. **S. cataphractum,** *A. Cunn. Herb.* A diffuse shrub or undershrub, the under side of the leaves usually sprinkled with stellate hairs, the whole plant otherwise glabrous or nearly so except the corolla. Prickles straight, rather slender, very numerous on the stems, foliage, inflorescence, and calyxes. Leaves petiolate, sinuate-lobed or deeply pinnatifid, with broad or narrow sinuate lobes, the whole leaf 2 to 4 in. long. Flowers monoecious, the males in pedunculate racemes, the females on solitary lateral pedicels. Flowering calyx not seen. Corolla violet, about ¾ in. diameter, scarcely lobed, tomentose outside. Anthers short, very obtuse. Fruiting pedicel thickened upwards, 1 in. long or more. Berry ¾ in. diameter, enclosed in the enlarged densely prickly calyx. Seeds large and black.

N. Australia. Bat Island and Regent river, N.W. coast, *A. Cunningham* (with linear-lanceolate leaf-lobes); Montague Sound, *A. Cunningham* (with broad less deeply lobed leaves). The specimens are all in fruit only. I describe the flowers from Cunningham's notes and from a drawing of a plant formerly raised in Kew Gardens from his seeds. He distinguished the broad-leaved form as a species under the name of *S. pectinatum.*

34. **S. pungetium,** *R. Br. Prod.* 446. A diffuse herb, the branches foliage and inflorescence sprinkled with stellate hairs, without any glandular pubescence. Prickles slender but not very long, rather numerous on the branches, leaves, inflorescence, and calyxes. Leaves petiolate, from broadly ovate to almost oblong, irregularly sinuate-lobed, with short and broad but acute and sinuately toothed lobes, green on both sides, the larger leaves 2 to 4 in. long. Flowers lateral, solitary or 2 together, each on a slender pedicel without any common peduncle. Calyx 3 to 4 lines long, with narrow acuminate lobes, slightly enlarged when in fruit. Corolla of a bluish-violet, about ¾ in. diameter, with rather broad and short triangular lobes.—Dun. in DC. Prod. xiii. part i. 295.

Queensland. Bowen river, *Bowman.*
N. S. Wales. Port Jackson, *Banks and Solander, R. Brown;* Illawarra, *Back-house.*
Victoria. Broadribb river, *F. Mueller.*

This species resembles some forms of *S. armatum,* with which F. Mueller is disposed to unite it; but, besides the indumentum, the inflorescence pointed out by Brown appears to be constant. In *S. armatum,* when the raceme is reduced to 2 flowers, if one pedicel is sessile on the stem the other is always raised on a peduncle.

35. **S. eremophilum,** *F. Muell. in Linnæa,* xxv. 432. A perennial or undershrub, either small and diffuse or tall and erect, the branches foliage and inflorescence hirsute with stellate hairs but scarcely tomentose. Prickles rigid and rather long on the stem, leaves, and calyxes. Leaves petiolate, broadly ovate, obtuse, undulate and broadly sinuate-lobed, green on both sides, scarcely above 1 in. long in the specimens seen. Racemes short and few-flowered, the pedicels at length above ½ in. long. Calyx broadly campanulate, about 3 lines long at the time of flowering, divided to below the middle into broadly lanceolate membranous lobes, much enlarged after flowering. Corolla about ¾ in. diameter, with broad acute lobes not reaching to the middle. Anthers tapering upwards. Berry globular, almost covered by the enlarged slightly prickly calyx.

N. S. Wales. Macquarrie river, *Bowman.*

S. Australia. Clayey somewhat saline pastures, Flinders Range, and between Rocky river and Rocky Creek, *F. Mueller.*

From each station I have seen only a single small specimen in Herb. F. Mueller.

36. **S. campanulatum,** *R. Br. Prod.* 446. A coarse erect herb (or undershrub?) of 2 to 3 ft., the branches foliage and inflorescence more or less hirsute with stellate or simple hairs mixed with a glandular pubescence. Prickles straight, rather slender, numerous on the stem, leaves and inflorescence, few and small on the calyxes. Leaves petiolate, ovate, sinuate-lobed, with short broad angular or sinuate lobes or rarely more deeply pinnatifid, green on both sides, 2 to 4 in. long. Flowers few, in loose lateral racemes, the pedicels at length above 1 in. long and distant along the common peduncle. Calyx 4 to 5 lines long, with subulate-acuminate lobes, enlarged in fruit and deeply divided into lanceolate acuminate segments. Corolla violet or blue, broadly campanulate or sometimes almost rotate, but always less open than in other Australian species, about 1 in. diameter, very shortly and broadly lobed. Anthers but slightly tapering upwards. Berry globular, ¾ to 1 in. diameter, surrounded by but not completely enclosed in the enlarged prickly calyx.—Dun. in DC. Prod. xiii. part i. 297 ; Bot. Mag. t. 3672.

Queensland. Araucaria Ranges, Burnet river, *Leichhardt* (apparently this species, but the specimen imperfect).

N. S. Wales. Port Jackson and Grose river, *R. Brown ;* Curreejong, *Woolls ;* New England, *C. Stuart ;* Clarence river, *Beckler.*

37. **S. adenophorum,** *F. Muell. Fragm.* ii. 162. An erect perennial, sometimes under 1 ft. and from that to 2 ft. high, the branches foliage and inflorescence hirsute with simple or stellate hairs mixed with a glandular pubescence. Prickles slender, rather numerous on the branches, leaves, inflorescence, and calyxes. Leaves petiolate, ovate, sinuately lobed or pinnatifid with rather obtuse sinuate lobes, green on both sides, 2 to 4 in. long. Flowers rather small, pale blue or white, in loose racemes on rather long peduncles. Calyx 3 to 4 lines long at the time of flowering, deeply divided into narrow subulate-acuminate lobes, enlarged in fruit and the points then very long. Corolla glabrous or with a very few stellate hairs outside, about ½ in. diameter, deeply divided into narrow lobes. Anthers rather long, tapering upwards. Berry whitish, globular, scarcely ½ in. diameter.

Queensland. Barren hills between the Mackenzie and Dawson rivers, *F. Mueller* (with most of the leaves rather deeply lobed) ; Rockingham Bay, *Dallachy* (leaves mostly sinuate-toothed or shortly lobed).

38. **S. cinereum,** *R. Br. Prod.* 446. An erect undershrub, the branches and inflorescence stellate-tomentose. Prickles slender, numerous on the branches, leaves, inflorescence, and calyxes. Leaves petiolate, ovate or ovate-lanceolate, mostly acuminate, more or less deeply sinuate-lobed or pinnatifid, 2 to 5 in. long, green above and glabrous or sprinkled with a few minute stellate hairs, white underneath with a soft stellate often floccose tomentum. Flowers blue, rather large, in pedunculate racemes often as long as the leaves, the pedicels at first short, much elongated in fruit. Calyx 3 to 4 lines long, somewhat enlarged after flowering, densely prickly, with acu-

minate lobes.   Corolly nearly 1 in. diameter, with short broad acute lobes.
Berry globular, ¾ to 1 in. diameter.—Dun. in DC. Prod. xiii. part i. 294.

N. S. Wales.   Grose river, *R. Brown ;* Hunter's and Mackenzie rivers and Whin-stone rocks on the skirts of Liverpool Plains, *Fraser, A. Cunningham ;* Nepean river, *Fraser* (with the calyx less prickly) ; Gwydir river, *Leichhardt ;* near Bathurst, *Woolls.*

S. elegans, Dun. Syn. Sol. 28, and in DC. Prod. xiii. part i. 335, from the detailed description, does not appear to me to differ from *S. cinereum.*

S. semiarmatum has sometimes the aspect of this species, but is readily known by the small calyx with short teeth and without prickles, and by the more numerous smaller flowers.

39. **S. lacunarium,** *F. Muell. in Trans. Phil. Soc. Vict.* i. 18, *and in Hook. Kew Journ.* viii. 166.   A small perennial or undershrub, most of the specimens under 6 in. and none exceeding 1 ft., the branches inflorescence and under side of the leaves hoary with a minute stellate tomentum.   Prickles rather numerous, usually red, on the branches, leaves, and calyxes.   Leaves deeply pinnatifid, with distant oblong and short or narrow and long lobes, all very obtuse and entire or sinuate, the whole leaf 1 to 3 in. long, the upper surface glabrous or sprinkled with a few minute stellate hairs.   Flowers not numerous, in loose pedunculate racemes often as long as the leaves.   Calyx campanulate, about 2 lines long, with short broad lobes.   Corolla rather above ½ in. diameter, tomentose outside, deeply divided into acute lobes.   Anthers tapering upwards.   Berry globular, yellow, surrounded by the slightly enlarged calyx.

N. S. Wales.   Desert of the Murray and Darling, *F. Mueller, Victorian Expedition,* and others.

40. **S. petrophilum,** *F. Muell. in Linnæa,* xxv. 433.   A low spreading shrub or undershrub, the branches foliage and inflorescence covered with a dense soft stellate tomentum, often yellowish or rusty.   Prickles rather slender, on the branches, leaves, and calyxes.   Leaves ovate lanceolate or oblong, obtuse, much undulate and sinuate-lobed, thick and soft, mostly ¾ to 1 in long.   Flowers large, bluish, in terminal or lateral racemes, the pedicels at first very short and under ½ in. when in fruit.   Calyx about 4 lines long, deeply divided into narrow lobes, each with a very prominent midrib terminating in the point, and after flowering the scarcely enlarged calyx is often almost reduced to the 5 linear ribs.   Corolla fully 1 in. diameter, with short and broad lobes.   Anthers slightly tapering upwards.   Berry depressed-globular, under ½ in. diameter.

N. S. Wales.   Mutanie Ranges, *Victorian Expedition.*
S. Australia.   Dry rocky wastes about Lake Torrens, *F. Mueller ;* Flinders Range, *Howitt's Expedition ;* in the N.W. interior, *M'Douall Stuart ;* also probably this species, head of Spencer's Gulf, *R. Brown* (without flowers).

41. **S. diversiflorum,** *F. Muell. Fragm.* vi. 146.   A straggling shrub or undershrub of 1 ft. or more, the branches foliage and inflorescence thickly covered with stellate hairs, not usually so soft as in the allied species, but sometimes floccose on the young leaves.   Prickles very small or rarely long, few or numerous on the branches and leaves, often entirely wanting on the male flowers, longer and more dense on the fertile and fruiting calyx.

Leaves deeply pinnatifid, with oblong very obtuse entire or sinuate lobes, the whole leaf 1 to 2 in. long. Flowers in lateral racemes, often as long as the leaves, but as in several allied species usually sterile except the lowest one of each raceme, which is on a longer pedicel proceeding from the base of the peduncle. Calyx about 3 lines long, with lanceolate acuminate lobes, without prominent keels or midribs, enlarged and very prickly round the growing fruit. Corolla (about ¾ in. diameter?) with short broad lobes. Anthers tapering upwards. Fruit only seen young.

N. Australia. Upper Victoria river, *F. Mueller;* La Grange Bay, N.W. coast, *Marten;* Port Walcott, *Harper.*

The monœcious character upon which the specific name was founded is common to several of the following species as well as to the American and Asiatic group of *Melongenæ*, but appears to exist in a less degree in some other groups so as to be scarcely available, in the present state of our acquaintance with the genus, as a sectional distinction.

42. **S. carduiforme,** *F. Muell. Fragm.* ii. 163. An erect herb of 1 to 3 ft. of a pale glaucous green, covered with a stellate tomentum rather loose on the branches, very short and not dense on both sides of the leaves. Prickles rather slender but long and very numerous on the branches, leaves, and especially on the female calyxes. Leaves narrow, irregularly pinnatifid, with rather narrow obtuse entire or sinuate lobes, the whole leaf 3 to 4 in. long. Sterile flowers numerous, in dense racemes on long lateral peduncles. Calyx at the time of flowering about 3 lines long, campanulate, with broad lobes. Corolla not large. Fertile flowers probably solitary on lateral peduncles, which are still very short in fruit. Fruiting calyx large, globular, very densely armed with long rigid prickles, enclosing a globular berry of ½ in. or more.

N. Australia. Sandy and rocky banks of Nicholson river, Gulf of Carpentaria, *F. Mueller.*

43. **S. melanospermum,** *F. Muell. Fragm.* ii. 163. An erect shrub or undershrub of 1 to 3 (or 4?) ft., the branches foliage and inflorescence densely and softly stellate-tomentose. Prickles not very long, straight, rather numerous on the stem, few on the leaves, more abundant and stouter on the calyxes. Leaves petiolate, ovate, scarcely acute, thick and soft, 1½ to 2½ in. long, rather deeply sinuate-lobed, with broad very obtuse lobes. Flowers not seen. Fruiting pedicels solitary, lateral, above 1 in. long. Berry yellow, globular, at least 1 in. diameter, surrounded by the large, almost membranous broadly lobed calyx, armed with stout prickles, at first closely appressed and almost covering the fruit, at length reflexed. Seeds large and black as in *S. cataphractum.*

N. Australia. Abel Tasman river, *F. Mueller.* Like *S. cataphractum,* this is probably monœcious and of the *Melongena* group.

44. **S. horridum,** *Dun. Syn. Sol.* 28, *and in DC. Prod.* xiii. *part* i. 296. Branches foliage and inflorescence very copiously woolly-hirsute with long loose stellate hairs of a yellowish or rusty colour. Prickles long straight and very numerous on the stem and leaves, usually rather smaller on the calyxes. Leaves on long petioles, ovate or ovate-lanceolate, acute or scarcely obtuse, entire or sinuate and often much undulate, 1 to 2 in. long. Pedicels in the

specimens seen solitary and lateral. Calyx about 3 lines long at the time of flowering, with narrow lobes, much enlarged afterwards and more deeply divided into ovate-lanceolate acuminate lobes. Corolla under ¾ in. diameter, with short broad lobes. Anthers tapering upwards. Berry large, globular, the enlarged calyx spreading under it and very prickly, but the prickles smaller than those of the stem and leaves.

N. Australia, *Baudin's Expedition* (*Herb. Banks*), and apparently the same species, with rather broader more undulate leaves, Depuech Island, N.W. coast, *Bynoe.*

The species is, according to Dunal, also in Timor.

45. **S. echinatum,** *R. Br. Prod.* 447. An erect or diffuse undershrub, the branches foliage and inflorescence very densely and softly stellate-tomentose, often velvety or floccose. Prickles slender, rather small on the stems, few or none on the leaves, more abundant and longer on the calyxes. Leaves on rather long petioles, ovate oblong or lanceolate, obtuse, entire or slightly sinuate, very thick and soft, mostly 1 to 2 in. long. Racemes lateral, loose, the common peduncle elongated. Flowering calyx 2 to 3 lines long, with short lanceolate lobes, very tomentose, with small prickles ; when in fruit globular, membranous, very prickly, about ¾ in. diameter, completely enclosing the globular berry, the broad triangular lobes almost meeting over it. Corolla very tomentose, ½ to ¾ in. diameter, shortly and broadly lobed.—Dun. in DC. Prod. xiii. part i. 297.

N. Australia. N.W. coast, *Bynoe;* Upper Victoria river, *F. Mueller;* islands of the Gulf of Carpentaria, *R. Brown.*

According to R. Brown the berry is almost 4-celled (*subquadrilocularis*), which I have been unable to verify in our specimens. In two flowers that I examined I found the ovary 2-celled only, but with slight indications of transverse spurious dissepiments, which may probably grow out as the fruit enlarges.

46. **S. lasiophyllum,** *Dun. in Poir. Dict. Suppl.* iii. 764, *not of Syn. Sol.* A stout rigid shrub or undershrub attaining 2 or 3 ft., the branches foliage and inflorescence very densely and softly stellate-tomentose. Prickles very slender and not long, few or numerous on the branches, few or none on the leaves, rarely wanting on the young calyx and always present on the fruiting one. Leaves very shortly petiolate, from ovate-oblong to nearly orbicular, very obtuse, entire or scarcely sinuate, very thick and soft, rarely 2 in. long and often under 1 in. Flowers few, large, in short dense racemes, the peduncle and pedicels thick and soft. Calyx very thick and woolly, with short thick narrow lobes, 3 to 4 lines long when in flower, enlarged afterwards. Corolla 1 to 1½ in. diameter, very shortly and broadly lobed, the lobes generally with a short point. Anthers tapering at the end. Ovary 2-celled. Berry ovoid-globular, almost enclosed in the calyx, which is then globular and above ½ in. diameter.—*S. eriophyllum,* Dun. Syn. Sol. 30, and in DC. Prod. xiii. part i. 300 ; F. Muell. Fragm. vi. 145.

W. Australia. Sharks' Bay, *Milne;* Murchison river, *Oldfield.* The specimen of Baudin's in Herb. Mus. Brit. (probably from the W. not the E. coast), referred to by Dunal, has the leaves rather more sinuate than Drummond's. Dunal does not say for what reason he adopted Kunth's unpublished name of *S. lasiophyllum* for a S. American species, when his own, for the present species, had already been published by Poiret for three years.

Var. ? *crassissimum.* Leaves broad, almost orbicular, cordate at the base, 2 to 3 in.

diameter, excessively thick. Racemes longer; flowers and calyx the same but more woolly.

N. Australia. N.W. coast, *Bynoe.*

47. **S. ellipticum,** *R. Br. Prod.* 446. A shrub or undershrub, either very low and spreading or taller and erect, the branches foliage and inflorescence covered with a dense stellate tomentum, sometimes very thick soft and velvety or floccose, sometimes shorter and closer. Prickles slender, few or numerous on the stems and calyxes, few or none on the leaves. Leaves petiolate, from broadly ovate to ovate-lanceolate or oblong, obtuse, entire or slightly sinuate or undulate, rounded or cordate at the base, mostly 1 to nearly 3 in. long. Flowers in lateral racemes, often longer than the leaves, the pedicels usually short. Calyx-tube at the time of flowering 1 to 2 lines long, the lobes or teeth narrow and thick, from very short to fully twice as long as the tube; after flowering the calyx much enlarged and dividing into broad lobes with short or long narrow points. Corolla violet, with short broad lobes, apparently varying in size from about ½ to above ¾ in. diameter. Ovary 2-celled. Berry globular, surrounded by but not enclosed in the enlarged calyx.—Dun. in DC. Prod. xiii. part i. 298; *S. lithophilum,* F. Muell. in Linnæa, xxv. 434.

N. Australia. Hammersley Range, N.W. coast, *Maitland Brown;* Sea Range and Sturt's Creek, *F. Mueller.*

Queensland. Broad Sound, *R. Brown;* near Peak Range, *Leichhardt;* Suttor desert, Dawson and Mackenzie rivers, *F. Mueller;* Rockhampton, *O'Shanesy, Dallachy;* Suttor and Connor rivers and Nerkool Creek, *Bowman;* Flinders river, *Sutherland;* Maranoa river, *Mitchell;* Armadillo, *Barton.*

N. S. Wales. Peele's Range, *Fraser;* Darling river, *E. Giles;* thence to the Barrier Range, *Victorian Expedition;* Mount Murchison, *Bonney.*

S. Australia. Flinders Range and Cudnaka, *F. Mueller;* near Spencer's Gulf, *Warburton;* in the interior, *M'Douall Stuart;* Cooper's River, *A. C. Gregory;* Lake Gillies, *Burkitt.*

W. Australia, *Drummond, n.* 87; Murchison river, *Oldfield.*

Var. *pannifolium,* A. Cunn. Tomentum ferruginous, very copious, almost woolly. Stems very prickly but not the leaves.—Cambridge Gulf, N.W. coast, *A. Cunningham.*

48. **S. quadriloculatum,** *F. Muell. Fragm.* ii. 161. An undershrub attaining several feet, the branches foliage and inflorescence densely and softly stellate-tomentose. Prickles straight, rather slender, long or short, few or many on the stems and calyxes, few or none on the leaves. Leaves petiolate, ovate or ovate-lanceolate, acute or acuminate, entire, very unequal at the base, thick and soft, 2 to 4 or even 5 in. long. Flowers small and numerous, in long lateral racemes (the upper ones sterile?), the pedicels rather short. Calyx-tube campanulate, under 2 lines long, with short or long soft points or lobes, enlarged after flowering and dividing into broad lobes with narrow points. Corolla above ½ in. diameter, rather deeply lobed. Ovary 4-celled. Berry globular, surrounded by but not enclosed in the enlarged calyx, but not seen quite ripe.

N. Australia. Upper Victoria river and Nicholson river, Gulf of Carpentaria, *F. Mueller;* in the interior, lat. 22°, *M'Douall Stuart.*

49. **S. phlomoides,** *A. Cunn. Herb.* An undershrub or shrub, either

low and prostrate or erect and attaining 1½ to 2 ft., very densely and softly woolly with stellate hairs. Prickles slender, few or numerous, as in *S. quadri-loculatum.* Leaves as in that species ovate or ovate-lanceolate, mostly acuminate, 3 to 4 in. long, but thicker and softer. Flowers large, numerous, in long racemes, on very short thick pedicels. Calyx nearly ½ in. long when in flower, rather deeply divided into narrow thick lobes, much enlarged afterwards, divided under the fruit into broad acuminate lobes at least 1 in. long and very open. Corolla 1 to 1¼ in. diameter, with short broad lobes. Berry depressed, globular, above 1 in. diameter.

N. Australia. Enderby island, N.W. coast, *A. Cunningham ;* Hammersley Range, *Maitland Brown.* I have been unable to examine the ovary, having found it destroyed by insects in the flowers I opened.

50. **S. Cunninghamii,** *Benth.* An undershrub or shrub, from under 1 ft. to 4 or 5 ft. high, the branches foliage and inflorescence covered with a stellate tomentum, sometimes thick and floccose especially on the under side of the leaves, shorter and harsher on the upper side. Prickles few and small except on the calyx-tube. Leaves petiolate, lanceolate or almost ovate, rather obtuse, entire, 2 to 3 in. long, rather thick and soft. Pedicels in all the specimens seen 1-flowered, lateral, 1 to 1½ in. long. Calyx-tube globular, not 3 lines long, densely armed with long prickles ; lobes linear, almost terete, fully twice as long as the tube. Corolla large with acute or acuminate lobes, not well open in our specimens, but evidently above 1½ in. diameter. Fruit not seen.

N. Australia. Cygnet Bay, N.W. coast, *A. Cunningham, Bynoe ;* Glenelg district, *Marten.* The flowers are probably monœcious, as in the *Melongenas,* and, if so, the male or sterile flowers may be, as in other species, racemose and less prickly.

*2. **NICANDRA,** Gærtn.

Calyx of 5 distinct broadly cordate segments or sepals, becoming much enlarged and inflated in fruit. Corolla campanulate, with 5 broad short lobes, folded (and perhaps also slightly imbricated) in the bud. Anthers short, opening longitudinally. Ovary 3- to 5-celled. Fruit a berry, enclosed in the enlarged calyx. Embryo curved in a fleshy albumen.—An erect annual, with the habit and foliage nearly of *Physalis.*

The genus is limited to a single species.

*1. **N. physalodes,** *Gærtn. ; Dun. in DC. Prod.* xiii. *part* i. 434. An erect glabrous annual or biennial, attaining sometimes 5 or 6 ft., but usually smaller. Leaves petiolate, ovate, irregularly sinuate or coarsely toothed or lobed, 3 or 4 in. long or sometimes larger. Flowers pale blue, solitary, on short pedicels in the upper axils, forming a terminal leafy raceme. Calyx-segments at the time of flowering a little more than ½ in. long and herbaceous, when in fruit above 1 in. long, thin and much veined, closely connivent, forming a vesicular calyx with very prominent angles. Corolla nearly 1 in. long. Berry globular.

N. S. Wales. A native of S. America, which has established itself as a weed in several parts of the warmer regions of the Old World, and has been received as such from various parts of N. S. Wales (*Herb. F. Mueller* and others).

3. PHYSALIS, Linn.

Calyx 5-toothed or 5-lobed, inflated after flowering. Corolla broadly campanulate or nearly rotate, 5-angled, folded in the bud. Anthers short, opening longitudinally. Ovary 2-celled. Berry globular, enclosed in the inflated calyx. Embryo circular or spiral round the fleshy albumen.—Herbs either annual or with a perennial stock. Leaves often in pairs. Flowers solitary, usually small, on axillary or lateral pedicels.

A genus rather numerous in America, of which two or three species, including the Australian ones, extend over the warmer regions of the Old World.

Stock perennial, the whole plant softly pubescent . . . . 1. *P. peruviana.*
Annual, sparingly pubescent. Flowers very small . . . . 2. *P. minima.*

*1. **P. peruviana,** *Linn.; Dun. in DC. Prod.* xiii. *part* i. 440. A herbaceous perennial of 1 to 2 ft., softly pubescent or tomentose with simple hairs. Leaves petiolate, broadly ovate, acuminate, entire or slightly sinuate-toothed, mostly cordate at the base, 2 to 3 in. long. Pedicels short, rarely ½ in. long even in fruit, recurved after flowering. Calyx when in flower about 3 lines long, with narrow lobes as long as the tube. Corolla rather above ½ in. diameter, pale yellow with purple spots in the centre. Fruiting calyx vesicular, with connivent teeth, 1 to 1½ in. long, reticulate with the principal veins prominent but not so angular as in *P. minima.* Berry globular, yellow. —Nees in Pl. Preiss. i. 344 ; *P. pubescens,* R. Br. Prod. 447, and of Linn. Herb. but not of Linn. Spec. Pl. ; *P. edulis,* Sims, Bot. Mag. t. 1068.

N. Australia. Sturt river, *F. Mueller.*
Queensland. Brisbane river, *F. Mueller.*
N. S. Wales. Very common in the colony, *R. Brown ;* Clarence river, *Beckler.*
S. Australia. Near Adelaide, *Blandowsky.*
W. Australia. Cape Leschenault, *Oldfield.*

The species is of South American origin, and perhaps really indigenous in the islands of the Pacific, but long since cultivated for its berries, said to be edible, and established as a weed in several tropical countries, and therefore perhaps introduced only in Australia.

2. **P. minima,** *Linn.; Dun. in DC. Prod.* xiii. *part* i. 445. An erect annual of about 1 ft., with spreading branches, more or less pubescent with scattered simple hairs. Leaves petiolate, ovate, acute or acuminate, irregularly sinuate-toothed or rarely entire, thin and membranous, mostly 2 to 3 in. long. Flowers very small, on filiform pedicels sometimes very short, sometimes above ½ in. long. Calyx when in flower scarcely 1½ lines long, with short acuminate teeth. Corolla about twice as long as the calyx, pale yellow, the centre often purple. Fruiting calyx about 1 in. long, vesicular, with 5 prominent angles and acuminate connivent teeth. Berry globular.—*P. parviflora,* R. Br. Prod. 447 ; Dun. in DC. Prod. xiii. part i. 444, with some other supposed species enumerated by Dunal.

N. Australia. Victoria river, *F. Mueller.*
Queensland. Broad Sound and Keppel Bay, *R. Brown ;* Moreton Bay, *Leichhardt, F. Mueller ;* Rockhampton, *O'Shanesy, Dallachy;* Nerkool and Crocodile Creeks, *Bowman ;* Rockingham Bay, *Dallachy ;* Port Molle, *M'Gillivray.*

The species is dispersed over tropical America, Asia, and Africa, and very common in E. India.

4. LYCIUM, Linn.

Calyx with 5, rarely 4 teeth, often ultimately dividing into 3 to 5 lobes. Corolla more or less funnel-shaped, the tube expanding into a campanulate 5- rarely 4-lobed limb, the lobes imbricate in the bud. Stamens usually unequal, longer or shorter than the corolla; anthers opening longitudinally. Ovary 2-celled. Berry ovoid or globular. Embryo curved or semicircular, in a fleshy embryo.—Shrubs, usually glabrous, the branchlets often spinescent. Leaves entire, usually small, often clustered on the old nodes. Flowers pedicellate, solitary or several together at the ends of the branchlets or in the clusters of leaves.

The genus is widely spread over the temperate and subtropical regions of the world, especially numerous in S. America and S. Africa. The only Australian species is endemic.

1. **L. australe,** *F. Muell. in Trans. Phil. Soc. Vict.* i. 20, and *Fragm.* i. 83. A scrubby spreading glabrous shrub of 2 to 3 ft., the smaller branchlets often degenerating into spines. Leaves clustered at the old nodes. obovate spathulate or oblong, obtuse, thick and fleshy, not ¼ in. long. Flowers usually solitary at the nodes, on short recurved pedicels. Calyx scarcely 1 line long, with minute teeth. Corolla white (*F. Muell.*), about 5 lines long, the tube rather slender, gradually dilated upwards, with 5 rarely 4 ovate obtuse lobes of about 1 line in length. Filaments inserted near the base of the tube, the longest nearly as long as the corolla, hairy to about the middle.

N. S. Wales. Desert of the Murray and Darling, *F. Mueller, Hergott.*
S. Australia. Subsaline pastures on the Murray, *Behr, F. Mueller.*

The species has entirely the aspect of some of the small-leaved S. African ones.

L. chinense, Mill. ; Dun. in DC. Prod. xiii. part i. 510, which includes *L. vulgare*, Dun. l. c. 509 ; Miers, Illustr. ii. 120. t. 70, a species much planted and now naturalized in various parts of Europe and Asia, is also sent as an introduced plant from Port Phillip by F. Mueller, and is in Leichhardt's collection. It is a tall glabrous shrub, with long weak, recurved or pendulous branches. Leaves oblanceolate to obovate, ½ to 1 in. long or even longer. Corolla with a very short tube and deeply-lobed campanulate limb, the lobes about 3 lines long. Stamens exserted.

5. ANTHOTROCHE, Endl.

Calyx broadly campanulate, 5- rarely 6-lobed. Corolla broadly campanulate, with 5 rarely 6 lobes, induplicate in the bud. Anthers reniform, 1-celled (the 2 cells completely confluent and opening in a single slit), turned outwards in the bud. Ovary 2-celled. Capsule small, smooth, opening in 4 valves. Embryo curved, in a fleshy albumen.—Shrubs, more or less tomentose with plumose or stellate hairs. Leaves rather small, obtuse, entire. Flowers solitary in the axils of the floral leaves.

The genus is endemic in Australia.

Flowers sessile, crowded on the branches, loosely woolly. Old leaves
  nearly glabrous . . . . 1. *A. pannosa.*
Flowers pedicellate, distant. Tomentum of the whole plant close or
  rather loose, usually rusty . . . . 2. *A. Walcottii.*

1. **A. pannosa,** *Endl. Nov. Stirp. Dec.* 7. An apparently erect shrub, the branches rather stout, terete, covered as well as the young shoots and calyxes with a woolly tomentum, consisting chiefly of plumose hairs mixed

2 H 2

with a few stellate ones.   Leaves very shortly petiolate or almost sessile, ovate obovate or oblong, very obtuse, thick, $\frac{1}{2}$ to 1 in. long, nearly glabrous when full grown.   Flowers sessile, solitary in the axils of small leaves, which are usually crowded in the axils of the upper leaves and ends of the branches. Calyx about $2\frac{1}{2}$ lines long, deeply divided into lanceolate lobes.   Corolla dull purple, about 4 or 5 lines diameter, the lobes broad, obtuse, as long as or longer than the tube, reflexed when fully open, very pubescent outside. Filaments dilated and pubescent at the base, and closed over the ovary, then filiform and recurved, rather shorter than the corolla.   Capsule enclosed in the persistent calyx, quite glabrous.   Seeds usually very few, and sometimes only a single one ripening; testa reticulate, rugose, slightly coriaceous.—A. DC. Prod. xiii. part i. 676; Miers, Illustr. ii. App. 36. t. 86.

W. Australia, *Roe (Endl. l. c.), Drummond.*

2. **A. Walcottii,** *F. Muell. Fragm.* i. 123.   An erect much-branched shrub of 1 to $1\frac{1}{2}$ ft., covered with a dense tomentum, sometimes very short and close, sometimes looser and almost floccose, usually rust-coloured, consisting, as in *A. pannosa,* of plumose hairs, persistent on the leaves as well as on the rest of the plant.   Leaves ovate obovate or orbicular, contracted into a petiole more prominent than in *A. pannosa,* very obtuse, $\frac{1}{2}$ to 1 in. long or in some specimens all under $\frac{1}{2}$ in.   Flowers scattered along the branches, solitary in the axils, on pedicels at least half as long as the leaves, smaller than in *A. pannosa,* but otherwise similar.   Calyx about 2 lines long, with obtuse lobes.   Corolla dark purple.   Filaments very hispid at the base.

W. Australia.   Murchison river, *Oldfield, Drummond;* Dirk Hartog's Island, *Milne.*

6. DATURA, Linn.

Calyx tubular, circumsciss near the base after flowering.   Corolla funnel-shaped, with a long tube and a broad 5-angled or 5-toothed limb, folded in the bud.   Ovary 2-celled, each cell incompletely divided into two.   Fruit an ovoid or globular capsule, opening in 4 short valves, and usually beset with prickles.   Embryo curved round a fleshy albumen.—Tall coarse herbs, or, in S. American species, shrubs or soft-wooded trees.   Leaves alternate, often in pairs.   Flowers solitary, terminal or lateral, usually very large.

A small genus, chiefly American, with two or three species equally common in, and perhaps indigenous to, the Old World.   The only Australian species is endemic, at least in its Australian form.

1. **D. Leichhardtii,** *F. Muell. in Trans. Phil. Soc. Vict.* i. 20.   An erect annual of 1 to 3 ft., sparingly pubescent.   Leaves petiolate, ovate, acute or shortly acuminate, irregularly sinuate-toothed or lobed, mostly 3 to 4 in. long.   Flowers of a pale yellowish-white, on short peduncles either terminal or in the forks, and recurved after flowering.   Calyx scarcely above 1 in. long.   Corolla about twice as long as the calyx, the angles produced into short points.   Capsule reflexed, globular, about 1 in. diameter, very prickly, resting on the broadly expanded persistent base of the calyx.—*D. alba,* F. Muell. Fragm. vi. 144, but scarcely of Nees.

N. Australia.   Gulf of Carpentaria, *Landsborough;* Ashburton river, *Walcott.*

Queensland. Gilbert river, *F. Mueller;* Comet river, *Leichhardt;* Suttor river, *Dorsay;* Rockingham Bay, *O'Shanesy;* Armadillo, *Barton.*

The very common East Indian *D. alba,* Nees, or *D. Metel,* Roxb., has the flowers fully twice as large as *D. Leichhardtii,* and the leaves more entire. The Australian plant has more the aspect of the common *D. Stramonium* or of *D. ferox,* with the small flowers of the latter, but differs from both in the reflexed capsule.

D. Tatula, Linn. Sp. Pl. 256 ; Sweet, Brit. Fl. Gard. t. 83, regarded by Dunal and most authors as a variety of *D. Stramonium* with blue flowers, but whose claims to be retained as a species have been recently again brought forward by Naudin's hybridizing experiments, has appeared in Australia as an introduced weed.

7. NICOTIANA, Linn.

Calyx campanulate, 5-lobed, persistent. Corolla with a cylindrical tube, the limb more or less spreading, 5-lobed, induplicate or folded in the bud. Stamens 5, included in the tube, often unequal ; anthers 2-celled, opening longitudinally. Ovary 2-celled ; stigma broadly 2-lobed. Fruit a capsule opening in 2 bifid valves parallel to the dissepiment which remains attached to the axis. Seeds numerous. Embryo slightly curved, in a fleshy albumen. —Herbs usually erect and coarse. Leaves alternate, entire. Flowers white greenish-yellow or dull-red, in terminal racemes often branching into very loose panicle-like cymes.

The genus is entirely American, with the exception of the single Australian species, which, however, is scarcely to be distinguished from a S. American one, and of one nearly allied to it from the S. Pacific islands. Some species, long cultivated under the name of *Tobacco,* have become almost naturalized in the warmer regions of the Old World, but we have as yet seen no Australian specimens. F. Mueller's collection contains, however, as an escape from gardens, a specimen of *N. glauca,* Grah. in Bot. Mag. t. 2837, a perfectly glabrous glaucous species, with rather slender tubular flowers of a greenish yellow, with a very small limb.

1. **N. suaveolens,** *Lehm. Hist. Nicot.* 43. An erect annual or biennial of 1 to 2 ft., more or less pubescent or villous and usually viscid. Lower leaves on long petioles, ovate or spathulate, the upper ones usually narrow and sessile although contracted at the base, but exceedingly variable, sometimes all cordate and the upper small ones clasping the stem, sometimes all narrow with very few on the stem, the petiole in some specimens dilated at the base, and stem-clasping or shortly decurrent. Flowers sweet-scented, especially at night, of a pure white or greenish outside, in loose terminal racemes often branching into irregular panicles, and exceedingly variable in size, on short or long pedicels. Bracts usually small and linear or none under the upper pedicels, but sometimes all larger and leafy. Calyx varying from 3 to 6 lines long, the lobes usually very narrow and as long as the tube. Corolla-tube slender or broad, varying from $\frac{1}{2}$ in. to 2 in. in length, usually slightly swollen under the throat ; limb spreading flat, from $\frac{1}{2}$ to 1 in. diameter, the lobes short and broad, emarginate obtuse or almost acute, the 2 upper ones usually rather smaller than the others. Filaments adnate high up ; anthers ovate or oblong, 4 usually at the throat of the corolla, the fifth much lower down. Capsule ovate, slightly acuminate, rather shorter than the calyx-lobes. Seeds very small and numerous.—Dun. in DC. Prod. xiii. part i. 565 ; *N. undulata,* Vent. Jard. Malm. t. 10 ; Bot. Mag. t. 673 ; R. Br. Prod. 447 ; *N. Australasiæ,* R. Br. in Tuck. Cong. App. 472, Misc. Works, ed. Benn. i.

158 ; *N. rotundifolia,* Lindl. Bot. Reg. 1838, Misc. 59 ; *N. fastigiata,* Nees in Pl. Preiss. i. 343.

N. Australia. N.W. coast, *Bynoe* (with broad cordate leaves) ; Nichol Bay, *Gregory's Expedition* (with small narrow leaves).

Queensland. Rockhampton and Rockingham Bay, *Dallachy;* Bowen river, *Bowman ;* in the interior, *Mitchell ;* Curriwillighi and Armadillo, *Dalton.*

N. S. Wales. Port Jackson to the Blue Mountains, *R. Brown* and others ; Macleay and Hastings rivers, *Beckler ;* southward to Kiama, *Harvey ;* in the interior from Lachlan and Darling rivers to the Barrier Range, *Victorian and other Expeditions.*

Victoria. Port Phillip, *Gunn ;* Murray river, *F. Mueller;* near Ballarat, *H. W. Locker.*

S. Australia. Head of Spencer's Gulf, *R. Brown ;* Murray river, *F. Mueller;* Torrens river, *Whittaker* (very hirsute, with broad leaves and decurrent auriculate petioles); Lake Gillies, *Burkett ;* abundant at Wills Creek, *Howitt's Expedition.*

W. Australia, *Drummond ;* Murchison and Blackwood rivers, *Oldfield.*

I cannot readily distinguish the species from the Chilian *N. acuminata,* Grah., which is perhaps again the same as *N. angustifolia,* Ruiz and Pav., from the same country. In Australia it varies exceedingly both in foliage and flowers, the most marked forms I have seen are the following :—

Var. *parviflora.* Corolla much under 1 in. long. Leaf-petioles sometimes auriculate, sometimes not. Panicle large and loose.—Queensland and northern part of N. S. Wales.

Var. *longiflora.* Corolla-tube at least 2 in. long. Leaves various.—In the interior of Queensland and N. S. Wales.

Var. *cordifolia.* Leaves almost all cordate. Calyx large with broad lobes. Corolla of the common size (1 to 1½ in. long).—N.W. coast, *Bynoe.*

N. Neesii, Lehm. in Pl. Preiss. i. 344, from W. Australia, which I have not seen, is probably a variety of the same species, remarkable for the stem, especially near the base, being densely covered with white wool.

ORDER LXXXIII. SCROPHULARINEÆ.

Flowers irregular or seldom nearly regular. Sepals 5, either free or more frequently united in a toothed or lobed calyx. Corolla usually 2-lipped, but sometimes nearly regular, with 5 or rarely 4 or more than 5 lobes, more or less imbricate, and in one tribe folded in the bud. Stamens usually 2 or 4, in pairs, inserted in the tube and alternating with the lower lobes of the corolla ; the fifth stamen, between the 2 upper lobes, usually deficient or rudimentary or sterile, very rarely perfect ; anthers 2-celled or 1-celled by the confluence of the cells or by the abortion of one of them, the cells opening longitudinally. Ovary 2-celled, with several ovules in each cell, attached to a single placenta in the centre of the dissepiment. Style simple with a 2-lobed or rarely entire stigma. Fruit a 2-celled capsule or very rarely an indehiscent berry. Seeds with more or less of albumen, the testa usually reticulate or tubercular-rugose, sometimes crustaceous. Embryo straight or rarely curved.—Herbs or rarely shrubs or small trees. Leaves usually opposite (or verticillate) in the lower part of the plant, alternate higher up, but sometimes all alternate or all opposite, without stipules. Flowers in terminal racemes or cymes, or the lower ones, rarely all, axillary. Bracts small or none besides the floral leaves, bracteoles very rare.

A large Order widely dispersed over every part of the globe. Of the thirty Australian genera, fourteen belong to the tropical Asiatic flora, several of them extending into Africa, and a few species occurring also (probably introduced) in S. America, five are tropical, both in the New and the Old World, five are chiefly American and Andine or extratropical, of which two are also represented in the mountains of tropical Asia, two belong chiefly to the extratropical flora of both hemispheres, and only four are endemic in Australia.

The Order is closely allied to *Solaneæ*, differing chiefly in that irregularity of flower which connects it with *Bignoniaceæ*, *Acanthaceæ*, *Verbenaceæ*, and others of the personate or bilabiate group, and which is evidenced in *Scrophularineæ*, either by the didynamy of the stamens or by the bilabiate æstivation of the corolla, or, in most cases, by both characters.

SUBORDER I. **Salpiglossideæ.**—*Corolla 5-lobed, the lobes induplicate or folded in the bud. Stamens in the Australian genera 4, didynamous. Inflorescence centrifugal, (often irregular in the Australian genera).*

Fruit a berry. Anthers 1-celled . . . . 1. DUBOISIA.
Fruit a capsule. Anthers 1- or 2-celled . . . . 2. ANTHOCERCIS.

SUBORDER II. **Antirrhinidæ.**—*Corolla 5-lobed or 2-lipped, imbricate in the bud, the upper lip or 2 upper lobes outside. Inflorescence centripetal or, in genera not Australian, compound. (Æstivation uncertain in some of the minute-flowered Limselleæ.)*

TRIBE *. **Verbasceæ.**—*Corolla rotate. Stamens declinate. Anthers 1-celled. Erect coarse herbs with alternate leaves.*

Stamens 5 . . . . *. VERBASCUM.
Stamens 4 . . . . *. CELSIA.

TRIBE *. **Antirrhineæ.**—*Corolla tubular at the base, the tube produced into a spur or protuberance. Stamens ascending, included in the tube. Capsule opening in pores or detached opercula. Lower leaves or all opposite.*

Corolla spurred (prostrate pubescent annual) . . . . *. LINARIA.

TRIBE. **Gratioleæ.** — *Corolla tubular at the base, neither spurred nor gibbous. Stamens shorter than the corolla, ascending. Capsule opening in 2 or 4 valves, or very rarely indehiscent.*

SUBTRIBE 1. **Eugratioleæ.**—*Leaves, at least the lower ones, opposite. Stamens all inserted in the tube and (in the Australian genera) entirely included. Capsule (in the Australian genera) opening loculicidally in 2 entire or bifid valves or 4-valved, or septicidal with bifid valves.*

Stamens 4, all perfect.
Calyx tubular, 5-angled, 5-toothed. Anther-cells contiguous . 3. MIMULUS.
Calyx campanulate, 5-lobed. Anther-cells contiguous . . . 4. MAZUS.
Calyx divided to the base or nearly so. Anther-cells more or
    less stipitate or separated from each other.
    Dissepiment of the capsule splitting and forming the in-
        flexed margins of the valves, leaving the two placentas
        free and separate.
        Anthers of the longer stamens 1-celled. . . . . . 5. ADENOSMA.
        Anthers all 2-celled . . . . 6. STEMODIA.
    Dissepiment of the capsule splitting, but leaving the placentas
        consolidated in a single column . . . . 7. MORGANIA.
    Dissepiment of the capsule remaining entire, at least at the
        base, and forming wings to the placental column . . . 8. LIMNOPHILA.
Calyx divided to the base, the outer segment much broader than
    the others. Anther-cells contiguous . . . . 9. HERPESTIS.
Stamens 2 perfect, the 2 lower ones reduced to filiform staminodia
    or entirely deficient.
Calyx divided to the base. Capsule 4-valved, leaving the pla-
    centas consolidated in a single column . . . . . . 10. GRATIOLA.
Calyx campanulate, 5-lobed. Capsule 2-valved, bearing the
    separate placentas in their centre . . . . 11. DOPATRIUM.

SUBTRIBE 2. **Lindernieæ.**—*Stem-leaves opposite. Upper stamens perfect and included in the tube. Lower stamens inserted in the throat, either reduced to staminodia or with long arched filaments with an angle or small lobe or appendage near the base, the anthers contiguous or cohering under the upper lip. Capsule opening in 2 entire valves parallel to the dissepiment.*

Perfect stamens 4.
Calyx deeply divided into herbaceous segments, dilated and
imbricate at the base. Flowers large. Appendage to the
lower stamens broad and flat . . . . 12. ARTANEMA.
Calyx deeply divided into linear segments, sometimes cohering
in a 5-toothed calyx (not angular as in *Torenia*). Appendage
to the lower stamens linear . . . . 13. VANDELLIA.
Perfect stamens 2. Staminodia 2. Calyx divided to the base.
Staminodia acute with an angle tooth or lobe near the base.
Capsule globular or broadly ovoid . . . . 14. ILYSANTHES.
Staminodia linear and obtuse, entire. Capsule oblong or linear. 15. BONNAYA.

SUBTRIBE 3. **Limoselleæ.**—*Small creeping or prostrate herbs with opposite or clustered leaves. Corolla (minute) with a short tube and 5 nearly equal lobes (æstivation variable ?). Anthers 1-celled. Capsule various.*

Calyx 5-toothed. Stamens 2. Leaves opposite.
Capsule indehiscent or bursting irregularly or obscurely 4-
valved . . . . 16. PEPLIDIUM.
Capsule loculicidally 2-valved . . . . 17. MICROCARPÆA.
Calyx obtusely 3- or 4-lobed. Stamens 2 or 4. Capsule loculici-
dally 2-valved. Leaves opposite . . . . 18. GLOSSOSTIGMA.
Calyx 5-toothed. Stamens 4. Leaves clustered or alternate.
Capsule opening in 2 valves parallel to the dissepiment . . . 19. LIMOSELLA.

SUBORDER III. **Rhinanthideæ.**—*Corolla either with 4, 5 (or rarely more in genera not Australian) spreading lobes, variously imbricate in the bud, the upper ones very rarely outside, or 2-lipped with the upper lip inside. Inflorescence centripetal or very rarely in genera not Australian compound.*

Leaves all alternate. Calyx 5-cleft to the base. Corolla deeply
5-lobed. Stamens 4 or 5. Anthers 2-celled, sagittate . . . 20. CAPRARIA.
Leaves, at least the lower ones, opposite.
Corolla rotate, 4-lobed. Stamens 4. Anthers equally 2-celled,
sagittate . . . . 21. SCOPARIA.
Corolla rotate or with a distinct tube, 4-lobed. Stamens 2, exserted; anthers with confluent cells, not mucronate . . . 22. VERONICA.
Corolla with a distinct tube or broadly campanulate, lobes 5,
nearly equal. Stamens 4.
Calyx deeply 5-lobed. Anthers with confluent cells, not mu-
cronate . . . . 23. OURISIA.
Calyx 5-toothed or shortly 5-lobed. Anthers with one large
scarcely mucronate cell and one stipitate empty cell . . 24. SOPUBIA.
Calyx herbaceous, split on one side. Anther-cells unequal,
mucronate or awned . . . . 25. CENTRANTHERA.
Corolla with a distinct tube, often long, the limb 2-lipped.
Corolla-tube slender. Stamens 4. Anthers with 1 vertical
usually linear obtuse cell.
Corolla-tube straight. Capsule obtuse . . . . . . 26. BUCHNERA.
Corolla-tube bent above the middle. Capsule obtuse . . 27. STRIGA.
Corolla-tube elongated. Capsule acuminate, the beak often
oblique . . . . 28. RHAMPHICARPA.
Corolla-tube slender. Stamens 2. Anthers with one large
mucronate or aristate cell . . . . 29. HEMIARRHENA.
Corolla-tube not very slender. Stamens 4. Anther-cells
equal, both mucronate . . . . 30. EUPHRASIA.

The introduced plants belonging to the genera, marked above with the asterisk *, are the following, all European weeds :—

Verbascum Blattaria, Linn ; Benth. in DC. Prod. x. 230. An erect coarse simple or scarcely branched biennial of 2 to 3 ft., either glabrous or slightly glandular-pubescent in the upper part. Leaves alternate, oblong, coarsely toothed or sinuate, the lower ones petiolate, the upper ones sessile and sometimes slightly decurrent. Flowers yellow or rarely white, in a long loose simple raceme, on pedicels of 3 to 6 lines. Calyx deeply 5-cleft. Corolla rotate, with 5 broad rounded lobes. Stamens 5, declinate, the filaments woolly with purple hairs; anthers 1-celled. Capsule 2-valved, with numerous small seeds.—N. S. Wales, Victoria, and S. Australia. *V. virgatum*, With.; Benth. in DC. Prod. x. 229, perhaps a variety of *V. Blattaria*, differs in the greater abundance of the glandular pubescence and in the pedicels of the flowers very short, usually from 2 to 6 together within each bract.—Victoria.

Celsia cretica, Linn.; Benth. in DC. Prod. x. 244. An erect biennial with much the habit of *Verbascum Blattaria*, pubescent and more or less viscid. Lower and radical leaves lyrate-pinnatifid, upper ones cordate and stem-clasping. Flowers yellow, larger than in *Verbascum Blattaria*, sessile within each bract, in a long terminal loose spike. Calyx divided into 5 broad serrate segments. Corolla rotate. Stamens 4, declinate, the 2 upper ones with woolly filaments and short reniform anthers, the 2 lower with much longer glabrous filaments and linear adnate anthers. Capsule 2-valved.—Naturalized on Buchan river and Plenty Creek, Victoria, *F. Mueller*.

Linaria Elatine, Mill.; Benth. in DC. Prod. x. 268. A prostrate hairy annual with slender stems. Leaves alternate or the lower ones opposite, nearly sessile, ovate and mostly angular or hastate at the base. Flowers small, solitary on slender pedicels in the axils of the upper small leaves. Calyx divided into 5 lanceolate segments. Corolla yellowish, the tube produced into a straight spur at the base, the throat closed by a projecting *palate*, the upper lip 2-lobed and purplish, the lower lip 3-lobed. Capsule opening on each side by the falling off of a circular valve-like operculum.—Established in cultivated places about Paramatta, *Woolls*.

SUBORDER I. SALPIGLOSSIDEÆ.—Corolla 5-lobed, the lobes more or less induplicate or folded in the bud and sometimes also slightly imbricate, the 2 upper ones (those next to the main axis of inflorescence) outside. Embryo often slightly curved. Inflorescence centrifugal.

As already observed in De Candolle's ' Prodromus,' this suborder might almost equally well be referred to *Solaneæ*, to several genera of which it is closely allied, but the stamens are constantly didynamous, with the fifth upper one reduced to a sterile staminodium or more frequently to a mere rudiment or entirely wanting. Miers proposes to unite it with several *Solaneæ* in an intermediate Order, Atropaceæ, but that appears to me rather to increase the difficulty of giving definite distinctive characters without establishing a more natural distribution. He considers, moreover, the two following Australian genera as forming with *Anthotroche* a very distinct tribe, remarkable for its reniform extrorse anthers, but unilocular reniform anthers occur in many Scrophularineous genera of other tribes, and, if examined in the bud, they are decidedly extrorse in several species at least of *Petunia* and *Nierembergia*. *Anthotroche* also, notwithstanding a similarity in anthers, differs essentially from *Duboisia* and *Anthocercis* in inflorescence and in the perfect regularity of the flowers and equal development of the five stamens.

1. DUBOISIA, R. Br.

Calyx 5-toothed. Corolla ovate-campanulate, the lobes broad, induplicate in the bud. Stamens 4, didynamous, included in the tube, the upper ones the longest, the fifth uppermost one reduced to a minute rudiment; anthers reniform, turned outwards at least when fully out, the cells confluent at the apex. Stigma slightly dilated and 2-lobed. Fruit an indehiscent berry. Seeds few, curved, with a crustaceous tubercular-rugose testa; embryo curved, the albumen not copious.—Small glabrous tree. Leaves alternate, entire. Flowers small, in terminal centrifugal panicles.

The genus is, as far as known, limited to a single species extending from E. Australia to New Caledonia, but it is not improbable that *Anthocercis Leichhardtii*, of which the fruit is unknown, may prove to be a second *Duboisia*.

1. **D. myoporoides,** *R. Br. Prod.* 448. A tall shrub or small tree, quite glabrous. Leaves alternate, from obovate-oblong to oblong-lanceolate, obtuse or rarely acute, entire, contracted into a petiole, 2 to 4 in. long. Panicles terminal, sometimes leafy at the base, usually much branched, broadly pyramidal or corymbose. Bracts minute. Calyx broadly campanulate, with broad obtuse teeth. Corolla about 2 lines long, white or pale lilac, the lobes rather short and obtuse. Stamens included in the tube. Berry small, nearly globular.—Endl. Iconogr. t. 77; Benth. in DC. Prod. x. 191; Miers, Illustr. t. 87; *Notelæa ligustrina,* Sieb. Pl. Exs.

Queensland. Brisbane river, Moreton Bay, *Fraser, F. Mueller;* Rockingham Bay, *Dallachy.*

N. S. Wales. Port Jackson to the Blue Mountains, *R. Brown, Sieber, n.* 259, and many others; Sydney woods, Paris Exhibition, 1857, *M'Arthur, n.* 81; Hastings and Clarence rivers, *Beckler;* Port Macquarrie, *Fraser;* Richmond river, *Henderson;* southward to Illawarra, *A. Cunningham, Ralston.*

The species is also in New Caledonia.

2. **ANTHOCERCIS,** Labill.

(Cyphanthera, *Miers;* Eadesia, *F. Muell.*)

Calyx 5-toothed or 5-lobed. Corolla-tube campanulate, shortly contracted at the base; lobes 5, spreading, nearly equal or the 2 upper rather shorter or longer than the others, all induplicate in the bud and the 2 upper slightly overlapping the lateral ones. Stamens 4, didynamous, included in the tube, with occasionally a small rudiment of the uppermost fifth one. Anthers 1- or 2-celled, turned outwards in the bud. Stigmatic lobes very short, rather broad. Capsule oblong ovoid or globular, opening in 2 entire or bifid valves. Seeds usually somewhat curved, with a reticulate crustaceous testa. Embryo straight or slightly curved, in a copious albumen.—Shrubs, sometimes almost arborescent, glabrous glandular-pubescent or hoary with a stellate tomentum. Leaves entire or rarely toothed, often rather thick. Peduncles 1- to 3-flowered, irregularly arranged in terminal racemes or panicles often leafy. Bracts very small or none. Corolla white or yellow, the tube usually streaked inside with purple or green.

The genus is limited to Australia. As a whole it is a very natural one, immediately connected with none except *Duboisia*, from which it differs solely in the capsular fruit. The two sections are very readily distinguished by a constant and absolute character, but appear to be too artificial to be conveniently adopted as genera as proposed by Miers. The anthers in the one are those of *Petunia*, in the other 1-celled as in *Duboisia*.

SECT. I. **Euanthocercis.**—*Anthers 2-celled, the cells not confluent. Plants glabrous or glandular-pubescent, without stellate hairs.*

Plant quite glabrous (often viscid).
Leaves obovate, conspicuously glandular-dotted. Corolla large,
  white . . . . 1. *A. viscosa.*

Leaves obovate or oblong-cuneate, not dotted. Corolla yellow,
  with narrow lobes . . . . 2. *A. littorea.*
Leaves linear, usually few and small. Corolla rather small, with
  narrow linear lobes.
  Branches virgate, not spinescent . . . . 3. *A. gracilis.*
  Branches intricate, with numerous divaricate spines (reduced
    branchlets) . . . . 4. *A. genistoides.*
Upper part of the plant, or at least the pedicels and calyx, glandular-
  pubescent.
  Corolla-lobes acute, longer than the tube.
    Leaves shortly cuneate, glandular-pubescent. Branches intri-
      cate, rigid, spinescent . . . . 5. *A. anisantha.*
    Leaves linear-cuneate, glabrous or nearly so. Branches intricate.
      Internodes distant. Branchlets occasionally spinescent . . 6. *A. intricata.*
      Leaves rather crowded. No spines . . . . 7. *A. arborea.*
    Leaves narrow-linear, almost acute, glandular-pubescent.
      Branches virgate or paniculate . . . . 8. *A. angustifolia.*
  Corolla-lobes obtuse, shorter than the tube. Leaves oblong-linear
    or oblanceolate . . . . 9. *A. fasciculata.*

SECT. II **Cyphanthera.**—*Anthers 1-celled. Plants glabrous glandular-pubescent
or stellate-tomentose.*

Branches and leaves glandular-pubescent. Pedicels solitary, terminal
  or leaf-opposed. Corolla-lobes broad and obtuse.
  Much-branched small shrub. Leaves under 1 line long . . . . 10. *A. microphylla.*
  Diffuse undershrub. Leaves 2 to 6 lines long . . . . : . . . 11. *A. myosotidea.*
Branches and leaves pubescent or tomentose with stellate hairs. Co-
  rolla-lobes narrow, acute.
  Leaves under ½ in. long, scabrous-pubescent. Pedicels filiform,
    solitary, terminal or leaf-opposed . . . . 12. *A. scabrella.*
  Leaves ¼ to ½ in. long, tomentose. Pedicels short, often several
    together, terminal or in the upper axils . . . . 13. *A. albicans.*
  Leaves ¾ to 1½ in. long, tomentose. Pedicels short, often several
    together, terminal or in the upper axils . . . . 14. *A. tasmanica.*
Leaves glabrous, young branches minutely tomentose or almost gla-
  brous. Leaves mostly under 2 in. long. Flowers in irregular
  leafy racemes or narrow panicles.
  Leaves oblong or lanceolate. Corolla-lobes narrow, acute . . . 15. *A. Eadesii.*
  Leaves linear. Corolla-lobes broad, obtuse . . . . 16. *A. racemosa.*
Leaves and branches quite glabrous. Leaves 2 to 4 in. long. Flowers
  paniculate.
  Leaves linear. Corolla-lobes broad, obtuse . . . . 17. *A. Hopwoodii.*
  Leaves oblong or lanceolate. Corolla-lobes narrow, acute . . . 18. *A. Leichhardtii.*

SECT. I. EUANTHOCERCIS.—Anthers 2-celled, the cells not confluent.
Plants glabrous or glandular-pubescent, without stellate hairs. Capsule
usually ovoid oblong or acuminate. Species all western, one of them extending into S. Australia.

1. **A. viscosa,** *R. Br. Prod.* 448. An erect shrub, usually of 6 to 8 ft.,
but attaining sometimes 20 ft. (*F. Mueller*), glabrous in all our specimens
and more or less viscid. Leaves broadly obovate, entire or minutely scabrous-denticulate especially when dry, contracted into a short petiole, rather
thick, marked with conspicuous glandular dots, mostly 1½ to 2½ in. long.
Peduncles 1- to 3-flowered, shorter than or exceeding the leaves. Flowers
white, much larger than in any other species. Calyx-tube about 2 lines

long; lobes lanceolate, longer than the tube.   Corolla-tube about ¾ in. long,
streaked inside with green; lobes ovate to lanceolate, very spreading, about
as long as the tube, but variable.   Anthers 2-celled.   Capsule acuminate,
slightly exceeding the calyx, the valves entire or shortly split at the end.—
Benth. in DC. Prod. x. 191; Bartl. in Pl. Preiss. i. 341 ; F. Muell. Fragm.
vi. 143 ; Bot. Mag. t. 2961 ; Bot. Reg. t. 1624; Maund, Botanist, t. 59;
Miers, Illustr. t. 82 ; *A. littorea,* Endl. Iconogr. t. 68, not of Labill. (the
denticulations of the leaves much exaggerated).

W. Australia. King George's Sound, *R. Brown, Drummond, n.* 498, *Preiss, n.* 1963,
and many others.   Said to be one of the poison plants (*F. Mueller*).

2. **A. littorea,** *Labill. Pl. Nov. Holl.* ii. 19. *t.* 158.   A glabrous often
slightly viscid shrub of from 2 to 8 ft.   Leaves from oblong-cuneate to obovate, but usually narrower and smaller than in *A. viscosa,* more sessile, and
without the glandular dots of that species, quite entire or the lower ones
marked with a few prominent teeth, mostly ¾ to 1½ in. long, usually rather
thick.   Flowers yellow, often numerous, on slender but short pedicels, forming at the ends of the branches irregular leafy racemes or narrow panicles
often more or less developed into terminal branching leafless panicles of 1 ft.
or more.   Calyx-tube scarcely 1 line long; lobes narrow, acute, shorter or
longer than the tube.   Corolla variable in size, the tube usually 3 to 4 lines
long, streaked inside with purple; lobes narrow, acute, from rather longer
than the tube to twice as long.   Anthers 2-celled.   Capsule narrow, acuminate, often ½ in. long, the valves usually entire.—R. Br. Prod. 448 ; Benth.
in DC. Prod. x. 191 ; Lehm. in Pl. Preiss. ii. 237 ; Bot. Reg. t. 212;
Sweet, Fl. Austral. t. 17 ; Maund, Botanist, t. 102 ; Miers, Illustr. t. 83.

W. Australia. King George's Sound, *Labillardière, R. Brown, F. Mueller,* and
others ; eastward to Goose Island Bay, *R. Brown,* and to Cape Arid and Point Malcolm,
Maxwell; Swan River, *Fraser, Drummond, 1st Coll., Preiss, n.* 1473, 1474; Murchison
river, *Oldfield.*

A. ilicifolia, Hook. Bot. Mag. under n. 2961 and t. 4200; Benth. in DC. Prod. x. 192;
Lehm. Pl. Preiss. ii. 237; Miers, Illustr. t. 83, from Swan River, has the leaves more
toothed, the panicle more developed, and the flowers rather smaller than usual, but in all
these respects the southern as well as the Swan River specimens are very variable.

A. glabella, Miers, Illustr. ii. App. 26, from Swan River, *Gilbert, n.* 126, appears to
me to be a narrow-leaved form of *A. littorea,* with very long narrow corolla-lobes.

3. **A. gracilis,** *Benth. in DC. Prod.* x. 192.   Stems apparently tall,
erect, slender, virgate and paniculately branched.   Lower leaves unknown,
upper ones very narrow, linear, a few at the base of the branches ½ to 1 in.
long, otherwise all very small and distant.   Flowers not very numerous, on
filiform pedicels.   Calyx-tube about 1 line long; lobes narrow, as long as
the tube.   Corolla-tube under 3 lines long, the narrow base longer in proportion to the broad part than in most species; lobes filiform, much longer
than the tube.   Anthers 2-celled.   Capsule acuminate.—Miers, Illustr. t.
83.

W. Australia. Swan River, *Drummond, 1st Coll.*

4. **A. genistoides,** *Miers, Illust.* ii. *App.* 27. *t.* 83.   An erect glabrous
shrub of 3 or 4 ft., with numerous intricate flexuose branches, the smaller

branchlets reduced to rigid divaricate spines. Leaves very small, linear, often reduced to small scales and rarely ½ in. long. Pedicels solitary or 2 or 3 together in the axils of the spines, filiform but short. Calyx-tube about 1 line long, small or rarely as long as the tube. Corolla white (*F. Mueller*), the tube about 2 lines long, the lobes narrow, acute, longer than the tube. Anthers 2-celled. Capsule 2 to 3 lines long, oblong but scarcely acuminate. —*A. spinescens*, F. Muell. Fragm. i. 122, vi. 143.

W. Australia, *Drummond, n.* 86; Kalgan river, *F. Mueller;* Thomas river and W. Mount Barren, *Maxwell;* Murchison river, *Oldfield.*

There is in the Muellerian herbarium a single small specimen from Murchison river, *Oldfield,* closely allied to *A. gracilis* and to *A. genistoides,* with the unarmed branches and linear leaves of the former, and the flowers of the latter, except that the corolla-lobes are contorted-imbricate in the bud; but that may possibly be an abnormal monstrosity.

5. **A. anisantha,** *Endl. in Ann. Wien. Mus.* ii. 201. An erect rigid intricately-branched shrub, more or less glandular-pubescent, the smaller branchlets reduced to divaricate spines as in *A. genistoides,* but stouter. Leaves small, from linear-cuneate and about ½ in. long, to almost obovate and about ¼ in., or often reduced to minute scales. Flowers on short pedicels, solitary or 2 or 3 together in the axils of the spines. Calyx glandular-hirsute, the tube about 1 line long, the teeth either short or as long as the tube. Corolla-tube broad, about 2 lines long; lobes narrow, acute, longer than the tube. Anthers 2-celled.—Benth. in DC. Prod. x. 192; F. Muell. Fragm. vi. 143.

S. Australia. Boston Island, *Wilhelmi;* Lake Gillies, *Burkitt.*
W. Australia, *Drummond.*

The above are all single specimens in Herb. F. Muell. I have not seen the typical specimen from W. Australia, *Roe,* but Endlicher's description agrees perfectly with Drummond's plant, except that I do not find the corolla-lobes more unequal than in several other species.

6. **A. intricata,** *F. Muell. Fragm.* i. 211. A shrub of 10 to 12 ft., closely allied to *A. arborea,* with divaricate flexuose glabrous branches, the smaller ones occasionally spinescent. Leaves often clustered at the nodes, and similar to those of *A. arborea,* but smaller and fewer. Pedicels short, solitary or clustered, glandular-pubescent as well as the calyxes. Flowers entirely of *A. arborea.* Fruit not seen.

W. Australia. Murchison river, *Oldfield.* Probably a slightly spinescent variety of *A. arborea;* the specimens of both forms insufficient for their proper definition.

7. **A. arborea,** *F. Muell. Fragm.* i. 212. A shrub or small tree, attaining 12 ft., with an erect trunk of 4 to 6 ft. (*Oldfield*), glabrous except the inflorescence, which is glandular-pubescent, the branches flexuose and intricate, but without spines in the specimens seen. Leaves usually clustered at the nodes or scattered on the short flowering branchlets, linear-oblong, obtuse, entire, narrowed into a petiole, rather thin, mostly nearly 1 in. long. Flowers whitish, on slender pedicels, in short cymes or almost solitary on the branchlets. Calyx glandular-hirsute, the tube under 1 line long, the teeth narrow, usually longer than the tube. Corolla-tube 2 to 2½ lines long, the lobes narrow, acute, 2 or 3 times as long as the tube. Anthers 2-celled. Capsule narrow, 4 to 5 lines long, the valves bifid at the end.

W. Australia. Murchison river, *Oldfield.*

8. **A. angustifolia,** *F. Muell. in Trans. Phil. Soc. Vict.* i. 21, *and in Hook. Kew Journ.* viii. 202. A glandular-pubescent unarmed shrub. Leaves linear, acute or almost obtuse, entire, contracted at the base, mostly ¾ to 1½ in. long. Flowers white, few, terminal or leaf-opposed, on rather slender pedicels. Calyx glandular-hirsute, the tube about 1 line long, the lobes narrow, usually longer than the tube. Corolla-tube ¼ in. long, yellowish inside; lobes narrow, acute, at least twice as long as the tube. Anthers 2-celled. Capsule ovoid, exceeding the calyx, but not seen perfect.

S. Australia. Stony glens near Mount Lofty and on the Torrens river, *F. Mueller.*

9. **A. fasciculata,** *F. Muell. Fragm.* i. 122. A shrub of about 3 ft., the branches and foliage viscid but scarcely pubescent, the inflorescence glandular-pubescent. Leaves oblong linear or oblanceolate, obtuse, entire, narrowed into a petiole, rather thick, ¾ to 1½ in. long, the upper ones much reduced. Flowers shortly pedicellate, in terminal almost leafless panicles. Calyx glandular-pubescent, the tube about 1 line long, the teeth rather shorter than the tube and obtuse. Corolla-tube fully 2 lines long; lobes obtuse, shorter than the tube. Anthers 2-celled. Capsule 4 to 6 lines long, acuminate, the valves scarcely bifid, diverging at the end.

W. Australia. Phillips river, *Maxwell.* Some specimens closely resemble the larger-leaved specimens of *A. myosotidea,* but are readily distinguished by the 2-celled anthers.

SECT. II. CYPHANTHERA.—Anthers 1-celled, reniform. Plants glabrous, glandular-pubescent or stellate-tomentose. Capsule short, globular or ovoid, the valves bifid.—*Cyphanthera,* Miers; *Eadesia,* F. Muell.

10. **A. microphylla,** *F. Muell. Fragm.* i. 179; vi. 143. A small intricately branched shrub, viscid with a minute glandular pubescence, the older leaves and branches becoming glabrous, the smaller branchlets slender and rigid but not spinescent. Leaves sessile, ovate or oblong, obtuse, entire, rather thick, ½ to 1 line long. Pedicels terminal or opposite the upper leaves, 1 to 2 lines long, glandular-pubescent. Calyx glandular-pubescent, not exceeding 1 line including the obtuse teeth or lobes. Corolla-tube 1½ lines long or rather more, the upper portion broadly campanulate, the lobes broad, obtuse, shorter than the tube. Anthers 1-celled. Capsule nearly globular, as long as the calyx. Seeds few.—*Cyphanthera microphylla,* Miers, Illust. ii. App. 33. t. 85.

W. Australia, *Drummond (5th Coll. ?), n.* 177; Salt river, *Maxwell.*

11. **A. myosotidea,** *F. Muell. in Trans. Phil. Soc. Vict.* i. 20, *in Hook. Kew Journ.* viii. 202, *and Fragm.* vi. 143. A low diffuse undershrub, the branches and foliage pubescent and viscid with short glandular hairs, without stellate tomentum. Leaves sessile, from oval-oblong to linear, from under ¼ in. long when broad to above ½ in. when narrow, obtuse, entire, rather thick, the margins recurved. Pedicels rather short, solitary, terminal or opposite the upper leaves, usually reflexed after flowering. Calyx glandular-pubescent, the tube about 1 line long; teeth narrow, usually as long as the tube. Corolla white, varying in size, the tube from under 2 lines to nearly 2½ lines,

very open at the top ; lobes broad, obtuse, about as long as the tube. Anthers 1-celled. Capsule shorter than or scarcely exceeding the calyx, ovoid-globular; valves bifid. Seeds few.—*A. amblyantha,* F. Muell. Fragm. i. 179.

Victoria. Grampians and Wimmera, *Dallachy.*
S. Australia. Gravelly sand ridges on the Murray, *F. Mueller ;* Tattara County, *Woods.*

12. **A. scabrella,** *Benth. in DC. Prod.* x. 192. A spreading shrub, with the foliage nearly of *A. albicans,* but the branches much more slender, the smaller ones almost filiform, and scabrous as well as the leaves with small stellate hairs, but not tomentose. Leaves ovate or oblong, obtuse, entire, very shortly petiolate, from ¼ in. long when broad to ½ in. when narrow. Pedicels mostly filiform, solitary, terminal or opposite the upper leaves. Calyx and corolla-tube of *A. albicans,* but the acute corolla-lobes appear to be longer and more slender than in that species. Anthers 1-celled. Capsule not seen.—*Cyphanthera scabrella,* Miers, Illustr. ii. App. 32. t. 85.

N. S. Wales. Nepean river, *R Cunningham.*

13. **A. albicans,** *A. Cunn. in Field, N. S. Wales,* 335, t. 2. An erect much-branched shrub of 2 or 3 ft., the branches and foliage densely covered with a stellate tomentum, rather loose and almost floccose in the typical form. Leaves ovate or oblong, very obtuse, entire, the margins often recurved, sessile or very shortly petiolate, from ¼ in. long when broad to about ½ in. when narrow. Flowers shortly petiolate, usually 2 or 3 together in the upper axils, forming sometimes narrow leafy panicles. Calyx-tube under 1 line long, glabrous or loosely stellate-tomentose in the typical form, the teeth narrow, shorter or longer than the tube. Corolla white, the tube about 2 lines long, streaked with purple inside ; lobes very narrow, acute, longer than the tube. Anthers 1-celled. Capsule small, globular, the valves bifid. Seeds few.—Benth. in DC. Prod. x. 192 ; Sweet, Fl. Austral. t. 16 ; *Cyphanthera albicans,* Miers, Illustr. ii. App. 31. t. 84, and *C. ovalifolia,* Miers, l. c. 33. t. 85.

N. S. Wales. Pine Hills near Bathurst, *A. Cunningham,* also *Backhouse ;* near Cassilis, *C. Moore.*

Var. *tomentosa,* Benth. l. c. Tomentum close and white, covering the calyxes as well as the rest of the plant.—*Cyphanthera tomentosa,* Miers, Illust. ii. App. 32. t. 85.—Peel's Range, *A. Cunningham.*

14. **A. tasmanica,** *Hook. f. Fl. Tasm.* i. 289. *t.* 92. An erect shrub, attaining 10 to 12 ft., the branches and foliage covered with a short stellate tomentum. Leaves oblong or lanceolate, obtuse, entire, the margins usually recurved, contracted into a short petiole, mostly ¾ to 1½ in. or when luxuriant 2 in. long. Flowers of a yellowish white, on short petioles, 2 or 3 together in the axils of small floral leaves and crowded at the ends of the branchlets. Calyx tomentose, the tube about 1 line long ; lobes or teeth acuminate, usually about as long as the tube. Corolla-tube 2 to 2½ lines long, streaked with purple inside ; lobes lanceolate, acute, at least as long as the tube. Anthers 1-celled. Capsule small, nearly globular ; valves bifid. Seeds few.—*Cyphanthera tasmanica,* Miers, Illust. ii. App. 30. t. 84.

Tasmania. Kelvedon, Great Swan Port, amongst gum-trees, *Backhouse ;* not un-common on the E. coast, *Gunn ;* Cygnet river, *C. Stuart.* Differs from *A. Eadesii* chiefly in the tomentum.

15. **A. Eadesii,** *F. Muell. Fragm.* ii. 139. An erect shrub of 3 or 4 ft., the branches and young foliage slightly hoary with a minute pubescence scarcely stellate and sometimes slightly glandular, the adult leaves and stems usually glabrous. Leaves oblong or lanceolate, obtuse, entire, contracted into a short petiole or the upper ones sessile, the larger ones 3 to 4 in. long, but those of the flowering branches under 2 in. Flowers nearly white, irre-gularly and loosely cymose in the upper axils, forming terminal oblong leafy panicles, the pedicels and calyxes slightly glandular-pubescent. Calyx-tube above 1 line long, the teeth rather obtuse, rarely as long as the tube and often very short. Corolla-tube $2\frac{1}{2}$ lines long.; lobes oblong, acute, rather shorter than the tube. Anthers 1-celled. Capsule small, nearly globular. Seeds few.— *Cyphanthera frondosa* and *C. cuneata,* Miers, Illust. ii. App. 29 and 31. t. 84 ; *Eadesia anthocercidea,* F. Muell. in Trans. Phil. Inst. Vict. ii. 72.

N. S. Wales. Near Camden, *M'Arthur.*
Victoria. Ranges near Mount Zero, Grampians, *Wilhelmi ;* summit of Mount Ara-piles and Wimmera, *Dallachy.*
S. Australia. Tattiara country, *Woods.*

16. **A. racemosa,** *F. Muell. Fragm.* i. 211. A shrub of 4 to 6 ft., glabrous except the very young shoots, which are occasionally white with a minute stellate tomentum. Leaves sessile, the lower ones lanceolate, the upper ones linear, obtuse, thick, flat or with recurved margins, $\frac{1}{2}$ to $1\frac{1}{2}$ in. long. Flowers 2 or 3 together in the axils, or the upper ones forming irregular ra-cemes. Pedicels rather slender. Calyx-tube scarcely 1 line long; lobes linear, nearly as long as or rather longer than the tube. Corolla-tube nearly 2 lines long; lobes obtuse, about as long as or even longer than the tube. Anthers 1-celled. Capsule small, ovoid, the valves deeply bifid. Seeds few.

W. Australia. Murchison river, *Oldfield.* The specimens very imperfect.

17. **A. (?) Hopwoodii,** *F. Muell. Fragm.* ii. 138. A glabrous tree or shrub. Leaves narrow-linear, acutely acuminate, with the point often re-curved, entire, rather thick, narrowed into a short petiole, 2 to 4 in. long. Flowers in short terminal cymes or leafy pyramidal panicles. Bracts mi-nute. Calyx small, broadly campanulate, with obtuse teeth. Corolla-tube campanulate, 3 to $3\frac{1}{2}$ lines long ; lobes broad, very obtuse, shorter than the tube. Anthers 1-celled. Fruit unknown.

N. S. Wales. Darling river, very rare, *Victorian Expedition.*
W. Australia, *Drummond* (with rather smaller flowers than in the N. S. Wales spe-cimens).

18. **A. (?) Leichhardtii,** *F. Muell. Fragm.* vi. 142. A glabrous shrub (or tree ?), with the foliage and inflorescence of *Duboisia myoporoides.* Leaves oblong-lanceolate, rather obtuse, entire, narrowed into a petiole, quite flat, 2 to 4 in. long. Panicles terminal, somewhat leafy at the base, broadly pyramidal or corymbose. Bracts very small. Pedicels short. Calyx small,

broadly campanulate, with short broad teeth.   Corolla-tube nearly 2 lines
long, ovate-campanulate; lobes narrow, acuminate, rather longer than the
tube.   Anthers 1-celled.   Fruit unknown.

Queensland, *Leichhardt,* the precise locality unknown.   The specimens might be mistaken for those of *Duboisia myoporoides,* were it not for the narrow acute corolla-lobes ; and,
as in the case of *A. Hopwoodii,* until the fruit shall have been observed it is in some measure uncertain whether it should be referred to *Anthocercis* or to *Duboisia.*

SUBORDER II. ANTIRRHINIDEÆ.—Corolla 5-lobed or 2-lipped, imbricate
in the bud, the upper lip or 2 upper lobes outside.   Inflorescence centripetal
or, in genera not Australian, compound, the primary peduncles centripetal
but branching into centrifugal cymes.

In the great majority of genera the difference in æstivation between the *Antirrhinideæ*
and *Rhinanthideæ* is well marked and easily ascertained.   It is only in some of the minute-
flowered *Limoselleæ,* and a very few non-Australian small genera of *Sibthorpieæ* that the
æstivation is uncertain and perhaps variable.   The dehiscence of the capsule in *Antirrhinideæ* is very variable, and in a very few non-Australian genera, the fruit is a berry.

TRIBE GRATIOLEÆ.—Corolla tubular at the base, neither spurred nor
gibbous.   Stamens shorter than the corolla, ascending, didynamous or reduced to two.   Capsule opening in 2 or 4 valves or very rarely indehiscent.

SUBTRIBE I. EUGRATIOLEÆ.—Leaves, at least the lower ones, opposite.
Stamens all inserted in the tube and, in the Australian genera, entirely included.   Capsule, in the Australian genera, opening loculicidally in 2 entire
or 2-fid valves, or 4-valved, or septicidal with 2-fid valves.

3. MIMULUS, Linn.

(Uvedalia, *R. Br.*)

Calyx tubular, with 5 prominent angles, ending in 5 small teeth.   Corolla
tubular at the base, the upper lip erect or spreading, 2-lobed ; the lower lip
spreading, 3-lobed, usually with 2 protuberances at its base in the throat ;
all the lobes broad and rounded.   Stamens 4 in pairs ; anthers all perfect,
2-celled, but the cells often confluent at the top.   Style with 2 ovate nearly
equal stigmatic laminæ.   Capsule scarcely furrowed, opening loculicidally in
2 valves which sometimes split along the dissepiment ; leaving an entire or
bifid central column bearing the placentas.   Seeds small, numerous.—Erect
or prostrate herbs.   Leaves opposite.   Flowers solitary on axillary pedicels,
without bracteoles, the upper ones forming sometimes a terminal raceme.

The genus is widely dispersed over the temperate regions of N. and S. America, as well
as along the range of the Andes, more sparingly in Eastern Asia, the mountains of tropical
Asia and in S. Africa.   Of the four Australian species, one is closely allied to, if not identical with, a common one in Asia and Africa, another extends to New Zealand, the remaining
two are endemic.

Stems ascending or erect, not much branched except at the base.   Plant
   glabrous.
Annual (?), very slender and weak, with small linear-lanceolate distant leaves . . . . 1. *M. Uvedaliæ.*
Perennial.   Leaves oblong or lanceolate, obtuse . . . . . . 2. *M. gracilis.*
2 I

Stems prostrate or creeping, much branched.
　Plant glabrous. Leaves rather thick, ovate or oblong . . . . . 3. *M. repens.*
　Plant more or less pubescent. Leaves very small, narrow-oblong.
　　Corolla-tube long . . . . 4. *M. prostratus.*

1. **M. Uvedaliæ,** *Benth. in DC. Prod.* x. 369. Apparently annual and quite glabrous, the stems very slender and weak, slightly branched, under 1 ft. long. Radical leaves rosulate, ovate, but very soon disappearing; stem-leaves small and distant, linear-lanceolate, acute or scarcely obtuse, entire, stem-clasping, rarely exceeding ¼ in. Pedicels in the upper axils slender, 1 to 2 in. long or more. Calyx 2 to 2½ lines long, the teeth very small. Corolla about twice as long as the calyx, pale blue with a yellow throat (*Soland. MSS.*). Capsule oblong, shorter than the calyx, the valves readily splitting. —*Uvedalia linearis*, R. Br. Prod. 440.

Queensland. Endeavour river, *Banks and Solander, A. Cunningham ;* Broad Sound and Shoalwater Bay, *R. Brown.*

Var. *lutea.* Corolla yellow (*F. Mueller*), the dried specimens absolutely undistinguishable from the typical form.—*M. debilis*, F. Muell. in Trans. Phil. Inst. Vict. iii. 62.

N. Australia. Swamps at the source of the Macarthur river, Providence Hill and Macadam Range, growing with *Uvedalia linearis*, F. Mueller.

2. **M. gracilis,** *R. Br. Prod.* 439. Quite glabrous. Stems from a perennial somewhat creeping rhizome, erect, usually about 6 in. and rarely in the Australian specimens nearly 1 ft. high, not much branched except at the base. Leaves linear-oblong to oblong-lanceolate, obtuse, entire, in some specimens attaining 1 in., but in others all under ½ in. long. Pedicels sometimes scarcely longer than the leaves, but often attaining 1 to 2 in. Calyx about 2½ lines long, with short acute teeth. Corolla violet purple or blue, the tube shortly exceeding the calyx or rarely half as long again, the lobes very broad, those of the lower lip retuse, all minutely ciliolate. Capsule enclosed in the calyx, oblong, the valves readily splitting.—Benth. in DC. Prod. x. 369 ; *M. pusillus*, Benth. l. c.

Queensland. Broad Sound, *R. Brown ;* Dawson river, *F. Mueller ;* Rockhampton, *O'Shanesy ;* Curriwillighi, *Dalton ;* Warwick, *Beckler.*

N. S. Wales. Hunter's River, *R. Brown ;* Blue Mountains, *A. Cunningham, Woolls,* and others ; New England, *C. Stuart, C. Moore,* and others ; towards Bathurst, *A. Cunningham ;* Murray and Darling rivers and Monument Creek, *Victorian Expedition:*

Victoria. Station Peak and Avoca river, *F. Mueller ;* Wimmera, *Dallachy.*

The species is also widely spread over hilly regions in Asia and Africa, but there represented chiefly by a luxuriant variety larger in all its parts, which I had originally published under the name of *M. strictus*, and from which the description of *M. gracilis* in the ' Prodromus ' is chiefly taken. The common form in Australia is the smaller one which I had considered as a distinct species under the name of *M. pusillus*, but some of the luxuriant Queensland specimens come very near to the Asiatic ones.

3. **M. repens,** *R. Br. Prod.* 439. A small glabrous prostrate perennial, creeping and rooting at the joints. Leaves sessile or scarcely petiolate, sometimes stem-clasping, from broadly ovate to oblong, obtuse, rather thick, often all under 2 lines long and rarely exceeding 3 lines. Flowers few, the pedicel often shorter than the leaves at the time of flowering, but lengthening considerably afterwards. Calyx scarcely 2 lines long, truncate, with small distant teeth. Corolla blue often yellow in the centre, the tube not 3 lines

long, dilated upwards, the lobes all broad and as long as the tube, the upper ones not much shorter than the lower. Capsule nearly globular, about 2 lines diameter, the valves readily splitting.—Benth. in DC. Prod. x. 373 ; Hook. f. Fl. Tasm. i. 290 ; Bot. Mag. t. 5423.

N. S. Wales. Manly Beach, *Woolls;* Blue Mountains and Illawarra, *A. Cunningham;* Darling river, *Neilson.*

Victoria. Swamps on the Murray, about Melbourne, etc., *F. Mueller* and others ; Portland, *Allitt.*

Tasmania. Port Dalrymple, *R. Brown ;* common in saline situations, on muddy banks of rivers, etc., *J. D. Hooker.*

S. Australia. Near Kaiserstuhl, towards Mount Remarkable, *F. Mueller.*

W. Australia. Murchison river, *Oldfield, Drummond, 6th coll., n.* 129 (apparently this species, but the specimens not good).

The species is also in New Zealand. The habit and foliage is often that of smaller specimens of *Herpestis Monnieria,* but the calyx and corolla are very different.

4. **M. prostratus,** *Benth. in DC. Prod.* x. 373. A small diffuse or prostrate much-branched perennial, more slender than *M. repens,* and not so frequently rooting at the joints, the whole plant rarely exceeding 2 or 3 in., the branches and peduncles and often the foliage also pubescent and sometimes slightly glandular. Leaves sessile, narrow-oblong, obtuse, entire, 1 to 2 or rarely 3 lines long, resembling those rather of *M. gracilis* than of *M. repens.* Pedicels filiform, usually longer than the leaves, and sometimes ¾ in. long. Calyx scarcely 2 lines long, with short acute teeth. Corolla-tube at least twice as long as the calyx, and more slender than in the other Australian species. Capsule oblong, shorter than the calyx, the valves usually entire.

Queensland. Bokhara scrub, *Leichhardt* (with rather long leaves, attaining 3 lines, but the specimens imperfect).

N. S. Wales. Lachlan river, *A. Cunningham ;* Murray and Darling rivers, *Dallachy, Victorian Expedition ;* Mount Murchison, *Bonney.*

S. Australia. S. of Wills' Creek, *Howitt's Expedition.*

F. Muell. Fragm. vi. 103, from my character in the 'Prodromus' (in which I had omitted the pubescence, having overlooked it in Cunningham's imperfect specimens), unites this with *M. repens,* the specimens were, however, referred in his collections to *M. gracilis.* It appears to me to be perfectly distinct from *M. repens* in its foliage and capsule, from *M. gracilis* in its dwarf prostrate habit, and from both in the pubescence, of which there is no trace in any other Australian species, as well as in the longer and more slender tube of the corolla.

4. MAZUS, Lour.

Calyx broadly campanulate, 5-lobed. Corolla with the upper lip erect, ovate, shortly bifid ; the lower lip much larger, spreading, broadly 3-lobed, with 2 slight protuberances at its base in the throat. Stamens 4, all fertile ; anther-cells contiguous, at length divaricate. Style with 2 ovate equal stigmatic laminæ. Capsule globular or compressed, obtuse, opening loculicidally in 2 entire valves.—Low herbs. Lower leaves opposite, the upper ones alternate, or all nearly rosulate. Flowers in terminal one-sided racemes or solitary.

The genus comprises a small number of tropical and east Asiatic species, besides the Australian one, which only extends to New Zealand.

1. **M. pumilio,** *R. Br. Prod.* 439. A small perennial, with a creeping rhizome. Stems very short or scarcely any besides the peduncle. Leaves forming an erect tuft or spreading rosette, from obovate and not ½ in. long to oblong and above 2 in. long, all obtuse, irregularly sinuate-toothed or rarely entire, contracted into a petiole, sprinkled with a few hairs on the upper surface, glabrous or nearly so underneath, rarely glabrous on both sides. Scapes or peduncles leafless, usually exceeding the leaves, bearing either a single flower or a loose raceme of very few flowers on long pedicels. Bracts very few and minute, scattered more frequently on the pedicels than on the peduncle, and often entirely wanting. Calyx about 2 lines long, the lobes narrow, shorter than the tube, enlarged and more deeply lobed after flowering. Corolla-tube scarcely exceeding the calyx ; lobes of the lower lip longer than the tube. Capsule enclosed in the calyx.—Benth. in DC. Prod. x. 375 ; Hook. f. Fl. Tasm. i. 290 ; Endl. Iconogr. t. 102 ; Hook. Ic. Pl. t. 567 (the flowers too small).

N. S. Wales. Hastings and Clarence rivers, *Beckler ;* Archer's Station, *Leichhardt.*
Victoria. Ovens river, Plenty Ranges, Dandenong Mountains, Wilson's Promontory, *F. Mueller ;* Fitzroy river, *Robertson ;* Portland, *Allitt.*
Tasmania. Port Dalrymple, *R. Brown ;* common in wet places, *J. D. Hooker ;* King's Island, *R. Brown* (a dwarf form with leaves of ½ in. and short 1-flowered scapes).

The species is also in New Zealand, and comes near to some of the smaller forms of the common Asiatic *M. rugosus,* but besides the difference in habit and foliage, the calyx is narrow and much less open.

5. ADENOSMA, R. Br.

(Pterostigma, *Benth.*)

Calyx divided to the base into 5 segments or sepals, the upper one larger. Corolla tubular at the base, the upper lip erect, entire or notched, the lower one spreading, 3-lobed. Stamens 4, in pairs ; anthers of the lower pair 1-celled (by the abortion of the other cell), of the upper pair 2-celled, with the cells separate and rather distant, or rarely 1-celled. Style dilated at the end into 2 short stigmatic lobes, and more or less winged below the lobes. Capsule acuminate, opening septicidally in 2 bifid valves or in 4 valves, the placentas of the 2 carpels completely separating at maturity. Seeds numerous, small, striate and reticulate.—Glandular-pubescent or villous herbs, usually strong-scented and turning black in drying. Leaves opposite. Flowers solitary in the upper axils, the upper ones often forming terminal spikes or heads. Bracteoles 2, linear, close under the calyx.

The genus consists of a very few tropical Asiatic plants. Of the two Australian species, one has a wide range in the Archipelago and some parts of India, the other is endemic. It differs from the section *Adenosmoides* of *Stemodia,* chiefly in the abortion of one cell of the upper anthers.

Stems erect. Leaves shortly petiolate . . . . 1. *A. cærulea.*
Stems decumbent. Leaves on long petioles . . . . .2. *A. Muelleri.*

1. **A. cærulea,** *R. Br. Prod.* 443. An erect, simple or branched, rather coarse annual, from under 1 ft. to nearly 2 ft. high, glandular-pubescent or villous all over, strongly scented. Leaves shortly petiolate ovate and scarcely exceeding 1 in. in the typical form, ovate-lanceolate and above 2 in. long in

some Asiatic specimens, the floral ones gradually smaller and passing into sessile bracts not exceeding the calyxes. Flowers blue, very shortly pedicellate, the lower ones axillary and distant, the upper ones forming a more or less compact terminal raceme. Bracteoles shorter than the calyx. Calyx very villous, oblique and slightly curved, 3 to 4 lines long, the upper segment lanceolate, the lower ones linear. Corolla 5 to 6 lines long, the lips nearly equal, shorter than the tube. Lower stamens nearly as long as the corolla, with 1-celled anthers; upper ones shorter, the anthers with 2 cells separated by a broad thick connectivum. Capsule acuminate, rather hard, somewhat incurved, about as long as the calyx.—*Pterostigma villosum*, Benth. Scroph. Ind., and in DC. Prod. x. 380.

Queensland. Endeavour Bay, *Banks and Solander;* Point Lookout, *R. Brown.*

Although I had originally considered this genus to be the true *Adenosma* of Brown, I was subsequently misled by some specimens sent to me by A. Cunningham, as having been identified with the Banksian plant, and which proved to be inseparable from *Stemodia*. The examination, however, which I have now made of Brown's and Banks's typical specimens, shows that they belong in fact to my *Pterostigma villosum*, and that Cunningham's plant is very different, both in station and in character. The name of *Adenosma* must, therefore, be now given to the whole of my genus *Pterostigma*.

2. **A. Muelleri,** *Benth.* Apparently annual though rather hard, diffuse or decumbent, branched, glandular-villous and viscid, strongly scented, even when dry. Leaves on rather long petioles, ovate or ovate-oblong, obtuse, crenulate, rugose, ½ to 1 in. long, the floral ones passing into the small sessile uppermost ones. Flowers rather large, sessile or very shortly pedicellate. Calyx glandular-villous, above 3 lines long when in fruit, the segments lanceolate, the upper one nearly twice as broad as the others. Corolla dark-coloured, the tube about as long as the calyx, the lips probably as long; the upper one broad and entire, the lower one 3-lobed. Anthers of the longer stamens with only one cell, the other one abortive or rudimentary, those of the shorter stamens 2-celled. Capsule acuminate, as long as the calyx.—*Stemodia odoratissima*, F. Muell. Herb.

N. Australia. Macarthur River, *F. Mueller.*

6. STEMODIA, Linn.

Calyx divided to the base into 5 segments or sepals, all equal or the upper one scarcely larger. Corolla tubular at the base, the upper lip broad, entire or notched, the lower one spreading, 3-lobed. Stamens 4, in pairs; anthers 2-celled with the cells quite separate, usually stipitate. Style dilated at the summit into 2 stigmatic lobes or rarely entire, not winged. Capsule globular, ovate or acuminate, opening septicidally in 2 usually 2-fid valves or in 4 valves, the placentas of the two carpels completely separating at maturity (at least in the Australian section). Seeds numerous, small, striate and usually reticulate.—Herbs, rarely undershrubs, more or less glandular-pubescent or villous and often strong-scented. Leaves opposite or in whorls of 3 or 4. Flowers solitary in the axils, the upper ones often forming terminal spikes. Bracteoles usually 2, linear, close under the calyx.

The genus is chiefly from tropical and southern extratropical America, represented by two species in tropical Asia and Africa.   Of the four Australian species, one is the common Asiatic one, the other three endemic.   They all belong to the section to which I had given the name of *Adenosma*, in the belief that it included Brown's genus of that name; but, as that now proves to be my *Pterostigma*, the present section, differing from it only by the anthers having all 2 perfect cells, may take the name of *Adenosmoides*.

Leaves mostly lanceolate or oblong, sessile and stem-clasping or a few of
    the lowest rarely petiolate.   Stems erect or ascending.
  Flowers sessile or very shortly pedicellate.
    Corolla (about 3 lines long) shortly exceeding the calyx . . . .   1. *S. lythrifolia.*
    Corolla (about 6 to 7 lines long) twice as long as the calyx . . .   2. *S. grossa.*
  Flowers on pedicels longer than the calyx . . . .   3. *S. viscosa.*
Leaves mostly ovate on long petioles.   Stems decumbent. (Corolla 3 to
    4 lines long ?)   . . . .   4. *S. debilis.*

1. **S. lythrifolia,** *F. Muell. in Herb. Hook.*   A hard erect slightly-branched herb attaining 1 to 2 ft., very softly villous all over, almost woolly, and sometimes slightly viscid.   Leaves ovate-lanceolate, oblong or lanceolate, serrate or almost entire, narrowed below the middle but usually dilated and stem-clasping at the base, soft and rugose, the larger ones 1 to 2 in. long, the lowest sometimes more distinctly petiolate, the floral ones small and ovate passing into entire bracts.   Flowers small, sessile in the upper axils, the uppermost forming a compact spike with the ovate bracts almost imbricate in 4 rows and scarcely exceeding the calyxes.   Calyx glandular-pubescent, about 2 lines long, the segments narrow-lanceolate, acute, rather unequal.   Corolla shortly exceeding the calyx, the upper lip broad, truncate or slightly notched.   Anthers all 2-celled.   Capsule hard, acuminate, not exceeding the calyx.—*Stemodia cærulea,* Benth. in DC. Prod. x. 381, as to A. Cunningham's plant but not R. Brown's synonym.

N. Australia.   Common in the rocky islands of the N.W. coast, *A. Cunningham, Bynoe;*   Upper Victoria river, *F. Mueller;*   islands of the Gulf of Carpentaria, *R. Brown* (not inserted in Brown's Prodromus).

Var. ? *tenuior.*   Less woolly, the leaves broader and more membranous, the floral ones all toothed and not imbricate.—York Sound, N.W. coast, *A. Cunningham.*   Perhaps a distinct species, but the specimens too imperfect to determine.

2. **S. grossa,** *Benth.*   A stout erect hard perennial or undershrub of 1 to 2 ft., glandular-villous all over, and strongly scented when fresh.   Leaves mostly in whorls of three, ovate oblong or lanceolate, acutely toothed, the lower ones contracted below the middle, dilated and stem-clasping at the base, the larger ones above 1 in. long, the floral ones gradually smaller and more ovate.   Flowers large for the genus, sessile in the upper axils, forming a terminal interrupted leafy spike.   Calyx glandular-villous, about 3 lines long, the segments lanceolate, nearly equal.   Corolla dark-coloured, at least 6 or 7 lines long, the tube broad, hairy inside, the upper lip very broad, entire, as long as the tube, the lower lip of the same length, with ovate obtuse lobes.   Anthers all 2-celled.   Capsule acuminate, about 2 lines long.

N. Australia.   Desert Island of the N.W. coast, *Bynoe;*   Nichol Bay, *Walcott.*

3. **S. viscosa,** *Roxb. Pl. Corom.* ii. 33. *t.* 163.   A perennial with ascending or erect not much-branched stems from under 6 in. to above 1 ft. high, the whole plant pubescent or villous, viscid and scented.   Leaves opposite or in whorls of three, the lower ones often ovate and contracted into a

petiole, the upper ones or nearly all lanceolate, acute, serrate, often dilated and stem-clasping at the base, the larger ones 1 to 2 in. long but often all under 1 in., the upper floral ones gradually smaller but usually distant. Flowers axillary, on pedicels always longer than the calyx and sometimes exceeding the leaves. Calyx usually about 2 lines long, the segments narrow, nearly equal or one larger. Corolla at least twice as long as the calyx, the upper lip very broad, entire or slightly notched, the lower with 3 ovate very obtuse lobes. Anthers all 2-celled. Capsule acuminate, as long as the calyx.—Benth. in DC. Prod. x. 381.

N. Australia. Victoria river, *F. Mueller*; Gulf of Carpentaria, *Landsborough.*
W. Australia. Murchison river, *Oldfield, Drummond, 6th Coll. n.* 127.

The species is common in East India, and I can find no difference in the above Australian specimens.

Var. ? *grandiflora.* A coarser plant, with the flowers ¾ in. long and very broad.—Murchison river, *Oldfield, Drummond, 6th Coll. n.* 128.

4. **S. debilis,** *Benth.* Apparently annual and diffuse or decumbent as in *Adenosma Muelleri,* but much more slender, loosely hairy and slightly glandular. Leaves on long slender petioles, ovate, toothed, membranous, glandular dotted, ½ to ¾ in. long, the upper floral ones smaller, more sessile, narrower and approximate but not imbricate. Flowers small, sessile or very shortly pedicellate. Calyx-segments lanceolate, acuminate, the upper one often 3 lines long, the others usually much smaller. Corolla shortly exceeding the calyx, but those expanded in the specimen not perfect; in the bud the upper lip broad and entire as in the preceding species. Anthers all 2-celled. Capsule acuminate, as long as the calyx.

W. Australia. Victoria river, *F. Mueller.*

7. MORGANIA, R. Br.

Calyx divided to the base into 5 narrow segments, all equal or nearly so. Corolla tubular at the base, the upper lip broad, entire or shortly 2-lobed, the lower one spreading, 3-lobed. Stamens 4, in pairs; anthers 2-celled, with the cells quite separate and somewhat stipitate. Style deflected at the summit, with 2 short spathulate stigmatic lobes, scarcely winged at the bend. Capsule ovoid oblong or shortly acuminate, opening septicidally in 2 2-fid or in 4 valves, leaving the placentas united in a single column in the centre. Seeds numerous, small, striate.—Herbs either glabrous or slightly pubescent. Leaves opposite or rarely in whorls of 3, narrow. Flowers solitary in the axils, sessile or pedicellate, with small linear bracts close under the calyx.

The genus is limited to Australia. The four forms here admitted as species are very closely allied to each other, and are all united by F. Mueller, Fragm. vi. 104, with the Australian specimens of *Stemodia viscosa* into one species, transferred to *Limnophila* under the name of *L. Morgania,* and there is no doubt but that the three genera *Stemodia, Morgania,* and *Limnophila* are so closely connected that they might almost equally well be considered as sections of one genus characterized chiefly by the stamens. Still the differences in the capsule between *Stemodia* (sect. *Adenosma*) *Morgania* and *Limnophila* prove, upon examination, to be rather more definite than I had thought when working up *Morgania* for the 'Prodromus' on insufficient materials. In *Stemodia* (*Adenosma*) the carpels separate completely each one carrying off its own placenta; in *Morgania* the dissepiment splits, forming inflexed margins to the valves, but the two placentas remain consolidated in a single central column; in *Limnophila* the margins of the valves are still somewhat inflexed, but a considerable portion of the dissepiment remains entire, detached from the valves, and bearing the placentas on its face: the dehiscence in the first two is septicidal, in the third partially septifragal. The American section *Diamoste* of *Stemodia* has, however, much more the dehiscence of *Morgania;* and if the latter intermediate genus is to be united with one of the two others, it is rather with *Stemodia* than with *Limnophila.* The three, however, as now constituted, are natural groups easily recognized.

Flowers sessile or the pedicels rarely as long as the calyx. Corolla-lips
  as long as the tube, the upper one entire.
  Plant glabrous or nearly so . . . . 1. *M. floribunda.*
  Plant hoary-pubescent . . . . 2. *M. pubescens.*
Pedicels mostly longer than the calyx.
  Flowers rather large, the lips shorter than the tube, the upper one
    entire . . . . 3. *M. glabra.*
  Flowers small, the lips as long as the tube, the upper one shortly 2-
    lobed . . . . 4. *M. parviflora.*

1. **M. floribunda,** *Benth. in Mitch. Trop. Austr.* 384. Stems from a perennial stock erect, usually taller less branched and more rigid than in *M. glabra,* glabrous or nearly so and often glaucous. Leaves linear or linear-lanceolate, entire or with few teeth, from about ½ in. to above 1 in. long. Flowers (blue?) almost sessile or on pedicels usually very short or rarely as long as the calyx, often appearing clustered with small leaves in the axils owing to the partial development of axillary branchlets. Calyx rather larger than in *M. glabra* and corolla the same size, but much more deeply cleft, the tube scarcely exceeding the calyx and the lips as long as the tube, the upper one broad truncate and entire as in *M. glabra.* Capsule shortly acuminate.

Queensland. Rockhampton and Keppel Bay, *Thozet;* Crocodile Creek, *Bowman;* Balonne and Narran rivers, *Mitchell.*
N. S. Wales. Macquarrie river, *Mitchell;* Murray and Darling rivers, *Victorian Expedition, Dallachy,* and others; Mount Murchison, *Giles.*
Victoria. Wimmera, *Dallachy.*
S. Australia, *Behr;* Holdfast Bay, *F. Mueller,* towards Spencer's Gulf, *Warburton;* Torrens river, *Whittaker;* Wills' Creek, Cooper's Creek, etc., *Howitt's Expedition.*
W. Australia. Murchison river, *Oldfield, Drummond, 6th Coll. n.* 126.

2. **M. pubescens,** *R. Br. Prod.* 441. Very nearly allied to *M. floribunda,* and, as far as I can ascertain in the few specimens seen, with the same nearly sessile flowers calyx and corolla, but the whole plant hoary with a short soft pubescence.—Benth. in DC. Prod. x. 385; Endl. Iconogr. t. 103.

N. Australia. Roper river, *F. Mueller.*
Queensland. Comet river, *Leichhardt;* Broad Sound, *R. Brown.*

3. **M. glabra,** *R. Br. Prod.* 441. Stems from a perennial stock erect, usually branched, rather slender, ½ to 1 ft. high, glabrous or with a minute almost granular pubescence on the upper parts and flowers. Leaves sessile, linear or linear-lanceolate, entire or with very few small teeth, ½ to 1 in. long. Flowers in the upper axils, on slender pedicels, sometimes short at first but at length much longer than the calyx. Calyx not 2 lines long, deeply divided into narrow segments. Corolla above ½ in. long, the tube twice as long as the calyx, the lips broad, the upper one truncate, the lower 3-lobed, both

much shorter than the tube. Capsule shortly acuminate.—Benth. in DC. Prod. x. 385.

N. Australia. Upper Roper river and Alligator Point, *F. Mueller;* Albert river, *Henne;* Gulf of Carpentaria, *Landsborough.*

Queensland. Broad Sound, *R. Brown;* estuary of the Burdekin, *F. Mueller;* Fitzroy river, *Bowman;* Barcoo river, *Mitchell;* Curriwillighi, *Dalton.*

N. S. Wales. Plains of the Gwydir, *Mitchell;* between the Darling and Cooper's Creek, *Neilson;* Ballandool river, *Locker.*

4. **M. parviflora,** *Benth.* Stems from a perennial almost woody stock erect, paniculately branched, 6 in. to above 1 ft. high, glabrous or slightly pubescent. Leaves very few, small and distant, all linear, a few of the largest $\frac{1}{2}$ to $\frac{3}{4}$ in. long, but mostly reduced to small scales. Flowers much smaller than in the other species, on short rigid pedicels. Calyx about $1\frac{1}{2}$ lines long, glandular-pubescent. Corolla scarcely above 3 lines long, the lips about as long as the tube, the upper one shortly 2-lobed, the lower one 3-lobed to about the middle; anthers of the longer stamens smaller than those of the shorter ones, but all 2-celled. Capsule $1\frac{1}{2}$ lines long, scarcely acuminate.

N. Australia. Arnhem's Land, *F. Mueller.*

8. LIMNOPHILA, R. Br.

Calyx divided to the base or below the middle into 5 narrow segments, all equal or nearly so. Corolla tubular at the base, the upper lip broad, entire, notched or shortly 2-lobed, the lower one spreading, 3-lobed. Stamens 4, in pairs; anthers 2-celled, with the cells quite separate and somewhat stipitate. Style deflected at the summit, with 2 short flat stigmatic lobes, scarcely winged at the bend. Capsule broadly ovoid or oblong, usually obtuse, opening in 4 valves, leaving the dissepiment entire at least at the base, bearing the placentas on its faces, thus forming as it were two wings to the undivided placental column. Seeds numerous, small, striate and transversely reticulate. —Herbs usually growing in marshes or shallow water, glabrous or slightly pubescent, usually scented and marked with pellucid dots. Leaves opposite or whorled, toothed or deeply cut, the submerged ones in some species divided into numerous capillary segments. Flowers solitary in the axils, the upper ones sometimes forming a terminal raceme. Bracteoles linear, close under the calyx.

A considerable genus, chiefly tropical, and limited to the Old World. The four Australian species are all widely spread in tropical Asia, and one at least extends into Africa.

Lower leaves (or all) deeply divided. Flowers pedicellate. Calyx-
segments 1-nerved . . . . 1. *L. gratioloides.*
Leaves all undivided.
Flowers pedicellate. Calyx-segments several-nerved, striate.
     Glabrous . . . . 2. *L. punctata.*
     Pubescent or villous . . . . 3. *L. hirsuta.*
Flowers sessile. Calyx-segments united at the base, slightly striate   4. *L. serrata.*

1. **L. gratioloides,** *R. Br. Prod.* 442. Stems from a creeping base, ascending or erect, usually about 6 in. high, but sometimes very short decumbent and branched, or drawn up into simple stems of 1 to 2 ft., the whole plant glabrous. Leaves mostly opposite, but the lower ones usually divided to the base into narrow toothed or pinnatifid segments so as to appear

whorled, and when under water cut up into numerous capillary segments or lobes; the upper ones sometimes, or very rarely nearly all, undivided, sessile, linear or lanceolate and slightly toothed, all under 1 in. long and usually about ½ in. Pedicels in the upper axils longer than the calyx and usually exceeding the leaves. Bracteoles small. Calyx usually under 2 lines long at the time of flowering, the segments lanceolate, acuminate, broad at the base especially after flowering, membranous and 1-nerved. Corolla blue, with the centre yellow inside, about 5 or 6 lines long, the tube exceeding the calyx, the lips broad and shorter than the tube, the upper one shortly 2-lobed. Anthers cohering in pairs. Capsule broad and obtuse.—Benth. in DC. Prod. x. 389, with the synonyms quoted (except the reference to Gaudichaud's plate in Freyc. Voy. t. 57. f .1, which is evidently *L. sessiliflora*); F. Muell. Fragm. vi. 104.

N. Australia. Gulf of Carpentaria, *F. Mueller.*
Queensland. Broad Sound, *R. Brown, Bowman;* Port Denison, *Fitzalan;* Rockingham Bay, *Dallachy ;* Rockhampton, *O'Shanesy.*

The species is widely dispersed over tropical Asia and Africa. The flowers are variously described by Australian collectors as yellow pink or red.

2. **L. punctata,** *Blume; Benth. in DC. Prod.* x. 388. Stems ascending or erect, usually taller and stouter than in *L. gratioloides*, often above 1 ft. high, the whole plant glabrous. Leaves opposite or rarely in whorls of 3, sessile and stem-clasping, oblong-lanceolate, minutely serrate, 1 to 1½ or even 2 in. long, the upper ones smaller, the larger ones sometimes rugose. Flowers violet-blue, 7 to 8 lines long, on pedicels scarcely shorter than the floral leaves, the upper ones sometimes forming a very loose leafy raceme. Calyx-segments lanceolate, subulate-acuminate, often above 3 lines long, striate with 5 to 7 prominent nerves. Corolla-lips shorter than the tube, the upper one very broad and retuse than scarcely lobed. Anthers cohering in pairs. Capsule oblong, shorter than the calyx.

Queensland. Wide Bay, *Bidwill;* Rockingham Bay, *Dallachy.* Common in the Indian Archipelago, also in Ceylon.

3. **L. hirsuta,** *Benth. in DC. Prod.* x. 388. Very nearly allied to *L. punctata*, with the same undivided leaves, striate calyx, and rather large blue-violet corolla ; but it is usually a smaller plant, the leaves more frequently in whorls of 3, the pedicels shorter, and the stems, pedicels, and calyx always, and usually the leaves also, pubescent or hirsute.

N. Australia. Arnhem's Land, *F. Mueller.* The precise station uncertain, as his labels of this and *L. gratioloides* have been mixed both in the Muellerian and in the Hookerian herbaria. The species has an extensive range in tropical Asia.

4. **L. serrata,** *Gaudich. in Freyc. Voy.* 448. *t.* 57. Decumbent or erect, not much branched and quite glabrous, the stems usually slender, ½ to 1 ft. long. Leaves ovate oblong or lanceolate, obtuse, obtusely-serrulate, the lower ones contracted at the base, the upper ones with a broader base, all stem-clasping, under 1 in. long. Flowers closely sessile in the upper axils, mostly distant, but the upper ones sometimes crowded into a short terminal leafy spike. Bracteoles small, linear. Calyx thinner than in the two preceding species, not exceeding 2 lines, the segments subulate-acuminate,

slightly striate and connected at the base into a short tube. Corolla rather slender, nearly twice as long as the calyx, the lips not half as long as the tube, the upper one broad, slightly notched, the lower of 3 broad rounded lobes. Anthers slightly cohering in pairs. Capsule ovoid, the persistent dissepiment broad.—Benth. in DC. Prod. x. 387.

N. Australia. Victoria, Upper Roper, and Fitzmaurice rivers, and swamps near Providence Hill, *F. Mueller.* The species is dispersed over the Indian Archipelago, and extends to the Pacific Islands, and if, as is probable, *L. conferta*, Benth. l. c., is but a variety of the same, it is also in Ceylon and several parts of E. India.

9. HERPESTIS, Gærtn. f.

Calyx divided to the base into 5 distinct sepals, the outer one much broader than the others. Corolla tubular at the base, the upper lip erect or spreading, notched or 2-lobed, the lower lip spreading, 3-lobed, or sometimes the 5 lobes nearly equal. Stamens 4, in pairs, the anthers all perfect, 2-celled, the cells contiguous. Style dilated at the summit, concave or slightly 2-lobed. Capsule opening loculicidally in 2 often bifid valves or in 4 valves, leaving the placentas on a free central column or dissepiment. Seeds numerous, usually striate and transversely reticulate.—Glabrous or rarely pubescent herbs. Leaves opposite, entire, toothed or in some non-Australian aquatic species the submerged ones cut into numerous capillary segments. Flowers axillary, or, in species not Australian, in a terminal raceme. Bracteoles under the calyx only in a very few species.

A considerable tropical and subtropical genus, chiefly American, with a few species natives of the Old World. Of the two Australian species, one is common over nearly the whole area of the genus, the other extends over tropical Asia and Africa.

Erect. Leaves narrow. Flowers numerous, on short pedicels . . . 1. *H. floribunda.*
Procumbent or creeping. Leaves small, obovate or oblong, rather thick.
Flowers few, on long pedicels . . . . 2. *H. Monnieria.*

1. **H. floribunda,** *R. Br. Prod.* 442. Apparently annual, erect and branching, rarely above 1 ft. high, quite glabrous. Leaves lanceolate or linear-lanceolate, rather obtuse, entire, narrowed to the base, 1-nerved, rarely above 1 in. long. Pedicels slender but usually shorter than the petiole, often 3 together in each axil owing to the partial development of an axillary branchlet. Bracteoles very small, a little below the calyx. Calyx scarcely above 1½ lines long at the time of flowering, 2 to 3 lines when in fruit, the segments thin, at first herbaceous, at length membranous and reticulate, the outer one broadly ovate, the 2 next narrow-ovate, the 2 innermost almost linear. Corolla scarcely exceeding the calyx. Capsule ovoid-globular, shorter than the calyx, opening in 4 valves.—Benth. in DC. Prod. x. 400.

N. Australia. Victoria river, *F. Mueller;* South Goulburn Island, *A. Cunningham.*
Queensland. Shoalwater Bay, *R. Brown;* Burdekin river, *Bowman.*
The species extends over tropical Asia and Africa.

2. **H. Monnieria,** *H. B. and K.; Benth. in DC. Prod.* x. 400. A low creeping or procumbent glabrous leafy annual (or perennial ?). Leaves obovate or oblong, rarely above ½ in. long, rather thick, entire or crenate, without prominent veins or obscurely 1- or 3-nerved. Flowers few, pale blue or almost white, on pedicels usually rather longer than the leaves, with 2 small

bracteoles under the calyx. Calyx about 2 lines long or 3 lines when in fruit, the outer sepal oval, the others ovate-lanceolate or lanceolate. Corolla-tube scarcely so long as the calyx, the 5 lobes spreading, broad, as long as the tube, the 2 upper ones rather smaller and less deeply separated than the others. Capsule ovoid, shorter than the calyx, opening loculicidally in 2 valves, which at length separate from the dissepiment and sometimes split into 2.—Bot. Mag. t. 2557.

Queensland. Moreton Island, *M'Gillivray;* Burnett river, *F. Mueller;* Nerkool Creek, *Bowman.*

N. S. Wales. Paramatta, *Woolls.*

The species is one of the commonest marsh plants in the tropical and subtropical regions of both the New and the Old World, and has been described under a great variety of names, as detailed in the above-quoted ' Prodromus.'

10. GRATIOLA, Linn.

Calyx divided to the base into 5 nearly equal segments or sepals. Corolla tubular at the base, the upper lip broad and entire or shortly 2-lobed, the lower 3-lobed. Stamens 2 perfect, with the anthers connivent, the cells parallel and distinct but contiguous; the lower stamens reduced to slender staminodia or entirely wanting. Style dilated and deflected at the summit, entire or with 2 flat lobes. Capsule 4-valved, leaving a single columnar placenta bordered by a portion of the dissepiment. Seeds small, striate and transversely reticulate.—Erect or procumbent herbs, glabrous or glandular-pubescent. Leaves opposite, undivided. Flowers axillary, sessile or pedicellate, with a pair of bracteoles close under the calyx.

The species are not numerous, dispersed over the temperate and subtropical regions of both hemispheres. Of the three Australian species, one extends to New Zealand and extra-tropical South America, another to New Zealand only ; the third appears to be endemic, but is closely allied to a common N. American one.

Flowers pedunculate. Stems erect. Leaves lanceolate. Staminodia
  none . . . . 1. *G. pedunculata.*
Flowers sessile. Stems erect. Leaves ovate or lanceolate. Staminodia filiform (often exceedingly slender) . . . . 2. *G. peruviana.*
Stems procumbent or creeping. Flowers shortly pedicellate. Leaves
  oblong or obovate. Staminodia filiform . . . . 3. *G. nana.*

1. **G. pedunculata,** *R. Br. Prod.* 435. Stems from a shortly decumbent or sometimes creeping base, erect or ascending, scarcely branched, ½ to 1 ft. high, the whole plant minutely viscid-pubescent or rarely glabrous. Leaves lanceolate or oblong, bordered by a few teeth or nearly entire, the lower ones often contracted at the base but mostly stem-clasping, the larger ones sometimes above 1 in. long, but generally smaller. Pedicels shorter or sometimes longer than the leaves, rarely shorter than the calyx. Bracteoles linear, sometimes as long as the calyx. Calyx-segments linear-lanceolate, acute, rather unequal, 2 to 2½ lines long. Corolla white, yellowish inside, at least twice as long as the calyx, the lips short and broad, the upper one very shortly 2-lobed. Anthers of the perfect stamens cohering, the cells parallel and transverse, the lower stamens entirely wanting. Capsule ovoid-globular, rather obtuse, often slightly exceeding the calyx.—Benth. in DC. Prod. x. 403.

Queensland. Burnett river, *F. Mueller ;* Brisbane river, Moreton Bay, *A. Cunningham ;* Rockhampton, *O'Shanesy.*

N. S. Wales. Port Jackson, *R. Brown* and others ; New England, *C. Stuart ;* Richmond river, *C. Moore ;* Darling Downs, *Law.*

Victoria. Avoca and Murray rivers, Tambo, Forest Creek, *F. Mueller.*

W. Australia, *Drummond, n.* 82, *and 4th Coll. n.* 158 (a glabrous form, with the pedicels usually shorter and the corolla smaller).

G. virginiana, from North America, is scarcely to be distinguished from this species by a more branching habit, the leaves more narrowed at the base, and the capsule not exceeding the calyx.

2. **G. peruviana,** *Linn. ; Benth. in DC. Prod.* x. 403. Stems from a procumbent or creeping base, often rooting at the lower nodes, ascending or erect, 6 in. to 1 ft. high, the whole plant quite glabrous or viscid-pubescent. Leaves sessile and stem-clasping, from ovate to lanceolate, obtuse or acute, serrate or almost entire, usually 3-nerved especially when broad, ½ to 1 in. long. Flowers sessile or nearly so in the upper axils, larger than in *G. pedunculata.* Calyx 2 to 3 lines long or even longer when in fruit, the segments linear-lanceolate, acuminate. Corolla 6 to 7 lines long, the lips broad, much shorter than the tube, the upper one notched. Anthers connivent, almost cohering, with transverse parallel cells. Staminodia filiform, with minute globular heads, sometimes short and so slender as to be very difficult to find, sometimes more elongated. Capsule ovoid-globular, rather obtuse, membranous.—*G. pubescens,* R. Br. Prod. 435 ; Benth. in DC. Prod. x. 404 ; Bartl. in Pl. Preiss. i. 342 (the narrow-leaved pubescent form) ; *G. latifolia,* R. Br. l.c. ; Benth. l.c. 403 ; Hook. f. Fl. Tasm. i. 291 (the broad-leaved glabrous form) ; *G. glabra,* Walp. Rep. iii. 287 (given by mistake as a name of Brown's).

Queensland. Moreton Bay, *Fitzalan* (with rather broad glabrous leaves).

N. S. Wales. Port Jackson to the Blue Mountains, *R. Brown, Woolls,* and others ; Hastings and Clarence rivers, *Beckler ;* Illawarra, *A. Cunningham* (all glabrous or nearly so, with broad or rather narrow leaves) ; Macleay river, *Beckler* (with broad, very pubescent leaves).

Victoria. About Melbourne, *Adamson, F. Mueller ;* near Portland, *Robertson, Allitt ;* Victoria Range and Station Peak, *F. Mueller* (all glabrous, with broad or sometimes narrow leaves) ; Snowy River, Dandenong Ranges, marshes on the Murray, *F. Mueller* (with narrow, viscid-pubescent leaves).

Tasmania. Port Dalrymple, *R. Brown ;* common everywhere in wet gravelly places, etc., *J. D. Hooker* (usually with broad leaves, and glabrous or nearly so) ; Jacke's Plain, and rocks in the Meandee near Cheshunt, *Archer* (pubescent, with narrow leaves).

S. Australia. Torrens river, Gulf of St. Vincent, *F. Mueller ;* Kangaroo Island, *Waterhouse* (mostly with broad leaves, and nearly glabrous).

W. Australia. King George's Sound, *R. Brown, Preiss, n.* 2326 *and* 2331 ; Karri Dale, *Walcott ;* granite rocks, Mount Melville, *F. Mueller* (all viscid-pubescent, with narrow leaves).

Var. *pumila.* Plant of 2 or 3 in., with small narrow leaves, glabrous or viscid-pubescent. —*G. pumila,* F. Muell. in Linnæa, xxv. 431.—Port Jackson and Port Dalrymple, *R. Brown ;* Victoria, *F. Mueller.*

The species is also in New Zealand and in extratropical S. America, where it is usually narrow-leaved and nearly glabrous, as represented by J. A. Schmidt in Mart. Fl. Bras. Scroph. t. 49, but occasionally very viscid-pubescent, and more rarely broad-leaved.

3. **G. nana,** *Benth. in DC. Prod.* x. 404. A dwarf procumbent or creeping much-branched plant, glabrous or minutely viscid-pubescent, not rising above a few inches from the ground, with something of the habit of

Herpestis Monnieria. Leaves oblong or obovate, narrowed at the base, very obtuse, rather thick, entire or obscurely toothed, 3 to 4 lines long. Flowers few, rather small, on short pedicels. Calyx glandular-pubescent, the segments rather obtuse, about 2 lines long. Corolla about 4 lines long. Anthers connivent, with transverse parallel cells ; staminodia filiform, with minute heads, rather long. Capsule broadly ovoid.—Hook. f. Fl. Tasm. i. 291.

Victoria. Highest part of the Australian Alps, *F. Mueller (Herb. F. Muell.).*

Tasmania. Sandy and marshy banks of rivers, etc., in alpine situations, Marlborough, Hampshire Hills, and Arthur's Lakes, *Milligan, Gunn ;* South Port, *C. Stuart ;* Recherche Bay, *Oldfield.*

The species is also in New Zealand.

11. DOPATRIUM, Hamilt.

Calyx campanulate, 5-lobed. Corolla tubular at the base, with the lips spreading, the upper one 2-lobed, the lower larger and 3-lobed. Stamens, 2 upper ones perfect, included in the tube ; anther-cells distinct and parallel, 2 lower reduced to minute filiform staminodia. Style with 2 flat stigmatic lobes. Capsule opening loculicidally in 2 entire or rarely bifid valves, bearing in their centre the separate placentas.—Slender glabrous herbs. Leaves opposite, chiefly at the base of the stem, the others usually minute and few. Pedicels filiform. Bracteoles none.

A genus with very few species, inhabitants of marshy or rich moist places in tropical Asia and Africa. The only Australian species is a common one in India.

The ovary and capsule of this genus, not quite correctly described in the ' Prodromus,' and still more inaccurately figured in Wight's plate of *D. lobelioides* (Ic. t. 859), differ from those of all other *Gratioleæ* in being scarcely perfectly 2-celled. The broad flat placentas are at right angles to the dissepiment, and although their inner faces are contiguous and bear no ovules or seeds, yet they do not cohere, the ovules and seeds being very numerous on their backs or outer faces, turned towards the walls of the cavity.

1. **D. junceum,** *Hamilt. ; Benth. in DC. Prod.* x. 407. A glabrous erect annual, branching chiefly at the base, sometimes scarcely above 2 or 3 in. high, but when luxuriant its slender stems attain 1 ft. Lower and radical leaves oblong, obtuse, entire, contracted at the base and often above $\frac{1}{2}$ in. long, the others small, sessile, ovate, the upper ones few and distant and scarcely 1 line long. Flowers in the upper axils usually short but sometimes nearly $\frac{1}{4}$ in. long. Calyx scarcely $\frac{3}{4}$ line long, divided to about the middle into narrow obtuse lobes. Corolla-tube about $1\frac{1}{2}$ lines long, the throat very open, the upper lip 2-lobed, the lower very broadly 3-lobed, as long as the tube. Capsule globular, scarcely 1 line diameter.—*Gratiola juncea,* Roxb. Pl. Corom. ii. t. 129.

Queensland. Rockhampton, *O'Shanesy.* Common in E. India.

SUBTRIBE II. LINDERNIEÆ.—Stem-leaves opposite. Stamens, 2 upper ones inserted in the tube and usually included in it and perfect, the anthers approximate or cohering, the cells contiguous, often divaricate and sometimes confluent into one ; 2 lower ones inserted in (or adnate to) the throat, either reduced to club-shaped linear or 2-fid staminodia, or when perfect with long arched filaments (short in the European *Lindernia*), with an angle or lobe near the base, the anthers cohering under the upper lip of the corolla, the cells usually divaricate and often confluent.   Capsule opening in 2 membranous entire valves, parallel to the broad thin dissepiment.

Like the majority of *Eugratioleæ*, the species are chiefly abundant in marshes or rich moist soils, and are more or less glandular-dotted.   The principal genera are closely allied, and often distinguished chiefly by characters derived from the stamens, which are difficult to ascertain in dried specimens owing to the delicacy of the corollas, but said to be readily seen in living plants.   F. Mueller proposes the reuniting several of them under the name of *Lindernia* or of *Vandellia*, but that would entail the regarding the whole subtribe as a genus, of which the present genera would be sections.   This would appear to me to be more a nominal than a real change, involving all the inconvenience of a great addition to the synonymy without any corresponding advantage.

12. ARTANEMA, Don.

Calyx deeply divided into 5 herbaceous acuminate segments, dilated and much imbricate at the base.   Corolla with a long broad tube, the upper lip broad and notched, the lower 3-lobed.   Perfect stamens 4, the anthers cohering in pairs with divaricate cells, the upper stamens included in the tube; filaments of the lower ones adnate almost to the throat, long and arched, with a broad appendage near the base.   Style with 2 flat stigmatic lobes.   Capsule globular, opening in 2 thin valves parallel to the broad thin dissepiment. —Erect herbs.   Leaves opposite.   Flowers large, pedicellate in the axils of small bract-like floral leaves, without bracteoles.

The genus consists of only three closely allied species, one of them Australian, the other two from tropical Asia.   It only differs from *Vandellia* in the larger flowers, broader calyx-segments, and in the shape of the staminal appendage.

1. **A. fimbriatum,** *Don in Sweet Brit. Fl. Gard. ser.* 2. *t.* 234.   An erect rather coarse annual (or sometimes perennial?) of 1 to 2 ft., the angles of the stems and upper surface of the leaves scabrous, otherwise glabrous. Lower leaves petiolate and ovate-lanceolate, upper ones more sessile and lanceolate, sometimes very narrow, the larger ones 2 to 4 in. long, all more or less serrate or rarely entire.   Flowers violet, in distant pairs, forming very loose terminal racemes, the floral leaves reduced to small bracts.   Pedicels ½ to 1 in. long.   Calyx-segments 3 to 4 lines long, the points usually recurved. Corolla above 1 in. long; lobes broad and rounded, minutely and irregularly crenulate or jagged.   Appendage of the lower filaments broad, rounded, and scale-like.   Capsule 3 to 4 lines diameter.—Benth. in DC. Prod. x. 408 ; *Torenia fimbriata*, Grah. in Edinb. New Phil. Journ. xi. 379 ; *T. scabra*, Grah. in Bot. Mag. t. 3104.

Queensland.   Brisbane river, Moreton Bay, *Fraser, F. Mueller*, and others ; Wide Bay, *Bidwill* (with narrow leaves) ; Rockingham Bay, *Dallachy*.

N. S. Wales.   Hastings and Clarence rivers, *Beckler ;* Richmond river, *Henderson*.

13. VANDELLIA, Linn.

Calyx either divided to the base into 5 narrow segments, or the segments more or less cohering into a short 5-toothed calyx (not folded and prominently angled, as in *Torenia*).   Corolla tubular at the base, the upper lip erect, shortly 2-lobed, the lower lip larger, spreading, 3-lobed.   Perfect stamens 4, the anthers cohering in pairs, with divaricate cells, the upper stamens included in the tube; filaments of the lower ones adnate to the throat, long and arched, with an angle tooth or linear lobe near the base. Style with 2 flat stigmatic lobes. Capsule globular oblong or linear, opening in 2 entire valves parallel to the thin dissepiment.—Slender herbs, erect or diffuse and much-branched, glabrous or pubescent. Leaves opposite, undivided. Flowers opposite, or alternate by the abortion of one of each pair, axillary or in terminal racemes, the racemes sometimes contracted into umbels, without bracteoles.

The genus comprises a considerable number of species, mostly common weeds in the tropical and subtropical regions of the Old World, two of them being also found in S. America. Of the five Australian species, one is the commonest over the whole range of the genus; the others appear to be all endemic.

Calyx-segments united in a 5-toothed calyx, at least at the time of
  flowering. Stems diffuse. Leaves ovate . . . . 1. *V. crustacea.*
Calyx-segments separate from the first.
  Leaves ovate, chiefly near the base of the stem.
    Plant pubescent or hirsute . . . . 2. *V. pubescens.*
    Plant glabrous.
      Corolla-tube rather longer than the calyx . . . . . . 3. *V. alsinoides.*
      Corolla-tube fully twice as long as the calyx . . . . . 4. *V. scapigera.*
  Leaves linear-subulate, few and mostly small . . . . 5. *V. subulata.*

1. **V. crustacea,** *Benth. Scroph. Ind. and in DC. Prod.* x. 413. A diffuse much-branched annual, glabrous or with a very few small scattered hairs, usually not exceeding 6 in. but attaining nearly 1 ft. when very luxuriant. Leaves shortly petiolate, ovate, broadly crenate or almost entire, sometimes almost cordate at the base, from under ½ in. to about ¾ in. long. Pedicels usually ½ to nearly 1 in. long, axillary or forming loose leafy racemes. Calyx about 2 lines long, membranous and 5-toothed, with 5 scarcely prominent nerves at the time of flowering, often splitting into 5 segments when the flowering is over. Corolla scarcely twice as long as the calyx. Capsule ovoid or almost oblong, shorter than or as long as the calyx.—Wight, Ic. t. 863; *Capraria crustacea,* Linn., and the numerous synonyms quoted in DC. Prod. as above; *Torenia flaccida* and *T. scabra,* R. Br. Prod. 440; *V. Brownii,* Benth. in DC. Prod. x. 413.

N. Australia. Islands of the Gulf of Carpentaria, *R. Brown.*
Queensland. Endeavour river, *Banks and Solander;* Wide Bay,*'Bidwill;* Port Denison, *Fitzalan;* Rockingham Bay, *Dallachy;* Burdekin river and Broad Sound, *Bowman;* Rockhampton, *O'Shanesy.*
The species is very common in tropical Asia, extending into tropical Africa and America.

2. **V. pubescens,** *Benth. in DC. Prod.* x. 415. Stems branching at the base, decumbent or erect, rarely exceeding 6 in., pubescent or hirsute as well as the foliage. Leaves chiefly crowded at the base of the stems, ovate, obtuse, entire, contracted into a short petiole, often 1 in. long, with 1 or 2 pairs of small sessile leaves higher up, the floral ones very small. Pedicels few, in pairs, ½ to 1 in. long, pubescent, reflexed after flowering. Calyx-segments very narrow, about 1½ lines long. Corolla-tube at least 2 lines long. Capsule broadly ovoid, obtuse.

N. Australia. Port Essington, *Armstrong.*

3. **V. alsinoides,** *Benth. in DC. Prod.* x. 415.   A slender branching erect or diffuse annual, usually glabrous and not exceeding 6 in.   Stem-leaves chiefly in the lower part of the plant, very shortly petiolate, broadly ovate or nearly orbicular, angular-toothed, thin and membranous, the larger ones 6 to 8 lines diameter, the floral ones very small or reduced to small bracts.   Flowers small, on slender pedicels of ½ to 1 in. reflexed after flowering, and usually one only to each pair of floral leaves.   Calyx about 1½ lines long, divided to the base into linear-subulate segments.   Corolla-tube rather longer than the calyx, the upper lip short, the lower not so long as the tube. Appendage of the lower filaments linear, glandular.   Capsule ovoid-oblong, rather longer than the calyx.—*Lindernia alsinoides*, R. Br. Prod. 441 ; *Tittmannia alsinoides*, Spreng. Syst. ii. 800 ; *Ilyogeton alsinoides*, Endl. in Walp. Rep. iii. 297.

Queensland. Facing Island, *R. Brown ;* Wide Bay, *Bidwill ;* Lizard Island, *M'Gillivray ;* Moreton Bay, *C. Stuart ;* Rockhampton, *O'Shanesy.*

4. **V. scapigera,** *Benth. in DC. Prod.* x. 415.   Stems branching at the base, ascending or erect, very slender, 6 in. high or more, the whole plant glabrous.   Leaves chiefly collected at the base of the stems, ovate, entire, under ½ in. long, the lowest broader and contracted into a short petiole, and 1 or 2 pairs higher up quite sessile, the floral ones very small and distant. Pedicels slender, 1 to 2 lines long, usually 1 only to each pair of floral leaves. Calyx divided to the base into linear segments scarcely 1 line long when in flower, rather longer afterwards.   Corolla-tube about 2 lines long.   Capsule ovoid-oblong, as long as the calyx.—*Lindernia scapigera*, R. Br. Prod. 441 ; *Tittmannia scapigera*, Spreng. Syst. ii. 800 ; *Ilyogeton scapigerum*. Endl. in Walp. Rep. iii. 297.

N. Australia. Islands of the Gulf of Carpentaria, *R. Brown ;* near Macadam Range, *F. Mueller.* Very near *V. alsinoides,* but more slender, the calyx smaller and the corolla larger.

5. **V. subulata,** *Benth. in DC. Prod.* x. 415.   Stems numerous, slender, erect, glabrous, usually branched, often attaining 1 ft.   Leaves linear-subulate, entire, the lower ones often ½ in. long, the upper ones few and small, the floral ones setaceous.   Pedicels filiform, often above 1 in. long, usually one only to each pair of floral leaves, but the uppermost often collected in a cluster or almost an umbel, with many minute floral leaves at their base.   Calyx divided into linear-subulate segments, scarcely above 1 line long. · Corolla fully 5 lines long.   Capsule ovoid-oblong.—*Lindernia subulata*, R. Br. Prod. 441 ; *Tittmannia subulata*, Spreng. Syst. ii. 801; *Ilyogeton subulatum*, Endl. in Walp. Rep. iii. 297.

N. Australia. Elsey's River, *F. Mueller* (specimens past flower and somewhat doubtful) ; Port Essington, *Armstrong.*
Queensland. Endeavour river, *Banks and Solander, A. Cunningham.*

14. ILYSANTHES, Rafin.

Calyx divided to the base into 5 narrow segments.   Corolla tubular at the base, the upper lip erect, shortly 2-lobed, the lower larger, spreading, 3-lobed. Perfect stamens 2, included in the tube, the anthers cohering, with divaricate cells, the lower pair reduced to staminodia adnate to the throat, thence usually projecting and 2-lobed, one lobe ascending, acute, filiform or reduced to a short tooth, the other obtuse and glandular or reduced to an angle. Style with 2 flat stigmatic lobes. Capsule globular ovoid or shortly oblong, opening in 2 entire valves parallel to the thin dissepiment.—Glabrous slender annuals. Leaves opposite. Flowers on slender pedicels, axillary or in terminal loose racemes, without bracteoles.

There are several species dispersed over the warmer regions of Asia, Africa, and America, extending into more temperate North America and South Africa. The only Australian species appears to be endemic. The genus differs from *Vandellia* and *Lindernia* in the abortion of the lower stamens, from *Bonnaya* chiefly in the short capsule.

1. **I. lobelioides,** *Benth.* A glabrous erect very slender annual, attaining about 6 in. and scarcely branched. Leaves few, near the base of the stem, ovate obovate or oblong, entire, narrowed into a short petiole and only 3 or 4 lines long; and 1 or 2 pairs of minute distant narrow sessile leaves higher up the stem, the floral ones reduced to minute bracts. Flowers in a short loose terminal raceme, on slender pedicels of $\frac{1}{2}$ to 1 in., opposite or one only to each pair of bracts. Calyx-segments linear-lanceolate, $1\frac{1}{2}$ lines long. Corolla-tube above 3 lines long, the lower lip much shorter than the tube, the upper one still shorter. Staminodia very shortly ascending, acute, the glandular lobe reduced to a prominent angle near its base. Capsule broadly ovate, obtuse, about as long as the calyx.—*Vandellia lobelioides*, F. Muell. in Trans. Phil. Inst. Vict. iii. 61.

N. Australia. Victoria Range, *F. Mueller.* In the 'Fragmenta,' vi. 102, F. Mueller refers this to *Vandellia scapigera*, which, however, besides the difference in foliage, has always 4 perfect stamens.

15. BONNAYA, Link and Otto.

Calyx divided to the base or nearly so into 5 narrow segments. Corolla tubular at the base, the upper lip erect, shortly 2-lobed, the lower larger, spreading, 3-lobed. Perfect stamens 2, included in the tube, the anthers cohering, with divaricate cells, the lower pair reduced to staminodia adnate to the throat, the ends either scarcely prominent or linear, entire obtuse and glandular. Style with 2 small flat stigmatic lobes. Capsule linear, longer than the calyx, opening in 2 entire valves parallel to the thin dissepiment.— Annuals usually glabrous. Leaves opposite. Flowers axillary or in terminal racemes, without bracteoles.

A small genus, spread over tropical and subtropical Asia. Of the two Australian species, one is the one most common in Asia, the other appears to be endemic.

Stems diffuse, leafy. Staminodia free at the end, erect linear and
obtuse . . . . 1. *B. veronicæfolia.*
Stems erect, the leaves few at the base of the stem. Staminodia
wholly adnate . . . . 2. *B. clausa.*

1. **B. veronicæfolia,** *Spreng.; Benth. in DC. Prod.* x. 421. A glabrous annual, much-branched, diffuse and rooting at the lower nodes, the flowering branches often ascending to 6 in. or more. Leaves sessile or narrowed into a short stem-clasping petiole, oblong-lanceolate or almost linear, the lower ones 1 to $1\frac{1}{2}$ in. long, entire or serrate, and often rather thick.

Flowers in terminal racemes, on spreading stiff pedicels of 3 to 6 lines, which are usually angular and thickened upwards, the subtending floral leaves reduced to minute bracts. Calyx narrow, 1 to $1\frac{3}{4}$ lines long, the segments sometimes united at the base. Corolla about twice as long as the calyx. Capsule linear, nearly $\frac{1}{2}$ in. long.—*B. verbenæfolia,* Benth. in DC. Prod. x. 421, with the numerous synonyms given under both names; *Gratiola veronicæfolia,* Roxb. Pl. Corom. t. 154; *Lindernia veronicifolia,* F. Muell. Fragm. vi. 101.

Queensland, *Bowman ;* Rockhampton, *O'Shanesy.* Common in tropical Asia, extending northwards to Loochoo.

2. **B. clausa,** *F. Muell. in Herb. Hook.* A glabrous erect annual, 6 in. to above 1 ft. high, slender and scarcely branched. Leaves few at the base of the stem, petiolate, broadly ovate, entire or obscurely toothed, under $\frac{1}{2}$ in. long, the stems otherwise leafless except the minute floral leaves or bracts, and sometimes a single pair of minute leaves lower down. Flowers in irregular racemes, occupying nearly the upper half of the stem, the pedicels rarely above $\frac{1}{4}$ in. long. Calyx-segments linear, about 1 line long. Corolla-tube slender and twice as long as the calyx, " the throat closed," the lower lip spreading, very broad, 3-lobed, with the middle lobe notched, longer than the tube, the upper lip shorter and slightly notched. Anthers cohering, with divaricate cells confluent so as to appear 1-celled; staminodia totally adnate, forming prominent ridges in the throat of the corolla. Capsule oblong-linear, about 2 lines long.—*Vandellia clausa,* F. Muell. in Trans. Phil. Inst. Vict. iii. 60; *Lindernia clausa,* F. Muell. Fragm. vi. 102.

N. Australia. Sand plains, Victoria river, *F. Mueller.* There are very few corollas on the specimens, and their texture is so delicate that I was unable to verify all the particulars described by F. Mueller in the only one I could examine.

SUBTRIBE III. LIMOSELLEÆ.—Small creeping or prostrate herbs with opposite or clustered leaves. Corolla usually minute, with a short tube and 5 lobes nearly equal or one or two rather larger than the others, the æstivation apparently variable. Anthers 1-celled.

The little plants here collected together are evidently nearly allied to each other, although formerly, from differences imperfectly observed in their æstivation, I had placed them in different tribes. They are very difficult to examine in the dried state, and some are often mixed in collections with one another or with *Montia fontana* and *Elatine americana.*

16. **PEPLIDIUM,** Delile.

Calyx tubular, 5-angled, 5-toothed or shortly 5-lobed. Corolla with a short tube and 5 nearly equal lobes. Stamens 2, the filaments somewhat dilated at the base; anthers 1-celled (by the confluence of 2 divaricate cells ?). Ovary completely 2-celled. Style short, dilated upwards into a broad spathulate lamina curved over the stamens. Capsule globular or ovoid, indehiscent or irregularly bursting (or sometimes 4-valved ?).—Small creeping or prostrate herbs. Leaves opposite. Flowers very small, axillary, without bracteoles.

The genus is limited to the two Australian species, of which one is widely diffused over the warmer regions of Asia and Africa, the other is endemic. The genus ought, perhaps, to be reunited with *Microcarpæa*, in which Smith had placed the common species. The anthers appear to have been erroneously described as bilocular.

Flowers sessile or nearly so. Capsule globular, obtuse . . . . . . 1. *P. humifusum.*
Flowers distinctly pedicellate. Capsule ovoid, acute . . . . . . 2. *P. Muelleri.*

1. **P. humifusum,** *Delile ; Benth. in DC. Prod.* x. 422. A dwarf prostrate glabrous plant, creeping and rooting at the nodes, sometimes forming dense tufts of 2 or 3 in. diameter, sometimes spreading to a considerable extent. Leaves ovate obovate or orbicular, obtuse, entire, contracted into a short petiole, rather thick especially when small, $\frac{1}{4}$ to $\frac{1}{2}$ in. long or rarely rather larger (in very wet situations?), the short petioles of each pair connected by their membranous margins. Flowers sessile or nearly so in the axils. Calyx scarcely above 1 line at the time of flowering, with 5 prominent angles or folds and membranous between them, the teeth short and obtuse. Corolla-tube rather shorter than the calyx, the lobes very short and rounded. Filaments rather thick, especially towards the base, angularly incurved. Capsule globular, large for the plant, very obtuse, enclosed in the distended calyx, about $1\frac{1}{2}$ lines diameter, membranous and indehiscent or at length bursting irregularly towards the base.—*Microcarpæa cochlearifolia,* Sm.; Hook. Bot. Misc. iii. 95. t. suppl. 29, and other synonyms quoted in the 'Prodromus.'

N. Australia. Upper Victoria river, *F. Mueller.*
Queensland. Rockhampton, *O'Shanesy ;* Cape river, *Bowman ;* between the Darling and the Lachlan rivers, *Burkitt.*

The species extends over the greater part of tropical and subtropical Asia and Africa.

2. **P. Muelleri,** *Benth.* Stems procumbent, much firmer than in *P. humifusum,* and not rooting at the nodes, glabrous or sparingly scabrous-pubescent. Leaves petiolate, ovate or obovate, very obtuse, entire, rather thick, 4 to 8 lines long. Flowers usually 2 together in each axil, on pedicels of 1 to 2 lines. Calyx tubular, $1\frac{1}{2}$ lines long, 5-angled, with obtuse teeth. Corolla-tube nearly as long as the calyx ; lobes oval-oblong, at least half as long as the tube, with 2 very prominent ridges (rudiments of staminodia?) in the throat opposite the sinus of the lower lobes, which are entirely wanting in *P. humifusum.* Filaments scarcely curved. Capsule ovoid, acute, readily opening in 2 or 4 valves, although not quite ripe in our specimens.

N. Australia. Upper Victoria river, *F. Mueller.* Several specimens of this are in the Hookerian herbarium, sent by F. Mueller as a large-leaved variety of *P. humifusum ;* but, besides the foliage, the pedicellate and longer flowers, the shape of the corolla, the stamens and the fruit appear to me to be quite different from those of *P. humifusum,* which is remarkably constant in its character throughout its very extended range.

17. MICROCARPÆA, R. Br.

Calyx tubular, 5-angled, 5-toothed. Corolla with a short tube and 5 nearly equal lobes (the 2 upper more united, the lowest rather larger). Stamens 2 ; filaments filiform ; anthers 1-celled (by the confluence of 2 divaricate cells). Ovary completely 2-celled. Style short, dilated upwards into a broad spathulate lamina curved over the stamens. Capsule ovoid, included in the calyx, opening loculicidally in 2 entire valves, leaving the transverse dissepiment free.—Small creeping herb. Leaves opposite. Flowers very small, axillary, without bracteoles.

The genus, as now constituted, is limited to the single Australian species, which extends into tropical Asia. If, however, the dehiscence of the capsule be neglected, it might include *Peplidium*, and even *Glossostigma* might be added as a section, differing chiefly in the calyx.

1. **M. muscosa,** *R. Br. Prod.* 436. A dwarf slender intricately-branched prostrate plant, creeping and rooting at the nodes, nearly glabrous or the margins of the leaves, angles of the stems and calyxes ciliate with small rigid hairs. Leaves sessile, linear, narrow-oblong or linear-lanceolate, obtuse, entire, under 2 lines long. Flowers all but sessile in the axils, usually one only to each pair of leaves. Calyx ¾ line long, prominently angled, with 5 acute ciliate teeth. Corolla-tube shorter than the calyx and the lobes very shortly exceeding it. Stamens nearly as long as the corolla. Capsule much shorter than the calyx.—Benth. in DC. Prod. x. 433.

N. Australia. Near Macadam Range, *F. Mueller.*
Queensland. Shoalwater Bay, *R. Brown.*

18. GLOSSOSTIGMA, Arn.

(Tricholoma, *Benth.*)

Calyx campanulate, obtusely 3- or 4-lobed, the upper lobes sometimes slightly notched. Corolla very small, with a short tube and 5 nearly equal lobes (the 2 upper more united, the lowest rather larger). Stamens 2 or 4; filaments filiform; anthers 1-celled (by the confluence of 2 diverging or divaricate cells). Style short, dilated upwards into a broad spathulate lamina curved over the stamens in the bud. Capsule globular or ovoid, included in the calyx, opening loculicidally in 2 entire valves, leaving the placental column free.—Small creeping herbs. Leaves opposite but often clustered at the nodes. Flowers very small, on axillary pedicels, without bracteoles.

The genus is apparently limited to the three Australian species, of which one extends to tropical Asia and Africa, another to New Zealand, and the third is endemic. It differs from *Microcarpæa* in the calyx, from *Limosella* in the opposite leaves, in the calyx, style, ovary and capsule. F. Mueller has, however, (in his herbarium as well as in Fragm. vi. 104,) united the three species under the name of *Limosella Drummondii.*

Stamens 2 (**Glossostigma**) . . . . 1. *G. spathulatum.*
Stamens 4 (**Tricholoma**).
   Calyx usually 3-lobed. Stamens as long as or exceeding the
     very short corolla-lobes . . . . 2. *G. Drummondii.*
   Calyx 4-lobed. Stamens much shorter than the ovate fringed
     corolla-lobes . . . . 3. *G. elatinoides.*

1. **G. spathulatum,** *Arn.; Benth. in DC. Prod.* x. 426. A very slender and minute intricately-branched glabrous plant, creeping and rooting at the nodes. Leaves linear-spathulate, obtuse, entire, 1 to 2 lines long, but usually tapering into a much longer petiole. Pedicels slender, scarcely exceeding the leaves. Calyx scarcely above ½ line long, 3-lobed. Corolla scarcely exceeding the calyx, with very small blue entire lobes. Stamens 2, nearly as long as the corolla. Capsule not exceeding the calyx, opening loculicidally in 2 valves.—*Microcarpæa spathulata,* Hook. Bot. Misc. ii. 101. t. suppl. 4.

Queensland. Rockhampton, *O'Shanesy*, who observes that the numerous little blue flowers look like tiny drops of dew. The species is dispersed over tropical Asia and Africa.

2. **G. Drummondii,** *Benth. in DC. Prod.* x. 426. A minute glabrous plant, creeping and rooting at the nodes like *G. spathulatum.* Leaves linear-spathulate or oblong, entire, 1 to 2 lines long, but narrowed into a slender petiole sometimes much longer than the lamina. Pedicels usually longer than the leaves. Calyx scarcely above ½ line long, 3-lobed as in *G. spathulatum,* one lobe often broader than the others. Corolla slightly exceeding the calyx, with short rounded lobes not fringed. Stamens 4, as long as or sometimes longer than the corolla. Capsule nearly globular, not exceeding the calyx, opening loculicidally in 2 valves.

W. Australia, *Drummond, n.* 19, 109, *4th Coll. n.* 111 (mixed with sessile depressed-globular fruits probable of an *Elatine* and with other minute plants) ; Murchison river, *Oldfield ;* foot of the Stirling Range, *F. Mueller* (luxuriant specimens with long petioles and pedicels, and mixed with *Limosella*).

3. **G. elatinoides,** *Benth. in Hook. Fl. N. Zeal.* i. 189. A small glabrous intricately-branched moss-like plant, creeping and rooting at the nodes, but often rather longer and more leafy than the two preceding species. Leaves linear-spathulate or oblong, obtuse, entire, rarely above 2 lines long, narrowed into a petiole as long as the lamina or shorter. Pedicels shorter than the leaves. Calyx ¾ lines long, with 4 short broad very obtuse lobes. Corolla-tube nearly as long as the calyx ; lobes ovate, much longer than in *G. Drummondii,* though still very small, the lower one rather larger than the others, all fringed with minute cilia. Stamens 4, shorter than the corolla. Ovary 2-celled. Capsule not seen ripe.—Hook. f. Fl. Tasm. i. 292 ; *Tricholoma elatinoides,* Benth. in DC. Prod. x. 426.

N. S. Wales. Glendon, *Leichhardt ;* Tumbarumba, *W. P. Ball.*
Victoria. Goulburn, Broken, Latrobe, Yarra, and Murray rivers, *F. Mueller.*
Tasmania. Banks of the Esk, near Launceston, *Gunn.*

The species is also in New Zealand. I have searched in vain our rather numerous specimens without succeeding in finding a single ripe capsule to show its dehiscence.

19. LIMOSELLA, Linn.

Calyx campanulate, 5-toothed or lobed. Corolla broadly campanulate or almost rotate, with 5 nearly equal lobes. Stamens 4. Anthers 1-celled (by the confluence of 2 divaricate cells). Ovary 2-celled at the base only. Style short, thickened at the end. Capsule globular, membranous, scarcely dehiscent or opening in 2 valves parallel to the very incomplete dissepiment. —Small herbs, tufted creeping or floating. Leaves clustered or alternate on short barren shoots. Peduncles usually very short, clustered with the leaves, without bracteoles. Flowers in the common species very small, larger in some S. African ones.

Besides the Australian species, which appears to be the same as the one which spreads over the northern hemisphere and the whole of western America, there are one or two from S. Africa with much larger flowers and broader leaves.

1. **L. aquatica,** *Linn. ; Benth. in DC. Prod.* i. 426. A glabrous an-

nual, forming little tufts of 1 or 2 in. diameter, and occasionally emitting creeping shoots terminating in another tuft or rarely short barren branches with alternate leaves. Leaves chiefly clustered in the tufts, almost linear in the common Australian form, more oblong in Europe and Asia but variable in both countries, obtuse and entire, $\frac{1}{4}$ to $\frac{1}{2}$ in. long, besides a petiole often twice as long. Flowers clustered with the leaves on very short pedicels. Calyx about 1 line long. Corolla very shortly exceeding the calyx, the lobes shortly ovate. Capsule ovoid-globular, exceeding the calyx when perfect.— Hook. f. Fl. Tasm. i. 292; *L. tenuifolia,* Nutt.; Benth. in DC. Prod. x. 427; *L. australis,* R. Br. Prod. 443.

N. S. Wales. Near Mudgee, *Woolls.*

Victoria. Avoca river, Station Peak, *F. Mueller;* near Portland, *Allitt.*

Tasmania. Port Dalrymple and Kent's Island, Bass's Straits, *R. Brown;* probably common in marshy situations, though frequently overlooked, *J. D. Hooker.*

S. Australia. Kangaroo Island, *R. Brown;* Light river and Mount Remarkable, *F. Mueller.*

W. Australia. Gordon river, *Oldfield;* foot of Stirling Range, mixed with *Glossostigma Drummondii,* F. Mueller.

The species extends nearly the whole length of western America with the same usually narrow rarely broader leaves as in Australia, and over a great part of Europe and temperate Asia usually with rather broader leaves.

SUBORDER III. RHINANTHIDEÆ.—Corolla either with 4, 5 (or rarely more in genera not Australian) spreading lobes, variously imbricated in the bud, the upper ones very rarely outside, or 2-lipped with the upper lip inside. Inflorescence centripetal or very rarely in genera not Australian compound.

It is only in the first two genera and a few non-Australian *Sibthorpieæ* that the æstivation is doubtful or variable; in all the rest of the suborder the upper lip or lobes are invariably inside in the bud.

20. CAPRARIA, Linn.

Calyx divided to the base into 5 equal segments. Corolla broadly campanulate, divided to below the middle into 5 nearly equal lobes, imbricate in the bud. Stamens 4 or rarely 5, shorter than the corolla; anthers sagittate, the cells confluent at the top. Style thickened at the end, the stigma obtuse, with 2 diverging lobes at the base. Capsule ovate, obtuse, opening loculicidally in 2 valves at length 2-fid and leaving a free placentiferous column. Seeds numerous, small, with a reticulate testa.—Perennials or undershrubs. Leaves alternate, serrate. Pedicels axillary, usually 2 together, without bracteoles.

The genus consists of a very few American species. The only Australian one, if a true congener, appears to be endemic.

1. **C. calycina,** *A. Gray in Proc. Amer. Acad.* vi. 49. Low and glabrous. Leaves lanceolate or linear, with few coarse divaricate teeth near the base. Flowers solitary in the axils, on pedicels of 3 or 4 lines. Calyx-segments leafy, 4 lines long when in flower, $\frac{1}{2}$ in. when in fruit, sometimes slightly denticulate. Corolla not exceeding the calyx. Stamens 4. Stigma emarginate.

N. S. Wales. Hunter's River, *American Exploring Expedition.* This plant is only known to me from A. Gray's character, from which the above is taken. We have no specimen, and it has not, as yet, turned up in any other collection.

21. SCOPARIA, Linn.

Calyx divided to the base into 4 or 5 segments. Corolla rotate, 4-lobed, hairy at the throat, the lobes imbricate in the bud. Stamens 4; anthers sagittate. Style slightly club-shaped at the top, truncate or emarginate. Capsule opening septicidally in 2 entire valves, leaving the placental column free.—Much-branched herbs or low undershrubs. Leaves opposite or whorled. Pedicels axillary, usually 2 together, without bracteoles.

The genus consists of but few species, all South American, including the Australian one, which is now a common weed in almost all tropical regions.

1. **S. dulcis,** *Linn.; Benth. in DC. Prod.* x. 431. A much-branched glabrous annual (or sometimes perennial?), erect or decumbent at the base, 1 to 3 ft. high. Leaves usually in whorls of 3, oblong-lanceolate or the upper ones linear in the Australian specimens, the lower ones broader, in some American ones dentate, narrowed into a petiole often rather long, the lamina varying from $\frac{1}{2}$ to $1\frac{1}{2}$ in. Flowers numerous, small, white, on filiform pedicels of 2 to 4 lines. Calyx-segments 4, ovate-oblong, about 1 line long. Corolla about 3 lines diameter. Capsule rather longer than the calyx.— R. Br. Prod. 443.

N. Australia. Gulf of Carpentaria, *F. Mueller.*
Queensland. Broad Sound, *R. Brown, Bowman;* Shoalwater Bay, *R. Brown;* Lizard Island, *M'Gillivray;* Nerkool Creek, *Bowman;* Rockingham Bay, *Dallachy;* Rockhampton, *Dallachy, O'Shanesy.*

The species is supposed to be of American origin, now a common tropical weed.

22. VERONICA, Linn.

Calyx deeply divided into 4 or rarely 5 segments. Corolla either rotate or with a distinct tube and spreading limb; lobes 4 or very rarely 5, imbricate in the bud, the lateral ones or one of them outside. Stamens 2, inserted in the tube and exserted from it; anthers with confluent cells, without points or awns. Style filiform, with an undivided somewhat capitate stigma. Capsule compressed or turgid, furrowed on each side, either septicidally dehiscent with the placentas separating or loculicidally dehiscent with the valves remaining adherent to the undivided placental column, or separating from it and septicidally bifid. Seeds ovate or orbicular, compressed, attached by the inner flat concave or slightly convex surface, the outer surface more or less convex.—Herbs undershrubs or shrubs. Leaves opposite or rarely the upper ones alternate, the floral leaves or bracts always alternate. Flowers blue pink or white, solitary in the axils of the floral leaves or bracts, without or very rarely with bracteoles, forming usually terminal or axillary racemes.

A large genus, abundant in the temperate and colder regions of the northern hemisphere, in New Zealand, and the Antarctic regions, ascending to great elevations and high latitudes, with a very few tropical species, and those chiefly in mountain regions or descending along streams. Of the fifteen Australian species, one is a common American weed, probably introduced in Australia; another is equally common in the temperate regions of the New and the Old World, as well as in tropical mountains, and may be indigenous in Australia; a third is apparently the same as a New Zealand species; the remaining twelve are endemic.

SECT. I. **Hebe.**—*Evergreen shrubs or densely tufted or tall and erect herbs. Leaves all opposite. Flowers in axillary racemes, very rarely reduced to single flowers. Capsule more or less turgid and septicidally dividing when ripe, at least at the top.*

Densely tufted dwarf perennial, with small decussate leaves. Flowers
sessile, solitary, with 2 bracteoles . . . . 1. *V. densifolia.*
Erect much-branched shrubs. Racemes short, loose, in terminal co-
rymbose leafy panicles.
　Leaves rather crowded, ovate to lanceolate, under ½ in. long . . 2. *V. formosa.*
　Leaves linear, ¾ to 1¼ in. long, usually distant . . . . 3. *V. decorosa.*
Stems from a perennial base tall, simple or nearly so. Racemes
elongated, many-flowered.
　Leaves ovate, stem-clasping and mostly connate, entire or rarely
　　toothed, glaucous . . . . 4. *V. perfoliata.*
　Leaves broadly lanceolate, serrate . . . . 5. *V. Derwentia.*
　Leaves linear or linear-lanceolate, entire or serrate . . . . . 6. *V. arenaria.*

SECT. II. **Chamædrys.**—*Herbs from a perennial usually creeping rootstock, diffuse, ascending or erect. Leaves all opposite. Flowers in axillary racemes. Capsule compressed, the valves not separating from the placental columns.*

Leaves deeply divided into linear segments. Stems tall . . . . 7. *V. nivea.*
Leaves toothed or entire.
　Leaves narrow-lanceolate, entire or rarely toothed, mostly sessile . 8. *V. gracilis.*
　Leaves ovate-lanceolate, acutely toothed, sessile or scarcely petio-
　　late. Stems glabrous or minutely pubescent. Flowers small . 9. *V. arguta.*
　Leaves ovate or ovate-lanceolate, toothed, shortly petiolate. Stems
　　glabrous pubescent or hirsute. Flowers large . . . . . . 10. *V. distans.*
Leaves more petiolate, rounded truncate or cordate at the base.
　Leaves broadly ovate, mostly ½ to 1 in. long.
　　Stems hirsute with long hairs, long and procumbent or short
　　　and erect. Calyx-segments large, obtuse, ciliate . . . . 11. *V. calycina.*
　　Stems slender, shortly pubescent, long and procumbent, rarely
　　　short and erect. Calyx-segments rather acute . . . . 12. *V. plebeia.*
　Leaves ovate-lanceolate, 1 to 3 in. long. Stems erect, often
　　tall, loosely pubescent or hirsute . . . . 13. *V. notabilis.*

SECT. III. **Veronicastrum.**—*Annual or perennial herbs, usually decumbent or small. Stem-leaves opposite, passing into the alternate floral leaves or bracts. Racemes or spikes terminal, simple, the lower bracts like the stem-leaves. Capsule as in* Chamædrys.

Plants perennial, decumbent, and rooting at the base. Flowers distinctly pedicellate . . . . 14. *V. serpyllifolia.*
Annual. Flowers sessile or nearly so . . . . 15. *V. peregrina.*

All the Australian species, except *V. densifolia*, have the corolla rotate or nearly so, with a very short tube, and none have bracteoles except the same *V. densifolia*; the bracts subtending the pedicels are small and narrow in all except the section *Veronicastrum.* The several species of the section *Chamædrys*, with the exception of *V. nivea*, appear connected by so many intermediate forms that their delimitation is, as here given, very unsatisfactory.

SECT. I. HEBE.—Evergreen shrubs or densely tufted or tall and erect perennial herbs. Leaves all opposite. Flowers in axillary racemes, reduced in *V. densifolia* to single flowers. Capsule more or less turgid and septicidally dividing when ripe, at least at the top, where it is then more or less 4-valved.

4. **V. densifolia,** *F. Muell. Fragm.* ii. 137 *and Lithogr. t.* 63. A small densely tufted much-branched prostrate or shortly creeping perennial,

the short ascending branches not above 1 in. high.   Leaves densely crowded and decussate, entirely covering the branches, ovate, very obtuse and thick, keeled underneath, under 2 lines long, minutely ciliate at the base, otherwise glabrous.   Flowers sessile in the uppermost axils, with a pair of oblong bracteoles at their base shorter than the calyx.   Calyx about 3 lines long, divided to the middle into 5 equal obtuse lobes, ciliolate and glandular-pubescent.   Corolla, when apparently normal, with a distinct tube of $1\frac{1}{2}$ lines and 5 oblong nearly equal lobes of about 2 lines, but in most of the flowers 1, 2 or 3 of the lobes are very broad or there is an additional sixth lobe inside.   Capsule "shorter than the calyx, $1\frac{1}{2}$ lines long, obcordate, pubescent in the notch."—*Pœderota densifolia,* F. Muell. in Hook. Kew Journ. viii. 202, and in Trans. Phil. Soc. Vict. i. 107.

N. S. Wales.   Summits of Mount Kosciusko on the Victorian frontier, *F. Mueller.*

Victoria.   Highest summits of the Munyong Mountains at an elevation of 6000 to 6500 ft., *F. Mueller (Herb. Hook.).*

In habit this is allied to the N. Zealand *V. tetragona* and its allies; the multiplication of the calycine and corolla-lobes is like that of the N. Zealand genus or section *Pygmea,* Hook. f.; the inflorescence is peculiar.   The specimens are not numerous, and I could only analyse two flowers; one was regular with the stamens perfect, the other had some of the corolla-lobes enlarged, with a sixth inner one as figured by F. Mueller; but there I found one of the anthers enlarged and probably sterile, and the other entirely replaced by the sixth corolla-lobe.

2. **V. formosa,** *R. Br. Prod.* 434.   A beautiful evergreen corymbosely branched shrub, attaining 2 to 3 or 4 ft., glabrous except a short pubescence decurrent from the margins of the leaves on opposite sides of the stem. Leaves rather crowded, oval-oblong or lanceolate, entire or very rarely obscurely toothed, thick, often recurved, usually about $\frac{1}{4}$ in. long, but from that to $\frac{1}{2}$ in. when narrow.   Flowers pale lilac, in short loose racemes in the upper axils, forming terminal leafy corymbs.   Calyx 1 to $1\frac{1}{2}$ lines long, deeply divided into 5 nearly equal lobes or one smaller than the others.   Corolla-lobes at least 3 lines long.   Capsule oblong, acute or obtuse, considerably longer than the calyx, turgid at the base and readily septicidal.—Benth. in DC. Prod. x. 462; Hook. f. Fl. Tasm. i. 293; *V. diosmæfolia,* Knowles and Westc. Fl. Cab. iii. 65. t. 106, not of A. Cunn.

Tasmania.   Port Dalrymple and Mount Wellington, *R. Brown ;* common on rocky hills in various parts of the island, *J. D. Hooker.*

3. **V. decorosa,** *F. Muell. in Linnæa,* xxv. 430.   An erect branching shrub of several feet, with minute pubescent lines decurrent from the margins of the leaves, otherwise glabrous.   Leaves sessile, linear, entire or rarely toothed, $\frac{3}{4}$ to $1\frac{1}{2}$ in. long.   Flowers white or pink with dark streaks, in rather loose racemes in the upper axils, rarely twice as long as the leaves, and forming, in good specimens, handsome corymbose leafy panicles.   Pedicels longer than the calyx.   Calyx-segments acute, 2 to $2\frac{1}{2}$ or rarely 3 lines long. Corolla-lobes fully 4 lines long, the upper one broader and the lower one narrower than the others.   Capsule turgid, very obtuse and slightly notched, as broad as or broader than long and much shorter than the calyx, but not quite ripe in the specimens.

S. Australia.   Rocky valleys of the Flinders Range, from Mount Remarkable to Mount Brown, *F. Mueller,* Mount Searl, *Warburton.*

F. Mueller (Fragm. vi. 102) reduces this to *V. arenaria*, A. Cunn., but that wɛs probably without actual comparison of specimens, for Cunningham's plant has tall simple ɹerbaceous stems with long racemes and short pedicels, and is more nearly allied to *V. Derʋentia.*

4. **V. perfoliata,** *R. Br. Prod.* 434. Stems from a perennial or shortly shrubby base erect, but often flexuose, simple or slightly branched, attaining several feet, the whole plant glabrous and usually glaucous. Leaves stem-clasping and often more or less connate by their broad bases, ovate or ovate-lanceolate, acuminate or acute, quite entire or with a few prominent teeth, 1 to 2 or rarely 3 in. long. Flowers of a bluish violet streaked with purple, in long slender racemes in the upper axils. Calyx-segments 4, linear, rather unequal, 1½ to 2 lines long. Corolla-lobes 2½ to above 3 lines long, rather unequal, nearly rotate but obscurely 2-lipped as in *V. Derwentia.* Capsule ovoid or oblong, turgid at the base, readily septicidal.—Benth. in DC. Prod. x. 463; Bot. Mag. t. 1936; Bot. Reg. t. 1930; Lodd. Bot. Cab t. 781; *V. imperfoliata*, Benth. in DC. Prod. x. 463.

N. S. Wales. Blue Mountains, *R. Brown*, *A. Cunningham*, and others; on the Murrumbidgee, *M'Arthur;* Mount Mitchell, *Beckler;* southward to Twofold Bay, *F. Mueller.*

Victoria. Forest Creek, Fuller's Range, Ovens and Broken rivers, Mount M'Ivor, Grampians, etc., usually indicating auriferous regions, *F. Mueller.*

5. **V. Derwentia,** *Andr. Bot. Rep. t.* 531. Stems from a perennial base erect, simple, 2 to 3 ft. high, glabrous as well as the foliage except a few cilia at the junction of the leaves, and sometimes a slight pubescence in 2 decurrent lines on the stem, or the inflorescence shortly pubescent Leaves sessile, broadly lanceolate, acuminate, serrate, attaining 3 or 4 in. Flowers pale blue or white, rather crowded, in racemes often 6 to 8 in. long in the upper axils. Calyx about 1½ lines long, divided to below the middle into 4 lanceolate or almost linear lobes, with usually a small upper fifth lobe. Co-rolla-lobes rather broad, acute, nearly 3 lines long, not very unequal but obscurely arranged in 2 lips. Capsule ovoid or oblong, obtuse or acute, turgid at the base, exceeding the calyx, readily septicidal.—*V. labiata*, R. Br. Prod. 434; Benth. in DC. Prod. x. 463; Hook. f. Fl. Tasm. i. 293; Bot. Mag. t. 1660, and 3461.

N. S. Wales. Blue Mountains, Macquarrie river, and to the west of Bathurst, *A. Cunningham* ; Tweed river, *C. Moore ;* Clarence river, *Beckler ;* Mount Lindsay, *W. Hill;* and southward to Twofold Bay, *A. Cunningham;* Maneroo plains, *Lhotzky.*

Victoria. Port Phillip, *R. Brown ;* Loddon river, Creswick Creek, Mount Disappoint-ment, Grampians, *F. Mueller ;* Ballarook forest, *Whan ;* mouth of the Glenelg, *Allitt.*

Tasmania. Port Dalrymple, *R. Brown ;* abundant in many places, especially in the northern and central parts of the island, *J. D. Hooker.*

S. Australia. Near Adelaide, *Blandowsky ,* Bugle Range, *F. Mueller.*

R. Brown does not state for what reason he rejected Andrews's older name, which he quotes as given by Littlejohn, probably from private information. This, however, can scarcely be recognized, as it does not appear to have been previously published, ɲor is Little-john referred to by Andrews in the Repository.

6. **V. arenaria,** *A. Cunn. ; Benth. in DC. Prod.* x. 463. Stems from a perennial (or suffrutescent ?) base, erect, simple or nearly so, 1 to 2 ft. high, glabrous as well as the foliage. Leaves sessile, linear or rarely linear-lanceo-

late, entire or with a few prominent teeth, rather thick, 1 to 2 in. long. Flowers in rather slender virgate racemes of ½ to 1 ft. in the upper axils, the pedicels very short. Calyx-segments very narrow, about 1 line long when in flower, but lengthening to 2 lines. Corolla-lobes acute, 3 to 4 lines long. Capsule oval-oblong, emarginate, often exceeding the calyx, turgid and septicidal when quite ripe.—*V. pulchra,* G. Don in Loud. Hort. Brit. 7 ; *V. dianthifolia,* A. Cunn. in Loud. l. c. 467.

N. S. Wales. Arid sandy flats in the plains of Daby on the Cugeegong river, *A. Cunningham.*

Sᴇᴄᴛ. II. Cʜᴀᴍᴇᴅʀʏs.—Herbs, from a perennial usually creeping rootstock, diffuse ascending or erect. Leaves all opposite. Flowers in axillary racemes. Capsule compressed, opening loculicidally on the margin, the valves not separating from the narrow placental column.

7. **V. nivea,** *Lindl. Bot. Reg.* 1842, *Misc.* 42. Stems from a perennial probably creeping rootstock, ascending or erect, ½ to 1½ ft. high, the whole plant glabrous or the inflorescence minutely pubescent. Leaves pinnately divided into linear entire or toothed or pinnatifid segments. Flowers in rather dense racemes of 2 or 3 in., terminal or in the upper axils, the pedicels short. Calyx-segments lanceolate, unequal, 1 to 1½ lines long. Corolla-lobes obtuse, not 2 lines long. Capsule compressed, broadly obcordate, longer than the calyx, opening loculicidally along the margin, the valves remaining attached to the placental column in the centre.—Benth. in DC. Prod. x. 471 ; Hook. f. Fl. Tasm. i. 294.

Victoria. Mount Latrobe, Baw-Baw Mountains, Mount Wellington in Gipps' Land, *F. Mueller.*
Tasmania. In alpine situations, rather local, *J. D. Hooker;* Mount Wellington, *Gunn;* Western Mountains and Lake Arthur, *Lawrence.*

8. **V. gracilis,** *R. Br. Prod.* 435. Stems from a creeping rootstock ascending or erect, simple or slightly branched, rarely above 6 in. high and sometimes not above 2 in., glabrous as well as the whole plant, or with a line of hairs decurrent on each side from the margins of the leaves. Leaves sessile or very shortly petiolate, lanceolate or linear, acute, entire or rarely with very few prominent teeth, ½ to 1 in. long, the floral ones shorter. Racemes in the upper axils loose but short and almost corymbose, on peduncles longer than the leaves, the pedicels slender, as long as or longer than the calyxes. Calyx-segments lanceolate, acute, from 2 to above 3 lines long. Corolla-lobes broad, rounded, scarcely exceeding the calyx. Capsule broad, half as long as the calyx, slightly notched, somewhat glandular-pubescent, compressed, but not seen quite ripe.—Benth. in DC. Prod. x. 478 ; Hook. f. Fl. Tasm. i. 295.

N. S. Wales. Argyle county, *M'Arthur* (the Port Jackson station given in the 'Prodromus' was probably a mistake in Herb. Lambert).
Victoria. Glenelg, Yarra, and Macalister rivers and Maroka valley, at an altitude of 4000 ft., *F. Mueller;* Creswick Range, *Whan.*
Tasmania. Port Dalrymple, *R. Brown;* moist places, common in many parts of the island, *J. D. Hooker.*
S. Australia. Onkaparinga river, *F. Mueller.*

9. **V. arguta,** *R. Br. Prod.* 435. Stems from a creeping rootstock ascending or erect, very slender, slightly pubescent. Leaves nearly sessile or the lower ones shortly petiolate, ovate or ovate-lanceolate, acute and acutely toothed, truncate or almost cordate at the base, ½ to ¾ in. long. Racemes in the upper axils almost filiform, with few small distant flowers on slender pedicels. Calyx scarcely above 1 line long when in flower and not 2 lines when in fruit. Corolla-lobes broad, obtuse, about 2 lines long. Capsule broad, but not seen ripe.

N. S. Wales. Grose river, *R. Brown.* This may possibly prove to be a slender small-flowered form of *V. distans*, or a broad-leaved variety of *V. gracilis*, to both of which it appears to me to be nearer allied than to *V. plebeia*.

10. **V. distans,** *R. Br. Prod.* 435. Stems from a creeping rootstock ascending or erect, simple or branched, rarely above 6 in. high, glabrous pubescent or rarely hirsute, the hairs usually in opposite lines but sometimes almost round the stem. Leaves sessile or shortly petiolate, from ovate to lanceolate, coarsely toothed or very rarely nearly entire, ½ to 1 in. long, glabrous or sprinkled with a few hairs underneath. Flowers white streaked with lilac, rather large, often only 2 or 3 and never numerous, in rather loose pedunculate racemes in the upper axils, appearing often at first terminal, and often only 1 or 2 racemes to the stem. Calyx-segments usually broad, 2 to 3 lines long, acute or obtuse. Corolla larger than in *V. gracilis*, the broad round lobes at least 4 lines long in many specimens. Capsule broadly obcordate, as long as the calyx, opening loculicidally, the valves adhering to the placental column. Seeds slightly incurved, closely packed.—Benth. in DC. Prod. x. 478; Hook. f. Fl. Tasm. i. 294; *V. Drummondii*, Benth. in DC. Prod. x. 478; *V. Hildebrandii*, F. Muell. in Trans. Phil. Soc. Vict. i. 49, and in Hook. Kew Journ. viii. 202.

S. Australia. Limestone cliffs on Lake Alexandrina, along the Coorong and near Spencer's Gulf, *F. Mueller ;* Lake Hamilton, *Wilhelmi.*

W. Australia, *Drummond, 1st coll. ;* King George's Sound, *R. Brown.* These, as well as the South Australian forms, nearly glabrous and the leaves often rather thick.

Var. ? *pubescens.* More pubescent, the leaves rather more distinctly petiolate, and the calyx-segments more obtuse, almost intermediate between *V. distans* and *V. calycina.*—*V. Novæ-Hollandiæ*, Poir. Dict. viii. 526 ?.

Tasmania. Recherche Bay, *Labillardière, C. Stuart ;* common on the sand hills near Circular Head, *Gunn.*

11. **V. calycina,** *R. Br. Prod.* 435. Stems from a creeping rootstock either procumbent, spreading to a considerable extent and rooting at the lower nodes, or some of the flowering ones usually ascending or erect, from a few inches to nearly 1 ft. long, more or less hirsute, the hairs usually rather long and in 2 opposite rows but sometimes nearly all round the stem. Leaves more petiolate than in the preceding species, broadly ovate, coarsely crenate-toothed, rounded truncate or cordate at the base, from under ½ in. to 1 in. or rarely rather more in length and often almost as broad, the floral ones smaller and sometimes more sessile and narrower. Flowers in the ascending stems in pedunculate few-flowered rather loose axillary racemes, or on the procumbent stems almost reduced to clusters, the pedicels long with a very short common peduncle. Calyx-segments broadly ovate, obtuse,

ciliate, usually about 2 lines long when in flower but soon enlarged and sometimes twice that size and thin. Corolla-lobes obtuse, either scarcely exceeding the calyx or twice as long. Capsule compressed, broadly obcordate or truncate, shorter than the calyx.—Benth. in DC. Prod. x. 477 ; Hook. f. Fl. Tasm. i. 294; *V. stolonifera,* Lehm. Del. Sem. Hort. Hamb. ·1842, and in Pl. Preiss. i. 342 ; Benth. in DC. Prod. x. 477 and 490 (from the character given); *V. cycnorum,* Miq. in Pl. Preiss. i. 342 (from the character given) ; *V. Gunnii,* Benth. in DC. Prod. x. 477.

Queensland. Burnett river, *F. Mueller* (apparently the same, but the specimens not sufficient).

N. S. Wales. Hastings river, *Beckler.*

Victoria. Port Phillip, *R. Brown; Fitzroy river, Robertson ;* Loddon river, Bunip Creek, Buffalo and Dandenong ranges, Grampians, Wilson's Promontory, *F. Mueller;* Little river, *Fullagar ;* Creswick Creek, *Whan.*

Tasmania. Port Dalrymple and Derwent river, *R. Brown ;* common in rich soil throughout the colony, *J. D. Hooker.*

S. Australia. Rivoli Bay, *F. Mueller ;* Kangaroo island, *Waterhouse.*

W. Australia, *Drummond, n.* 99, 215, *4th coll. n.* 159 (with very large calyxes) ; Kalgan river, *Oldfield, F. Mueller.*

Var. ? *longifolia.* Leaves narrow-ovate or ovate-lanceolate, 1 in. long or rather more.— Hampshire Hills, Tasmania, *Gunn.* Included by J. D. Hooker, Fl. Tasm. i. 295, among the forms of *V. arguta,* but with neither the stature nor the long acute leaves of the *V notabilis,* still less is it the true *V. arguta,* Br., the whole species, however, although well-marked in its common typical form, varies occasionally so as to make it difficult to give any absolute character to distinguish it on the one hand from *V. distans,* and on the other from some forms of *V. plebeia.* The most northern stations may require further confirmation.

12. **V. plebeia,** *R. Br. Prod.* 435. Stems from a creeping rootstock procumbent, elongated and much more slender than in the other species, sometimes several feet long, occasionally rooting at the nodes, rarely emitting a tuft of erect branches of a few inches, usually minutely pubescent, without the long hairs of *V. calycina.* Leaves on rather long petioles, broadly ovate sometimes almost deltoid, deeply acutely and irregularly toothed, truncate or broadly cordate at the base, from under ½ in. to about 1 in. long. Racemes as in *V. calycina,* sometimes rather slender pedunculate and 2 or even 3 in. long with the pedicels not much longer than the calyx, sometimes almost reduced to clusters of 2 or 3 flowers on long pedicels with a very short common peduncle. Calyx-segments about 2 lines long when in flower, and rarely above 3 when in fruit, rather acute and minutely ciliolate. Corolla not much longer than the calyx. Capsule shorter than the calyx, compressed, nearly orbicular, not at all or only very slightly emarginate.—Benth. in DC. Prod. x. 478 ; *V. deltoidea,* Spreng. Syst. Cur. Post. 17.

Queensland. Brisbane river, Moreton Bay, *F. Mueller, C. Stuart ;* Maranoa river, *Mitchell.*

N. S. Wales. Port Jackson to the Blue Mountains, *R. Brown, Sieber, n.* 483, and others ; northward to Hastings and Clarence rivers, *Beckler ;* New England, *C. Stuart;* southward to the island of Tallaburga, *Mapleston.*

Victoria. Low bushy hills on the Yarra, Bunip Creek, Tambo river in Gipps' Land, *F. Mueller.*

The New Zealand *V. elongata,* Benth. (*V. calycina,* A. Cunn. in Bot. Mag. under n. 3461), does not appear to be really distinct from *V. plebeia.*

13. **V. notabilis,** *F. Muell. Herb.* Stems from a creeping or decumbent base, ascending or erect, 1 ft. high or more, often much stouter than in the preceding species, loosely pubescent or hirsute. Leaves petiolate, ovate-lanceolate or lanceolate, acute and acutely toothed, 1 to 3 in. long. Racemes in the upper axils loose, 3 to 8 in. long, the pedicels usually longer than the calyx. Calyx-segments rather acute, 2 lines long when in flower, lengthening to 3 lines in fruit. Corolla not much exceeding the calyx, but not seen very perfect. Capsule shorter than the calyx, broad, truncate or slightly notched.

N. S. Wales. Grose river, *R. Brown;* Clarence river, *Beckler;* near Berwick, *Robinson;* Illawarra, *A. Cunningham.*

Victoria. Shady places, Dandenong Ranges, and Sealer's Cove, rare, *F. Mueller.*

Tasmania. St. Patrick's River, *Gunn.*

This species, which had been determined by A. Cunningham to be the *V. arguta* of Brown, and was included under that name by myself in the 'Prodromus' and by Hooker in the 'Tasmanian Flora,' proves to be very different from Brown's plant, and apparently as distinct a species as any of the *Chamædrys* group in Australia except *V. nivea.*

SECT. III. VERONICASTRUM.—Annual or perennial herbs, usually decumbent or small. Stem-leaves opposite, passing into the alternate floral leaves or bracts. Racemes or spikes terminal, simple, the lower bracts leafy like the stem-leaves. Capsule compressed, opening loculicidally on the edges, the valves cohering in the centre to the narrow placental column.

14. **V. serpyllifolia,** *Linn.; Benth. in DC. Prod.* x. 482. A perennial with shortly creeping very much branched stems, forming a small flat dense leafy tuft, the flowering branches ascending from 2 in. to nearly ½ ft., the whole plant minutely pubescent or nearly glabrous. Lower leaves shortly petiolate, the upper ones sessile or nearly so, ovate, obtuse, slightly crenate, rarely exceeding ½ in. Flowers very small, of a pale blue or white with darker streaks, on pedicels of 1 to 1½ lines or rarely nearly sessile, in a simple terminal raceme or spike, the subtending bracts, especially the lower ones, rather large and leaf-like and passing into the stem leaves. Calyx but little more than 1 line long at the time of flowering, somewhat enlarged in fruit. Corolla scarcely exceeding the calyx. Capsule broad, compressed, often rather deeply notched.

N. S. Wales. New England, *C. Stuart.*

Victoria. Snowy and Upper Mitta Mitta rivers, Munyong Mountains, and others of the Australian Alps at an elevation of 4000 to 5000 ft., *F. Mueller.*

The species is common in the temperate and colder regions of both the northern and southern hemispheres ascending to high latitudes and great elevations, and also in mountain ranges within the tropics.

15. **V. peregrina,** *Linn.; Benth. in DC. Prod.* x. 482. An annual with erect or ascending stems, simple or branching at the base, glabrous or minutely glandular-pubescent, usually about 6 in. high, but lengthening occasionally to 1 ft. Radical and lowest leaves petiolate and ovate but soon dying off, the others sessile, oblong or linear, entire or serrate, rarely exceeding ½ in., passing into smaller alternate linear floral leaves or bracts. Flowers small, pale blue or white, sessile in the axils of the floral leaves or bracts,

forming a terminal interrupted leafy spike. Calyx-segments oblong, but little
more than 1 line long, slightly enlarged after flowering. Corolla not exceeding the calyx. Capsule about as broad as long, compressed, slightly notched,
about 1½ lines diameter. Seeds very small.

N. S. Wales. Between the Lachlan and Darling rivers, *Burkitt.*
Victoria. Near Geelong, Forest Creek, Rocky river, *F. Mueller.*
Tasmania. South Esk river, *C. Stuart.*
W. Australia, *Drummond, n.* 443.

The species is common in extratropical America, rather less abundant within the tropics,
and appears here and there in the Old World introduced from America. It is believed also
to have been introduced only in all the above Australian localities.

23. OURISIA, Comm.

Calyx 5-lobed or 5-cleft. Corolla more or less oblique or curved, the tube
very short or elongated ; lobes 5, flat, imbricate in the bud, one of the lateral
ones outside. Stamens 4, not exserted ; anthers reniform, not mucronate,
with confluent cells. Style filiform, with a capitate stigma. Capsule loculicidally 2-valved, the valves entire, carrying off the placentas along their
centre. Seeds several, with a loose reticulate testa.—Perennial herbs, the
stock often woody. Leaves opposite, sometimes all or nearly all radical.
Flowers either solitary in the axils, or forming a raceme sometimes contracted to an umbel, on a scape-like peduncle, without bracteoles.

The genus comprises a considerable number of species from the Andes of S. America and
New Zealand. The only Australian one appears to be endemic in Tasmania.

1. **O. integrifolia,** *R. Br. Prod.* 439. A small glabrous perennial,
with a creeping stock, rooting at the nodes. Radical leaves on the stock or
short barren shoots, ovate obovate or nearly orbicular, obtuse, entire, rather
thick, ¼ to nearly ½ in. long, narrowed into a petiole often as long as the
lamina. Flowering stems erect, simple and 1-flowered or slightly branched,
2 to 3 in. high, bearing 1 or 2 pairs or whorls of 3 small sessile oblong-
linear leaves, the flowers on long pedicels above the last pair. Calyx about
3 lines long, divided to much below the middle into oblong segments often
minutely ciliate. Corolla nearly ½ in. long, broadly and obliquely campanulate, tapering into a very short tube, the lobes all obtuse and rather longer
than the entire part. Capsule ovate, about as long as the calyx.—Benth. in
DC. Prod. x. 493 ; Hook. f. Fl. Tasm. i. 295.

Tasmania. Mount Wellington, *R. Brown, Gunn,* and others ; not uncommon by
alpine rivulets in shaded places, as Mount Wellington, the Western Mountains, etc., *J. D.
Hooker.*

24. SOPUBIA, Hamilt.

Calyx campanulate, with 5 teeth or lobes, valvate in the bud. Corolla
broadly campanulate, nearly rotate or tapering into a short tube, with 5 flat
spreading lobes nearly equal. Stamens 4, the anthers cohering in pairs, each
with one ovate scarcely mucronate perfect cell and one small stipitate empty
cell. Style thickened and slightly flattened towards the end. Capsule ovate
or oblong, truncate or notched, opening loculicidally in 2 entire or at length
bifid valves.—Erect scabrous herbs, drying black. Leaves narrow, often

divided, opposite or the upper ones alternate.—Flowers yellow purple or pink, in terminal racemes or spikes, with a pair of bracteoles on the pedicel.

A small genus, dispersed over tropical Asia and Africa. The only Australian species is one of the Asiatic ones. The species are probably all parasitical.

1. **S. trifida,** *Hamilt.; Benth. in DC. Prod.* x. 522. An erect rigid scabrous slightly branched annual of 1 to 3 ft. Leaves narrow linear, the lower ones on the main stem often 3-fid, the upper ones and those of the side branches entire, ½ to 1 in. long with smaller ones often clustered in the axils, the upper ones alternate. Flowers usually distant, forming a very loose terminal leafy raceme. Pedicels at first short, at length ½ in. long. Bracteoles linear, close under the calyx. Calyx 2 to 2½ lines long, with triangular acute lobes as long as the tube, woolly inside. Corolla with a very short tube, almost rotate, about ¼ in. diameter, yellow with a purple centre or all purple. Capsule truncate, as long as the calyx.

Queensland? In Leichhardt's collection without the precise station (*Herb. F. Mueller*). The species has a wide range in the hilly districts of India, extending to Ceylon, and (in a slight variety) to Madagascar.

25. CENTRANTHERA, R. Br.

Calyx compressed, obliquely acute, split down the lower edge, entire or 2- to 5-toothed at the top. Corolla with a curved tube dilated upwards, the limb spreading, with 5 broad lobes nearly equal or obscurely 2-lipped. Stamens 4, included in the tube; anthers in pairs, the cells transverse, with an awn-like point at the end, one cell usually smaller than the other or empty. Style with a lanceolate flattened stigmatic end. Capsule obtuse, opening loculicidally in 2 entire valves. Seeds minute, testa loose, reticulate; albumen scanty.—Scabrous herbs. Leaves opposite or the upper ones alternate. Flowers almost sessile, axillary or in interrupted terminal spikes with small bracteoles.

The genus consists of a few tropical Asiatic species, including the only Australian one. They are probably several of them if not all parasites.

1. **C. hispida,** *R. Br. Prod.* 438. A stiff erect annual, simple or with spreading branches, 6 in. to 1 ft. high or rarely more, very scabrous with minute hairs or tubercles. Leaves mostly linear, entire, the longer ones 1 to 1½ in. long, the upper ones much smaller. Flowers nearly sessile in the upper axils, alternate and distant. Calyx herbaceous, 3 to 4 lines long. Corolla ¾ to 1 in. long, variously said to be pink purple or yellow. One cell of each anther much narrower than the other, with a long point. Capsule ovoid-globose.—*Wall. Pl. As. Rar.* t. 45; *Benth. in DC. Prod.* x. 525.

N. Australia. Alluvial flats near Fish river, Glenelg district, N.W. coast, *Marten;* Victoria river and moist grassy flats, Arnhem's Land, *F. Mueller.*

Queensland. Endeavour river, *Banks and Solander;* Brisbane river, Moreton Bay *W. Hill;* Rockhampton and Rockingham Bay, *Dallachy.*

N. S. Wales. Richmond river, *Beckler.*

The species is widely distributed over tropical Asia, from Ceylon and the Peninsula to the Archipelago and northward to the Himalaya and S. China.

26. BUCHNERA, Linn.

Calyx tubular, obscurely nerved, shortly 5-toothed. Corolla-tube slender, straight or slightly curved, the limb with 5 almost equal obovate or oblong spreading lobes, the 2 upper ones inside in the bud. Stamens 4, in pairs, included in the tube; anthers 1-celled, vertical. Style club-shaped at the top, entire. Capsule straight, not acuminate, opening loculicidally in 2 entire valves.—Stiff erect herbs, usually drying black. Lower leaves opposite, the upper ones alternate. Flowers sessile, forming terminal dense or interrupted spikes, with a pair of bracteoles under the calyx.

The genus is widely dispersed over the tropical and subtropical regions of Asia, Africa, and America. The limits of the species are exceedingly difficult to determine, and the Australian ones may be considered either as all endemic or nearly so, or all except *B. tetragona* may be referred as varieties to a single species common in tropical Asia and Africa and very near to a common American one.

Flowers in short dense 4-sided spikes, the imbricate bracts very broad
and as long as the calyx . . . . 1. *B. tetragona.*
Flowers in slender interrupted spikes, the bracts either narrow or
much shorter than the calyx.
Radical and lower leaves broad, rosulate; upper ones narrow,
acute. Corolla glabrous . . . . 2. *B. urticifolia.*
Leaves all narrow, the lower ones oblong, the upper ones linear,
mostly acute. Corolla glabrous.
Corolla-tube 3 to 4 lines long . . . . 3. *B. linearis.*
Corolla-tube not 2 lines long . . . . 4. *B. tenella.*
Leaves all narrow and obtuse, usually hoary, the lower ones
oblong.
Stems simple. Corolla glabrous outside . . . . 5. *B. gracilis.*
Stems branching. Corolla pubescent or hispid outside . . . 6. *B. ramosissima.*

1. **B. tetragona,** *R. Br. Prod.* 437. Erect tall and stout, some specimens simple and fully 2 ft. high, others smaller and branched, and all quite glabrous. Lower leaves ovate or oblong, obtuse, coarsely and irregularly sinuate-toothed, narrowed into a short broad petiole and sometimes 3 to 4 in. long, upper ones lanceolate and sometimes all under 2 in. Spikes usually 3 to 5 together, almost sessile within the last pair of leaves, very thick and 1½ to 2 in. long, the flowers densely imbricate in 4 rows, each one sessile within a bract 2 to 3 lines long, much broader than long, very shortly acuminate in the middle. Bracteoles narrow, complicated, acuminate, as long as the calyx. Calyx 2½ lines long, not at all or scarcely compressed, the lobes narrow, very acute, nearly as long as the tube. Corolla-tube slender, nearly 4 lines long, the lobes broad, nearly equal, spreading to 3 or 4 lines diameter. Capsule oblong, rather longer than the calyx.—Benth. in DC. Prod. x. 495.

N. Australia. Port Essington, *Armstrong.*
Queensland. Endeavour river, *Banks and Solander.*

Allied to the E. Indian *B. tetrasticha*, but readily distinguished by the smoothness of the whole plant as well as by the calyxes and bracts.

2. **B. urticifolia,** *R. Br. Prod.* 437. Scabrous-pubescent or nearly glabrous. Stems erect and simple or branching and slightly decumbent at the base, rather slender, often above 1 ft. high. Radical and lower leaves almost rosulate at the base of the stem, obovate or broadly oblong, usually

sessile, obtuse, entire or slightly sinuate-toothed, 1 to 1½ or rarely 2 in. long ; stem-leaves narrower, the upper ones linear or linear-lanceolate, acute. Flowers purplish or nearly white, in slender interrupted terminal spikes. Bracts mostly ovate, acute, ciliate, about half as long as the calyx or the lower ones longer and narrower ; bracteoles similar, but smaller. Calyx narrow, rarely 2 lines long, the teeth acute. Corolla glabrous outside, the tube slender, not twice as long as the calyx. Capsule oblong, obtuse, either equal to or rather exceeding the calyx.—Benth. in DC. Prod. x. 496 ; Endl. Iconogr. t. 78.

N. Australia. Victoria river, *F. Mueller ;* Glenelg district, N.W. coast, *Marten.*
Queensland. Common along the coast, *R. Brown* and others; from Cape York, *Daemel,* to Moreton Bay, *F. Mueller.*

The common E. Indian *B. hispida* differs chiefly in being much more hirsute. The African *B. leptostachya* can scarcely be distinguished from some forms of the species, which might indeed include, as slight varieties, the following four.

3. **B. linearis,** *R. Br. Prod.* 437. Scabrous-pubescent. Stems erect, simple or slightly branched, often exceeding 1 ft., the upper leaves linear and acute as in *B. urticifolia,* and sometimes the lower ones scarcely broader, but usually those near the base of the stem are oblong, obtuse, often obscurely toothed, narrowed into a petiole and not sessile nor rosulate. Flowers and fruit the same as in *B. urticifolia,* or rather larger.—Benth. in DC. Prod. x. 497.

N. Australia. Islands of the Gulf of Carpentaria, *R. Brown ;* S. Goulburn Island, *A. Cunningham ;* Port Essington, *Armstrong;* Victoria river and near Macadam Range, *F. Mueller ;* King's Ponds, in the interior, *M'Douall Stuart's Expedition.*

Var. *asperata.* *B. asperata,* R. Br. Prod. 438 ; Benth. in DC. Prod. x. 496, appears to be a rather larger, coarser, and more scabrous form of the same species.

Queensland. Bustard Bay and Bay of Inlets, *Banks and Solander.*

4. **B. tenella,** *R. Br. Prod.* 437. More slender than the other species, simple or branched, often 1 ft. high or more, the foliage and lower part of the plant sparingly hirsute, the upper part often quite glabrous. Leaves all narrow and mostly narrow-linear and acute. Flowers " yellowish-brown," smaller than in *B. linearis* and *B. urticifolia,* but otherwise similar, the corolla glabrous outside, the tube not 2 lines long.—Benth. in DC. Prod. x. 497.

N. Australia. South Goulburn Island, *A. Cunningham ;* head of Victoria river, *F. Mueller ;* islands of the Gulf of Carpentaria, *R. Brown.*
Queensland. Endeavour river, *Banks and Solander ;* Facing Island, *R. Brown.*

5. **B. gracilis,** *R. Br. Prod.* 437. Very near *B. ramosissima,* with the same somewhat hoary indumentum and narrow obtuse leaves, but the stem slender, erect, usually simple or branching at the base only, and the corolla-tube glabrous outside or very rarely sprinkled with a few hairs at the top of the tube.—Benth. in DC. Prod. x. 497.

N. S. Wales. Port Jackson, *R. Brown.* The only specimens hitherto detected so far south.

6. **B. ramosissima,** *R. Br. Prod.* 438. Erect or decumbent at the base, more branching than the other species and usually more hoary with a short scabrous pubescence, sometimes under 6 in. but often attaining 1 ft or

more. Lower leaves oblong, obtuse, narrowed into a short petiole, ¾ to 1¼ in. long; upper ones linear but almost always obtuse, and all usually quite entire. Bracts and bracteoles usually narrow and short. Calyx 2 to 3 lines long, with acute teeth. Corolla-tube more or less exserted, always pubescent or hispid outside, especially at the top, the lobes narrow, about 1½ lines long. Capsule about as long as the calyx.—Benth. in DC. Prod. x. 496.

N. Australia. Hunter's River, York Sound, N.W. coast, *A. Cunningham* (a large variety, attaining 2 ft. or more).

Queensland. Thirsty Sound, *R. Brown;* Port Denison, *Fitzalan;* Gracemere and near Rockhampton, *Bowman.*

Var. ? *parviflora.* Corolla much smaller, slightly pubescent outside.—*B. pubescens,* Benth. in DC. Prod. x. 496.—Endeavour river, *A. Cunningham.*

27. STRIGA, Lour.

Calyx tubular-campanulate, with prominent nerves, 5-toothed or 5-lobed. Corolla-tube slender, abruptly bent at or above the middle, the limb 2-lipped, the upper lip emarginate or 2-lobed, innermost in the bud, the lower 3-lobed. Stamens 4, in pairs, included in the tube; anthers vertical, 1-celled. Style club-shaped at the top, entire. Capsule straight, not acuminate, opening loculicidally in 2 valves.—Rigid erect annuals, usually scabrous and drying black. Lower leaves opposite, upper ones alternate, sometimes, in species not Australian, all reduced to small scales. Flowers sessile, usually forming terminal interrupted spikes.

A genus of several species, dispersed over the tropical regions of the Old World, and all probably parasites on roots. Of the four Australian species, one is a common one in tropical Asia; the other three, closely allied to each other, may be all endemic. The characters by which several of the species are distinguished, those especially which are derived from the size and proportions of the corolla, are very difficult to observe correctly in dried specimens, and appear often to be very variable

Calyx with 10 equally prominent ribs . . . . 1. *S. hirsuta.*
Calyx with 5 prominent ribs, smooth between them or rarely here and
   there an obscure vein.
   Corolla scarcely ¼ in. long, the upper lip more than half as long as the
     lower . . . . 2. *S. parviflora.*
   Corolla above ½ in. long, the upper lip more than half as long as the
     lower . . . . 3. *S. multiflora.*
   Corolla nearly or fully ¾ in. long, the upper lip less than half as long
     as the lower . . . . 4. *S. curviflora.*

1. S. hirsuta, *Benth. in DC. Prod.* x. 502. An erect, scabrous or pubescent, simple or slightly branched annual, usually about 6 in. high, and not always drying so black as the other species. Leaves linear or the lower ones lanceolate. Flowers yellow red or white, in terminal interrupted spikes, the lower ones distant. Calyx variable in size, usually 2 to 2½ lines long, with 10 very prominent scabrous or hispid nerves, one of them very rarely here and there divided, the furrows between them very narrow. Corolla-tube glabrous, 4 to 5 lines long, bent near the top; the upper lip much shorter than the lower one.—*Campuleia coccinea,* Hook. Exot. Fl. t. 203.

Queensland. Burdekin river, *Bowman.* Frequent in tropical Asia, extending westward into Africa, eastward to the Archipelago, and northward to S. China.

2. S. parviflora, *Benth. in Comp. Bot. Mag. and in DC. Prod.* x. 501.

A very scabrous, erect, simple or slightly branched annual of 6 to 9 in. Leaves linear, usually short, the floral ones very narrow. Flowers small (blue ?), in more or less interrupted terminal spikes. Calyx 1 to 1½ lines long, with 5 very scabrous and prominent ribs, and smooth between them or here and there with an imperfect row of minute prickles. Corolla scarcely 3 lines long, the tube bent near the top, the lobes all very short, but the upper lip more than half as long as the lower one. Capsule broad.—*Buchnera parvi-flora,* R. Br. Prod. 438.

Queensland. Keppel Bay, *R. Brown;* Peak Range, *Leichhardt;* Broad Sound, Suttor and Bowen rivers, Nerkool Creek, Gracemere, *Bowman.*

3. **S. multiflora,** *Benth. in Comp. Bot. Mag. and in DC. Prod.* x. 501. Nearly allied to *S. parviflora* and to *S. curviflora,* and in some respects inter-mediate between the two, with a similar calyx but different corolla. Stems erect and usually branched, often above 1 ft. high. Leaves linear, often above 1 in. long, the floral ones small and narrow. Flowers usually nume-rous (blue or purple ?). Corolla glabrous glandular or pubescent, interme-diate in size between those of *S. parviflora* and *S. curviflora,* but in some spe-cimens fully as large as in the latter, the upper lip shortly and broadly 2-lobed, more than half as long as the lower lip.

N. Australia. Victoria river and Sturt's Creek, *F. Mueller;* on all the islands to the westward of Goulburn island, *A. Cunningham;* Port Essington, *Armstrong;* Camden Har-bour, Glenelg district, N.W. coast, *Marten* (with remarkably large flowers).

I have now some doubts whether the Philippine Island and Molucca plant I referred to this species in the 'Prodromus' be really the same.

4. **S. curviflora,** *Benth. in Comp. Bot. Mag. and in DC. Prod.* x. 501. Usually a much taller and stouter plant than *S. parviflora,* many of the spe-cimens above 1 ft. high, simple and slightly branched and very scabrous. Leaves linear, the lower ones above 1 in. high. Flowers (blue or purple ?) in terminal interrupted spikes. Calyx 3 lines long or more, with long subu-late-acuminate teeth, the tube with 5 prominent scabrous ribs, and smooth between them. Corolla pubescent, the tube 4 to 5 lines long, bent near the top, the lobes of the lower lip 3 to 4 lines long, the upper lip slightly notched, only 1 to 1½ lines long, usually somewhat recurved.—*Buchnera curviflora,* R. Br. Prod. 438.

N. Australia. Islands of the Gulf of Carpentaria, *R. Brown;* N.W. coast, *Bynoe.*
Queensland. Endeavour river, *Banks and Solander;* Rockhampton, *O'Shanesy;* Cape York, *Daemel.*

28. RHAMPHICARPA, Benth.

Calyx campanulate, 5-lobed. Corolla-tube long and slender, straight or slightly curved; lobes 5, obovate, nearly equal or the 2 upper (inside in the bud) rather smaller. Stamens 4, in pairs; anthers 1-celled, vertical, obtuse. Capsule ovate, compressed or turgid, acuminate, with a straight or oblique beak, opening loculicidally in 2 valves.—Erect branching glabrous herbs, drying black and perhaps parasitical. Lower leaves opposite, upper ones alternate, entire or the lower ones pinnately divided. Flowers in terminal racemes, usually without bracteoles.

A small genus, chiefly African, with one Asiatic species, the same as the Australian one.

1. R. longiflora, *Benth. in Comp. Bot. Mag and in DC. Prod.* x. 504.

An erect slender but rigid branching annual, more or less scabrous, from under 6 in. to nearly 1 ft. high.   Leaves pinnately divided into linear-subulate segments, rather short and distant, or sometimes again toothed or pinnate, the whole leaf usually above 1 in. long.   Flowers in the upper axils, on pedicels of ½ to 1 in., without bracteoles.   Calyx broadly campanulate, 2 to 3 lines long, the lobes ending in fine points.   Corolla-tube slender, about 1 in. long when perfect, with a campanulate throat, the lobes broad, varying in size, but always 2 or 3 times shorter than the tube.   Capsule ovate, acuminate, without prominent margins, the beak nearly straight or somewhat oblique in the Australian form.—*R. fistulosa,* Benth. in DC. Prod. x. 504.

N. Australia, *F. Mueller* (imperfect specimens in Herb. Hook.); lat. 17° 58′, *M'Douall Stuart's Expedition* (imperfect specimens in Herb. F. Muell.)   These Australian specimens seem to connect the African *R. fistulosa,* which has usually the capsule bordered by a raised nerve or wing, but the beak straight, with the Asiatic *R. longiflora* (Wight, Ic. f. 1415), which has not the raised nerve, but the beak of the capsule more or less oblique or recurved.   Neither character appears, however, to be quite constant, and the foliage and flowers are the same in all.

29. HEMIARRHENA, Benth.

Calyx deeply divided into narrow obtuse segments.   Corolla tubular at the base, the throat dilated, the upper lip erect, narrow, concave, entire, the lower one longer, spreading, divided into 3 narrow lobes folded over the upper lip in the bud.   Stamens 2, without any rudiment of the upper pair; filaments arched; anthers connivent under the upper lip but free, each with one pendulous cell, with a fine rigid point or awn at the end, opening longitudinally from the base to near the end.   Style filiform, slightly dilated at the end, entire.   Capsule ovoid, opening in 2 entire thin valves, parallel to the thin dissepiment.   Seeds numerous, striate and reticulate, like those of *Gratioleæ.*— Slender perennial.   Leaves opposite rosulate or clustered at the base of the stem.   Flowers in short terminal racemes, without bracteoles.

The genus is limited to a single species endemic in Australia, and singularly exceptional in whichever of the great suborders it is placed.   The form and æstivation of the corolla and aristate anthers, so decidedly those of Euphrasieæ, are absolutely unknown in Antirrhinideæ, whilst the capsule and seeds, exactly those of Lindernieæ, are as foreign to any of the genera hitherto known in Euphrasieæ, or indeed in any but a very doubtful one of the whole suborder of Rhinanthideæ.

1. H. plantaginea, *Benth.*   Stems from a thick perennial almost

woody stem, erect, very slender, simple, often above 1 ft. long, quite glabrous.   Leaves in few pairs at the base of the stem, almost rosulate, very shortly petiolate, ovate or broadly oblong, obtuse, entire, glabrous, ½ to 1 in. long, and sometimes 1 or 2 pairs of minute scale-like sessile leaves higher up the stem.   Flowers densely crowded in a short oblong terminal raceme, with sometimes a branch proceeding from the base bearing a second raceme.   Pedicels very short, glandular-pubescent, in the axils of minute bracts.   Calyx-segments above 1 line long, membranous, with a dark-coloured midrib and a few large glands on each side.   Corolla-tube slender, about 3 lines long, the throat dilated, the upper lip scarcely above 1 line long, the lobes of the lip longer, the whole corolla of a delicate texture and veined like that of *Euphrasia*. Capsule obtuse, not exceeding the calyx.—*Vandellia plantaginea*, F. Muell. in Trans. Vict. Inst. iii. 62 ; *Lindernia plantaginea*, F. Muell. Fragm. vi. 102.

N. Australia. Mount King, Glenelg district, N.W. coast, *Marten ;* between Providence Hill and M'Adam Range, *F. Mueller ;* Arnhem's Land, *M'Douall Stuart's Expedition*.

30. EUPHRASIA, Linn.

Calyx tubular or campanulate, 4-lobed. Corolla tubular at the base, 2-lipped, the upper lip concave or hood-shaped, with 2 broad spreading lobes ; the lower lip spreading, 3-lobed, overlapping the upper ones in the bud. Stamens 4 in pairs ; anthers 2-celled, connivent under the upper lip of the corolla, the cells mucronate, often hairy. Style slightly dilated at the end, the stigma obtuse, entire or with a small upper lobe. Capsule oblong, compressed, opening loculicidally in 2 valves. Seeds oblong, striate.—Herbs either annual or perennial and branching at the base, believed to be often partially parasitical on roots. Leaves opposite, toothed or lobed. Flowers sessile or nearly so in short and dense or long and interrupted terminal spikes, the floral leaves or bracts usually more acute than the stem-leaves. Bracteoles none. Anther-cells equal in all the Australian species, unequal in some others.

The genus comprises a small number of very variable species distributed over the temperate and colder regions of the northern hemisphere, extratropical South America, and New Zealand. Of the eight Australian forms here admitted as species, one is also in New Zealand and Fuegia, another is very near a New Zealand one, and the remainder appear to be endemic, but some of them are scarcely more than marked varieties.

Perennials, branching at the base with ascending or erect stems.
  Leaves ovate or broadly oblong, very pubescent and rugose. Flowers
    very large. Stems tall . . . . .1. *E. speciosa*.
  Leaves narrow, pubescent or glabrous. Flowers variable in size, the
    spikes usually interrupted . . . . 2. *E. collina*.
  Lower leaves small obovate or broadly cuneate. Flowers usually in
    compact spikes. Alpine species.
    Anthers very hairy. Stems usually 4 to 8 in. . . . . . . 3. *E. alpina*.
    Anthers glabrous or shortly hairy along the suture. Stems usually
      2 to 3 in. 4. *E. striata*.
Pubescent erect annuals (usually above 6 in.), not drying so black as
  the perennials, and the corolla-lobes not so broad. Anthers very
  hairy.
  Flowers yellow. Teeth of the upper leaves rather acute . . . . 5. *E. scabra*.
  Flowers white or purplish. Teeth of the upper leaves very acute . 6. *E. arguta*.
Dwarf annuals (under 4 in.). Anthers glabrous or minutely or very
  sparingly ciliate.
  Glabrous or nearly so. Leaves digitately 4- to 8-lobed, the lobes
    acute or cuspidate . . . . 7. *E. cuspidata*.
  Glandular-pubescent. Leaves pinnatifid or 3-lobed, the lobes obtuse. 8. *E. antarctica*.

1. **E. speciosa,** *R. Br. Prod.* 437. Stems stout, erect or ascending, often exceeding 1 ft., very scabrous-pubescent as well as the foliage. Leaves sessile, ovate or broadly oblong, obtusely and coarsely toothed, undulate and very rugose, 4 to 8 lines long, the floral ones broad and deeply crenate. Flowering spike at first dense, afterwards interrupted, the flowers large.

Calyx 2 to 3 lines long, pubescent, the lobes broad, dilated and very obtuse or rarely almost acute. Corolla pubescent, the tube much dilated upwards, the lower lip large and broad with the middle lobe notched.

N. S. Wales. Port Jackson, *R. Brown.*
Victoria. Forest Creek, *F. Mueller.*

F. Mueller may be right in considering this as a remarkably large-flowered variety of the following species, and certainly there are some of the more pubescent and vigorous specimens of the variety *paludosa*, which come near to the *E. speciosa*, and which in the ' Prodromus' I had referred to that species. But Brown's specimens show larger flowers than any others, except those above-quoted from F. Mueller, with broader more sessile and much more rugose leaves.

2. **E. collina,** *R. Br. Prod.* 436. Stems from a hard usually if not always perennial much-branched base, ascending or erect, from 6 in. to above 1 ft. high, glabrous or pubescent as well as the foliage, the inflorescence usually more or less glandular-pubescent. Leaves sessile or the lower ones narrowed into a short petiole, from oblong to linear-cuneate, obtuse and obtusely toothed at the end only or more frequently to near the base, usually $\frac{1}{4}$ to $\frac{1}{2}$ in. long, but larger in luxuriant specimens ; the floral ones smaller broader and less toothed, the upper ones often entire (rarely cuneate and more deeply toothed?). Flowers purple bluish or white, sometimes mixed with yellow, rarely quite yellow, in terminal spikes, usually long and interrupted, with the flowers in distant pairs, at least when the flowering is advanced, rarely compact but occasionally remaining so even in fruit. Calyx usually about 2 lines long at first and lengthening to 3 lines, but very variable, the lobes acute or obtuse, equal to or shorter than the tube. Corolla-tube exserted, the throat broad, the lobes large but scarcely so long as the tube, the middle lower one emarginate, the others very obtuse or retuse, the whole corolla varying from $\frac{1}{2}$ to $\frac{3}{4}$ in. in length. Anthers hirsute. Capsule exceeding the calyx, shortly mucronate or rather acute.—Benth. in DC. Prod. x. 553 ; Hook. f. Fl. Tasm. i. 296 ; *E. tetragona,* R. Br. Prod. 436 ; Benth. l. c. ; Bartl. in Pl. Preiss. i. 343 ; *E. multicaulis,* Benth. l. c. ; Hook. f. Fl. Tasm. i. 297.

N. S. Wales. Port Jackson to the Blue Mountains, *R. Brown, Sieber, n.* 183, 507, and many others ; northward to Hastings river, *Beckler ;* and New England, *C. Stuart ;* southward to Twofold Bay, *F. Mueller ;* westward to the Lachlan, *A. Cunningham* and others (all chiefly the var. *paludosa* and other large forms).
Victoria. Abundant from the Glenelg to Gipps' Land, Wimmera, and the Grampians, ascending to 4500 ft:, *F. Mueller* and others.
Tasmania. Derwent river, *R. Brown ;* common on dry hilly situations as well as in marshy ground, *J. D. Hooker.*
S. Australia. Memory Cove, *R. Brown ;* around St. Vincent's and Spencer's Gulfs, *F. Mueller* and others ; and (large varieties approaching *E. speciosa*) Mount Rous, *Wilhelmi ;* Flinders and Lofty Ranges, *F. Mueller.*
W. Australia. King George's Sound, *R. Brown* and others, *Preiss, n.* 2338, and eastward to Cape le Grand and Esperance Bay, *Maxwell.*

Var. *paludosa.* Tall and often pubescent. Leaves nearly of *E. speciosa*, but much narrower. Flowers usually distant in long interrupted spikes, purple white or sometimes yellow (*R. Brown, Woolls*).—*E. paludosa,* R. Br. Prod. 436.—In marshes chiefly in N. S. Wales, but including a few of the larger Victorian and S. Australian specimens.

The preceding *E. speciosa* and the following *E. alpina*, and even *E. striata*, are not separated from *E. collina* by any very marked characters, and F. Mueller (Fragm. v. 88) unites them all under the name of *E. Brownii.* It does not appear necessary, however, to discard Brown's names *E. collina* or *E. speciosa,* either of which might, without inconvenience, be applied to the collective species. This has the appearance of being generally, if not always, perennial, but probably not of long duration, and sometimes evidently flowering the first year, but its mode of growth requires further observation of the living plant. *E. tetragona,* Br., from King George's Sound, is certainly one of the common forms of *E. collina,* the stems are but very obscurely angular or compressed in the origina' typical specimens. *E. multicaulis,* Benth., appears to be the typical *E. collina,* Br. The whole series are closely allied to the New Zealand *E. cuneata,* Forst.

3. **E. alpina,** *R. Br. Prod.* 436. A perennial, branching at the base, with the habit of the smaller specimens of *E. collina,* usually glabrous or very minutely pubescent, the stems ascending to 6- to 8 in. or rarely taller. Leaves obovate to oblong-cuneate, narrowed at the base, very obtuse, with few very obtuse teeth, rather thick, 2 to 3 lines long or the upper narrower ones in luxuriant specimens 4 to 5 lines long. Flowers rather large, white or bluish purple with darker streaks, usually in short compact spikes, rarely more distant in interrupted spikes, the floral leaves or bracts broadly cuneate and crenate. Anthers very hairy.—Benth. in DC. Prod. x. 553; Hook. f. Fl. Tasm. i. 296 (partly); *E. diemenica,* Spreng. Syst. ii. 777.

N. S. Wales. Mount Kosciusko, near the Victorian frontier, *F. Mueller.*
Victoria. Summits of the Cobberas and Munyong mountains, *F. Mueller.*
Tasmania. Mount Wellington, *R. Brown, Gunn,* and others, Western Mountains, *Gunn, C. Stuart.*

This may be an alpine form of *E. collina,* differing from the glabrous varieties of that species in its short broad lower leaves, the inflorescence usually more compact with broader more cuneate floral leaves. The variety *humilis* of the 'Prodromus' is *E. striata;* the var. *angustifolia* must be reduced to *E. collina.*

4. **E. striata,** *R. Br. Prod.* 436. A perennial branching at the base, with the habit of *E. alpina,* but smaller, the ascending or erect flowering stems usually only 2 or 3 in. high and very rarely exceeding 6 in., glabrous except two decurrent lines of pubescence, and the inflorescence sometimes glandular-pubescent. Leaves small, obovate or oblong, crenate-toothed, the floral ones cuneate, usually broad and toothed or almost digitate at the end. Flowers in short compact spikes, usually white or pale coloured streaked with red or purple. Anthers either quite glabrous or shortly or sparingly hairy along the line of dehiscence.—Benth. in DC. Prod. x. 554; Hook. f. Fl. Tasm. i. 297; *E. alpina,* var. *humilis,* Benth. l. c. 553.

Victoria. Summits of the Baw-Baw, Cobberas, and Munyong mountains, *F. Mueller.*
Tasmania. Summit of Mount Wellington, *R. Brown, A. Cunningham,* and others; Western Mountains, *Archer;* Birch's Inlet, Macquarrie harbour, *Milligan;* South Port, *C. Stuart.*

5. **E. scabra,** *R. Br. Prod.* 437. An erect, rigid, simple or branched annual of ¼ to 1 ft., scabrous-pubescent and not drying so black as *E. collina* and its allies. Leaves oblong-lanceolate, with a few teeth more prominent and less obtuse than in *E. collina,* and the upper ones often acute, the stem-leaves rarely above ½ in. long, the floral ones narrow and more entire, the uppermost linear. Flowers yellow and scarcely or not at all streaked, in terminal spikes at first dense but at length often long and interrupted. Calyx narrow, glandular-pubescent, the lobes almost acute. Corolla-tube ex-

ceeding the calyx and dilated at the top, but not so much so as in *E. collina*, and the lobes very much shorter and entire, the whole corolla usually about ½ in. long. Anthers very hairy.—Benth. in DC. Prod. x. 554 ; Hook. f. Fl. Tasm. i. 297 ; Bartl. in Pl. Preiss. i. 343.

N. S. Wales. Port Jackson to the Blue Mountains, *A. Cunningham, Sieber, n.* 490, *Woolls ;* grassy lands north of Bathurst, *A. Cunningham ;* New England, *C. Stuart ;* Mount Mitchell, *Beckler.*

Victoria. Glenelg river, *Robertson, Allitt ;* Port Phillip, *Gunn ;* thence to the lower part of the Australian Alps, *F. Mueller ;* Upper Murray river, *Bull ;* Creswick, *Whan.*

Tasmania. Port Dalrymple, *R. Brown ;* abundant about Circular Head, *Gunn ;* Cheshunt, *Archer.*

S. Australia. Around St. Vincent's Gulf, *F. Mueller* and others.

W. Australia. King George's Sound and adjoining districts, *Drummond, n.* 14, 244, *Preiss, n.* 2337, *Oldfield, F. Mueller ;* eastward to Esperance Bay and Cape Knobb, *Maxwell.*

6. **E. arguta,** *R. Br. Prod.* 437. An erect branching annual of ½ to 1½ ft., pubescent like *E. scabra*, but not usually so scabrous. Leaves oblong-lanceolate, usually deeply toothed, the lower ones like those of *E. scabra*, the upper ones with more acute often numerous teeth or lobes, and the floral ones usually but not always with long subulate points to the lobes. Flowers in long spikes at length interrupted, very near those of *E. scabra*, but the corolla, with a bluish tint when dry, is described (by R. Brown and A. Cunningham) as white with a yellowish throat, and the throat is rather broader and the lobes rather larger than in *E. scabra*, though less so than in *E. collina*, the lobes entire as in *E. scabra*.

N. S. Wales. Paterson's and Williams' rivers, *R. Brown ;* plains near Bathurst, *A. Cunningham ,* New England, *Leichhardt ;* Mudgee, *Woolls.*

Victoria. Plains of the Cobberas Mountains, *F. Mueller.*

This species is certainly very closely allied to *E. scabra,* although the specimens I had when describing for the ' Prodromus,' and upon which I united the two species, have proved not to have been correctly identified with Brown's plant.

7. **E. cuspidata,** *Hook. f. Fl. Tasm.* i. 298. An erect simple or scarcely branched annual, glabrous or rarely with slightly pubescent stems, drying very black, 2 to 4 in. high. Leaves broadly cuneate, digitately divided to near the middle into 4 to 8 acute or cuspidate flat lobes, the floral ones usually broader with as many or even more lobes or teeth. Flowers in short dense leafy spikes, shortly exceeding the floral leaves. Calyx 3 to 4 lines long in flower, 5 lines when in fruit, glandular-pubescent, the lobes acute, rather shorter than the tube. Corolla-tube scarcely so long as the calyx, the lobes of the lower lip emarginate, as long as the tube. Anthers very minutely ciliolate along the line of dehiscence of the cells or quite glabrous. Capsule oval-oblong, shorter than the calyx, obtuse or slightly notched. Seeds not numerous, the loose testa forming a wing round them.

Tasmania. Mount Sorrell, Macquarrie harbour, *Milligan ;* Western Mountains, *Archer ;* Mount Lapeyrouse, *Oldfield, C. Stuart.*

8. **E. antarctica,** *Benth. in DC. Prod.* x. 555. An erect or diffuse branching glandular-pubescent annual, 1 to 2 in. or rarely 3 in. high. Leaves oblong, obtuse, pinnatifid, narrowed at the base or almost petiolate, ¼ to ½

in. long, the floral ones mostly 3-fid only. Flowers in the uppermost axils, sometimes not exceeding the floral leaves, sometimes nearly twice as long. Calyx 2 to nearly 3 lines long, the lobes obtuse, shorter than the tube.   Corolla-tube about as long as the calyx, the lower lip as long as the tube, the lobes emarginate.   Anthers either quite glabrous or very sparingly hirsute along the line of dehiscence of the cells.   Capsule oval-oblong, as long as the calyx, slightly notched.   Seeds few.—*E. alsa*, F. Muell. in Trans. Phil. Soc. Vict. i. 107, and in Hook. Kew Journ. viii. 203.

Victoria.   Wet gravelly places on the summits of the Munyong Mountains, at an elevation of 6000 ft., *F. Mueller.*

The species is also in New Zealand, Fuegia, and S. Chili.   As in New Zealand, some of the dwarf specimens of ½ to 1 in. have the flowers, and especially the calyxes, much smaller.

Order LXXXIV. LENTIBULARIEÆ.

Calyx free, with 2 to 5 segments lobes or teeth.   Corolla irregular, the tube usually projecting into a spur or pouch at the base, the limb 2-lipped. Stamens 2, included in the tube and inserted at its base.   Anthers 1-celled. Ovary superior, 1-celled, with several ovules attached to a free central placenta.   Style short, with a 2-lipped stigma.   Fruit a capsule.   Seeds small, often minute, the testa usually reticulate, without albumen.   Embryo with very short cotyledons or apparently undivided.—Herbs either aquatic or growing in marshes or wet places.   Leaves radical or floating or none. Flowers solitary or several in a raceme, on leafless radical or terminal scapes or peduncles.

The Order, comprising but very few genera, is dispersed over the greater part of the globe. Of the two Australian genera, one, the principal one of the Order, occupies its whole area ; the other, dismembered from it by a purely artificial character, is endemic.

Calyx of 2 opposite segments . . . . 1. Utricularia.
Calyx of 4 segments in pairs, the inner ones lateral . . . . . 2. Polypompholyx.

1. UTRICULARIA, Linn.

Calyx deeply divided into 2 lobes or segments.   Corolla with a spur at the base rarely reduced to a small protuberance, the mouth of the tube usually closed or nearly so by a convex palate, the upper lip erect, broad, entire, sinuate or 2-lobed, the lower usually longer and broader, entire or 3-lobed (rarely 2-lobed or 4-lobed by the suppression or division of the middle lobe), with the lobes reflexed or the whole lip spreading horizontally, with a convex palate at the base, often bearing a small 3-lobed protuberance.   Capsule globular, opening in 2 valves.—Herbs either floating with submerged root-like leaves divided into capillary segments and interspersed with little vesicles or bladders full of air, or marsh plants either leafless or with entire radical leaves.   Peduncles or scapes radical or axillary.   Flowers solitary or in a raceme, alternate or opposite, with a small scale-like bract under each pedicel and sometimes 1 or very few similar minute scales on the scape below the flowers.   Bracteoles in many species 2 at the base of the pedicels, but very minute and often concealed within the bract.

A considerable genus, dispersed over nearly the whole globe except the extreme north and south, and especially numerous within the tropics both in the New and the Old World, several species having a very wide range. Of the 20 Australian species, five are the same as tropical Asiatic ones; another extends to New Zealand; the remaining 14 appear to be endemic; but one or two of the minute ones require further comparison from better specimens with Indian species. The corollas, from which some of the chief specific characters are drawn, are indeed of so delicate a nature that it is exceedingly difficult to ascertain their precise form in dried specimens; and it is to be feared that, in several of the following descriptions, slight inaccuracies may have crept in, which will have to be corrected chiefly from the examination of living plants.

§ 1. **Natantes.**—*Stems floating. Leaves submerged, divided into capillary segments, mostly interspersed with bladders. Flowers yellow, on axillary peduncles.*

Peduncles bearing about the middle a cluster or false whorl of oblong or
linear vesicles . . . . 1. *U. stellaris.*
Peduncles without vesicles.
Pedicels thickened and reflexed after flowering. Corolla rather large.
Style about 1 line long . . . . 2. *U. flexuosa.*
Peduncles filiform, pedicels slender and erect in fruit. Corolla small.
Stigma almost sessile. Seeds winged . . . . 3. *U. exoleta.*

§ 2. **Limosæ.**—*Plants growing in mud (sometimes under water) with erect leafless scapes. Leaves radical, often accompanied by filaments of which some bear utricles, or no leaves at all at the time of flowering.*

Bracts not produced below their insertion, always alternate.
Flowers yellow or white, 1 or 2 on short filiform scapes.
Corolla (white) upper lip emarginate, lower shortly and broadly
3-lobed . . . . 4. *U. albiflora.*
Corolla (yellow?) upper lip entire, lower with 1 broad lobe and
2 lateral narrow ones . . . . 5. *U. pygmæa.*
Flowers yellow, several in a raceme.
Pedicels very short, erect in fruit, not winged.
Spur horizontal. Palate spotted with red . . . . 6. *U. fulva.*
Spur descending . . . . 7. *U. chrysantha.*
Pedicels as long as the calyx, reflexed, and more or less winged in
fruit. Spur descending . . . . 8. *U. bifida.*
Flowers blue or white, several in a raceme. Pedicels short, not
winged. Calyx-segments rather acute . . . . 9. *U. cyanea.*
Flowers small, purple. Calyx-segments very obtuse.
Flowers distant, on very short pedicels. Scapes 1 to 4 in. . . . 10. *U. lateriflora.*
Flowers solitary. Scapes about 1 in.
Corolla upper lip broad. Spur as long as the lower lip . . . 11. *U. simplex.*
Corolla upper lip narrow. Spur half as long as the lower lip . 12. *U. monanthos.*
Bracts produced below their insertion into a small free appendage.
Flowers purple (or deep blue?).
Pedicels opposite, in 1, 2 or 3 pairs or in whorls of three. Corolla
lower lip large, semicircular.
Scapes erect, slender. Eastern species . . . . 13. *U. dichotoma.*
Scapes twining. Western species . . . . 14. *U. volubilis.*
Scapes 1-flowered. Bracts opposite or in whorls of three.
Spur shorter than the large semicircular lower lip.
Upper lip entire or shortly 2-lobed. Eastern species . . . . 13. *U. dichotoma.*
Upper lip deeply 2-lobed. Western species . . . . 15. *U. Hookeri.*
Spur longer than the lower lip.
Flowers under ¼ in. long. Lower lip shortly 3-lobed . . . . 16. *U. violacea.*
Flowers ½ to 1 in. long. Lower lip deeply 3-lobed . . . . 17. *U. Menziesii.*
Flowers racemose, alternate as well as the bracts. (Flowers blue?).
Pedicels as long as or longer than the calyx. Lower lip 2 lobed.

Flowers ¼ in. long or more. Spur longer than the lower lip . 18. *U. biloba.*
Flower scarcely 2 lines long. Spur shorter than the lower lip . 19. *U. limosa.*
Pedicels scarcely any, Lower lip entire . . . . 20. *U. Baueri.*

U. barbata, R. Br. Prod. 432 ; A. DC. Prod. viii. 16, from Queensland, *Banks and So-lander,* is unknown to me. There is no specimen in the Banksian or in Brown's herbarium that I can identify with it, nor indeed any in which I can discover the palate to be hairy or bearded as described.

U. compressa, R. Br. Prod. 431 ; A. DC. Prod. viii. 15, from Queensland, *Banks and Solander,* cannot now be identified. The character given will apply to several species, and no specimen is preserved either in the Banksian or in Brown's herbarium.

§ 1. *Stems floating. Leaves submerged, divided into capillary segments.*

1. **U. stellaris,** *Linn. f. ; A. DC. Prod.* viii. 3. Stems floating, branched, extending to a considerable length. Submerged leaves root-like, branching into numerous capillary segments interspersed with little globular vesicles. Peduncles slender, 2 to 4 in. long in the Australian specimens but longer in some Indian ones, bearing at some distance below the raceme a cluster or almost a whorl of 3 to 5 oblong or narrow vesicles each about ½ in. long, tapering at both ends, and bearing a few short simple or branched capillary segments. Flowers several, yellow, on pedicels of 2 to 6 lines, which are slender at the time of flowering, often thickened under the fruit, and then spreading or reflexed. Calyx-segments ovate and about 1 line long in flower, broad and 2 lines diameter in fruit. Corolla upper lip ovate or rounded, obtuse, longer than the calyx, lower lip nearly orbicular, scarcely longer than the upper one, truncate or slightly 3-toothed ; spur turned upwards under the lower lip and about its length. Capsule nearly as long as the calyx, membranous. Seeds peltate, with an angular margin.—Oliv. in Journ. Linn. Soc. iii. 174 ; F. Muell. Fragm. vi. 161 ; Wight, Ic. t. 1567 (not good).

N. Australia. Nicholson, Robinson, and Gilbert rivers, Gulf of Carpentaria. *F. Mueller.* Common in tropical Asia, extending also into tropical Africa.

2. **U. flexuosa,** *Vahl ; Oliv. in Journ. Linn. Soc.* iii. 175. Submerged floating stems extending sometimes to several feet. Leaves all submerged and root-like, branching into numerous capillary segments, interspersed with little globular vesicles or utricles. Peduncles usually 3 to 6 in. long, without vesicles, bearing a raceme of 3 to 6 yellow flowers. Pedicels erect and slender at the time of flowering, usually reflexed and thickened upwards when in fruit. Calyx-segments rather unequal, enlarged when in fruit. Corolla fully 5 lines across, the upper lip ovate, entire or slightly emarginate, the lower nearly reniform, the palate marked with brown veins ; spur obtuse, shorter than the lower lip. Style about 1 line long. Capsule nearly 3 lines diameter.—*U. fasciculata,* Roxb. ; Wight, Ic. t. 1568 ; A. DC. Prod. viii. 7 ; *U. australis,* R. Br. Prod. 430 ; A. DC. Prod. viii. 6 ; Lehm. Pl. Preiss. i. 338 ; Hook. f. Fl. Tasm. i. 298 ; F. Muell. Fragm. vi. 161.

N. Australia. Robinson river, Gulf of Carpentaria and near Providence Hill, *F. Mueller.*
Queensland. Burnett and Brisbane rivers, *F. Mueller ;* Midge Creek and Burdekin river, *Bowman ;* near Rockhampton, *O'Shanesy.*
N. S. Wales. Paramatta, *R. Brown.*

Victoria. Lagoons on the Yarra, Goulburn river, and near Omeo, *F. Mueller.*
Tasmania. Derwent river, *R. Brown ;* pools near the S. Esk river, near Launceston, *Gunn.*
W. Australia. Avon river, *Preiss, n.* 1875. These, as well as some of the specimens from the other colonies, are without flower, and therefore in some measure doubtful ;
but those from Omeo, as well as Brown's and several of the northern ones, are in flower or
fruit, and agree perfectly with those from tropical Asia, where the species has a wide range.

3. **U. exoleta,** *R. Br. Prod.* 430. Nearly allied to *U. flexuosa,* but
very much smaller and more slender. Floating submerged stems capillary,
with exceedingly fine filiform leaves, not much divided, interspersed with
minute vesicles, the largest of which are scarcely $\frac{1}{2}$ line diameter, and in
marshy places the linear-filiform leaves are nearly entire. Scapes filiform, 1
to 3 in. long or rarely more, bearing 1, 2 or 3 small yellow flowers. Bracts
very obtuse, not produced below their insertion. Pedicels slender and erect
in fruit as well as in flower. Calyx-segments broad, very obtuse, $\frac{1}{2}$ to $\frac{3}{4}$ line
long in flower, slightly enlarged under the fruit but not exceeding the capsule.
Corolla not above 3 lines long to the end of the spur and sometimes scarcely
2 lines, the lips nearly equal, both broad ; spur narrow-conical, obtuse, horizontal or turned upwards, as long as or longer than the lower lip. Stigma
sessile. Capsule membranous. Seeds peltate, bordered by a thin irregular
wing.—A. DC. Prod. viii. 7 ; F. Muell. Fragm. vi. 162 ; *U. diantha,* Rœm.
and Schult. ; Oliv. in Journ. Linn. Soc. iii. 176, not of A. DC. ; Wight, Ic.
t. 1569.

N. Australia. Victoria river and swamps near sea range, *F. Mueller.*
Queensland. Gracemere, *Bowman ;* near Rockhampton, *O'Shanesy.*
N. S. Wales. Nepean river, *R. Brown.*
The species has a wide range in tropical Asia.

§ 2. *Scapes erect. Leaves radical or none.*

4. **U. albiflora,** *R. Br. Prod.* 431. Scapes filiform, $\frac{1}{2}$ to 1 in. long,
bearing a single small white flower. Leaves none at the time of flowering.
Bract minute, not produced at the base. Pedicel very short. Calyx-seg-
ments $\frac{1}{2}$ line long at the time of flowering, $\frac{3}{4}$ line when in fruit. Corolla
not 2 lines long, the upper lip small, emarginate, lower lip broad, shortly 3-
lobed ; spur descending, as long as the lower lip.—A. DC. Prod. viii. 15.

Queensland. Endeavour river, *Banks and Solander.* Perhaps a white-flowered variety
of *U. pygmæa,* but both of these minute species require further investigation from better
specimens.

5. **U. pygmæa,** *R. Br.? Prod.* 432. Scapes filiform, 1 to 3 in. high.
Leaves radical, narrow-linear or none. Flowers small, yellow, solitary or 2
distant ones on rather long filiform pedicels. Bracts obtuse, not produced at
the base. Calyx-segments obtuse, about 1 line long. Corolla : upper lip
not twice as long as the calyx, obovate or orbicular, entire ; lower lip larger,
the middle lobe convex, fully 2 lines across, " the lateral lobes linear, divaricate ;" spur ascending, as long as the lower lip. Capsule membranous, the
fruiting pedicel not reflexed.—A. DC. Prod. viii. 16.

Queensland. Cape Grafton, *Banks and Solander ;* Brisbane river, *W. Hill.*
The Banksian specimens are minute, and the colour of the flower is not given ; their identity with Hill's yellow-flowered specimens (determined by F. Mueller) is therefore in some
measure uncertain.

6. **U. fulva,** *F. Muell. in. Trans. Phil. Inst. Vict.* iii. 63. Scapes simple or slightly branched, ½ to 1 ft. high, more rigid than in *U. chrysantha,* the flowering portion flexuose. Radical leaves none at the time cf flowering. Flowers yellow, rather distant, often numerous, almost sessile or the pedicel rarely 1 line long and erect in fruit. Bracts very obtuse, not produced at the base. Calyx-segments obtuse, above 1 line long. Corolla : upper lip 2 or 3 times as long as the calyx, broad and broadly 2-lobed ; lower lip very broad, obscurely 3-lobed, the very convex palate spotted with red ; spur rather slender, horizontal or ascending, nearly or quite as long as the lower lip. Capsule about 1 line diameter, slightly crustaceous. Seeds very small, ovoid.

N. Australia. Around stagnant waters near Macadam Range, *F. Mueller ;* Strangways river, *M'Douall Stuart.*

7. **U. chrysantha,** *R. Br. Prod.* 432. Scapes slender but tall, often exceeding 1 ft. Leaves usually none at the time of flowering. Flowers yellow, usually numerous but distant, rarely forming a more compact spike. Pedicels exceedingly short, erect in fruit and then not exceeding 1 line. Bracts very small, usually acute, not produced at the base. Calyx-segments ovate-lanceolate, about 1 line long. Corolla : upper lip much longer than the calyx, ovate-orbicular, entire ; lower lip rather longer, broad, very convex, the margins reflexed and 4-lobed (3-lobed with the middle lobe 2-fid) ; spur descending, straight or slightly curved, as long as the upper lip, the whole corolla usually about ½ in. long. Capsule globular, almost crustaceous, about 1 line diameter. Seeds minute.—A. DC. Prod. viii. 18.

N. Australia. Regent river, Brunswick Bay, N.W. coast, *A. Cunningham ;* Glenelg river, N.W. coast, *Marten ;* Victoria river, *F. Mueller ;* Port Essington, *Armstrong.*
Queensland. Point Lookout, *Banks and Solander ;* Rockingham Bay, *Dallachy.*

U. flava, R. Br. Prod. 432 ; A. DC. Prod. viii. 18, from Endeavour river, *Banks and Solander,* appears to me to be a slender variety of *U. chrysantha,* with the middle lobe of the lower lip nearly entire.

8. **U. bifida,** *Lam. ; Oliv. in Journ. Linn. Soc.* iii. 182. Scapes slender, usually about 6 in. high. Radical leaves small, linear or spathulate, very rare at the time of flowering. Flowers yellow, usually from 6 to 8 on the scape, rather distant. Pedicels 1 to 2 lines long, reflexed when in fruit and winged by the decurrent calyx-segments. Bracts acute, not produced at the base. Calyx-segments orbicular, obtuse, scarcely above 1 line long in flower, twice as large and decurrent when in fruit. Corolla : upper lip obovate, longer than the calyx, the lower lip broad, convex, longer than the upper one ; spur conical, descending, as long as the upper lip. Capsule membranous. Seeds small, ovoid, reticulate.—*U. diantha,* A. DC. Prod. viii. 21, not of Rœm. and Schult.

Queensland. Rockingham Bay, *Dallachy.* Extends over tropical Asia.

9. **U. cyanea,** *R. Br. Prod.* 431. Scapes slender, rarely exceeding 6 in. Leaves radical, linear or slightly spathulate, ¼ to ½ in. long, or sometimes very narrow and more than twice that length, but usually disappearing before the flowering. Flowers blue, sometimes very pale or white, several on the scape but distant. Pedicels shorter than or scarcely exceeding the

calyx, ascending or erect in fruit and not at all or scarcely winged. Bracts acute or acuminate, not produced at the base. Calyx-segments about 2 lines long when in flower and scarcely enlarged in fruit, acute or scarcely obtuse. Corolla : upper lip shorter than the calyx or scarcely exceeding it, obovate ; lower lip rather longer, broader and very convex ; spur descending, conical, rather obtuse, as long as the lower lip. Capsule membranous. Seeds nearly globular, appearing tuberculate or almost muricate when dry, reticulate only when soaked.—A. DC. Prod. viii. 15 ; *U. graminifolia,* R. Br. Prod. 432, but scarcely of Vahl.

N. Australia. Fitzmaurice river, *F. Mueller.*
Queensland. Endeavour river, *Banks and Solander ;* Brisbane river, *F. Mueller* (with leaves 1 in. long).
N. S. Wales. Grose river, *R. Brown ;* Manly Beach swamp, *Woolls.*

Var. *alba.* Corolla white, the calyx not quite so large in proportion at the time of flowering.—Rockingham Bay, *Dallachy.*

U. affinis, Wight, Ic. t. 1580 ; Oliv. in Journ. Linn. Soc. iii. 178, from the Indian Peninsula, appears to be, as suggested by Oliver, the same plant, differing from the common Asiatic *U. cærulea,* Linn. (which is probably the true *V. graminifolia,* Vahl,) in the shorter pedicels, usually less acute calyxes, and perhaps a few other characters of no great importance.

10. **U. lateriflora,** *R. Br. Prod.* 431. Scapes slender, sometimes filiform, but often rather rigid and drying black, 1 to 4 in. high. Leaves radical, spathulate, very rarely present at the time of flowering. Flowers small, purple, usually only 2 to 4, distant and very shortly pedicellate or almost sessile. Bracts obtuse, not produced at the base. Calyx not 1 line long. Corolla : upper lip oblong or linear, truncate or emarginate, shortly exceeding the calyx ; lower lip broader than long, 3 to 4 lines across, entire or obscurely crenate, the sides reflexed ; spur conical, nearly as long as the lower lip or sometimes rather longer. Capsule membranous. Seeds small, ovoid.—A. DC. Prod. viii. 15 ; Hook. f. Fl. Tasm. i. 299.

N. S. Wales. Near Sydney, frequent, *R. Brown.*
Tasmania. King's Island, *R. Brown ;* Rocky Cape and near Franklin river, *Gunn ;* lagoons, Brisbane Bay, Macquarrie Harbour, *Milligan ;* Western Mountains, *Archer ;* South Port, *C. Stuart.*

U. parviflora, R. Br. Prod. 431 ; A. DC. Prod. viii. 15, appears to me to be a slight variety of *U. lateriflora,* with long filiform scapes and rather smaller flowers.

11. **U. simplex,** *R. Br. Prod.* 431. Scapes filiform, about 1 in. long in the specimens seen, simple and 1-flowered. Leaves radical, linear-cuneate, very few or none at the time of flowering. Bract not produced at the base. Pedicel short. Calyx-segments broad, obtuse, about ¾ line long. Corolla : upper lip broadly obovate or rounded, the lower larger, broader than long, 3 to 4 lines across, entire or obscurely crenate, the sides reflexed ; spur ascending, flattened, about as long as the lower lip.—A. DC. Prod. viii. 15.

W. Australia. Moist heaths, King George's Sound, *R. Brown ;* swamps at the base of Mount Melville, *F. Mueller.*

12. **U. monanthos,** *Hook. f. Fl. Tasm.* i. 299. A little plant, closely resembling *U. simplex,* with the same filiform, simple, 1-flowered scapes of

about 1 in., and bracts and calyx the same, but the upper lip of the corolla very narrowly obovate, emarginate, much shorter than the broad, semicircular, scarcely notched lower lip, and the spur not above half as long as the lower lip.

Tasmania. Wet sandy ground near Arthur's Lakes, *Gunn.* Also in New Zealand. United by F. Mueller (Fragm. vi. 162) with *U. simplex,* but probably without actual comparison of specimens.

13. **U. dichotoma,** *Labill. Pl. Nov. Holl.* i. 11. *t.* 8. Scapes slender, from a few inches to above 1 ft. high. Leaves radical, petiolate, from almost ovate and 1 line long to linear or oblong, 3 to 4 lines long (or in a few abnormal specimens more than 1 in. and very narrow), accompanied by filiform fibres, some of them bearing small fringed utricles, the leaves sometimes disappearing before the flowering. Flowers purple or lilac, opposite in 1 or 2 pairs or whorls of 3 at the end of the stem, rarely reduced to a single terminal flower. Pedicels filiform, at first very short but lengthening to from $\frac{1}{4}$ to $\frac{1}{2}$ in. under the fruit. Bracts always opposite or in threes (even when the flower is solitary), small and narrow, very shortly produced below their insertion. Calyx-segments usually about $1\frac{1}{2}$ lines long, broad and obtuse. Corolla : upper lip small, broadly ovate or obovate, obtuse or obscurely 2-lobed ; lower lip horizontal, broadly semicircular, $\frac{1}{2}$ to $\frac{3}{4}$ in. across, the palate with a small 3-lobed prominence ; spur descending, obtuse, much shorter than the lower lip and sometimes very short. Capsule membranous. Seeds small, ovoid.—A. DC. Prod. viii. 14 ; Hook. f. Fl. Tasm. i. 299 ; F. Muell. Fragm. vi. 161 ; *U. speciosa,* R. Br. Prod. 430 ; *U. oppositiflora,* R. Br. l. c. ; A. DC. Prod. viii. 14.

N. S. Wales. Port Jackson, *R. Brown* and others ; New England, *C Stuart ;* near Goulburn, *Woolls.*

Victoria. Wendu Vale, *Robertson ;* Port Phillip, near Brighton, Station Peak, Grampians, etc., *F. Mueller ;* Skipton, *Whan.*

Tasmania. In pools and wet soil, abundant, ascending to 3500 ft., *J. D. Hooker.*

Var. *uniflora.* Flowers smaller, mostly solitary.—*U. uniflora,* R. Br. Prod. 431 ; A. DC. Prod. viii. 14 ; Hook. f. Fl. Tasm. i. 299. Generally mixed with the 2 or more-flowered specimens, but chiefly in N. S. Wales. The specimens distinguished by Brown as his three species appear to me to differ only in the number and size of the flowers, which are always variable in all the sets of *U. dichotoma* which I have seen.

14. **U. volubilis,** *R. Br. Prod.* 430. Scapes rather slender, twining sometimes to the length of 3 ft. or more, with a radical tuft of filaments, some of them bearing rather large utricles (often 2 lines diameter), but no leaves seen. Flowers large, purple, opposite in one or two pairs or rarely reduced to a single terminal flower. Pedicels $\frac{1}{2}$ to 1 in. Bracts opposite, narrow, produced below their insertion into an appendage sometimes nearly as long as the upper part. Calyx-segments thin, coloured, very obtuse, nearly 2 lines long. Corolla : upper lip scarcely twice as long as the calyx, broadly obovate or almost orbicular, retuse ; lower lip horizontal, semicircular, entire, $\frac{3}{4}$ to 1 in. across, the palate yellowish, with a small 3-lobed protuberance at its base ; spur descending, short and very obtuse.—A. DC. Prod. viii. 14 ; Lehm. Pl. Preiss. i. 339 ; F. Muell. Fragm. vi. 160.

W. Australia. Swamps, King George's Sound and adjoining districts, *R. Brown, Preiss, n.* 1922, *F. Mueller.*

15. **U. Hookeri,** *Lehm. Nov. Stirp. Pug.* viii. 47, *and Pl. Preiss.* i. 339. Scapes slender, 3 to 5 in. high, bearing a single terminal purple flower. Leaves linear, often $\frac{1}{2}$ in. long, not expanded into an obtuse lamina, but acute though flat, accompanied by filiform fibres, some of them bearing small utricles. Bracts opposite or 3 in a whorl, shortly produced below their insertion. Pedicel 2 to 4 lines long. Calyx-segments ovate, obtuse, coloured, about $1\frac{1}{2}$ lines long, usually unequal. Corolla : upper lip much contracted at the base, deeply divided into 2 oblong diverging lobes about 2 lines long ; lower lip almost reniform, entire or broadly 3-lobed, $\frac{1}{2}$ to $\frac{3}{4}$ in. across, the palate with a small 3- or 5-lobed protuberance at the base; spur obtuse, shorter than the lower lip. Capsule membranous.—*U. inæqualis,* A. DC. Prod. viii. 666 ; *U. linearifolia,* Benj. in Linnæa, xx. 306 (partly) ; *U. latilabiata,* Benj. l. c. 315.

W. Australia. Swan River, *Drummond, n.* 128, 508, *Preiss, n.* 1918; Tone and Vasse rivers, *Oldfield.*

Benjamin has, in the Hookerian herbarium, named some specimens *U. latilabiata,* others *U. uniflora* and *U. linearifolia ;* his character of the latter is a compound of two or three species, but appears to have been taken chiefly from Drummond's specimens n. 508 of *U. Hookeri.*

U. similis, Lehm. Nov. Stirp. Pug. viii. 46, and Pl. Preiss. i. 339, and *U. Preissii,* A. DC. Prod. viii. 666, are both founded on Preiss's specimens n. 1919, from Swan River, which I have not seen. From the descriptions of the two authors (which do not quite agree with each other) the species only appears to differ from *U. Hookeri* in the lower lip of the corolla more deeply 3-lobed, a character in which *U. Hookeri* and the allied species are variable. Neither author describes the bracts as produced below their insertion, but Lehmann states them to be opposite, which, as far as known, only occurs in the groups of *U. dichotoma,* where they are always more or less produced. A. DC. describes a single 3-fid bract,—a mistake arising probably from the 3 whorled bracts being so closely packed as to appear united.

16. **U. violacea,** *R. Br. Prod.* 431. Scapes filiform, 1 to 2 in. high, bearing a single small terminal purple flower. Leaves linear or slightly spathulate, rarely almost ovate, obtuse, very small, often accompanied by filaments, a few of them bearing small utricles. Bracts opposite, oblong, obtuse, produced below their insertion. Pedicel short. Calyx-segments very obtuse, rather unequal, about $\frac{3}{4}$ line long in flower, often $1\frac{1}{2}$ lines in fruit. Corolla not 3 lines long to the end of the spur ; upper lip shortly exceeding the calyx, obtuse, more or less but usually very shortly 2- or 3-lobed or almost entire ; lower lip twice as long as the upper lip, broader than long, crenately 3-lobed ; spur conical, horizontal, nearly twice as long as the lower lip. Capsule membranous. Seeds minute.—A. DC. Prod. viii. 15 ; *U. perminuta,* F. Muell. Fragm. vi. 160.

W. Australia, *Drummond, n.* 86 *and* 213 ; King George's Sound, *R. Brown ;* Mount Melville, near King George's Sound, *F. Mueller.* Drummond's specimens in the Hookerian herbarium were confounded by Benjamin with *Polypompholyx tenella,* which resembles it in size, but is very readily distinguished by the bracts and calyx.

17. **U. Menziesii,** *R. Br. Prod.* 431. Scapes filiform, 1 to 2 in. high, bearing a single large terminal purple flower, remarkable for its long spur. Leaves linear-spathulate, 1 to 2 lines long, on long petioles, more abundant at the time of flowering than in most species, accompanied by numerous transparent filaments, a few of which bear small utricles. Bracts opposite or

in whorls of three, linear-oblong, shortly produced below their insertion. Pedicel 2 to 4 lines long. Calyx-segments obtuse, coloured, nearly 2 lines long. Corolla: upper lip cuneate or obovate, retuse, not twice as long as the calyx and reflexed over it; lower lip larger, broad, entire or crenate, about 4 lines across; spur somewhat curved, very obtuse, ½ to ¾ in. long.—A. DC. Prod. viii. 15; Lehm. Pl. Preiss. i. 339; *U. macroceras,* A. DC. Prod. viii. 666.

W. Australia. King George's Sound, *Menzies,* and thence to Swan River, *Drummond, 1st Coll., Preiss, n.* 1917, *Harvey, Oldfield.*

18. **U. biloba,** *R. Br. Prod.* 432. Scapes slender, simple or slightly branched, 6 to 9 in. high, usually with several small scales, which as well as the bracts are alternate and produced below their insertion, mostly acute. Leaves very small and rare at the time of flowering. Flowers blue (*R. Brown*), dark when dry, several in a raceme, on filiform pedicels usually as long as or rather longer than the calyx. Calyx-segments about 1 line long when in flower and but slightly enlarged afterwards, very broad and obtuse. Corolla: upper lip very short, ovate, 2-lobed, with the sides reflexed; lower lip broader than long, broadly 2-lobed; spur conical, obtuse, horizontal or descending, longer than the lower lip, the whole corolla 3 or 4 lines long.—A. DC. Prod. viii. 24.

N. S. Wales. Port Jackson, *R. Brown, Backhouse.* This and the following two species usually turn black in drying, which is not the case with *U. cyanea U. lateriflora* sometimes turns black, but it is readily distinguished by the bracts.

19. **U. limosa,** *R. Br. Prod.* 432. Scapes filiform, simple or branched, 6 to 10 in. high. Leaves none at the time of flowering. Flowers (blue?) in a long loose raceme, all alternate. Bracts narrow, much produced below their insertion, acute at both ends. Pedicels filiform, 1 to 2 lines long. Calyx-segments orbicular, obtuse, about ½ line long when in flower, slightly enlarged afterwards. Corolla scarcely above 2 lines long, the upper lip short, ovate, entire; lower lip much larger, broad, deeply 2-lobed; spur descending, shorter than the lower lip. Capsule small.—A. DC. Prod. viii. 24.

Queensland. Endeavour river, *Banks and Solander.* Very near *U. biloba,* but very much more slender, the flowers much smaller, the pedicels longer, and the spur shorter.

20. **U. Baueri,** *R. Br.? Prod.* 431. Scapes slender but rather long, simple or slightly branched, more rigid than in *U. biloba,* bearing usually several scales below the inflorescences, which, like the bracts, are all alternate, narrow, produced below their insertion, acuminate and very acute at both ends. Flowers (blue?) almost sessile in short terminal spikes, with rarely the lower flower at some distance below the spike. Calyx-segments obtuse, small. Corolla: upper lip short, narrow-ovate, erect, entire; lower lip much larger (about 3 lines across), broader than long, apparently quite entire with the sides reflexed; spur straight, horizontal, considerably longer than the lower lip.—A. DC. Prod. viii. 15.

Queensland. Endeavour river, *Banks;* Shoalwater Bay, *R. Brown.*
N. S. Wales. Port Jackson, *Bauer?*

U. Baueri was described by Brown from a drawing made by Bauer of a plant of which no specimen was preserved; but in his notes he says he believes it to be the same as one of which he had a specimen before him, which there is now no certain means of identifying.

The character, however, agrees well with the specimens described above, which are named
by Solander in the Banksian herbarium *U. juncea,* and with one in Brown's own herbarium
labelled *U. obscura,* neither of which names are taken up by Brown.

2. POLYPOMPHOLYX, Lehm.

(Tetralobus, *A. DC.*)

Characters and habit of *Utricularia,* except that the calyx, besides the two
fore-and-aft segments of that genus, has two additional inner lateral segments
alternating with them.—Marsh plants with radical leaves, leafless scapes and
racemose or solitary pink flowers, the bracts alternate and not produced at
the base.

The genus is limited to the two W. Australian species.

Scapes several inches to above 1 ft. high. Flowers rather large, the spur
   not half as long as the lower lip . . . . 1. *P. multifida.*
Scapes usually under 3 in. Flowers small, the spur usually as long as
   the lower lip . . . . 2. *P. tenella.*

1. **P. multifida,** *F. Muell. Fragm.* vi. 162. Scapes sometimes filiform,
under 6 in. high, bearing only 2 or even a single pink flower, sometimes
stouter, above 1 ft. high, with a loose raceme of 5 or 6 flowers. Leaves
radical, linear-spathulate, accompanied by filiform fibres, some of them bearing
utricles, but often all disappearing before the flowering. Bracts minute, obtuse. Pedicels usually distant, filiform, varying in length, erect or spreading
in fruit. Calyx, outer segments rather unequal broad and obtuse, a little
more than 1 line long in flower, more or less enlarged in fruit ; inner ones
similar but considerably shorter. Corolla : upper lip scarcely exceeding the
calyx, deeply divided into 2 narrow usually acuminate lobes ; lower lip large
(from under ½ in. to fully ¾ in. across according to the size of the flower),
more or less deeply divided into 3 obtuse retuse or bifid lobes, the palate
marked with a small digitately 5-lobed protuberance ; spur obtuse, not half
so long as the lower lip. Capsule membranous. Seeds very numerous, peltate.—*Utricularia multifida,* R. Br. Prod. 432 ; A. DC. Prod. viii. 18 ;
Benth. in Hueg. Enum. 82 ; *U. latiloba,* Benth. l. c. ; *Polypompholyx
Endlicheri,* Lehm. Nov. Stirp. Pug. viii. 48, and Pl. Preiss. i. 340 ; *P.
latiloba,* Lehm. ll. cc. 49 and 341 ; *Tetralobus Preissii,* A. DC. Prod. viii.
667.

W. Australia. King George's Sound and adjoining districts, *Menzies, R. Brown,
Preiss, n.* 1921, and many others ; Vasse and Tone rivers, *Oldfield ;* Swan River, *Huegel,
Drummond, n.* 509 (also 507 in Herb. Hook., but probably a mistake), *Preiss, n.* 1923 (the
latter specimen not seen.
The specimens in the Hookerian Herbarium are variously named by Benjamin, *P. lati-
loba,* Lehm., *U. uniflora,* Br., *U. oppositiflora,* Br., and *U. linearifolia,* Benj.

2. **P. tenella,** *Lehm. Nov. Stirp. Pug.* viii. 50, *and Pl. Preiss.* i. 341.
Closely allied to *P. multifida,* differing chiefly in its small size and longer
spur. Scapes filiform, 1 to 2 or rarely 3 in. high, bearing 1 or 2 small pink
flowers. Calyx as in *P. multifida,* but smaller. Corolla with the short
upper lip deeply divided into acuminate lobes and the lower with 3 retuse
lobes, as in that species, but scarcely 3 lines across, and the spur usually as

long as the lower lip.—Hook. f. Fl. Tasm. i. 300; *Utricularia tenella*, R.
Br. Prod. 432; A. DC. Prod. viii. 16; *Tetralobus pusillus*, A. DC. Prod. viii.
667; *P. exigua*, F. Muell. in Trans. Phil. Soc. Vict. i. 50, in Hook. Kew
Journ. viii. 203, and Lithogr. t. 64.

Victoria. Near Melbourne, *Adamson;* mossy, peaty, or boggy places in the Gram-
pians, Serra and Victoria ranges, *F. Mueller.*
Tasmania. Flinders Island, *Gunn.*
S. Australia. Echunga, *F. Mueller.*
W. Australia. King George's Sound, *R. Brown;* summit of Mount Melville, *F.
Mueller;* Swan River, *Preiss, n.* 1920 (not seen), *Drummond, n.* 85, 507.

ORDER LXXXV. **OROBANCHACEÆ.**

Flowers irregular. Sepals 4 or 5, united in a variously split calyx. Corolla tubular or campanulate, usually curved or oblique; the limb more or
less 2-lipped; the upper lip erect or spreading, emarginate or 2-lobed; the
lower lip spreading, 3-lobed. Stamens 4, in pairs, inserted in the tube.
Anthers 2-celled, the cells usually but not always pointed or awned, opening
longitudinally. Ovary superior, 1-celled, with 2 (very rarely 3) double or
bifid placentas, or 4 distinct placentas, more or less protruding from the sides
into the cavity, but not united in the axis. Ovules several, usually very numerous. Style simple, with a capitate or 2-lobed stigma. Capsule 2-valved.
Seeds small, with a minute embryo and abundant albumen.—Leafless herbs,
not green, parasites on roots. Stems usually thick, the leaves replaced by
alternate scales or bracts of the colour of the rest of the plant. Flowers
solitary in the axils of the bracts, usually forming terminal spikes or
racemes.

An Order not very numerous in species, but widely distributed over nearly the whole
globe, except the extreme north and south, and much more abundant in the northern than
in the southern hemisphere. The only Australian genus is the principal one of the Order,
though almost limited to the northern hemisphere.

1. **OROBANCHE,** Linn.

Calyx divided to the base on the upper side, and often also on the lower
side, so as to form 2 lateral sepals, either entire or 2-cleft, either distinct from
each other or more or less connected at the base on the lower side, and
sometimes connected also on the upper side by the intervention of a small
fifth lobe, and always pointed. Habit and other characters those of the
Order.

The principal genus of the Order, abundant in the northern hemisphere in the Old World,
less so in North America, and a very few of the European species have also appeared in the
southern hemisphere, and amongst them the only Australian one.

1. **O. cernua,** *Læfl.; Reut. in DC. Prod.* xi. 32. Stems stout, erect,
simple, from about 6 in. to above 1 ft. high, of a pale brown colour more or
less tinged with blue and loosely pubescent; the scales ovate, the lower ones
ovate, the upper ones acute or acuminate. Flowers of a lurid bluish purple,
pale or whitish towards the base, in a rather dense spike, occupying about

one-third of the stem. Bracts acuminate, acute, shorter than the corolla. Sepals in the Australian specimens 2, entire, lanceolate, with long points, nearly as long as the bracts. Corolla tubular, incurved, about ¾ in. long, glabrous or minutely glandular-pubescent towards the top; upper lip very concave, with 2 short broad lobes, not ciliate; lower lip divided into 3 ovate shortly acuminate spreading lobes. Filaments glabrous; anthers not mucronate. Style glabrous, with short very thick stigmatic lobes.

Victoria. Black Forest on *Senecio lautus,* and Cape Grant, *F. Mueller;* Murray river, *Dallachy.*

S. Australia. Near Cudnaka, *F. Mueller.*

W. Australia, *Drummond, n.* 185, 198; Swan River, *Oldfield;* Flinders Bay, *Collie.*

The species is an inhabitant of the Mediterranean region of the northern hemisphere, where it is found on several species of *Artemisia,* and extends to E. India. Its introduction into Australia is not easily accounted for.

Order LXXXVI. GESNERIACEÆ.

(Cyrtandraceæ, *DC. Prod.*)

Flowers usually irregular. Calyx with 5 teeth lobes or distinct sepals. Corolla with a long or short tube, the limb 2-lipped or of 5 spreading lobes, imbricate in the bud. Stamens 2 or 4, in pairs, inserted in the tube, with the addition sometimes of a fifth barren one or staminodium. Anthers 2-celled or 1-celled by the confluence of the two, the cells opening longitudinally. Ovary superior or more or less inferior, 1-celled, with 2 parietal entire or lobed placentas, protruding more or less into the cavity, but not united in the axis. Ovules numerous. Style simple, with an entire or lobed stigma. Fruit a berry or capsule. Seeds small, numerous, with or without albumen. Embryo straight.—Herbs or rarely shrubs or climbers. Leaves opposite or whorled.

A considerable Order, chiefly tropical, with a very few species from more temperate climates. Of the two Australian genera, one is endemic, the other is Asiatic, extending into China beyond the tropics, and westward to the Seychelles islands. Both belong to the tribe of *Cyrtandreæ,* characterized by a superior ovary and by the seeds containing little or no albumen. This tribe is limited to the Old World, with the exception of a very few American species, whilst the other two tribes, *Gesnerieæ* and *Beslerieæ,* are exclusively American. The Order differs from *Scrophularineæ* and *Bignoniaceæ* chiefly in the parietal placentation of the ovary.

Woody climber or epiphyte. Fruit globular, slightly pulpy, indehiscent . . 1. FIELDIA.
Herb with radical leaves. Fruit a spirally twisted linear capsule . . . . 2. BÆA.

1. FIELDIA, A. Cunn.

Calyx divided to the base into 5 segments. Corolla tubular, the limb 5-lobed, somewhat 2-lipped. Stamens 4, didynamous; anther-cells parallel, distinct. Stigma 2-lobed. Fruit slightly pulpy, indehiscent.—Woody climber or epiphyte. Leaves opposite, unequal. Pedicels axillary, 1-flowered, with herbaceous bracteoles under the calyx.

The genus is limited to a single species, endemic in Australia.

1. **F. australis,** *A. Cunn. in Field, N. S. Wales,* 364. *t.* 2. A tall climbing shrub clinging to the trunks of large trees by adventitious roots, the branches foliage and inflorescence hirsute with articulate hairs. Leaves very unequal in each pair, the larger one obovate elliptical or oblong, coarsely toothed, narrowed into a short petiole, 1¼ to 3 in. long; the opposite one sessile, ovate, ¼ to ½ in. long or rarely half as long as the larger one. Flowers of a greenish yellow, pendulous from axillary pedicels, shorter than the larger leaf. Bracteoles herbaceous, thin, ovate-lanceolate, acuminate, often ½ in. long. Calyx-segments lanceolate, acuminate, about as long as the bracteoles. Corolla-tube nearly cylindrical or slightly enlarged upwards, above 1 in. long; lobes very short and broad, nearly equal. Stamens inserted near the base of the tube and nearly as long as the corolla, the filaments dilated especially in the lower part; anther-cells obtuse; staminodium between the upper stamens small and slender. Placentas of the ovary 2-lobed, densely covered with exceedingly numerous ovules. Fruit about as long as the calyx, the pericarp membranous. Seeds exceedingly numerous, oblong, minutely striate. Embryo straight, surrounded by scanty albumen.—DC. Prod. ix. 286; Hook. Exot. Fl. t. 232; Bot. Mag. t. 5089; *Basileophyta Friderici Augusti,* F. Muell. First Gen. Rep. 16.

N. S. Wales. Generally diffused in the moist shady woods of the Blue Mountains, *A. Cunningham* and others; Shoalhaven, *Backhouse;* Illawarra, *A. Cunningham, Shepherd.*

Victoria. Sealer's Cove and Streletzky Range, Gipps' Land, *F. Mueller.*

2. BÆA, Commers.

Calyx divided to the base into 5 segments. Corolla with a short broadly campanulate tube, the limb somewhat 2-lipped, the upper lip 2-lobed, the lower 3-lobed, the lobes all flat and spreading. Stamens 2, shorter than the corolla; anther-cells diverging or divaricate, confluent at the apex into a single cell; staminodia usually 3, very small. Stigma 2-lobed. Capsule linear, spirally twisted, splitting usually into 4 valves. Seeds minute.— Herbs with a perennial stock and radical leaves, or in species not Australian a developed stem and opposite leaves. Peduncles or scapes axillary, bearing usually a dichotomous or umbellately branched panicle of flowers, without bracteoles.

The genus comprises a very few Asiatic species, one of them extratropical, and one from the Seychelles Islands. The only Australian one is endemic.

1. **B. hygroscopica,** *F. Muell. Fragm.* iv. 146. A perennial with a short thick woolly stock. Leaves radical, rosulate, broadly ovate or orbicular, crenate, sessile or contracted into a short broad petiole, thick soft and rugose, densely clothed with long woolly hairs, the larger ones 4 to 5 in. long, but usually half that size. Scapes 4 to 8 in. high, bearing a loose umbellately branched panicle of rather numerous deep blue flowers, the inflorescence glabrous or sprinkled with a few hairs. Bracts few and minute. Calyx-segments linear-oblong, about 1½ lines long. Corolla-tube broad and not above 1 line long, upper lip of 2 orbicular lobes about 2 lines diameter, lobes of the lower lip smaller. Filaments thickly clavate, longer than the anthers;

anther-cells quite divaricate, forming a single narrow-oblong cell. Stigma of 2 short broad lobes. Capsule from ¾ to above 1 in. long.

Queensland. Rockingham Bay, *W. Hill, Dallachy.* At first sight closely resembles the N. Chinese *B. hygrometrica,* Br. (*Dorcoceras,* Bunge), which has the same foliage, but the scape in the Chinese plant is less divided, the corolla-tube much larger (that figured in Deless. Ic. v. t. 95, is an imperfectly developed bud), and the anthers reniform on short filaments.

Order LXXXVII. **BIGNONIACEÆ.**

Flowers irregular. Calyx tubular or campanulate, truncate toothed or laterally split. Corolla-tube elongated or rarely short and campanulate ; lobes 5, spreading, often arranged in 2 lips, variously imbricate or rarely induplicate-valvate in the bud. Stamens 2 or 4, in pairs, inserted in the tube, the fifth staminodium usually small, rarely wanting. Anthers 2-celled, the cells opening longitudinally. Ovary usually 2-celled, with 2 distinct placentas in each cell attached to the dissepiments, and either contiguous or separated by a considerable interval, or, in some genera not Australian, the dissepiment discontinued between the placentas, and the ovary then 1-celled ; ovules several, often numerous to each placenta. Style filiform, with 2 short stigmatic lobes. Fruit a capsule, often elongated, opening loculicidally or septifragally in 2 valves, leaving the dissepiment free. Seeds transverse, usually flattened and bordered by a membranous wing, without albumen. Embryo straight or rarely curved ; cotyledons flat or fleshy ; radicle next the hilum. —Trees shrubs or woody climbers, very rarely (in species not Australian) herbs. Leaves opposite or rarely scattered, compound or rarely simple, without stipules. Flowers solitary in the axils or more frequently paniculate.

An Order almost entirely tropical, and most abundant in South America, with a few Asiatic and African species. Of the four Australian genera, two extend at least to tropical Asia, another is perhaps a congener of a New Caledonian plant, the fourth appears to be endemic. But the Order is at present in a state of great confusion, and not the less so that it has been partially elaborated by different botanists, who entertain very different views as to the theoretical structure of the ovary. The limits of the genera must therefore remain very uncertain until a satisfactory rearrangement of the whole shall have been laid before us. All the Australian genera belong to De Candolle's second subtribe *Catalpeæ,* in which the dehiscence of the capsule is loculicidal, the dissepiment being transverse, that is, attached (before maturity) to the centres of the valves.

Stamens included in the corolla-tube.
  Calyx truncate or slightly toothed. Woody climbers. . . . . 1. TECOMA.
  Calyx spathaceous, split longitudinally. Erect trees . . . . . 2. SPATHODEA.
Stamens exserted, longer than the corolla.
  Corolla-tube elongated ; lobes induplicate-valvate. Woody climb-
    er, with compound leaves . . . . 3. HAUSSMANNIA.
  Corolla-tube campanulate ; lobes imbricate. Erect tree, with
    simple whorled leaves . . . . 4. DIPLANTHERA.

1. **TECOMA**, Juss.

Calyx truncate or shortly 5-toothed. Corolla tubular, the lobes spreading, nearly equal, obscurely 2-lipped or oblique. Stamens 4, in pairs, included in the tube ; anther-cells diverging or divaricate. Style with 2 short ovate stigmatic lobes. Ovules numerous, in several rows on each placenta. Capsule (oblong in the Australian species) opening loculicidally in 2 very concave valves, the dissepiment transverse with relation to the valves, and not laterally dilated. Seeds overlaying each other in several rows, flat, broadly winged.—Tall woody climbers. Leaves opposite, pinnate. Flowers in terminal panicles. Bracts minute ; bracteoles none.

The genus is at present in a state of too much uncertainty to fix its geographical limits. The two Australian species are endemic, and with some botanists alone constitute the genus *Pandorea.* The two typical *Tecomæ* are West Indian and South Africa, and many other more or less allied species from tropical and northern America, from Japan, E. India, and S. Africa had been included in it by De Candolle and others, but have been recently again separated from it.

Panicles loose. Corolla-tube under 1 in. long, the lobes less than
　half as long . . . . 1. *T. australis.*
Panicles compact, corymbose. Corolla-tube above 1 in., the lobes more
　than half as long . . . . 2. *T. jasminoides.*

1. **T. australis,** *R. Br. Prod.* 471. A tall woody glabrous climber, with more or less twining branches. Leaflets usually 5 to 9, ovate-oblong ovate-lanceolate or almost linear, entire or here and there coarsely crenate, from under 1 in. to nearly 3 in. long, but exceedingly variable, all small or all large, sometimes, especially on barren shoots, all coarsely toothed, and then occasionally all very small and much more numerous. Flowers of a yellowish-white, tinged inside with purple or red, in loose terminal panicles, leafy at the base, the primary and often the secondary branches opposite, the ultimate inflorescence cymose or racemose. Calyx smooth, 1 to 1½ lines long. Corolla-tube from about ½ to ¾ in. long, slightly curved and dilated upwards ; lobes broad, not one-third as long as the tube, the 2 upper rather smaller with purple or red spots or streaks at their base, the throat bearded inside under the lower lip. Capsule 1½ to 3 in. long, usually acute at both ends, the valves hard and very concave. Seeds very flat, obovate, surrounded by a broad wing.— DC. Prod. ix. 225 ; Maund, Botanist, t. 8 ; *Bignonia Pandorea,* Vent. Jard. Malm. t. 43 ; Andr. Bot. Rep. t. 86 ; Bot. Mag. t. 865 ; *B. meonantha,* Link, Enum. Hort. Berol. ii. 130 ; *Tecoma meonantha,* G. Don, Gen. Syst. iv. 224 ; *T. Oxleyi* and *T. floribunda,* A. Cunn. in DC. Prod. ix. 225 ; *T. diversifolia,* G. Don, Gen. Syst. iv. 224 ; DC. Prod. ix. 225 ; *T. ochroxantha,* Kunth and Bouché, Ind. Sem. Hort. Berol. 1847, 12 (according to the character given and Seemann's verification).

N. Australia. Macdonnel Ranges in the interior, *M'Douall Stuart.*
Queensland. Thirsty Sound, Keppel Bay, etc., *R. Brown ;* evidently abundant in numerous localities, sent by many collectors, from Cape York, *Daemel,* to Moreton Bay, *A. Cunningham, F. Mueller,* and others ; in the interior to the Mantuan Downs, *Mitchell.*
N. S. Wales. Port Jackson to the Blue Mountains, frequent, *R. Brown, Sieber, n.* 265, and many others ; northward to Clarence river, *Beckler ;* New England, *C. Stuart ;* southward to Illawarra, *A. Cunningham,* and Twofold Bay, *F. Mueller ;* in the interior to St. George's and Peel's Ranges, *A. Cunningham ;* Lachlan and Darling rivers, *L. Morton ;* Lord Howe's Island, frequent, *Milne.*
Victoria. Sealer's Cove, Dandenong and Buffalo Ranges, *F. Mueller.*

2. **T. jasminoides,** *Lindl. Bot. Reg. t.* 2002. A tall glabrous woody climber, resembling the luxuriant specimens of *T. australis,* but with much

larger flowers. Leaflets usually 5 or 7, ovate and acuminate or ovate-lanceolate, 1 to 2 in. long, all entire and, as far as hitherto observed, not presenting the remarkable variations of *T. australis.* Flowers white, streaked with red in the throat, in compact terminal corymbose panicles. Calyx smooth, fully 3 lines long. Corolla-tube above 1 in. long, much more dilated upwards than in *T. australis,* the lobes very broad, more than half as long as the tube, the throat scarcely bearded inside or marked with 2 decurrent lines of short hairs. Fruit of *T. australis,* the seeds rather broader, almost obcordate, the wing either entirely surrounding them or chiefly on the two sides. —DC. Prod. ix. 225 ; Bot. Mag. t. 4004.

Queensland. Brisbane river, Moreton Bay, *A. Cunningham, F. Mueller* ; Ipswich, *Nernst.*

N. S. Wales. Richmond river, *Henderson ;* Clarence river (*Beckler ?*).

2. SPATHODEA, Beauv.

(*Sect. or gen.* Dolichandra *or* Dolichandrone, *Fenzl.*)

Calyx spathaceous, herbaceous, acuminate, oblique, split on the upper edge. Corolla tubular, the lobes spreading, nearly equal, obscurely 2-lipped or oblique, imbricate in the bud. Stamens 4, in pairs, included in the tube, with a small fifth staminodium ; anther-cells parallel. Ovules crowded on the placentas but almost in a single row. Capsule linear, elongated, flattened or nearly terete, opening loculicidally in 2 concave or nearly flat valves, the dissepiment transverse with relation to the valves, but so much laterally dilated between the placentas as to appear flattened and parallel to the valves. Seeds in a single row to each placenta, flat, broadly winged on each side, but scarcely overlapping each other.—Small trees. Leaves scattered or irregularly whorled, entire or pinnate. Flowers in terminal racemes. Bracts minute ; bracteoles none.

Like *Tecoma,* the genus is at present in a state of great confusion, and no two botanists are agreed as to its limits. The Australian species are certainly congeners of the Asiatic *S. crispa,* retained in the genus by *Bureau,* but it is as yet very uncertain how many of the American and African species (among the latter of which are the two species of which one must be considered as typical) are to be associated with them.

Leaves ovate, simple, mostly scattered . . . . 1. *S. alternifolia.*
Leaves simple or pinnate, mostly whorled, lanceolate or linear as well
as their leaflets . . . . 2. *S. heterophylla.*
Leaves pinnate ; leaflets long, terete, almost filiform . . . . . . 3. *S. filiformis.*

1. **S. alternifolia,** *R. Br. Prod.* 472. A tree, evidently allied to *S. heterophylla,* and a variety only according to Seemann, but the few specimens known insufficient for determining the point. Leaves scattered, alternate or irregularly opposite, simple, ovate or broadly ovate-lanceolate, acuminate, very coriaceous, obliquely veined, narrowed into a long petiole, no pinnate ones occurring on any of the specimens known. Flowers unknown. Capsule as in *S. heterophylla.*—DC. Prod. ix. 209.

Queensland. Endeavour river, *Banks and Solander ;* Burdekin river, *F. Mueller.*

2. **S. heterophylla,** *R. Br. Prod.* 472. A scrubby tree of 10 to 15 ft., with a rugged bark, quite glabrous. Leaves crowded on the young shoots,

mostly in whorls of 3, simple or pinnate with 3 to 7 leaflets, varying from oblong-lanceolate to linear, from 1 to 3 in. long, the simple leaves usually lanceolate, from 1½ to 5 in. long and narrowed into the petiole without articulation, both leaves and leaflets thickly coriaceous with very oblique veins. Flowers white, very fragrant, in short terminal simple racemes, the pedicels ½ to 1 in. long. Calyx nearly 1 in. long. Corolla-tube slender, 1½ in. long, dilated only at the top; lobes nearly ½ in. diameter, broadly rounded with the margins undulate and crisped. Hypogynous disk thick and fleshy, the margin forming a short ring round the base of the ovary. Capsule from a few in. to above 1 ft. long, compressed (or nearly terete when fresh?); valves slightly concave; dilatations of the dissepiment rather thick and corky, almost reaching the margins of the valves. Seeds transversely oblong, the wing on each side as long as the seed itself.—DC. Prod. ix. 207; *Dolichandrone heterophylla*, F. Muell. Fragm. iv. 149.

N. Australia. Islands of the Gulf of Carpentaria, *R. Brown, Henne;* Victoria river, *F. Mueller;* Careening Bay, N.W. coast, *A. Cunningham;* Roebuck Bay, N.W. coast, *Marten;* King's Ponds, in the interior, *M'Douall Stuart.*
Queensland. Rockingham Bay, *W. Hill, Dallachy.*

Seemann, in adopting the genus *Dolichandrone*, Fenzl, for this and the following *Spathodeas* (Ann. Nat. Hist. ser. 3, x. 31), places them in a section distinguished from *Tecoma* as having the " Capsula *marginicida;* septum valvis oppositum." That must, however, be giving a different meaning to the ordinary one of the term *marginicidal,* for the dehiscence is certainly loculicidal in all the three Australian species. The septum, although apparently parallel to the valves as above explained, is really transverse as correctly stated by Seemann, which is incompatible with a marginicidal dehiscence in the ordinary acceptation of the term.

3. **S. filiformis,** *DC. Prod.* ix. 209. A small tree, quite glabrous. Leaves scattered or, in some specimens, irregularly opposite, pinnate; leaflets in few distant pairs, very narrowly linear-terete, almost filiform as well as the rhachis, 6 to 10 in. long in some specimens, half that length and more crowded in others, occasionally only 3 to the leaf (or rarely the leaves simple?). Flowers entirely like those of *S. heterophylla,* but the racemes shorter and the pedicels usually longer. Capsule above 1 ft. long, terete according to A. Cunningham, appearing somewhat compressed when dry, the structure and seeds as in *S. heterophylla.*—*Dolichandrone filiformis,* Fenzl; F. Muell Fragm. iv. 149.

N. Australia. Copeland Island, *A. Cunningham;* Victoria river, *F. Mueller.*

3. HAUSSMANNIA, F. Muell.

Calyx campanulate, truncate or minutely 5-toothed. Corolla tubular, incurved, dilated upwards; lobes 5, nearly equal, obscurely arranged in 2 lips induplicate-valvate in the bud. Stamens 4, inserted in the tube, longer than the corolla, with a fifth small staminodium; anther-cells diverging or divaricate. Hypogynous disk cupular, completely enclosing the ovary. Ovary short, slightly compressed, the dissepiment transverse. Ovules numerous, in several rows in each placenta. Style with 2 ovate stigmatic lobes. Fruit unknown.—Woody climber. Leaves opposite, compound. Flowers in short racemes. Bracts minute; bracteoles none.

The genus, as far as at present known, is limited to the single Australian species, and is very remarkable for the æstivation of the corolla. I do not see the affinity with the genus (or section of *Tecoma*) *Campsis,* suggested by F. Mueller.

1. **H. jucunda,** *F. Muell. Fragm.* iv. 148. A tall glabrous woody climber.   Leaflets 3, digitate and articulate at the end of a petiole of 1 to 2 in., each leaflet oval or elliptical, shortly acuminate, entire, membranous, penniveined, narrowed into a short petiolule, 2 to 4 in. long, and occasionally the central leaflet confluent with one of the lateral ones, but no simple leaves in the specimen seen.   Flowers "purple," in short racemes in the axils of the terminal pair of leaflets.   Pedicels 2 to 3 lines long.   Calyx 2 to 3 lines long.   Corolla-tube about 1 in. long ; lobes ovate, not 1 line long, hairy inside.   Stamens hairy at their insertion below the middle of the tube, shortly exceeding the corolla-lobes.   Hypogynous disk above 1 line long.—*Campsis Haussmannii,* F. Muell. l. c.

Queensland.   Seaview Range, Rockingham Bay, *Dallachy.*

4. DIPLANTHERA, R. Br.

(Bulweria, *F. Muell.*)

Calyx campanulate, with 5 equal lobes, valvate and connivent in the bud, and often cohering in 2 lips or in pairs after the calyx is open.   Corolla with a broad campanulate tube and 5 broad nearly equal lobes, imbricate in the bud.   Stamens 4, without the fifth staminodium, exserted, involute in the bud ; anthers with 2 linear distinct cells, parallel in the bud, at length divaricate.   Ovary 2-celled ; ovules very numerous, crowded in several rows on 2 distinct but approximate placentas in each cell.   Style long, with 2 oval, flat, stigmatic lobes.   Capsule oblong-fusiform, opening loculicidally in 2 woody valves leaving the placenta free.   Seeds very flat, with a broad transparent wing.—Tree.   Leaves simple, whorled or opposite.   Flowers yellow, in terminal panicles.

The genus is limited to the single Australian species, unless it should include, as suggested by Seemann, the New Caledonian *Deplanchea,* which is, no doubt, closely allied in foliage and inflorescence.   Our specimens of the latter have not the flowers in a sufficiently perfect state for examination, but Bureau describes the fifth sterile stamen or staminodium as present, and the two placentas of each cell of the ovary as distant from each other, leaving a bare space between them ; and the fruit being unknown, it remains yet to be determined whether the two would be most appropriately considered as congeners or not.   The fruit of *Diplanthera* closely resembles that of the Australian *Tecomas.*

1. **D. tetraphylla,** *R. Br. Prod.* 449.   A moderate-sized or sometimes lofty tree, with a soft wood and spongy bark ; the thick branchlets, under side of the leaves, and inflorescence covered with a thick soft tomentum, often assuming a golden or bronzed hue, and consisting of single or clustered but scarcely stellate hairs.   Leaves crowded at the ends of the branches, in whorls of 4 or the first leaves of young shoots opposite, on short petioles, ovate, obtuse, entire, 1 to 2 ft. long or those immediately under the panicle 6 to 8 in., the upper surface glabrous or slightly scabrous.   Flowers yellow, in a dense terminal panicle, nearly sessile above the last leaves, the primary branches whorled, each one dichotomously branched, with a flower shortly

pedicellate in each fork. Bracts linear, minute. Calyx coriaceous, about ½ in. long, the lobes acute, as long as the tube. Corolla-tube shortly exceeding the calyx, the lobes as long as the tube, broadly rounded. Stamens and style exceeding the corolla by an inch or more, very divergent. Hypogynous disk rather thick. Capsule (only seen open with the valves detached) 2 to 3 in. long, the valves hard and woody, smooth inside with a longitudinal line probably where the dissepiment was attached, the placenta-bearing dissepiment not broad and rather thick. Seeds apparently ripe, but the embryo not perfect in those examined.—F. Muell. in Seem. Journ. Bot. v. 212; *Bulweria nobilissima* or *Tecomella Bulweri*, F. Muell. Fragm. iv. 147; *Deplanchea Bulwerii*, F. Muell. Fragm. v. 72.

Queensland. Endeavour river, *Banks and Solander*; Cape York, *M'Gillivray, Daemel;* Rockingham Bay, *Dallachy.*

F. Mueller describes the upper leaves as 2-foliolate. I have in vain sought for these bi-foliolate leaves in his own as well as in the Hookerian herbarium. He was probably misled by the young shoots in some of the upper axils bearing only a pair of leaves, but always with the terminal bud between them.

Order LXXXVIII. ACANTHACEÆ.

Flowers irregular. Calyx more or less deeply divided into 5 lobes segments or distinct sepals, the upper one often smaller and sometimes wanting or the 2 lowest united into one. Corolla with a long or short tube, the limb either 2-lipped or of 5 spreading lobes, contorted or otherwise imbricate in the bud or expanded into a single lower lip. Stamens inserted in the tube, 4 in pairs or 2 only, the upper ones then reduced to staminodia or entirely wanting. Anthers 2-celled or 1-celled by the abortion of the other cell. Ovary superior, 2-celled, with 2 or more ovules or rarely a single one in each cell. Style simple, usually subulate, with an entire or 2-lobed stigma, the lobes not dilated and the upper one often reduced to a small tooth. Capsule opening loculicidally in 2 valves, usually elastically recurved and bearing the placentas along their centre. Seeds usually flat, attached to hooked processes from the dissepiment called *retinaculæ*, or in the first two tribes the seeds globular and resting on cup-shaped dilatations or mere papillæ, sometimes almost inconspicuous. Albumen none. Embryo usually curved.—Herbs shrubs or rarely trees. Leaves opposite, entire or rarely toothed, or in a few species not Australian lobed. Flowers axillary or terminal, in spikes racemes or clusters, more or less bracteate, the primary inflorescence centripetal, the secondary sometimes dichotomous and centrifugal. Bracteoles rarely wanting and sometimes large and leafy.

A large Order, diffused over both the New and the Old World, chiefly within the tropics, a very few species occurring in more temperate regions, either in the northern or the southern hemisphere. Of the eleven Australian genera, ten are Asiatic, most of them extending into Africa, and several are also American. The station of the previously known species of the eleventh genus (*Graptophyllum*) is uncertain, probably the Eastern Archipelago.

In the delimitation of the genera of this Order I have endeavoured to follow the views of Dr. T. Anderson, who has elaborated the African and E. Indian species with great care and success, but unfortunately his detailed generic characters have not yet been published; and in the very concise synoptic enumeration in the 'Journal of the Linnean Society' a few errors, probably clerical, render the distinctions upon which some of his groups are established rather difficult to make out. His arrangement has, however, dispelled much of the obscurity in which that of the ' Prodromus ' by Nees had been involved.

TRIBE I. **Thunbergieæ.**—*Corolla-limb with* 5 *nearly equal spreading lobes, contorted in the bud. Seeds globular, resting on a cup-shaped expansion of the placenta.*

Stem twining. Calyx an entire or toothed ring concealed within
 the bracteoles. Capsule beaked . . . . 1. THUNBERGIA.

TRIBE II. **Nelsonieæ.**—*Corolla-limb of* 5 *nearly equal lobes or* 2-*lipped, imbricate in the bud, the upper lobes or lip outside. Seeds globular. Retinacula none or reduced to minute papillæ.*

Corolla-lobes nearly equal. Stamens 2 . . . . 2. NELSONIA.
Corolla-lobes nearly equal. Stamens 4 . . . . 3. EBERMAIERA.

TRIBE III. **Ruellieæ.**—*Corolla-lobes* 5, *nearly equal or* 2-*lipped, contorted in the bud. Seeds flat. Retinacula prominent.*

Corolla 2-lipped. Stamens 4. Ovules several in each cell . . . 4. HYGROPHILA.
Corolla-lobes nearly equal, spreading. Stamens 4. Ovules 6 or
 more in each cell . . . . 5. RUELLIA.

TRIBE IV. **Justicieæ.**—*Corolla* 2-*lipped or* 1-*lipped by the reduction of the upper one or rarely nearly equally* 5-*lobed, the lobes variously imbricate but not contorted in the bud. Seeds flat. Retinacula prominent.*

Corolla with one broad flat (lower) lip. Stamens 4. Leaves
 usually prickly-toothed . . . . 6. ACANTHUS.
Corolla 2-lipped. Stamens 2.
 Bracts small or not enclosing the flowers.
  Anther-cells attached one higher up than the other. Upper
   corolla lip erect . . . . 7. JUSTICIA.
  Anther-cells equal and parallel. Upper corolla-lip incurved.
   Flowers red . . . . 8. GRAPTOPHYLLUM.
 Bracts in 2 pairs, forming an involucre enclosing 1 to 3 flowers.
  Inner involucral bracts broad and appressed. Outer ones
   spreading subulate or spinous. Anthers 2-celled . . . 9. DICLIPTERA.
  Involucre cylindrical, the 2 pairs of bracts nearly equal and
   usually connate, one pair within the other. Anthers 1-
   celled . . . . 10. HYPOESTES.
Corolla-lobes nearly equal, spreading. Stamens 2 . . . . . 11. ERANTHEMUM.

TRIBE I. THUNBERGIEÆ.—Corolla-limb with 5 nearly equal spreading lobes, contorted in the bud. Seeds nearly globular, resting on a cup-shaped expansion of the placenta.

* 1. **THUNBERGIA,** Linn. f.

Calyx reduced to an entire or many-toothed ring and concealed within 2 large bracteoles. Corolla-lobes 5, nearly equal, spreading, contorted in the bud. Stamens 4, included in the tube; anther-cells parallel. Ovules 2 in each cell of the ovary. Capsule globose and seed-bearing at the base, terminating in a flattened beak. Seeds globular, hollowed out on the inner face, and inserted on a cupular (sometimes very small) expansion of the placenta. —Twining or rarely dwarf and prostrate herbs. Flowers in axillary pedicels or in terminal racemes.

The genus is limited to tropical Asia and Africa and southern Africa. The following species is probably introduced only in Australia.

* 1. **T. alata,** *Boj. in Sims, Bot. Mag. t.* 2591. A herbaceous softly pubescent or villous twiner. Leaves broadly angular-cordate, on rather long petioles which are always more or less winged. Flowers pale orange or in one variety white, with the tube purple inside, on axillary pedicels shorter than the leaves. Bracteoles herbaceous, ovate-lanceolate or ovate, ½ to ¾ in. long. Calyx reduced to a ring of from 10 to 12 small acute teeth. Corolla-tube shortly exceeding the bracteoles with 5 rounded spreading nearly equal lobes.—Nees in DC. Prod. xi. 58; Hook. Exot. Fl. t. 177; Bot. Mag. t. 3512.

Queensland. Rockhampton, *Thozet.* A native of the S.E. coast of Africa or of the Mascarene Islands, long since cultivated in Indian as well as European gardens, and probably introduced only in Australia.

Tribe II. NELSONIEÆ.—Corolla-limb with 5 nearly equal lobes or 2-lipped, the upper lobes or the upper lip outside in the bud. Seeds small, globular, the retinacula reduced to minute papillæ or quite inconspicuous.

2. NELSONIA, R. Br.

Calyx of 4 distinct segments, the lowest 2-fid. Corolla-lobes 5, nearly equal, the 2 upper ones outside in the bud. Stamens 2, included in the corolla-tube; anther-cells distinct, divergent; no staminodia. Ovules rather numerous; stigmatic lobes of the style unequal. Capsule 2-celled from the base, terminating in a seedless beak. Seeds small, globular, resting on minute scarcely conspicuous papillæ.—Diffuse herb. Flowers small, sessile in terminal leafy spikes.

The genus appears to be limited to a single species, a common tropical weed.

1. **N. campestris,** *R. Br. Prod.* 481. A diffuse or prostrate herb, the slender stems much branched and extending sometimes to above 1 ft., clothed as well as the foliage with long soft hairs which are often white and silky on the young shoots and inflorescence. Leaves oblong or elliptical, narrowed into a short petiole or rarely broadly ovate or almost orbicular, rarely exceeding ¾ in. except the radical and lowest which are sometimes much longer, the floral ones sessile, ovate, acute, 3 to 4 lines long, crowded or almost imbricate in short terminal spikes. Flowers nearly sessile, not exceeding the floral leaves. Calyx about 2 lines long, the upper and lower segments rather broader than the others, the lowest from minutely 2-toothed to deeply 2-lobed. Corolla-tube about as long as the calyx, the lobes rounded, 2 upper ones nearly 1½ lines long, the 3 lower rather smaller. Capsule oblong-linear, not exceeding the floral leaves.— Endl. Icongr. t. 79; *N. rotundifolia,* R. Br. l. c.; *N. tomentosa,* Dietr.; T. Anders. in Journ. Linn. Soc. ix. 450; the whole five species of *Nelsonia* and their numerous synonyms given by Nees in DC. Prod. xi. 65 to 67.

N. Australia. Islands of the Gulf of Carpentaria, *R. Brown;* between Victoria and Fitzmaurice rivers, *F. Mueller;* Albert river, *Henne.*

Queensland. Endeavour river, *Banks and Solander;* Rockhampton, *Dallachy.*

The species is a common tropical weed in Asia and Africa, and is already abundant in several parts of tropical America. The name *N. tomentosa* was attributed by Nees to

Willd. Sp. Pl. ed. 2. This is a mistake. Willdenow never published a second edition of his 'Species Plantarum,' and never knew the genus *Nelsonia.* The specific name *tomentosa* was first given by Roxburgh, and *Nelsonia tomentosa* is Dietrich's. R. Brown's very appropriate name, *N. campestris,* is much older than any of them.

3. EBERMAIERA, Nees.

Calyx divided to the base into 5 segments, the upper one broader than the others. Corolla-lobes 5, nearly equal, the 2 upper ones outside in the bud. Stamens 4, in pairs, included in the corolla-tube ; anthers ovate, transverse, 2-celled. Ovules numerous ; stigmatic lobes of the style unequal. Capsule oblong-linear, not beaked, 2-celled from the base. Seeds numerous, very small, nearly globular ; retinacula reduced to minute papillæ or quite inconspicuous.—Herbs. Flowers small, sessile in the axils of the floral leaves or bracts, forming terminal dense or interrupted leafy or leafless spikes.

The genus comprises a considerable number of species from tropical Asia and Africa, with a few American ones. The only Australian species is a common Asiatic one.

1. **E. glauca,** *Nees in DC. Prod.* xi. 73. Stems at first simple and erect, at length diffuse prostrate or creeping and rooting at the nodes, with ascending or erect branches of ½ to 1 ft., the whole plant slightly pubescent, the inflorescence often glandular. Leaves oblong-lanceolate or elliptical, obtuse, narrowed into a petiole, 1 to 2 in. long, the floral ones much smaller, mostly under ½ in. Flowers nearly sessile in the axils of the floral leaves, between 2 leafy bracteoles about as long as the calyx, forming rather long leafy spikes, terminal or sometimes also in the axils of the upper stem-leaves. Calyx upper segment lanceolate, 3 to 4 lines long, lower ones linear and rather shorter. Corolla-tube about as long as the calyx, slightly dilated upwards ; lobes short, obovate, obscurely 2-lipped. Capsule as long as the calyx.—T. Anders. in Journ. Linn. Soc. ix. 450 ; Wight, Ic. t. 1488.

N. Australia. Providence Hill and Macadam Range towards Fitzmaurice river, *F. Mueller.* Common in E. India and the Archipelago.

TRIBE III. RUELLIEÆ.—Corolla-limb with 5 nearly equal lobes contorted in the bud. Seeds flat, subtended by hooked retinacula.

4. HYGROPHILA, R. Br.

Calyx more or less deeply divided into 5 or rarely 4 lobes or segments. Corolla-limb 2-lipped, the upper lip 2-lobed, the lower 3-lobed, the lobes usually short and contorted in the bud. Stamens 4, in pairs, or in species not Australian only 2 perfect ; anthers erect, the cells parallel and equal. Style subulate, with a small upper tooth. Ovules several in each cell of the ovary. Capsule oblong or linear, 2-celled from the base. Seeds flat ; retinacula hooked.—Erect or decumbent herbs. Flowers sessile in axillary clusters.

A small genus, widely distributed over the tropical and subtropical regions of the New and the Old World. The only Australian species is a common Asiatic one.

1. **H. salicifolia,** *Nees in Wall. Pl. As. Rar. and in DC. Prod.* xi. 92. Stems erect or ascending, branched, from ½ to 1½ ft. high, glabrous or slightly pubescent with appressed hairs as well as the foliage. Leaves lan-

ceolate or almost linear, contracted into a short petiole, 3 or 4 in. long in stout specimens, half that size in others. Flowers usually 2 or 3 together in the axils of the stem-leaves, purple or pale blue (or yellow according to *Dallachy*). Bracteoles concave, acute, usually shorter than the calyx. Calyx pubescent, tubular, the lobes shorter than the entire part, the 2 lower ones often more united. Corolla-tube scarcely exceeding the calyx, upper lip 2-lobed, the lower lip 3-lobed, convex, with 2 lines of hairs decurrent from the sinus, the lobes all nearly equal, slightly contorted or almost valvate in the bud. Stamens inserted near the top of the tube. Capsule linear, about ½ in. long. Seeds about 6 to 8 in each cell.—T. Anders. in Journ. Linn. Soc. ix. 456, with the synonyms adduced ; *Ruellia salicifolia,* Vahl, Symb. iii. 84 ; *Hygrophila angustifolia,* R. Br. Prod. 479 ; Nees in DC. Prod. xi. 91.

N. Australia. Van Diemen's Gulf, N.W. coast, *A. Cunningham ;* Victoria river, *F. Mueller ;* Port Essington, *Armstrong.*

Queensland. Endeavour river, *Banks and Solander, R. Brown ;* Rockingham Bay, *Dallachy ;* Broad Sound, *Bowman ;* Beddome Creek, *Thozet ;* Moreton Bay, *W. Hill.*

It has appeared to me that the æstivation of the corolla-lobes is somewhat variable in *Hygrophila,* but the overlapping is often so slight as to make it difficult to ascertain it correctly from dried specimens.

5. RUELLIA, Linn.

(Dipteracanthus *and* Cryphiacanthus, *Nees.*)

Calyx more or less deeply divided into 5 lobes or segments. Corolla-lobes 5, nearly equal, spreading, contorted in the bud. Stamens 4, included in the corolla-tube; anther-cells parallel and equal. Ovules 6 or more in each cell of the ovary. Capsule oblong-linear or clavate, more or less contracted and seedless at the base, very rarely equally 2-celled throughout. Seeds flat ; retinacula hooked, often denticulate at the top.—Herbs or rarely shrubs. Flowers usually blue, mostly axillary, solitary or clustered, rarely in terminal or axillary spikes.

A considerable genus, distributed over the warmer regions of the New and the Old World. The Australian species are perhaps all endemic, although one of them is very closely allied to an E. Indian one.

SECT. I. **Dipteracanthus.**—*Bracteoles usually longer and broader than the calyx. Capsule contracted or flattened and seedless at the base.*

Corolla with a slender tube of 1 in., the broader portion or throat half
as long. Capsule ¾ in. Flowers pedunculate . . . . 1. *R. bracteata.*
Corolla-tube very shortly slender at the base, the throat much longer.
  Capsule ⅓ in. long. Flowers nearly sessile.
  Corolla-throat nearly 1 in. long . . . . 2. *R. primulacea.*
  Corolla-throat scarcely ½ in. long . . . . 3. *R. corynotheca.*

SECT. II. **Cryphiacanthus.**—*Bracteoles linear-subulate, shorter than the calyx or none. Capsule equally 2-celled from the base or nearly so.*

Flowers sessile in the axils or nearly so. Bracteoles shorter than the
  calyx . . . . 4. *R. australis.*
Flowers distant in axillary leafless spikes. Bracteoles very small . . 5. *R. spiciflora.*
Flowers solitary on elongated pedicels. Bracteoles none . . . . 6. *R. acaulis.*

SECT. I. DIPTERACANTHUS.—Bracteoles usually longer and broader than the calyx. Capsule contracted or flattened and seedless at the base.

1. **R. bracteata,** *R. Br. Prod.* 479. Stems usually simple, 6 in. to 1 ft. high, more or less hirsute as well as the foliage. Leaves shortly petiolate, oblong or elliptical, the larger ones above 2 in. long, the lowest small and obovate. Flowers blue, on axillary pedicels varying from $\frac{1}{2}$ in. to above 1 in. in length. Bracteoles herbaceous, oval-oblong, $\frac{1}{2}$ to $\frac{3}{4}$ in. long, enclosing the base of the flower. Calyx-segments narrow, 2 to 3 lines long. Corolla with a very slender straight tube of about 1 in., the campanulate broad part or throat nearly half that length, the lobes broad and rounded, the margins slightly crisped. Stamens inserted near the top of the slender tube and reaching to the top of the throat. Capsule nearly $\frac{3}{4}$ in. long, contracted into a broad flattened seedless base. Seeds in the upper part, 7 or 8 in each cell. —Endl. Iconogr. t. 104 (the corolla-lobes more crisped than in any of the specimens seen); *Dipteracanthus bracteatus,* Nees in DC. Prod. xi. 143.

N. Australia. Arnhem N. and S. Bay, *R. Brown.*
Queensland. Cape York, *M'Gillivray.*

The E. Indian *Ruellia suffruticosa,* Roxb., is evidently closely allied to if not identica with *R. bracteata.*

2. **R. primulacea,** *F. Muell. Herb.* A perennial, apparently with the habit of *R. corynotheca,* but larger and more villous. Leaves ovate, nearly sessile, $\frac{1}{2}$ to 1 in. long. Flowers large, blue, nearly sessile in the upper axils. Bracteoles herbaceous, oval or oblong, much longer than the calyx. Calyx-segments narrow, 2 to 3 lines long. Corolla with an exceedingly short narrow base, the remainder of the tube much dilated, above 1 in. long, forming a long broad rather oblique throat; lobes broad, scarcely half as long as the tube. Capsule oblong-clavate, about $\frac{1}{2}$ in. long, contracted and seedless at the base. Seeds about 6 in each cell.

Queensland. Burdekin river, *F. Mueller;* Selheim and Elliot rivers, *Bowman;* Barcoo river, *M'Douall Stuart* (the latter specimen doubtful, more villous, with narrower more petiolate leaves).

3. **R. corynotheca,** *F. Muell. Herb.* A perennial, usually shortly scabrous-pubescent, with rather slender decumbent or erect stems of 6 in. to nearly 1 ft. Leaves petiolate, mostly ovate and small, often under $\frac{1}{2}$ in. and rarely 1 in. long. Flowers nearly sessile in the upper axils. Bracteoles oblong-linear or oblong, contracted at the base, longer than the calyx. Calyx $1\frac{1}{2}$ to 2 lines long, the segments united at the base. Corolla-tube 3 to 4 lines long, gradually dilated upwards, the lobes fully half as long as the tube. Capsule clavate, about $\frac{1}{2}$ in. long, the lower portion contracted and seedless, 2 lines broad above the middle. Seeds about 4 in each cell, rather large, all attached very near the middle of the capsule.

Queensland. Burdekin river, *F. Mueller;* Suttor river, *Bowman, Dorsay.* Very near the E. Indian *R. patula,* Jacq., but the corolla-tube appears to be shorter and not so slender, and there may be a few other trifling differences.

SECT. II. CRYPHIACANTHUS.—Bracteoles linear-subulate, shorter than the calyx or none. Capsule equally 2-celled from the base or nearly so.

4. **R. australis,** *R. Br. Prod.* 479. A small perennial with erect or diffuse branching stems often under 6 in. but sometimes nearly 1 ft. long, hirsute as well as the foliage or nearly glabrous. Leaves from obovate or oblong and under ½ in. to oblong-lanceolate and 2 in. long, narrowed into a petiole. Flowers blue, axillary, sessile or very shortly pedicellate, with narrow bracteoles shorter than or very rarely as long as the calyx. Calyx-segments subulate-acuminate, 3 to 4 or even 5 lines long. Corolla-tube exceeding the calyx, gradually but considerably enlarged upwards; lobes spreading, more than half as long and sometimes nearly as long as the tube. Stamens inserted near the base of the tube and the anthers scarcely reaching above the middle. Capsule about ½ in. long, mucronate, linear, not enlarged upwards, and usually 2-celled from the base. Seeds about 6 in each cell.— Cav. Ic. vi. 62. t. 586 ; Nees in DC. Prod. xi. 151 as to Brown's synonym, but not the plant described ; *Cryphiacanthus australis,* Nees in DC. Prod. xi. 198.

Queensland. Bay of Inlets, *Banks and Solander;* Thirsty Sound and Keppel Bay, *R. Brown;* Cape York, *M'Gillivray;* Rockhampton, *Dallachy;* Warwick, *Beckler.*
N. S. Wales. Port Jackson, *R. Brown* and others ; Hastings river, *Beckler;* Liverpool plains, *C. Moore;* between the Darling and Cooper's Creek, *Neilson.*
Var. *scabra.* A coarse form. Leaves ovate or ovate-lanceolate, rigid and scabrous. Capsule rather larger.—Gilbert river, *F. Mueller;* Armadillo, *Barton.*
Var. *pumila.* Very small and nearly glabrous. Leaves mostly oblong.—*R. pumilio,* R. Br. Prod. 479 ; *Dipteracanthus pumilio,* Nees in DC. Prod. xi. 124.—Burdekin river, *F. Mueller;* Rockhampton, *O'Shanesy;* Mogill Scrub, *C. Stuart;* Darling Downs, *Law* Armadillo, *Barton;* Port Jackson, *R. Brown.*

The garden plant described by Nees as *R. australis,* and which on his authority (he having named it so in my herbarium) I described as such in Maund's 'Botanist,' t. 175, is the E. Indian *Hemigraphis elegans,* Nees. Some specimens of *R. australis* much more closely resemble the *Hemigraphis hirta,* T. Anders., or *Ruellia hirta,* Vahl.

5. **R. spiciflora,** *F. Muell. Herb.* Stems creeping and rooting at the base, ascending, under 1 ft. long in the specimens seen, loosely pubescent as well as the veins of the leaves underneath. Leaves petiolate, ovate or oblong, those of each pair very unequal, the larger one 1 to 2 in. long. Flowers distant, along slender axillary peduncles, forming interrupted spikes, each one sessile within a small linear-setaceous bract with still smaller bracteoles. Calyx-segments linear-setaceous. Corolla-tube rather broad, nearly ½ in. long, the lobes about half as long as the tube. Stamens short. Capsule nearly ½ in. long, apparently like that of *R. australis,* but not seen ripe.

Queensland. Archer's Creek, *Leichhardt.* The habit of the plant is that of *Asystasia gangetica,* T. Anders. (*A. coromandeliana,* Nees), but the contorted corolla-lobes and the stamens are those of *Ruellia australis.*

6. **R. acaulis,** *R. Br. Prod.* 479. A dwarf almost stemless hirsute perennial, with a short branching stock, rarely emitting a few rather longer decumbent stems. Leaves mostly radical or nearly so, petiolate, oval-elliptical to oblong, obtuse, from under 1 in. to nearly 2 in. long. Flowers on pedicels varying from about the length of the calyx to three times that length, and always longer than the fruit. Bracteoles none. Calyx-segments 4 to 5 lines long. Corolla-tube about ½ in. long, not much dilated, the lobes

2 N 2

about half as long as the tube. Stamens of *R. australis.* Capsule 6 to 8 lines long, 2-celled from the base.—Nees in DC. Prod. xi. 154.

Queensland. Bustard Bay, Bay of Inlets and Cape Grafton, *Banks and Solander;* Port Denison, *Fitzalan;* Burdekin river, *F. Mueller;* Suttor river, *Thozet;* Broad Sound, *Bowman;* Bogee river, *Dallachy.*

TRIBE IV. JUSTICIEÆ.—Corolla 2-lipped or 1-lipped by the reduction of the upper one, or rarely nearly equally 5-lobed, the lobes variously imbricate but not contorted in the bud. Seeds flat. Retinacula prominent, usually hooked.

6. ACANTHUS, Linn.

(Dilivaria, *Juss.*)

Calyx divided to the base into 4 distinct segments or sepals. Corolla with a very short tube ; the upper lip short and truncate or entirely wanting, the lower lip large entire or 3-lobed. Stamens 4, exserted ; anthers 1-celled, hirsute or ciliate. Ovules 2 in each cell of the ovary. Capsule 2-celled from the base. Seeds large, flat ; retinacula thick.—Herbs or shrubs. Leaves usually prickly-toothed. Flowers in bracteate spikes.

A small genus, spread over tropical Asia, Africa, and southern Europe. The only Australian species is a common maritime plant in tropical Asia.

1. **A. ilicifolius,** *Linn.; T. Anders. in Journ. Linn. Soc.* ix. 501. An erect glabrous shrub of several feet. Leaves sessile or nearly so, oval or broadly oblong, 4 to 8 in. long, coriaceous and shining when full grown, bordered with undulate prickly teeth or short lobes or rarely entire, with a pair of divaricate short prickles at their base in the place of stipules, sometimes 4 or 5 lines long, sometimes very short or entirely wanting. Spikes terminal or in the upper axils, 6 in. to 1 ft. long. Bracts ovate, acute or obtuse, often at least half as long as the calyx ; bracteoles similar but smaller, sometimes very small or wanting. Calyx-segments or sepals oblong, coriaceous, the 2 outer ones 6 to 8 lines long, the 2 inner ones smaller. Corolla upper lip exceedingly short truncate and coriaceous, the lower expanded into an obovate entire or shortly 3-lobed limb of above 1 in. Filaments hard, thick and shining, more than half as long as the lower lip. Anthers 3 to 4 lines long, very densely ciliate-hirsute. Capsule ¾ to 1 in. long, coriaceous, shining, very obtuse.—R. Br. Prod. 480 ; *Dilivaria ilicifolia,* Juss. ; Nees in DC. Prod. xi. 268 ; Wight, Ic. t. 459.

N. Australia. Gulf of Carpentaria, *R. Brown;* Albert river, *Henne.*

Queensland. Cape York, *Daemel;* Endeavour river, *Banks and Solander;* Cape Conway, *A. Cunningham;* Rockhampton, *O'Shanesy, Thozet.*

*A. ebracteatus,*Vahl, Symb. ii. 75. t. 40 ; R. Br. Prod. 480 (*Dilivaria ebracteata,* Juss. ; Nees in DC. Prod. xi. 269) ; is distinguished by almost all botanists as a species by the absence of bracts, and usually by the want of the stipular spines at the base of the leaves. Both the bracts and these spines are so very variable in size, that in the usually indifferent specimens in the collections before me, I am quite unable to ascertain whether there really are or not two distinct forms, all the other characters being precisely the same in both. A Malacca specimen answering to Vahl's figure, has smaller flowers than the common form, and the bracts very deciduous, but they are present under the buds on the young spikes. Brown's specimens of *A. ebracteatus* have the subtending bract, but the bracteoles very minute or deficient and the stipular spines wanting.

7. JUSTICIA, Linn.

(Rostellularia *and* Rhaphidospora, *Nees.*)

Calyx divided to the base into 5 or 4 segments. Corolla 2-lipped, the upper lip erect, concave, entire or notched, the lower convex or with a longitudinal fold and veined in the centre, 3-lobed. Stamens 2; anther-cells oblique, one attached higher up than the other, the lower one usually mucronate or spurred. Ovules 2 in each cell of the ovary. Style usually entire. Capsule contracted or compressed and seedless at the base. Seeds flat; retinacula obtuse.—Herbs or shrubs. Flowers solitary or in clusters or cymes, axillary or forming terminal spikes or panicles. Bracts various.

A large genus, widely distributed over the tropical and subtropical regions of the globe. Of the five Australian species, two are common tropical weeds in Asia, another is closely allied to, if not identical with an Asiatic one, the remaining two are, as far as is known, endemic.

Flowers (small) in dense terminal bracteate spikes.
Bracts linear or lanceolate, acute, hispid or ciliate, not bordered.　1. *J. procumbens.*
Bracts obtuse, bordered by a broad white margin . . . . .　2. *J. peploides.*
Flowers in axillary sessile clusters surrounded by a few broad obcordate bracts . . . .　3. *J. hygrophiloides.*
Flowers in pairs on axillary simple or forked peduncles. Bracts setaceous . . . .　4. *J. cavernarum.*
Flowers in a terminal dichotomous panicle . . . .　5. *J. eranthemoides.*

1. **J. procumbens,** *Linn.; T. Anders. in Journ. Linn. Soc.* ix. 511. A procumbent, prostrate or rarely erect annual, often extending to above 1 or 2 ft. when trailing, shorter when erect. Leaves usually oblong lanceolate or almost linear, the lower ones small and more ovate, sometimes all ovate from ¼ to 1 in. long or all narrow and then sometimes nearly 2 in. long; as well as the whole plant pubescent hirsute or nearly glabrous. Flowers pink or white, solitary under each bract, in terminal rather dense spikes of ¾ to above 1 in., with often 1 or 2 pairs of flowers at some distance below. Bracts linear-lanceolate or linear, acute, hirsute and ciliate, as long as the calyx; bracteoles rather smaller; calyx-segments 4 with occasionally a small fifth one, linear, 2 to 3 or rarely 4 lines long. Corolla-tube nearly as long as the calyx; upper lip short, erect; lower one broad, spreading, nearly as long as the tube. Lower anther-cell spurred and often empty. Capsule 3 to 4 lines long, the seedless base short.—*Rostellaria* (or *Rostellularia*) *procumbens,* Nees in Wall. Pl. As. Rar., and in DC. Prod. ix. 371; Wight, Ic. t. 1539; *Justicia juncea, J. media,* and *J. adscendens,* R. Br. Prod. 476; *Rostellularia media* and *R. juncea,* Nees in DC. Prod. ix. 374, 376; *R. pogonanthera,* F. Muell. in Linnæa, xxv. 431; besides the numerous synonyms adduced by T. Anderson, l. c.

N. Australia. Gulf of Carpentaria, *R. Brown;* Victoria and Fitzmaurice rivers, *F. Mueller;* Cooper's River, *A. C. Gregory.*
Queensland. Endeavour river, *Banks and Solander;* Thirsty Sound, Broad Sound, Keppel Bay, etc., *R. Brown,* and from very numerous stations both on the coast and in the interior by most of the subsequent collectors.
N. S. Wales. Hunter's River, *A. Cunningham;* Hastings and Clarence rivers, *Beckler;* from the Lachlan, *A. Cunningham;* and Darling river to the Barrier Range, *Victorian and other Expeditions.*

S. Australia. Near Akaba, *F. Mueller* ; Mount Serle, *Warburton* ; Purdie's Ponds, *Waterhouse;* Flinders Range, *Howitt's Expedition.*

The species is a very common and variable weed throughout tropical Asia, extending into Africa. The Australian forms are chiefly narrow-leaved, either erect (*J. juncea,* Br.) or procumbent (*J. adscendens,* Br.), but there are several others which at first sight look very distinct although connected by numerous intermediates. The most remarkable are one with very small ovate leaves, chiefly from Mount Serle, *Warburton,* and New England, *C. Stuart,* and one with rather large ovate nearly sessile leaves, more glabrous and tending to dry black, from Endeavour river, *Banks and Solander,* Rockingham Bay, *Dallachy,* and a few others. In one luxuriant specimen of F. Mueller's from Victoria river, some of the bracts show a slight tendency to white margins, although but a very distant approach to those of *J. peploides.*

2. **J. peploides,** *T. Anders. in Journ. Linn. Soc.* ix. 511. Nearly allied to *J. procumbens,* with similar procumbent ascending or rarely erect stems inflorescence and flowers. Leaves ovate oblong or broadly lanceolate, on longer petioles than in *J. procumbens,* and not so much contracted at the base, usually pubescent. Spikes cylindrical, compact, ¾ to 1½ in. long, very rarely interrupted at the base. Flowers white, smaller than in *J. procumbens.* Bracts and bracteoles obtuse, bordered by a broad white margin. Calyx-segments also somewhat membranous on the margin, but acute.—*Rostellaria peploides,* Nees, and other synonyms quoted by T. Anderson, l. c.

Queensland. Brisbane river, Moreton Bay, *A. Cunningham, F. Mueller, Henne, C. Stuart.* Also in E. India, but not so widely spread as *J. procumbens.*

3. **J. hygrophiloides,** *F. Muell. Fragm.* vi. 89. An erect densely-branched shrub, glabrous or minutely pubescent. Leaves ovate-lanceolate or lanceolate, obtuse, narrowed into a short petiole, 1 to 1½ or rarely 2 in. long. Flowers white or according to some labels yellow, in axillary clusters of 2 to 6, surrounded by 3 or 4 very broadly obcordate or 2-lobed obtuse leafy bracts, at least as long as the calyx and rather broader than long. Calyx-segments 5 or rarely 4, linear-subulate, about 3 lines long. Corolla under ½ in. long, the tube shorter than the calyx, the lips as long as the tube, the upper one concave and notched, the lower one broadly obovate, shortly divided into 3 broad nearly equal lobes. Stamens nearly as long as the upper lip, the lower anther-cell with a basal appendage or spur. Capsule oblong, obtuse, about as long as the calyx, the basal seedless part very variable in length, sometimes very short. Seeds 2 in each cell.

Queensland. Brisbane river, Moreton Bay, *F. Mueller, C. Stuart ;* Cape Conway, *A. Cunningham;* Rockingham Bay, *Dallachy;* Rockhampton, *Dallachy, O'Shanesy.*

4. **J. cavernarum,** *F. Muell. Fragm.* vi. 91. Branches slender, apparently decumbent or divaricate, shortly hirsute or pubescent as well as the foliage and inflorescence. Leaves petiolate, ovate, 1 to 1½ in. long. Peduncles in one axil of each pair of leaves, longer than the leaves, bearing at the end 2 sessile flowers, or forked with 2 sessile flowers at the end of each branch. Bracts and bracteoles setaceous, shorter than the calyx. Calyx-segments linear-setaceous, not 2 lines long. " Corolla small, pubescent outside, glabrous inside. Stamens 2 ; anthers 2-celled with the lower cell conspicuously spurred."

Queensland. Mountain caves near Rockhampton, *Thozet.* Described from fragmentary specimens in Herb. F. Mueller, which closely resemble the pubescent variety of the E. Indian *J. glabra,* Kœn. (*Rhaphidospora glabra,* Nees), but are not in a state to determine whether they really belong or not to that species.

5. **J. eranthemoides,** *F. Muell. Fragm.* vi. 90. Branches, veins of the under side of the leaves and inflorescence pubescent. Leaves petiolate, ovate-lanceolate or lanceolate, 2 to 2½ in. long. Flowers white, in terminal dichotomous corymbose panicles as long as the leaves. Bracts small, linear-subulate. Pedicels very short, without bracteoles. Calyx-segments linear-setaceous, about 2 lines long. Corolla 5 or 6 lines long, the tube nearly straight, dilated upwards ; lips rather shorter than the tube, the upper one narrow, erect, notched, innermost in the bud, lower lip broader, with the prominent longitudinal fold of the genus, 3-lobed to the middle, the middle lobe broader than the others and overlapping them in the bud. Anther-cells inserted one higher than the other as in the rest of the genus, but the lower one without any basal appendage. Capsule narrow, 5 or 6 lines long, contracted flattened and seedless at the base. Seeds 2 in each cell.

N. S. Wales. Tweed river, *C. Moore.* Described from a single small specimen in Herb. F. Mueller, very different from any species known to me, especially in inflorescence, which, however, comes nearest to that of the section *Rhaphidospora.*

8. GRAPTOPHYLLUM, Nees.

(Earlia, *F. Muell.*)

Calyx divided to the base into 5 segments. Corolla-tube incurved, the limb 2-lipped, the upper lip concave, incurved, notched, the lower divided to the base into 3 nearly equal lobes. Stamens 2, ascending under the upper lip, anther-cells parallel, nearly equal, without basal appendages ; staminodia 2. Ovules 2 in each cell of the ovary. Capsule oblong-clavate, contracted into a solid seedless base. Seeds flat ; retinacula hooked.—Tall shrubs with glabrous shining leaves. Flowers red, in axillary or terminal clusters or short racemes. Bracts and bracteoles very small.

Besides the two Australian endemic species, there is only one known and commonly cultivated in tropical Asia, but of uncertain origin, probably from some of the early visited islands of the Eastern Archipelago. It is the *G. pictum,* Nees, known in our hothouses under the name of the *Caricature-plant,* and only differs in foliage from *G. ilicifolium.*

Leaves small oblong entire or minutely toothed . . . . 1. *G. Earlii.*
Leaves large, broadly ovate, prickly-toothed . . . . 2. *G. ilicifolium.*

1. **G. Earlii,** *F. Muell. Fragm.* vi. 87. A beautiful glabrous shrub or tree of 10 to 15 ft. (*Dallachy*). Leaves oblong-elliptical, acute or mucronulate, entire or with a few very small acute teeth, ¾ to 1½ in. long. Flowers of a rich red, solitary in the axils or in clusters of very few. Pedicels 2 to 3 lines long, with minute bracts and bracteoles at the base. Calyx-segments narrow, acute, about 2 lines long. Corolla-tube incurved and dilated into a broad oblique throat above ½ in. long ; upper lip incurved, much shorter than the tube, the lower lip rather shorter, equally divided into rather broad almost acute lobes. Filaments hairy at the base ; staminodia filiform. Capsule hard, almost woody, about ¾ in. long.—*Earlia excelsa,* F. Muell. Fragm. iii. 160 ; *Thyrsacanthus Earlii,* F. Muell. Fragm. vi. 87.

Queensland. Near Rockhampton, *Dallachy, Thozet.* *Thyrsacanthus* is a South American genus with a very different habit and corolla.

2. **G. ilicifolium,** *F. Muell. Herb.* A glabrous shrub of 10 to 15 ft. (*Nernst*). Leaves very shortly petiolate, broadly· ovate, obtuse or acute, bordered by irregular mucronate or prickly teeth, 3 to 4 in. long, coriaceous, much veined but very shining. Flowers of a rich red, in short dense clusters or racemes, axillary in our specimens. Pedicels short, with very small bracts and bracteoles at the base. Calyx-segments 3 to 4 lines long. Corolla-tube ¾ in. long, dilated into a broad throat but not so oblique nor so broad as in *G. Earlii,* the lips ½ in long, the upper one concave and incurved, the lower one divided to the base into 3 equal narrow lobes. Stamens as in *G. Earlii.* Capsule above 1 in. long.

Queensland. Mount Blackwood, Mackay district, *Nernst.* F. Mueller, Fragm. vi. 87, refers this as a variety to *G. Earlii.* It appears to me much nearer to *G. pictum,* of which it has the narrow-lobed corollas, and only differs, as far as I can ascertain, in foliage.

9. DICLIPTERA, Juss.

(Brochosiphon, *Nees.*)

Calyx deeply divided into 5 lobes or segments. Corolla-tube usually slender, dilated at the throat, the upper lip concave entire or notched, the lower broader nearly entire or 3-lobed, the middle lobe much broader than the others. Stamens 2, ascending under the upper lip; anthers 2-celled, the cells placed usually one higher than the other, but without any basal appendage. Ovules 2 in each cell. Capsule usually flat, shortly contracted and seedless at the base, the dissepiment separating from the valves when opened and turning upwards elastically with the retinacula. Seeds flat.— Herbs. Flowers 1 to 3 together, sessile within a flattened involucre of 2 bracts concealing the calyx, the involucres usually several in clusters or short cymes, in the axils of the floral leaves or forming terminal loose spikes or racemes, with usually 2 subulate or spinescent bracts outside the flat ones. Corolla, owing to the peculiar inflorescence, appearing frequently resupinate with relation to the main axis, the upper entire or 2-notched lip becoming the lowest.

A considerable genus dispersed over the tropical and subtropical regions of the New and the Old World. The two Australian species extend at least to Timor.

Involucral bracts orbicular, very flat, glabrous or glandular-pubescent, all in
   axillary clusters . . . . 1. *D. glabra.*
Involucral bracts ovate, aristate, convex, ciliate-hirsute on the upper side,
   the clusters forming terminal loose spikes or racemes . . . . . . 2. *D. spicata.*

1. **D. glabra,** *Dcne. Herb. Tim.* 55. A much-branched annual of 1 to 2 ft., glabrous or the foliage sprinkled with a few rather rigid hairs. Leaves lanceolate or almost linear, mostly acute, contracted into a very short petiole, 1 to 2 in. long. Involucres either 2 sessile in the axils or 4 in pairs on 2 very short peduncles or several in a more or less cymose but very dense cluster, the involucral bracts very broadly ovate or nearly orbicular, mucronate-acute, glabrous or glandular-pubescent and ciliate, flat and usually unequal, the larger one 3 to 6 lines diameter, and always with an outer pair of

rigid linear-subulate spreading or recurved outer bracts or spines. Flowers within the bracts solitary or rarely 2 or 3, with minute bracteoles. Calyx 1 to 1½ lines long, divided to below the middle into linear-lanceolate lobes. Corolla shortly exceeding the bracteoles, the lips nearly as long as the tube, the upper one ovate and notched, the lower one broad and 3-toothed. Capsule very small, flat, nearly orbicular, usually 2-seeded.—Nees in DC. Prod. xi. 476 ; *Brochosiphon australis*, Nees, l. c. 492 ; *Dicliptera armata*, F. Muell. Fragm. vi. 88.

N. Australia. Glenelg river, N.W. coast, *Marten ;* Upper Victoria river and Stirling Creek, *F. Mueller ;* S. Goulburn island, *A. Cunningham.* The specimers agree perfectly with Decaisne's character as well as with Cunningham's Timor specimens.

2. **D. spicata,** *Dcne. Herb. Tim.* 56. An erect paniculately branched annual of 1 to 2 ft., the stem and leaves glabrous or minutely pubescent. Leaves lanceolate or almost linear, very acute and mucronate, almost aristate, narrowed into a petiole, 1 to 2 in. long, the floral ones narrower and shorter. Involucres usually 3 on a common peduncle in the axil of each floral leaf and sometimes a second shorter peduncle in the same axil with a single involucre, the clusters of involucres numerous in terminal racemes leafy at the base, the upper floral leaves reduced to subulate bracts. Involucral bracts ovate, acute and aristate, the margins recurved, the upper or inner surface convex and hirsute, the larger one of each pair 3 to 4 lines long, with an external pair of subulate bracts. Corolla slender, shorter than the bracts, the lips as long as the tube. Stamens 2. Capsule clavate, ovate. Seeds 2, muricate. —Nees in DC. Prod. xi. 479 ; *D. racemifera*, F. Muell. Fragm. vi. 89.

Queensland. Cape York, *Daemel.* Also in Timor ; the typical specimens received from Decaisne, as well as others in the Banksian Herbarium, agree precisely with the Australian ones.

10. HYPOESTES, R. Br.

Calyx more or less deeply divided into 5 lobes or segments. Corolla with a slender tube, deeply 2-lipped, the upper lip narrow entire or rarely notched, the lower 3-lobed. Stamens 2, often nearly as long as the corolla ; anthers linear, 1-celled. Ovules 2 in each cell of the ovary. Style bifid at the top. Capsule compressed and seedless at the base, oblong or clavate. Seeds flat ; retinacula subulate.—Herbs shrubs or small trees. Flowers solitary or 2 or 3 together, within a cylindrical or clavate involucre of 2 pairs of bracts often united to the middle, the inner pair alternating with the outer, the involucres in axillary clusters or spikes or in terminal panicles.

The genus is dispersed over Africa and tropical Asia. The Australian varieties or species appear to be endemic, but require further comparison with some forms from the Eastern Archipelago of which we have very imperfect specimens.

1. **H. floribunda,** *R. Br. Prod.* 474. An erect branching perennial, attaining 2 or 3 ft. and usually glabrous except the minutely glandular-pubescent inflorescence. Leaves ovate-lanceolate or almost linear, acutely acuminate, contracted into a rather long petiole, usually thin and membranous and 2 to 4 in. long, but occasionally much larger. Involucres usually numerous in dense axillary clusters or racemes or loose terminal panicles each in-

volucre tubular, concrete, 2 to 4 lines long, 4-lobed to about the middle, the lobes acute, the 2 inner ones rather smaller. Flowers solitary in the involucre or rarely 2 or 3 together, but the accessary ones mostly rudimentary. Calyx very thin, divided to about the middle, much shorter than the involucre. Corolla slender, about ¾ in. long or rather larger, the lips as long as the tube, the upper one linear and entire, the lower one much broader, very shortly 3-lobed. Stamens nearly as long as the lips. Capsule rather narrow, 5 to 6 lines long.—Endl. Iconogr. t. 105; *H. laxiflora* and *H. floribunda* (partly), Nees in DC. Prod. xi. 508, 509.

The following forms of this very variable plant might be distributed according to the inflorescence into three principal varieties or perhaps species:—

1. *Densiflora.* Involucres mostly 2 to 3 lines long in short dense spikes or clusters chiefly axillary.

N. Australia. Lagrange Bay, N.W. coast, *Marten.*

Queensland. Moreton Bay, *A. Cunningham, F Mueller;* Rockhampton, *Thozet;* Edgecombe and Rockingham Bays, *Dallachy;* Nerkool Creek, *Bowman;* Port Denison *Fitzalan.* (All nearly glabrous.)

Var. *canescens.* Branches inflorescence and under side of the leaves hoary with a very minute pubescence.—Cape York, *Daemel.*

Var. *pubescens.* Rather densely clothed with a scabrous or a soft pubescence.—Wide Bay, *Bidwill;* Burdekin river, *Leichhardt;* N. coast of Arnhem's Land, *Kinley.*

N. S. Wales, Clarence river, *Beckler.*

2. *Paniculata.* Involucres usually 3 to 4 lines long, in elongated interrupted spikes, usually numerous in the upper axils, forming rather large terminal panicles.

N. Australia. South Goulburn Island, *A. Cunningham;* Cape Upstart, *Bynoe;* Port Essington, *Armstrong.*

Queensland. Shoalwater Bay, *R. Brown.*

Var. *angustifolia* Leaves narrow-lanceolate or almost linear.—Victoria and Fitzmaurice rivers, *F. Mueller.*

3. *Distans.* Stems long and decumbent. Involucre few and very distant along the branches of a very loose terminal panicle.

N. Australia. Hunter's River, N.W. coast, *A. Cunningham.*

R. Brown's specimens belong to the paniculate form, which is included by Nees in his *H. laxiflora* β, with some Javanese specimens which appear to me quite different. Nees's typical *H. laxiflora* has a remarkably dense inflorescence and long subulate-acuminate involucral bracts, and agrees much better with the Javanese plants determined by him to be *H. rosea,* Dcne., but not agreeing with Decaisne's character. Why he suppressed Decaisne's *H. rosea* to substitute a *H. rosea* of his own (p. 503) does not appear.

11. ERANTHEMUM, Linn.

Calyx deeply divided into 5 lobes or segments. Corolla-tube long and slender; limb spreading, 5-lobed, the lobes nearly equal, imbricate but not contorted in the bud. Stamens 2, inserted high up in the tube; anthers partially exserted, with 2 parallel and equal cells; staminodia 2, usually very small. Ovules usually 2 in each cell. Capsule oblong-clavate or linear, much contracted and seedless at the base. Seeds flat; retinacula curved.—Herbs undershrubs or shrubs. Flowers solitary or in little cymes of 3 to 5, sessile or very shortly pedunculate in the upper axils or more frequently forming terminal interrupted spikes with the floral leaves reduced to small bracts. Bracteoles very small or none.

A considerable genus, dispersed over the tropical and subtropical regions of the New as well as the Old World. The two Australian species appear to be both endemic.

Herb with a creeping rhizome and erect nearly simple stems. Flowers
in spikes . . . . 1. *E. variabile.*
Slender branching shrub. Flowers axillary . . . . 2. *E. tenellum.*

1. **E. variabile,** *R. Br. Prod.* 477. A perennial with a creeping
rhizome and ascending or erect simple or slightly branched stems varying
from a few inches to above 1 ft. high, the whole plant glabrous pubescent or
hirsute. Leaves petiolate, ovate oblong lanceolate or linear, 1 to 3 in. long.
Flowers white, solitary or rarely in little cymes of 3 or 5, in the axils of
bracts always very small and sometimes almost inconspicuous, forming
racemes or spikes, sometimes short and dense in the upper axils, but mostly
slender interrupted and terminal; pedicels very short. Calyx-segments
linear-setaceous, varying from under 2 to above 4 lines in length. Corolla
glabrous or pubescent, the slender straight tube 5 to 8 lines long ; lobes oblong, from under half the length of the tube to nearly its length. Capsule
about ½ in. long, the lower half contracted and seedless.—Nees in DC. Prod.
xi. 456.

Queensland. Brisbane river, Moreton Bay, *A. Cunningham, F. Mueller,* and others ;
Rockhampton and Rockingham Bay, *Dallachy ;* Nerkool Creek, Broad Sound, and Amity
Creek, *Bowman ;* Wide Bay, *Bidwill ;* Burdekin river, *F. Mueller ;* Cape York, *M'Gillivray.*
N. S. Wales. Port Jackson to the Blue Mountains, *R. Brown* and others ; Hastings
and Clarence rivers, *Beckler ;* Richmond river, *Fawcett ;* New England, *C. Stuart.*

Var. *molle.* Leaves rather large, ovate, thin, softly pubescent. Flowers distant in
slender leafless racemes.—Cape York, *Daemel ;* Rockhampton, *Dallachy.*

Var. *lineare.* Leaves narrow linear.—Gilbert river, *F. Mueller ;* Moreton Bay, *C.
Stuart.*

Var. ? *grandiflorum.* Corolla-tube above 1 in. long ; lobes above ½ in.—Lord Howe's
Island, *Milne,* a single specimen. Possibly a distinct species.

The plant figured in Paxt. Mag. xiii. 75 as *E. variabile* is a very different species, not
Australian.

2. **E. tenellum,** *Benth.* An erect shrub of 2 to 3 ft. with slender scabrous-pubescent or glabrous branches. Leaves petiolate, ovate or oblong,
those of each pair very unequal in size, the larger one sometimes rather
above 1 in. long but usually half that size, and its opposite one much
smaller. Flowers white, solitary in the axils, on short pedicels. Calyx-
segments linear-setaceous, under 2 lines long. Corolla like the smaller
form of *E. variabile.*

Queensland. Rockhampton, *Dallachy ;* Broad Sound, *Bowman.* Evidently nearly
allied to *E. tuberculatum,* Hook. Bot. Mag. t. 5405, but without the peculiar warts of that
species, the flowers smaller, and the leaves differently shaped.

ORDER LXXXIX. **PEDALINEÆ.**

(Sesameæ, *DC.*)

Flowers irregular. Calyx 5-lobed or divided into 5 segments. Corolla tubular ; lobes 5, spreading, often arranged in 2 lips, the lowest often rather larger
than the others, imbricate or rarely valvate in the bud. Stamens 4, didynamous,
with a small fifth staminodium, rarely only 2 perfect ; anthers 2-celled, the cells
opening longitudinally. Ovary of 2 or rarely 3 or 4 carpels, but divided, at
least after flowering, into twice as many cells. Ovules in each cell either several

superposed in a single row or rarely solitary.　Style filiform, with as many stigmatic lobes as carpels.　Fruit dry, hard and indehiscent or opening in valves.　Seeds with a thin testa.　Albumen scanty or none.　Embryo straight, with a very short radicle.—Herbs.　Leaves all or at least the lower ones opposite.　Flowers solitary in the axils of the floral leaves or bracts, the upper ones often forming a terminal raceme, with or without bracteoles.

A small Order, dispersed over the tropical and subtropical regions both of the New and the Old World, the only Australian genus extending to the Eastern Archipelago.　De Candolle's arrangement, including Sesameæ and Pedaleæ in one Order, appears to be far the most natural ; thus forming a small group connected on the one hand with Gesneriaceæ and Bignoniaceæ by their flowers, and on the other hand with Verbenaceæ by their ovary or fruit divided into twice as many cells as carpels.

1. JOSEPHINIA, Vent.

Calyx divided to the base into 5 segments.　Corolla tubular, the lobes spreading, short, the lowest rather larger than the others.　Stamens didynamous, included in the tube; anther-cells parallel, the connectivum usually tipped with a small gland.　Ovary of 4, 6 or 8 cells, each with 1 erect ovule; stigmatic lobes 2, 3 or 4.　Fruit hard and indehiscent, armed with conical prickles, shortly or not at all beaked.　Seeds 1 in each cell, oblong, erect.—Herbs with the habit of *Sesamum*.　Leaves opposite, entire, toothed or divided.　Flowers in the upper axils on short pedicels without bracteoles.

The genus extends into the Archipelago.　Of the three Australian species, one is also in the Archipelago, the two others are endemic.　The solitary erect ovules and seeds connect this genus with *Verbenaceæ*, but the habit and corollas are those of *Pedalineæ*.

Leaves glabrous or nearly so, quite entire.　Ovary usually 8-celled.
　Fruit with a small terete or conical beak . . . . 1. *J. grandiflora.*
Leaves pubescent underneath, the lower ones coarsely toothed.　Ovary
　usually 6-celled.　Fruit with a triangular truncate beak . . . . 2. *J. imperatricis.*
Plant densely villous.　Lower leaves divided into 3 distinct segments.
　Ovary usually 4-celled.　Fruit not beaked . . . . 3. *J. Eugeniæ.*

1. J. grandiflora, *R. Br. Prod.* 520.　Stems erect or diffuse (2 to 3 ft. high ?), glabrous or sprinkled with a few minute hairs.　Leaves petiolate, lanceolate or the lower ones ovate-lanceolate, all quite entire, 1½ to 3 in. long, glabrous or minutely and sparingly pubescent underneath.　Pedicels shorter than the petiole.　Calyx-segments narrow, acuminate, about 2 lines long, the upper one usually shorter.　Corolla at least 1 in. long, pubescent outside, the tube gibbous at the base on the upper side, gradually dilated upwards ; lobes broad, the 4 upper ones nearly equal, the lower twice as long and broader than the others.　Ovary in the flowers examined 8-celled ; stigmatic lobes 4 (sometimes 3 according to Endlicher's figure).　Fruit ovoid-globular, under ½ in. diameter, very hard, villous with short soft hairs, armed with thick conical very unequal prickles, the persistent thickened base of the style forming a cylindrical or slightly conical beak, sometimes very short, sometimes at least as long as the prickles.—Endl. Iconogr. t. 106.

Queensland.　Endeavour Straits, *R. Brown ;* Low Island, *R. Brown, Henne ;* islands of Howick's group, *F. Mueller ;* Three Isles and Lizard Island, *M'Gillivray ;* Pelican and Haggerstone Islands, *A. Cunningham.*

Decaisne (Herb. Tim. 76), followed by De Candolle (Prod. ix. 255) and by F. Mueller (Fragm. vi. 163), unites this with *J. imperatricis*, but probably without having had good specimens at his disposal, for, besides the indumentum, the foliage corolla and fruit appear to me to be different. Probably also neither of these authors had consulted Ventenat's plate, for they all copy the misquotation of its number, originating with Brown at a time when complete copies of the work may not have reached England. Endlicher in the above-quoted figure (from Bauer's drawings) has reversed the fruit (fig. *n*), so as to make the beak appear as a pedicel ; the seeds *o* are also reversed.

2. **J. imperatricis,** *Vent. Jard. Malm. t.* 67 (*not* 103). An erect herb of 2 or 3 ft. Leaves petiolate, the lower ones cordate-ovate, coarsely toothed, 3 or 4 in. long, the upper ones much smaller, broadly lanceolate, entire or nearly so, all minutely but rather densely pubescent underneath. Pedicels shorter than the petioles. Calyx-segments nearly equal. Corolla " with a short tube, much dilated upwards, the lower lobe not much larger than the others." Ovary in the specimens seen 6-celled. Fruit ovoid, villous, hard, armed with conical prickles as in *J. grandiflora*, but terminating in a thick obtuse or truncate, very prominently 3-angled beak, as long as or longer than the prickles.—R. Br. Prod. 520.

N. Australia. N.W. coast, *Baudin* (*Herb. Banks*). I have also a Timor specimen, sent by Decaisne, probably from the same Expedition, and a specimen in the Hookerian herbarium from Java sent by Miquel (probably *J. celebica*, Blume) appears to be the same species. In Ventenat's plate the fruit, fig. 4, is reversed, as in Endlicher's, the beak appearing as a stipes. I have not yet seen good corollas, but, besides Decaisne's description, the above quoted figure shows them to be much shorter and broader than in *J. grandiflora*.

3. **J. Eugeniæ,** *F. Muell. in Hook. Kew Journ.* ix. 370. *t.* 11. Stems procumbent ascending or erect, attaining 2 ft. or rather more, every part of the plant densely villous with articulate simple or branched hairs. Lower leaves on long petioles mostly divided into 3 petiolulate segments, the segments as well as the upper simple leaves oblong or lanceolate, coarsely toothed, 1 to 2 in. long ; the uppermost floral ones small, lanceolate and entire. Flowers pink, very small, on short pedicels in the upper axils. Calyx-segments linear, obtuse, very hispid, scarcely above 1 line long. Corolla about 3 lines long, hirsute outside, the 4 upper lobes short and broad, the lowest one larger. Ovary 4-celled ; stigmatic lobes 2, oblong-linear. Fruit ovoid, very hirsute, about ¼ in. long, the prickles small, often not exceeding the hairs, without any beak, but slightly raised ribs across the summit.

N. Australia. Upper Victoria river, *F. Mueller.*
Queensland. Suttor river, *Bowman.*
S. Australia. Near Cooper's Creek (*Herb. F. Mueller*).

The following references have been accidentally omitted :—
P. 44. Leschenaultia filiformis, add : F. Muell. Fragm. vi. t. 48.
P. 44. Leschenaultia agrostophylla, add : F. Muell. Fragm. vi. t. 47.
P. 46. Velleia connata, add : F. Muell. Lithogr. t. 49.
P. 340. Marsdenia leptophylla, add : *Bidaria leptophylla*, F. Muell. in Trans. Phil. Inst. Vict. iii. 60.
P. 344. Gymnema stenophyllum, add : *Bidaria erecta*, F. Muell. in Trans. Phil. Inst. Vict. iii. 59.

559

INDEX OF GENERA AND SPECIES.

The Synonyms and Species incidentally mentioned are printed in Italics.

2 o 2

END OF VOL. IV.

PRINTED BY TAYLOR AND CO.,
LITTLE QUEEN STREET, LINCOLN'S INN FIELDS.